Topological Aspects of Condensed Matter Physics

École de Physique des Houches

Session CIII, 4–29 August 2014

Topological Aspects of Condensed Matter Physics

Edited by

Claudio Chamon

Physics Department, Boston University, Boston, Massachusetts, 02215, USA

Mark O. Goerbig

Laboratoire de Physique des Solides, CNRS UMR 8502, Université Paris-Sud, Université Paris-Saclay F-91405, France

Roderich Moessner

Max-Planck-Institut für Physik komplexer Systeme, 01187 Dresden, Germany

Leticia F. Cugliandolo

Sorbornne Universités, Université Pierre et Marie Curie, Laboratoire de Physique Théorique et Hautes Energies, CNRS UMR 7589, Paris, France

OXFORD
UNIVERSITY PRESS

OXFORD
UNIVERSITY PRESS

Great Clarendon Street, Oxford, OX2 6DP,
United Kingdom

Oxford University Press is a department of the University of Oxford. It furthers the University's objective of excellence in research, scholarship, and education by publishing worldwide. Oxford is a registered trade mark of Oxford University Press in the UK and in certain other countries

First Edition published in 2017
Reprinted 2018

Published in the United States of America by Oxford University Press
198 Madison Avenue, New York, NY 10016, United States of America

British Library Cataloguing in Publication Data
Data available

Library of Congress Control Number: 2016943652

ISBN 978–0–19–878578–1

École de Physique des Houches

Service inter-universitaire commun
à l'Université Joseph Fourier de Grenoble
et à l'Institut National Polytechnique de Grenoble

Subventionné par l'Université Joseph Fourier de Grenoble,
le Centre National de la Recherche Scientifique,
le Commissariat à l'Énergie Atomique

Directeur:
Leticia F. Cugliandolo, Sorbonne Universités, Université Pierre et Marie Curie
Laboratoire de Physicque Théorique et Hautes Energies, Paris, France

Directeurs scientifiques de la session:
Claudio Chamon, Physics Department, Boston University, Boston, Massachusetts, USA
Mark O. Goerbig, Laboratoire de Physique des Solides, CNRS UMR 8502, Université Paris-Sud, France
Roderich Moessner, Max-Planck-Institut für Physik komplexer Systeme, Dresden, Germany
Leticia F. Cugliandolo, Sorbornne Universités, Université Pierre et Marie Curie, Laboratoire de Physique Thèorique et Hautes Energies, Paris, France

Previous Sessions

I	1951	Quantum mechanics. Quantum field theory
II	1952	Quantum mechanics. Statistical mechanics. Nuclear physics
III	1953	Quantum mechanics. Solid state physics. Statistical mechanics. Elementary particle physics
IV	1954	Quantum mechanics. Collision theory. Nucleon-nucleon interaction. Quantum electrodynamics
V	1955	Quantum mechanics. Non equilibrium phenomena. Nuclear reactions. Interaction of a nucleus with atomic and molecular fields
VI	1956	Quantum perturbation theory. Low temperature physics. Quantum theory of solids. Ferromagnetism
VII	1957	Scattering theory. Recent developments in field theory. Nuclear and strong interactions. Experiments in high energy physics
VIII	1958	The many body problem
IX	1959	The theory of neutral and ionized gases
X	1960	Elementary particles and dispersion relations
XI	1961	Low temperature physics
XII	1962	Geophysics; the earths environment
XIII	1963	Relativity groups and topology
XIV	1964	Quantum optics and electronics
XV	1965	High energy physics
XVI	1966	High energy astrophysics
XVII	1967	Many body physics
XVIII	1968	Nuclear physics
XIX	1969	Physical problems in biological systems
XX	1970	Statistical mechanics and quantum field theory
XXI	1971	Particle physics
XXII	1972	Plasma physics
XXIII	1972	Black holes
XXIV	1973	Fluids dynamics
XXV	1973	Molecular fluids
XXVI	1974	Atomic and molecular physics and the interstellar matter
XXVII	1975	Frontiers in laser spectroscopy
XXVIII	1975	Methods in field theory
XXIX	1976	Weak and electromagnetic interactions at high energy
XXX	1977	Nuclear physics with heavy ions and mesons

XXXI	1978	Ill condensed matter
XXXII	1979	Membranes and intercellular communication
XXXIII	1979	Physical cosmology
XXXIV	1980	Laser plasma interaction
XXXV	1980	Physics of defects
XXXVI	1981	Chaotic behavior of deterministic systems
XXXVII	1981	Gauge theories in high energy physics
XXXVIII	1982	New trends in atomic physics
XXXIX	1982	Recent advances in field theory and statistical mechanics
XL	1983	Relativity, groups and topology
XLI	1983	Birth and infancy of stars
XLII	1984	Cellular and molecular aspects of developmental biology
XLIII	1984	Critical phenomena, random systems, gauge theories
XLIV	1985	Architecture of fundamental interactions at short distances
XLV	1985	Signal processing
XLVI	1986	Chance and matter
XLVII	1986	Astrophysical fluid dynamics
XLVIII	1988	Liquids at interfaces
XLIX	1988	Fields, strings and critical phenomena
L	1988	Oceanographic and geophysical tomography
LI	1989	Liquids, freezing and glass transition
LII	1989	Chaos and quantum physics
LIII	1990	Fundamental systems in quantum optics
LIV	1990	Supernovae
LV	1991	Particles in the nineties
LVI	1991	Strongly interacting fermions and high Tc superconductivity
LVII	1992	Gravitation and quantizations
LVIII	1992	Progress in picture processing
LIX	1993	Computational fluid dynamics
LX	1993	Cosmology and large scale structure
LXI	1994	Mesoscopic quantum physics
LXII	1994	Fluctuating geometries in statistical mechanics and quantum field theory
LXIII	1995	Quantum fluctuations
LXIV	1995	Quantum symmetries
LXV	1996	From cell to brain
LXVI	1996	Trends in nuclear physics, 100 years later
LXVII	1997	Modeling the earths climate and its variability
LXVIII	1997	Probing the Standard Model of particle interactions
LXIX	1998	Topological aspects of low dimensional systems
LXX	1998	Infrared space astronomy, today and tomorrow
LXXI	1999	The primordial universe

LXXII	1999	Coherent atomic matter waves
LXXIII	2000	Atomic clusters and nanoparticles
LXXIV	2000	New trends in turbulence
LXXV	2001	Physics of bio-molecules and cells
LXXVI	2001	Unity from duality: Gravity, gauge theory and strings
LXXVII	2002	Slow relaxations and nonequilibrium dynamics in condensed matter
LXXVIII	2002	Accretion discs, jets and high energy phenomena in astrophysics
LXXIX	2003	Quantum entanglement and information processing
LXXX	2003	Methods and models in neurophysics
LXXXI	2004	Nanophysics: Coherence and transport
LXXXII	2004	Multiple aspects of DNA and RNA
LXXXIII	2005	Mathematical statistical physics
LXXXIV	2005	Particle physics beyond the Standard Model
LXXXV	2006	Complex systems
LXXXVI	2006	Particle physics and cosmology: the fabric of spacetime
LXXXVII	2007	String theory and the real world: from particle physics to astrophysics
LXXXVIII	2007	Dynamos
LXXXIX	2008	Exact methods in low-dimensional statistical physics and quantum computing
XC	2008	Long-range interacting systems
XCI	2009	Ultracold gases and quantum information
XCII	2009	New trends in the physics and mechanics of biological systems
XCIII	2009	Modern perspectives in lattice QCD: quantum field theory and high performance computing
XCIV	2010	Many-body physics with ultra-cold gases
XCV	2010	Quantum theory from small to large scales
XCVI	2011	Quantum machines: measurement control of engineered quantum systems
XCVII	2011	Theoretical physics to face the challenge of LHC
Special Issue	2012	Advanced data assimilation for geosciences
XCVIII	2012	Soft interfaces
XCIX	2012	Strongly interacting quantum systems out of equilibrium
C	2013	Post-Planck cosmology
CI	2013	Quantum Optics and Nanophotonics
Special Issue		Statistical Physics, Optimization, Inference and Message-Passing Algoritms

Publishers

- Session VIII: Dunod, Wiley, Methuen
- Sessions IX and X: Herman, Wiley
- Session XI: Gordon and Breach, Presses Universitaires
- Sessions XII–XXV: Gordon and Breach
- Sessions XXVI–LXVIII: North Holland
- Session LXIX–LXXVIII: EDP Sciences, Springer
- Session LXXIX–LXXXVIII: Elsevier
- Session LXXXIX– : Oxford University Press

Preface

Topological condensed matter physics is a recent arrival among the disciplines of modern physics, with a distinctive and substantive nature. Its roots, however, reach far back, to Dirac's magnetic monopoles and strings of the early 1930s and Skyrme's topologically non-trivial solutions in the nonlinear sigma models of nuclear physics of the 1960s. In the 1970s, the influx of topology into physics came in parallel in different subdisciplines: mathematical physics, high-energy physics, and condensed matter physics.

Much of the current importance of topological condensed matter physics derives from exciting developments in the last half-century. These include, in the 1970s, Wegner's farsighted analysis of lattice gauge theories and the foundation of the study of what is now known as order beyond the Landau–Ginzburg–Wilson paradigm; the work of Kosterlitz and Thouless on the eponymous phase; Anderson's pioneering work on an alternative type of magnetic state of matter, the resonating valence bond liquids; and the seminal studies by Jackiw and Rebbi and by Su, Schrieffer, and Heeger of solitons and topological zero modes with concomitant charge fractionalization in one-dimensional systems.

In the 1980s, there was an outburst of developments in which topology took centre stage in condensed matter physics, many inspired by the experimental discovery of the integer quantum Hall effect by von Klitzing and of the fractional effect by Tsui, Störmer, and Gossard. This was the time when the foundational concepts of Berry's phase and Thouless's pumps arose, when Haldane added a topological term to field theories of spin chains, and when Laughlin's work on the fractional quantum Hall effect lifted charge fractionalization to two-spatial dimensions and provided a setting where anyon statistics could appear. The fractional quantum Hall effect also opened the door for topological field theories to enter condensed matter physics, along with protected edge states and the notion of topological order. These concepts and their relation with quantum Hall physics were at the centre of an earlier Les Houches school, 'Topological Aspects of Low-Dimensional Systems', organized in 1998.

The field has since advanced rapidly, grown explosively, and diversified greatly. We now have a zoo of topological phenomena—the quantum spin Hall effect, topological insulators, Coulomb spin liquids, and non-Abelian anyonic statistics and their potential application in topological quantum computing, to name but a few—as well as an increasingly sophisticated set of concepts and methods underpinning their understanding.

Our aim for this Les Houches Summer School was to present an overview of this field, along with a sense of its origins and its place on the map of advances in fundamental physics. The school comprised a set of basic lectures (Part 1) aimed at a pedagogical introduction of the fundamental concepts, together with more advanced lectures (Part 2) covering individual topics at the forefront of today's research in condensed matter physics.

This school was, to its date, the most sought after of the Les Houches Schools in terms of number of applicants, which we read as evidence of the interest and relevance of the topic to modern condensed matter physics. Its popularity forced us to reject such a large number of suitable applicants that we decided to video-record the lectures, which can now be viewed at the school's website http://topo-houches.pks.mpg.de.

We thus hope that this book, along with these video recordings, will provide access to the knowledge shared at the school to audiences from all corners of the globe—an earnest way to express our intentions, even if geometrically imprecise!

The book starts with two conceptual presentations, by J. Moore and by B. A. Bernevig (with T. Neupert). The former introduces topological band theory, topological field theories, and Berry phases. The latter introduces the concepts of topological superconductivity and of fractional excitations along with their description within the framework of category theory. These are followed by the lecture notes of J. Chalker, which are oriented towards spin systems, more specifically spin liquids and frustrated magnetism.

The physics of the quantum Hall effect, which may be viewed as a paradigm of topological condensed matter physics, was covered in several advanced lectures. A modern view of the quantum Hall effect is provided by N. Regnault, within the framework of entanglement spectra and advanced numerical techniques. Historical and experimental aspects of this field were covered in a short series of seminars by K. von Klitzing, which, along with other experimental seminars, are available in video form on the school's website (http://topo-houches.pks.mpg.de).

The second part of the book is concerned with the advanced and more specialized lectures. F. Wegner's presentation provides a broad view on lattice gauge theories, harking back to the very origins of the field in the duality construction of phase transitions not associated with a local order parameter.

Generalizations of topological insulators due to interaction effects are the topic of A. Vishwanath's contribution. The topological tenfold-way classification of insulators and superconductors, as well as a stability analysis of one-dimensional edge channels at the boundary of two-dimensional Abelian topological phases, are presented by C. Mudry, who also gives an introduction to Abelian bosonization.

One-dimensional lattice systems are covered by F. Pollmann's lecture notes, where state-of-the art numerical techniques are introduced, such as matrix product states and their descendants. To conclude the study of one-dimensional topological systems, F. von Oppen expands on the topic of topological superconductivity in quantum wires and quantum chains.

The lecture notes by D. Carpentier are concerned with the transport properties of the $(D-1)$-dimensional surface states that accompany many topological phases living in D dimensions. A particular aspect of quantum Hall systems, namely quantum Hall ferromagnetism and its charged topological spin textures, skyrmions, is presented within the context of a modern approach by B. Douçot.

C. Castelnovo covers out-of-equilibrium properties in spin systems and kinetically constrained models. The topical lectures conclude with those by A. Niemi on topology and the physics of proteins and their folding.

All in all, we felt an acute need for a written account of the state of the field, much of the introductory material on which is scattered among review articles in the literature. The lecture notes collated in this volume, along with their oral deliveries, hopefully present a reasonably complete and concise introduction to the burgeoning field of topological condensed matter physics.

The school and the realization of this book would not have been possible without generous financial support from the continuous-formation programme 'École thématique' of CNRS and from the Franco-German University (DFH-UFA). Financial support from the European Spallation Source should also be acknowledged. We would furthermore like to thank all the participants—lecturers and students—for their stimulating interactions and scientific discussions, the outcome of which is also reflected in the present lecture notes. These interactions were favoured by the extremely pleasant environment of the Physics Centre at Les Houches, and we would therefore also like to acknowledge its administrative staff. Finally, we would like to most warmly thank Titus Neupert for the portraits of the various speakers, which figure at the beginning of their respective lecture notes.

As we were finalising these lecture notes for publication, the Nobel Prize 2016 was awarded to Duncan Haldane, Michael Kosterlitz and David Thouless "for theoretical discoveries of topological phase transitions and topological phases of matter", and the Buckley Prize 2017 of the American Physical Society to Alexei Kitaev and Xiao-gang Wen "for theories of topological order and its consequences in a broad range of physical systems". We hope that this recognition will further motivate students of all ages to learn about, and start contributing to, this exciting field of physics.

Claudio Chamon,
Mark O. Goerbig,
Roderich Moessner,
Leticia F. Cugliandolo

Hard Rock

Contents

List of Participants

Organizers

CHAMON Claudio
Boston University, USA

GOERBIG Mark-Oliver
CNRS Université Paris Sud, Orsay, France

MOESSNER Roderich
Max Planck Institute, Dresden, Germany

Lecturers

BERNEVIG Andrei
Princeton University, USA

CARPENTIER David
CNRS ENS, Lyon, France

CASTELNOVO Claudio
University of Cambridge, UK

CHALKER John T.
University of Oxford, UK

DIL Hugo
PSI and EPFL, Lausanne, Switzerland

DOUCOT Benoît
CNRS Université Pierre et Marie Curie, Paris, France

GAULIN Bruce
McMaster University, Hamilton, Canada

KIM Philip
Columbia University, New York, USA

LEVITOV Leonid
MIT, Cambridge, USA

MOLENKAMP Laurens
Würzburg University, Germany

MOORE Joel
University of California Berkeley, USA

MUDRY Christopher
Paul Scherrer Institute, Villengen, Switzerland

NIEMI Antti
Uppsala University, Sweden

POLLMANN Frank
Max Planck Institute, Dresden, Germany

REGNAULT Nicolas
CNRS ENS, Paris, France; Princeton University, USA

STERN Ady
Weizmann Institute Rehovot Israel

TAKAGI Hidenori
Max Planck Institute, Stuttgart, Germany

VISHWANATH Ashvin
University of California Berkeley, USA

VON KLITZING Klaus
Max Planck Institute Stuttgart, Germany

VON OPPEN Felix
Freie Universität Berlin, Germany

WEGNER Franz
Heidelberg University, Germany

WEN Xiao-Gang
MIT, Cambridge, USA

Participants

AASEN David
California Institute of Technology, Pasadena, USA

ASMAR Mahmoud M.
Ohio University, Athens, USA

BAXEVANIS Benjamin
University of Hamburg, Germany

BEGUE Frédéric
IRSAMC, Toulouse, France

BELOPOLSKI Ilya
Princeton University, USA

COOK Ashley
University of Toronto, Canada

DASSONNEVILLE Bastien
Leibniz Institute IFW, Dresden, Germany

de COSTER George Joseph
University of Oregon, USA

FRUCHART Michel
ENS, Lyon, France

GANESH JAYA Sreejith
NORDITA, Stockholm, Sweden

GAZIT Snir
Israel Institute of Technology, Haifa, Israel

GERAEDTS Scott
California Institute, Pasadena, USA

GROMOV Andrey
Stony Brook University, USA

GRUSDT Fabian
University of Kaiserslautern, Germany

HALASZ Gabor
Rudolf Peierls Centre, Oxford, UK

HARPER Fenner
Rudolf Peierls Centre, Oxford, UK

HUANG Po-Hao
Boston University, USA

IADECOLA Thomas
Boston University, USA

JIAN Chaoming
Stanford University, USA

JONES Nicholas G.
University of Bristol, UK

KESELMAN Anna
Weizmann Institute, Rehovot, Israel

KOURTIS Stefanos
Leibniz Institute, Dresden, Germany

LIU Tianhan
UPMC, Ecole Polytechnique, Palaiseau, France

LUNDGREN Rex
University of Texas, Austin, USA

MORAMPUDI Siddhardh
Max Planck Institute, Dresden, Germany

MOTRUK Johannes
Max Planck Institute, Dresden, Germany

NAKOSAI Sho
University of Tokyo, Japan

NEUPERT Titus
Princeton University, USA

PENG Yang
Freie Universität Berlin, Germany

PETRESCU Tudor-Alexandru
Yale University, New Haven, USA and Ecole Polytechnique/CNRS Palaiseau, France

PETROVA Olga
Max Planck Institute, Dresden, Germany

PONTE Pedro
Perimeter Institute, Waterloo, Canada

POTASZ Pawel
Wroclaw University, Poland

QUEIROZ Raquel
Max Planck Institute, Stuttgart, Germany

QUELLE Anton
Utrecht University, Netherlands

QUITO Victor
University of Campinas, Brazil

RAOUX Arnaud
ENS Paris and Université Paris Sud, France

REHN Jorge Armando
Max Planck Institute, Dresden, Germany

RIEDER Maria-Theresa
University of Berlin, Germany

RISCHAU Willem
Ecole Polytechnique, Palaiseau, France

SANTOS Raul
Weizmann Institute, Rehovot, Israel

SBIERSKI Bjoern
University of Berlin, Germany

STICLET Doru Cristian
Université de Bordeaux, France

SÜSSTRUNK Roman
ETH, Zürich, Switzerland

TABERT Calvin
University of Guelph, Canada

TANG Evelyn
MIT, Cambridge, USA

TEKER Aykut
Sabanci University, Istanbul, Turkey

THAKURATHI Manisha
Indian Institute of Science, Bangalore, India

VAJNA Szabolcs
Budapest University, Hungary

VAYRYNEN Jukka
Yale University, New Haven, USA

WITCZACK-KREMPA William
Perimeter Institute, Waterloo, Canada

WU Yinghai
Pennsylvania University, University Park, USA

YOSHIMURA Yukinori
Hiroshima University, Japan

ZHANG Junyi
ENS, Paris, France

Part I

Basic Lectures

1
An introduction to topological phases of electrons

Joel E. Moore

University of California, Berkeley and
Lawrence Berkeley National Laboratory, USA

Chapter Contents

1.1 Introduction

These lectures seek to present a coherent picture of some key aspects of topological insulators and the quantum Hall effect. Rather than aiming for completeness or historical accuracy, the goal is to show that a few important ideas, such as the Berry phase and the Chern and Chern–Simons differential forms, occur repeatedly and serve as links between superficially different areas of physics. Non-interacting topological phases, electrical polarization, and some transport phenomena in metals can all be understood in a unified framework as consequences of Abelian and non-Abelian Berry phases. The fractional quantum Hall effect is then discussed as an example of topological order, and we introduce its description by the (Abelian) Chern–Simons topological field theory.

Some effort has been made to avoid duplicating the material covered in other Les Houches lectures, both past and present. Readers seeking alternative approaches and a comprehensive list of references are encouraged to consult the many review articles on topological insulators [16, 17, 35] and the recent book by Bernevig [4]. For the fractional quantum Hall effect, our treatment parallels closely the review article of Wen [40], and the Les Houches notes of Girvin [14] provide an overview of the topic that includes more physical background than we provide here.

As part of our goal is to explain the topological invariants that underlie various topological phases, we start in Section 1.2 with some preliminaries: a few examples of the two kinds of topology (homotopy and cohomology) that appear most frequently in condensed matter physics and a derivation of the Berry phase formula for adiabatic transport. No claims of rigour or completeness are made, and the book of Nakahara [30] is a good place to start learning more; readers focused on physics content should feel free to skip this section and refer back to it as necessary.

Section 1.3 introduces non-interacting topological phases of electrons (the integer quantum Hall effect and topological insulators are two important examples) and related phenomena, focusing on topological aspects. The concept of the Berry phase links these topological phases to important physical observables such as electrical polarization and the magnetoelectric effect.

Section 1.4 discusses topological phases in interacting systems, using a field-theory approach that starts with an example of topological terms in a conventional field theory (the Haldane gap in spin systems) and then moves on to the purely topological description of the fractional quantum Hall effect via Chern–Simons theory. Many additional steps in describing Chern–Simons field theory properly can be found in the Les Houches notes of Dunne [7]. While neither the interacting nor the non-interacting topological phases are discussed in full detail, it is hoped that this way of presenting them gives some practical understanding of what it means for an electronic state of matter to be 'topological'.

1.2 Basic concepts

1.2.1 Mathematical preliminaries

1.2.1.1 An intuitive example of global geometry and topology: Gauss–Bonnet

You may have heard a topologist described as 'a mathematician who can't tell the difference between a donut and a coffee cup'. As an example of the connections between geometry and topology, we start by discussing an integral that will help us classify two-dimensional (2D) compact manifolds (surfaces without boundaries) embedded smoothly in three dimensions. The integral we construct is 'topologically invariant' in that if one such surface can be smoothly deformed into another, then the two will have the same value of the integral. The integral can't tell the difference between the surface of a coffee cup and that of a donut, but it can tell that the surface of a donut (a torus) is different from a sphere. Such global integrals of geometrical quantities are a common origin of topological quantities in physics.

We start with a bit of local geometry. Given our 2D surface in 3D, we can choose coordinates at any point on the surface so that the $(x, y, z = 0)$ plane is tangent to the surface, which can locally be specified by a single function $z(x, y)$. We choose $(x = 0, y = 0)$ to be the given point, so $z(0, 0) = 0$. The tangency condition is

$$\left.\frac{\partial z}{\partial x}\right|_{(0,0)} = \left.\frac{\partial z}{\partial y}\right|_{(0,0)} = 0. \tag{1.1}$$

Hence we can approximate z locally from its second derivatives:

$$z(x, y) \approx \frac{1}{2} \begin{pmatrix} x & y \end{pmatrix} \begin{pmatrix} \frac{\partial^2 z}{\partial x^2} & \frac{\partial^2 z}{\partial x\, \partial y} \\ \frac{\partial^2 z}{\partial y\, \partial x} & \frac{\partial^2 z}{\partial y^2} \end{pmatrix} \begin{pmatrix} x \\ y \end{pmatrix}. \tag{1.2}$$

The 'Hessian matrix' that appears here is real and symmetric. It can be diagonalized and has two real eigenvalues λ_1, λ_2, corresponding to two orthogonal eigendirections in the (x, y) plane. The geometric interpretation of these eigenvalues is simple: their magnitude is an inverse radius of curvature, and their sign tells us whether the surface is curving towards or away from the positive z direction in our coordinate system. To see why the first is true, suppose that we carried out the same process for a circle of radius r tangent to the x axis at the origin. Parametrize the circle by an angle θ that is 0 at the origin and traces the circle counterclockwise, i.e.

$$x = r \sin\theta, \qquad y = r(1 - \cos\theta). \tag{1.3}$$

Near the origin, we have

$$y = r\left[1 - \cos\left(\sin^{-1}\frac{x}{r}\right)\right] = r - \left(1 - \frac{x^2}{2r^2}\right) = \frac{x^2}{2r}, \tag{1.4}$$

which corresponds to an eigenvalue $\lambda = 1/r$ of the matrix in (1.2).

Going back to the Hessian, its determinant (the product of its eigenvalues, $\lambda_1\lambda_2$) is called the Gaussian curvature and has a remarkable geometric significance. First, consider a sphere of radius r, which at every point has $\lambda_1 = \lambda_2 = 1/r$. Then we can integrate the Gaussian curvature over the sphere's surface:

$$\int_{S^2} \lambda_1\lambda_2\, dA = \frac{4\pi r^2}{r^2} = 4\pi. \tag{1.5}$$

Beyond simply being independent of radius, this integral actually gives the same value for any compact manifold that can be smoothly deformed to a sphere.

However, we can easily find a compact manifold with a different value for the integral. Consider the torus made by revolving the circle in (1.3), with $r = 1$, around the axis of symmetry $x = t, y = -1, z = 0$, with $-\infty < t < \infty$. To compute the Gaussian curvature at each point, we sketch the calculation of the eigenvalues of the Hessian as follows. One eigenvalue is around the smaller circle, with radius of curvature r: $\lambda_1 = 1/r = 1$. Then the second eigenvalue must correspond to the perpendicular direction, which has a radius of curvature that depends on the angle θ around the smaller circle (we keep $\theta = 0$ to indicate the point closest to the axis of symmetry). The distance from the axis of symmetry is $2 - \cos\theta$, so we might have guessed $\lambda_2 = (2 - \cos\theta)^{-1}$, but there is an additional factor of $\cos\theta$ that appears because of the difference in direction between the surface normal and this curvature. So our guess is that

$$\lambda_2 = -\frac{\cos\theta}{2 - \cos\theta}. \tag{1.6}$$

As a check and to understand the sign, note that this predicts a radius of curvature 1 at the origin and other points closest to the symmetry axis, with a negative sign in the eigenvalue indicating that this curvature is in an opposite sense as that described by λ_1. At the top, the radius of curvature is 3 and in the same sense as that described by λ_1, and on the sides, λ_2 vanishes because the direction of curvature is orthogonal to the tangent vector.

Now we compute the curvature integral. With ϕ the angle around the symmetry axis, the curvature integral is

$$\begin{aligned}\int_{T^2} \lambda_1\lambda_2\, dA &= \int_0^{2\pi} d\theta \int_0^{2\pi} (2 - \cos\theta)\, d\phi\, \lambda_1\lambda_2 \\ &= \int_0^{2\pi} d\theta \int_0^{2\pi} d\phi\, (-\cos\theta) = 0.\end{aligned} \tag{1.7}$$

Again this zero answer is generic to any surface that can be smoothly deformed to the torus. The general result (the Gauss–Bonnet formula) of which the above are examples is

$$\int_S \lambda_1\lambda_2\, dA = 2\pi\chi = 2\pi(2 - g), \tag{1.8}$$

where χ is a topological invariant known as the Euler characteristic and g is the genus, essentially the number of 'holes' in the surface.[1] For a compact manifold with boundaries, the Euler characteristic becomes $2 - 2g - b$, where b is the number of

[1] A good question is why we write the Euler characteristic as $2 - 2g$ rather than $1 - g$; one way to motivate this is by considering polygonal approximations to the surface. The discrete Euler characteristic $V - E + F$, where V, E, and F count vertices, edges, and faces, respectively, is equal to χ. For example, the five Platonic solids all have $V - E + F = 2$.

boundaries: one can check this by noting that by cutting a torus, one can produce two discs (by slicing a bagel) or alternatively a cylinder with two boundaries (by slicing a Bundt cake).

More generally, we will encounter several examples where a topological invariant is expressed as an integral over a local quantity with a geometric significance. We now turn to a simpler example in order to allow us to introduce some basic concepts of algebraic topology.

1.2.1.2 *Invariant integrals along paths in two dimensions: exact forms*

As our first example of a topological property, let's ask about making line integrals along paths (not path integrals in the physics sense, where the path itself is integrated over[2]) that are nearly independent of the precise path: they will turn out to depend in some cases on topological properties (homotopy or cohomology). We will assume throughout these notes, unless otherwise specified, that all functions are smooth (i.e. $\mathbb{C}^\infty$, meaning derivatives of all orders exist).

First, suppose that we deal with paths on some open set U in the two-dimensional plane $\mathbb{R}^2$. (An open set is one for which some neighbourhood of each point in the set is also in the set.) We consider a smooth path $(u(t), v(t))$, where $0 \leq t \leq 1$ and the endpoints may be different[3]. Now let $f(x, y) = (p(x, y), q(x, y))$ be a two-dimensional vector field that lets us compute line integrals of this path:

$$W = \int_0^1 dt\, p\frac{du}{dt} + q\frac{dv}{dt}\, dt, \tag{1.9}$$

where p and q are evaluated at $(x(t), y(t))$.

Mathematical note In more fancy language, f is a differential form, a '1-form' to be precise. All that this means is that f is something we can use to form integrals over paths that are linear and probe the tangent vector of the path. Another way to state this, with which you may be more familiar, is that the tangent vector to a path, which we call a 'vector', transforms naturally in the opposite way to the gradient of a function, which we call a 'covector'. To convince yourself that this is true, think about how both transform under a linear transformation on the underlying space. We will say a bit more about such forms in a moment.

Our first goal is to show that the following three statements are equivalent:

(a) W depends only on the endpoints $(u(0), v(0))$ and $(u(1), v(1))$;
(b) $W = 0$ for any closed path;
(c) f is the gradient of a function g: $(p, q) = (\partial_x g, \partial_y g)$.

[2] There are additional topological properties that emerge ('quantum topology') when integrals over paths and other structures are incorporated; actually the Chern-Simons field theory that we discuss in Section 1.4 was an important tool in the history of that field.

[3] To make these results more precise, we should provide for adding one path to another by requiring only piecewise-smooth paths and require that u and v be smooth in an open set including $t \in [0, 1]$. For additional rigour, see the first few chapters of Fulton's book on algebraic topology [13].

The formal language used for (c) is that f is an *exact form*: $f = dg$ is the differential of a 0-form (a smooth function) g.

Note that (c) obviously implies (a) and (b), since then $W = g(u(1), v(1)) - g(u(0), v(0))$. To show that (b) implies (a), suppose (b) is true and (a) is not. Then there are two paths γ_1, γ_2 that have different integrals but the same endpoints. Form a new path γ so that, as t goes from 0 to $\frac{1}{2}$, γ_1 is traced, and then as t goes from $\frac{1}{2}$ to 1, γ_2 is traced opposite to its original direction (now you can see why piecewise-smooth paths are needed if one wants to be rigorous). Then this integral is non-zero, which contradicts (b).

It remains to show that (a) implies (c). Define $g(x, y)$ as equal to 0 at $(0, 0)$, or some other reference point in U if U does not include the origin. Everywhere else, set g equal to the W obtained by integrating over an arbitrary path from $(0, 0)$ to the final point, which by (a) is path-independent. (If U is not connected, then carry out this process on each connected component.) We will show that $\partial_x g = p$, and the same logic then implies $\partial_y g = q$. We need to compute

$$\partial_x g = \lim_{\Delta x \to 0} \frac{g(x + \Delta x, y) - g(x, y)}{\Delta x}. \tag{1.10}$$

We can obtain g by any path we like, so let's take an arbitrary path to define $g(x, y)$, then add a short horizontal segment to that path to define the path for $g(x + \Delta x, y)$. The value of the integral along this extra horizontal segment converges to $p(x, y)(\Delta x)$, as needed.

It turns out that the above case is simple because the plane we started with is 'topologically trivial'. Before proceeding to look at a non-trivial example, let us state one requirement on f that is satisfied whenever f is exact ($f = dg$). The fact that partial derivatives commute means that, with $f = dg = (p, q)$, $\partial_y p = \partial_x q$. We can come up with an elegant notation for this property by expanding our knowledge of differential forms.

Before, we obtained a 1-form f as the differential of a scalar g by defining

$$f = dg = \partial_x g \, dx + \partial_y g \, dy. \tag{1.11}$$

Note that we now include the differential elements dx, dy in the definition of f and that 1-forms form a real vector space (spanned by dx, dy): we can add them and multiply them by scalars. To obtain a 2-form as the differential of a 1-form, we repeat the process: writing $f = f_i \, dx_i$ (with $x_1 = x, x_2 = y, f_1 = p, f_2 = q$),

$$df = \sum_j \frac{\partial f_i}{\partial x_j} dx_j \wedge dx_i. \tag{1.12}$$

where the $\wedge$ product between differential forms satisfies the rule $dx_i \wedge dx_j = -dx_j \wedge dx_i$, which implies that if any coordinate appears twice, then we get zero: $dx \wedge dx = 0$. For some intuition about why this anticommutation property is important, note that in our 2D example,

$$df = (\partial_x f_y - \partial_y f_x) \, dx \wedge dy, \tag{1.13}$$

so that the function appearing in df is just the curl of the 2D vector field represented by f. So our statement about partial derivatives commuting is just the statement that if $f = dg$, then $df = 0$, or that the curl of a gradient is zero. We label any 1-form satisfying $df = 0$ a *closed form.* While every exact form is also closed, we will see that not every closed form is exact, with profound consequences.

1.2.1.3 Topologically invariant integrals along paths: closed forms

As an example of non-trivial topology, we would now like to come up with an example where integrals over paths are only path-independent in a limited 'topological' sense: the integral is the same for any two paths that are *homotopic*, one of the fundamental concepts of topology (to be defined in a moment). Basically, two paths are homotopic if one can be smoothly deformed into another. Consider the vector field

$$f = (p, q) = \left(-\frac{y}{x^2 + y^2}, \frac{x}{x^2 + y^2} \right) = \frac{-y\,dx + x\,dy}{x^2 + y^2}, \tag{1.14}$$

where in the second step we have used our 1-form notation. This vector field is well defined everywhere except the origin. This 1-form looks locally like the differential of $g = \tan^{-1}(y/x)$ (which just measures the angle in polar coordinates), but that function can only be defined smoothly on some open sets. For example, in a disc around the origin, the 2π ambiguity of the inverse tangent prevents us from defining g globally.

So, if we have a path that lies entirely in a region where g can be defined, then the integral of this 1-form over the path will give the change in angle between the starting point and end point, $g(u(1), v(1)) - g(u(0), v(0))$. What about other types of paths, for example, paths in $\mathbb{R}^2 \setminus (0, 0)$, the 2D plane with the origin omitted, that circle the origin and return to the starting point? We can still integrate using the 1-form f, even if it is not the gradient of a scalar function g, and will obtain the value $2\pi n$, where n is the 'winding number', a signed integer that describes how many times the closed path $(u(t), v(t))$ circles the origin as t goes from 0 to 1.

Now this winding number does not change as we make a small change in the closed path, as long as the path remains in $\mathbb{R}^2 \setminus (0, 0)$. What mathematical property of f guarantees this? We have seen that any exact 1-form (the differential of a scalar function) is also closed. While f is not exact, we can see that it is closed:

$$\begin{aligned} df &= \left(\partial_x \frac{x}{x^2 + y^2} \right) dx \wedge dy + \left(\partial_y \frac{-y}{x^2 + y^2} \right) dy \wedge dx \\ &= \frac{2 - 2}{x^2 + y^2} dx \wedge dy = 0. \end{aligned} \tag{1.15}$$

In other words, $(-y, x)/(x^2 + y^2)$ is curl-free ('irrotational'), while $(-y, x)$ has constant non-zero curl. Now suppose that we are given two paths γ_1 and γ_2 that differ by going in different ways around some small patch dA in which the 1-form remains defined. The difference in the integral of f over these two paths is then the integral of df over the enclosed surface by Stokes's theorem, which is zero if f is a closed form.

So we conclude that if f is a closed form, then the path integral is path-independent if we move the path through a region where f is always defined. For an exact form, the integral is completely path-independent. In the case of $\mathbb{R}^2 \setminus (0,0)$, the 1-form in (1.14) is locally but not completely path-independent. The set of closed forms and the set of exact forms are both vector spaces (we can add and multiply by scalars), which are typically infinite-dimensional, although their quotient as vector spaces is typically finite-dimensional. (The quotient of a vector space A by a vector space B is the vector space that identifies any two elements of A that differ only by an element of B.) A basic object in 'cohomology' is the first de Rham cohomology group (a vector space is by definition a group under addition):

$$H^1(M) = \frac{\text{closed 1-forms on } M}{\text{exact 1-forms on } M} = \frac{Z^1(M)}{B^1(M)}. \tag{1.16}$$

If you wonder why the prefix 'co-' appears in 'cohomology', there is a dual theory of linear combinations of curves, etc., called homology, in which the differential operator in de Rham cohomology is replaced by the boundary operator. However, while it may be equally more basic mathematically, homology seems to crop up less frequently in physics.

In this introductory discussion, we will focus on cohomology with real coefficients. The first and second Chern numbers defined later and applied to topological phases are related to elements of the even cohomology groups with *integer* coefficients $H^{2k}(M, \mathbb{Z})$. An even simpler object is the zeroth de Rham cohomology group. To understand this, realize that a closed 0-form is one whose gradient is zero, i.e. one that is constant on each connected component of U. There are no (-1)-forms and hence no exact 0-forms. So the zeroth group is just $\mathbb{R}^n$, where n is the number of connected components.

We can show that $H^1 = \mathbb{R}$ for the unit circle S^1 using the angle form f in (1.14), by showing that this form (more precisely, its equivalence class up to exact forms) provides a basis for H^1. Given some other form f', we use the unit-circle path, parametrized by an angle θ going from zero to 2π, to define

$$c = \frac{\displaystyle\int_0^{2\pi} f'}{\displaystyle\int_0^{2\pi} f}. \tag{1.17}$$

Now $f' - cf$ integrates to zero. We can define a function g via

$$g(\theta) = \int_0^{\theta} (f' - cf). \tag{1.18}$$

Now g is well defined and periodic because of how we defined c, and $f' = cf + dg$, which means that f' and cf are in the same equivalence class, because dg is an exact form. We say that f' and f are cohomologous because they differ by an exact form. So cf, $c \in \mathbb{R}$, generates H^1, and $H^1(S^1)$ is isomorphic to $\mathbb{R}$. With a little more work, one can show that $\mathbb{R}^2 \setminus (0,0)$ also has $H^1 = \mathbb{R}$.

Actually we can connect the results of this subsection to those of the previous one: a general expression for the Euler characteristic is

$$\chi(M) = \sum_i (-1)^i \dim H^i(M) = \sum_i (-1)^i \dim \frac{Z^i(M)}{B^i(M)}. \tag{1.19}$$

The dimension of the ith cohomology group is called the ith Betti number. To be pedantic, the Betti numbers are defined for homology rather than cohomology, but we can use a duality relationship. There is a compact way to express the idea of cohomology and homology that will let us introduce some notation and terminology. If Ω_r is the vector space of r-forms, and C_r is the dual space of r-chains, then the action of the boundary operator and the differential is as follows:

$$\longleftarrow C_r \underset{[\partial_{r+1}]}{\longleftarrow} C_{r+1} \underset{[\partial_{r+2}]}{\longleftarrow} C_{r+2} \longleftarrow, \tag{1.20}$$

$$\longrightarrow \Omega_r \underset{[d_{r+1}]}{\longrightarrow} \Omega_{r+1} \underset{[d_{r+2}]}{\longrightarrow} \Omega_{r+2} \longrightarrow. \tag{1.21}$$

The rth cohomology group is the quotient $\ker d_{r+1}/\operatorname{im} d_r$, and the rth homology group is $\ker \partial_r/\operatorname{im} \partial_{r+1}$.

The duality relationship is provided by Stokes's theorem. Recall that this theorem relates the integral of a form over a boundary to the integral of the differential of the form over the interior. In terms of the linear operator (f, c) that evaluates the form f on the chain c, we have the compact expression

$$(f, \partial c) = (df, c). \tag{1.22}$$

Now we move on to a different type of topology that is perhaps more intuitive and will be useful for our first physics challenge: how to classify defects in ordered systems.

1.2.1.4 Homotopy

What if we did not want to deal with smooth functions and calculus? An even more basic type of topology is homotopy theory, which can be defined without reference to calculus, differential forms, etc. Suppose that we are given a continuous map from $[0, 1]$ to a manifold M such that 0 and 1 get mapped to the same point; we can think of this as a closed curve on M. We say that two such curves γ_1, γ_2 are homotopic if there is a continuous function (a homotopy) f from $[0, 1] \times [0, 1]$ to M that satisfies

$$f(x, 0) = \gamma_1(x), \qquad f(x, 1) = \gamma_2(x). \tag{1.23}$$

Intuitively, f describes how to smoothly distort γ_1 to γ_2. Now, homotopy is an equivalence relation and hence defines equivalence classes: $[\gamma_1]$ is the set of all paths homotopic to γ_1. Furthermore, concatenation of paths (i.e. tracing one after the other) defines a natural group structure on these equivalence classes: the inverse of any path can be obtained by tracing it in the opposite direction.[4] We conclude that the equivalence classes of closed paths form a group $\pi_1(M)$, called the fundamental group or first homotopy group. Higher homotopy groups $\pi_n(M)$ are obtained by considering mappings from the n-sphere S^n to M in the same way.

The homotopy groups of a manifold are not independent of the cohomology groups: for example, if $\pi_1(M)$ is trivial, then so is the first de Rham group. The cohomology groups are always Abelian; in general, the first de Rham group *with integer coefficients* is the Abelianization of π_1 (which need not be Abelian, although higher homotopy groups are). If you are interested in further details, the result of Hurewicz gives a relationship between higher cohomology and homotopy groups. The examples above of $\mathbb{R}^2 \setminus (0,0)$ and S^1 both have $\pi_1(M) = \mathbb{Z}$: there is an integer-valued winding number that we can use to classify paths, and this winding number can be computed by the angle form given above. So our 2D examples already contains the two types of topology that occur most frequently in physics: cohomology and homotopy. We will return to homotopy in much more detail in a moment, when we explain how it can be used to classify topological defects such as vortices in broken-symmetry phases.

1.2.1.5 Application of homotopy to topological defects in symmetry-breaking phases

As a direct physical application of homotopy theory, consider the notion of a 'vortex' in an ordered phase such as a superfluid. Such a configuration has a core where there is no order, but far away from the core the system is always locally in an ordered state. However, *which* ordered state the system is in varies smoothly as we move around the vortex core. For a 2D defect with a point core, such as a vortex of the 2D XY model, the vortex core is enclosed by a large real-space circle S^1, and as we move around this circle we have a map from S^1 to S^1, where the first circle is real space and the second circle reflects that the 'order-parameter manifold' of distinct ordered configurations of the XY model is also a circle.

The mathematical classification of topological defects has been carried out for a variety of systems. Vortex-like defects (defects that can be circled by a loop) are related to the group $\pi_1(M)$, where M is the manifold of degenerate values of the order parameter once its magnitude has been set (for example, S^1 for XY and S^2 for the Heisenberg model, where S^d is the unit sphere in $d+1$ dimensions). $\pi_1(M)$ is known as the first homotopy group and is the group of equivalence classes of mappings from S^1 to M: for example, the mappings from S^1 to S^1 are characterized by an integer winding number $n \in \mathbb{Z}$, so $\pi_1(S^1) = \mathbb{Z}$, while $\pi_1(S^2) = 0$ (the group with one element), since any loop on the sphere is contractible to a point.

In other words, $\pi_1(M)$ gives the set of equivalence classes up to smooth deformations of closed paths on M. Multiplication of equivalence classes in the group is defined by concatenation of paths. The second homotopy group $\pi_2(M)$ classifies mappings from S^2 to M, and describes defects circled by a sphere, such as pointlike defects

[4] To be precise, one should define homotopy with reference to a particular point where paths start and end; for a symmetric space where all points are basically equivalent, this is unnecessary.

in 3D. For example, $\pi_2(S^2)$ is non-zero, and there are stable point defect configurations of Heisenberg spins (known descriptively as 'hedgehogs') but not of XY spins. There can also be topological configurations that are not 'defects' but are not homotopic to the identity: the most famous example is the skyrmion configuration of a uniaxial ferromagnet in 2D, where all spins at infinity point in the same direction and the spin direction moves in the plane in such a way as to cover the sphere once. Shankar's monopoles and other defect-free configurations in 3D are related to the group π_3.

There is a considerable technology built up for the calculation of homotopy groups of general order-parameter manifolds $M = G/H$, whose elements are cosets of the residual symmetry group H, i.e. any symmetries that survive in the ordered phase, in the high-temperature symmetry group G. For example, for a uniaxially ordered Heisenberg ferromagnet, $G = SO(2)$ and $H = SO(3)$, so indeed $M = S^2$ as anticipated earlier. The advent of complicated ordered states in systems such as liquid crystals and spinor condensates stimulated the development of the techniques described in the *Review of Modern Physics* article by Mermin [25].

1.2.2 Berry phases in quantum mechanics

We now turn to a beautiful geometric property of quantum mechanics that was understood relatively recently: the geometric or Berry phase. The connection to the Gauss–Bonnet theorem we mentioned earlier is as follows. Curvature is a property of Riemannian manifolds, which have a (real) inner product defined on the tangent space at each point.[5] This inner product varies smoothly from point to point, which allows us to define a number of important concepts, including parallel transport and curvature.

Frequently in quantum mechanics, we have, instead of a tangent space, a Hilbert space (including a Hermitian inner product) that varies smoothly from point to point in parameter space. Hence one can think of the Berry-phase objects we are about to define as really quite similar to curvature and related properties on Riemannian manifolds, except that the Berry phase does not come from the intrinsic geometry of the manifold of parameters but rather is related to how the attached Hilbert space evolves as parameters change.

An important result from undergraduate quantum mechanics is the 'adiabatic approximation'. Suppose that a system is prepared in a non-degenerate eigenstate of a time-dependent Hamiltonian H. For later reference, we will write H as a function of some parameters λ_i that depend on time: $H(t) = H(\lambda_1(t), \lambda_2(t), \ldots)$. If the eigenstate remains non-degenerate, then the adiabatic approximation is the result that if the Hamiltonian changes slowly in time (how slowly depends primarily on the energy gap between adjacent eigenstates), then there are no transitions between eigenstates.

This approximation, even when correct, only gives part of the story: it describes the probability to remain in the eigenstate that evolved from the initial eigenstate, but

[5] The combination of a differentiable manifold and its tangent space at each point is the 'tangent bundle', the simplest example of a vector bundle, an attachment of a vector space to each point of a manifold.

there is actually non-trivial information in the *phase* of the final state as well. This result may seem quite surprising because the overall phase in quantum mechanics is in general independent of observable quantities. However, the Berry phase from an adiabatic evolution is observable: for example, one system can be taken around a closed path in parameter space, while another system initially identical to the first can be taken around a different path, or the null path; an interference experiment on the final states will reveal the Berry phase. The first example of this type of geometric phase in physics was found more than 50 years ago by Pancharatnam in an optical example, but Berry's classic paper of 1984 was the first to explain the concept in its full generality.

Berry's result for a closed path is relatively simple to state, but some careful thought is required to understand and derive it. In moving a system adiabatically around a closed path in parameter space, the final wavefunction is in the same eigenstate as the initial one (again, under the assumptions of the adiabatic approximation as stated above), but its phase has changed:

$$|\psi(t_f)\rangle = \exp\left[-\frac{i}{\hbar}\oint_{t_i}^{t_f} E(t')\,dt'\right] e^{i\gamma}|\psi(t_i)\rangle. \tag{1.24}$$

Here $E(t')$ means the corresponding eigenvalue of the Hamiltonian at that time, and γ is the Berry phase, expressed as an integral over a path in *parameter* space with no time dependence:

$$\gamma = i\int \langle\tilde{\psi}(\lambda_i)|\nabla_{\boldsymbol{\lambda}}|\tilde{\psi}(\lambda_i)\rangle \cdot d\boldsymbol{\lambda} \tag{1.25}$$

Note that there are two different wavefunctions ψ and $\tilde{\psi}$ in the above formulas. $\psi(t)$ has a time argument and means the wavefunction of the system at that time. The 'reference wavefunction' $\tilde{\psi}(\lambda_i)$ has a parameter argument and indicates the wavefunction we have chosen of the Hamiltonian for that value of the parameters, which we assume to be smoothly varying[6] A key assumption of the following derivation is that there is some smooth choice of the $\tilde{\psi}(\lambda_i)$ throughout a surface in parameter space with the loop as boundary.

For an open path, we need to describe the phase of the wavefunction relative to this reference set, so the expression becomes more complicated (for the closed path, we could simply compare the initial and final wavefunctions, without needing the reference set at these points). We will show that, assuming $\psi(t_i) = \tilde{\psi}(\lambda(t_i))$ so that

[6] A smooth choice of reference wavefunctions is always possible locally but not possible globally, as in the example of a spin-$\frac{1}{2}$ particle moving in a Zeeman magnetic field. Computing the Berry phase in this example is a nice exercise for the reader, and also physically useful: in making a path integral for a quantum spin, one needs to include a term that reflects the Berry phase of the path in time (either real or imaginary), and we will use this in Section 1.4. A pedagogical derivation of this path integral including the Berry phase is in the book of Auerbach [2].

the initial wavefunction is equal to the reference state at the corresponding value of parameters,

$$\langle\tilde{\psi}(\lambda_i(t))|\psi(t)\rangle = \exp\left[-\frac{i}{\hbar}\int_0^t E(t')\,dt'\right]e^{i\gamma} \equiv e^{i\theta(t)}, \tag{1.26}$$

i.e. the Berry phase appears as an extra contribution, beyond the expected contribution related to the energy, when comparing the actual time-dependent evolved state $\psi(t)$ with the reference state at the same point in parameter space $\lambda_i(t)$. We write $\theta(t)$ for the total phase including both energetic and Berry contributions. We can take the time derivative using the time-dependent Schrödinger equation

$$i\hbar\frac{\partial\psi}{\partial t} = H(t)\psi. \tag{1.27}$$

The first two quantities in (1.26) agree initially from our choice of the phase of $\psi(t_i)$. The time derivative of the leftmost is

$$\langle\tilde{\psi}(\lambda_i(t))|\,\frac{-iE(t)}{\hbar}\,|\psi(t)\rangle + \frac{d\lambda_j}{dt}\langle\partial_{\lambda_j}\tilde{\psi}(\lambda_i(t))|\psi(t)\rangle, \tag{1.28}$$

Since $e^{i\theta(t)} = \langle\psi(\lambda_i(t))|\psi(t)\rangle$, this gives

$$\begin{aligned} i\partial_t\theta(t) &= i\left(\frac{d}{dt}e^{i\theta(t)}\right)(-ie^{-i\theta(t)}) \\ &= \left[\frac{-iE(t)}{\hbar}\langle\tilde{\psi}(\lambda_i(t))| + \frac{d\lambda_j}{dt}\langle\partial_{\lambda_j}|\tilde{\psi}(\lambda_i(t))|\right]|\psi(t)\rangle\langle\psi(t)|\tilde{\psi}(\lambda_i(t))\rangle, \end{aligned} \tag{1.29}$$

and this is satisfied if we set (note that for E we do not need to distinguish between time and λ dependence)

$$\partial_t\theta(t) = -\frac{E(t)}{\hbar} - i\frac{d\lambda_j}{dt}\langle\partial_{\lambda_j}\tilde{\psi}(\lambda_i(t))|\tilde{\psi}(\lambda_i(t))\rangle, \tag{1.30}$$

which is our desired conclusion. We have used the fact that ψ and $\tilde{\psi}$ differ only by a phase factor, since they describe the same non-degenerate state, to eliminate $|\psi\rangle\langle\psi|$.

The 'Berry connection' or 'Berry vector potential' $A_j = i\langle\psi(\lambda_i)|\partial_{\lambda_j}\psi(\lambda_i)\rangle$ is real, which follows from noting that $\partial_{\lambda_j}\langle\tilde{\psi}((\lambda_i)|\tilde{\psi}(\lambda_i)\rangle = 0$ by constancy of normalization. It is required for a non-zero Berry phase that H evolve in such a way that the wavefunction changes by more than just a phase, so that that the evolution of the wavefunction is more than just a simple phase factor, even though the actual rate of change in H drops out and only the path taken by H enters the Berry phase.

Now one can ask whether the Berry connection $\mathbf{A}$ is independent of how we chose the reference wavefunctions (in this case, the $U(1)$ degree of freedom in the wavefunction at each λ). While for an open path the Berry phase is clearly not phase-independent, it is so for a closed path—for exactly the same reasons as a closed line integral of $\mathbf{A}$ is gauge-independent in electrodynamics: we can integrate the 'Berry flux' or 'Berry curvature' $\epsilon_{ij}\partial_i A_j$ (which you can check is phase-independent, just like $F_{\mu\nu}$ in electrodynamics) on the surface bounded by the path. Alternatively, we can note that a phase change changes A by the gradient of a scalar, so that on a closed loop there is no change.

Independent of Berry's work and at roughly the same time, condensed matter physicists such as Thouless were realizing that Berry phases of wavefunctions on the Brillouin zone have the same mathematical structure as gauge fields in parameter space, even though there is no longer a notion of time evolution. The Berry vector potential $\mathbf{A}$ is a way to compare or 'connect' the Hilbert spaces at neighbouring points in parameter space. The gauge-invariant or nearly gauge-invariant quantities constructed from $\mathbf{A}$ and its derivatives control a variety of physical quantities. For the specific case of wavefunctions on the Brillouin zone, we will see that $\mathbf{A}$ is intimately related to the location of the wavefunctions within the unit cell in real space.

To get some geometric intuition for what the Berry phase means in general, we explain why the Berry connection A is called a connection, and the flux F is sometimes called a curvature. A connection is a way to compare vector spaces that are attached to different points of a manifold, forming a 'vector bundle'. In our case, there is a one-dimensional complex vector space attached at each point in parameter space, spanned by the local eigenstate. The inner product lets us compare vectors at the same point in parameter space, but the Berry connection appears when we try to compare two vectors from slightly different points.

An example we used above of a real vector bundle is the 'tangent bundle' to a Riemannian manifold (say, a sphere), made up of tangent vectors at each point, which have a dot product corresponding to the inner product in quantum mechanics. The connection in this case, which gives rise to 'parallel transport' of tangent vectors, is related to the same curvature that we previously discussed with the Gauss–Bonnet theorem. Consider an aeroplane moving around the surface of the Earth and carrying a gyroscope that is fixed to lie in the tangent plane to the Earth's surface (i.e. free to rotate around the normal axis to the tangent plane). If the aeroplane follows a great circle, then it will appear to be going straight ahead to a passenger on board, and the gyroscope will not rotate relative to the plane's axis.

However, if the aeroplane follows a line of latitude other than the equator, or any other path that is not a 'geodesic' (see a differential geometry text for details), it will feel constantly as though it is turning, and the gyroscope will appear to rotate relative to the aeroplane's direction. After going around a closed path in the aeroplane, the gyroscope may have rotated compared with a stationary gyroscope (the same physics that underlies Foucault's pendulum). As an exercise, you can work out that the total angle of rotation in circling a line of latitude is $2\pi\sin(\phi)$, where ϕ is the latitude. At the equator this gives no rotation, while at the north pole it gives a 2π rotation. This is a geometrical version of the same idea of holonomy (failure of a gyroscope to return to its initial direction) that underlies the Berry phase.

Note that a vector potential in a gauge theory and the associated Wilson loop are also examples of the concept of holonomy in a (now complex) vector bundle. The $U(1)$ Berry phase described above generalizes immediately to a non-Abelian $U(N)$ Berry phase when there are degenerate states or the energy differences between states are irrelevant, which has some important applications in condensed matter physics that have only recently been discovered. Our primary mathematical objects later in this chapter will be properties of the wavefunctions on the Brillouin zone, which form a Hermitian bundle (a smoothly varying Hilbert space) on the d-dimensional torus.

One reason for introducing the idea of cohomology above was to give a sense of the mathematical structures hiding in the background of the simple calculations we do: to pick one example, the integral physicists do to calculate the Chern number, which determines the contribution of a filled 2D band to the quantum Hall effect, would be viewed by a mathematician as using the first Chern form to classify smooth complex line bundles on the Brillouin zone, and the group of line bundles under tensor products is isomorphic to the second cohomology class with integer coefficients. However, our hope is that the physical examples we discuss will be readily comprehensible even for readers not terribly excited about algebraic technology.

1.3 Topological phases: Thouless phases arising from Berry phases

The integer quantum Hall effect (IQHE) has the remarkable property that, even at finite temperature in a disordered material, a transport quantity is quantized to remarkable precision: the transverse (a.k.a. Hall) conductivity is $\sigma_{xy} = ne^2/h$, where n is integral to 1 part in 10^9. This quantization results because the transport is determined by a topological invariant, as stated most clearly in the work of Thouless and collaborators. Consequently, we use the term 'Thouless phases' for phases where a response function is determined by a topological invariant.

In the cases we discuss, including the recently discovered 'topological insulators' and quantum spin Hall effect (QSHE), this topological invariant results from integration of an underlying Berry phase. It turns out that the Berry phase can be rather important even when it is not part of a topological invariant. In crystalline solids, the electrical polarization, the anomalous Hall effect, and the magnetoelectric polarizability all derive from Berry phases of the Bloch electron states, which are introduced in the following subsection. We will avoid the conventional textbook presentation of the IQHE in terms of Landau levels of a continuum electron. As we will use Landau levels when we discuss the fractional quantum Hall effect later, readers who are unfamiliar with the IQHE may wish to learn the standard treatment (see e.g. [14]) and compare it with the approach using Bloch electrons below. The connection between the two can be made precise in the limit of small flux per unit cell, when a flat magnetic Bloch band becomes equivalent to a Landau level.

1.3.1 Bloch states

One of the cornerstones of the theory of crystalline solids is Bloch's theorem for electrons in a periodic potential. We will demonstrate this in the following form: given a potential invariant under a set of lattice vectors $\mathbf{R}$, $V(\mathbf{r}+\mathbf{R}) = V(\mathbf{r})$, the electronic eigenstates can be labelled by a 'crystal momentum' $\mathbf{k}$ and written in the form

$$\psi_{\mathbf{k}}(\mathbf{r}) = e^{i\mathbf{k}\cdot\mathbf{r}} u_{\mathbf{k}}(\mathbf{r}), \tag{1.31}$$

where the function u has the periodicity of the lattice. Note that the crystal momentum $\mathbf{k}$ is only defined up to addition of reciprocal lattice vectors, i.e. vectors whose dot product with any of the original lattice vectors is a multiple of 2π.

We give a quick proof of Bloch's theorem in one spatial dimension, then consider the Berry phase of the resulting wavefunctions. A standard fact from quantum mechanics tells us that, given two Hermitian operators that commute, we can find a basis of simultaneous wavefunctions. In the problem at hand, we have a non-Hermitian operator (lattice translations by the lattice spacing a: $(T\psi)(x) = \psi(x+a)$) that commutes with the Hamiltonian. It turns out that only one of the two operators needs to be Hermitian for simultaneous eigenstates to exist, and therefore we can find wavefunctions that are energy eigenstates and satisfy

$$(T\psi)(x) = \lambda\psi(x). \tag{1.32}$$

Now if the magnitude of λ is not 1, repeated application of this formula will give a wavefunction that either blows up at spatial positive infinity or negative infinity. We would like to find wavefunctions that can extend throughout an infinite solid with bounded probability density, and hence require $|\lambda| = 1$. From this, it follows that $\lambda = e^{i\theta}$, and we define $k = \theta/a$, where we need to specify an interval of width 2π to uniquely define θ, say $[-\pi, \pi)$. In other words, k is ambiguous by addition of a multiple of $2\pi/a$, as expected. So we have shown

$$\psi_k(x+a) = e^{ika}\psi_k(x). \tag{1.33}$$

The last step is to define $u_k(x) = \psi_k(x)e^{-ikx}$; then (1.33) shows that u_k is periodic with period a, and $\psi_k(x) = e^{ikx}u_k(x)$. [7]

While the energetics of Bloch wavefunctions underlies many properties of solids, there is also Berry-phase physics arising from the dependence of u_k on k that was understood only rather recently. Note that, even though this is one-dimensional, there is a non-trivial 'closed loop' in the parameter k that can be defined because of the periodicity of the 'Brillouin zone' $k \in [-\pi/a, \pi/a)$:

$$\gamma = \oint_{-\pi/a}^{\pi/a} \langle u_k | i\partial_k | u_k \rangle \, dk. \tag{1.34}$$

How are we to interpret this Berry phase physically, and is it even gauge-invariant? We will derive it from scratch below, but an intuitive clue is provided if we make the replacement $i\partial_k$ by x, as would be appropriate if we consider the action on a plane wave. This suggests, correctly, that the Berry phase may have something to do with the spatial location of the electrons, but evaluating the position operator in a Bloch state gives an ill-defined answer; for this real-space approach to work, we would need to introduce localized 'Wannier orbitals' in place of the extended Bloch states.

[7] Readers interested in more information and the 3D case can consult any solid state physics text, e.g. Ashcroft and Mermin [1].

Another clue to what the phase γ might mean physically is provided by asking if it is gauge-invariant. Before, gauge-invariance resulted from assuming that the wavefunction could be continuously defined on the interior of the closed path. Here we have a closed path on a noncontractible manifold; the path in the integral winds around the Brillouin zone, which has the topology of the circle. What happens to the Berry phase if we introduce a phase change $\phi(k)$ in the wavefunctions, $|u_k\rangle \to e^{-i\phi(k)}|u_k\rangle$, with $\phi(\pi/a) = \phi(-\pi/a) + 2\pi n, n \in \mathbb{Z}$? Under this transformation, the integral shifts as

$$\gamma \to \gamma + \oint_{-\pi/a}^{\pi/a} (\partial_k \phi)\, dk = \gamma + 2\pi n. \tag{1.35}$$

So, redefinition of the wavefunctions shifts the Berry phase. This corresponds to changing the polarization by a multiple of the 'polarization quantum', which in one dimension is just the electron charge. (In higher dimensions, the polarization quantum is one electron charge per transverse unit-cell area.) Physically, the ambiguity of polarization corresponds to the following idea: given a system with a certain bulk unit cell, there is an ambiguity in how that system is terminated and how much surface charge is at the boundary; adding an integer number of charges to one allowed termination gives another allowed termination [37]. The Berry phase is not gauge-invariant, but any fractional part it had in units of a *is* gauge-invariant. However, the above calculation suggests that, to obtain a gauge-invariant quantity, we need to consider a two-dimensional crystal rather than a one-dimensional one. Then integrating the Berry curvature, rather than the Berry connection, has to give a well-defined gauge-invariant quantity.

We will give a physical interpretation of γ in the next section as a 1D polarization by relating changes in γ to electrical currents. (A generalization of this Berry phase is remarkably useful for the theory of polarization in real, 3D materials.) In the next section, we will understand how this 1D example is related to the 2D IQHE. Historically, the understanding of Berry phases in the latter came first, in a paper by Thouless, Kohmoto, den Nijs, and Nightingale [39]. They found that when a lattice is put in a commensurate magnetic field (one with rational flux per unit cell, in units of the flux quantum so that Bloch's theorem applies), each occupied band j contributes an integer

$$n_j = \frac{i}{2\pi} \int dk_x\, dk_y\, \left(\langle \partial_{k_x} u_j | \partial_{k_y} u_j \rangle - \langle \partial_{k_y} u_j | \partial_{k_x} u_j \rangle \right) \tag{1.36}$$

to the total Hall conductance:

$$\sigma_{xy} = \frac{e^2}{h} \sum_j n_j. \tag{1.37}$$

Now we derive this topological quantity (the 'Chern number', expressed as an integral over the Berry flux, which is the curl of the Berry connection $A^j = i\langle u_j | \nabla_k u_j \rangle$) for the case of 1D polarization, then explain its mathematical significance.

1.3.2 1D polarization and 2D IQHE

We start with the question of 1D polarization mentioned earlier. More precisely, we attempt to compute the change in polarization by computing the integral of current through a bulk unit cell under an adiabatic change:

$$\Delta P = \int_0^1 d\lambda \, \frac{dP}{d\lambda} = \int_{t_0}^{t_1} dt \, \frac{dP}{d\lambda}\frac{d\lambda}{dt} = \int_{t_0}^{t_1} j(t)\, dt. \tag{1.38}$$

In writing this formula, we are assuming implicitly that there will be some definition of dP in terms of a parameter λ of the bulk Hamiltonian. Our treatment will follow that of Resta [37], but with a few more mathematical details in the derivation. (We write q for 1D momentum and k_x, k_y for 2D momenta in the following.) We will use Bloch's theorem in the following form: the periodic single-particle orbitals $u_n(q, r)$ are eigenstates of

$$H(q, \lambda) = \frac{1}{2m}(p + \hbar q)^2 + V^{(\lambda)}(r). \tag{1.39}$$

The current operator is

$$j(q) = ev(q) = \frac{ie}{\hbar}[H(q, \lambda), r] = \frac{e}{m}(p + \hbar q) = \frac{e}{\hbar}\partial_q H(q, \lambda). \tag{1.40}$$

The current at any fixed λ in the ground state is zero, but changing λ adiabatically in time drives a current that generates the change in polarization. To compute this current, we need to use the first correction to the adiabatic theorem (cf. the quantum mechanics book of Messiah [26]). Following Thouless, we choose locally a gauge in which the Berry phase is zero (this can only be done locally and is only meaningful if we obtain a gauge-invariant answer for the instantaneous current) and write for the many-body wavefunction

$$|\psi(t)\rangle = \exp\left[-\frac{i}{\hbar}\int^t E_0(t')\, dt'\right]\left[|\psi_0(t)\rangle + i\hbar \sum_{j\neq 0} |\psi_j(t)\rangle (E_j - E_0)^{-1} \langle \psi_j(t)|\dot{\psi}_0(t)\rangle\right]. \tag{1.41}$$

Here $E_i(t)$ are the local eigenvalues and $|\psi_j(t)\rangle$ a local basis of reference states. The first term is just the adiabatic expression we derived before, but with the Berry phase eliminated with a phase rotation to ensure $\langle \psi_0(t)|\dot{\psi}_0(t)\rangle = 0$.

We want to use the above expression to write the expectation value of the current. The ground state must differ from the excited state by a single action of the (one-body) current operator, which promotes one valence electron (i.e. an electron in an occupied state) to a conduction electron. Using the one-particle states, we get

$$\frac{dP}{d\lambda} = 2\hbar e \,\mathrm{Im} \sum_{v,c} \int \frac{dq}{2\pi} \frac{\langle u_v(q)|v(q)|u_c(q)\rangle\langle u_c(q)|\partial_\lambda u_v(q)\rangle}{E_c(q) - E_v(q)}. \tag{1.42}$$

For example, we wrote

$$\langle \psi_j(t)|\dot{\psi}_0(t)\rangle = \sum_{v,c} \langle u_c|\partial_\lambda u_v\rangle \frac{d\lambda}{dt}. \tag{1.43}$$

This sum involves both valence and conduction states. For simplicity, we assume a single valence state in the following. We can rewrite the sum simply in terms of the valence state using the first-order time-independent perturbation theory expression for the wavefunction change under a perturbation Hamiltonian $H' = dq\,\partial_q H$:

$$|\partial_q u_j(q)\rangle = \sum_{j\neq j'} |u_{j'}(q)\rangle \frac{\langle u_{j'}(q)|\partial_q H(q,\lambda)|u_j(q)\rangle}{E_j(q) - E_{j'}(q)}. \tag{1.44}$$

Using this and $v(q) = \hbar^{-1}\partial_q H(q,\lambda)$, we obtain

$$\begin{aligned}\frac{dP}{d\lambda} &= 2\hbar e \,\mathrm{Im} \sum_c \int \frac{dq}{2\pi} \frac{\langle u_v(q)|v(q)|u_c(q)\rangle\langle u_c(q)|\partial_\lambda u_v(q)\rangle}{E_c(q) - E_v(q)} \\ &= 2e\,\mathrm{Im} \int \frac{dq}{2\pi} \langle \partial_q u_v(q)|\partial_\lambda u_v(q)\rangle.\end{aligned} \tag{1.45}$$

We can convert this to a change in polarization under a finite change in parameter λ:

$$\Delta P = 2e\,\mathrm{Im} \int_0^1 d\lambda \int \frac{dq}{2\pi} \langle \partial_q u_v(q)|\partial_\lambda u_v(q)\rangle. \tag{1.46}$$

The last expression is in 2D and involves the same type of integrand (a Berry flux) as in the 2D TKNN formula (1.36). However, in the polarization case, there does not need to be any periodicity in the parameter λ. If this parameter is periodic, so that $\lambda = 0$ and $\lambda = 1$ describe the same system, then the total current run in a closed cycle that returns to the original Hamiltonian must be an integer number of charges, consistent with quantization of the TKNN integer in the IQHE.

If we define polarization via the Berry connection,

$$P = ie \int \frac{dq}{2\pi} \langle u_v(q)|\partial_q u_v(q)\rangle, \tag{1.47}$$

so that its derivative with respect to λ will give the result above with the Berry flux, we note that a change of gauge changes P by an integer multiple of the charge e. Only the fractional part of P is gauge-independent. Note that u and the Bloch Hamiltonian are not generally periodic with the Brillouin zone, even though the energy levels and Berry curvature are. This does not affect the calculation of topological invariants, but it does matter for evaluating the polarization via (1.47). The relationship between polarization in 1D, which has an integer ambiguity, and the IQHE in 2D, which has an integer quantization, is the simplest example of the relationship between Chern–Simons forms in odd dimension and Chern forms in even dimension. We will turn soon to the mathematical properties of these differential forms, which in the case above (and others to be discussed) came from the Berry phases of a band structure.

1.3.3 Interactions and disorder: the flux trick

One might worry whether the TKNN integer defined in (1.36) is specific to non-interacting electrons in perfect crystals. An elegant way to generalize the definition physically, while keeping the same mathematical structure, was developed by Niu, Thouless, and Wu [31]. This definition also makes somewhat clearer, together with our polarization calculation above, why this invariant should describe σ_{xy}. First, note that from the formula for the Bloch Hamiltonian in the polarization calculation above, we can reinterpret the crystal momentum q as a parameter describing a flux threaded through a unit cell of size a: the boundary conditions are periodic up to a phase $e^{iqa} = e^{ie\Phi/\hbar c}$. We will start by reinterpreting the non-interacting case in terms of such fluxes, then move to the interacting case.

The setup is loosely similar to the Laughlin argument for quantization in the IQHE. Consider adiabatically pumping a flux Φ_x though one circle of a toroidal system, in the direction associated with the periodicity $x \to x + L_x, y \to y$. The change in this flux in time generates an electric field pointing in the $\hat{\mathbf{x}}$ direction. Treating this flux as a parameter of the crystal Hamiltonian, we compute the resulting change in $\hat{\mathbf{y}}$ polarization, which is related to the y current density:

$$\frac{dP_y}{dt} = j_y = \frac{dP_y}{d\Phi_x}\frac{d\Phi_x}{dt} = \frac{dP_y}{d\Phi_x}(cE_x L_x). \tag{1.48}$$

We are going to treat the polarization P_y as an integral over y flux but keep Φ_x as a parameter. Then [32]

$$P_y(\Phi_x) = \frac{ie}{2\pi}\int d\Phi_y \, \langle u|\partial_{\Phi_y} u\rangle \tag{1.49}$$

and we see that polarization now has units of charge per length, as expected. In particular, the polarization quantum in the y direction is now one electronic charge per L_x. The last step to obtain the quantization is to assume that we are justified in averaging j_y over the flux:

$$\langle j_y\rangle = \left\langle \frac{dP_y}{d\Phi_x}\right\rangle (cE_x L_x) \to \frac{\Delta P_y}{\Delta\Phi_x}(cE_x L_x), \tag{1.50}$$

where Δ means the change over a single flux quantum, $\Delta\Phi_x = hc/e$. So the averaged current is determined by how many y-polarization quanta change in the periodic adiabatic process of increasing the x flux by hc/e:

$$\langle j_y\rangle = \frac{e}{hc}\frac{ne}{L_x}(cE_x L_x) = \frac{ne^2}{h}E_x. \tag{1.51}$$

The integer n follows from noting that computing $dP_y/d\Phi_x$ and then integrating $d\Phi_x$ gives just the expression for the TKNN integer (1.36), now in terms of fluxes.

1.3.4 TKNN integers, Chern numbers, and homotopy

In this subsection, we will give several different ways to understand the TKNN integer or Chern number described above. First, a useful trick for many purposes is to define the Berry flux and first Chern number in a manifestly gauge-invariant way, using projection operators. For the case of a single non-degenerate band, define $P_j = |u_j\rangle\langle u_j|$ at each point of the Brillouin zone. This projection operator is clearly invariant under $U(1)$ transformations of u_j. The Chern number can be obtained as

$$n_j = \frac{i}{2\pi}\int \mathrm{Tr}\left[dP_j \wedge P_j\, dP_j\right], \tag{1.52}$$

where $\wedge$ is the wedge product and $dP_j = \partial_{k_x} P_j\, dk_x + \partial_{k_y} P_j\, dk_y$ is a differential form where the coefficients are operators. (Note that the wedge product in the above formula acts only on dk_x and dk_y.) It is a straightforward exercise to verify that this reproduces the TKNN definition (1.36).

Then the generalization to degenerate bands, for example, is naturally studied by using the gauge- and basis-invariant projection operator $P_{ij} = |u_i\rangle\langle u_i| + |u_j\rangle\langle u_j|$ onto the subspace spanned by $|u_i\rangle$ and $|u_j\rangle$: the index of this operator gives the total Chern number of bands i and j. In general, when two bands come together, only their total Chern number is defined. The total Chern number of all bands in a finite-dimensional band structure (i.e. a finite number of bands) is argued to be zero below. Often one is interested in the total Chern number of all occupied bands because this describes the IQHE through the TKNN formula; because of this zero-sum rule, the total Chern number of all *unoccupied* bands must be equal and opposite.

In the remainder of this subsection, we use a powerful homotopy argument of Avron, Seiler, and Simon to show indirectly that there is one Chern number per band, but with a 'zero-sum rule' that all the Chern numbers add up to zero. We will not calculate the Chern number directly, but rather the homotopy groups of Bloch Hamiltonians. To get some intuition for the result, we first consider the example of a non-degenerate two-band band structure, then give the general result, which is an application of the 'exact sequence of a fibration'.

The Bloch Hamiltonian for a two-band non-degenerate band structure can be written in terms of the Pauli matrices and the two-by-two identity as

$$H(k_x, k_y) = a_0(k_x, k_y)\mathbf{1} + a_1(k_x, k_y)\sigma_x + a_2(k_x, k_y)\sigma_y + a_3(k_x, k_y)\sigma_z. \tag{1.53}$$

The non-degeneracy constraint is that a_1, a_2, and a_3 are not all simultaneously zero. Now we first argue that a_0 is only a shift in the energy levels and has no topological significance, i.e. it can be smoothly taken to zero without a phase transition. Similarly, we can deform the other a functions to describe a unit vector on $\mathbb{Z}_2$: just as the punctured plane $\mathbb{R}^2 \setminus (0,0)$ can be taken to the circle, we are taking punctured 3-space to the 2-sphere via

$$(a_1, a_2, a_3) \to \frac{(a_1, a_2, a_3)}{\sqrt{{a_1}^2 + {a_2}^2 + {a_3}^2}} \tag{1.54}$$

at each point in k-space.

Now we have a map from T^2 to S^2. We need to use one somewhat deep fact: under some assumptions, if $\pi_1(M) = 0$ for some target space M, then maps from the torus $T^2 \to M$ are contractible to maps from the sphere $S^2 \to M$. Intuitively, this is because the images of the non-contractible circles of the torus, which make it different from the sphere, can be contracted on M. By this logic, the two-band non-degenerate band structure in 2D is characterized by a single integer, which can be viewed as the Chern number of the occupied band.

What is the Chern number, intuitively? For simplicity let's consider maps from S^2 to the non-degenerate two-band Hamiltonians described above. One picture is in terms of $\pi_2(S^2)$. An alternative picture is that a non-zero Chern number is an 'obstruction' to globally defining wavefunctions, in the following sense. F, the first Chern form, is a 2-form. Let's consider a constant non-zero F, which for the case $S^2 \to S^2$ can be viewed as the field of a monopole located at the centre of the target sphere. *Locally*, it is possible to find wavefunctions giving a vector potential A with $F = dA$, but not *globally*. (There has to be a 'Dirac string' passing through the surface of the sphere somewhere.) In other words, states with non-zero Chern number have Chern forms that are non-trivial elements of the second cohomology class: they are closed 2-forms that are not globally exact.

The one subtle thing about this two-band model is that there is a non-trivial invariant in *three* spatial dimensions, since $\pi_3(S^2) = \mathbb{Z}$ (the 'Hopf invariant'). In other words, even if the Chern numbers for the three 2D planes in this 3D structure are zero, there can still be an integer-valued invariant.[8] This map is familiar to physicists from the fact that the Pauli matrices can be used to map a normalized complex 2-component spinor, i.e. an element of S^3, to a real unit vector, i.e. an element of S^2: $n^i = \mathbf{z}^\dagger \sigma^i \mathbf{z}$. This 'Hopf map' is an example of a map that cannot be deformed to the trivial (constant) map. The Hopf invariant does not generalize naturally to more than two bands, but the Chern number does, as we now see.

Now we consider the case of a non-degenerate 2D band structure with multiple bands, which we study using a method of Avron, Seiler, and Simon [3]. By the same argument as in the two-band case, we would like to understand π_1 and π_2 of the target space $H_{n\times n}$, non-degenerate $n \times n$ Hermitian matrices. As before, we will find that π_1 is zero, so maps from T^2 are equivalent to maps from S^2, but the latter will be quite non-trivial. We first diagonalize H at each point in k-space:

$$H(k) = U(k)D(k)U^{-1}(k). \tag{1.55}$$

Here $U(k)$ is unitary and $D(k)$ is real diagonal and non-degenerate. We can smoothly distort D everywhere in the Brillouin zone to a reference matrix with eigenvalues $1, 2, \ldots$ because of the non-degeneracy: if we plot the jth eigenvalue of D as a function of k_x and k_y, then this distortion corresponds to smoothing out ripples in this plot to obtain a constant plane.

[8] The nature of this fourth invariant changes when the Chern numbers are non-zero, as shown by Pontryagin in 1941: it becomes an element of a finite group rather than of the integers.

The non-trivial topology is contained in $U(k)$. The key is to note that $U(k)$ in the above is ambiguous: right multiplication by any diagonal unitary matrix, an element of $DU(N)$, will give the same $H(k)$. So we need to understand the topology of $M = U(N)/DU(N) = SU(N)/SDU(N)$, where $SDU(N)$ means diagonal unitary matrices with determinant 1. We can compute π_2 of this quotient by using the exact sequence of a fibration and the following facts: $\pi_2(SU(N)) = \pi_1(SU(N)) = 0$ for $N \geq 2$. These imply that $\pi_2(M) \cong \pi_1(SDU(N)) = \mathbb{Z}^{n-1}$, i.e. $n-1$ copies of the integers. This follows from viewing $SDU(N)$ as N circles connected only by the requirement that the determinant be 1. Similarly, we obtain $\pi_1(M) = 0$. We interpret these $n-1$ integers that arise in homotopy theory as just the Chern numbers of the bands, together with a constraint that the Chern numbers sum to zero.

1.3.5 Time-reversal invariance in Fermi systems

Now we jump to 2004–2005, when it was noted that imposing time-reversal symmetry in 2D electronic systems leads to new topological invariants. While non-zero Chern numbers cannot be realized with time-reversal invariance, the zero-Chern-number class gets subdivided into two pieces: 'ordinary' insulators that do not in general have an edge state, and a QSHE or 'topological insulator' where a bulk topological invariant forces an edge state. The topological invariant is not an integer here but rather a two-valued or $\mathbb{Z}_2$ invariant.

The idea that triggered this development started from considering two copies of the quantum Hall effect, one for spin-up electrons and one for spin-down, with opposite effective magnetic fields for the two spins. This combination, studied early on by Murakami, Nagaosa, and Zhang [28], for example, is time-reversal-invariant because acting with the time-reversal operator T changes both the magnetic field direction and the spin. Note that in a model such as this, S_z is a conserved quantum number even though $SU(2)$ (spin-rotation invariance) is clearly broken, as up and down spins behave differently. Heuristically, think of the spin–orbit coupling as arising from intra-atomic terms like $\mathbf{L} \cdot \mathbf{S}$, and consider specifically $L_z S_z$. For an electron of fixed spin, this coupling to the orbital motion described by L_z is just like the coupling in a constant magnetic field, since the orbital motion L_z generates a magnetic dipole moment. In the simplest case of a Chern number $+1$ state of up electrons and a Chern number -1 state of down electrons, the edge will have counterpropagating modes: for example, up spin moves clockwise along the edge and down spin moves counterclockwise. This turns out to be not a bad caricature of the quantum spin Hall phase in a more realistic system: one can tell by symmetry arguments that it will have no quantum Hall effect (i.e. $\alpha_c = 0$ in $J_i = \alpha_c \epsilon_{ijk} E_j B_k$), but it will have a spin Hall effect

$$J_j^i = \alpha_s \epsilon_{ijk} E_k, \tag{1.56}$$

where α_c and α_s are some numerical constants and J_j^i is a spin current (a current of angular momentum i in spatial direction j.[9] The appearance of the electric field rather

[9] There are some challenges that arise in trying to define a spin current in a realistic physical system, chiefly because spin is not a conserved quantity. Spin currents are certainly real and measurable in

than the magnetic field in the quantum spin Hall equation results from the goal of having a potentially dissipationless current equation. If dissipation provides no 'arrow of time', then both sides should transform in the same way under the time-reversal operation, which fixes the field on the right side to be E rather than B.

As an example of this 'two copies of the IQHE' generated by spin–orbit coupling, consider the model of graphene introduced by Kane and Mele [19]. This is a tight-binding model for independent electrons on the honeycomb lattice (Fig. 1.1). The spin-independent part of the Hamiltonian consists of a nearest-neighbour hopping, which alone would give a semimetallic spectrum with Dirac nodes at certain points in the 2D Brillouin zone, plus a staggered sublattice potential whose effect is to introduce a gap:

$$H_0 = t \sum_{\langle ij \rangle \sigma} c_{i\sigma}^\dagger c_{j\sigma} + \lambda_v \sum_{i\sigma} \xi_i c_{i\sigma}^\dagger c_{i\sigma}. \tag{1.57}$$

Here $\langle ij \rangle$ denotes nearest-neighbour pairs of sites, σ is a spin index, ξ_i alternates sign between sublattices of the honeycomb, and t and λ_v are parameters.

The insulator created by increasing λ_v is an unremarkable band insulator. However, the symmetries of graphene also permit an 'intrinsic' spin–orbit coupling of the form

$$H_{SO} = i\lambda_{SO} \sum_{\langle\langle ij \rangle\rangle \sigma_1 \sigma_2} \nu_{ij} c_{i\sigma_1}^\dagger s_{\sigma_1\sigma_2}^z c_{j\sigma_2}. \tag{1.58}$$

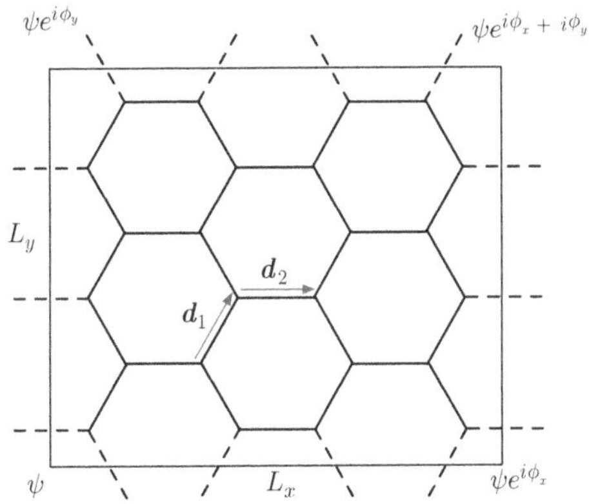

Fig. 1.1 The honeycomb lattice on which the tight-binding Hamiltonian resides. For the two sites depicted, the factor ν_{ij} of equation (1.58) is $\nu_{ij} = -1$. The phases $\phi_{x,y}$ describe twisted boundary conditions that are used in the text to give a pumping definition of the $\mathbb{Z}_2$ invariant. (Reprinted with permission from [9]. Copyright 2007 by the American Physical Society.)

various situations, but the fundamental definition we give of the quantum spin Hall phase will actually be in terms of charge; 'two-dimensional topological insulator' is another term for the same phase.

Here $\nu_{ij} = (2/\sqrt{3})\hat{\boldsymbol{d}}_1 \times \hat{\boldsymbol{d}}_2 = \pm 1$, where i and j are next-nearest neighbours and $\hat{\boldsymbol{d}}_1$ and $\hat{\boldsymbol{d}}_2$ are unit vectors along the two bonds that connect i to j. Including this type of spin–orbit coupling alone would not be a realistic model. For example, the Hamiltonian $H_0 + H_{SO}$ conserves s^z, the distinguished component of electron spin, and reduces for fixed spin (up or down) to Haldane's model [15]. Generic spin–orbit coupling in solids should not conserve any component of electron spin.

This model with S_z conservation is mathematically treatable using the Chern number above, since it just reduces to two copies of the IQHE. It is therefore not all that interesting, in addition to not being very physical, because of the requirement of S_z conservation. In particular, the stability of the phase is dependent on a subtle property of spin-$\frac{1}{2}$ particles (here we use the terms spin-$\frac{1}{2}$ and Fermi interchangeably). The surprise is that the quantum spin Hall phase survives, with interesting modifications, once we allow more realistic spin–orbit coupling, as long as time-reversal symmetry remains unbroken.

The time-reversal operator T acts differently in Fermi and Bose systems, or, more precisely, in half-integer versus integer spin systems. Kramers showed long ago that the square of the time-reversal operator is connected to a 2π rotation, which implies that

$$T^2 = (-1)^{2S}, \tag{1.59}$$

where S is the total spin quantum number of a state: half-integer-spin systems pick up a minus sign under two time-reversal operations.

An immediate consequence of this is the existence of 'Kramers pairs': every eigenstate of a time-reversal-invariant spin-$\frac{1}{2}$ system is at least twofold-degenerate. We will argue this perturbatively, by showing that a time-reversal-invariant perturbation H' cannot mix members of a Kramers pair (a state ψ and its time-reversal conjugate $\phi = T\psi$). To see this, note that

$$\langle T\psi|H'|\psi\rangle = \langle T\psi|H'|T^2\psi\rangle = -\langle T\psi|H'|\psi\rangle = 0, \tag{1.60}$$

where in the first step we have used the antiunitarity of T and the time-reversal symmetry of H' and in the second step the fact that $T^2 = -1$, while the last step is just to note that if $x = -x$, then $x = 0$.

Combining Kramers pairs with what is known about the edge state, we can say a bit about why a odd–even or $\mathbb{Z}_2$ invariant might be physical here. If there is only a single Kramers pair of edge states and we consider low-energy elastic scattering, then a right-moving excitation can only backscatter into its time-reversal conjugate, which is forbidden by the Kramers result above if the perturbation inducing scattering is time-reversal-invariant. However, if we have two Kramers pairs of edge modes, then a right-mover can backscatter to the left-mover that is *not* its time-reversal conjugate. This process will, in general, eliminate these two Kramers pairs from the low-energy theory.

Our general belief based on this argument is that a system with an even number of Kramers pairs will, under time-reversal-invariant backscattering, localize in pairs down to zero Kramers pairs, while a system with an odd number of Kramers pairs will wind up with a single stable Kramers pair. Additional support for this odd–even argument will be provided by our next approach. We would like, rather than just trying to understand whether the edge is stable, to predict from bulk properties whether the edge will have an even or odd number of Kramers pairs. Since deriving the bulk–edge correspondence directly is quite difficult, what we will show is that, starting from the bulk T-invariant system, there are two topological classes. These correspond in the example above (of separated up and down spins) to paired IQHE states with even or odd Chern number for one spin. Then the known connection between Chern number and number of edge states is good evidence for the statements above about Kramers pairs of edge modes.

A direct Abelian Berry-phase approach for the 2D $\mathbb{Z}_2$ invariant is provided in Appendix 1.A, along with an introduction to Wess–Zumino (WZ) terms in $(1+1)$-dimensional field theory and a physical interpretation of the invariant in terms of pumping cycles. The common aspect between these two is that in both cases the 'physical' manifold (either the 2-sphere in the WZ case, or the 2-torus in the QSHE case) is extended in a certain way, with the proviso that the resulting physics must be independent of the precise nature of the extension. When we later go to 3D, it turns out that there is a very nice 3D non-Abelian Berry-phase expression for the 3D $\mathbb{Z}_2$ invariant; while in practice it is harder to compute than the original expression based on applying the 2D invariant, it is much more elegant mathematically, so we will focus on that. Actually, for practical calculations, a very important simplification for the case of inversion symmetry (in both $d = 2$ and $d = 3$) was made by Fu and Kane [12]: the topological invariant is determined by the product of eigenvalues of the inversion operator at the 2^d time-reversal-symmetric points of the Brillouin zone.

1.3.6 Experimental status of 2D insulating systems

This completes our discussion of 1D and 2D insulating systems. The 2D topological insulator was observed by a transport measurement in (Hg, Cd)Te quantum wells [21], following theoretical predictions [5]. A simplified description of this experiment is that it observed, in zero magnetic field, a two-terminal conductance $2e^2/h$, consistent with the expected conductance e^2/h for each edge if each edge has a single mode, with no spin degeneracy. More recent work has observed some of the predicted spin transport signatures as well, although, as expected, the amount of spin transported for a given applied voltage is not quantized, unlike the amount of charge.

In the next subsection, we start with the 3D topological insulator and its remarkable surface and magnetoelectric properties. We then turn to metallic systems in order to understand another consequence of Berry phases of Bloch electrons.

1.3.7 3D band structure invariants and topological insulators

We will give a very quick introduction to the band structure invariants that allowed generalization of the previous discussion of topological insulators to 3D. However, most of our discussion of the 3D topological insulator will be in terms of emergent properties that are difficult to perceive directly from the bulk band structure invariant. We start by asking to what extent the 2D IQHE can be generalized to 3D. A generalization of the previous homotopy argument [3] can be used to show that there are three Chern numbers per band in 3D, associated with the xy, yz, and xz planes of the Brillouin zone. A more physical way to view this is that a 3D integer quantum Hall system consists of a single Chern number and a reciprocal lattice vector that describes the 'stacking' of integer quantum Hall layers. The edge of this 3D IQHE is quite interesting: it can form a 2D chiral metal, as the chiral modes from each IQHE combine and point in the same direction.

Consider the Brillouin zone of a 3D time-reversal-invariant material. Our approach will be to build on our understanding of the 2D case: concentrating on a single band pair, there is a $\mathbb{Z}_2$ topological invariant defined in the 2D problem with time-reversal invariance. Taking the Brillouin zone to be a torus, there are two inequivalent xy planes that are distinguished from others by the way time-reversal acts: the $k_z = 0$ and $k_z = \pm\pi/a$ planes are taken to themselves by time reversal (note that $\pm\pi/a$ are equivalent because of the periodic boundary conditions). These special planes are essentially copies of the 2D problem, and we can label them by $\mathbb{Z}_2$ invariants $z_0 = \pm 1$, $z_{\pm 1} = \pm 1$, where $+1$ denotes 'even Chern parity' or ordinary 2D insulator and -1 denotes 'odd Chern parity' or topological 2D insulator. Other xy planes are not constrained by time reversal and hence do not have to have a $\mathbb{Z}_2$ invariant.

The most interesting 3D topological insulator phase (the 'strong topological insulator') results when the z_0 and $z_{\pm 1}$ planes are in different 2D classes. This can occur if, moving in the z direction between these two planes, one has a series of 2D problems that interpolate between ordinary and topological insulators by breaking time reversal invariance. We will concentrate on this type of 3D topological insulator here. Another way to make a 3D topological insulator is to stack 2D topological insulators, but considering the edge of such a system shows that it will not be very stable: since two 'odd' edges combine to make an 'even' edge, which is unstable in the presence of T-invariant backscattering, we call such a stacked system a 'weak topological insulator'.

Above we found two xy planes with 2D $\mathbb{Z}_2$ invariants. By the same logic, we could identify four other such invariants: x_0, $x_{\pm 1}$, y_0, $y_{\pm 1}$. However, not all six of these invariants are independent: some geometry (an exercise for the reader) shows that there are two relations, reducing the number of independent invariants to four:[10]

$$x_0 x_{\pm 1} = y_0 y_{\pm 1} = z_0 z_{\pm 1}. \tag{1.61}$$

We can take these four invariants in 3D as $(x_0, y_0, z_0, x_0 x_{\pm 1})$, where the first three describe layered 'weak' topological insulators and the last describes the genuinely 3D invariant that distinguishes 'strong' topological insulators. (Note that one material can hence be both a strong and a weak topological insulator by this definition.)

[10] *Sketch of geometry:* To establish the first of these equalities, consider evaluating the Fu-Kane 2D formula on the four effective Brillouin zones (EBZs, see Section 1.A.2) described by the four invariants x_0, x_{+1}, y_0, y_{+1}. These define a torus, on whose interior the Chern 2-form F is well defined. Arranging the four invariants so that all have the same orientation, the A terms drop out, and the F integral vanishes since the torus can be shrunk to a loop. In other words, for some gauge choice, the difference $x_0 - x_{+1}$ is equal to $y_0 - y_{+1}$.

Alternatively, the 'axion electrodynamics' field theory in the next subsection can be viewed as suggesting that there should be only one genuinely 3D $\mathbb{Z}_2$ invariant.

For example, the strong topological insulator cannot be realized in any model with S_z conservation, while, as explained earlier, a useful example of the 2D topological insulator (a.k.a. QSHE) can be obtained from combining IQHE phases of up and down electrons. The impossibility of making a strong topological insulator with S_z conservation follows from noting that all planes normal to z have the same Chern number, since the Chern number is a topological invariant whether or not the plane is preserved by time- reversal. In particular, the $k_z = 0$ and $k_z = \pm\pi/a$ phases have the same Chern number for up electrons, say, which means that these two planes are either both 2D ordinary or both 2D topological insulators.

While the above argument is valid and useful for connecting the 3D topological insulators to the 2D case, it doesn't give much insight into what sort of gapless surface states we should expect at the surface of a strong topological insulator. The answer can be obtained by other means (some properties can be found via the field-theory approach given in the next section): the spin-resolved Fermi surface encloses an odd number of Dirac points. In the simplest case of a single Dirac point, believed to be realized in Bi_2Se_3, the surface state can be pictured as 'one-quarter of graphene'. Graphene, a single layer of carbon atoms that form a honeycomb lattice, has two Dirac points and two spin states at each k; spin–orbit coupling is quite weak, since carbon is a relatively light element. The surface state of a 3D topological insulator can have a single Dirac point and a single spin state at each k. As in the edge of the 2D topological insulator, time-reversal invariance implies that the spin state at k must be the T conjugate of the spin state at $-k$.

1.3.8 Axion electrodynamics, second Chern number, and magnetoelectric polarizability

The 3D topological insulator turns out to be connected to a basic electromagnetic property of solids. We know that in an insulating solid, Maxwell's equations can be modified because the dielectric constant ϵ and magnetic permeability μ need not take their vacuum values. Another effect is that solids can generate the electromagnetic term

$$\Delta\mathcal{L}_{EM} = \frac{\theta e^2}{2\pi h}\mathbf{E}\cdot\mathbf{B} = \frac{\theta e^2}{16\pi h}\epsilon^{\alpha\beta\gamma\delta}F_{\alpha\beta}F_{\gamma\delta}. \tag{1.62}$$

This term describes a magnetoelectric polarizability: an applied electrical field generates a magnetic dipole, and vice versa. An essential feature of the above 'axion electrodynamics' theory (cf. Wilczek [42]) is that when the axion field $\theta(\boldsymbol{x}, t)$ is constant, it plays no role in electrodynamics; this follows because θ couples to a total derivative, $\epsilon^{\alpha\beta\gamma\delta}F_{\alpha\beta}F_{\gamma\delta} = 2\epsilon^{\alpha\beta\gamma\delta}\partial_\alpha(A_\beta F_{\gamma\delta})$ (here we have used that F is closed, i.e. $dF = 0$), and so does not modify the equations of motion. However, the presence of the axion field can have profound consequences at surfaces and interfaces, where gradients in $\theta(\mathbf{x})$ appear.

A bit of work shows that at a surface where θ changes, there is a surface quantum Hall layer of magnitude

$$\sigma_{xy} = \frac{e^2(\Delta\theta)}{2\pi h}. \tag{1.63}$$

(This can be obtained by moving the derivative from one of the A fields to act on θ, leading to a Chern–Simons term for the electromagnetic field at the surface. The connection between Chern–Simons terms and the quantum Hall effect will be a major subject of the last part of this chapter.) The magnetoelectric polarizability described above can be obtained from these layers: for example, an applied electric field generates circulating surface currents, which in turn generate a magnetic dipole moment. In a sense, σ_{xy} is what accumulates at surfaces because of the magnetoelectric polarizability, in the same way as charge is what accumulates at surfaces because of ordinary polarization.

We are jumping ahead a bit in writing θ as an angle: we will see that, like polarization, θ is only well defined as a bulk property modulo 2π (for an alternative picture showing why θ is periodic that is more appropriate for electroweak symmetry breaking, see [42]). The integer multiple of 2π is only specified once we specify a particular way to make the boundary. How does this connect to the 3D topological insulator? At first glance, $\theta = 0$ in any time-reversal-invariant system, since $\theta \to -\theta$ under time reversal. However, as θ is periodic, $\theta = \pi$ also works, since $-\theta$ and θ are equivalent because of the periodicity, and is inequivalent to $\theta = 0$.

Here we will not give a microscopic derivation of how θ includes, for a band structure of non-interacting electrons, a part that is an integral of the Chern–Simons form:

$$\theta = \frac{1}{2\pi}\int_{\mathrm{BZ}} d^3k\, \epsilon_{ijk}\, \mathrm{Tr}\left[\mathcal{A}_i\partial_j\mathcal{A}_k - i\frac{2}{3}\mathcal{A}_i\mathcal{A}_j\mathcal{A}_k\right], \tag{1.64}$$

which can be done by imitating our previous derivation of the polarization formula [10, 34]. In general, unlike for the electrical polarization, there are additional non-geometrical contributions as well [11, 24]. Instead, we will focus on understanding the physical and mathematical meaning of the Chern–Simons form that constitutes the integrand, chiefly by discussing analogies with our previous treatment of polarization in 1D and the IQHE in 2D. These analogies are summarized in Table 1.1.

Throughout this subsection,

$$\mathcal{F}_{ij} = \partial_i\mathcal{A}_j - \partial_j\mathcal{A}_i - i[\mathcal{A}_i, \mathcal{A}_j] \tag{1.65}$$

is the (generally non-Abelian) Berry curvature tensor ($\mathcal{A}_\lambda = i\langle u|\partial_\lambda|u\rangle$), and the trace and commutator refer to band indices. We can gain an understanding of the above Chern–Simons form $K = \mathrm{Tr}[\mathcal{A}_i\partial_j\mathcal{A}_k - i\frac{2}{3}\mathcal{A}_i\mathcal{A}_j\mathcal{A}_k]$ by starting from the second Chern form $\mathrm{Tr}[\mathcal{F}\wedge\mathcal{F}]$; the relationship between the two is

$$dK = \mathrm{Tr}[\mathcal{F}\wedge\mathcal{F}], \tag{1.66}$$

Table 1.1 Comparison of Berry-phase theories of polarization and magnetoelectric polarizability

	POLARIZATION	MAGNETOELECTRIC POLARIZABILITY
d_{min}	1	3
Observable	$\mathbf{P} = \dfrac{\partial \langle H \rangle}{\partial E}$	$M_{ij} = \dfrac{\partial \langle H \rangle}{\partial E_i \, \partial B_j} = \dfrac{\delta_{ij} \theta e^2}{2\pi h}$
Quantum	$\Delta \mathbf{P} = \dfrac{e\mathbf{R}}{\Omega}$	$\Delta M = \dfrac{e^2}{h}$
Surface	$q = (\mathbf{P}_1 - \mathbf{P}_2) \cdot \hat{\mathbf{n}}$	$\sigma_{xy} = M_1 - M_2$
Electromagnetic coupling	$\mathbf{P} \cdot \mathbf{E}$	$M\mathbf{E} \cdot \mathbf{B}$
Chern–Simons form	$\mathcal{A}_i$	$\epsilon_{ijk} \left(\mathcal{A}_i \mathcal{F}_{jk} + i\frac{1}{3} \mathcal{A}_i \mathcal{A}_j \mathcal{A}_k \right)$
Chern form	$\epsilon_{ij} \partial_i \mathcal{A}_j$	$\epsilon_{ijkl} \mathcal{F}_{ij} \mathcal{F}_{kl}$

just as $\mathcal{A}$ is related to the first Chern form: $d(\operatorname{Tr} \mathcal{A}) = \operatorname{Tr} \mathcal{F}$. These relationships hold locally (this is known as Poincaré's lemma—that, given a closed form, it is *locally* an exact form) but not globally, unless the first or second Chern form generates the trivial cohomology class. For example, we saw that the existence of a non-zero first Chern number on the sphere prevented us from finding globally defined wavefunctions that would give an $\mathcal{A}$ with $d\mathcal{A} = \mathcal{F}$. We are assuming in even writing the Chern–Simons formula for θ that the ordinary Chern numbers are zero, so that an $\mathcal{A}$ can be defined in the 3D Brillouin zone. We would run into trouble if we assumed that an $\mathcal{A}$ could be defined in the 4D Brillouin zone if the *first or second* Chern number were non-zero. Note that the electromagnetic action above is just the second Chern form of the (Abelian) electromagnetic field.

The second Chern form is closed and hence generates an element of the de Rham cohomology we studied earlier. There are higher Chern forms as well: the key is that symmetric polynomials can be used to construct closed forms, by the antisymmetry properties of the exterior derivative. In physics, we typically keep the manifold fixed (in our Brillouin zone examples, it is usually a torus T^n), and are interested in classifying different fibre bundles on the manifold. In mathematical language, we want to use a properly normalized cohomology form to compute a homotopy invariant (i.e. with respect to changing the connection, not the manifold). This is exactly what we did with the Chern number in the IQHE, which was argued to compute certain integer-valued homotopy π_2 invariants of non-degenerate Hermitian matrices.

More precisely, we saw that the $U(1)$ gauge dependence of polarization was connected to the homotopy group $\pi_1(U(1)) = \mathbb{Z}$, but that this is connected also to the existence of integer-valued Chern numbers, which we explained in terms of π_2. (These statements are not as inconsistent as they might seem, because our calculation of

π_2 came down to π_1 of the diagonal unitary group.) We can understand the second Chern and Chern–Simons form similarly, using the homotopy invariants π_3 (gauge transformation in $d = 3$) and π_4 (quantized state in $d = 4$). The Chern–Simons integral for θ given above, in the non-Abelian case, has a $2\pi n$ ambiguity under gauge transformations, and this ambiguity counts the integer-valued homotopy invariant

$$\pi_3(SU(N)) = \mathbb{Z}, \quad N \geq 2. \tag{1.67}$$

In other words, there are 'large' (non-null-homotopic) gauge transformations. Note that the Abelian Chern–Simons integral is completely gauge-invariant, consistent with $\pi_3(U(1)) = 0$.

The quantized state in $d = 4$ was originally discussed in the context of time-reversal-symmetric systems. The set $\mathcal{Q}$ has one integer-valued π_4 invariant for each band pair, with a zero-sum rule. These invariants survive even once T is broken, but realizing the non-zero value requires that two bands touch somewhere in the 4D Brillouin zone. In this sense, the '4D quantum Hall effect' is a property of how pairs of bands interact with each other, rather than of individual bands. Even if this 4D QHE is not directly measurable, it is mathematically connected to the 3D magnetoelectric polarizability in the same way as 1D polarization and the 2D IQHE are connected.

The above Chern–Simons formula for θ works, in general, only for a non-interacting electron system. This is not true for the first Chern formula for the IQHE, or the polarization formula, so what is different here? The key is to remember that the 3D Chern formula behaves very differently in the Abelian and non-Abelian cases; for example, in the Abelian case, θ is no longer periodic as the integral is fully gauge-invariant. Taking the ground state many-body wavefunction and inserting it into the Chern–Simons formula is not guaranteed to give the same result as using the multiple one-particle wavefunctions.

However, we can give a many-body understanding of θ that clarifies the geometric reason for its periodicity even in a many-particle system. Consider evaluating dP/dB by applying the 3D polarization formula

$$P_i = e \int_{BZ} \frac{d^3k}{(2\pi)^3} \operatorname{Tr} \mathcal{A}_i \, . \tag{1.68}$$

to a rectangular-prism unit cell. The minimum magnetic field normal to one of the faces that can be applied to the cell without destroying the periodicity is one flux quantum per unit cell, or a field strength $h/(e\Omega)$, where Ω is the area of that face. The ambiguity of polarization (1.68) in this direction is one charge per transverse unit cell area, i.e. e/Ω. Then the ambiguity in dP/dB is

$$\Delta \frac{P_x}{B_x} = \frac{e/\Omega}{h/(e\Omega)} = \frac{e^2}{h} = 2\pi \frac{e^2}{2\pi h}. \tag{1.69}$$

So the periodicity of 2π in θ is really a consequence of the geometry of polarization, and is independent of the single-electron assumption that leads to the microscopic Chern–Simons formula.

1.3.9 Anomalous Hall effect and Karplus–Luttinger anomalous velocity

Our previous examples of Berry phases in solids have concentrated on insulators, but one of the most direct probes of the Berry phase of Bloch electrons is found in metals that break time-reversal symmetry. The breaking of T allows a non-zero transverse conductivity σ_{xy} to exist along with the metallic diagonal conductivity σ_{xx}. This 'anomalous Hall effect' (AHE) can originate from several different microscopic processes. The most interesting from a geometric point of view is the intrinsic AHE that results from Berry phases of a time-reversal-breaking band structure when the Fermi level is in the middle of a band. We will not attempt to discuss this interesting physics here but refer the reader to a comprehensive review by Nagaosa et al. [29] and note that there are an increasing number of other examples of Berry-phase effects in metals. There are similar effects related to the orbital moment of Bloch electrons, which is similar in some ways but not purely geometric since it arises from both the wavefunctions and the Hamiltonian, unlike the Berry phase, which is purely a wavefunction property.

1.4 Introduction to topological order

Now we consider strongly interacting topological phases, defined as those that cannot be understood in terms of free particles. In contrast, the IQHE and topological insulators can be understood in terms of free particles, although these phases are stable in the sense that they survive over a finite region of interaction strength until a phase transition occurs. Our main tool will be quantum field theory, which is a powerful language to describe the long-wavelength physics of interacting systems. After giving some microscopic motivation from the fractional quantum Hall effect (FQHE), we give a first example of field theory applied to spin chains as an example of how an analysis of topological terms in a simple field theory led to a clear experimental prediction (the 'Haldane gap') regarding antiferromagnetic integer-spin Heisenberg chains.

We then return to the quantum Hall effect and develop Abelian Chern–Simons theory, an example of a truly topological field theory. Although it is written in terms of one or more $U(1)$ gauge fields, similar to ordinary electromagnetism, its behaviour is strikingly different from that of the conventional field theories with which the reader may already be familiar. In lieu of a microscopic derivation, which has been carried out but is somewhat tedious, we show that it unifies properties such as ground-state degeneracy, braiding statistics, and edge excitations. We will follow increasingly standard parlance and use the term 'topological order' specifically for phases of matter described by a non-trivial topological field theory, hence having ground-state degeneracy, fractional statistics, etc. Thus the IQHE, which is certainly a topological phase of matter and well described by the Abelian Chern–Simons theory given below with $k = 1$, does not have topological order in the sense introduced by Wen.

1.4.1 FQHE background

We give quickly some standard background on the FQHE in order to motivate the Chern–Simons field theory introduced below. The goal of that field theory is to give a compact universal description of the key features of the topological order in quantum Hall states, similar in spirit to the Ginzburg–Landau field theory of symmetry-breaking phases. Most of this material is standard and can be found in a number of edited volumes and textbooks on the quantum Hall effects [6, 18, 33].

Our discussion centres on the Laughlin wavefunction for 2D electrons ($z_j = x_j + iy_j$ describes the jth electron, $j = 1, \dots, N$)

$$\Psi_m = \left[\prod_{i<j}(z_i - z_j)^m\right] \exp\left(-\sum_i |z_i|^2/4\ell^2\right). \tag{1.70}$$

The magnetic length is $\ell = \sqrt{\hbar c/eB}$ and the wavefunction is not normalized. This wavefunction clearly can be expanded over the single-electron lowest-Landau-level wavefunctions in the rotational gauge,

$$\psi_m = z^m e^{-|z|^2/4\ell^2}. \tag{1.71}$$

where $m = 0, 1, \dots$ labels angular momentum. For $m = 1$, the Laughlin state is just a Slater determinant for the filled lowest Landau level, but for higher m, it is believed not to be a sum of any finite number of Slater determinants in the $N \to \infty$ limit.

This wavefunction can be justified using the pseudopotential approach introduced by Haldane: it is the maximum-density zero-energy state of a repulsive interaction that vanishes for relative angular momentum greater than or equal to m. We checked that its density is $\nu = 1/m$ by looking at the degree of the polynomial factor, which is directly related to $\langle r^2 \rangle$, and argued that it contains 'quasihole' excitations of charge $-q/m$, where q is the charge of the electrons. The wavefunction for a quasihole at z_0 is

$$\Psi_{\text{quasihole}} = \left[\prod_i (z_i - z_0)\right] \Psi_m. \tag{1.72}$$

The fractional charge can be understood by noting that m copies of the extra factor here would lead to a wavefunction with an electron at z_0, but without treating z_0 as an electron coordinate; in other words, a wavefunction with a 'hole' added at z_0. It has edge states that at first glance are loosely similar to those in the filled Landau level.

1.4.2 Topological terms in field theories: the Haldane gap and Wess–Zumino–Witten models

As a warm-up for fully topological field theories, we give an example of how topological terms can have profound consequences in 'ordinary' field theories (i.e. theories without gauge fields). By a topological term, we mean loosely one whose value in any specific configuration (a path in the path integral) is a topological invariant, so that the set of all paths can be divided into topological sectors by the value of the topological term. A famous example of this phenomenon found by Haldane led to the first understanding of the gapped spin-1 Heisenberg antiferromagnet in one spatial dimension, which has recently been interpreted as a symmetry-protected topological phase of interacting particles because it can be smoothly connected to a trivial phase by breaking spatial symmetries such as inversion. We will focus on topological terms that appear in nonlinear σ-models, which, despite their unwieldy name, are a very basic type of field theory for systems in or near an ordered phase breaking a continuous symmetry.

We first present Haldane's example (following closely the treatment of Auerbach [2]), and then discuss a different kind of topological term that appears in Wess–Zumino–Witten models, again in one spatial dimension; details of the latter are provided in Appendix 1.A. The nonlinear σ-model (NLSM) is an example of an effective theory, a simplified description of the low-energy degrees of freedom of a complicated system. Ginzburg–Landau theory is another such effective theory, and one use of the NLSM is as a further simplification of Ginzburg–Landau theory where we have thrown away the 'hard' or 'massive' fluctuations of the magnitude of the order parameter, keeping only the 'soft' or 'massless' fluctuations within the order-parameter manifold.

For definiteness, we consider a d-dimensional XY model, which would be described in Ginzburg–Landau theory by a 2-component real or 1-component complex order parameter. The mean-field physics in the ordered phase as a function of the order parameter is as follows: the order-parameter manifold of symmetry-related ground states is a circle, and we can expect that fluctuations along this circle are 'soft' in the sense of requiring little energy (since this is a flat direction of the energy), while those perpendicular to the circle are more costly. This order-parameter manifold is the same as that considered in the discussion of topological defects in Section 1.2, where defects were classified using maps from surfaces enclosing the defect in real space to the order-parameter manifold. At low temperature, we might expect that a reasonable description of the system is therefore obtained just from fluctuations of the order parameter's direction, leading to a functional integral for the coarse-grained classical partition function:

$$Z_{\mathrm{NLSM}} = \int \mathcal{D}\theta(x) \exp\left[-\beta c \int \frac{(\nabla\theta)^2}{2}\, d^d x\right]. \tag{1.73}$$

Here c is a coupling constant with units of energy if $d = 2$; one could estimate c simply from the coupling strength in a lattice XY model. The NLSM it is called nonlinear because the circle is defined by a hard constraint on the $\hat{\mathbf{n}}$ field, which in more complicated target manifolds such as the sphere leads to interaction (i.e. nonlinear) terms in the fields obtained in a perturbative expansion; it is called a σ-model because of its first appearance in particle physics.

For a quantum-mechanical model at zero temperature, we might expect on general grounds that imaginary time will become an extra dimension in any Euclidean path-integral representation of the partition function, in the same way as the Dirac–Feynman path integral for quantum mechanics involves integration of the Lagrangian over time (a $(0+1)$-dimensional theory). Now we will obtain a NLSM for a quantum-mechanical problem in one spatial dimension. Heuristically, we might expect an NLSM to be a reasonable description for a quantum model that is 'close to' having symmetry-breaking order.

Our approach is to derive a connection between the low-energy, long-wavelength degrees of freedom of the spin path integral of the Heisenberg antiferromagnet. This process is known as Haldane's mapping in the context of spin systems: we will use it to show that there is a topological term present for half-integer spin but not for integer spin, which is believed to explain the different behaviour seen numerically and experimentally in these two cases.

First we look for a more general way of writing the Berry -phase term for a spin that results from setting up a coherent-state path integral for spin. In order to make a path integral, we should set up an integral over 'classical' trajectories—what is the classical trajectory of a spin? One answer is to use the overcomplete basis of coherent states for the spin-S Hilbert space [2], which are labelled by a unit vector $\hat{\mathbf{\Omega}}$. As S increases, the spin wavefunction becomes more and more concentrated around $\hat{\mathbf{\Omega}}$:

$$\omega[\hat{\mathbf{\Omega}}] = -\int_0^\beta d\tau \dot{\phi} \cos\theta. \tag{1.74}$$

For a closed path on the sphere, this corresponds to the signed spherical area enclosed by the path. An overall ambiguity of $\pm 4\pi$ in this area does not affect the physics, since the area ω appears in the path-integral action with a coefficient $-iS$. For a many-spin system, the full action is

$$S[\hat{\mathbf{\Omega}}] = -iS\sum_i \omega[\hat{\mathbf{\Omega}}_i] + \int_0^\beta d\tau \, \frac{S^2 J}{2} \sum_{ij} \hat{\mathbf{\Omega}}_i \cdot \hat{\mathbf{\Omega}}_j. \tag{1.75}$$

For now, we return to a single spin to set up an improved way of writing the Berry-phase term.

Let the vector potential $\mathbf{A}(\hat{\mathbf{\Omega}})$ be assumed to have the property that its line integral over a closed orbit on the sphere should give the area enclosed by the orbit:

$$\omega = \int_0^\beta d\tau \, \mathbf{A}(\hat{\mathbf{\Omega}}) \cdot \dot{\hat{\mathbf{\Omega}}}. \tag{1.76}$$

Then Stokes's theorem fixes curl $\mathbf{A}$ to be the magnetic field of a magnetic monopole (a vector with uniform outward component):

$$\nabla \times \mathbf{A} \cdot \hat{\mathbf{\Omega}} = \epsilon^{\alpha\beta\gamma} \frac{\partial A_\beta}{\partial \hat{\Omega}_\alpha} \hat{\Omega}^\gamma = 1. \tag{1.77}$$

Two explicit examples to check that this can be done are

$$\mathbf{A}^a = -\frac{\cos\theta}{\sin\theta}\hat{\phi}, \qquad \mathbf{A}^b = \frac{1-\cos\theta}{\sin\theta}\hat{\phi}. \tag{1.78}$$

Clearly, $\mathbf{A}^a$ has singularities at the north and south poles, while $\mathbf{A}^b$ has a singularity only at the south pole.[11]

Now we can use this representation to write concisely the variation of the Berry-phase term under a small variation in the path from imaginary time 0 to imaginary time t. Suppose that we want to calculate

$$\begin{aligned}\delta\omega[\hat{\boldsymbol{\Omega}}] &= \int_0^t dt'\,\delta(\mathbf{A}\cdot\dot{\hat{\boldsymbol{\Omega}}}) \\ &= \int_0^t dt' \left(\frac{\partial A^\alpha}{\partial\hat{\Omega}^\beta}\,\delta\hat{\Omega}^\beta\,\dot{\hat{\Omega}}^\alpha + A^\alpha \frac{d}{dt}\delta\hat{\Omega}^\alpha\right)\end{aligned} \tag{1.79}$$

under a small variation of the path $\delta\hat{\Omega}$ that is assumed to keep the endpoints fixed. Now subtract $(\partial A^\alpha/\partial\hat{\Omega}^\beta)\dot{\hat{\Omega}}^\beta\,\delta\hat{\Omega}^\alpha$ from the first term and add it to the second, to get

$$\begin{aligned}\delta\omega[\hat{\boldsymbol{\Omega}}] &= \int_0^t dt'\,\frac{\partial A^\alpha}{\partial\hat{\Omega}^\beta}\epsilon^{\alpha\beta\gamma}(\dot{\hat{\boldsymbol{\Omega}}}\times\delta\hat{\boldsymbol{\Omega}})_\gamma + \int_0^t dt'\left(A^\alpha\frac{d}{dt}\delta\hat{\Omega}^\alpha + \frac{\partial A^\alpha}{\partial\hat{\Omega}^\beta}\dot{\hat{\Omega}}^\beta\,\delta\hat{\Omega}^\alpha\right) \\ &= \int_0^t dt'\,\hat{\boldsymbol{\Omega}}\cdot(\dot{\hat{\boldsymbol{\Omega}}}\times\delta\hat{\boldsymbol{\Omega}}) + \int_0^t dt'\,\frac{d}{dt'}(\mathbf{A}\cdot\delta\hat{\boldsymbol{\Omega}}) \\ &= \int_0^t dt'\,\hat{\boldsymbol{\Omega}}\cdot(\dot{\hat{\boldsymbol{\Omega}}}\times\delta\hat{\boldsymbol{\Omega}}).\end{aligned} \tag{1.80}$$

Here we have used the condition (1.77) and also, in rewriting the first term, the fact that the quantity in parentheses $(\dot{\hat{\boldsymbol{\Omega}}}\times\delta\hat{\boldsymbol{\Omega}}) \parallel \hat{\boldsymbol{\Omega}}$ because of the constant length of the vector $\hat{\boldsymbol{\Omega}}$.

Now, after this prelude, we are ready to rewrite the full path integral for the many-spin Heisenberg model. The first step is to write the spin $\hat{\boldsymbol{\Omega}}_i$ in terms of two continuous fields of spacetime $\hat{\mathbf{n}}$ and $\mathbf{L}$:

$$\hat{\boldsymbol{\Omega}}_i = \eta_i\hat{\mathbf{n}}(\mathbf{x}_i)\sqrt{1-\left|\frac{\mathbf{L}(\mathbf{x}_i)}{S}\right|^2} + \frac{\mathbf{L}(\mathbf{x}_i)}{S}. \tag{1.81}$$

Here η_i alternates between sublattices, $\hat{\mathbf{n}}(\mathbf{x})$ is a unit vector field, sometimes referred to as the Néel field, and $\mathbf{L}$ is constrained to be perpendicular to $\hat{\mathbf{n}}$. Hence a constant value of $\hat{\mathbf{n}}$ corresponds to a classical Néel state. It seems like we have greatly increased the degrees of freedom by this rewriting; what we do now is restrict the allowed Fourier components of the new fields (i.e. the Brillouin zone) to small momenta in such a way that the total number of degrees of freedom is unchanged (cf. Auerbach [2] for details). The spirit of this approximation is that we are interested in long-lengthscale

[11] Actually, $\mathbf{A}^b$ is a good representation of the field of a Dirac monopole: a singular flux ('Dirac string') enters through the south pole and then goes out uniformly over the rest of the sphere. A small circle around the south pole contains flux 4π, which contributes $4\pi S$ to the action, but recall that this winds up giving zero physical contribution to the path integral.

physics, so details on the scale of the lattice spacing are unimportant. It turns out that we assume slow variations in $\hat{\mathbf{n}}$ but only that $|\mathbf{L}| \ll S$, i.e. that $\mathbf{L}$ is small but not necessarily slowly varying. We now expand the path integral in powers of $|\mathbf{L}|/S$.

A pair of spins gives a contribution

$$\hat{\mathbf{\Omega}}_i \cdot \hat{\mathbf{\Omega}}_j \approx \eta_i\eta_j - \frac{1}{2}\eta_i\eta_j(\hat{\mathbf{n}}_i - \hat{\mathbf{n}}_j)^2 \\ + \frac{1}{S^2}\left[\mathbf{L}_i \cdot \mathbf{L}_j - \frac{1}{2}\eta_i\eta_j(\mathbf{L}_i^2 + \mathbf{L}_j^2)\right] + \frac{1}{S}(\eta_j\mathbf{L}_i \cdot \hat{\mathbf{n}}_j + \eta_i\mathbf{L}_j \cdot \hat{\mathbf{n}}_i) + \dots \quad (1.82)$$

Here the neglected terms are of order $|\mathbf{L}|^2(\hat{\mathbf{n}}_i - \hat{\mathbf{n}}_j)$ or smaller. In the first term, use a Taylor expansion to convert differences of the Néel field into derivatives and keep only the leading contribution. You can show that the cross-terms (those with both $\mathbf{L}$ and $\hat{\mathbf{n}}$) vanish by the symmetry of the Heisenberg Hamiltonian. The term with two $\mathbf{L}$ factors we rewrite below in Fourier space, where it is much simpler and where we will be able to 'integrate it out'.

What we are left with, after going from the lattice to integrals using

$$\sum_i F_i \to a^{-d} \int d^d x \sum_i \delta(\mathbf{x} - \mathbf{x}_i)F(x), \quad (1.83)$$

is the continuum representation

$$H = E_0 + \frac{1}{2}\int d^d x \left[\rho_s \sum_l |\partial_l \hat{\mathbf{n}}|^2 + \int d^d x' \left(\mathbf{L}_x \cdot \chi^{-1}_{xx'} \cdot \mathbf{L}_{x'}\right)\right]. \quad (1.84)$$

Here E_0 is the classical energy

$$E_0 = \frac{S^2}{2}\sum_{ij} J_{ij}\eta_i\eta_j. \quad (1.85)$$

In the first term, the spin stiffness is

$$\rho_s = -\frac{S^2}{2dNa^d}\sum_{ij} J_{ij}\eta_i\eta_j|\mathbf{x}_i - \mathbf{x}_j|^2. \quad (1.86)$$

The second or 'canting' term in Fourier space is simply

$$\int \frac{d^d q}{(2\pi)^d}\frac{\mathbf{L}_\mathbf{q} \cdot \mathbf{L}_{-\mathbf{q}}}{J(\mathbf{q}) - J(\pi, \pi, \dots)}, \qquad J(\mathbf{q}) = \sum_j e^{i\mathbf{q}\cdot(\mathbf{x}_i - \mathbf{x}_j)} J_{ij}. \quad (1.87)$$

Now we just need to rewrite the geometric phase

$$-iS\sum_i \omega_i = -iS\int_0^\beta d\tau \sum_i \mathbf{A}(\hat{\mathbf{\Omega}}_i) \cdot \dot{\hat{\mathbf{\Omega}}}_i. \quad (1.88)$$

Assume that the vector potential is chosen so that $\mathbf{A}(\hat{\mathbf{\Omega}}) = \mathbf{A}(-\hat{\mathbf{\Omega}})$, as works for one of the examples above. Now expanding in terms of the new fields,

$$\begin{aligned} -iS\sum_i \omega_i &= -iS\sum_i \eta_i \omega[\hat{\mathbf{n}}_i + \eta_i \mathbf{L}_i/S] \\ &= -iS\sum_i \eta_i \omega\left[\hat{\mathbf{n}}_i + \frac{\delta\omega}{\delta\hat{\mathbf{n}}_i}\cdot(\mathbf{L}_i/S)\right] \\ &= -i\Upsilon - i\int_0^\beta d\tau\,\sum_i(\hat{\mathbf{n}}_i \times \partial_\tau\hat{\mathbf{n}}_i \cdot \mathbf{L}_i). \end{aligned} \tag{1.89}$$

In the last line, we have used our earlier formula for the variation of ω, and defined

$$\Upsilon = S\sum_i \eta_i \omega[\hat{\mathbf{n}}(\mathbf{x}_i)], \tag{1.90}$$

switching to the spatial continuum limit.

Now our goal is going to be to combine the classical and geometric terms in order to obtain a simple long-wavelength action. The key step is to note that the second term in (1.89) couples one power of L to a combination of n fields. So integrating out the L degrees of freedom (a Gaussian integral) will give rise to the following: considering only the terms involving $\mathbf{L}$ and doing the integral in Fourier space, we get (ignoring an unimportant overall constant)

$$Z_L \propto \int \mathcal{D}\hat{\mathbf{n}} \exp\left\{-\frac{1}{2}\int d\tau\,\frac{d^dq}{(2\pi)^d}[J(\mathbf{q}) - J(\pi,\pi,\ldots)](\hat{\mathbf{n}}\times\partial_\tau\hat{\mathbf{n}})_{\mathbf{q}}\cdot(\hat{\mathbf{n}}\times\partial_\tau\hat{\mathbf{n}})_{-\mathbf{q}}\right\}. \tag{1.91}$$

We can simplify this much further: for long wavelengths, we approximate $\chi(\mathbf{q}) \approx \chi(0)$ and use

$$|\hat{\mathbf{n}}\times\partial_\tau\hat{\mathbf{n}}|^2 = |\partial_\tau\hat{\mathbf{n}}|^2 \tag{1.92}$$

from the constant length of $\hat{\mathbf{n}}$, to get just the following (the real-space constant $\chi_0 = a^{-d}\chi(0) = a^{-d}[J(0,0,\ldots) - J(\pi,\pi,\ldots)]$):

$$Z_L = \int \mathcal{D}\hat{\mathbf{n}} \exp\left(-\frac{1}{2}\int_0^\beta d\tau \int d^dx\,\chi_0|\partial_\tau\hat{\mathbf{n}}|^2\right). \tag{1.93}$$

So, putting it all together, we have

$$Z \propto \int \mathcal{D}\hat{\mathbf{n}}\, e^{i\Upsilon} \exp\left[-\frac{1}{2}\int_0^\beta d\tau \int d^dx\,\left(\chi_0|\partial_\tau\hat{\mathbf{n}}|^2 + \rho_s|\partial_{x^\alpha}\hat{\mathbf{n}}|^2\right)\right]. \tag{1.94}$$

This now looks much more symmetric between space and time; if desired, one can just rescale time to make the theory look like it lives in an isotropic $(d+1)$-dimensional space. This gives

$$Z \propto \int \mathcal{D}\hat{\mathbf{n}}\, e^{i\Upsilon} \exp\left(-\int d^{d+1}x\, \mathcal{L}_{\mathrm{NLSM}}\right), \qquad \mathcal{L}_{\mathrm{NLSM}} = \sum_{\alpha=1}^{d+1} \frac{\partial_{x^\alpha}\hat{\mathbf{n}} \cdot \partial_{x^\alpha}\hat{\mathbf{n}}}{2}. \tag{1.95}$$

This NLSM is the simplest field theory of maps from the space $\mathcal{R}^{d+1}$ to the unit sphere. We still need to say a bit about the topological term Υ: in one spatial dimension, this term fundamentally modifies the physics, for reasons we shall see. We expand it for slowly varying $\hat{\mathbf{n}}(x)$: recall that Υ is defined to include factors η_i, so

$$\begin{aligned}\Upsilon^{d=1} &= -S\sum_i \{\omega[\hat{\mathbf{n}}(x_{2i})] - \omega[\hat{\mathbf{n}}(x_{2i-1})]\} \\ &= \frac{S}{2}\int \frac{dx}{a}\frac{\delta\omega}{\delta\hat{\mathbf{n}}} \cdot \partial_x\hat{\mathbf{n}}a = 2\pi S\Theta[\hat{\mathbf{n}}(x,\tau)].\end{aligned} \tag{1.96}$$

Here Θ comes from using our previous variation form for the variation $d\omega$:

$$\Theta = \frac{1}{4\pi}\int d\tau \int dx\, (\hat{\mathbf{n}} \times \partial_\tau\hat{\mathbf{n}} \cdot \partial_x\hat{\mathbf{n}}). \tag{1.97}$$

This form is known as the Pontryagin index, which is a topological invariant like a winding number. It is an integer and is constant under smooth deformations of $\hat{\mathbf{n}}$. Essentially, it measures the number of times the map from $(-L/2, L/2) \times (0, \beta)$ 'wraps' the sphere S^2. You can easily construct examples with $\Theta = 0$ (the constant map) and $\Theta = 1$ (spherical projection). If you want a sense of why it is a topological invariant (which is not that hard to show), imagine that someone gives you a sphere wrapped with paper. The paper can't be 'contracted to a point' without tearing, unlike a loop drawn on the sphere. So maps $S^1 \to S^2$ are all contractible, while maps $S^2 \to S^2$ are classified by the Pontryagin index.

The important thing to note is that the coefficient in front of this integer is only $2\pi S$, so that there will be a difference between integer and half-integer spins. For integer spin the topological term doesn't do anything, while for half-integer spins, there is interference between terms with odd or even values of the Pontryagin index. So Haldane's mapping explains (with a few approximations along the way!) the profound difference between integer and half-integer spins in one dimension, later confirmed experimentally and numerically. It is actually easier just to solve the spin-$\frac{1}{2}$ chain using the Bethe ansatz than to explicitly solve its continuum theory with Berry phases, although a proof has been given that the latter is indeed gapless. The Lieb–Schultz–Mattis theorem discussed by Chalker in Chapter 3 of this volume provides a general reason why the half-integer-spin case is gapless. Experimental results by neutron scattering confirm the existence of a gap and also the existence of spin-$\frac{1}{2}$ edge states at the end of chains, which are discussed by Regnault in Chapter 4 of this volume .

So we have seen how the unusual geometry of spin space, in the path-integral representation, gives rise to a profound difference between integer-spin and half-integer spin chains. We can connect the above result to the exact solution by Affleck, Kennedy, Lieb, and Tasaki (AKLT) of a spin-1 chain with additional biquadratic interactions:

$$H = J \sum_i \left[S_i \cdot S_j + \alpha (S_i \cdot S_j)^2 \right] \tag{1.98}$$

with $J > 0$ and $\alpha = \frac{1}{3}$. This value of α is special in that the two terms on each bond act as projectors onto the total spin-0 and spin-1 subspaces of the two spins, with equal weight. The full phase diagram of the bilinear–biquadratic phase diagram from numerical density-matrix renormalization group studies has been computed [22]. We note that the Haldane problem of the purely bilinear chain is in the same gapped phase as the AKLT solution, but that there are other phases as well, and there are also parameter values for which the system is gapless. As the last part of our discussion of topological terms for now, we explain the existence of two critical points with labels $SU(3)_1$ and $SU(2)_2$ in the phase diagram found in [22], which combine a Lie group with a subscripted integer known as the level. These points are examples of field theories with both conformal invariance and Lie group symmetry known as Wess–Zumino-Witten (WZW) models.

The NLSM for the XY model in (1.73) can be written in a different way if we think about the order-parameter manifold (the circle) as the Lie group $U(1)$. Writing $g = e^{i\theta}$, we note that $\partial_i \theta = g^{-1} \partial_i g$, so

$$Z_{\text{NLSM}} = \int \mathcal{D}\theta(x) \exp\left[\beta c \int d^2x \sum_i \tfrac{1}{2} (g^{-1} \partial_i g)^2 \right]. \tag{1.99}$$

In taking the trace here, we are looking ahead to a generalization. There is not a Lie group structure on the sphere, but we might be tempted to generalize to other Lie groups, for example by taking $g \in U(N)$ or $SU(N)$. Then $g^{-1}\partial_i g$ is an element of the Lie algebra, which has an inner product known as the Killing form; for $SU(N)$, $\mathcal{K}(X, Y) = 2N \operatorname{Tr}(XY)$. Generalizing the kinetic term that is the only term in the action above to the Lie algebra is straightforward.

However, it turns out that the low-energy physics of this generalization with just the resulting term is quite different from the $U(1)$ case. As for the NLSM into the sphere, the manifolds of unitary groups are curved once we go beyond the circle, leading to interactions that result in a mass gap. If we want instead to obtain a gapless model with Lie group symmetry, we must add an additional topological term first written down by Wess and Zumino. This term is quite unusual in that it requires extending the manifold on which the theory lives into an extra dimension. Assume $N > 1$ in what follows, and pick $g \in SU(N)$ for definiteness. Let us compactify the two-dimensional space into S^2 as for the Haldane chain above. Given a configuration of the Lie group field g on the surface of a sphere, we can always find a way to smoothly deform that configuration to the constant configuration, since $\pi_2(SU(N))$ is trivial.

We will keep writing the generalized models in Euclidean space, although their primary relevance is to quantum models in one spatial dimension. The action of the WZW model in the usual notation is then (see Appendix 1.A for a physics motivation)

$$\begin{aligned} S = &- \frac{k}{8\pi} \int_{S^2} d^2x \, \mathcal{K}(g^{-1}\partial^\mu g, g^{-1}\partial_\mu g) \\ &- \frac{k}{24\pi} \int_{B^3} d^3y \, \epsilon^{ijk} \mathcal{K}(g^{-1}\partial_i g, [g^{-1}\partial_j g, g^{-1}\partial_k g]) \, . \end{aligned} \tag{1.100}$$

The meaning of upper and lower indices in the first term is that the metric of spacetime appears. In the second term, in contrast, the ϵ term appears instead of the metric, a sign that the term is topological in the sense of being metric-independent. In the second term, we have chosen a continuation of the field g into the interior B^3 of the sphere S^2. While, as mentioned above, those continuations certainly exist, we should check to make sure that the physics is independent of precisely which continuation we chose.

This independence is related to another topological fact about $SU(N)$. Consider two different continuations from S^2 into B^3. Actually, as a simpler example, consider two different continuations from S^1 into B^2. We could combine those into a field configuration on S^2, where one continuation gives the northern hemisphere and the other gives the southern hemisphere. In the same way, combining our two continuations from S^2 to B^3 gives a field configuration on S^3. Since $\pi_3(SU(N)) = \mathbb{Z}$, there are integer-valued classes of such configurations, and in fact the Wess–Zumino term is defined so as to compute this topological invariant Z: more precisely, the difference of the above integral for two different continuations into the bulk is k times $2\pi n$, where $n \in \mathbb{Z}$ measures the topological invariant of the map $S^3 \to S^3$ resulting from combining the two continuations as described above.

When we put this action into a quantum path integral, it therefore leads to a quantization of the level k to *integers*. $SU(2)_k$ with $k = 2$ can be viewed as a different representation of the same symmetry as the $SU(2)_1$ realized in the spin-$\frac{1}{2}$ Heisenberg chain, in the same way as the spins on one site are in different representations of ordinary $SU(2)$. The full demonstration that the model is gapless is beyond our present scope, but at least we have a topological understanding of why the Wess–Zumino term is a natural quantity to consider. One way to tell apart the gapless points associated with different levels or Lie groups is by computing the central charge c, a measure of how many degrees of freedom are gapless at the critical point, in units where one free boson gives $c = 1$. The WZW model for Lie group g at level k has central charge

$$c = \frac{k \dim SU(N)}{k + n}, \tag{1.101}$$

where $\dim SU(N) = N(N-1)$. Hence, $SU(3)_1$ has $c = 2$, $SU(2)_2$ has $c = \frac{3}{2}$, and $SU(2)_1$ has $c = 1$, consistent with its bosonized representation as a single boson.

1.4.3 Topologically ordered phases: the FQHE

1.4.3.1 Chern–Simons theory I: flux attachment and statistics change

We will now start the process of developing a more abstract description of the FQHE that will help us understand what type of order it has. For example, this will define precisely what it means to say that the physical state is adiabatically connected to the Laughlin wavefunction. Our main tool will be Chern–Simons theory; we briefly encountered the Chern–Simons term of the electromagnetic gauge potential when we discussed quantum Hall layers at the surface of the strong topological insulator, and we will come to that in a moment. However, a more fundamental use of Chern–Simons theory is to describe the internal degrees of freedom of the quantum Hall liquid. In other words, we will have both an 'internal' Chern–Simons theory describing the quantum Hall liquid and a Chern–Simons term induced in the electromagnetic action.

Since that sounds complicated, let's start by understanding why a Chern–Simons theory might be useful. To begin, we come up with a picture for the Laughlin state by noting that since the filled lowest Landau level has one magnetic flux quantum per electron, the Laughlin state at $m = 3$ (i.e. $\nu = \frac{1}{3}$) has three flux quanta per electron. To get a picture of how the Laughlin state is connected to the $\nu = 1$ state, we imagine attaching two of these flux quanta to each electron. The resulting 'composite fermion' still has fermionic statistics, by the following counting. Interchanging two electrons gives a -1 factor. The Aharonov–Bohm factor from moving an electron all the way around a flux quantum is $+1$, but in this exchange process, each electron moves only halfway around the flux quanta attached to the other electron. So when one of these objects is exchanged with another, the wavefunction picks up three factors of -1 and the statistics is still fermionic.

These composite fermions now can form the integer quantum Hall state in the remaining field of one flux quantum per composite fermion, leading to a $\nu = \frac{1}{3}$ incompressible state in terms of the original electrons. More generally, the phase picked up by a particle of charge q moving completely around a flux Φ is

$$e^{i\theta} = e^{iq\Phi/\hbar c}. \tag{1.102}$$

We will now see how the Chern–Simons term lets us carry out a 'flux attachment' related to the above composite fermion idea: in fact, by attaching three flux quanta rather than two to each electron, we would obtain bosons moving in zero applied field, and the Laughlin state can be viewed as a Bose–Einstein condensate of these 'composite bosons' [36, 44].[12]

The Abelian Chern–Simons theory we will study is described by the following Lagrangian density in (2 + 1)-dimensional Minkowski spacetime:

$$\mathcal{L} = 2\gamma\epsilon^{\mu\nu\lambda}a_\mu\partial_\nu a_\lambda + a_\mu j^\mu, \tag{1.103}$$

[12] One feature of the composite fermion picture that is preferable to the composite boson picture is that the former is naturally described as 'topological order', while the latter would lead to a picture of the phase in terms of the symmetry-breaking order of a Bose–Einstein condensate.

where γ is a numerical constant that we will interpret later, a is the Chern–Simons gauge field, and j is a conserved current describing the particles of the theory. Under a gauge transformation $a_\mu \to a_\mu + \partial_\mu \chi$, the Chern–Simons term (the first one) transforms as

$$\epsilon^{\mu\nu\lambda} a_\mu \partial_\nu a_\lambda \to \epsilon^{\mu\nu\lambda} a_\mu \partial_\nu a_\lambda + \epsilon^{\mu\nu\lambda} \partial_\mu \chi \partial_\nu a_\lambda, \tag{1.104}$$

where the term with two derivatives of χ drops out by antisymmetry. The new term can be written as

$$\delta S = 2\gamma \int d^2x\, dt\, \epsilon^{\mu\nu\lambda} \partial_\mu (\chi \partial_\nu a_\lambda), \tag{1.105}$$

where again the term with two derivatives acting on a gives zero by antisymmetry. So, *if we can neglect the boundary*, the Abelian Chern–Simons term is gauge-invariant. (As we discussed previously in the context of magnetoelectric polarizability, the non-Abelian Chern–Simons term is not gauge-invariant, because 'large' (non-null-homotopic) gauge transformations change the integral; this is related to the third homotopy group of $SU(N)$.) Later on. we will actually consider a system with a boundary and see how the boundary term leads to physically important effects.

Consider the equation of motion obtained by varying this action. We get

$$4\gamma \epsilon^{\mu\nu\lambda} \partial_n u a_\lambda = -j^\mu. \tag{1.106}$$

where the 4 rather than a 2 appears because the Chern–Simons term has non-zero derivative with respect to both a and ∂a. For a particle sitting at rest, the spatial components of the current vanish, but there must be a flux: writing in components, we have

$$\int d^2x\, (\partial_1 a_2 - \partial_2 a_1) = -\frac{1}{4\gamma} \int d^2x\, j^0. \tag{1.107}$$

Hence a charged particle in the theory gains a flux of the a field (since the left term is just the integral of a magnetic field). If the charge is localized, then the flux is localized as well.

What good is this? Well, we know that when one charged particle with respect to the a field moves around another, it will now pick up an Aharonov–Bohm phase from the attached flux in addition to any statistics factor. The additional statistics factor is

$$\theta = \frac{1}{8\gamma}, \tag{1.108}$$

where the $\frac{1}{2}$ here results because the particles only move halfway around each other in an exchange. In other words, if we started with $\theta = 0$ bosonic particles but added a $\gamma = 1/8\pi$ Chern–Simons term, we would obtain fermions, and vice versa. But so far nothing constrains γ, suggesting that in 2D, 'braiding' statistics is not constrained to be bosonic or fermionic. Particles in 2D that are neither bosonic nor fermionic are known as 'anyons'.

Why is two spatial dimensions so special? It turns out that an argument about why generalized statistics are possible for point particles in two spatial dimensions but not higher dimensions was given long ago by Leinaas and Myrheim [23]. The key observation is that an exchange path that takes one particle around another and back to its original location is not smoothly contractible in 2D without having the particles pass through each other, while in higher dimensions, such a path is contractible. The consequence of this is that in 2D, phase factors are not defined just for permutations of the particles but rather for any 'braiding'.[13]

1.4.3.2 Chern–Simons theory II: integrating out gauge fields and coupling to electromagnetism

Aside from the composite fermion/composite boson pictures, why might the Chern–Simons theory with Lagrangian density given by (1.103) describe quantum Hall states? Without working through a detailed derivation starting from non-relativistic quantum mechanics of many interacting electrons in a magnetic field (which is still not all that rigorous—for a discussion, see Zee's lecture notes [43]) we can note the following. A conserved electromagnetic current in $2+1$ dimensions can always be written as the curl of a gauge field:

$$J^\mu = \frac{1}{2\pi}\epsilon^{\mu\nu\lambda}\partial_\nu a_\lambda \tag{1.109}$$

(note that this electromagnetic current might in general be distinct from the particle current above). Here a is automatically a gauge field since the $U(1)$ gauge transformation does not modify the current. Gauge invariance forbids the mass term $a^\mu a_\mu$, so the lowest-dimension possible term is the Chern–Simons term, which we write for future use with a normalization different from that used above:

$$\mathcal{L}_{CS} = \frac{k}{4\pi}\epsilon^{\mu\nu\lambda}a_\mu\partial_\nu a_\lambda. \tag{1.110}$$

The point of the new normalization $k = 8\pi\gamma$ compared with (1.103) is that the boson–fermion statistics transformation above now corresponds just to $k = 1$. We will argue later that k should be an integer for the electron to appear somewhere in the spectrum of excitations of the theory.

Does this term need to appear? No—for example, in a system that has P or T symmetry, it cannot appear. However, if it does appear, then, since there is only one spatial derivative, it dominates the Maxwell term at large distances. Effectively, we define the quantum Hall phase as one in which $\mathcal{L}_{CS}$ appears in the low-energy Lagrangian; for example, this is true in both the Laughlin state and the physical state with Coulomb interactions, even though the overlap between those two ground-state wavefunctions is presumably zero in the thermodynamic limit.

[13] Even non-Abelian statistics are possible if there are multiple ground states: the phase factor associated with a particular braid is then a matrix acting on the set of ground states, and two such matrices need not commute.

What if we added the $a_\mu J^\mu$ coupling and integrated out the gauge field? Well, the main reason not to do so is that we then obtain a nonlocal current–current coupling. Since the original action is quadratic in the fields, this integration is not too difficult, but an alternative, equivalent way to do it is to solve for a in terms of J. Given a general Lagrangian

$$\mathcal{L} = \phi \mathcal{Q} \phi + \phi J, \tag{1.111}$$

where $\mathcal{Q}$ denotes some operator, we have the formal equation of motion from varying ϕ,

$$2\mathcal{Q}\phi = -J, \tag{1.112}$$

which is solved by

$$\phi = \frac{-1}{2\mathcal{Q}} J. \tag{1.113}$$

Then substituting this into the Lagrangian (and ignoring some subtleties about ordering of operators), we obtain

$$\mathcal{L} = \frac{1}{4} J \frac{1}{\mathcal{Q}} J - J \frac{1}{2\mathcal{Q}} J = -J \frac{1}{4\mathcal{Q}} J. \tag{1.114}$$

So, for the Chern–Simons term, we need to define the inverse of the operator $\epsilon^{\mu\nu\lambda}\partial_\nu$ that appears between the a fields. This is a bit subtle because there is a zero mode of the original operator, related to gauge invariance: for any smooth function g, $\epsilon^{\mu\nu\lambda}\partial_\nu(\partial_\lambda g) = 0$. To define the inverse, we fix the Lorentz gauge $\partial_\mu a_\mu = 0$. In this gauge, we look for an inverse using

$$(\epsilon^{\mu\nu\lambda}\partial_\nu)(\epsilon^{\lambda\alpha\beta}\partial_\alpha a_\beta) = \epsilon^{\mu\nu\lambda}\epsilon^{\lambda\alpha\beta}(\partial_\nu \partial_\alpha a_\beta). \tag{1.115}$$

We can combine the ϵ tensors by noting that $\epsilon^{\mu\nu\lambda} = \epsilon^{\lambda\mu\nu}$, so there are two types of non-zero terms in the above: either $\mu = \alpha$ and $\nu = \beta$ or vice versa, with a minus sign in the second case. From the first type of term, we obtain $\partial_\alpha(\partial_\beta a_\beta)$, which is zero by our gauge choice. From the second type, we obtain

$$-\partial_\nu^2 a_\mu. \tag{1.116}$$

So the inverse of the operator appearing in the Chern–Simons term in this gauge is $-\epsilon^{\mu\nu\lambda}\partial_\nu/\partial^2$, and the Lagrangian (1.103) with the gauge field integrated out is just

$$\mathcal{L} = \frac{1}{8\gamma} j_\mu \left(\frac{\epsilon^{\mu\nu\lambda}\partial_\nu}{\partial^2} \right) j_\lambda. \tag{1.117}$$

Aside from showing another interesting difference between the Chern–Simons term and the Maxwell term, we can use this inverse to couple the Chern–Simons theory to an external electromagnetic gauge potential $\mathcal{A}_\mu$. We will set $e = \hbar = 1$ except when otherwise noted. We do not include the Maxwell term to give this field dynamics, but rather view it as an imposed field *beyond the magnetic field producing the phase*. For example, we could use this additional field to add an electrical field, and we should find a Hall response. Let's try this:

$$\begin{aligned}\mathcal{L} &= \frac{k}{4\pi}\epsilon^{\mu\nu\lambda}a_\mu\partial_\nu a_\lambda - \frac{1}{2\pi}\epsilon^{\mu\nu\lambda}A_\mu\partial_\nu a_\lambda \\ &= \frac{k}{4\pi}\epsilon^{\mu\nu\lambda}a_\mu\partial_\nu a_\lambda - \frac{1}{2\pi}\epsilon^{\mu\nu\lambda}a_\mu\partial_\nu A_\lambda,\end{aligned} \tag{1.118}$$

where in the second step we have dropped a boundary term and used the antisymmetry property of the ϵ tensor. Note that to obtain the second term, we have just rewritten $A_\mu J^\mu$ using (1.109).

Now we can integrate out a_μ using equation (1.117) above, recalling $\gamma = k/8\pi$, and obtain

$$\begin{aligned}\mathcal{L}_{\text{eff}} &= \frac{\pi}{k}J_\mu\epsilon^{\mu\nu\lambda}\partial_\nu\frac{1}{\partial^2}J_\lambda \\ &= \frac{1}{4\pi k}\epsilon^{\mu\alpha\beta}\partial_\alpha A_\beta\epsilon^{\mu\nu\lambda}\partial_\nu\frac{1}{\partial^2}\epsilon^{\lambda\gamma\delta}\partial_\gamma A_\delta.,\end{aligned} \tag{1.119}$$

where in the second step we have used the rewritten Lagrangian in (1.118) to identify $J^\mu = (1/2\pi)\epsilon^{\mu\nu\lambda}\partial_\nu A_\lambda$. As above, the non-zero possibilities are $\alpha = \nu$ and $\beta = \lambda$ $(+1)$ or vice versa (-1), and also $\gamma = \mu$ and $\delta = \nu$ $(+1)$ or vice versa (-1). Working through these, we are left with the $\gamma = \nu$ and $\delta = \mu$ terms:

$$\mathcal{L}_{\text{eff}} = \frac{1}{4\pi k}\epsilon^{\mu\nu\lambda}A_\mu\partial_\nu A_\lambda. \tag{1.120}$$

This is the *electromagnetic* Chern–Simons term. The electromagnetic current is obtained by varying A:

$$J^\mu = -\frac{\delta\mathcal{L}_{\text{eff}}}{\delta A_\mu} = \frac{1}{2\pi k}\epsilon^{\mu\nu\lambda}\partial_\nu A_\lambda. \tag{1.121}$$

where the factor of 2 is obtained because the variation can act on either A.

We can see immediately that this predicts a Hall effect: in response to an electrical field along x, we obtain a current along y. What about the factor $1/2\pi$? That is here just so that the response, once we restore factors of e and $\hbar$, is

$$\sigma_{xy} = \frac{e^2}{(2\pi)k\hbar} = \frac{1}{k}\frac{e^2}{h}. \tag{1.122}$$

Here we get a clue about the physical significance of k. Another clue is to consider the electromagnetic charge J^0 induced by a change in the magnetic field δB (i.e. an additional field beyond the one producing the FQHE):

$$J^0 = \delta n = \frac{1}{2\pi k}\delta B. \tag{1.123}$$

where we have written $J^0 = \delta n$ to indicate that this electromagnetic density describes the change in electron density from the ground state without the additional field. For the IQHE, a change of one flux quantum corresponds to one additional electron, while we can see that the $k = 3$ Chern–Simons theory predicts a change in density $e/3$, consistent with the quasihole and quasiparticle excitations.

To summarize what we have learned so far, we now see that Chern–Simons theory predicts a connection between the Hall quantum, the statistics of quasiparticles in the theory (from the previous section), and the effective density induced by a local change in the magnetic field. Here, a 'quasiparticle', which we will discuss later, means whatever particle couples to the Chern–Simons theory as in Section 1.4.3.1.1, which need not be an electron.

1.4.3.3 Chern–Simons theory III: topological aspects and gapless edge excitations

One obvious respect in which the Chern–Simons theory is topological is that, because ϵ rather than the metric tensor g was used to raise the indices, there is no dependence on the metric. In Zee's language, it describes a world without rulers or clocks. Since the stress–energy tensor in a relativistic theory is determined by varying the Lagrangian with respect to the metric, the stress–energy tensor is identically zero.

How can a theory be interesting if all its states have zero energy, as in the pure Chern–Simons theory? Well, one interesting fact is that the number of zero-energy states is dependent on the manifold where the theory is defined. We will not try to compute this in general but will solve the theory for the case of the torus. It is quite surprising that we can solve this $(2+1)$-dimensional field theory exactly; the key will be that there are very few physical degrees of freedom once the $U(1)$ gauge invariance is taken into account.

We wish to solve the pure Chern–Simons theory with action

$$\mathcal{L}_{CS} = \frac{k}{4\pi}\epsilon^{\mu\nu\lambda} a_\mu \partial_\nu a_\lambda \tag{1.124}$$

on the manifold $\mathbb{R}(\text{time}) \times T^2(\text{space})$. The gauge invariance is under $a_\mu \to a_\mu + \partial_\mu \chi$, with χ an arbitrary scalar function. Given an arbitrary configuration of the gauge field a_μ, we first fix $a_0 = 0$ by the gauge transformation $a_\mu \to a_\mu + \partial_\mu \chi$ with $\chi = -\int a_0\, dt$. The Lagrangian is then

$$\mathcal{L} = -\frac{k}{4\pi}\epsilon^{ij} a_i \dot{a}_j, \tag{1.125}$$

where $i, j = 1, 2$. The equation of motion from varying the original Lagrangian with respect to a_0 now gives a constraint

$$\epsilon_{ij}\partial_i a_j = 0. \tag{1.126}$$

There is still some gauge invariance remaining in a_1, a_2: we can add a purely spatially dependent χ, so that a_0 remains 0, to make $\partial_i a_i = 0$ (an exercise for the reader).

Then $(a_i(t), a_j(t))$ have zero spatial derivatives and hence are purely functions of time. The Lagrangian (1.125) is now just the minimal coupling of a particle moving in a position-dependent vector potential; thinking of (a_1, a_2) as the coordinates of a particle moving in the plane, and noting that a constant magnetic field can be described by the vector potential $(\frac{1}{2}By, -\frac{1}{2}Bx) = (\frac{1}{2}Ba_2, -\frac{1}{2}Ba_1)$, we see that this is the interaction term of a particle in a constant magnetic field.

So far, using gauge invariance, we can reduce the degrees of freedom from a $(2+1)$-dimensional field theory to the path integral for the quantum mechanics of a particle moving in 2D. There is one last bit of gauge invariance we need to use. This will reduce the space on which our particle moves, which so far is $\mathbb{R}^2$ because the gauge fields are non-compact, to the torus T^2 on which the theory is defined. We consider a gauge transformation of the form $a_j \to a_j - iu^{-1}\partial_j u$, where u is purely a function of space. Note that if we can write $u = e^{i\theta}$, then this becomes a conventional gauge transformation $a_j \to a_j + \partial_j\theta$. This gauge transformation will not break the previous two gauge constraints if $\nabla^2\theta = 0$.

However, the periodicity of the torus means that we might not be able to define θ periodically, even if u is defined globally and the gauge transformation is indeed periodic. Taking the torus to be $L_1 \times L_2$, the following θ has zero Laplacian everywhere and gives rise to a periodic u and hence a periodic gauge transformation, even if θ is not itself periodic:

$$\theta = \frac{2\pi n_1 x}{L_1} + \frac{2\pi n_2 y}{L_2}. \tag{1.127}$$

The effect of this gauge transformation is that we can shift the particle's trajectory by an arbitrary constant integer multiple of L_1 in the x direction and L_2 in the y direction. To make the torus equivalent to the unit torus, we can rescale $a_i(t) = (2\pi/L_i)q_i(t)$. So, finally, we have shown

$$\begin{aligned} S &= \int d^2x\, dt\, \frac{k}{4\pi}\epsilon^{\mu\nu\lambda}a_\mu\partial_\nu a_\lambda \\ &= -\frac{kL_1L_2}{4\pi}\int dt\, \frac{(2\pi)^2}{L_1L_2}\epsilon^{ij}q_i\dot{q}_j. \end{aligned} \tag{1.128}$$

Here one L_1L_2 factor is from the spatial integrals and one is from the change of variable from a_i to q_i. We still haven't done anything quantum-mechanical to solve the path integral. However, we can temporarily add a term $m\dot{q}_i^2/2$ to the Lagrangian and recognize it as the path integral for a particle moving on the torus in a constant magnetic field. The gauge potential is $A_i = k\pi\epsilon_{ij}q_j$, which corresponds to a magnetic field $B = 2\pi k$ (this factor of 2 always appears in the rotational gauge). This is in our theorist's units with $\hbar = e = 1$; it means that there are a total of k flux quanta through the torus.

The limit we care about for pure Chern–Simons theory is $m \to 0$, which takes all states not in the lowest Landau level to infinite energy. This makes sense because in a topological theory there can be no energy scale—the states either have some constant energy (the lowest Landau level here), which can be taken to zero, or infinite energy (the other Landau levels here). A quick calculation shows that there are exactly k states in the lowest Landau level on the torus pierced by k flux quanta; note that the 'shift' of one extra level on the sphere is absent. For example, the lowest Landau level with one flux quantum through the *sphere* corresponds to the coherent-state path integral for a $s = \frac{1}{2}$ particle (see [2]), with two degenerate states.

The conclusion is that the parameter k also controls the ground-state degeneracy on the torus. An argument by Wen and Niu [41] (regrettably, direct calculation seems to be more difficult) shows that the general degeneracy of the pure Abelian Chern–Simons theory on a 2-manifold of genus g is k^g. So, for a topological theory, the physical content of the model is determined not just by explicit parameters in the action, such as k, but also by the topology of the manifold where the theory is defined. In this sense, topological theories are sensitive to global or 'long-ranged' properties, even though the theory is massive/gapped. (Of course, in the pure Chern–Simons theory, there is no notion of length, so the distinction between local and global doesn't mean much, but adding a Maxwell term or something like that would not modify the long-distance properties—it would just mean that the other Landau levels were no longer at infinite energy.)

1.4.3.4 Bulk–edge correspondence

We noted above that the Chern–Simons term has different gauge-invariance properties from the Maxwell term: in particular, in a system with a boundary, it is not gauge-invariant by itself, because the boundary term we found above need not vanish. Our last goal in this section is to see that this gauge invariance leads to the free massless chiral boson theory at the edge:

$$S_{\text{edge}} = \frac{k}{4\pi} \int dt\, dx\, (\partial_t + v\partial_x)\phi\partial_x\phi. \tag{1.129}$$

Here k is exactly the same integer coefficient as in the bulk Chern–Simons theory, while v is a non-universal velocity that depends on the confining potential and other details. Note that the kinetic term here is 'topological' in the sense that it does not contribute to the Hamiltonian, because it is first-order in time. The second term is not topological and hence shouldn't be directly obtainable from the bulk theory.

The theory of the bulk and boundary is certainly invariant under 'restricted' gauge transformations that vanish at the boundary: $a_\mu \to a_\mu + \partial_\mu \chi$ with $\chi = 0$ on the boundary. From (1.105), the boundary term vanishes if $\chi = 0$ there. This constraint means that degrees of freedom that were previously gauge degrees of freedom now become dynamical degrees of freedom. We will revisit this idea later.

To start, choose the gauge condition $a_0 = 0$ as in Section 1.4.3.3 and again use the equation of motion for a_0 as a constraint.[14] Then $\epsilon^{ij} a_j = 0$ and we can write $a_i = \partial_i \phi$. Substituting this into the bulk Chern–Simons Lagrangian, we have

[14] Here and before, we are assuming that the Jacobians from our gauge-fixings and changes of variables are trivial. That this is the case is argued in Elitzur et al. [8]. Another nice discussion in this paper is how, for the non-Abelian case, the bulk can be understood as providing the Wess–Zumino term that keeps the edge theory gapless.

$$\begin{aligned} S &= -\frac{k}{4\pi}\int \epsilon^{ij} a_i \partial_0 a_j \, d^2x\, dt \\ &= -\frac{k}{4\pi}\int (\partial_x\phi\partial_0\partial_y\phi - \partial_y\phi\partial_0\partial_x\phi)\, d^2x\, dt \\ &= -\frac{k}{4\pi}\int [\partial_x(\phi\partial_0\partial_y\phi) - \partial_y(\phi\partial_0\partial_x\phi)]\, d^2x\, dt \\ &= -\frac{k}{4\pi}\int (\nabla\times\mathbf{v})_z\, d^2x\, dt = -\frac{k}{4\pi}\int \mathbf{v}\cdot d\mathbf{l}\, dt, \end{aligned} \tag{1.130}$$

where $\mathbf{v}$ is the vector field

$$\mathbf{v} = (\phi\partial_0\partial_x\phi, \phi\partial_0\partial_y\phi). \tag{1.131}$$

(You might wonder why this doesn't let us transform the action simply to zero in the case of the torus studied in Section 1.4.3.3. The reason is that when using Stokes's theorem in the final line, we have assumed the disk topology—since the torus has non-trivial topology, we are not allowed to use Stokes's theorem to obtain zero, cf. Section 1.2.1.) So at the boundary, which we will assume to run along x for compactness, the resulting action is, after an integration by parts,

$$S_{\text{edge}} = \frac{k}{4\pi}\int \partial_t\phi\, \partial_x\phi\, dx\, dt. \tag{1.132}$$

We're now almost done—this predicts a 'topological' edge theory determined by the bulk physics, and this edge theory is topological in that the Hamiltonian is zero. However, in order to obtain an accurate physical description, we need to include non-universal, non-topological physics arising from the details of how the Hall droplet is confined. One approach to this is to start from a hydrodynamical theory of the edge and then recognize one term in that theory as S_{edge} above. The other term in that theory is a non-universal velocity term, and the combined action is

$$S_{\text{edge}} = \frac{k}{4\pi}\int (\partial_t\phi - v\partial_x\phi)\, \partial_x\phi\, dx\, dt. \tag{1.133}$$

Here the non-universal parameter v clearly has units of a velocity, and in the correlation functions of the theory discussed below indeed appears as a velocity. The Hamiltonian density is

$$\mathcal{H} = \frac{kv}{4\pi}(\partial_x\phi)^2. \tag{1.134}$$

Note that for the Hamiltonian to be positive-definite, the product kv needs to be positive: in other words, the sign of the velocity is determined by the bulk parameter k even thought the magnitude is not, and the edge is indeed chiral. (The density at the edge is found from the hydrodynamical argument to be proportional to $\partial_x\phi/2\pi$, so the above interaction term corresponds to a short-ranged density–density interaction; as usual, we will neglect the differences that arise if the long-ranged Coulomb interaction is retained instead.)

1.4.3.5 Chern–Simons theory IV: connecting edge theory to observables

We give a quick overview of how the above theory leads to detailed predictions of several edge properties. The general approach to treating 1D electronic systems via free-boson theories is known as 'bosonization', and is the subject of several books (e.g. [38]). While we will not calculate the main results in detail, it turns out that there is a close similarity between the 1D free (chiral or non-chiral) boson Lagrangian and the theory of the algebraic phase of the XY model studied previously.

The reason such a connection exists is simple: the Euclidean version of the non-chiral version of the above free-boson theory is just the 2D Gaussian theory. However, we know from the study of the XY model that subtleties such as the Berezinskii–Kosterlitz–Thouless transition arise when the variable appearing in the Gaussian theory is taken to be periodic, as when it describes an angular variable in that model. One of the surprising results we found was a power-law phase with continuously variable exponents: the correlations of spin operators $S_x + iS_y = e^{i\theta}$ go as a power law with the coefficient depending on the prefactor of the Gaussian.

The connection between the edge theory above and physical quantities is that the electron correlation function is represented in the bosonized theory as $e^{ik\phi}$: effectively, ϕ describes a single quasiparticle and k quasiparticles make up the electron. The electron propagator in momentum space is likewise here found to have an exponent that depends on k: there is a factor of k^2 from the k's in the electron operator, and a factor of k^{-1} from the quasiparticle propagator since k appears as a coefficient in the Lagrangian. The result is

$$G(q,\omega) \propto \frac{(vq+\omega)^{k-1}}{vq-\omega}. \tag{1.135}$$

This describes an electron density of states $N(\omega) \propto |\omega|^{k-1}$, and this exponent can be measured in tunnelling exponents: $dI/dV \propto V^{k-1}$. As a sanity check, the $k=1$ case describes a constant density of states and the predicted conduction is Ohmic: $I \propto V$.

Experimental agreement is reasonable but hardly perfect; at $\nu = \frac{1}{3}$, the observed tunnelling exponent $I \propto V^{\alpha}$ satisfies $\alpha \approx 2.7$, which is far from the Ohmic value ($\alpha = 1$) but reasonably close to the predicted value $\alpha = 3$. The tunnelling exponent also does not appear to be perfectly constant when one is on a Hall plateau, as the theory would predict. Other measurements include 'noise' measurements that attempt to see the quasiparticle charge directly and, in recent years, interferometry measurements that try to check more subtle aspects of the theory.

In closing, we comment briefly on the generalization of the above Chern–Simons and edge theories to more complicated (but still Abelian) quantum Hall states. These states, as suggested by the hierarchy picture, have multiple types of 'particles', and two particles can have non-trivial statistics whether or not they belong to the same species. These statistics are defined by a universal integer 'K matrix' that can be taken as a fundamental aspect of the topological order in the state. (Information must also be provided about the allowed quasiparticle types.) The resulting Chern–Simons theory is

$$\mathcal{L} = \frac{1}{4\pi} K^{IJ} a^I_\mu \partial_\nu a^J_\lambda. \tag{1.136}$$

This effective theory works for all but a few of the many quantum Hall states observed in experiment. It is hoped that this chapter has clarified how a few geometrical structures (the Berry phase, Chern and Chern–Simons forms, etc.) underlie a remarkable diversity of physics.

1.A Topological invariants in 2D with time-reversal invariance

The point of this appendix is to provide additional details on the topological invariants in the 2D case, which are more difficult to write down explicitly in terms of the Berry phase than in the 3D case. One way of doing this is educational since it builds on the classic work of Wess and Zumino that was alluded to earlier in the discussion of spin chains.

1.A.1 An interlude: Wess–Zumino terms in 1D nonlinear σ-models

A mathematical strategy similar to what we will need for the QSHE was developed by Wess and Zumino in the context of $(1+1)$-dimensional field theory. The free boson is described by the action

$$S_0 = -\frac{K}{2}\int_{\mathbb{R}^2} (\nabla\phi)^2, \tag{1.137}$$

which for a compact boson field ϕ is the nonlinear sigma model into the circle S^1, which is the manifold of the Lie group $U(1)$. The direct generalization of this to a more complicated Lie group such as $SU(N)$ is written as

$$S_0 = -\frac{k}{8\pi}\int_{S^2} \mathcal{K}(g^{-1}\partial^\mu g, g^{-1}\partial_\mu g)\,, \tag{1.138}$$

where we have compactified the plane to the sphere, changed the prefactor, and written the interaction in terms of the 'Killing form' $\mathcal{K}$ on the Lie algebra associated with g. (This Killing form is a symmetric bilinear form that, in the $U(1)$ case above, is just the identity matrix.) Unfortunately, this action behaves quite differently from the $U(1)$ case: it does not describe a critical theory (in particle physics language, it develops a mass).

To fix this problem, Wess and Zumino wrote a term that is quite remarkable:

$$S_{WZ} = -\frac{2\pi k}{48\pi^2}\int_{B^3} \epsilon_{\mu\nu\lambda}\mathcal{K}(g^{-1}\partial_\mu g, [g^{-1}\partial_\nu g, g^{-1}\partial_\lambda g])\,. \tag{1.139}$$

Even writing this term depends on being able to take an original configuration of g on the sphere S^2 and extend it into the sphere's interior B^3. (We will not show here that this term accomplishes the desired purpose, just that it is topologically well defined.) At least one contraction into the ball exists, because $\pi_2(G) = 0$. Different contractions exist, because $\pi_3(G) = \mathbb{Z}$, and the coefficient of the second term is chosen so that, if k (the 'level' of the resulting WZW theory) is an integer, the different topological classes differ by a multiple of $2\pi i$ in the action, so that the path integral is independent of what contraction is chosen. The reason that $\pi_3(G)$ is relevant here is that two different contractions into the interior B^3 can be joined together at their common boundary to form a 3-sphere, in the same way as two disks with the same boundary can be joined together to form the top and bottom hemispheres of a 2-sphere.

1.A.2 Topological invariants in time-reversal-invariant Fermi systems

The main subtlety in finding a topological invariant for time-reversal-invariant band structures will be in keeping track of the time-reversal requirements. We introduce $\mathcal{Q}$ as the space of time-reversal-invariant Bloch Hamiltonians. This is a subset of the space of Bloch Hamiltonians with at most pairwise degeneracies (the generalization of the non-degenerate case we described above; we need to allow pairwise degeneracies because bands come in Kramers-degenerate pairs). In general, a $\mathcal{T}$-invariant system need not have Bloch Hamiltonians in $\mathcal{Q}$ except at the four special points where $k = -k$. The homotopy groups of $\mathcal{Q}$ follow from similar methods to those used above: $\pi_1 = \pi_2 = \pi_3 = 0$, $\pi_4 = \mathbb{Z}$. $\mathcal{T}$-invariance requires an even number of bands $2n$, so $\mathcal{Q}$ consists of $2n \times 2n$ Hermitian matrices for which H commutes with Θ, the representation of $\mathcal{T}$ in the Bloch Hilbert space:

$$\Theta H(k)\Theta^{-1} = H(-k). \tag{1.140}$$

Our goal in this subsection is to give a geometric derivation of a formula, first obtained by Fu and Kane [12] via a different approach, for the $\mathbb{Z}_2$ topological invariant in terms of the Berry phase of Bloch functions:

$$D = \frac{1}{2\pi}\left[\oint_{\partial(\text{EBZ})} d\mathbf{k}\cdot\boldsymbol{\mathcal{A}} - \int_{\text{EBZ}} d^2k\,\mathcal{F}\right] \quad \text{mod } 2. \tag{1.141}$$

Here EBZ stands for 'effective Brillouin zone' [27], which describes one half of the Brillouin zone together with appropriate boundary conditions. Since the Brillouin zone is a torus (see Fig. 1.2(a)), the EBZ can be viewed as a cylinder, and its boundary $\partial(\text{EBZ})$ as two circles, as in Fig. 1.2(b). While $\mathcal{F}$ is gauge-invariant, $\boldsymbol{\mathcal{A}}$ is not, and different (time-reversal-invariant) gauges, in a sense made precise below, can change the boundary integral by an even amount. The formula (1.141) was not the first definition of the 2D $\mathbb{Z}_2$ invariant, as the original Kane–Mele paper [20] gave a definition based on counting of zeros of the 'Pfaffian bundle' of wavefunctions. However, (1.141) is both easier to connect to the IQHE and easier to implement numerically.

The way to understand this integral is as follows. Since the EBZ has boundaries, unlike the torus, there is no obvious way to define Chern integers for it; put another way, the $\mathcal{F}$ integral above is not guaranteed to be an integer. However, given a map from the EBZ to Bloch Hamiltonians, we can imitate the Wess–Zumino approach above and consider 'contracting' or 'extending' the map to be one defined on the sphere (Fig. 1.3), by finding a smooth way to take all elements on the boundary to some constant element $\mathcal{Q}_0 \in \mathcal{Q}$. The geometric interpretation of the line integrals of $\mathcal{A}$ in (1.141) is that these are the integrals of $\mathcal{F}$ over the boundaries, and the requirement

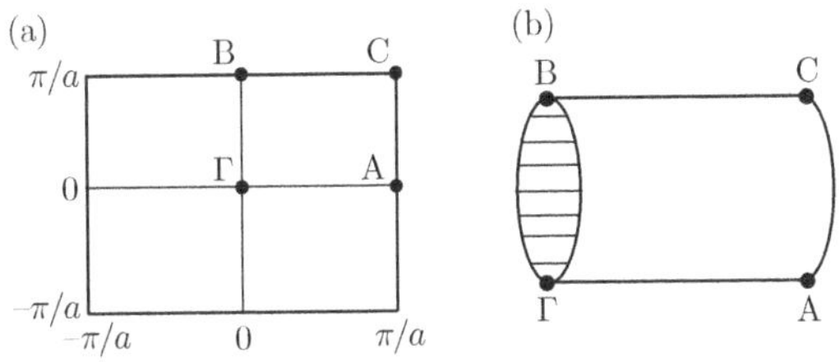

Fig. 1.2 (a) A 2D Brillouin zone; note that any such Brillouin zone, including that for graphene, can be smoothly deformed into a torus. The labelled points are time-reversal-invariant momenta. (b) The effective Brillouin zone (EBZ). The horizontal lines on the boundary circles ∂(EBZ) connect time-reversal-conjugate points, where the Hamiltonians are related by time reversal and therefore cannot be specified independently. (Reprinted with permission from [27]. Copyright 2007 by the American Physical Society.)

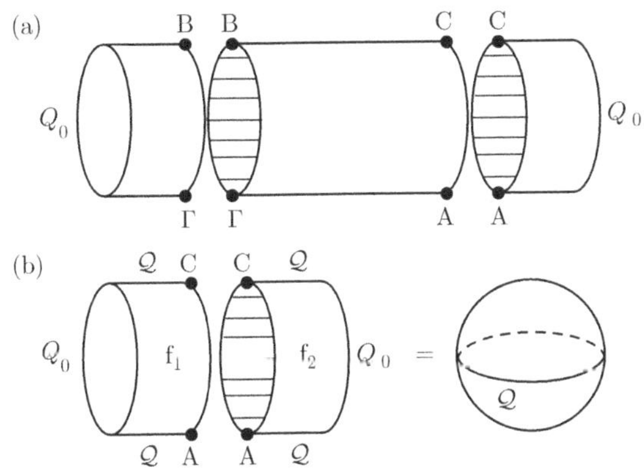

Fig. 1.3 (a) Contracting the extended Brillouin zone to a sphere. (b) Two contractions can be combined to give a mapping from the sphere, but this sphere has a special property: points in the northern hemisphere are conjugate under $\mathcal{T}$ to those in the southern, in such a way that overall every band pair's Chern number must be even. (Reprinted with permission from [27]. Copyright 2007 by the American Physical Society.)

on the gauge used to define the two $\mathcal{A}$ integrals is that each extends smoothly in the associated cap. The condition on the cap is that each vertical slice satisfy the same time-reversal-invariance condition as an EBZ boundary; this means that a cap can alternatively be viewed as a way to smoothly deform the boundary to a constant while maintaining the time-reversal condition at each step.

The two mathematical steps, as in the Wess–Zumino term, are showing that such contractions always exist and that the invariant D in (1.141) is invariant under the choice of contraction. The first step is rather straightforward and follows from $\pi_1(\mathcal{H}) = \pi_1(\mathcal{Q}) = 0$. The second step is more subtle and gives an understanding of why only a $\mathbb{Z}_2$ invariant or 'Chern parity' survives, rather than an integer-valued invariant as in the IQHE. We can combine two different contractions of the same boundary into a sphere, and the Chern number of each band pair on this sphere gives the difference between the Chern numbers of the band pair obtained using the two contractions (Fig. 1.3).

The next step is to show that the Chern number of any band pair on the sphere is even. To accomplish this, we note that Chern number is a homotopy invariant and that it is possible to deform the Bloch Hamiltonians on the sphere so that the equator is the constant element $\mathcal{Q}_0$ (here the equator came from the time-reversal-invariant elements at the top and bottom of each allowed boundary circle.) The possibility of deforming the equator follows from $\pi_1(\mathcal{Q}) = 0$, and the equivalence of different ways of deforming the equator follows from $\pi_2(\mathcal{Q}) = 0$. Then the sphere can be separated into two spheres, related by time reversal, and the Chern numbers of the two spheres are equal, so the total Chern number is zero.

The above argument establishes that the two values of the $\mathbb{Z}_2$ invariant are related to even or odd Chern number of a band pair on half the Brillouin zone. Note that the lack of an integer-valued invariant means, for example, that we can smoothly go from an S_z-conserved model with $\nu = 1$ for spin $\uparrow$, $\nu = -1$ for spin $\downarrow$ to a model with $\nu = \pm 3$ by breaking S_z conservation in between. This can be viewed as justification for the physical argument given above in terms of edge states annihilating in pairs, once we define the $\mathbb{Z}_2$ invariant for disordered systems in the following subsection.

1.A.3 Pumping interpretation of $\mathbb{Z}_2$ invariant

We expect that, as for the IQHE, it should be possible to reinterpret the $\mathbb{Z}_2$ invariant as an invariant that describes the response of a finite toroidal system to some perturbation. In the IQHE, the response is the amount of charge that is pumped around one circle of the torus as a 2π flux (i.e. a flux hc/e) is pumped adiabatically through the other circle.[15] Here, the response will again be a pumped charge, but the cyclic process that pumps the charge is more subtle.

Instead of inserting a 2π flux through a circle of the toroidal system, we insert a π flux, adiabatically (Fig. 1.4); this is consistent with the part of D in (1.141) that is obtained by integration over half the Brillouin zone. However, while a π flux is compatible with T invariance, it is physically distinct from zero flux, and hence this process is not a closed cycle. We need to find some way to return the system to its initial conditions. We allow this return process to be anything that does not close the gap, but require that the Hamiltonians in the return process *not* break time-reversal symmetry. Since the forward process, insertion of a π flux, definitely breaks time-reversal symmetry, this means that the whole closed cycle is a non-trivial loop in Hamiltonian space. The $\mathbb{Z}_2$ invariant then describes whether the charge pumped by this closed cycle through the other circle of the torus is an odd or even multiple of the

[15] A previous pumping definition that involves a π flux but considers pumping of '$\mathbb{Z}_2$' from one boundary to another of a large cylinder was given by Fu and Kane [12].

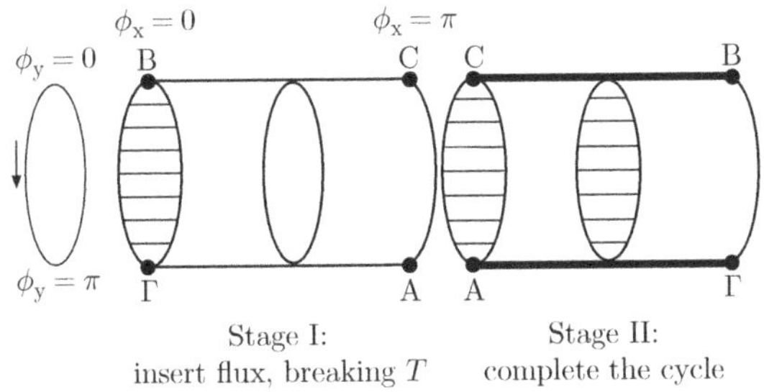

Fig. 1.4 Graphical representation of charge pumping cycle for Chern parities. The first stage takes place as the flux ϕ_x increases adiabatically from 0 to π. In the second stage, the Hamiltonian at $(\phi_x = \pi, \phi_y)$ is adiabatically transported through the space of Hamiltonians to return to the Hamiltonian at $(\phi_x = 0, \phi_y)$. The difference between the second stage and the first is that at every step of the second stage, the Hamiltonians obey the time-reversal conditions required at $\phi_x = 0$ or $\phi_x = \pi$. The bold lines indicate paths along which all Hamiltonians are time-reversal-invariant, and the disk with horizontal lines indicates, as before, how pairs of points in the second stage are related by time reversal. (Reprinted with permission from [9]. Copyright 2007 by the American Physical Society.)

electron charge; while the precise charge pumped depends on how the cycle is closed, the parity of the pumped charge (i.e. whether it is odd or even) does not.

This time-reversal-invariant closure is one way to understand the physical origin of the $\mathcal{A}$ integrals in (1.141), although here, by requiring a closed cycle, we have effectively closed the EBZ to a torus rather than a sphere. One weakness of the above pumping definition, compared with the IQHE, is that obtaining the $\mathbb{Z}_2$ invariant depends on Fermi statistics, so that this definition cannot be directly applied to the many-body wavefunction as in the IQHE case. The problem of a pumping-like definition of the 3D topological insulator for many electrons is solved (aside from possible exotic states with fractional charge) by the 'axion electrodynamics' discussed in Section 1.3.8.

References

[1] N. W. Ashcroft and N. D. Mermin. *Solid State Physics.* Holt, Rinehart, and Winston, New York, 1976.

[2] A. Auerbach. *Interacting Electrons and Quantum Magnetism.* Springer-Verlag, New York, 1994.

[3] J. E. Avron, R. Seiler, and B. Simon. Homotopy and quantization in condensed matter physics. *Phys. Rev. Lett.*, **51**, 51 (1983).

[4] B. A, Bernevig with T. L. Hughes. *Topological Insulators and Topological Superconductors.* Princeton University Press, Princeton, NJ, 2013.

[5] B. A. Bernevig, T. L. Hughes, and S.-C. Zhang. Quantum spin Hall effect and topological phase transition in HgTe quantum wells. *Science*, **314**, 1757 (2006).

[6] S. Das Sarma and A. Pinczuk (eds.). *Perspectives in Quantum Hall Effects: Novel Quantum Liquids in Low-Dimensional Semiconductor Structures*. Wiley, New York, 1997.

[7] G. V. Dunne. Aspects of Chern–Simons theory. In A. Comtet, T. Jolicoeur, S. Ouvry, and F. David (eds.). *Topological Aspects of Low Dimensional Systems. Proceedings of Les Houches Summer School, Session LXIX, 7–31 July 1998*, p. 177. Springer-Verlag, Berlin and Les Editions de Physique, Les Ulis, 1999.

[8] S. Elitzur, G. Moore, A. Schwimmer, and N. Seiberg. Remarks on the canonical quantization of the Chern–Simons–Witten theory. *Nucl. Phys. B*, **326**, 108 (1989).

[9] A. M. Essin and J. E. Moore. Topological insulators beyond the Brillouin zone via Chern parity. *Phys. Rev. B*, **76**, 165307 (2007).

[10] A. M. Essin, J. E. Moore, and D. Vanderbilt. Magnetoelectric polarizability and axion electrodynamics in crystalline insulators. *Phys. Rev. Lett.*, **102**, 146805 (2009).

[11] A. M. Essin, A. M. Turner, J. E. Moore, and D. Vanderbilt. Orbital magnetoelectric coupling in band insulators. *Phys. Rev. B*, **81**, 205104 (2010).

[12] L. Fu and C. L. Kane. Topological insulators with inversion symmetry. *Phys. Rev. B*, **76**, 045302 (2007).

[13] W. Fulton. *Algebraic Topology: A First Course*. Springer-Verlag, New York, 1995.

[14] S. M. Girvin. The quantum Hall effect: novel excitations and broken symmetries. In A. Comtet, T. Jolicoeur, S. Ouvry, and F. David (eds.). *Topological Aspects of Low Dimensional Systems. Proceedings of Les Houches Summer School, Session LXIX, 7–31 July 1998*, p. 53. Springer-Verlag, Berlin and Les Editions de Physique, Les Ulis, 1999.

[15] F. D. M. Haldane. Model for a quantum Hall effect without Landau levels: condensed-matter realization of the 'parity anomaly'. *Phys. Rev. Lett.*, **61**, 2015 (1988).

[16] M. Z. Hasan and C. L. Kane. Topological insulators. *Rev. Mod. Phys.*, **82**, 3045 (2010).

[17] M. Z. Hasan and J. E. Moore. Three-dimensional topological insulators. *Annu. Rev. Condens. Matter Phys.*, **2**, 55 (2011).

[18] J. K. Jain. *Composite Fermions*. Cambridge University Press, Cambridge, 2007.

[19] C. L. Kane and E. J. Mele. Quantum spin Hall effect in graphene. *Phys. Rev. Lett.*, **95**, 146802 (2005).

[20] C. L. Kane and E. J. Mele. Z_2 topological order and the quantum spin Hall effect. *Phys. Rev. Lett.*, **95**, 146802 (2005).

[21] M. Koenig, S. Wiedmann, C. Bruene, A. Roth, H. Buhmann, L. W. Molenkamp, X.-L. Qi, and S.-C. Zhang. Quantum spin Hall insulator state in HgTe quantum wells. *Science*, **318**, 766 (2007).

[22] A. Läuchli, G. Schmid, and S. Trebst. Spin nematics correlations in bilinear–biquadratic $s = 1$ spin chains. *Phys. Rev. B*, **74**, 144426 (2006).

[23] J. M. Leinaas, and J. Myrheim. On the theory of identical particles. *Nuovo Cim. B*, **37**, 1 (1977).

[24] A. Malashevich, I. Souza, S. Coh, and D. Vanderbilt. Theory of orbital magnetoelectric response. *New J. Phys.*, **12**, 053032 (2010).

[25] N. D. Mermin. The topological theory of defects in ordered media. *Rev. Mod. Phys.*, **51**, 591 (1979).

[26] A. Messiah. *Quantum Mechanics.* Wiley, New York, 1958 (republished Dover, New York, 2014).

[27] J. E. Moore and L. Balents. Topological invariants of time-reversal-invariant band structures. *Phys. Rev. B*, **75**, 121306(R) (2007).

[28] S. Murakami, N. Nagaosa, and S.-C. Zhang. Dissipationless spin current at room temperature. *Science*, **301**, 1348 (2003).

[29] N. Nagaosa, J. Sinova, S. Onoda, A. H. MacDonald, and N. P. Ong. Anomalous Hall effect. *Rev. Mod. Phys.*, **82**, 1539 (2010).

[30] M. Nakahara. *Geometry, Topology and Physics*, 2nd edn. Taylor & Francis, New York, 2003.

[31] Q. Niu, D. J. Thouless, and Y.-S. Wu. Quantized Hall conductance as a topological invariant. *Phys. Rev. B*, **31**, 3372 (1985).

[32] G. Ortiz and R. M. Martin. Macroscopic polarization as a geometric quantum phase: many-body formulation. *Phys. Rev. B*, **49**, 14202 (1994).

[33] R. E. Prange and S. M. Girvin (eds.). *The Quantum Hall Effect.* Springer-Verlag, New York, 1990.

[34] X.-L. Qi, T. L. Hughes, and S.-C. Zhang. Topological field theory of time-reversal invariant insulators. *Phys. Rev. B*, **78**, 195424 (2008).

[35] X.-L. Qi and S.-C. Zhang. Topological insulators and superconductors. *Rev. Mod. Phys.*, **83**, 1057 (2011).

[36] N. Read. Order parameter and Ginzburg–Landau theory for the fractional quantum Hall effect. *Phys. Rev. Lett.*, **62**, 86 (1989).

[37] R. Resta. Theory of the electric polarization in crystals. *Ferroelectrics*, **136**, 51 (1992).

[38] M. Stone (ed.). *Bosonization.* World Scientific, Singapore, 1994.

[39] D. J. Thouless, M. Kohmoto, M. P. Nightingale, and M. den Nijs. Quantized Hall conductance in a two-dimensional periodic potential. *Phys. Rev. Lett.*, **49**, 405 (1982).

[40] X.-G. Wen. Theory of the edge states in fractional quantum Hall effects. *Int. J. Mod. Phys. B*, **6**, 1711 (1992).

[41] X. G. Wen and Q. Niu. Ground-state degeneracy of the fractional quantum Hall states in the presence of a random potential and on high-genus Riemann surfaces. *Phys. Rev. B*, **41**, 9377 (1990).

[42] F. Wilczek. Two applications of axion electrodynamics. *Phys. Rev. Lett.*, **58**, 1799 (1987).

[43] A. Zee. Quantum Hall fluids. In H. B. Geyer (ed.). *Field Theory, Topology and Condensed Matter Physics. Proceedings of the Ninth Chris Engelbrecht Summer School in Theoretical Physics Held at Storms River Mouth, Tsitsikamma National Park, South Africa, 17–28 January 1994*, p. 99. Lecture Notes in Physics, Vol. 456. Springer-Verlag, Berlin, 1995.

[44] S. C. Zhang, T. H. Hansson, and S. Kivelson. Effective-field-theory model for the fractional quantum Hall effect. *Phys. Rev. Lett.*, **62**, 82 (1989).

2
Topological superconductors and category theory

Andrei Bernevig and Titus Neupert

Department of Physics,
Princeton University,
Princeton, NJ 08544,
USA
Princeton Center for Theoretical Science,
Princeton University,
Princeton, NJ 08544,
USA

Topological Aspects of Condensed Matter Physics. First Edition. C. Chamon et al.

Chapter Contents

Colour figures. For those figures in this chapter that use colour, please see the version of these lecture notes at http://topo-houches.pks.mpg.de. and arXiv: 1506.05805 [cond-mat.str-el]. These figures are indicated by '[Colour online]' at the start of the caption.

Preface

We give a pedagogical introduction to topologically ordered states of matter, with the aim of familiarizing the reader with their axiomatic topological quantum field theory description. We introduce basic non-interacting topological phases of matter protected by symmetries, including the Su–Schrieffer–Heeger model and the one-dimensional p-wave superconductor. The defining properties of topologically ordered states are illustrated explicitly using the toric code and—on a more abstract level—Kitaev's 16-fold classification of two-dimensional topological superconductors. Subsequently, we present a short review of category theory as an axiomatic description of topological order in two dimensions. Equipped with this structure, we revisit Kitaev's 16-fold way.

Thee lectures on which this chapter is based were also presented in part at:

- XVIII Training Course in the Physics of Strongly Correlated Systems, 6–17 October 2014, International Institute for Advanced Scientific Studies, Vietri sul Mare, Italy
- 7th School on Mathematical Physics, 'Topological Quantum Matter: From Theory to Applications', 25–29 May 2015, Universidad de los Andes, Bogotá, Colombia

2.1 Introduction to topological phases in condensed matter

2.1.1 The notion of topology

In these lectures, we will learn how to categorize and characterize some phases of matter that have topological attributes. A topological property of a phase, such as boundary modes (in an open geometry), topological response functions, or the character of its excitations, is described by a set of quantized numbers, related to so-called topological invariants of the phase. The quantization immediately implies that topological properties are *universal* (they can be used to label the topological phase) and in some sense *protected*, because they cannot change smoothly when infinitesimal perturbations are added. Topological properties, in the sense that we want to discuss them here, can only be defined for

- spectrally *gapped* ground states on a manifold without boundary of
- *local* Hamiltonians at
- *zero temperature.*

The spectral gap allows us to define an equivalence class of states, i.e. a phase, with the help of the adiabatic theorem. Two gapped ground states are in the same phase if there exists an adiabatic interpolation between their respective Hamiltonians, such that the spectral gap above the ground state as well as the locality are preserved for all Hamiltonians along the interpolation.

Often it is useful to further modify these rules to define topological phases that are subject to symmetry constraints. We refer to topological states as being protected/enriched by a symmetry group G if the Hamiltonian has a symmetry G and only G-preserving interpolations are allowed. Since the G-preserving interpolations are a subset of all local interpolations, it is clear that symmetries make a topological classification of Hamiltonians more refined.

The locality of a Hamiltonian is required to guarantee the quantization of topological response functions and to distinguish topological characterizations depending on the dimensionality of space. If we were not to impose locality, any system could in essence be zero-dimensional and there would be no notion of boundary states (which are localized over short distances) or point-like and line-like excitations, etc.

Equipped with this definition of a topological phase, the exploration of topological states of matter above all poses a classification problem. We would like to know how many phases of quantum systems exist that can be distinguished by their topological properties. We would like to obtain such a classification while imposing any symmetry G that is physically relevant, such as time-reversal symmetry, space-group or point-group symmetries of a crystal, particle number conservation, etc. To identify the right mathematical tools that allow for such a classification and to guarantee its completeness is a subject of ongoing research. Here, we shall focus on aspects of this classification problem, which are well established and understood.

Most fundamental is a distinction between two types of topological states of matter: those with *intrinsic (long-range entangled) topological order* [1] and those without. This notion is also core to the structure of these lectures. In this section, we only discuss phases without intrinsic topological order, while the following two sections are devoted to states with intrinsic topological order. A definition of intrinsic topological order can be based on several equivalent characterizations of such a phase, of which we give three:

- *Topological ground-state degeneracy.* On a manifold without boundary, the degeneracy of gapped topologically degenerate ground states depends on the topological properties of the manifold. There are no topologically degenerate ground states if the system is defined on a sphere. The matrix elements of any local operator taken between two distinct topologically degenerate ground states vanishes.
- *Fractionalized excitations.* There exist low-energy excitations that are point-like (in two dimensions (2D) or above) or line-like (in three dimensions (3D) or above). These excitations carry a fractional quantum number (e.g. a fractional charge) as compared with the microscopic degrees of freedom that enter the Hamiltonian and are deconfined and dynamical (i.e. free to move in the low-energy excited states).
- *Topological entanglement entropy.* The entanglement entropy between two parts of a system that is in a gapped zero-temperature ground state typically scales with the size of the line or surface that separates the two regions ('area-law entanglement'). Topologically ordered, long-range entangled states have a universal subleading correction to this scaling that is characteristic of the type of topological order.

(Note that these statements, like many universal properties we discuss, are only strictly true in the thermodynamic limit of infinite system size. For example, in a finite system, the ground-state degeneracy is lifted by an amount that scales exponentially with system size.) As fractionalized excitations in the above sense may only exist in two or higher dimensions, intrinsic topological order cannot be found in one-dimensional

(1D) phases of matter. Further, for intrinsic topological order to occur, interactions are needed in the system.

Examples of topologically non-trivial phases (both with and without intrinsic topological order) exist in the absence of any symmetry. However, most phases without intrinsic topological order belong to the so-called symmetry-protected topological (SPT) phases. In these cases, the topology is protected by a symmetry. These phases almost always possess topologically protected boundary modes when defined on a manifold with boundary, except if the boundary itself breaks the protecting symmetry (as could be the case with inversion symmetry, for example).

In contrast, phases with intrinsic topological order are not necessarily equipped with boundary modes, even if the boundary of the manifold preserves the defining symmetries of the phase. If the definition of a phase with intrinsic topological order relies on symmetries, it is called a symmetry-enriched topological (SET) phase.

An alternative characterization of topological properties of a phase uses the entanglement between different subsystems. While we opt not to touch upon this concept here, we want to make contact with the ensuing terminology: All phases with intrinsic topological order are called long-range entangled (LRE). The term 'short-range entangled (SRE) phase' is often used synonymously with 'no intrinsic topological order'. (Some authors also count 2D phases with non-vanishing thermal Hall conductivity, such as the $p + \mathrm{i}p$ superconductors, but no intrinsic topological order unless gauged as LRE.)

In these lectures, we will encounter two classifications of a subset of topological phases. The following subsection introduces the complete classification of non-interacting fermionic Hamiltonians with certain symmetries (which have no intrinsic topological order). Section 2.3 is concerned with the unified description of 2D phases with intrinsic topological order in the absence of any symmetries.

2.1.2 Classification of non-interacting fermion Hamiltonians: the 10-fold way

We have stated that SPT order in SRE states manifests itself via the presence of gapless boundary states in an open geometry. In fact, there exists a intimate connection between the topological character of the gapped bulk state and its boundary modes. The latter are protected against local perturbations on the boundary that (i) preserve the bulk symmetry and (ii) induce no intrinsic topological order or spontaneous symmetry breaking in the boundary modes. This bulk–boundary correspondence can be used to classify SPT phases. Two short-range entangled phases with the same symmetries belong to a different topological class, if the interface between the two phases hosts a state in the bulk gap and this state cannot be moved into the continuum of excited states by any local perturbation that obeys (i) and (ii). Equivalently, to change the topological attribute of a gapped bulk state via any smooth changes in the Hamiltonian, the bulk energy gap has to close and reopen.

Schnyder et al. [2]use this bulk–boundary correspondence to classify all non-interacting fermionic Hamiltonians. For the topological phases that they discuss, two fundamental symmetries, particle–hole symmetry (PHS) and time-reversal symmetry (TRS), are considered. In the following, we will review the essential results of this classification [2–4].

2.1.2.1 *Classification with respect to time-reversal and particle–hole symmetry*

Symmetries in quantum mechanics are operators that have to preserve the absolute value of the scalar product of any two vectors in Hilbert space. They can thus be either unitary operators, preserving the scalar product, or antiunitary operators, turning the scalar product into its complex conjugate (up to a phase). For a unitary operator to be a symmetry of a given Hamiltonian H, the operator has to commute with H. Consequently, the Hamiltonian can be block-diagonalized, where each block acts on one eigenspace of the unitary symmetry. If H has a unitary symmetry, we block-diagonalize it and then consider the topological properties of each block individually. This way, we do not have to include unitary symmetries (except for the product of TRS and PHS and the omnipresent particle number conservation) in our further considerations, as we will not focus on the burgeoning field of crystalline topological insulators.

A fundamental antiunitary operator in quantum mechanics is the reversal of time $\mathcal{T}$. Let us begin by recalling its elementary properties. If a given Hamiltonian H is TRS, that is,

$$\mathcal{T} H \mathcal{T}^{-1} = +H, \tag{2.1a}$$

then the time-evolution operator at time t should be mapped to the time-evolution operator at $-t$ by the operator $\mathcal{T}$:

$$\begin{aligned} \mathcal{T} \mathrm{e}^{-\mathrm{i}tH} \mathcal{T}^{-1} &= \mathrm{e}^{-\mathcal{T}\mathrm{i}\mathcal{T}^{-1}tH} \\ &= \mathrm{e}^{-\mathrm{i}(-t)H}. \end{aligned} \tag{2.1b}$$

We conclude that the reversal of time is indeed an antiunitary operator $\mathcal{T}\mathrm{i}\mathcal{T}^{-1} = -\mathrm{i}$. It can be represented as $\mathcal{T} = T\mathcal{K}$, where $\mathcal{K}$ denotes complex conjugation and T is a unitary operator. Applying the reversal of time twice on any state must return the same state up to an overall phase factor $\mathrm{e}^{\mathrm{i}\phi}$:

$$\begin{aligned} &\mathrm{e}^{\mathrm{i}\phi} \stackrel{!}{=} \mathcal{T}^2 = T(T^\mathsf{T})^{-1} \\ &\Rightarrow \quad T = \mathrm{e}^{\mathrm{i}\phi} T^\mathsf{T}, \quad T^\mathsf{T} = \mathrm{e}^{\mathrm{i}\phi} T. \end{aligned} \tag{2.1c}$$

Inserting the last two equations into one another, we obtain $T = \mathrm{e}^{2\mathrm{i}\phi}T$; that is, $\mathrm{e}^{2\mathrm{i}\phi}$ has to equal $+1$. We conclude that the time-reversal operator squares either to $+1$ or to -1:

$$\mathcal{T}^2 = +1, \qquad \mathcal{T}^2 = -1. \tag{2.1d}$$

The second fundamental antiunitary symmetry considered here is charge conjugation $\mathcal{P}$. Its most important incarnation in solid state physics is found in the theory of superconductivity. In an Andreev reflection process, an electron-like quasiparticle that enters a superconductor is reflected as a hole-like quasiparticle. The charge difference between incident and reflected states is accounted for by adding one Cooper pair to the superconducting condensate. In the mean-field theory of superconductivity, the energies of the electron-like state and the hole-like state are equal in magnitude and opposite in sign, giving rise to the PHS. In this case, rather than being a fundamental physical symmetry of the system like TRS is, PHS emerges owing to a redundancy in the mean-field description. We define a (single-particle) Hamiltonian H to be PHS if

$$\mathcal{P}H\mathcal{P}^{-1} = +H. \tag{2.1e}$$

In order to also reverse the sign of charge, $\mathcal{P}$ has to turn the minimal coupling $\mathbf{p} - \mathrm{i}e\mathbf{A}$ into $\mathbf{p} + \mathrm{i}e\mathbf{A}$, where $\mathbf{p}$ is the momentum operator and $\mathbf{A}$ is the electromagnetic gauge potential. This is achieved by demanding $\mathcal{P}\mathrm{i}\mathcal{P}^{-1} = -\mathrm{i}$. We conclude that $\mathcal{P}$ is indeed an antiunitary operator that can be decomposed as $\mathcal{P} = P\mathcal{K}$, where P is a unitary operator. As a consequence, the reasoning of (2.1c) also applies to $\mathcal{P}$ and we conclude that the charge conjugation operator squares either to $+1$ or to -1:

$$\mathcal{P}^2 = +1, \qquad \mathcal{P}^2 = -1. \tag{2.1f}$$

In the case where the operators $\mathcal{T}$ and $\mathcal{P}$ are both symmetries of H, their product is also a symmetry of H. We call this product chiral transformation $C := \mathcal{T}\mathcal{P}$. It is a unitary operator. The Hamiltonian H transforms under the chiral symmetry as

$$CHC^{-1} = +H. \tag{2.1g}$$

(It is important to note that both $\mathcal{P}$ and C anticommute rather than commute with the single-particle first-quantized Hamiltonian $\mathcal{H}_{\alpha,\alpha'}$ that we will introduce below.) Observe that a Hamiltonian can have a chiral symmetry, even if it possesses neither PHS nor TRS. We can now enumerate all combinations of the symmetries $\mathcal{P}$, $\mathcal{T}$, and C that a Hamiltonian can obey, accounting for the different signs of $\mathcal{T}^2$ and $\mathcal{P}^2$. There are in total 10 such symmetry classes, listed in Table 2.1. The main result of Schnyder et al. [2] is to establish how many distinct phases with protected edge modes exist on the $(d-1)$-dimensional boundary of a phase in d dimensions. We find three possible cases: if there is only one (topologically trivial) phase, then the entry $\varnothing$ is found in Table 2.1; if there are exactly two distinct phases (one trivial and one topological phase), then $\mathbb{Z}_2$ is listed; finally, if there exists a distinct topological phase for every integer, then $\mathbb{Z}$ is listed.

2.1.2.2 *Flatband Hamiltonians and homotopy groups*

There are several approaches to obtain the entries $\mathbb{Z}_2$ and $\mathbb{Z}$ in Table 2.1. For one, the theory of Anderson localization can be employed to determine in which spatial dimensions' boundaries can host localization-protected states (the topological surface states) under a given symmetry. This was done by Schnyder et al. [2]. Kitaev [4], on the other hand, derived the table using the algebraic structure of Clifford algebras in the various dimensions and symmetry classes. In mathematics, this goes under the name K-theory.

Table 2.1 Symmetry classes of non-interacting fermionic Hamiltonians from [3] and [4]. The columns contain, from left to right: Cartan's name for the symmetry class; the squares of the time-reversal operator, the particle–hole operator, and the chiral operator (∅ means that the symmetry is not present); the group of topological phases to which a Hamiltonian with the respective symmetry can belong for the dimensions $d = 1, \ldots, 8$ of space. The first two rows are called 'complex classes', while the lower eight rows are the 'real classes'. The homotopy groups of the former show a periodicity with period 2 in d, while those of the latter have a period 8 in d (Bott periodicity)

	$\mathcal{T}^2$	$\mathcal{P}^2$	$\mathcal{C}^2$	$d =$	1	2	3	4	5	6	7	8
A	∅	∅	∅		∅	$\mathbb{Z}$	∅	$\mathbb{Z}$	∅	$\mathbb{Z}$	∅	$\mathbb{Z}$
AIII	∅	∅	+		$\mathbb{Z}$	∅	$\mathbb{Z}$	∅	$\mathbb{Z}$	∅	$\mathbb{Z}$	∅
AII	−	∅	∅		∅	$\mathbb{Z}_2$	$\mathbb{Z}_2$	$\mathbb{Z}$	∅	∅	∅	$\mathbb{Z}$
DIII	−	+	+		$\mathbb{Z}_2$	$\mathbb{Z}_2$	$\mathbb{Z}$	∅	∅	∅	$\mathbb{Z}$	∅
D	∅	+	∅		$\mathbb{Z}_2$	$\mathbb{Z}$	∅	∅	∅	$\mathbb{Z}$	∅	$\mathbb{Z}_2$
BDI	+	+	+		$\mathbb{Z}$	∅	∅	∅	$\mathbb{Z}$	∅	$\mathbb{Z}_2$	$\mathbb{Z}_2$
AI	+	∅	∅		∅	∅	∅	$\mathbb{Z}$	∅	$\mathbb{Z}_2$	$\mathbb{Z}_2$	$\mathbb{Z}$
CI	+	−	+		∅	∅	$\mathbb{Z}$	∅	$\mathbb{Z}_2$	$\mathbb{Z}_2$	$\mathbb{Z}$	∅
C	∅	−	∅		∅	$\mathbb{Z}$	∅	$\mathbb{Z}_2$	$\mathbb{Z}_2$	$\mathbb{Z}$	∅	∅
CII	−	−	+		$\mathbb{Z}$	∅	$\mathbb{Z}_2$	$\mathbb{Z}_2$	$\mathbb{Z}$	∅	∅	∅

Here, we want to give a flavor of the mathematical structure behind the table by considering two examples. To keep matters simple, we shall restrict ourselves to the situation where the system is translationally invariant and periodic boundary conditions are imposed. In second quantization, the Hamiltonian H has the Bloch representation

$$H = \int \mathrm{d}^d\mathbf{k}\, \psi^\dagger_\alpha(\mathbf{k}) \mathcal{H}_{\alpha,\alpha'}(\mathbf{k}) \psi_{\alpha'}(\mathbf{k}), \tag{2.2a}$$

where $\psi^\dagger_\alpha(\mathbf{k})$ creates a fermion of flavor $\alpha = 1, \ldots, N$ at momentum $\mathbf{k}$ in the Brillouin zone (BZ) and the summation over α and α' is implicit. The flavour index may represent orbital, spin, or sublattice degrees of freedom. Energy bands are obtained by diagonalizing the $N \times N$ matrix $\mathcal{H}(\mathbf{k})$ at every momentum $\mathbf{k} \in \mathrm{BZ}$ with the aid of a unitary transformation $U(\mathbf{k})$:

$$U^\dagger(\mathbf{k})\mathcal{H}(\mathbf{k})U(\mathbf{k}) = \mathrm{diag}\left[\varepsilon_{m+n}(\mathbf{k}), \ldots, \varepsilon_{n+1}(\mathbf{k}), \varepsilon_n(\mathbf{k}), \ldots, \varepsilon_1(\mathbf{k})\right], \tag{2.2b}$$

where the energies are arranged in descending order on the right-hand side and $n, m \in \mathbb{Z}$ such that $n + m = N$. So as to start from an insulating ground state, we assume that there exists an energy gap between the bands n and $n+1$ and that the chemical potential μ lies in this gap:

$$\varepsilon_n(\mathbf{k}) < \mu < \varepsilon_{n+1}(\mathbf{k}) \quad \forall \mathbf{k} \in \mathrm{BZ}. \tag{2.3}$$

The presence of the gap allows us to adiabatically deform the Bloch Hamiltonian $\mathcal{H}(\mathbf{k})$ to the flatband Hamiltonian

$$\mathcal{Q}(\mathbf{k}) := U(\mathbf{k}) \begin{pmatrix} \mathbb{1}_m & 0 \\ 0 & -\mathbb{1}_n \end{pmatrix} U^{\dagger}(\mathbf{k}) \tag{2.4a}$$

that assigns the energy -1 and $+1$ to all states in the bands below and above the gap, respectively. This deformation preserves the eigenstates, but removes the non-universal information about energy bands from the Hamiltonian.

In other words, the degenerate eigenspaces of the eigenvalues ± 1 of $\mathcal{Q}(\mathbf{k})$ reflect the partitioning of the single-particle Hilbert space introduced by the spectral gap in the spectrum of $\mathcal{H}(\mathbf{k})$. The degeneracy of its eigenspaces equips $\mathcal{Q}(\mathbf{k})$ with an extra $\mathrm{U}(n) \times \mathrm{U}(m)$ gauge symmetry: while the $(n+m) \times (n+m)$ matrix $U(\mathbf{k})$ of Bloch eigenvectors that diagonalizes $\mathcal{Q}(\mathbf{k})$ is an element of $\mathrm{U}(n+m)$ for every $\mathbf{k} \in \mathrm{BZ}$, we are free to change the basis for its lower and upper bands by a $\mathrm{U}(n)$ and $\mathrm{U}(m)$ transformation, respectively. Hence $\mathcal{Q}(\mathbf{k})$ is an element of the space $C_0 := \mathrm{U}(n+m)/[\mathrm{U}(n) \times \mathrm{U}(m)]$ defining a map

$$\mathcal{Q}: \quad \mathrm{BZ} \to C_0. \tag{2.5}$$

The group of topologically distinct maps $\mathcal{Q}$, or, equivalently, the number of topologically distinct Hamiltonians $\mathcal{H}$, is given by the homotopy group

$$\pi_d(C_0) \tag{2.6}$$

for any dimension d of the BZ. (The homotopy group is the group of equivalence classes of maps from the d-dimensional *sphere* to a target space, in this case C_0. Even though the BZ is a d-dimensional *torus*, it turns out that this difference between torus and sphere does not affect the classification as discussed here.)

For example, in $d=2$, we have $\pi_2(C_0) = \mathbb{Z}$. A physical example of a family of Hamiltonians that exhausts the topological sectors of this group is found in the integer quantum Hall effect. The incompressible ground state with $r \in \mathbb{N}$ filled Landau levels is topologically distinct from the ground state with $\mathbb{N} \ni r' \neq r$ filled Landau levels. Two different patches of space with r and r' filled Landau levels have $|r-r'|$ gapless edge modes running at their interface, reflecting the bulk–boundary correspondence of the topological phases. In contrast, $\pi_3(C_0) = \mathbb{Z}_1$ renders all non-interacting fermionic Hamiltonians in 3D space topologically equivalent to the vacuum, if no further symmetries besides the $\mathrm{U}(1)$ charge conservation are imposed.

As a second example, let us discuss a Hamiltonian that has only chiral symmetry and hence belongs to the symmetry class AIII. The chiral symmetry implies a spectral symmetry of $\mathcal{H}(\mathbf{k})$. If gapped, $\mathcal{H}(\mathbf{k})$ must have an even number of bands $N = 2n$, $n \in \mathbb{Z}$. When represented in the eigenbasis of the chiral symmetry operator C, the spectrally flattened Hamiltonian $\mathcal{Q}(\mathbf{k})$ and the chiral symmetry operator have the representations

$$\mathcal{Q}(\mathbf{k}) = \begin{pmatrix} 0 & q(\mathbf{k}) \\ q^{\dagger}(\mathbf{k}) & 0 \end{pmatrix}, \qquad C = \begin{pmatrix} \mathbb{1}_n & 0 \\ 0 & -\mathbb{1}_n \end{pmatrix}, \tag{2.7a}$$

respectively. From $\mathcal{Q}(\mathbf{k})^2 = 1$, one concludes that $q(\mathbf{k})$ can be an arbitrary unitary matrix. We are thus led to consider the homotopy group $\pi_d(C_1)$ of the mapping

$$q: \quad \mathrm{BZ} \to C_1 = \mathrm{U}(n). \tag{2.7b}$$

For example, in $d = 1$ spatial dimensions $\pi_3(C_1) = \mathbb{Z}$. A tight-binding model with non-trivial topology that belongs to this symmetry class will be discussed in Section 2.1.3.

With these examples, we have discussed the two complex classes A and AIII. In the real classes, which have at least one antiunitary symmetry, it is harder to obtain the constraints on the spectrally flattened Hamiltonian $\mathcal{Q}(\mathbf{k})$. The origin of this complication is that the antiunitary operators representing time-reversal and particle–hole symmetry relate $\mathcal{Q}(\mathbf{k})$ and $\mathcal{Q}(-\mathbf{k})$ rather than acting locally in momentum space.

2.1.2.3 Topological invariants

Given a gapped non-interacting fermionic Hamiltonian with certain symmetry properties in d-dimensional space, one can use Table 2.1 to conclude whether the system can potentially be in a topological phase. However, to understand in which topological sector the system is, we have to do more work. To obtain this information, one computes topological invariants or topological quantum numbers of the ground state. Such invariants are automatically numbers in the group of possible topological phases ($\mathbb{Z}$ or $\mathbb{Z}_2$). For many of them, a variety of different-looking but equivalent representations are known.

To give concrete examples, we shall discuss the invariants for all $\mathbb{Z}$ topological phases found in Table 2.1. These are called Chern numbers in the symmetry classes without chiral symmetry and winding numbers in the classes with chiral symmetry.

In physics, topological attributes refer to *global* properties of a physical system that is made out of *local* degrees of freedom and might only have local, i.e. short-ranged, correlations. The distinction between global and local properties parallels the distinction between topology and geometry in mathematics, where the former refers to global structure, while the latter refers to the local structure of objects. In differential geometry, a bridge between topology and geometry is given by the Gauss–Bonnet theorem. It states that for compact 2D Riemannian manifolds M without boundary, the integral over the Gaussian curvature $F(\mathbf{x})$ of the manifold is (i) integer and (ii) a topological invariant

$$2(1-g) = \frac{1}{2\pi}\int_M \mathrm{d}^2\mathbf{x}\, F(\mathbf{x}). \tag{2.8}$$

Here, g is the genus of M; for example, $g = 0$ for a 2D sphere and $g = 1$ for a 2D torus. The Gaussian curvature $F(\mathbf{x})$ can be defined as follows. Attach to every point on M the tangent plane, a 2D vector space. Take some vector from the tangent plane at a given point on M and parallel-transport it around an infinitesimal closed loop on M. The angle mismatch of the vector before and after the transport is proportional to the Gaussian curvature enclosed in the loop.

In the physical systems that we want to describe, the manifold M is the BZ and the analogue of the tangent plane on M is a space spanned by the Bloch states of the occupied bands at a given momentum $\mathbf{k} \in \mathrm{BZ}$. The Gaussian curvature of differential geometry is now generalized to a curvature form, called the Berry curvature F. In our case, it is given by an $n \times n$ matrix of differential forms that is defined via the Berry connection A as

$$\mathrm{F} := F_{ij}(\mathbf{k})\, \mathrm{d}k_i \wedge \mathrm{d}k_j \tag{2.9a}$$

$$F_{ij}(\mathbf{k}) := \partial_i A_j(\mathbf{k}) - \partial_j A_i(\mathbf{k}) + [A_i(\mathbf{k}), A_j(\mathbf{k})], \qquad i,j = 1,\ldots,d, \tag{2.9b}$$

$$\mathrm{A} := A_i(\mathbf{k})\, \mathrm{d}k_i, \tag{2.9c}$$

$$A_i^{(ab)}(\mathbf{k}) := \sum_{\alpha=1}^{N} U^\dagger_{a\alpha}(\mathbf{k}) \partial_i U_{\alpha b}(\mathbf{k}), \qquad a,b = 1,\ldots,n, \quad i = 1,\ldots,d. \tag{2.9d}$$

(Two different conventions for the Berry connection are commonly used: it is either purely real or purely imaginary. Here we choose the latter option.) The unitary transformation $U(\mathbf{k})$ that diagonalizes the Hamiltonian was defined in (2.2b), both $A_i(\mathbf{k})$ and $F_{ij}(\mathbf{k})$ are $n \times n$ matrices, we write $\partial_i \equiv \partial/\partial k_i$, and the sum over repeated spatial coordinate components i, j is implicit.

Under a local $\mathrm{U}(n)$ gauge transformation in momentum space that acts on the states of the lower bands and is parametrized by the $n \times n$ matrix $G(\mathbf{k})$,

$$U_{\alpha a}(\mathbf{k}) \to U_{\alpha b}(\mathbf{k}) G_{ba}(\mathbf{k}), \qquad \alpha = 1,\ldots,N, \quad a = 1,\ldots,n, \tag{2.10a}$$

the Berry connection A changes as

$$\mathrm{A} \to G^\dagger \mathrm{A} G + G^\dagger \mathrm{d}G, \tag{2.10b}$$

while the Berry curvature F changes covariantly,

$$\mathrm{F} \to G^\dagger \mathrm{F} G, \tag{2.10c}$$

leaving its trace invariant.

Chern numbers For the spatial dimension $d = 2$, the generalization of the Gauss–Bonnet theorem (2.8) in algebraic topology was found by Chern to be

$$\begin{aligned} 2\mathsf{C}^{(1)} &:= \frac{\mathrm{i}}{2\pi} \int_{\mathrm{BZ}} \mathrm{tr}\, \mathrm{F} \\ &= 2 \frac{\mathrm{i}}{2\pi} \int_{\mathrm{BZ}} \mathrm{d}^2\mathbf{k}\, \mathrm{tr}\, F_{12}. \end{aligned} \tag{2.11}$$

This defines a gauge-invariant quantity, the first Chern number $\mathsf{C}^{(1)}$. Remarkably, $\mathsf{C}^{(1)}$ can take only integer values. In order to obtain a topological invariant for any even dimension $d = 2s$ of space, we can use the sth power of the local Berry curvature form F (using the wedge product) to build a gauge invariant d-form that can be integrated over the BZ to obtain a scalar. On taking the trace, this scalar is invariant under the gauge transformation (2.10a) and defines the sth Chern number:

$$2\mathsf{C}^{(s)} := \frac{1}{s!}\left(\frac{\mathrm{i}}{2\pi}\right)^{s}\int\limits_{\mathrm{BZ}} \mathrm{tr}[\mathrm{F}^{s}], \tag{2.12}$$

where $\mathrm{F}^{s} = \mathrm{F}\wedge\cdots\wedge\mathrm{F}$. As with the case $s = 1$ that we have exemplified above, $\mathsf{C}^{(s)}$ is an integer for any $s = 1, 2, \ldots$

From inspection of Table 2.1 we see that symmetry classes without chiral symmetry may have integer topological invariants $\mathbb{Z}$ only when the dimension d of space is even. In fact, all the integer invariants of these classes are given by the Chern number $\mathsf{C}^{(d/2)}$ of the respective dimension.

Winding numbers Let us now consider systems with chiral symmetry C. To construct their topological invariants as a natural extension of the above, we consider a different representation of the Chern numbers $\mathsf{C}^{(s)}$. In terms of the flatland projector Hamiltonian $\mathcal{Q}(\mathbf{k})$ that was defined in (2.4a), we can write

$$\mathsf{C}^{(s)} \propto \varepsilon_{i_1\ldots i_d}\int\limits_{\mathrm{BZ}} \mathrm{d}^d\mathbf{k}\,\mathrm{tr}\big[\mathcal{Q}(\mathbf{k})\partial_{i_1}\mathcal{Q}(\mathbf{k})\cdots\partial_{i_d}\mathcal{Q}(\mathbf{k})\big], \qquad d = 2s. \tag{2.13}$$

The form of (2.13) allows us to interpret $\mathsf{C}^{(s)}$ as the *winding number* of the unitary transformation $\mathcal{Q}(\mathbf{k})$ over the compact BZ. One verifies that $\mathsf{C}^{(s)} = 0$ for symmetry classes with chiral symmetry by inserting $CC^{\dagger}$ at some point in the expression and anticommuting C with all $\mathcal{Q}$, using the cyclicity of the trace. After $2s + 1$ anticommutations, we are back to the original expression up to an overall minus sign and have found $\mathsf{C}^{(s)} = -\mathsf{C}^{(s)}$. Hence, all systems with chiral symmetry have vanishing Chern numbers.

In odd dimensions of space, we can define an alternative topological invariant for systems with chiral symmetry by modifying (2.13) and using the chiral operator C:

$$\begin{aligned}\mathsf{W}^{(s)} &:= \frac{(-1)^{s}s!}{2(2s+1)!}\left(\frac{\mathrm{i}}{2\pi}\right)^{s+1}\varepsilon_{i_1\ldots i_d}\int\limits_{\mathrm{BZ}} \mathrm{d}^d\mathbf{k}\,\mathrm{tr}\big[C\mathcal{Q}(\mathbf{k})\partial_{i_1}\mathcal{Q}(\mathbf{k})\cdots\partial_{i_d}\mathcal{Q}(\mathbf{k})\big]\\ &= \frac{(-1)^{s}s!}{(2s+1)!}\left(\frac{\mathrm{i}}{2\pi}\right)^{s+1}\varepsilon_{i_1\ldots i_d}\int\limits_{\mathrm{BZ}} \mathrm{d}^d\mathbf{k}\,\mathrm{tr}\big[q^{\dagger}(\mathbf{k})\partial_{i_1}q(\mathbf{k})\partial_{i_2}q^{\dagger}(\mathbf{k})\cdots\partial_{i_d}q(\mathbf{k})\big],\\ &\qquad d = 2s+1.\end{aligned} \tag{2.14}$$

Upon anticommuting the chiral operator C once with all matrices $\mathcal{Q}$ and using the cyclicity of the trace, one finds that the expression for $\mathsf{W}^{(s)}$ vanishes for even dimensions. The second line of (2.14) allows us to interpret $\mathsf{W}^{(s)}$ as the *winding number* of the unitary off-diagonal part $q(\mathbf{k})$ of the chiral Hamiltonian that was defined in (2.7a). With (2.14), we have given topological invariants for all entries $\mathbb{Z}$ in odd dimensions d in Table 2.1.

In summary, we have now given explicit formulas for the topological invariants for all entries $\mathbb{Z}$ in Table 2.1 for systems with translational invariance. It is important to remember that the classification of Table 2.1 is restricted to systems without interactions. If interactions are allowed, that neither spontaneously nor explicitly break the defining symmetry of a symmetry class, one of two things can happen: (i) two phases that are distinguished by a non-interacting invariant like $\mathsf{W}^{(0)}$ might—sometimes but not always—be connected adiabatically (i.e. without a closing of the spectral gap) by turning on strong interactions; (ii) interactions can enrich the classification of Table 2.1 by inducing new phases with topological response functions that are distinct from those of the non-interacting phases. We will given an example of scenario (i) in Section 2.1.5.

Besides, interactions can strongly modify the topological boundary modes of the non-interacting systems to the extent that they can be gapped without breaking the protective symmetries, but at the expense of introducing topological order on the boundary.

2.1.3 The Su–Schrieffer–Heeger model

The first example of a topological band insulator that we consider here is also the simplest: The Su–Schrieffer–Heeger model [5] describes a 1D chain of atoms with one (spinless) electronic orbital each at half-filling. The model was originally proposed to describe the electronic structure of polyacetylene. This 1D organic molecule features a Peierls instability by which the hopping integral between consecutive sites is alternating between strong and weak. This enlarges the unit cell to contain two sites A and B. The second-quantized mean-field Hamiltonian reads

$$H = t\sum_{i=1}^{N}[(1-\delta)c^{\dagger}_{A,i}c_{B,i} + (1+\delta)c^{\dagger}_{B,i}c_{A,i+1} + \text{h.c.}]. \tag{2.15}$$

Here, $c^{\dagger}_{A,i}$ and $c^{\dagger}_{B,i}$ create an electron in the ith unit cell on sublattices A and B, respectively. If we identify $i = N+1 \equiv 1$, periodic boundary conditions are implemented. The corresponding Bloch Hamiltonian

$$H = t\sum_{k\in\text{BZ}}\sum_{\alpha=A,B} c^{\dagger}_{\alpha,k}h_{\alpha\beta,k}c_{\beta,k}, \tag{2.16a}$$

$$h_k = \begin{pmatrix} 0 & (1-\delta)+(1+\delta)\mathrm{e}^{-\mathrm{i}k} \\ (1-\delta)+(1+\delta)\mathrm{e}^{\mathrm{i}k} & 0 \end{pmatrix} \tag{2.16b}$$

$$= \sigma_x[(1-\delta)+(1+\delta)\cos k] + \sigma_y(1+\delta)\sin k, \tag{2.16c}$$

where σ_x and σ_y are the first two Pauli matrices acting on the sublattice index, t is the nearest-neighbour hopping integral, and δ is a dimensionless parametrization of the strong–weak dimerization of bonds.

We observe that the Hamiltonian (2.16) has time-reversal symmetry $\mathcal{T} = \mathcal{K}$, chiral symmetry $C = \sigma_z$, and thus also particle–hole symmetry $\mathcal{P} = \sigma_z\mathcal{K}$. This places it in class BDI of Table 2.1 with a $\mathbb{Z}$ topological characterization. Observe that breaking the time-reversal symmetry would not alter the topological properties, as long as the chiral symmetry was intact. The model would then belong to class AIII, which also features a $\mathbb{Z}$ classification. Hence, it is the chiral symmetry that is crucial to protecting the topological properties of the Hamiltonian (2.16). Notice that generic longer-range hopping (between sites of the same sublattice) breaks the chiral symmetry.

What are the different topological sectors that can be accessed by tuning the parameter δ in the Su–Schrieffer–Heeger model? We observe that the dispersion

$$\varepsilon_k^2 = 2[(1+\delta^2) + (1-\delta^2)\cos k] \tag{2.17}$$

is gapless for $\delta = 0$, hinting that this is the boundary between two distinct phases $\delta > 0$ and $\delta < 0$. As we are interested in understanding the topological properties of these phases, we can analyse them for any convenient value of the parameter δ and then conclude that they are the same in the entire phase by adiabaticity. We consider the Hamiltonian (2.15) with *open* boundary conditions and choose the following representative parameters:

- $\delta = +1$: The operators $c_{1,A}^\dagger$ and $c_{N,B}^\dagger$ do not appear in the Hamiltonian for the open chain. Hence, there exists a state at either end of the open chain that can be occupied or unoccupied at no cost of energy. Thus, either end of the chain supports a localized topological end state (see Fig. 2.1). Away from $\delta = +1$, as long as $\delta > 0$, the end states start to overlap and split apart in energy by an amount that is exponentially small in the length N of the chain. We can back up this observation by evaluating the topological invariant (2.14) for this phase. The off-diagonal projector is $q_k = \mathrm{e}^{-\mathrm{i}k}$, and its winding number evaluates to

$$\mathsf{W}^{(0)} = \frac{\mathrm{i}}{2\pi}\int \mathrm{d}k\, \mathrm{e}^{\mathrm{i}k}(-\mathrm{i})\mathrm{e}^{-\mathrm{i}k} = 1. \tag{2.18}$$

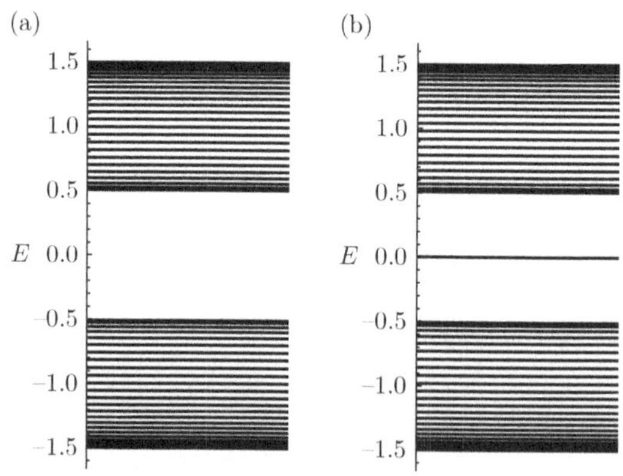

Fig. 2.1 Energy spectra for the Su–Schrieffer–Heeger model with open boundary conditions (a) in the trivial phase and (b) in the non-trivial topological phase with a zero-energy mode on each boundary point.

- $\delta = -1$: In this case, strong bonds form between the two sites in every unit cell and no topological end states appear. Correspondingly, as the off-diagonal projector $q_k = 1$ is independent of k, we conclude that the winding number vanishes: $\mathsf{W}^{(0)} = 0$.

One can visualize the winding number of a two-band Hamiltonian that has the form $h_k = \mathbf{d}_k \cdot \boldsymbol{\sigma}$ in the following way. If the Hamiltonian has chiral symmetry, we can choose this symmetry to be represented by $C = \sigma_z$ without loss of generality. Then $\mathbf{d}_k$ has to lie in the x–y plane for every k and may not be zero if the phase is gapped. The winding number $\mathsf{W}^{(0)}$ measures how often $\mathbf{d}_k$ winds around the origin in the x–y plane as k changes from 0 to 2π.

Besides the topological end states, the Su–Schrieffer–Heeger model also features topological domain wall states between a region with $\delta > 0$ and $\delta < 0$. Such topological midgap modes have to appear pairwise in any periodic geometry. As the system is considered at half-filling, each of these modes binds half an electron charge. This is an example of charge fractionalization at topological defects. It is important to remember that these defects are not dynamical, but are rigidly fixed external perturbations. Therefore, this form of fractionalization is not related to intrinsic topological order.

2.1.4 The 1D p-wave superconductor

In the Su–Schrieffer–Heeger model, particle–hole symmetry (and with it chiral symmetry) is in some sense fine-tuned, as it is lost if generic longer-range hoppings are considered. In superconductors, particle–hole symmetry arises more naturally as a symmetry that is inherent in the redundant description of mean-field Bogoliubov–de Gennes Hamiltonians.

Here, we want to consider the simplest model for a topological superconductor that has been studied by Kitaev [6]. The setup is again a 1D chain with one orbital for spinless fermion on each site. Superconductivity is encoded in pairing terms $c_i^\dagger c_{i+1}^\dagger$ that do not conserve particle number. The Hamiltonian is given by

$$H = \sum_{i=1}^{N} \big[- t(c_i^\dagger c_{i+1} + c_{i+1}^\dagger c_i) - \mu c_i^\dagger c_i + \Delta c_{i+1}^\dagger c_i^\dagger + \Delta^* c_i c_{i+1} \big]. \tag{2.19}$$

Here, μ is the chemical potential and Δ is the superconducting order parameter, which we will decompose into its amplitude $|\Delta|$ and complex phase θ, that is, $\Delta = |\Delta| \mathrm{e}^{\mathrm{i}\theta}$.

The fermionic operators $c_i^\dagger$ satisfy the algebra

$$\{c_i^\dagger, c_j\} = \delta_{i,j}, \tag{2.20}$$

with all other anticommutators vanishing. We can chose to trade the operators $c_i^\dagger$ and c_i on every site i for two other operators a_i and b_i that are defined by

$$a_i = \mathrm{e}^{-\mathrm{i}\theta/2} c_i + \mathrm{e}^{\mathrm{i}\theta/2} c_i^\dagger, \qquad b_i = \frac{1}{\mathrm{i}}(\mathrm{e}^{-\mathrm{i}\theta/2} c_i - \mathrm{e}^{\mathrm{i}\theta/2} c_i^\dagger). \tag{2.21}$$

These so-called Majorana operators obey the algebra

$$\{a_i, a_j\} = \{b_i, b_j\} = 2\delta_{ij}, \qquad \{a_i, b_j\} = 0 \qquad \forall i, j. \tag{2.22}$$

In particular, they square to 1,

$$a_i^2 = b_i^2 = 1, \tag{2.23}$$

and are self-conjugate,

$$a_i^\dagger = a_i, \qquad b_i^\dagger = b_i. \tag{2.24}$$

In fact, we can always break up a complex fermion operator on a lattice site into its real and imaginary Majorana components, though this may not always be a useful representation. As an aside, note that the Majorana anticommutation relation in (2.22) is the same as that of the generators of a Clifford algebra where the generators all square to +1. Thus, mathematically, one can think of the operators a_i (or b_i) as matrices forming by themselves the representation of Clifford algebra generators.

When rewritten in terms of the Majorana operators, the Hamiltonian (2.19) takes (up to a constant) the form

$$H = \frac{\mathrm{i}}{2}\sum_{i=1}^{N}[-\mu a_i\, b_i + (t+|\Delta|)b_i\, a_{i+1} + (-t+|\Delta|)a_i\, b_{i+1}]. \tag{2.25}$$

After imposing periodic boundary conditions, it is again convenient to study the system in momentum space. When defining the Fourier transform of the Majorana operators $a_i = \sum_i \mathrm{e}^{\mathrm{i}ki} a_k$, we note that the the self-conjugate property (2.24) that is local in position space translates into $a_k^\dagger = a_{-k}$ in momentum space (and likewise for the b_k). The momentum-space representation of the Hamiltonian is

$$H = \sum_{k\in\mathrm{BZ}}\sum_{\alpha=A,B} (a_k \;\; b_k)\, h_k \begin{pmatrix} a_{-k} \\ b_{-k} \end{pmatrix} \tag{2.26a}$$

$$h_k = \begin{pmatrix} 0 & -\frac{\mathrm{i}\mu}{2} + it\cos k + |\Delta|\sin k \\ \frac{\mathrm{i}\mu}{2} - it\cos k + |\Delta|\sin k & 0 \end{pmatrix} \tag{2.26b}$$

$$= \sigma_x |\Delta| \sin k + \sigma_y \left(\frac{\mu}{2} - t\cos k\right). \tag{2.26c}$$

While this Bloch Hamiltonian is formally very similar to that of the Su–Schrieffer–Heeger model (2.16), we have to keep in mind that it acts on entirely different single-particle degrees of freedom, namely in the space of Majorana operators instead of complex fermionic operators. As with the case of the Su–Schrieffer–Heeger model, the Hamiltonian (2.26) has a time-reversal symmetry $\mathcal{T} = \sigma_z\mathcal{K}$ and a particle–hole symmetry $P = \mathcal{K}$, which combine to give the chiral symmetry $C = \sigma_z$. Hence, it belongs to symmetry class BDI as well. For the topological properties that we explore below, only the particle–hole symmetry is crucial. If time-reversal symmetry is broken, the model changes to symmetry class D, which still supports a $\mathbb{Z}_2$ topological grading.

To determine its topological phases, we notice that Hamiltonian (2.26) is gapped except for $|t| = |\mu/2|$. We specialize again to convenient parameter values on either side of this potential topological phase transition:

- $\mu = 0$, $|\Delta| = t$. The Bloch matrix h_k takes exactly the same form as that of the Su–Schrieffer–Heeger model (2.16) for the parameter choice $\delta = +1$. We conclude that the Hamiltonian (2.26) is in a topological phase. The Hamiltonian reduces to

$$H = \mathrm{i}t \sum_j b_j a_{j+1}. \tag{2.27}$$

  A pictorial representation of this Hamiltonian is shown in Fig. 2.2(b). With open boundary conditions, it is clear that the Majorana operators a_1 and b_N are not coupled to the rest of the chain and are 'unpaired'. In this limit, the existence of two Majorana zero modes localized on the ends of the chain is manifest.
- $\Delta = t = 0$, $\mu < 0$. This is the topologically trivial phase. The Hamiltonian is independent of k and we conclude that the winding number vanishes $\mathsf{W}^{(0)} = 0$. In this case, the Hamiltonian reduces to

$$H = -\mu \frac{\mathrm{i}}{2} \sum_j a_j b_j. \tag{2.28}$$

  In its ground state, the Majorana operators on each physical site are coupled but the Majorana operators between each physical site are decoupled. In terms of the physical complex fermions, it is the ground state with either all sites occupied or all sites empty. A representation of this Hamiltonian is shown in Fig. 2.2(a). The Hamiltonian in the physical-site basis is in the atomic limit, which is another way

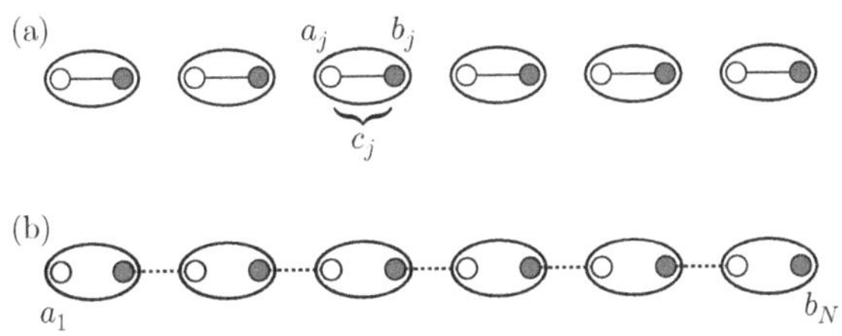

Fig. 2.2 [Colour online] Schematic illustration of the lattice p-wave superconductor Hamiltonian in (a) the trivial limit and (b) the non-trivial limit. The empty [white] and filled [red] circles represent the Majorana fermions making up each physical site (oval). The fermion operator on each physical site (c_j) is split up into two Majorana operators (a_j and b_j). In the non-trivial phase, the unpaired Majorana fermion states at the end of the chain are labelled with a_1 and b_N. These are the states that are continuously connected to the zero modes in the non-trivial topological superconductor phase.

to see that the ground state is trivial. If the chain has open boundary conditions, there will be no low-energy states on the end of the chain if the boundaries are cut between *physical* sites. That is, we are not allowed to pick boundary conditions where a physical complex fermionic site is cut in half.

These two limits give the simplest representations of the trivial and non-trivial phases. By tuning away from these limits, the Hamiltonian will have some mixture of couplings between Majorana operators on the same physical site, and operators between physical sites. However, since the two Majorana modes are localized at different ends of a gapped chain, the coupling between them will be exponentially small in the length of the wire and they will remain at zero energy. In fact, in the non-trivial phase, the zero modes will not be destroyed until the bulk gap closes at a critical point.

It is important to note that these zero modes count to a different many-body ground state degeneracy than the end modes of the Su–Schrieffer–Heeger model. The difference is rooted in the fact that one cannot build a fermionic Fock space out of an odd number of Majorana modes, because they are linear combinations of particles and holes. Rather, we can define a *single* fermionic operator out of *both* Majorana end modes a_1 and b_N as $c^\dagger := a_1 + ib_N$. The Hilbert space we can build out of a_1 and b_N is hence inherently non-local. This *non-local* state can be either occupied or empty, giving rise to a twofold-degenerate ground state of the chain with two open ends. (In contrast, the topological Su–Schrieffer–Heeger chain has a fourfold-degenerate ground state with two open ends, because it has one fermionic mode on each end.) The Majorana chain thus displays a different form of fractionalization than the Su–Schrieffer–Heeger chain. For the latter, we observed that the topological end modes carry fractional charge. In the Majorana chain, the end modes are a fractionalization of a fermionic mode into a superposition of particle and hole (and no longer have a well-defined charge), but the states $|0\rangle$ (with $c|0\rangle = 0$) and $c^\dagger|0\rangle$ do have distinct fermion parity. The non-local fermionic mode formed by two Majorana end modes is envisioned to work as a qubit (a quantum-mechanical two-level system) that stores quantum information (its state) in a way that is protected against local noise and decoherence.

2.1.5 Reduction of the 10-fold way classification by interactions: $\mathbb{Z} \to \mathbb{Z}_8$ in class BDI

When time-reversal symmetry $\mathcal{T} = \mathcal{K}$ is present, the model considered in Section 2.1.4 belongs to class BDI of the classification of non-interacting fermionic Hamiltonians in Table 2.1 with a $\mathbb{Z}$ topological characterization. We want to explore how interactions alter this classification, following a calculation by Fidkowski and Kitaev [8]. To this end, we consider a collection of n identical 1D topological Majorana chains in class BDI and only consider their Majorana end modes on one end, which we denote by $a_1, \dots, a_n$. We will take the point of view that if we can gap the edge, we can continue the bulk to a trivial state (insulator). This is not entirely a correct point of view in general (see 2D topologically ordered states such as the toric code discussed in Section 2.2.1), but works for our purposes. Given some integer n, we ask whether we can couple the Majorana modes locally on one end such that no gapless degrees of freedom are left on that end and the ground state with open boundary conditions becomes singly degenerate. To remain in class BDI, we only allow couplings that respect time-reversal symmetry. Let us first derive the action of $\mathcal{T}$ on the Majorana modes. The complex fermion operators are left invariant under time-reversal $\mathcal{T}c\mathcal{T}^{-1} = c$. Hence,

$$\begin{aligned} \mathcal{T}(a + \mathrm{i}b)\mathcal{T}^{-1} &= \mathcal{T}a\mathcal{T}^{-1} - \mathrm{i}\mathcal{T}b\mathcal{T}^{-1} \stackrel{!}{=} a + \mathrm{i}b \\ \Longrightarrow \quad \mathcal{T}a\mathcal{T}^{-1} &= a, \quad \mathcal{T}b\mathcal{T}^{-1} = -b. \end{aligned} \tag{2.29}$$

Thus, when acting on the modes localized on the left end of the wire (which transform like the a's), time-reversal symmetry leaves the Majorana operators invariant.

The most naive coupling term that would gap out two Majoranas is $\mathrm{i}a_1a_2$. This is because two Majoranas can form a local Hilbert space (unlike just one Majorana), and this local Hilbert space can be split unless some other symmetry prevents it from being split. However, time-reversal symmetry forbids these hybridization terms, since it sends $\mathrm{i}a_1a_2 \to -\mathrm{i}a_1a_2$. In spinful systems, another symmetry that can do this is $M\mathcal{T}$, where M is a mirror operator (which in spinful systems squares to -1 $M^2 = -1$) and $\mathcal{T}$ is the usual time-reversal operator $\mathcal{T}^2 = -1$, such that $(M\mathcal{T})^2 = M^2\mathcal{T}^2 = 1$ and hence $M\mathcal{T}$ acts like spinless time reversal [7]. Realizing that such a term is not allowed is the end of the story for non-interacting systems, yielding the classification $\mathbb{Z}$. Let's find out what interactions do to this system. The steps that we will now outline are summarized in Fig. 2.3.

We saw that two Majorana end states cannot be gapped: the only possible interacting or non-interacting Hamiltonian is $\mathrm{i}a_1a_2$. Three Majoranas clearly cannot be gapped either, since three is an odd number. Let us thus add two more Majorana end states into the mix. Any one-body term still is disallowed, but the term

$$H_{\text{int}} = a_1a_2a_3a_4 \tag{2.30}$$

can be present. We can now form two complex fermions, $c_1 = (a_1 + \mathrm{i}a_2)/\sqrt{2}$ and $c_2 = (a_3 + \mathrm{i}a_4)/\sqrt{2}$. In terms of these two fermions, the Hamiltonian reads

$$H_{\text{int}} = -\left(n_1 - \frac{1}{2}\right)\left(n_2 - \frac{1}{2}\right), \tag{2.31}$$

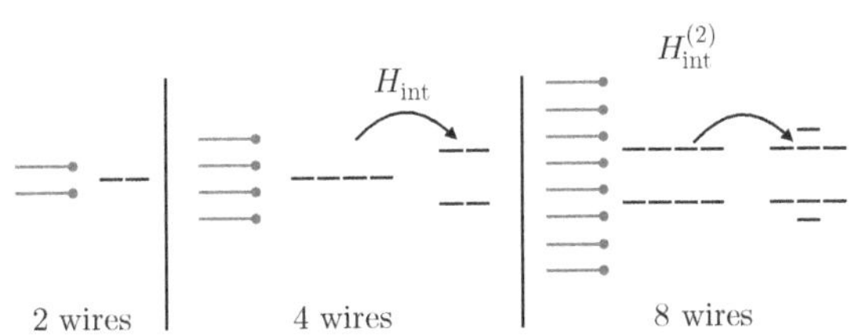

Fig. 2.3 Schematic illustration of the many-body energy levels for 2, 4, and 8 wires with Majorana end states as well as the (partial) lifting of their degeneracy by the Hamiltonians in (2.31) and (2.33).

where $n_1 = c_1^\dagger c_1$ and $n_2 = c_2^\dagger c_2$ are the occupation numbers. The Hamiltonian is diagonal in the eigenbasis $|n_1 n_2\rangle$ of the occupation number operators, and the states $|11\rangle, |00\rangle$ are degenerate at energy $-\frac{1}{4}$, while the states $|01\rangle, |10\rangle$ are degenerate at energy $+\frac{1}{4}$. The original non-interacting system of four Majorana fermions had a degeneracy of $2^2 = 4$. The interaction, however, has lifted this degeneracy, but not all the way to a single non-degenerate ground state. Irrespective of the sign of the interaction, it leaves the states doubly degenerate on one edge, and hence cannot be adiabatically continued to the trivial state of single degeneracy. However, if we add four more Majoranas wires so that we have $n = 8$ Majoranas, we can build an interaction that creates a singly degenerate ground state. We can understand this as follows. Add two interactions

$$H_{\text{int}}^{(1)} = -\alpha(a_1 a_2 a_3 a_4 + a_5 a_6 a_7 a_8). \tag{2.32}$$

These create two doublets, one in c_1, c_2 defined above, and one in $c_3 = (a_5 + ia_6)/\sqrt{2}, c_4 = (a_7 + ia_8)/\sqrt{2}$. We couple these doublets via the interaction

$$H_{\text{int}}^{(2)} = \sum_{i=x,y,z} \beta(c_1^\dagger \quad c_2^\dagger)\sigma_i \begin{pmatrix} c_1 \\ c_2 \end{pmatrix} (c_3^\dagger \quad c_4^\dagger)\sigma_i \begin{pmatrix} c_3 \\ c_4 \end{pmatrix}. \tag{2.33}$$

Representing each of the doublets as a spin-$\frac{1}{2}$ $\mathbf{S}$, this interaction is nothing but an $\mathbf{S} \cdot \mathbf{S}$ term. If we take $0 < \beta \ll \alpha$, then we can approximate the interaction β by its action within the two ground-state doublets. As such, this interaction creates a singlet and a triplet (in that doublet) and for the correct sign of β, we can put the singlet below the triplet, thereby creating a unique ground state

$$\frac{1}{\sqrt{2}} \left(|0110\rangle - |1001\rangle\right), \tag{2.34}$$

in terms of the occupation number states $|n_1 n_2 n_3 n_4\rangle$. This unique ground state can be adiabatically continued to the atomic limit. In this way the non-interacting $\mathbb{Z}$ classification of class BDI breaks down to $\mathbb{Z}_8$ if interactions are allowed.

2.2 Examples of topological order

So far, we have been concerned with symmetry-protected topological states and considered examples that were motivated by the topological classification of free-fermion Hamiltonians. The topological properties of these systems are manifest by the presence of protected boundary modes.

In this section, we want to familiarize ourselves with the concept of intrinsic topological order by ways of two examples. We will study the connections between different characterizations of topological order, such as fractionalized excitations in the bulk and the topological ground-state degeneracy. Our examples will be in 2D space, since topologically ordered states do not exist in 1D and are best understood in 2D. Our first example, the toric code, has Abelian anyon excitations, while the second example, the chiral p-wave superconductor, features non-Abelian anyons.

2.2.1 The toric code

The first example of a topologically ordered state is an exactly soluble model with vanishing correlation length. The significance of having zero correlation length is as follows. The correlation functions of local operators decay exponentially in gapped quantum ground states in 1D and 2D with a characteristic lengthscale given by the correlation length ξ [9]. In contrast, topological properties are encoded in quantized expectation values of non-local operators (e.g. the Hall conductivity) or the degeneracy of energy levels (e.g. the end states of the Su–Schrieffer–Heeger model). In finite systems, such quantizations and degeneracies are generically only exact up to corrections that are of order $e^{-L/\xi}$, where L is the linear system size. Models with zero correlation length are free from such exponential finite-size corrections and thus expose the topological features already for the smallest possible system sizes. The downside is that their Hamiltonians are rather contrived.

We define the toric code model [10] on a square lattice with a spin-$\frac{1}{2}$ degree of freedom on every *bond* j (see Fig. 2.4). The four spins that sit on the bonds emanating from a given site of the lattice are referred to as a star s. The four spins that sit on the bonds surrounding a square of the lattice are called a plaquette p. We define two sets of operators

$$A_s := \prod_{j \in s} \sigma_j^x, \tag{2.35a}$$

$$B_p := \prod_{j \in p} \sigma_j^z, \tag{2.35b}$$

that act on the spins of a given star s and plaquette p, respectively. Here, $\sigma_j^{x,z}$ are the respective Pauli matrices acting on the spin on bond j.

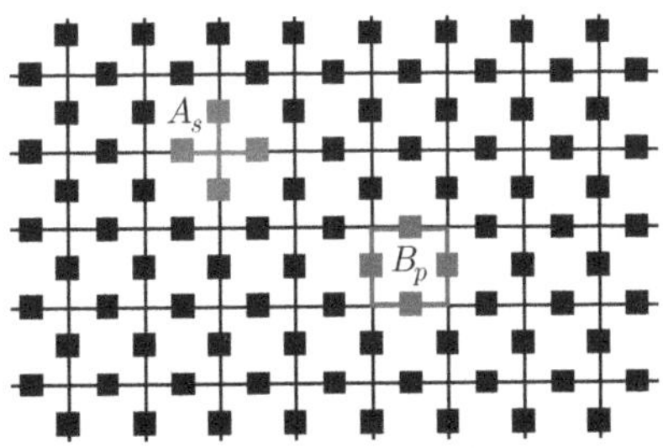

Fig. 2.4 [Colour online] The toric code model is defined on a square lattice with spin-$\frac{1}{2}$ degrees of freedom on every bond (black squares). The operator A_s acts with σ_x on all four spins on the bonds that are connected to a lattice site (a star s) [blue]. The operator B_p acts with σ_z on all four spins around a plaquette p [red].

These operators have two crucial properties that are often used to construct exactly soluble models for topological states of matter:

1. All of the A_s and B_p commute with each other. This is trivial for all cases except for the commutator of A_s with B_p if s and p have spins in common. However, any star shares with any plaquette an even number of spins (edges), so that commuting A_s with B_p involves commuting an even number of σ^z with σ^x, each of which comes with a minus sign.
2. The operators
$$\frac{1-B_p}{2}, \qquad \frac{1-A_s}{2} \tag{2.36}$$
are projectors. The former projects out plaquette states with an even number of spins polarized in the positive z direction. The latter projects out stars with an even number of spins in the positive x direction.

2.2.1.1 Ground states

The Hamiltonian is defined as a sum over these commuting projectors:

$$H = -J_{\mathrm{e}} \sum_{s} A_s - J_{\mathrm{m}} \sum_{p} B_p, \tag{2.37}$$

where the sums run over all stars s and plaquettes p of the lattice. Let us assume that both J_{e} and J_{m} are positive constants. Then, the ground state is given by a state in which all stars s and plaquettes p are in an eigenstate with eigenvalue $+1$ of A_s and B_p, respectively. (The fact that all A_s and B_p commute allows for such a state to exist, as we can diagonalize each of them separately.) Let us think about the ground state in the eigenbasis of the σ^x operators and represent by bold lines those bonds with spin up and and draw no lines along bonds with spin down. Then, A_s imposes on all spin configurations with non-zero amplitude in the ground state the constraint that an even number of bold lines meet at the star s. In other words, we can think of the bold lines as connected across the lattice and they may only form closed loops. Bold lines that end at some star ('open strings') are not allowed in the ground-state configurations; they are excited states. Having found out which spin configurations are allowed in the ground state, we need to determine their amplitudes. This can be inferred from the action of the B_p operators on these closed-loop configurations. The B_p flips all bonds around the plaquette p. Since $B_p^2 = 1$, given a spin configuration $|c\rangle$ in the σ^x-basis, we can write an eigenstate of B_p with eigenvalue 1 as

$$\frac{1}{\sqrt{2}} \left(|c\rangle + B_p|c\rangle\right), \tag{2.38}$$

for some fixed p. This reasoning can be extended to all plaquettes, and therefore we can write for the ground state

$$|\mathrm{GS}\rangle = \left(\prod_p \frac{1+B_p}{\sqrt{2}}\right)|c\rangle, \tag{2.39}$$

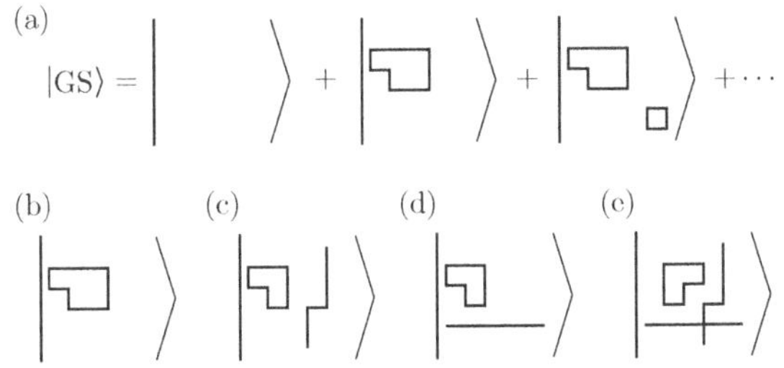

Fig. 2.5 Visualization of the toric code ground states on the torus. (a) The toric code ground state is the equal amplitude superposition of all closed-loop configurations. (b)–(e) Four base configurations $|c\rangle$ entering (2.39) that yield topologically distinct ground states on the torus.

where $|c\rangle$ is a closed-loop configuration; see Fig. 2.5(a). Is $|\mathrm{GS}\rangle$ independent of the choice of $|c\rangle$? In other words, is the ground state unique? We will see that the answer depends on the topological properties of the manifold on which the lattice is defined and thus reveals the topological order imprinted in $|\mathrm{GS}\rangle$.

To answer these questions, let us consider the system on two topologically distinct manifolds: the torus and the sphere. To obtain a torus, we consider a square lattice with $L_x \times L_y$ sites and impose periodic boundary conditions. This lattice hosts $2L_xL_y$ spins (2 per unit cell, since they are centred along the bonds). Thus, the Hilbert space of the model has dimension $2^{2L_xL_y}$. There are L_xL_y operators A_s and just as many B_p. Hence, together, they impose $2L_xL_y$ constraints on the ground state in this Hilbert space. However, not all of these constraints are independent. The relations

$$1 = \prod_s A_s, \tag{2.40a}$$

$$1 = \prod_p B_p \tag{2.40b}$$

make two of the constraints redundant, yielding $2L_xL_y - 2$ independent constraints. The ground-state degeneracy (GSD) is obtained as the quotient of the dimension of the Hilbert space and the dimension of the subspace modded out by the constraints:

$$\mathrm{GSD} = \frac{2^{2L_xL_y}}{2^{2L_xL_y-2}} = 4. \tag{2.41}$$

The four ground states on the torus are distinguished by having an even or an odd number of loops wrapping the torus in the x and y directions, respectively. Four configurations $|c\rangle$ that can be used to build the four degenerate ground states are shown in Fig. 2.5(b–e). This constitutes a set of 'topologically degenerate' ground states and is a hallmark of the topological order in the model.

Let us contrast this with the ground-state degeneracy on the sphere. Since we use a model with zero correlation length, we might as well use the smallest convenient lattice with the topology of a sphere. We consider the model (2.37) defined on the edges of a cube. The same counting as above yields that there are 12 degrees of freedom (the spins on the 12 edges), 8 constraints from the A_s operators defined on the corners, and 6 constraints from the B_p operators defined on the faces. Subtracting the 2 redundant constraints (2.40) yields $12 - (8 + 6 - 2) = 0$ remaining degrees of freedom. Hence, the model has a unique ground state on the sphere.

On a general manifold, we have

$$\mathrm{GSD} = 2^{\text{number of non-contractible loops}}. \tag{2.42}$$

An important property of the topologically degenerate ground states is that any local operator has vanishing off-diagonal matrix elements between them in the thermodynamic limit. Similarly, no local operator can be used to distinguish between the ground states. We can, however, define *non-local* operators that transform one topologically degenerate ground state into another and that distinguish the ground states by topological quantum numbers. (Notice that such operators may not appear in any physical Hamiltonian owing to their non-locality, and hence the degeneracy of the ground states is protected.) On the torus, we define two pairs of so-called Wilson loop operators as

$$W^{\mathrm{e}}_{x/y} := \prod_{j \in l^{\mathrm{e}}_{x/y}} \sigma^z_j, \tag{2.43a}$$

$$W^{\mathrm{m}}_{x/y} := \prod_{j \in l^{\mathrm{m}}_{x/y}} \sigma^x_j. \tag{2.43b}$$

Here, $l^{\mathrm{e}}_{x/y}$ are the sets of spins on bonds parallel to a straight line wrapping the torus once along the x and y directions, respectively. The $l^{\mathrm{m}}_{x/y}$ are the sets of spins on bonds perpendicular to a straight line that connects the centres of plaquettes and wraps the torus once along the x and y directions, respectively. We note that the $W^{\mathrm{e}}_{x/y}$ and $W^{\mathrm{m}}_{x/y}$ commute with all A_s and B_p,

$$[W^{\mathrm{e/m}}_{x/y}, A_s] = [W^{\mathrm{e/m}}_{x/y}, B_p] = 0, \tag{2.44}$$

and thus also with the Hamiltonian. Furthermore, they obey

$$W^{\mathrm{e}}_x W^{\mathrm{m}}_y = -W^{\mathrm{m}}_y W^{\mathrm{e}}_x. \tag{2.45}$$

This algebra must be realized in any eigenspace of the Hamiltonian. However, because of (2.45), it cannot be realized in a 1D subspace. We conclude that all eigenspaces of the Hamiltonian, including the ground state, must be degenerate. In the σ^x basis that we used above, $W^{\mathrm{m}}_{x/y}$ measures whether the number of loops wrapping the torus is even or odd in the x and y direction, respectively, giving four degenerate ground states. In contrast, $W^{\mathrm{e}}_{x/y}$ changes the number of loops wrapping the torus in the x and y directions between even and odd.

2.2.1.2 Topological excitations

To find the topological excitations of the system above the ground state, we ask which are the lowest-energy excitations that we can build. Excitations are a violation of the rule that all stars s are eigenstates of A_s and all plaquettes p are eigenstates of B_p. Let us first focus on star excitations, which we will call e. They appear as the endpoints of open strings, i.e. if the closed loop condition is violated. Since any string has two endpoints, the lowest excitation of this type is a pair of e. They can be created by acting on the ground state with the operator

$$W^{\mathrm{e}}_{l^{\mathrm{e}}} := \prod_{j \in l^{\mathrm{e}}} \sigma^z_j, \tag{2.46}$$

where l^{e} is a string of bonds connecting the two excitations e_1 and e_2 (see Fig. 2.6(a)). The state

$$|e_1, e_2\rangle := W^{\mathrm{e}}_{l^{\mathrm{e}}}|\mathrm{GS}\rangle \tag{2.47}$$

has energy $4J_{\mathrm{e}}$ above the ground state energy. Similarly, we can define an operator

$$W^{\mathrm{m}}_{l^{\mathrm{m}}} := \prod_{j \in l^{\mathrm{m}}} \sigma^x_j, \tag{2.48}$$

that creates a pair of plaquette defects m_1 and m_2 connected by the string l^{m} of perpendicular bonds; see Fig. 2.6(b). (Notice that the operator $W^{\mathrm{m}}_{l^{\mathrm{m}}}$ does not flip spins when the ground state is written in the σ^x basis. Rather, it gives weight $+1/-1$ to the different loop configurations in the ground state, depending on whether an even or an odd number of loops crosses l^{m}.) The state

$$|m_1, m_2\rangle := W^{\mathrm{m}}_{l^{\mathrm{m}}}|\mathrm{GS}\rangle \tag{2.49}$$

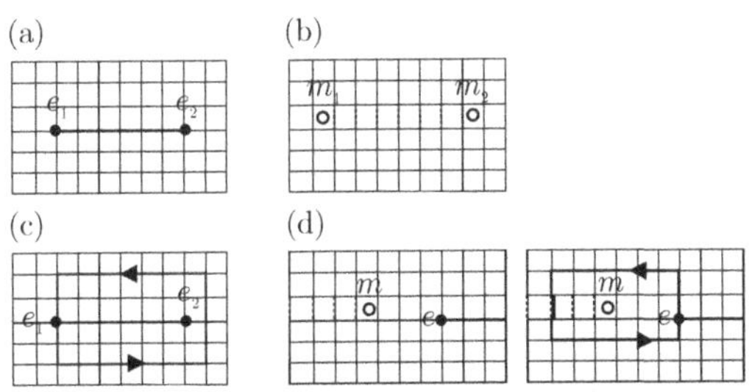

Fig. 2.6 Visualization of operations to compute the braiding statistics of toric code anyons. (a) Two e excitations above the ground state. (b) Two m excitations above the ground state. (c) Loop created by braiding e_1 around e_2. (c) Loop created by braiding e around m. A phase of -1 results for this process because there is a single bond on which both a σ^x operator (dotted line) and a σ^z operator (bold line) act.

has energy $4J_{\mathrm{m}}$ above the ground-state energy. Notice that the excited states $|e_1, e_2\rangle$ and $|m_1, m_2\rangle$ depend only on the positions of the excitations and not on the particular choice of string that connects them. Furthermore, the energy of the excited state is independent of the separation between the excitations. The excitations are thus 'deconfined', i.e. free to move independently of each other.

It is also possible to create a combined defect when a plaquette hosts an m excitation and one of its corners hosts an e excitation. We call this combined defect f and formalize the relation between these defects in a so-called fusion rule

$$e \times m = f. \tag{2.50a}$$

When two e-type excitations are moved to the same star, the loop l^{e} that connects them becomes a closed loop and the state returns to the ground state. For this, we write the fusion rule

$$e \times e = 1, \tag{2.50b}$$

where 1 stands for the ground state or vacuum. Similarly, moving two m-type excitations to the same plaquette creates a closed loop l^{m}, which can be absorbed in the ground state, i.e.

$$m \times m = 1. \tag{2.50c}$$

Superimposing the above processes yields the remaining fusion rules

$$m \times f = e, \qquad e \times f = m, \qquad f \times f = 1. \tag{2.50d}$$

It is now imperative to ask what type of quantum statistics these emergent excitations obey. We recall that quantum statistics are defined as the phase by which a state changes if two identical particles are exchanged. Rendering the exchange operation as an adiabatically slow evolution of the state, in three and higher dimensions only two types of statistics are allowed between point particles: that of bosons with phase $+1$ and that of fermions with phase -1. In 2D, richer possibilities exist and the exchange phase θ can be *any* complex number on the unit circle, opening the way for *any*ons. While the exchange is only defined for quantum particles of the same type, the double exchange (braiding) is well defined between any two deconfined anyons. We can compute the braiding phases of the anyons e, m, and f that appear in the toric code one by one. Let us start with the phase resulting from braiding e_1 with e_2. The initial state is $W^{\mathrm{e}}_{l^{\mathrm{e}}}|\mathrm{GS}\rangle$ depicted in Fig. 2.6(a). Moving e_1 around e_2 leaves a loop of flipped σ^x bonds around e_2; see Fig. 2.6(c). This loop is created by applying B_p to all plaquettes enclosed by the loop $l^{\mathrm{e}}_{e_1}$ along which e_1 moves. We can thus write the final state as

$$\begin{aligned}\left(\prod_{p\in l^{\mathrm{e}}_{e_1}} B_p\right) W^{\mathrm{e}}_{l^{\mathrm{e}}}|\mathrm{GS}\rangle &= W^{\mathrm{e}}_{l^{\mathrm{e}}}\left(\prod_{p\in l^{\mathrm{e}}_{e_1}} B_p\right)|\mathrm{GS}\rangle \\ &= W^{\mathrm{e}}_{l^{\mathrm{e}}}|\mathrm{GS}\rangle.\end{aligned} \tag{2.51}$$

Flipping the spins in a closed loop does not alter the ground state, since it is the equal amplitude of all loop configurations. We conclude that the braiding of two e particles gives no phase. Similar considerations can be used to conclude that the braiding of two m particles is trivial as well. In fact, not only the braiding, but also the exchange of two e particles and two m particles is trivial. (We have not shown that here.)

More interesting is the braiding of m with e. Let the initial state be $W^{\mathrm{m}}_{l^{\mathrm{m}}} W^{\mathrm{e}}_{l^{\mathrm{e}}} |\mathrm{GS}\rangle$ and move the e particle located on one end of the string $l^{\mathrm{e}}_{\mathrm{in}}$ around the magnetic particle m on one end of the string l^{m}. Again this is equivalent to applying B_p to all plaquettes enclosed by the path $l^{\mathrm{e}}_{\mathrm{e}}$ of the e particle, so that the final state is given by

$$
\begin{aligned}
\left(\prod_{p \in l^{\mathrm{e}}_{\mathrm{e}}} B_p\right) W^{\mathrm{m}}_{l^{\mathrm{m}}} W^{\mathrm{e}}_{l^{\mathrm{e}}} |\mathrm{GS}\rangle &= -\, W^{\mathrm{m}}_{l^{\mathrm{m}}} \left(\prod_{p \in l^{\mathrm{e}}_{\mathrm{e}}} B_p\right) W^{\mathrm{e}}_{l^{\mathrm{e}}} |\mathrm{GS}\rangle \\
&= -\, W^{\mathrm{m}}_{l^{\mathrm{m}}} W^{\mathrm{e}}_{l^{\mathrm{e}}} |\mathrm{GS}\rangle .
\end{aligned}
\tag{2.52}
$$

The product over B_p operators anticommutes with the path operator $W^{\mathrm{m}}_{l^{\mathrm{m}}}$, because there is a single bond on which a single σ^x and a single σ^z act at the crossing of l^{m} and l^{e}_{e}; see Fig. 2.6(d). As a result, the initial and final state differ by a -1, which is the braiding phase of e with m. Particles with this braiding phase are called (mutual) semions.

Notice that we have moved the particles on contractible loops only. If we create a pair of e or m particles, move one of them along a non-contractible loop on the torus, and annihilate the pair, we have effectively applied the operators $W^{\mathrm{e}}_{x/y}$ and $W^{\mathrm{m}}_{x/y}$ to the ground state (although, in the process, we have created finite-energy states). The operation of moving anyons on non-contractible loops thus allows us to operate on the manifold of topologically degenerate ground states. This exposes the intimate connection between the presence of fractionalized excitations and topological ground-state degeneracy in topologically ordered systems.

From the braiding relations of e and m, we can also determine the braiding and exchange relations of the composite particle f. This is most easily done in a pictorial way by representing the particle worldlines as moving upwards. For example, we represent the braiding relations of e and m as

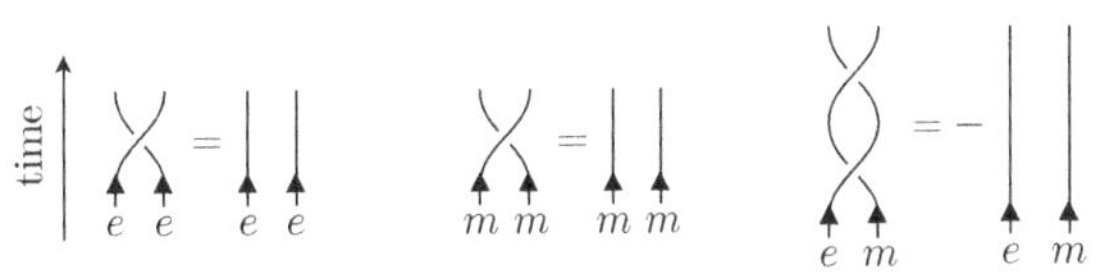

(2.53)

The exchange of two f, each of which is composed of one e and one m, is then

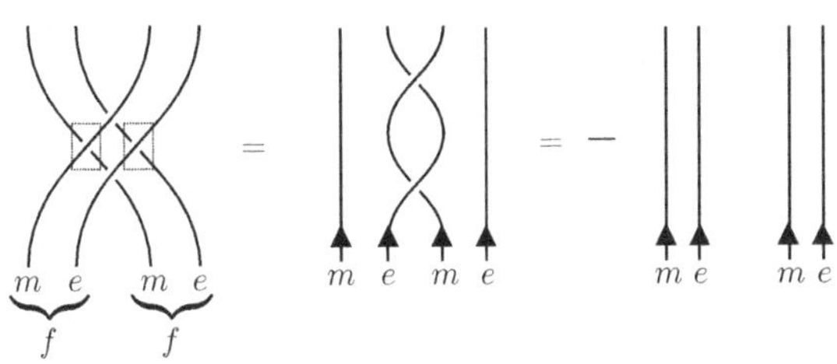

(2.54)

Notice that we have used (2.53) to manipulate the crossing in the dotted rectangles. Exchange of two f thus gives a phase -1, and we conclude that f is a fermion.

In summary, we have used the toric code model to illustrate topological ground-state degeneracy and emergent anyonic quasiparticles as hallmarks of topological order. We note that the toric code model does not support topologically protected edge states.

2.2.2 The 2D p-wave superconductor

The second example of a 2D system with anyonic excitations that we want to discuss here is the chiral p-wave superconductor. Unlike the toric code, owing to its chiral nature, it is a model with non-zero correlation length. The vortices of the chiral p-wave superconductor exhibit anyon excitations that have exotic non-Abelian statistics.[12–14] (The anyons in the toric code are Abelian—we will see below what that distinction refers to.) For the system to be topologically ordered, these vortices should appear as emergent, dynamical excitations. This requires that we treat the electromagnetic gauge field quantum-mechanically. (In fact, since the fermion number conservation is spontaneously broken down to the conservation of the fermion parity in the superconductor, the relevant gauge theory involves only a $\mathbb{Z}_2$ instead of a U(1) gauge field.) However, the topological properties that we want to discuss here can also be seen if we model the gauge field and vortices as static defects, rather than within a fluctuating $\mathbb{Z}_2$ gauge theory. This allows us to study a model very similar to the 'non-interacting" topological superconductor in 1D and still expose the non-Abelian statistics.

For pedagogy, we will use both lattice and continuum models of the chiral superconductor. We begin with the lattice Hamiltonian defined on a square lattice:

$$H = \sum_{m,n} \{ - t(c^\dagger_{m+1,n} c_{m,n} + c^\dagger_{m,n+1} c_{m,n} + \text{h.c.}) - (\mu - 4t) c^\dagger_{m,n} c_{m,n} + (\Delta c^\dagger_{m+1,n} c^\dagger_{m,n} + \mathrm{i}\Delta c^\dagger_{m,n+1} c^\dagger_{m,n} + \text{h.c.})\}. \tag{2.55}$$

The fermion operators $c_{m,n}$ annihilate fermions on the lattice site (m, n) and we are considering spinless (or equivalently spin-polarized) fermions. We set the lattice constant $a = 1$ for simplicity. The pairing amplitude is anisotropic and has an additional phase of i in the y direction compared with the pairing in the x direction. Because the pairing is not on-site, just as in the lattice version of the p-wave wire, the pairing terms will have momentum dependence. We can write this Hamiltonian in the Bogoliubov–de Gennes form and, assuming that Δ is translationally invariant, we can Fourier transform the lattice model to get

$$H_{\text{BdG}} = \frac{1}{2} \sum_{\mathbf{p}} \Psi^\dagger_{\mathbf{p}} \begin{pmatrix} \epsilon(p) & 2\mathrm{i}\Delta(\sin p_x + \mathrm{i}\sin p_y) \\ -2\mathrm{i}\Delta^*(\sin p_x - \mathrm{i}\sin p_y) & -\epsilon(p) \end{pmatrix} \Psi_{\mathbf{p}}, \tag{2.56}$$

where $\epsilon(p) = -2t(\cos p_x + \cos p_y) - (\mu - 4t)$ and $\Psi_\mathbf{p} = (c_\mathbf{p} \;\; c^\dagger_{-\mathbf{p}})^\mathsf{T}$. For convenience, we have shifted the chemical potential by the constant $4t$. As a quick aside, we note that the model takes a simple familiar form in the continuum limit ($\mathbf{p} \to 0$):

$$H^{(\text{cont})}_{\text{BdG}} = \frac{1}{2}\sum_\mathbf{p} \Psi^\dagger_\mathbf{p} \begin{pmatrix} \dfrac{p^2}{2m} - \mu & 2\text{i}\Delta(p_x + \text{i}p_y) \\ -2\text{i}\Delta^*(p_x - \text{i}p_y) & -\dfrac{p^2}{2m} + \mu \end{pmatrix} \Psi_\mathbf{p}, \tag{2.57}$$

where $m \equiv 1/2t$ and $p^2 = p_x^2 + p_y^2$. We see that the continuum limit has the characteristic $p_x + \text{i}p_y$ chiral form for the pairing potential. The quasiparticle spectrum of $H^{(\text{cont})}_{\text{BdG}}$ is $E_\pm = \pm\sqrt{(p^2/2m - \mu)^2 + 4|\Delta|^2 p^2}$, which, with a non-vanishing pairing amplitude, is gapped across the entire BZ as long as $\mu \neq 0$. This is unlike some other types of p-wave pairing terms (e.g. $\Delta(p) = \Delta p_x$), which can have gapless *nodal* points or lines in the BZ for $\mu > 0$. In fact, nodal superconductors, having gapless quasiparticle spectra, are not topological superconductors by definition (i.e. a bulk excitation gap does not exist).

We recognize the form of $H^{(\text{cont})}_{\text{BdG}}$ as a massive 2D Dirac Hamiltonian, and indeed (2.55) is just a lattice Dirac Hamiltonian, which is what we will consider first. In the first-quantized notation, the single-particle Hamiltonian for a superconductor is equivalent to that of an insulator with an additional particle–hole symmetry. It is thus placed in class D of Table 2.1 and admits a $\mathbb{Z}$ topological classification in 2D. Thus, we can classify the eigenstates of the Hamiltonian (2.55) by a Chern number—but, owing to the breaking of U(1) symmetry, the Chern number does not have the interpretation of Hall conductance. However, it is still a topological invariant.

We expect that H_{BdG} will exhibit several phases as a function of Δ and μ for a fixed $t > 0$. For simplicity let us set $t = \frac{1}{2}$ and make a gauge transformation $c_\mathbf{p} \to \text{e}^{\text{i}\theta/2} c_\mathbf{p}$, $c^\dagger_\mathbf{p} \to \text{e}^{-\text{i}\theta/2} c^\dagger_\mathbf{p}$, where $\Delta = |\Delta|\text{e}^{\text{i}\theta}$. The Bloch Hamiltonian for the lattice superconductor is then

$$\mathcal{H}_{\text{BdG}}(\mathbf{p}) = (2 - \mu - \cos p_x - \cos p_y)\,\sigma_z - 2|\Delta| \sin p_x \sigma_y - 2|\Delta| \sin p_y \sigma_x, \tag{2.58}$$

where σ_i, $i = x, y, z$, are the Pauli matrices in the particle/hole basis. Assuming $|\Delta| \neq 0$, this Hamiltonian has several fully gapped superconducting phases separated by gapless critical points. The quasiparticle spectrum for the lattice model is

$$E_\pm = \pm\sqrt{(2 - \mu - \cos p_x - \cos p_y)^2 + 4|\Delta|^2 \sin^2 p_x + 4|\Delta|^2 \sin^2 p_y} \tag{2.59}$$

and is gapped (under the assumption that $|\Delta| \neq 0$) unless the prefactors of all three Pauli matrices vanish simultaneously. As a function of (p_x, p_y, μ), we find three critical points. The first occurs at $(p_x, p_y, \mu) = (0, 0, 0)$. The second has two gap-closings in the BZ for the same value of μ: $(\pi, 0, 2)$ and $(0, \pi, 2)$. The third is again a singly degenerate point at $(\pi, \pi, 4)$. We will show that the phases for $\mu < 0$ and $\mu > 4$ are trivial superconductors, while those for $0 < \mu < 2$ and $2 < \mu < 4$ are topological superconductors with opposite chirality. In principle, one can define a Chern number topological invariant constructed from the eigenstates of the lower quasiparticle band to characterize the phases. We will show this calculation below, but first we make some physical arguments as to the nature of the phases, following the discussion in [11].

We will first consider the phase transition at $\mu = 0$. The low-energy physics for this transition occurs around $(p_x, p_y) = (0, 0)$ and so we can expand the lattice Hamiltonian around this point; this is nothing but (2.57). One way to test the character of the $\mu < 0$ and $\mu > 0$ phases is to make an interface between them. If we can find a continuous interpolation between these two regimes that is always gapped, then they are topologically equivalent phases of matter. If we cannot find such a continuously gapped interpolation, then they are topologically distinct. A simple geometry to study is a domain wall where $\mu = \mu(x)$ such that $\mu(x) = -\mu_0$ for $x < 0$ and $\mu(x) = +\mu_0$ for $x > 0$ for a positive constant μ_0. This is an interface that is translationally invariant along the y direction, and thus we can consider the momentum p_y as a good quantum number to simplify the calculation. What we will now show is that there exist gapless, propagating fermions bound to the interface that prevent us from continuously connecting the $\mu < 0$ phase to the $\mu > 0$ phase. This is one indication that the two phases represent topologically distinct classes.

The single-particle Hamiltonian in this geometry is

$$\mathcal{H}_{\mathrm{BdG}}(p_y) = \frac{1}{2}\begin{pmatrix} -\mu(x) & 2\mathrm{i}|\Delta|\left(-\mathrm{i}\dfrac{\mathrm{d}}{\mathrm{d}x} + \mathrm{i}p_y\right) \\ -2\mathrm{i}|\Delta|\left(-\mathrm{i}\dfrac{\mathrm{d}}{\mathrm{d}x} - \mathrm{i}p_y\right) & \mu(x) \end{pmatrix}, \tag{2.60}$$

where we have ignored the quadratic terms in p, and p_y is a constant parameter, not an operator. This is a quasi-1D Hamiltonian that can be solved for each value of p_y independently. We propose an ansatz for the gapless interface states:

$$|\psi_{p_y}(x, y)\rangle = \mathrm{e}^{\mathrm{i}p_y y} \exp\left(-\frac{1}{2|\Delta|}\int_0^x \mu(x')\,\mathrm{d}x'\right)|\phi_0\rangle \tag{2.61}$$

for a constant, normalized spinor $|\phi_0\rangle$. The secular equation for a zero-energy mode at $p_y = 0$ is

$$\mathcal{H}_{\mathrm{BdG}}(p_y)|\psi_0(x, y)\rangle = 0 \qquad \Longrightarrow \qquad \begin{pmatrix} -\mu(x) & -\mu(x) \\ \mu(x) & \mu(x) \end{pmatrix}|\phi_0\rangle = 0. \tag{2.62}$$

The constant spinor that is a solution of this equation is $|\phi_0\rangle = 1/\sqrt{2}\,(1, -1)^{\mathsf{T}}$. This form of the constant spinor immediately simplifies the solution of the problem at finite p_y. We see that the term proportional to p_y in (2.60) is $-2|\Delta|p_y\sigma_x$. Since $\sigma_x|\phi_0\rangle = -|\phi_0\rangle$, i.e. the solution $|\phi_0\rangle$ is an eigenstate of σ_x, we conclude that $|\psi_{p_y}(x, y)\rangle$ is an eigenstate of $\mathcal{H}_{\mathrm{BdG}}(p_y)$ with energy $E(p_y) = -2|\Delta|p_y$. Thus, we have found a normalizable bound-state solution at the interface of two regions with $\mu < 0$ and $\mu > 0$, respectively. This set of bound states, parametrized by the conserved quantum number p_y, is gapless and chiral, i.e. the group velocity of the quasiparticle dispersion is always negative and never changes sign (in this simplified model). The chirality is determined by the sign of the 'spectral' Chern number mentioned above, which we will calculate below.

These gapless edge states have quite remarkable properties and are not the same chiral complex fermions that propagate on the edge of integer quantum Hall states, but chiral real (Majorana) fermions. Using Clifford algebra representation theory, it can be shown that the so-called chiral Majorana (or Majorana–Weyl) fermions can only be found in spacetime dimensions $(8k+2)$, where $k = 0, 1, 2, \ldots$ Thus, we can only find chiral Majorana states in $(1+1)$ dimensions or in $(9+1)$ dimensions (or higher!). In condensed matter, we are stuck with $(1+1)$ dimensions, where we have now seen that they appear as the boundary states of chiral topological superconductors. The simplest interpretation of such chiral Majorana fermions is as half of a conventional chiral fermion, i.e. its real or imaginary part. To show this, we will consider the edge state of a Chern number 1 quantum Hall system for a single edge:

$$\mathcal{H}^{(\mathrm{QH})}_{\mathrm{edge}} = \hbar v \sum_p p \eta_p^\dagger \eta_p, \tag{2.63}$$

where p is the momentum along the edge. The fermion operators satisfy $\{\eta_p^\dagger, \eta_{p'}\} = \delta_{pp'}$. Similar to the discussion on the 1D superconducting wire, we can decompose these operators into their real and imaginary Majorana parts

$$\eta_p = \frac{1}{2}(\gamma_{1,p} + \mathrm{i}\gamma_{2,p}), \qquad \eta_p^\dagger = \frac{1}{2}(\gamma_{1,-p} - \mathrm{i}\gamma_{2,-p}), \tag{2.64}$$

where $\gamma_{a,p}$ $(a = 1, 2)$ are Majorana fermion operators satisfying

$$\gamma_{a,p}^\dagger = \gamma_{a,-p}, \qquad \{\gamma_{a,-p}, \gamma_{b,p'}\} = 2\delta_{ab}\delta_{pp'}. \tag{2.65}$$

The quantum Hall edge Hamiltonian now becomes

$$\begin{aligned}
\mathcal{H}^{(\mathrm{QH})}_{\mathrm{edge}} &= \hbar v \sum_{p\geq 0} p(\eta_p^\dagger \eta_p - \eta_{-p}^\dagger \eta_{-p}) \\
&= \frac{\hbar v}{4} \sum_{p\geq 0} p\{(\gamma_{1,-p} - \mathrm{i}\gamma_{2,-p})(\gamma_{1,p} + \mathrm{i}\gamma_{2,p}) - (\gamma_{1,p} - \mathrm{i}\gamma_{2,p})(\gamma_{1,-p} + \mathrm{i}\gamma_{2,-p})\} \\
&= \frac{\hbar v}{4} \sum_{p\geq 0} p(\gamma_{1,-p}\gamma_{1,p} + \gamma_{2,-p}\gamma_{2,p} - \gamma_{1,p}\gamma_{1,-p} - \gamma_{2,p}\gamma_{2,-p}) \\
&= \frac{\hbar v}{2} \sum_{p\geq 0} p(\gamma_{1,-p}\gamma_{1,p} + \gamma_{2,-p}\gamma_{2,p} - 2).
\end{aligned} \tag{2.66}$$

Thus,

$$\mathcal{H}^{(\mathrm{QH})}_{\mathrm{edge}} = \frac{\hbar v}{2} \sum_{p\geq 0} p(\gamma_{1,-p}\gamma_{1,p} + \gamma_{2,-p}\gamma_{2,p}), \tag{2.67}$$

up to a constant shift of the energy. This Hamiltonian is exactly two copies of a chiral Majorana Hamiltonian. The edge/domain-wall fermion Hamiltonian of the chiral p-wave superconductor will be

$$\mathcal{H}_{\text{edge}}^{(p-\text{wave})} = \frac{\hbar v}{2} \sum_{p \geq 0} p \gamma_{-p} \gamma_p. \tag{2.68}$$

Finding gapless states on a domain wall of μ is an indicator that the phases with $\mu > 0$ and $\mu < 0$ are distinct. If they were the same phase of matter, then we should be able to adiabatically connect these states continuously. However, we have shown a specific case of the more general result that any interface between a region with $\mu > 0$ and a region with $\mu < 0$ will have gapless states that generate a discontinuity in the interpolation between the two regions. The question remaining is whether $\mu > 0$ or $\mu < 0$ is non-trivial. The answer is that we have a trivial superconductor for $\mu < 0$ (adiabatically continued to $\mu \to -\infty$) and a topological superconductor for $\mu > 0$. Remember that for now we are only considering μ in the neighbourhood of 0 and using the continuum model expanded around $(p_x, p_y) = (0, 0)$. We will now define a bulk topological invariant for 2D superconductors that can distinguish the trivial superconductor state from the chiral topological superconductor state. For the spinless Bogoliubov–de Gennes Hamiltonian, which is of the form

$$H_{\text{BdG}} = \frac{1}{2} \sum_{\mathbf{p}} \Psi_{\mathbf{p}}^{\dagger} \left[\mathbf{d}(\mathbf{p}, \mu) \cdot \boldsymbol{\sigma} \right] \Psi_{\mathbf{p}}, \tag{2.69a}$$

$$\mathbf{d}(\mathbf{p}, \mu) = \left(-2|\Delta| p_y, -2|\Delta| p_x, p^2/2m - \mu \right), \tag{2.69b}$$

the topological invariant is the spectral Chern number defined in (2.11), which simplifies, for this Hamiltonian, to the winding number

$$\begin{aligned} \mathsf{C}^{(1)} &= \frac{1}{8\pi} \int \mathrm{d}^2\mathbf{p}\, \epsilon^{ij} \hat{\mathbf{d}} \cdot (\partial_{p_i} \hat{\mathbf{d}} \times \partial_{p_j} \hat{\mathbf{d}}) \\ &= \frac{1}{8\pi} \int \mathrm{d}^2\mathbf{p}\, \frac{\epsilon^{ij}}{|\mathbf{d}|^3} \mathbf{d} \cdot \left(\partial_{p_i} \mathbf{d} \times \partial_{p_j} \mathbf{d} \right). \end{aligned} \tag{2.70}$$

We defined the unit vector $\hat{\mathbf{d}} = \mathbf{d}/|\mathbf{d}|$, which is possible since $|\mathbf{d}| \neq 0$ owing to the existence of a gap. This integral has a special form and is equal to the degree of the mapping from momentum space onto the 2-sphere S^2 given by $\hat{d}_1^2 + \hat{d}_2^2 + \hat{d}_3^2 = 1$. As it stands, the degree of the mapping $\hat{\mathbf{d}} : \mathbb{R}^2 \to S^2$ is not well defined, because the domain is not compact, i.e. (p_x, p_y) is only restricted to lie in the Euclidean plane ($\mathbb{R}^2$). However, for our choice of the map $\hat{\mathbf{d}}$, we can define the winding number by choosing an equivalent, but compact, domain. To understand the necessary choice of domain, we can simply look at the explicit form of $\hat{\mathbf{d}}(\mathbf{p})$:

$$\hat{\mathbf{d}}(\mathbf{p}) = \frac{\left(-2|\Delta| p_y, -2|\Delta| p_x, p^2/2m - \mu \right)}{\sqrt{4|\Delta|^2 p^2 + (p^2/2m - \mu)^2}}. \tag{2.71}$$

We see that $\lim_{|\mathbf{p}|\to\infty} \hat{\mathbf{d}}(\mathbf{p}) = (0, 0, 1)$ and it does not depend on the direction in which we take the limit in the 2D plane. Because of the uniqueness of this limit, we are free to perform the *one-point compactification* of $\mathbb{R}^2$, which amounts to including the point at infinity in our domain. The topology of $\mathbb{R}^2 \cup \{\infty\}$ is the same as that of S^2 and thus we can consider the degree of our map from the compactified momentum space (S^2) to the unit $\hat{\mathbf{d}}$-vector space (S^2). Using the explicit form of the $\hat{\mathbf{d}}$-vector for this model, we find

$$\mathsf{C}^{(1)} = \frac{1}{\pi}\int \mathrm{d}^2\mathbf{p}\, \frac{|\Delta|^2(p^2/2m + \mu)}{[4|\Delta|^2 p^2 + (p^2/2m - \mu)^2]^{3/2}}. \tag{2.72}$$

The evaluation of this integral can be easily carried out numerically. The result is $\mathsf{C}^{(1)} = 0$ for $\mu < 0$ and $\mathsf{C}^{(1)} = 1$ for $\mu > 0$; i.e. there are two different phases separated by a quantum critical point at $\mu = 0$. Thus we have identified the phase that is in the chiral superconductor state to be $\mu > 0$.

2.2.2.1 Argument for the existence of Majorana bound states on vortices

A simple but rigorous argument can show us the presence of zero-energy bound states in the core of vortices in a superconductor. Assume we have a chiral ($p + ip$) superconductor in two geometries: a disc with an edge and a cylinder with two edges. Since it is a topological superconductor, the system will have chiral dispersing (Majorana) gapless modes along the edges. In Fig. 2.7, the spectra are plotted versus the momentum along the edge, and they are qualitatively very different in the two cases. For an edge of length L, the smallest difference between two momenta along the edge is $2\pi/L$. The energy difference between two levels is $v2\pi/L$, where v is the velocity of the edge mode.

In a single-particle superconducting Hamiltonian, the number of total single-particle eigenvalues is always even. This is clear from the fact that whatever the spinor of the non-superconducting Hamiltonian is, when superconductivity is added, we have a doubled spectrum, so that every energy state at $E > 0$ comes with a counterpart at

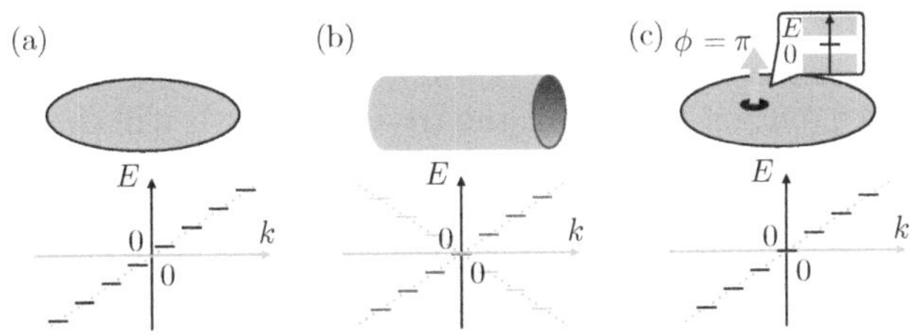

Fig. 2.7 [Colour online] Spectra of a chiral superconductor in different geometries: (a) disc; (b) cylinder; (c) disc with flux defect. The spectra of the chiral topological boundary modes are shown, including their finite-size quantization with level spacing $v2\pi/L$. If a π flux is inserted in the disc geometry (c), it binds an isolated zero-energy state. At the same time, a single zero-energy state appears on the edge.

energy $-E$. When labelled by momentum quantum number, for a system with just one edge, like the disc, there cannot be a single state at momentum $p = 0$ at energy $E = 0$. If such a state were there, the spectrum would contain an odd number of states. Hence the spectrum of the linearized edge mode cannot have a state at $E = 0$, $p = 0$ on the disc. The only way to introduce such a state is to have antiperiodic boundary conditions, with the spectrum of the edge being at momenta $\pi(2n+1)/L$, $n \in \mathbb{Z}$. On the cylinder, since two edges are present, periodic boundary conditions are allowed (as are antiperiodic, which can be obtained by threading a flux through the cylinder).

We now add a single vortex inside the disc, far away from the edge of the disc. What is the influence of the vortex on the edge? The vortex induces a phase 2π in units of the superconducting quantum $hc/2e$, which means that the phase of Δ changes by 2π and that of the electronic operators by π upon a full rotation around the edge. This implies that the antiperiodic boundary conditions on the edge without vortex change to periodic boundary conditions in the presence of the vortex. The spectrum on the edge is then translated by π/L compared with the case without the vortex, making it have an energy level at $p = 0, E = 0$. This would mean that the spectrum has an odd number of levels. However, this cannot be true, as we explained above, since the number of levels is always even. We are hence missing one unpaired level. Where is it? Since the only difference from the case with no vortex is the vortex itself, we draw the conclusion that the missing level is associated with the vortex, and is a bound state on the vortex. We also draw the conclusion that, since it is unpaired and really bound to the vortex, it has to rest exactly at $E = 0$, thereby showing that chiral superconductors have Majorana zero modes in their vortex core.

2.2.2.2 *Bound states on vortices in 2D chiral p-wave superconductors*

Let us explicitly show that a vortex in a chiral superconductor will contain a zero mode [12, 13, 15]. For this calculation, which is a variant of our calculation for the existence of a Majorana mode at the interface between a topological and a trivial superconductor, we follow the discussion in [11]. For this construction, consider a disc of radius R that has $\mu > 0$ surrounded by a region with $\mu < 0$ for $r > R$. We know from our previous discussion that there will be a single branch of chiral Majorana states localized near $r = R$, but no exact zero mode. If we take the limit $R \to 0$, this represents a vortex and all the low-energy modes on the interface will be pushed to higher energies. If we put a π flux inside the trivial region, it will change the boundary conditions such that even in the $R \to 0$ limit there will be a zero mode in the spectrum localized on the vortex.

Now let us take the Bogoliubov–de Gennes Hamiltonian in the Dirac limit ($m \to \infty$) and solve the Bogoliubov–de Gennes equations in the presence of a vortex located at $r = 0$ in the disc geometry in polar coordinates. Let $\Delta(r,\theta) = |\Delta(r)|e^{i\alpha(r)}$. The profile $|\Delta(r)|$ for a vortex will depend on the details of the model, but must vanish inside the vortex core region; for example, for an infinitely thin core, we just need $|\Delta(0)| = 0$. We take the phase $\alpha(\mathbf{r})$ to be equal to the polar angle at $\mathbf{r}$.

The first step in the solution of the bound state for this vortex profile is to gauge-transform the phase of $\Delta(r,\theta)$ into the fermion operators via $\Psi(\mathbf{r}) \to e^{i\alpha(\mathbf{r})/2}\Psi(\mathbf{r})$.

This has two effects: (i) it simplifies the solution of the Bogoliubov–de Gennes differential equations and (ii) it converts the boundary conditions of $\Psi(\mathbf{r})$ from periodic to antiperiodic around the vortex position $\mathbf{r} = 0$. In polar coordinates, the remaining single-particle Bogoliubov–de Gennes Hamiltonian is simply

$$\mathcal{H}_{\mathrm{BdG}} = \frac{1}{2}\begin{pmatrix} -\mu & 2|\Delta(r)|\mathrm{e}^{\mathrm{i}\theta}\left(\dfrac{\partial}{\partial r} + \dfrac{\mathrm{i}}{r}\dfrac{\partial}{\partial \theta}\right) \\ -2|\Delta(r)|\mathrm{e}^{-\mathrm{i}\theta}\left(\dfrac{\partial}{\partial r} - \dfrac{\mathrm{i}}{r}\dfrac{\partial}{\partial \theta}\right) & \mu \end{pmatrix}. \tag{2.73}$$

We want to solve $\mathcal{H}_{\mathrm{BdG}}\Psi = E\Psi = 0$, which we can do with the ansatz

$$\begin{aligned}\Psi_0(r,\theta) &= \frac{\mathrm{i}}{\sqrt{r}\mathcal{N}}\exp\left[-\frac{1}{2}\int_0^r \frac{\mu(r')}{|\Delta(r')|}\,\mathrm{d}r'\right]\begin{pmatrix} -\mathrm{e}^{\mathrm{i}\theta/2} \\ \mathrm{e}^{-\mathrm{i}\theta/2}\end{pmatrix} \\ &\equiv \mathrm{i}g(r)\begin{pmatrix} -\mathrm{e}^{\mathrm{i}\theta/2} \\ \mathrm{e}^{-\mathrm{i}\theta/2}\end{pmatrix},\end{aligned} \tag{2.74}$$

where $\mathcal{N}$ is a normalization constant. The function $g(r)$ is localized at the position of the vortex. We see that $\Psi_0(r, \theta + 2\pi) = -\Psi_0(r,\theta)$, as required. From an explicit check, one can see that $\mathcal{H}_{\mathrm{BdG}}\Psi_0(r,\theta) = 0$. The field operator that annihilates fermion quanta in this localized state is

$$\gamma = \int r\,\mathrm{d}r\,\mathrm{d}\theta\,\mathrm{i}g(r)[-\mathrm{e}^{\mathrm{i}\theta/2}c(r,\theta) + \mathrm{e}^{-\mathrm{i}\theta/2}c^\dagger(r,\theta)], \tag{2.75}$$

from which we can immediately see that $\gamma = \gamma^\dagger$. Thus, the vortex traps a single Majorana bound state at zero energy.

2.2.2.3 Non-Abelian statistics of vortices in chiral p-wave superconductors

We have just shown that on each vortex in a spinless chiral superconductor, there exists a single Majorana bound state. If we have a collection of $2N$ vortices that are well separated from each other, a low-energy subspace is generated which in the thermodynamic limit leads to a ground-state degeneracy of 2^N [16, 17]. For example, two vortices give a degeneracy of 2, which can be understood by combining the two localized Majorana bound states into a single complex fermion state, which can be occupied or unoccupied, akin to the end states of the superconducting wire. From $2N$ vortices, one can form N complex fermion states, giving a degeneracy of 2^N, which can be broken up into the subspace of 2^{N-1} states with even fermion parity and the 2^{N-1} states with odd fermion parity. As an aside, since we have operators that mutually anticommute and square to $+1$, we can define a Clifford algebra operator structure using the set of $2N$ γ_i.

To illustrate the statistical properties of the vortices under exchanges, we closely follow the work of Ivanov [14] and the discussion in [11]. Let us begin with a single pair of vortices which have localized Majorana operators γ_1 and γ_2 respectively and

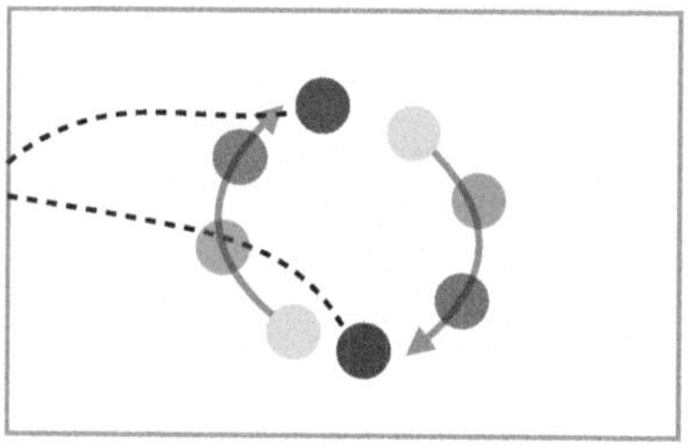

Fig. 2.8 [Colour online] Illustration of the exchange of two vortices in a chiral p-wave superconductor. The dashed lines represent branch cuts across which the phase of the superconducting order parameter jumps by 2π.

are assumed to be well separated. We imagine that we adiabatically move the vortices in order to exchange the two Majorana fermions. If we move them slowly enough, then the only outcome of exchanging the vortices is a unitary operator acting on the two degenerate states that make up the ground-state subspace. If we exchange the two vortices, then we have $\gamma_1 \to \gamma_2$ and $\gamma_2 \to \gamma_1$. However, if we look at Fig. 2.8, we immediately see there is a complication. In this figure, we have illustrated the exchange of two vortices, and the dashed lines represent branch cuts across which the phase of the superconductor order parameter jumps by 2π. Since our solution of the Majorana bound states used the gauge-transformed fermion operators, we see that the bound state on one vortex [red online], which passes through the branch cut of the other vortex [blue online], picks up an additional minus sign upon exchange. Thus the exchange of two vortices is effected by

$$\gamma_1 \to \gamma_2, \qquad \gamma_2 \to -\gamma_1. \tag{2.76}$$

In general, if we have $2N$ vortices, we can think of the different exchange operators $T_{ij}(\gamma_a)$, which, for our choice of conventions, send $\gamma_i \to \gamma_j$, $\gamma_j \to -\gamma_i$, and $\gamma_k \to \gamma_k$ for all $k \neq i, j$. We can construct a representation of this exchange process on the Hilbert space by finding a $\tau(T_{ij})$ such that $\tau(T_{ij})\gamma_a\tau^{-1}(T_{ij}) = T_{ij}(\gamma_a)$. Such a representation is given by

$$\tau(T_{ij}) = \exp\left(\frac{\pi}{4}\gamma_j\gamma_i\right) = \frac{1}{\sqrt{2}}\left(1 + \gamma_j\gamma_i\right). \tag{2.77}$$

Let us prove this by showing an explicit example for T_{12}, which will have the transformation given in (2.76):

$$\tau(T_{12})\gamma_1\tau^{-1}(T_{12}) = \frac{1}{2}\left(\gamma_1 - \gamma_1\gamma_2\gamma_1 + \gamma_2 - \gamma_1\right) = \gamma_2, \tag{2.78a}$$

$$\tau(T_{12})\gamma_2\tau^{-1}(T_{12}) = \frac{1}{2}\left(\gamma_2 - \gamma_1 + \gamma_2\gamma_1\gamma_2 - \gamma_2\right) = -\gamma_1, \tag{2.78b}$$

$$\tau(T_{12})\gamma_3\tau^{-1}(T_{12}) = \gamma_3\tau(T_{12})\tau^{-1}(T_{12}) = \gamma_3. \tag{2.78c}$$

Now that we have this representation, we can illustrate the non-Abelian statistics. We start with four vortices with Majorana operators $\gamma_1, \gamma_2, \gamma_3, \gamma_4$. To illustrate the action of the exchange operators on the fourfold-degenerate ground state space, we need to pair these Majorana operators into complex fermions

$$\begin{aligned} a &= \frac{1}{2}(\gamma_1 + \mathrm{i}\gamma_2), & a^\dagger &= \frac{1}{2}(\gamma_1 - \mathrm{i}\gamma_2), \\ b &= \frac{1}{2}(\gamma_3 + \mathrm{i}\gamma_4), & b^\dagger &= \frac{1}{2}(\gamma_3 - \mathrm{i}\gamma_4). \end{aligned} \tag{2.79}$$

The basis vectors of the ground-state subspace can now be written as

$$\{|0\rangle_a \otimes |0\rangle_b, |1\rangle_a \otimes |1\rangle_b, |1\rangle_a \otimes |0\rangle_b, |0\rangle_a \otimes |1\rangle_b\}, \tag{2.80}$$

where we have ordered the basis so that states of the same fermion parity are together. The notation $|n\rangle_{a,b}$ means $a^\dagger a|n\rangle_a = n|n\rangle_a$ and $b^\dagger b|n\rangle_b = n|n\rangle_b$. The set of statistical exchanges is generated by T_{12}, T_{23}, T_{34} and we want to understand how these exchanges act on the ground-state subspace. We can rewrite these three operators as

$$\tau(T_{12}) = \frac{1}{\sqrt{2}}(1 + \gamma_2\gamma_1) = \frac{1}{\sqrt{2}}\left[1 - \mathrm{i}(aa^\dagger - a^\dagger a)\right], \tag{2.81a}$$

$$\tau(T_{23}) = \frac{1}{\sqrt{2}}\left[1 - \mathrm{i}(ba - ba^\dagger + b^\dagger a - b^\dagger a^\dagger)\right], \tag{2.81b}$$

$$\tau(T_{34}) = \frac{1}{\sqrt{2}}\left[1 - \mathrm{i}(bb^\dagger - b^\dagger b)\right]. \tag{2.81c}$$

Taking matrix elements in our chosen ground-state basis (2.80), we find

$$\tau(T_{12}) = \frac{1}{\sqrt{2}}\begin{pmatrix} 1-\mathrm{i} & 0 & 0 & 0 \\ 0 & 1+\mathrm{i} & 0 & 0 \\ 0 & 0 & 1+\mathrm{i} & 0 \\ 0 & 0 & 0 & 1-\mathrm{i} \end{pmatrix}, \tag{2.82a}$$

$$\tau(T_{23}) = \frac{1}{\sqrt{2}}\begin{pmatrix} 1 & -\mathrm{i} & 0 & 0 \\ \mathrm{i} & 1 & 0 & 0 \\ 0 & 0 & 1 & -\mathrm{i} \\ 0 & 0 & \mathrm{i} & 1 \end{pmatrix}, \tag{2.82b}$$

$$\tau(T_{34}) = \frac{1}{\sqrt{2}}\begin{pmatrix} 1+\mathrm{i} & 0 & 0 & 0 \\ 0 & 1-\mathrm{i} & 0 & 0 \\ 0 & 0 & 1-\mathrm{i} & 0 \\ 0 & 0 & 0 & 1+\mathrm{i} \end{pmatrix}. \tag{2.82c}$$

We see that with our basis choice, T_{12} and T_{34} are Abelian phases acting on each state, while T_{23} exhibits non-trivial mixing terms between the states with the same fermion parity. Thus, the form of T_{23} represents non-Abelian statistics. Given an initial state $|\psi_{\mathrm{in}}\rangle = |0\rangle_a \otimes |0\rangle_b$, if we take vortex 2 around vortex 3, the final state is $|\psi_{\mathrm{f}}\rangle = \frac{1}{\sqrt{2}}(|0\rangle_a \otimes |0\rangle_b + \mathrm{i}|1\rangle_a \otimes |1\rangle_b)$. In principle, one must also keep track of the Berry phase contribution to the statistical phase. Here we have only considered the wavefunction monodromy; however, it can be proved that the Berry phase does not contribute in this case. The field of topological quantum computation is built on the idea that such exchange or braiding operations will lead to non-trivial quantum evolutions of the ground state that can be used for quantum computations.

2.2.2.4 The 16-fold way

We have now noticed that there are two characterizations of a topological superconductor, but they are seemingly different. First, the spectral Chern number is an integer $\mathsf{C}^{(1)} \in \mathbb{Z}$. Directly related to it is the number of chiral Majorana modes on the edge, which in turn is related to an experimental observable, the thermal conductivity on the edge. Hence the system has a $\mathbb{Z}$ index, which becomes obvious when an edge exists. We then saw that a $(p + \mathrm{i}p)$ superconductor (i.e. a topological superconductor with Chern number equal to 1) with a vortex threaded through it exhibits a Majorana zero-energy mode at the core of the vortex. A $(d + \mathrm{i}d)$ superconductor, with Chern number equal to 2, would exhibit two Majorana modes in the core of the vortex. However, those two Majorana modes would be unstable towards single-particle hybridization terms, which would push them away from zero energy, and leave the core of the vortex with no states in it. The generalization tells us that an even-Chern-number topological superconductor has no Majorana zero modes in the vortex, while an odd-Chern-number topological superconductor has one Majorana zero mode in its core. This shows that the defects (vortices) in a topological superconductor are classified by a $\mathbb{Z}_2$ number ($\mathsf{C}^{(1)} \mod 2$).

We now show that there is a third classification related to the idea of topological order [18]. In the absence of an edge and in the absence of vortex defects, there is a $\mathbb{Z}_{16}$ classification of topological superconductors indexed by $\mathsf{C}^{(1)} \mod 16$, which can be put on a solid basis by using the formalism of topological quantum field theory (TQFT) that we will introduce in Section 2.3. This shows that the edge–bulk correspondence needs revisiting—the bulk does not know if we add 16 edge modes or not, and hence that the edge contains more information than the bulk [19]. We first give a simple argument for the existence of a $\mathbb{Z}_{16}$ classification.

We ask how we can classify the system in the absence of an edge. One way would be to compute the phases that wavefunctions can acquire upon taking particles or quasiparticles around each other. However, the system is made out of electrons (it is a superconductor), so usually nothing special can happen to phases of electrons. The only 'special' excitation of the superconductor is a vortex, so we will look at the phase that two vortices acquire upon exchange. We can calculate this with an argument. Take two copies of the $(p + \mathrm{i}p)$ superconductor governed by the Hamiltonian

$$H = \frac{\mathrm{i}}{4} \sum_{j,k} A_{jk} (\gamma_{1,j}\gamma_{1,k} + \gamma_{2,j}\gamma_{2,k}), \tag{2.83}$$

written in terms of Majorana operators $\gamma_{1,j}$ for one copy and $\gamma_{2,j}$ for the other. These operators can be combined into an complex fermion $c_j = \frac{1}{2}(\gamma_{1,j} + \mathrm{i}\gamma_{2,j})$, in terms of which the Hamiltonian becomes

$$H = \mathrm{i} \sum_{j,k} A_{jk} c_j^\dagger c_k. \tag{2.84}$$

This Hamiltonian has a 'fake' U(1) symmetry given by our choice of A_{jk} for both Hamiltonians. (Since the system is gapped, we expect our universal conclusions to hold even when this symmetry is stripped away). Thus, the system is a quantum Hall state of Hall conductance $\mathsf{C}^{(1)}$ (in units of e^2/h) if each of the superconductors had Chern number $\mathsf{C}^{(1)}$. We now ask what happens when we thread a superconducting vortex $h/2e$, which is equal to π. Threading a flux 2π in a quantum Hall state of Chern number $\mathsf{C}^{(1)}$ pulls $\mathsf{C}^{(1)}$ electron charges to the vortex core through the Hall effect; hence a π flux pulls $\mathsf{C}^{(1)}/2$ electron charges towards the core. We then try to compute the phase acquired when a vortex is exchanged with another vortex. This is an exchange process, which is half a braid. A braid of two vortices is equivalent to $\mathsf{C}^{(1)}/2$ electrons braided with a π vortex, giving rise to a phase $\pi\mathsf{C}^{(1)}/2$ upon a braid, and $\pi\mathsf{C}^{(1)}/4$ under exchange. Since this is the phase for exchange of vortices in two exactly identical superimposed superconductors, the phase for exchange in one of them is half that, $\pi\mathsf{C}^{(1)}/8 = 2\pi\mathsf{C}^{(1)}/16$. This shows that the phase for vortex exchange is defined only mod 16.

We will show this more rigorously within the framework of TQFT that we will introduce axiomatically in Section 2.3. Before doing so, let us summarize what we have learned about the vortices in chiral superconductors with odd Chern number in a language that anticipates the formalism that we will introduce. We have seen that well-separated vortices hold a Majorana zero mode at their core. When these vortices come together, the two Majorana modes hybridize and split, giving rise to two states that differ by their fermion parity. Let us call the Bogoliubov–de Gennes vacuum 1, the Bogoliubov quasiparticle ψ, and the Majorana fermion of the vortex σ. We can then formalize the fusion of two vortices by writing down a fusion rule

$$\sigma \times \sigma = 1 + \psi, \tag{2.85}$$

which basically tells us that combining two Majoranas can go either go to a state with no fermion or to one with a fermion—the fermion parity (and density) would be different for the two states. Which one it is depends on the microscopics of the model. Hence, a quantum state of two Majoranas has to be described by another quantum number, which describes the 'fusion channel' of those two Majoranas—either the vacuum or the Bogoliubov quasiparticle. The fusion rule (2.85) allows for multiple fusion channels, unlike the fusion rules (2.50) that we deduced for the toric code. This difference is a manifestation of the fact that the Majoranas are non-Abelian anyons, while the toric code anyons are Abelian. When two Bogoliubov quasiparticles fuse, they condense (forming a Cooper pair) and go to the vacuum,

$$\psi \times \psi = 1, \tag{2.86}$$

while the fusion of a Bogoliubov and a Majorana quasiparticle basically creates another Majorana,

$$\psi \times \sigma = \sigma. \tag{2.87}$$

This can be rationalized by thinking of the complex Bogoliubov quasiparticle as made out of two Majoranas, which then couple to the third Majorana. The Hamiltonian is a 3×3 antisymmetric matrix that necessarily has a zero eigenvalue, which is another Majorana fermion coming as a result of the fusion.

2.3 Category theory

So far, we have tried to gain some intuition about topologically ordered phases of matter by ways of examples. In this section, we are going to define a framework that describes topological order in 2D space in a unified and axiomatic way. At the same time, our description strips away all non-universal details from the problem. A field theory with these properties is known as a topological field theory. It does not contain any information about energy scales of the problem.

The topological field theory that we study is based on the mathematical concepts of category theory [18, 20, 21]. We will, however, try to keep the description as light as possible. For our purpose, we can view category theory as a generalization of group theory that is based on the fusion rules between anyon species that we have already encountered in examples. Consistent implementation of fusion defines a fusion category. Subsequently, we can impose more structure on the fusion category that elevates it to a braiding category, or a braided tensor fusion category. Our presentation will follow [18, 21], while giving more examples of the use of the theory.

2.3.1 Fusion category

A fusion category is based on a finite number of topological sectors (also called anyons, topological charges, or simply particles), which we will label

$$a, b, c, \dots \tag{2.88}$$

For every charge a, there exists a unique conjugate charge or antiparticle, which we denote by $\bar{a}$. It is possible that an anyon is its own antiparticle, $a = \bar{a}$ (e.g. the Majorana σ). There exists a unique vacuum sector denoted by 1 (or sometimes by 0). The fusion category is defined by its fusion rules

$$a \times b = \sum_c N_{ab}^c c, \tag{2.89}$$

where $N_{ab}^c \in \mathbb{Z}_+$ are non-negative integers. We have already encountered two examples of fusion rules in Section 2.2, namely the toric code with charges 1, e, m, and f,

$$\begin{aligned} 1 \times e &= e, & 1 \times m &= m, & 1 \times f &= f, \\ e \times m &= f, & e \times f &= m, & m \times f &= e, \\ e \times e &= 1, & m \times m &= 1, & f \times f &= 1, \end{aligned} \tag{2.90}$$

and the so-called Ising anyon theory that we found realized by Majorana fermions in a chiral p-wave superconductor with charges 1, σ, and Ψ,

$$\begin{aligned} 1 \times \sigma &= \sigma, & 1 \times \psi &= \psi \\ \sigma \times \sigma &= 1 + \psi, & \sigma \times \psi &= \sigma, & \psi \times \psi &= 1. \end{aligned} \tag{2.91}$$

A principal difference between (2.90) and (2.91) is that the former has always only one fusion product on the right-hand side, while the fusion of two σ in the latter produces two outcomes. Hence, the $\times$-product in the toric code can still be thought of as a group operation, while this is not possible in the Ising theory. We will see that this distinction coincides with the notion of an Abelian theory (toric code) and a non-Abelian theory (Ising).

Does any choice of fusion rules, i.e. $N_{ab}^c \in \mathbb{Z}_+$, define a permissible fusion category? The answer to this question is negative, since we have to impose the following conditions on a fusion category:

- The fusion rules must be *commutative*:

$$a \times b = b \times a \qquad \Longrightarrow \qquad N_{ab}^c = N_{ba}^c. \tag{2.92}$$

- The fusion rules must be *associative*:

$$(a \times b) \times c = a \times (b \times c) \qquad \Longrightarrow \qquad \sum_m N_{ab}^m N_{mc}^n = \sum_m N_{am}^n N_{bc}^m. \tag{2.93}$$

  If we define the matrix N_a with matrix elements $(N_a)_{bc} = N_{ab}^c$, this relation becomes a vanishing commutator:

$$[N_a, N_c] = 0, \tag{2.94}$$

  which implies that all fusion matrices N_a are diagonalized by the same eigenvectors. We will exploit this fact later.
- Fusion with the identity leaves any anyon unchanged:

$$a \times 1 = a. \tag{2.95}$$

- The fusion product of a with its antiparticle $\bar{a}$ contains the vacuum with prefactor 1:

$$a \times \bar{a} = 1 + \sum_{c \neq 1} N_{a\bar{a}}^c c. \tag{2.96}$$

- There exists a solution to the consistency condition called the *pentagon equation*, which we will discuss below.

2.3.1.1 Diagrammatics

Before we turn to the pentagon equation, we want to introduce a diagrammatic language that will facilitate computations within the fusion and braiding categories. In this formalism, we denote anyon a travelling forwards in time as an upward-oriented line. It is the same as the associated antiparticle travelling backwards in time:

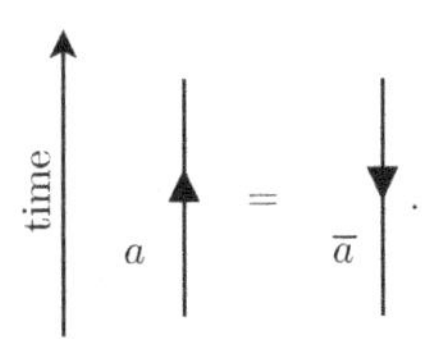

(2.97)

A diagram with open anyon worldlines at the top and bottom represents a state in a Hilbert space that depends on the number and types of open anyon worldlines. A diagram without open worldlines represents an amplitude or complex number. The simplest non-trivial Hilbert space is the *fusion space* V_{ab}^{c}. Its dimension is given by the number of ways that anyons a and b can fuse into c, that is, $\dim V_{ab}^{c} = N_{ab}^{c}$. A basis in V_{ab}^{c} is denoted by

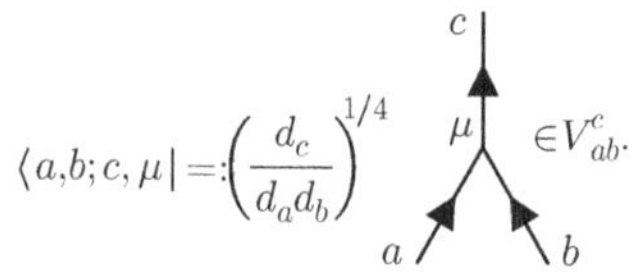

(2.98)

Here, $\mu = 1, \ldots, N_{ab}^{c}$ labels the fusion multiplicity and the real positive prefactor $(d_c/d_a\, d_b)^{1/4}$ should be understood as a normalization constant at this stage. We adopt the normalization of [21]. Likewise, we define the *splitting space* V_{c}^{ab} of the same dimension $\dim V_{c}^{ab} = N_{ab}^{c}$ and write a basis as

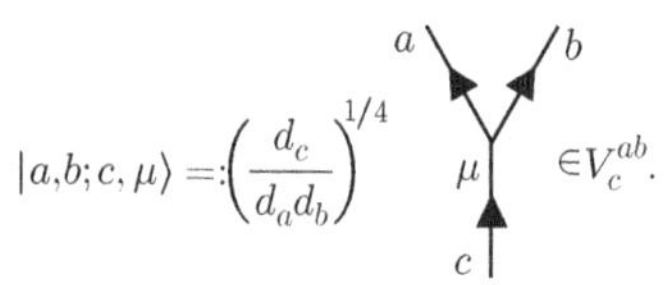

(2.99)

Two propagating particles a and b live in a vector space $V_{ab}^{ab} = \bigoplus_c V_{c}^{ab} \otimes V_{ab}^{c}$. The identity element I_{ab} in V_{ab}^{ab} is then represented by the *completeness* relation

$$I_{ab} = \sum_{c,\mu} |a, b; c, \mu\rangle\langle a, b; c, \mu|, \tag{2.100}$$

which we represent pictorially as

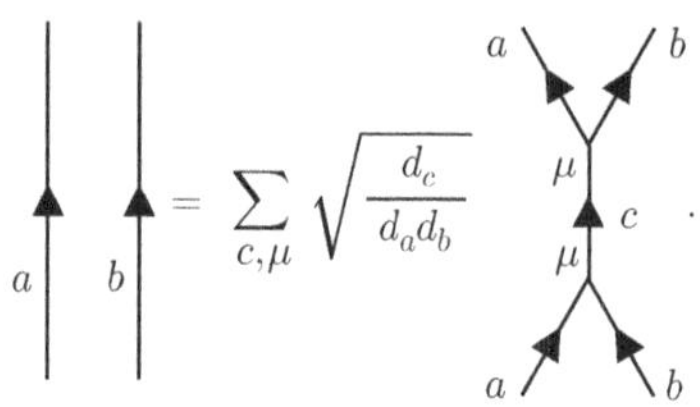

(2.101)

The basis vectors in V_{c}^{ab} and V_{ab}^{c} furthermore satisfy the *orthogonality* relation

$$\langle a, b; c, \mu | a, b; c', \mu'\rangle = \delta_{c,c'}\delta_{\mu\mu'}, \tag{2.102}$$

which we represent pictorially as

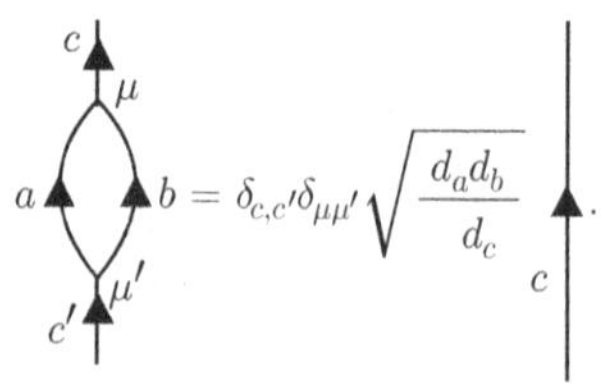

(2.103)

In particular, we can choose $c = 1$ (a dashed worldline) to obtain

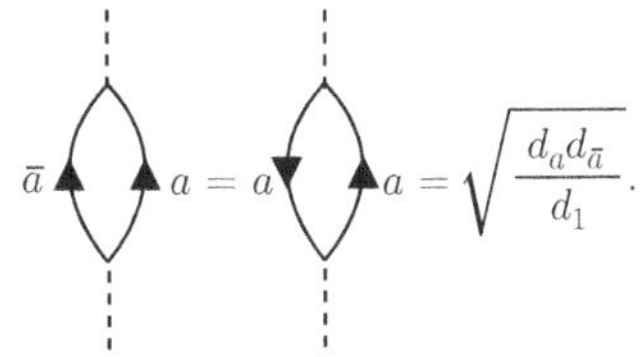

(2.104)

We can use this relation to determine the normalization constants d_a, which are called *quantum dimensions.* We have the freedom to choose $d_1 = 1$ and note that for all examples discussed here $d_a = d_{\bar{a}}$. It follows that

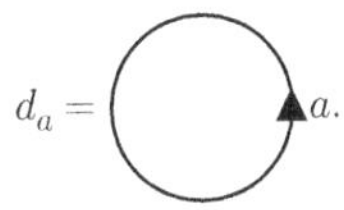

(2.105)

2.3.1.2 F-moves and the pentagon equation

As noted above, we need to impose a further consistency condition to complete the definition of a fusion category. For this, we generalize the notion of associativity that we imposed on the fusion coefficients by imposing associativity on the basis. The consistency equations basically say that observables only depend on the states of the particles at the beginning (the fusion channel) and at the end. Nothing in between can matter, up to phases and rotations in possibly degenerate spaces. We consider the Hilbert space of particle d splitting into three (not two) particles a, b, c:

$$V_d^{abc} = \sum_e V_e^{ab} \otimes V_d^{ec} = \sum_f V_d^{af} \otimes V_f^{bc}. \tag{2.106}$$

There is hence a unitary transformation F_d^{abc} ('F-move") between the two vector spaces,

$$|a, b; e, \alpha\rangle \otimes |e, c; d, \beta\rangle = \sum_{f,\mu,\nu} \left[F_d^{abc}\right]_{(e,\alpha,\beta),(f,\mu,\nu)} |b, c; f, \mu\rangle \otimes |a, f; d, \nu\rangle, \tag{2.107}$$

which reads diagrammatically

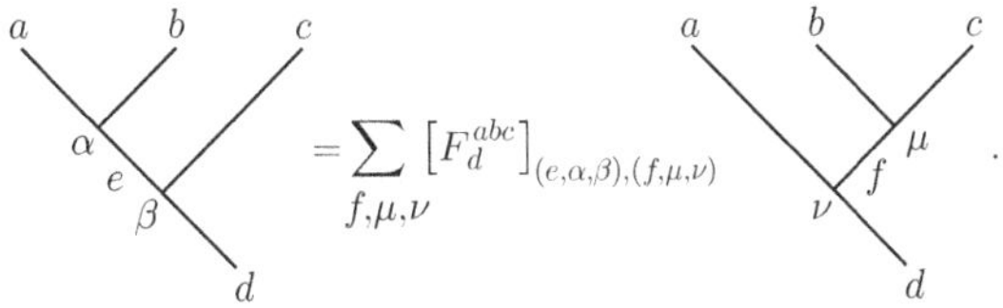

(2.108)

For a fusion category to be *unitary*, we require that $(F_d^{abc})^\dagger = (F_d^{abc})^{-1}$ (as a matrix). Notice that F_d^{abc} is trivial if any of $a, b, c = 1$.

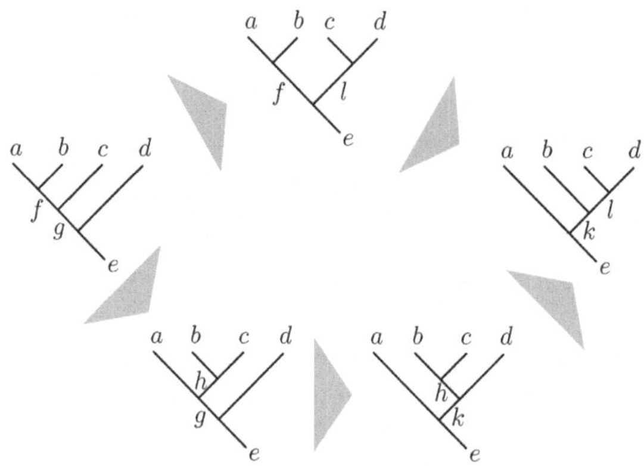

Fig. 2.9 The pentagon equation defines a consistency relation that must be imposed on a fusion category.

The pentagon equation is the consistency condition represented diagrammatically as in Fig. 2.9. It shows that there are two ways to build the same mapping between two vector spaces out of F-moves. Each distinct solution, up to gauge freedom, of the pentagon equation

$$\left[F_e^{fcd}\right]_{gl}\left[F_e^{abl}\right]_{fk}=\sum_h\left[F_g^{abc}\right]_{fh}\left[F_e^{ahd}\right]_{gk}\left[F_k^{bcd}\right]_{hl} \tag{2.109}$$

is a distinct fusion category (with the same fusion rules). Here and below, we have suppressed the Greek indices that correspond to the fusion multiplicities. We will focus on theories of multiplicity 1 only, i.e. all N_{ab}^c fusion coefficients will be either 0 or 1. As the F-moves relate different basis states, not all of them are gauge-invariant (see below). However, there are some F-symbols that are gauge-invariant. Those are related to invariants of the theory called Frobenius–Schur indicators.

2.3.1.3 Gauge freedom and its fixing

Consider a gauge transformation on the basis states $|a,b;c\rangle$:

$$|a,b;c\rangle'=u_c^{ab}|a,b;c\rangle. \tag{2.110}$$

We only consider the case without multiplicities, i.e. $N_c^{ab}=0,1$, for which $u_c^{ab}\in\mathbb{C}$ are scalars with $|u_c^{ab}|=1$. Likewise, if $N_f^{bc}=0,1$, the F-symbols are scalars with $|[F_d^{abc}]_{ef}|=1$. In view of the definition (2.107), the F-symbols are not invariant under the gauge transformation. They transform as

$$\left[F_d^{abc}\right]'_{ef}=\left[F_d^{abc}\right]_{ef}\frac{u_e^{ab}u_d^{ec}}{u_d^{af}u_f^{bc}}. \tag{2.111}$$

Furthermore, we need to set

$$u_c^{1b} = \delta_{bc}, \tag{2.112}$$

since the fusion of the identity particle can be added at any point in time to the worldline of any particle b without changing the state.

From this, we can conclude that the following F-symbols are gauge-invariant:

$$\left[F_d^{1bc}\right]_{ef} = \left[F_d^{1bc}\right]_{bd}, \qquad \left[F_d^{a1c}\right]_{ef} = \left[F_d^{a1c}\right]_{ac}, \qquad \left[F_d^{ab1}\right]_{ef} = \left[F_d^{ab1}\right]_{db}. \tag{2.113}$$

In a theory with no multiplicity, all these F-symbols are equal to unity because they represent identity maps from spaces $|b, c; d\rangle$ into the same space:

$$\left[F_d^{1bc}\right]_{ef} = \left[F_d^{1bc}\right]_{bd} = \left[F_d^{a1c}\right]_{ef} = \left[F_d^{a1c}\right]_{ac} = \left[F_d^{ab1}\right]_{ef} = \left[F_d^{ab1}\right]_{db} = 1. \tag{2.114}$$

2.3.1.4 Quantum dimensions and Frobenius–Schur indicators

Having completed the definition of a fusion category, we now explore its structure. First, we shall properly define the quantum dimension d_a of an anyon a, which has already entered several relations as a normalization factor. Physically, the definition of d_a can be obtained from imposing isotopy invariance, which means the ability to remove bends in particle worldlines. This should be possible as long as lines are not crossed and endpoints are not moved. Bending a line slightly (so that the line always flows upwards) is a trivial allowed move, but a complication arises when a line is bent so much that it acquires a turning point. The F-move associated with this type of bending is

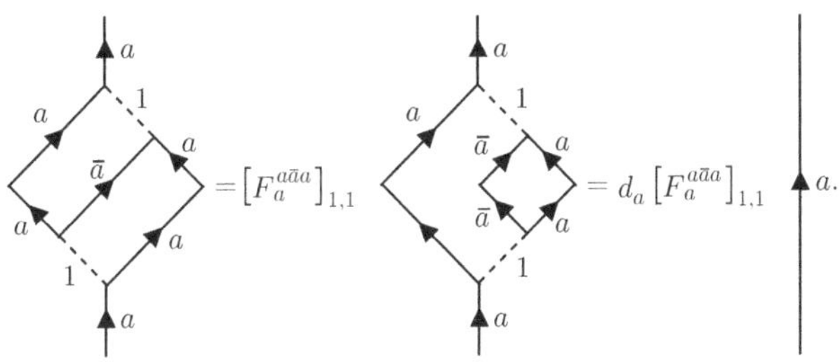

(2.115)

Notice that the symbol $[F_a^{a\bar{a}a}]_{11}$ is gauge-invariant. Hence its value is a topological invariant. Since we know that the line in the left diagram is isotopically equivalent to a line going up, it should be, up to a phase, equal to a line going up. We conclude that $[F_a^{a\bar{a}a}]_{11} = \chi_a/d_a$, where χ_a is a phase called the Frobenius–Schur indicator. If a is its own antiparticle, χ_a has to equal either $+1$ or -1. Since it can take different values, it is a topological invariant characterizing the fusion category. Interestingly, there are other Frobenius–Schur indicators that characterize the theory as topological invariants. For example, one on which we will not elaborate further is connected to the trivalent vertex [21].

We now know how to compute the quantum dimension through the F-symbols. However, there is another, easier, way of computing the quantum dimensions. This is obvious once the space of states has been endowed with a completeness and an orthonormality relation, which we gave in (2.101) and (2.103), respectively. One can use them to show the identity

$$d_a d_b = \sum_c N_{ab}^c d_c, \tag{2.116}$$

for

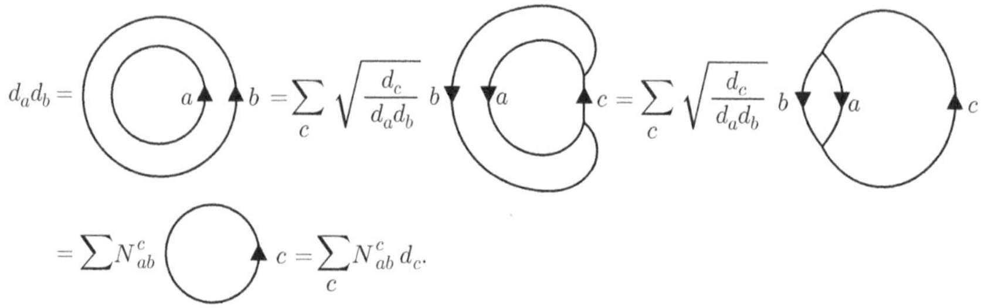

(2.117)

Equation (2.116) is key to understanding how the quantum dimensions follow from the fusion rules. It is again useful to render the fusion coefficients in a matrix form $(N_a)_{bc} = N_{ab}^c$ (b and c are the indices of the matrix N_a). Then, (2.116) is nothing but an eigenvalue equation for N_a. We see that d_a is an eigenvalue of N_a and its eigenvector is the vector that contains all quantum dimensions d_c. The existence of the real positive eigenvalue d_a is a highly non-trivial fact. The Perron–Frobenius theorem, proved by Oskar Perron (1907) and Georg Frobenius (1912), asserts, in its weak version, that a real square matrix with non-negative entries has a largest positive eigenvalue and that the corresponding (possibly degenerate) eigenvector has non-negative components. We would like to use it to show that the eigenvalue d_a is the largest eigenvalue of the matrix N_a. By assumption, the vector with entries d_c has only strictly positive components. Suppose we have another eigenvector v of N_a with non-negative components $v_c \geq 0$ and eigenvalue μ_a. Then the strict equality

$$\sum_c v_c d_c > 0 \tag{2.118}$$

holds since all d_c are strictly positive and at least one v_c is strictly positive as well. From

$$d_a \left(\sum_c d_c v_c \right) = \sum_{b,c} d_b N_{ab}^c v_b = \mu_a \left(\sum_c d_c v_c \right), \tag{2.119}$$

it thus follows that the eigenvalues d_a and μ_a are equal. In other words, *any* non-negative eigenvector of N_a has the same eigenvalue d_a. This includes the eigenvector of the largest eigenvalue of N_a, which is non-negative because of the Perron–Frobenius theorem. Hence, d_a is the largest eigenvalue of N_a.

Using the fact that d_a is the largest eigenvalue, we can now give a more physical interpretation of the quantum dimension. For that, consider the fusion of some anyon a with itself n times:

$$\underbrace{a \times a \times \cdots \times a}_{n} = \sum_{c_1, c_2, \ldots, c_{n-1}} (N_a)_{ac_1} (N_a)_{c_1 c_2} \times \cdots \times (N_a)_{c_{n-2} c_{n-1}} \, c_{n-1}. \tag{2.120}$$

The right-hand side contains the $(n-1)$th power of the matrix N_a. Hence, approximating N_a by its highest eigenvalue, we conclude the the dimension of the fusion space of n anyons of type a is dominated by the quantum dimension $\dim(\bigoplus_c V^c_{a \ldots a}) \sim d^n_a$ for large n. In other words, the quantum dimension tells us how fast the Hilbert space for a particle grows! Any non-Abelian particle has a quantum dimension strictly larger than unity (Abelian particles have quantum dimension unity).

2.3.1.5 *Examples*

Before moving on to impose more structure on the fusion category in order to obtain a braiding category, we shall briefly follow up on the two examples of semion TQFT and Ising TQFT.

Semion TQFT The simplest non-trivial TQFT has one particle s besides the identity and the semion fusion rules

$$s \times s = 1, \qquad 1 \times s = s. \tag{2.121}$$

The theory is Abelian, i.e. $d_s = 1$. Let us solve the pentagon equation for this theory. There is only one F-symbol, which is not entirely determined by gauge fixing alone, namely $[F^{sss}_s]_{11}$. We can deduce its allowed values from the pentagon equation

$$\left[F^{1ss}_1\right]_{s1} \left[F^{ss1}_1\right]_{1s} = [F^{sss}_s]_{11} \left[F^{s1s}_1\right]_{ss} [F^{sss}_s]_{11}, \tag{2.122}$$

which, using (2.114), yields the two possibilities

$$[F^{sss}_s]_{11} = \pm 1. \tag{2.123}$$

In fact, $[F^{sss}_s]_{11}$ is equal to the Frobenius–Schur indicator mentioned above and the two values ± 1 distinguish two different fusion categories. The choice $+1$ is trivial, while the -1 is what is commonly called the semion TQFT. For example, it is realized in the $\nu = \frac{1}{2}$ Laughlin state of bosons in the fractional quantum Hall effect.

Ising TQFT We gave the fusion rules of the non-Abelian Ising TQFT in (2.91). We now want to use (2.116) to compute the quantum dimensions of the anyons. For that, we note that the fusion matrix (in the basis $(1, \sigma, \psi)$) of the σ anyon is given by

$$N_\sigma = \begin{pmatrix} 0 & 1 & 0 \\ 1 & 0 & 1 \\ 0 & 1 & 0 \end{pmatrix}. \tag{2.124}$$

Its eigenvalues are given by $\pm\sqrt{2}$ and 0, the largest of which $d_\sigma = \sqrt{2}$ is the quantum dimension of the σ particle. Its corresponding eigenvector $(d_1, d_\sigma, d_\psi) = (1, \sqrt{2}, 1)$ indeed contains the quantum dimensions of all anyons. The quantum dimension $\sqrt{2}$ is compatible with our explicit calculation of the degeneracy resulting from Majorana bound states in the vortices of a p-wave superconductor. For example, two vortices with one Majorana state each gave rise to a degeneracy $\sqrt{2}^2 = 2$ of the state.

2.3.2 Braiding category

A bare fusion category has no means to relate the two fusion spaces V_c^{ab} and V_c^{ba}. The physical operation that corresponds to such a relation is an exchange between the particles a and b. In a braiding category, exchange is a map $R_{ab} : V_c^{ab} \to V_c^{ba}$. A double exchange is an automorphism in a given fusion space $R_{ba}R_{ab} : V_c^{ab} \to V_c^{ab}$. For fusion processes without multiplicities, for which the space is 1D, $R_{ba}R_{ab}$ is thus a phase (it is represented by a matrix if the multiplicity N_{ab}^c is larger than 1). Upon braiding, the state changes by flipping the particles, by convention,

$$R_{ab}|a, b; c, \mu\rangle = \sum_\nu [R_c^{ab}]_{\mu\nu}|b, a; c, \nu\rangle. \tag{2.125}$$

Diagrammatically, this operation ('R-move') is represented as

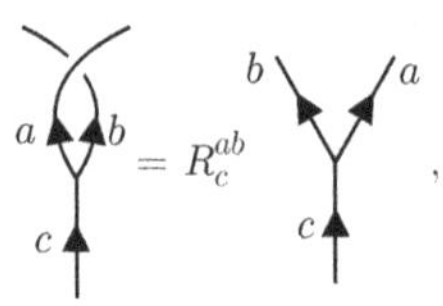

(2.126)

in the case without multiplicities.

For this braiding relation to define a consistent braiding category, the R- and F-symbols have to satisfy two consistency relations called hexagon equations. (The two hexagon equations differ only in the directions of the braids involved; one uses R, the other uses R^{-1}. For simplicity, we only present the equation for R here.) As with the pentagon equation for the F-symbols alone, the hexagon equation represents a way to mix F and R-moves to get between the same two diagrams in two different ways. Analytically, it reads

$$R_e^{ab}\left[F_d^{bac}\right]_{eg} R_g^{ac} = \left[F_d^{abc}\right]_{ef} R_d^{af}\left[F_d^{bca}\right]_{fg} \tag{2.127}$$

and they are shown diagrammatically in Fig. 2.10. The matrix R_c^{ab} is unitary, and if either of a or b is the identity particle, then the R matrix is unity (braiding with the vacuum is trivial):

$$R_c^{1b} = R_c^{a1} = 1. \tag{2.128}$$

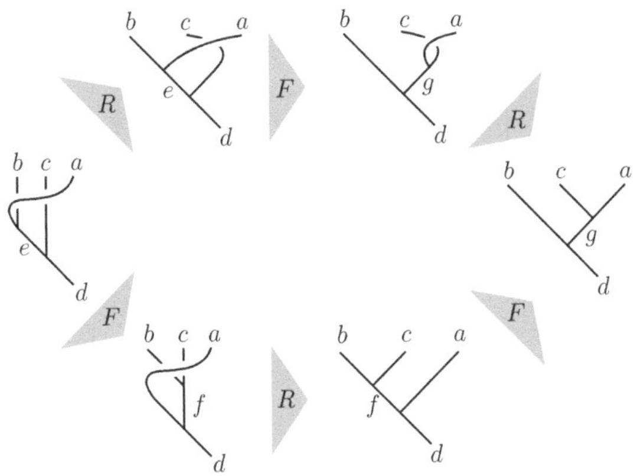

Fig. 2.10 The hexagon equation defines a consistency relation that has to be imposed on a braiding category.

2.3.2.1 Topological spin

Every quasiparticle type a in a braiding category carries another topological quantum number besides its quantum dimension, namely the complex-valued topological spin θ_a with $|\theta_a| = 1$. Physically, the topological spin can be interpreted as the phase resulting from a rotation of the particle around its own axis by 2π. It is -1 for fermions, but can be fractional for other particles. Diagrammatically, we can define θ_a as

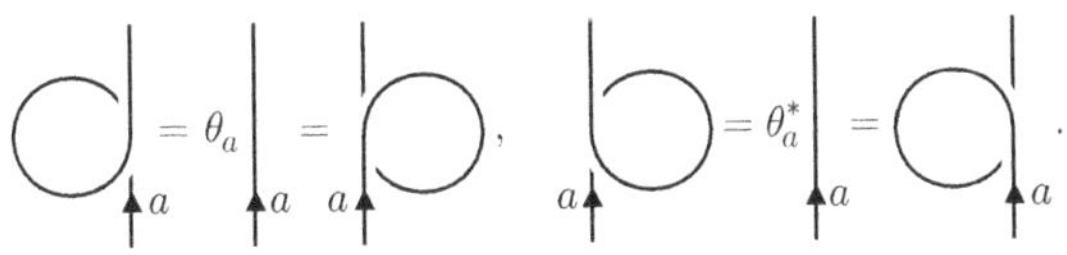

(2.129)

We bend the world line of a particle a, bring it down, then bring it back to itself, this time braided, then take it to infinity. This corresponds, up to a phase, to the configuration without bending.

To see how the topological spin is related to the R-symbols, we want to evaluate a diagram that represents an amplitude, i.e. is closed. For that, we simply connect up the open worldlines in (2.129) with an $\bar{a}$ worldline and obtain the definition

(2.130)

We evaluate this diagram by inserting an identity and applying an R-move (by the definition of the R-symbols, exchange is only defined between worldlines that are in a definite fusion channel):

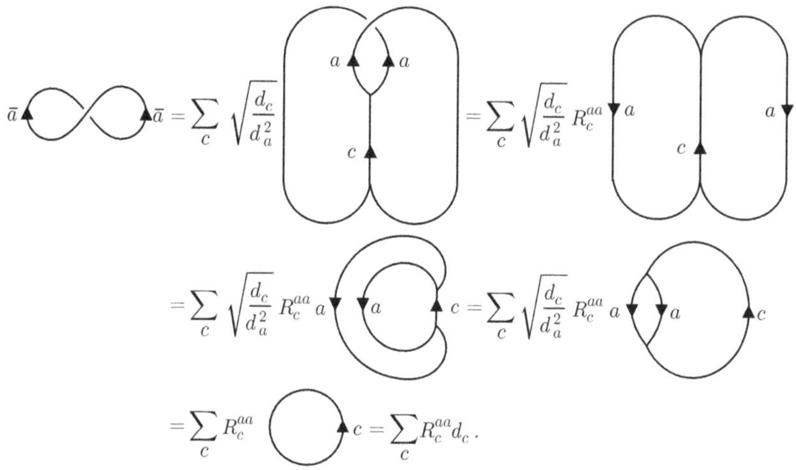

$$(2.131)$$

Combining (2.130) and (2.131), we obtain the definition of θ_a in terms of the R-symbol:

$$\theta_a := \frac{1}{d_a} \sum_c d_c \mathrm{Tr}[R_c^{aa}]. \tag{2.132}$$

Here, the matrix trace Tr is relevant for theories with multiplicities, in which case R_c^{aa} is a square matrix with dimension equal to the multiplicity of the fusion channel c of $a \times a$.

Another identity, which requires bending to be proved, is

$$R_1^{\bar{a}a} = \theta_a^* \chi_a \tag{2.133}$$

where χ_a is the Frobenius–Schur indicator.

2.3.2.2 Ribbon equation

We are now in good shape to prove an important identity relating topological spin to the R-matrices that is called the ribbon equation:

$$R_c^{ab} R_c^{ba} = \frac{\theta_c}{\theta_a \theta_b} I, \tag{2.134}$$

where I is the identity element in the space V_c^{ab}. Physically, it equates the operation of twisting each worldline in a splitting diagram by 2π with the operation of braiding the splitting products. We can prove the ribbon equation via diagrammatic manipulations. For that, we evaluate the same diagram in two different ways. For one, we can use the topological spin of the split particle c:

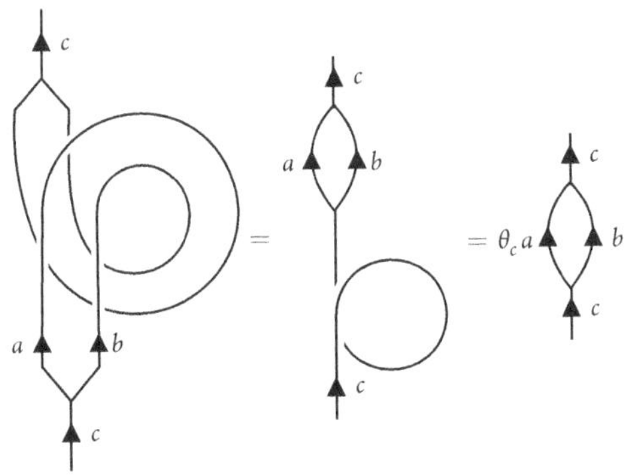

(2.135)

On the other hand, we can use a combination of topological spins and R-moves:

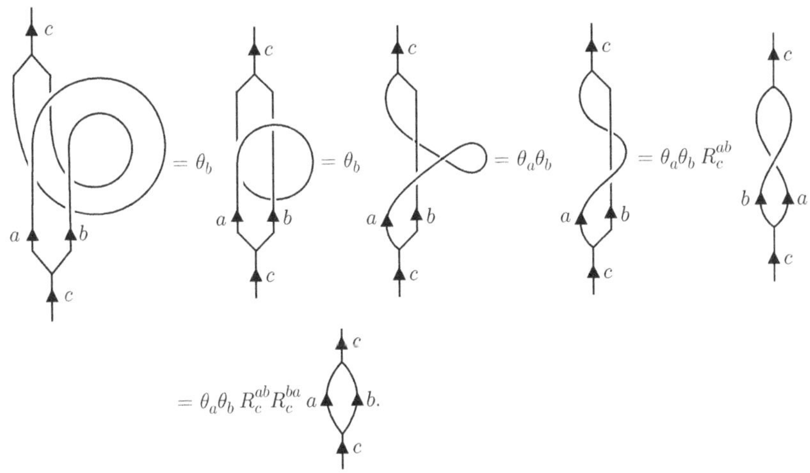

(2.136)

Together, (2.135) and (2.136) yield the ribbon equation (2.134).

2.3.2.3 *Vafa's theorem*

An important theorem relating the topological spin and the structure constants N_{ab}^{c} is Vafa's theorem [22]. It shows that the topological spin is a rational number. We will not derive it here, but point the reader to [18] for an easy derivation. Vafa's theorem proceeds by writing down in matrix form the two hexagon equations, one for R and one for R^{-1}, dividing them and using $R_c^{ab} R_c^{ba} = \theta_c/(\theta_a\theta_b)I$ to obtain

$$\prod_c \left(\frac{\theta_c}{\theta_a\theta_b}\right)^{N_{ab}^{c} N_{cd}^{e}} \prod_f \left(\frac{\theta_f}{\theta_a\theta_d}\right)^{N_{bf}^{e} N_{ad}^{f}} = \prod_r \left(\frac{\theta_e}{\theta_a\theta_r}\right)^{N_{bd}^{r} N_{ar}^{e}}. \tag{2.137}$$

As an example, we will later use Vafa's theorem to deduce the spin of the σ particle in the Ising TQFT.

2.3.3 Modular matrices

In the preceding subsections, we have constructed the fusion and braiding category from the F- and R-moves and deduced the universal data (quantum dimensions d_a and topological spins θ_a) that characterizes the anyons in this category.

For TQFT in physical systems, i.e. quantum liquid ground states of matter, it is important to address how the universal information about the TQFT can in general be accessed. We know that topological properties cannot be accessed by local measurements. In contrast, it turns out that there is a set of global measurements that can be used, namely the action of automorphisms on the manifold over which the system is defined. Automorphisms are transformations that map the manifold back to itself and form the so-called mapping class group of the manifold. Here, we will explore the most standard case, namely $(2+1)$-dimensional systems with periodic boundary conditions, in which case the manifold is a torus. Automorphisms on the torus form the modular group, which has two generators. The first generator S exchanges the two coordinate axes. The second generator T changes the angle between the coordinate axes.

A TQFT on the torus exhibits a topological ground-state degeneracy, where the number of ground states is equal to the number of anyons in the theory (including the identity). It is the representation of the S and T operations in this ground-state manifold that reveals information about the nature of the TQFT. We call these representations the S- and T-matrices. Instead of deriving the S- and T-matrices from the action of the respective transformations, we will simply give their definitions within the TQFT here and subsequently explore their properties (see [23] for a more complete discussion of the connection).

2.3.3.1 The S-matrix

In the TQFT, the S-matrix is defined diagrammatically as

$$S_{ab} := \frac{1}{D}$$ 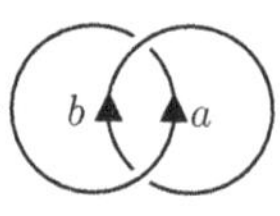

, (2.138)

where D is the total quantum dimension of the TQFT:

$$D = \sqrt{\sum_a d_a^2}. \tag{2.139}$$

(We choose the same convention for the diagrammatic definition of S_{ab} here as in [21], which is different from that chosen in [18].) We can relate it to the topological spins and fusion coefficients via the following diagrammatic manipulations:

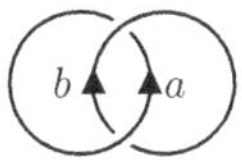

$= \sum_{c,\mu} \sqrt{\frac{d_c}{d_a d_b}}$
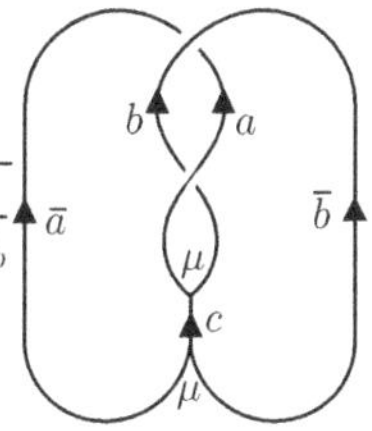

$= \sum_{c,\mu,\nu} \sqrt{\frac{d_c}{d_a d_b}}\, [R_c^{ab}]_{\mu\nu}\, [R_c^{ba}]_{\nu\mu}$
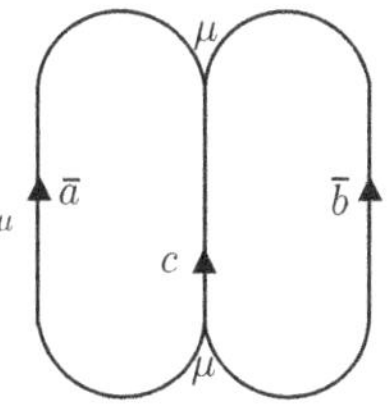

$= \sum_{c,\mu,\nu} [R_c^{ab}]_{\mu\nu}\, [R_c^{ba}]_{\nu\mu} \sqrt{\frac{d_c}{d_a d_b}}$
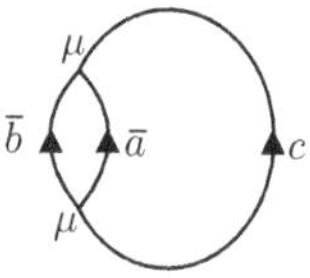

$$= \sum_{c} N_{ab}^{c} \mathrm{Tr}(R_c^{ab} R_c^{ba}) d_c . \tag{2.140}$$

From here, we use the ribbon equation (2.134) and (2.138) to arrive at the final expression for the S-matrix:

$$S_{ab} = \frac{1}{D} \sum_{c} N_{ab}^{c} \frac{\theta_c}{\theta_a \theta_b} d_c. \tag{2.141}$$

2.3.3.2 Verlinde formula

We now derive a fundamental formula in both TQFT and conformal field theory. This formula relates the S-matrix to the fusion coefficients. It allows us to find the set of braiding phases among the anyons, but not always the topological spin of every particle.

We start by showing that

$$S_{1x} = \frac{d_x}{D} . \tag{2.142}$$

To prove this, we first observe that

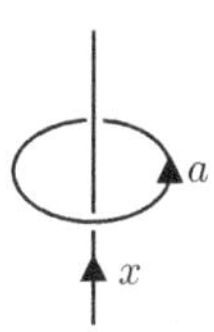

$= \frac{S_{ax}}{S_{1x}}$

(2.143)

holds, because we can close the x worldlines in this diagram to obtain amplitudes and an identity loop can be added without changing the value of the diagram. The relation (2.143) can be used to obtain the line of equalities

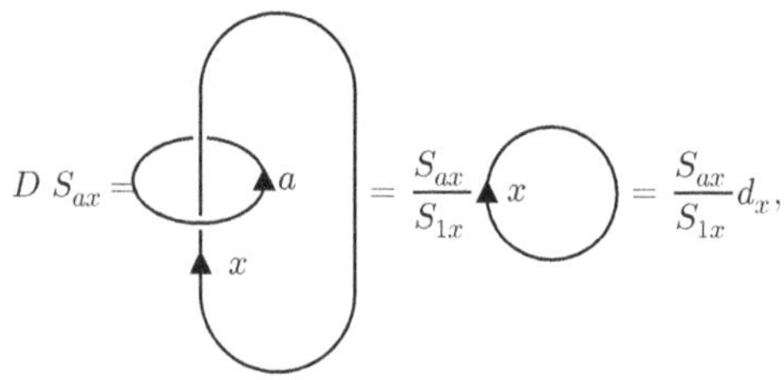

(2.144)

which yields (2.142). Equation (2.142) says that the first column of the S-matrix contains only positive numbers, the quantum dimensions of the theory.

The next step is to derive the all-important Verlinde formula. Staring from two copies of (2.143), we can perform the following set of diagrammatic manipulations:

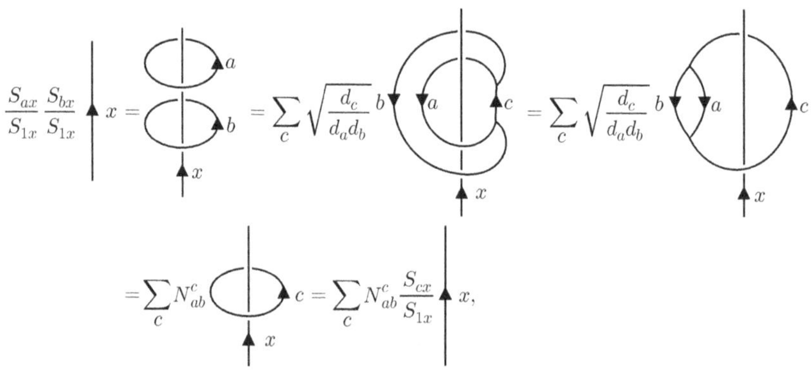

(2.145)

yielding the Verlinde formula

$$\sum_c N_{ab}^c S_{cx} = S_{bx} \frac{S_{ax}}{S_{1x}} \,. \tag{2.146}$$

This is a remarkable equality, which again takes the form of an eigenvalue equation for the fusion matrices N_a. We have already encountered an eigenvalue equation for N_a when solving for the quantum dimensions, i.e. the first column of the S-matrix. The Verlinde formula says that the S-matrix contains both the eigenvectors and the eigenvalues of the $(N_a)_{bc}$ matrices. If the S-matrix is unitary (and thus has an inverse), we call the theory a unitary modular category. In this case, we have a simple relation between the fusion coefficients and the S-matrix:

$$N_a = S D_a S^{-1}, \tag{2.147}$$

where we have defined the diagonal matrices

$$(D_a)_{mn} = S_{am}/S_{1m}\delta_{mn}. \tag{2.148}$$

This also implies that all the N_a matrices are commuting (which we already knew from (2.94)), since they are diagonalized by the same eigenvectors.

2.3.3.3 Obstruction for theories with multiplicities

In this subsection, we want to illustrate how we can use the structure of the braiding category to discard certain fusion categories as unphysical (or at least not unitary). As an example, we ask which theories are possible with only two particles, the identity 1 and a particle s. For s to have an inverse, all possible fusion rules

$$s \times s = 1 + m\, s \tag{2.149}$$

are labelled by a non-negative integer m. The case $m = 0$ is the semion TQFT of (2.121). For $m = 1$, we have the non-Abelian Fibonacci fusion rules.

We would like to answer the question regarding the values of $m > 1$ for which we can define a consistent modular braiding category. Observe that the fusion matrix

$$N_s = \begin{pmatrix} 0 & 1 \\ 1 & m \end{pmatrix} \tag{2.150}$$

yields the quantum dimension

$$d_s = \frac{m + \sqrt{m^2 + 4}}{2}. \tag{2.151}$$

From the definition of the S-matrix, we have

$$\begin{aligned} S_{s1} &= \frac{1}{D} \sum_c N_{s1}^c \frac{\theta_c}{\theta_s} d_c = \frac{1}{D} d_s = S_{1s}, \\ S_{11} &= \frac{1}{D}, \\ S_{ss} &= \frac{1}{D} \frac{1}{\theta_s^2} (1 + m\theta_s d_s). \end{aligned} \tag{2.152}$$

It is already possible to see that something will go wrong for large enough m if we demand a unitary theory, i.e. a theory with a unitary S-matrix. For a unitary matrix, all the matrix elements have to be less than or equal to 1. When m is large, d_s is proportional to m (and so is D), but the S_{ss} matrix element has an md_s/D that is proportional to m in the large-m limit. Hence, we forsake unitarity.

Lets us calculate the exact m at which unitarity breaks down. Imposing unitarity of the S-matrix yields

$$S^\dagger S = \frac{1}{D^2} \begin{pmatrix} D^2 & d_s \left(1 + \dfrac{1 + m\theta_s d_s}{\theta_s^2}\right) \\ d_s \left(1 + \dfrac{1 + m\theta_s^* d_s}{(\theta_s^*)^2}\right) & d_s^2 + (1 + m\theta_s^* d_s)(1 + m\theta_s d_s) \end{pmatrix}, \tag{2.153}$$

and we find

$$(1 + m\theta_s^* d_s)(1 + m\theta_s d_s) = 1, \tag{2.154}$$

which reduces to

$$\theta_s + \theta_s^* + md_s = 0, \tag{2.155}$$

where we have used the fact that the value of d_s is positive. Now, md_s is a positive number growing with m, while θ_s is a phase so the sum with its conjugate cannot be smaller than -2. Hence,

$$md_s \leq 2 \quad \Longrightarrow \quad m^2 + m\sqrt{m^2+4} \leq 4 \tag{2.156}$$

with the solutions $m = 0$ and $m = 1$. We can see that already $m = 2$ gives a left-hand side that is too large. We conclude that the semion and the Fibonacci TQFT are the only allowed modular unitary theories with one non-trivial anyon s.

2.3.3.4 The T-matrix

The T-matrix is diagonal and simply given by

$$T_{ab} = \theta_a \delta_{ab}. \tag{2.157}$$

2.3.4 Examples: the 16-fold way revisited

To motivate our study of TQFTs and as an example of topologically ordered phases, we have studied in Section 2.2.2.4 Kitaev's 16-fold way of classifying topological superconductors as gauge theories from their bulk properties. Equipped with a category theory understanding of TQFTs, we now want to revisit this classification, since it provides us with several examples of TQFTs. In particular, we want to characterize the theories from their S- and T-matrices.

2.3.4.1 Case: $\mathsf{C}^{(1)}$ odd

If we couple an odd number of layers of spinless chiral p-wave superconductors, the core of each vortex still carries an unpaired Majorana state. For that reason, the vortices will have Ising fusion rules (2.91) in this case. Using the Verlinde formula, we can compute the S-matrix in the basis $(1, \sigma, \psi)$:

$$S = \begin{pmatrix} 1 & \sqrt{2} & 1 \\ \sqrt{2} & 0 & -\sqrt{2} \\ 1 & -\sqrt{2} & 1 \end{pmatrix}. \tag{2.158}$$

If, in contrast, we compute the S-matrix via its definition (2.141), we find in particular for the (σ, σ) matrix element,

$$S_{\sigma\sigma} = \frac{1 + \theta_\psi}{2\theta_\sigma^2}. \tag{2.159}$$

For this to vanish, we conclude that $\theta_\psi = -1$, that is, ψ is a fermion. We cannot obtain θ_σ from similar relations using only the S-matrix. Rather, we can use Vafa's theorem (2.137) to deduce the spin of the σ particle. If we take $a = b = d = e = \sigma$, we find

$$\prod_c \left(\frac{\theta_c}{\theta_\sigma\theta_\sigma}\right)^{N^c_{\sigma\sigma}N^\sigma_{c\sigma}} \prod_f \left(\frac{\theta_f}{\theta_\sigma\theta_\sigma}\right)^{N^\sigma_{\sigma f}N^f_{\sigma\sigma}} = \prod_r \left(\frac{\theta_\sigma}{\theta_\sigma\theta_r}\right)^{N^r_{\sigma\sigma}N^\sigma_{\sigma r}}, \tag{2.160}$$

or

$$\frac{1}{\theta_\sigma^2}\frac{\theta_\psi}{\theta_\sigma^2}\frac{1}{\theta_\sigma^2}\frac{\theta_\psi}{\theta_\sigma^2} = \frac{1}{\theta_\psi}, \tag{2.161}$$

which gives

$$\theta_\sigma^8 = \theta_\psi^3. \tag{2.162}$$

Since $\theta_\psi = -1$, we have

$$\theta_\sigma^8 = \theta_\psi = -1, \tag{2.163}$$

i.e. the phase of the topological spin of the Majorana is an odd-integer multiple of $\frac{1}{16}$. We can thus discriminate 8 different TQFTs with Ising fusion rules by the values

$$\theta_\sigma = \mathrm{e}^{2\pi \mathrm{i} \mathsf{C}^{(1)}/16}, \tag{2.164}$$

where $\mathsf{C}^{(1)}$ is the (odd, as needed by (2.163)) Chern number or the number of stacked spinless chiral p-wave superconductors.

2.3.4.2 Case $\mathsf{C}^{(1)} = 2$ mod 4

If we stack $\mathsf{C}^{(1)} = 2 \bmod 4$ layers of chiral p-wave superconductors, then, as we have argued before, heuristically, the system can be described as one species of spinless Dirac fermions with Chern number $\tilde{\mathsf{C}}^{(1)} = \frac{1}{2}\mathsf{C}^{(1)}$. In this system, the 2π flux ψ binds odd-integer charge $\tilde{\mathsf{C}}^{(1)}$, that is, the 2π flux is a fermion ψ. However, unlike in the quantum Hall effect, the superconducting π or $-\pi$ fluxes are allowed topological excitations that bind half-integer charge. Let us denote the π flux with charge $\frac{1}{2}\tilde{\mathsf{C}}^{(1)}$ by a. Fusing such a π flux with the fermion (or 2π flux) gives another excitation, a $-\pi$ flux, with $\frac{3}{2}\tilde{\mathsf{C}}^{(1)}$ charge, that we call $\tilde{a}$. (We note that 4π flux is identified with zero flux, since this corresponds to a charge $2\tilde{\mathsf{C}}^{(1)}$ object, i.e. $\tilde{\mathsf{C}}^{(1)}$ Cooper pairs that can be absorbed by the condensate.) Two fluxes of either type a or $\tilde{a}$ thus fuse into ψ. This motivates the following fusion rules:

$$\begin{aligned} &a \times \tilde{a} = 1, \qquad && a \times \psi = \tilde{a}, \quad \tilde{a} \times \psi = a, \\ &a \times a = \tilde{a} \times \tilde{a} = \psi, \quad && \psi \times \psi = 1. \end{aligned} \tag{2.165}$$

These fusion rules are Abelian, so the quantum dimensions are $d_1 = d_a = d_{\tilde{a}} = d_\psi = 1$. Given the fusion matrices, we can compute the S-matrix as the matrix of their simultaneous eigenvectors. In the basis $(1, a, \tilde{a}, \psi)$, it can take one of two forms:

$$S^{(1)} = \frac{1}{2}\begin{pmatrix} 1 & 1 & 1 & 1 \\ 1 & -\mathrm{i} & \mathrm{i} & -1 \\ 1 & \mathrm{i} & -\mathrm{i} & -1 \\ 1 & -1 & -1 & 1 \end{pmatrix}, \qquad S^{(2)} = \frac{1}{2}\begin{pmatrix} 1 & 1 & 1 & 1 \\ 1 & \mathrm{i} & -\mathrm{i} & -1 \\ 1 & -\mathrm{i} & \mathrm{i} & -1 \\ 1 & -1 & -1 & 1 \end{pmatrix}. \tag{2.166}$$

From $S_{aa} = S_{\tilde{a}\tilde{a}} = \pm\frac{1}{2}\mathrm{i}$, we conclude

$$\theta_a^2 = \theta_{\tilde{a}}^2 = \mp\mathrm{i}, \tag{2.167}$$

while $S_{a\psi} = S_{\tilde{a}\psi} = -\frac{1}{2}$ gives

$$\theta_a = \theta_{\tilde{a}}, \tag{2.168}$$

with the help of (2.141). We conclude that there are four possible theories with topological spins

$$\theta_a = \theta_{\tilde{a}} = \mathrm{e}^{2\pi\mathrm{i}\mathsf{C}^{(1)}/16}. \tag{2.169}$$

2.3.4.3 Case $\mathsf{C}^{(1)} = 0 \bmod 4$

If we couple four layers of chiral p-wave superconductors, the π superconducting vortices (let us denote them by e) bind a full electron charge. However, the electrons exist also as free fermionic quasiparticles f in the theory. Hence, the fusion of e with f should yield a new excitation m that is a π vortex stripped of its charge. Together, e, m, and f obey the toric code fusion rules (2.90). The S-matrix of the theory in the basis $(1, e, m, f)$ can take one of two forms:

$$S^{(1)} = \frac{1}{2}\begin{pmatrix} 1 & 1 & 1 & 1 \\ 1 & -1 & 1 & -1 \\ 1 & 1 & -1 & -1 \\ 1 & -1 & -1 & 1 \end{pmatrix}, \qquad S^{(2)} = \frac{1}{2}\begin{pmatrix} 1 & 1 & 1 & 1 \\ 1 & 1 & -1 & -1 \\ 1 & -1 & 1 & -1 \\ 1 & -1 & -1 & 1 \end{pmatrix}. \tag{2.170}$$

If we assume that $\theta_f = -1$, we find from $S^{(1)}$ that $\theta_e^2 = \theta_m^2 = \theta_e\theta_m = -1$, while the theory with $S^{(2)}$ has $\theta_e^2 = \theta_m^2 = \theta_e\theta_m = 1$. In total, we have four possibilities

$$\theta_e = \theta_m = \mathrm{e}^{2\pi\mathrm{i}\mathsf{C}^{(1)}/16}, \tag{2.171}$$

with $\mathsf{C}^{(1)} = 0 \bmod 4$, as allowed theories. This concludes Kitaev's 16-fold way [18].

Acknowledgements

The authors were supported by NSF CAREER DMR-095242, ONR-N00014-11-1-0635, MURI-130-6082, MERSEC Grant, the Packard Foundation, and a Keck grant.

References

[1] X.-G. Wen, *Int. J. Mod. Phys. B* **4**, 239 (1990); *Adv. Phys.* **44**, 405 (1995).

[2] A. P. Schnyder, S. Ryu, A. Furusaki, and A. W. W. Ludwig, *Phys. Rev. B* **78**, 195125 (2008).

[3] S. Ryu, A. P. Schnyder, A. Furusaki, and A. W. W. Ludwig, *New J. Phys.* **12** (2010).

[4] A. Y. Kitaev, *AIP Conf. Proc.* **1134**, 22 (2009).

[5] W. P. Su, J. R. Schrieffer, and A. J. Heeger, *Phys. Rev. Lett.* **42**, 1698 (1979).

[6] A. Y. Kitaev, *Phys. Usp.* **44**, 131 (2001).

[7] C. Fang, M. J. Gilbert, and B. A. Bernevig, *Phys. Rev. Lett.* **112**, 106401 (2014)

[8] L. Fidkowski and A. Kitaev, *Phys. Rev. B* **81**, 134509 (2010).

[9] M. B. Hastings and T. Koma, *Commun. Math. Phys.* **265**, 781 (2006).

[10] A. Y. Kitaev, *Ann. Phys. (NY)* **303**, 2 (2003).

[11] B. A. Bernevig, with T. L. Hughes, *Topological Insulators and Topological Superconductors*, Princeton University Press, Princeton, NJ, 2013.

[12] G. Volovik, *JETP Lett.* **70**, 609 (1999).

[13] N. Read and D. Green, *Phys. Rev. B* **61**, 10267 (2000).

[14] D. A. Ivanov, *Phys. Rev. Lett.* **86**, 268 (2001).

[15] N. B. Kopnin and M. M. Salomaa, *Phys. Rev. B* **44**, 9667 (1991).

[16] G. Moore and N. Read, *Nucl. Phys. B* **360**, 362 (1991).

[17] C. Nayak and F. Wilczek, *Nucl. Phys. B* **479**, 529 (1996).

[18] A. Y. Kitaev, *Ann. Phys. (NY)* **321**, 2 (2006).

[19] J. Cano, M. Cheng, M. Mulligan, C. Nayak, E. Plamadeala, and J. Yard, *Phys. Rev. B* **89**, 115116 (2014).

[20] C. Nayak, S.H. Simon, A. Stern, M Freedman, and S. Das Sarma, *Rev. Mod. Phys.* **80**, 1083 (2008).

[21] P. Bonderson, *Non-Abelian Anyons and Interferometry*, PhD thesis, California Institute of Technology, http://thesis.library.caltech.edu/2447/2/thesis.pdf (2007).

[22] C. Vafa, *Phys. Lett. B* **206**, 421 (1988).

[23] M. Barkeshli, P. Bonderson, M. Cheng, and Z. Wang, arXiv:1410.4540 [cond-mat.str-el].

3 Spin liquids and frustrated magnetism

John T. Chalker

Theoretical Physics
Oxford University
1 Keble Road
Oxford OX1 3NP
UK

Topological Aspects of Condensed Matter Physics. First Edition. C. Chamon et al.

Chapter Contents

3.1 Introduction

Two important markers in the history of research on spin liquids and frustrated magnetism are Anderson's suggestion [1], over 40 years ago, of the resonating valence bond state as an alternative to Néel order, and Ramirez' influential review [2], some 20 years ago, of strongly frustrated magnets. There has been a tremendous amount of progress since then, but much remains to be done, especially in identifying experimental examples of spin liquids and understanding their properties. In these lecture notes, I aim to provide an introduction to the field that links our understanding of the classical statistical physics of these systems with approaches to their quantum mechanics. Short complementary introductions can be found in the articles by Lee [3] and by Balents [4]; more specialized reviews of various aspects of the field can be found in a recent book [5], and alternative approaches to the one taken in the present notes are outlined in Section 3.6.

The term *spin liquid* is presumably intended to draw an analogy between possible states of a magnet and the conventional three phases of matter, but this analogy fails to capture some of the most interesting features of spin liquids. More specifically, it is reasonable to see a paramagnet as being like a gas, since both states occur at high temperature and are essentially uncorrelated; and it is also appropriate to think of a Néel state as like a solid, in the sense that both have broken symmetry, characterized by a local order parameter, and occur at low temperature. But whereas classical fluids have only local correlations, we shall see that classical spin liquids may have a divergent correlation length and power-law correlations. And in place of the Fermi surface or Bose condensate of quantum fluids, quantum spin liquids may have topological order and fractionalized excitations.

3.1.1 Overview

We will be concerned with the statistical mechanics and quantum mechanics of models for magnetic degrees of freedom in Mott insulators. These have well-defined local moments, which we represent using simple spin Hamiltonians such as the Ising and Heisenberg models. The models are said to be frustrated if different contributions to the interaction energy have conflicting classical minima. The interest of frustration in these systems is that it acts to destabilize conventional ordered states. Classically, one sees this from a large contribution to ground-state degeneracy, which is *accidental* in the technical sense that it is not a consequence of symmetry.

The ideas of frustration and accidental degeneracy can be illustrated by considering a cluster of spins with antiferromagnetic interactions between all pairs in the cluster. To allow some generality, we take a cluster of q classical spins, represented using n-component unit vectors $\mathbf{S}_i$. The Hamiltonian

$$\mathcal{H} = J\sum_{\langle ij\rangle} \mathbf{S}_i \cdot \mathbf{S}_j = \frac{J}{2}|\mathbf{L}|^2 + \text{const}, \qquad \text{with} \quad \mathbf{L} = \sum_{i=1}^{q} \mathbf{S}_i \tag{3.1}$$

(where $\sum_{\langle ij\rangle}$ denotes a sum over pairs ij) is minimized in states for which the total magnetization $\mathbf{L}$ of the cluster is zero. When $q > 2$, no state can minimize the interaction energy $J\,\mathbf{S}_i \cdot \mathbf{S}_j$ of all pairs simultaneously. For example, a set of four Ising spins with nearest-neighbour antiferromagnetic interactions on a tetrahedron hence has a ground state in which two spins are up and two are down. This gives a ground-state degeneracy of six, rather than the generic value of two for Ising systems with time-reversal symmetry. Replacing these Ising spins with classical Heisenberg spins, the ground states (as illustrated in Fig. 3.1) have two internal degrees of freedom in addition to the three that arise from global rotations of any non-collinear spin arrangement.

Some of the most important lattices for the study of geometrically frustrated magnets can be constructed as corner-sharing arrangements of clusters: examples are the kagome and pyrochlore lattices, formed in this way from triangles or tetrahedra and shown in Fig. 3.2.

It is useful for orientation to discuss the some selected examples of geometrically frustrated magnetic materials, listed in Table 3.1, although we will not attempt any sort of survey. An important characterization is provided by the dependence of the magnetic susceptibility $\chi(T)$ on temperature T. At high temperatures, it obeys the Curie–Weiss law

$$\chi(T) \propto \frac{1}{T - \theta_{\mathrm{CW}}} \tag{3.2}$$

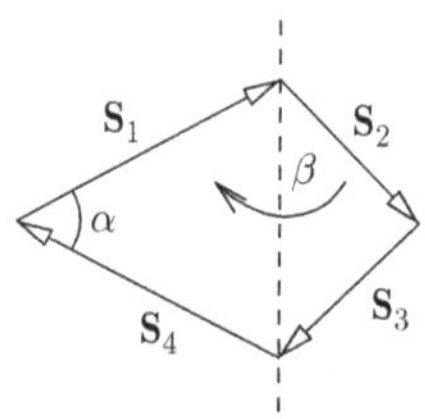

Fig. 3.1 Ground states of four antiferromagnetically coupled classical Heisenberg spins: the two accidental ground-state degrees of freedom are the angle α between spins $\mathbf{S}_1$ and $\mathbf{S}_4$, and the angle β between the plane containing $\mathbf{S}_1$ and $\mathbf{S}_4$ and that containing $\mathbf{S}_2$ and $\mathbf{S}_3$.

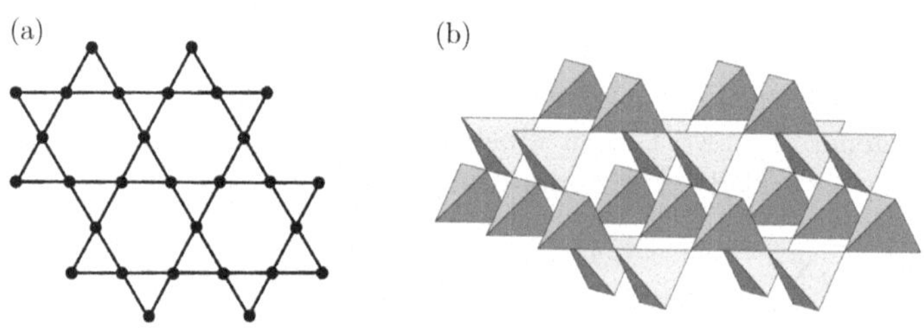

Fig. 3.2 Two lattices: (a) kagome; (b) pyrochlore. (Part (b) reprinted with permission from [6]. Copyright 2002 by the American Physical Society.)

Table 3.1 Selected examples of frustrated magnetic materials

SHORT NAME	MATERIAL	MAGNETIC LATTICE (SIZE OF MOMENTS)	SPIN MODEL (VALUE OF θ_{CW})
SCGO	$SrCr_{9-x}Ga_{3+x}O_9$	Pyrochlore slabs $\left(S = \frac{3}{2}\right)$	Heisenberg ($\theta_{CW} \approx -500\,K$)
Spin ice	$Ho_2Ti_2O_7$ $Dy_2Ti_2O_7$	Pyrochlore	Ising ($\theta_{CW} \approx +1.9\,K$)
Herbertsmithite	$ZnCu_3(OH)_6Cl_2$	Kagome $\left(S = \frac{1}{2}\right)$	Heisenberg ($\theta_{CW} \approx -300\,K$)
κ-ET	κ-(BEDT-TTF)$_2Cu_2(CN)_3$	Triangular $\left(S = \frac{1}{2}\right)$	Heisenberg ($\theta_{CW} \approx -400\,K$)

and the magnitude of the Curie–Weiss constant θ_{CW} (negative if exchange is antiferromagnetic) reflects the energy scale of exchange interactions. Without frustration, one expects ordering at a temperature scale set by $|\theta_{CW}|$. By contrast, highly frustrated systems remain in the paramagnetic phase to much lower temperatures, and in some cases to zero temperature. Their low-temperature fate may involve freezing or a structural, frustration-relieving transition at temperature T_c, and Ramirez introduced the ratio

$$f = -\frac{\theta_{CW}}{T_c} \tag{3.3}$$

as a simple measure for the degree of frustration. The state of a system in the temperature range $|\theta_{CW}| \gg T > T_c$, where spins are highly correlated but strongly fluctuating, was termed by Villain [7] a *cooperative paramagnet*. This is the spin-liquid state that we wish to characterize more thoroughly, at both the classical and the quantum levels.

These ideas are illustrated by the first material in our selection, SCGO. It has a frustration parameter $f \gtrsim 100$ [8] and magnetic neutron scattering shows strong spin correlations at low temperature (via a peak in scattering at intermediate wavevector) but no long-range order (from the absence of magnetic Bragg peaks) [9]. To obtain more detailed information about low-temperature spin correlations using neutron scattering requires single crystals. These are not available for SCGO, but for spin-ice materials, so-called *pinch-point* features in the diffuse scattering [10], which are sharp in reciprocal space, reveal long-range correlations in real space (see Section 3.4).

The other listed materials, herbertsmithite and κ-ET, are two of the best candidate quantum spin liquids. Neither shows signs of magnetic order, even at the lowest accessible temperatures. Moreover, in contrast to the sharp response from magnon excitations in an ordered magnet, single-crystal inelastic neutron scattering from herbertsmithite has structure broad in wavevector at all energies [11], as expected if the energy and momentum imparted by scattered neutrons are shared between multiple fractionalized excitations. Information on excitations can also be inferred from the dependence on temperature T of the heat capacity $C_{\rm p}$, which in κ-ET at low temperature fits the form $C_{\rm p} = \gamma T + \beta T^3$ [12]. While such behaviour is familiar in a metal, a contribution linear in T is remarkable in an insulator, suggesting the formation of a spinon Fermi surface in a system that does not have mobile charges.

3.1.2 Classical ground-state degeneracy

A first step in the discussion of classical frustrated magnets is to understand ground-state degeneracy. The character of this problem is different according to whether we treat discrete (e.g. Ising) or continuous (e.g. classical Heisenberg) spin variables, since we should count discrete ground states for the former and continuous degrees of freedom within the ground-state manifold for the latter. In either case, a signature of a highly frustrated system is that the number we obtain in this way is extensive, suggesting that within ground states there are local fluctuations that take place independently in different parts of a large sample.

For the discrete case, an illustrative example is provided by the nearest-neighbour Ising antiferromagnet on the pyrochlore lattice. Here an approximate but remarkably accurate estimate was provided by Pauling in the context of water ice [13] (see Section 3.4 for details of the link between water ice, spin ice, and the Ising antiferromagnet). As a start, note for a single tetrahedron that 6 states from a total of 16 are ground states. A pyrochlore Ising model consisting of $N_{\rm T}$ tetrahedra contains $N_{\rm S} = 2N_{\rm T}$ spins, since there are four spins per tetrahedron but each spin is shared between two tetrahedra. It therefore has $2^{2N_{\rm T}}$ states in total. Treating the restriction to ground states as if it were independent on each tetrahedron, the number of ground states for the entire system is then estimated to be

$$2^{2N_{\rm T}} \times \left(\frac{6}{16}\right)^{N_{\rm T}} = \left(\frac{3}{2}\right)^{N_{\rm T}} = \left(\frac{3}{2}\right)^{N_{\rm S}/2}, \tag{3.4}$$

and from this the ground-state entropy per spin is $\frac{1}{2}k_{\rm B} \ln \frac{3}{2}$. Measurements of the low-temperature entropy in spin ice, obtained from the magnetic contribution to the heat capacity, are in striking agreement with this estimate [14].

For models with spins that can be rotated continuously, ground-state degrees of freedom are the ones remaining after respecting all ground-state constraints. We can estimate their number D using an approach initiated in the context of mechanical systems by Maxwell [15–17]. Consider $N_{\rm S}$ classical n-component unit spins, constituting $F = N_{\rm S}(n-1)$ degrees of freedom. Suppose that these spins form a lattice of $N_{\rm C}$

corner-sharing, antiferromagnetically coupled, q-site clusters $\{\alpha\}$. Each cluster can be treated as in (3.1), and $N_S = qN_C/2$. Energy is minimized if the total magnetization $\mathbf{L}_\alpha$ of every cluster is zero: a set of $K = nN_C$ scalar constraints. If these constraints can be satisfied simultaneously and are linearly independent, then we have

$$D = F - K = \left[\frac{q}{2}(n-1) - n\right] N_C \,. \tag{3.5}$$

As an example, for Heisenberg spins ($n = 3$) on the pyrochlore lattice ($q = 4$), we get in this way the extensive result $F = N_C$.

3.1.3 Order by disorder

Extensive ground-state degeneracy, in either the discrete or the continuous sense, is characteristic of many highly frustrated systems and offers a potential route to understanding suppression of order. However, not being symmetry-protected, this degeneracy may be lifted by fluctuations, a phenomenon termed *order by disorder* [18, 19].

The issues at stake are illustrated schematically in Fig. 3.3, where we consider in phase space the configurations that are accessible at low temperature and lie close to the ground state. Introducing coordinates $\mathbf{x}$ on the ground-state manifold and $\mathbf{y}$ locally orthogonal to it, by integrating out small-amplitude fluctuations in $\mathbf{y}$, from an energy $\mathcal{H}(\mathbf{x}, \mathbf{y})$ at inverse temperature β one induces a probability distribution on the ground state of the form (before normalization)

$$\mathcal{Z}(\mathbf{x}) = \int \mathcal{D}\mathbf{y}\, \mathrm{e}^{-\beta\mathcal{H}(\mathbf{x},\mathbf{y})} \propto \prod_k \frac{k_B T}{\omega_k(\mathbf{x})} \,, \tag{3.6}$$

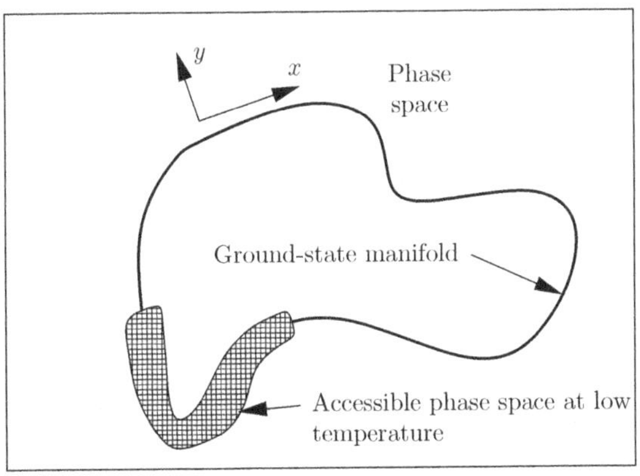

Fig. 3.3 Schematic view of phase space for a geometrically frustrated magnet: the ground-state manifold forms a high-dimensional subspace, and states accessible at low temperatures (partly marked by shading) lie close to it. Coordinates in phase space can be separated locally into ones ($\mathbf{x}$) within the ground-state subspace and others ($\mathbf{y}$) orthogonal to it.

where the right-hand expression follows from a harmonic approximation for the dependence of $\mathcal{H}(\mathbf{x}, \mathbf{y})$ on $\mathbf{y}$, under the assumption that the components of $\mathbf{y}$ consist of canonically conjugate pairs of generalized coordinates and momenta. Two alternatives now arise: this probability distribution may either represent a system that accesses all ground states at low temperature, or the probability density may be concentrated on a subset of ground states. For the latter, the states selected by fluctuations are ones with $\prod_k \omega_k(\mathbf{x})$ small. In practice, these are likely to be states that are ordered in some way, in which a subset of $\omega_k(\mathbf{x})$ vanish.

This point can be illustrated by a toy calculation for four antiferromagnetically coupled classical XY ($n = 2$) or Heisenberg ($n = 3$) spins, with (3.1) as the Hamiltonian. The existence of a soft mode for special ground states in which all four spins are collinear is demonstrated in Fig. 3.4. To examine the consequences of this soft mode, consider at inverse temperature β the thermal distribution $P(\theta)$ of the angle θ between a pair of these spins, defined via $\cos\theta = \mathbf{S}_1 \cdot \mathbf{S}_2$. Let

$$\mathcal{Z}(\theta) = \int \mathrm{d}\mathbf{S}_3\, \mathrm{d}\mathbf{S}_4\, \mathrm{e}^{-\beta\mathcal{H}}\,. \tag{3.7}$$

Taking into account the volume factors of $\mathrm{d}\theta$ and $\sin\theta\,\mathrm{d}\theta$ for $n = 2$ and $n = 3$, respectively, one has $P(\theta)\,\mathrm{d}\theta \propto \mathcal{Z}(\theta)\,\mathrm{d}\theta$ for $n = 2$ and $P(\theta)\,\mathrm{d}\theta \propto \mathcal{Z}(\theta)\sin\theta\,\mathrm{d}\theta$ for $n = 3$. It is straightforward to evaluate $\mathcal{Z}(\theta)$ at large β and so obtain

$$P(\theta) \propto \begin{cases} (\sin\theta)^{-1} & \text{XY}, \\ \sin\frac{1}{2}\theta & \text{Heisenberg}. \end{cases} \tag{3.8}$$

The non-integrable divergences of $P(\theta)$ at $\theta = 0$ and $\theta = \pi$ for XY spins (which are rounded at finite β) indicates that fluctuations select collinear spin configurations in

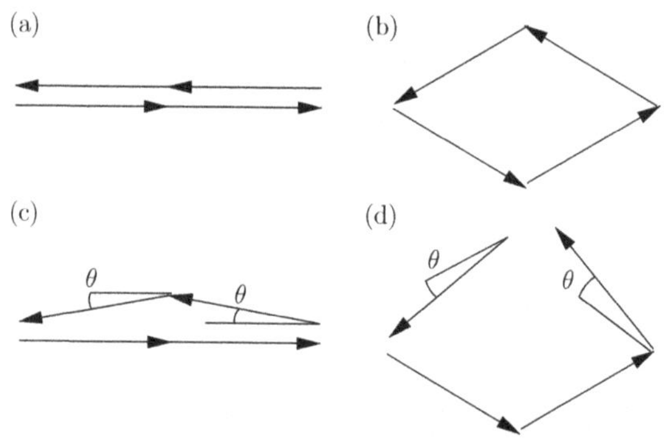

Fig. 3.4 An illustration of how soft modes arise in selected ground states. Consider two ground states (a) and (b) for four antiferromagnetically coupled spins and two states (c) and (d) obtained by rotating pairs of spins through small angles θ. The total magnetization varies differently with θ in the two cases. It is $\mathbf{L} \propto \theta^2$ in the example based on a collinear ground state, but $\mathbf{L} \propto \theta$ in the generic case. The energy cost of the excitation is hence $\mathcal{H} \propto \theta^4$ in the collinear case, but is generically $\mathcal{H} \propto \theta^2$.

the low-temperature limit. Conversely, Heisenberg spins sample all orientations even at arbitrarily temperature.

Moving from this toy problem to an extended lattice, there exists a catalogue of well-studied examples and a criterion for whether there is fluctuation-induced order. Instances of classical order by disorder include the kagome Heisenberg antiferromagnet (where coplanar spin configurations are selected [20]) and the pyrochlore XY antiferromagnet (with collinear order [17]), while a converse case is the pyrochlore Heisenberg antiferromagnet, which is thermally disordered at all temperatures [17]. A sufficient condition for order is that the ground-state probability distribution $\mathcal{Z}(\mathbf{x})$ has non-integrable divergences in the vicinity of the subset of configurations favoured by fluctuations, and one can assess whether that is the case by comparing the dimensionality of the full ground-state manifold with that of the soft subspace. For n-component spins on a lattice of corner-sharing clusters of q sites, order is expected if $n < (q+2)/(q-2)$ [17].

In Monte Carlo simulations of models with continuous degrees of freedom, the value of the low-temperature heat capacity C per spin provides a diagnostic for the presence of soft modes. A simple generalization of the equipartition principle shows that a dependence of energy $\mathcal{H}$ on mode coordinate θ with the form $\mathcal{H} \propto |\theta|^n$ implies a contribution to C from this mode of k_B/n. By this argument, one expects for unfrustrated classical Heisenberg models (with two degrees of freedom per spin) the value $C = k_\mathrm{B}$ at low temperature. In contrast, for the pyrochlore Heisenberg antiferromagnet, from (3.5) one-quarter of modes cost no energy. The remaining modes are conventional, with an energy cost quadratic in displacement, and so $C = \frac{3}{4}k_\mathrm{B}$ [17]. For the kagome Heisenberg antiferromagnet, one-sixth of the fluctuation modes from a coplanar ground state cost an energy quartic in displacement, while the remainder are quadratic, so that $C = \frac{11}{12}k_\mathrm{B}$ [20].

Quantum order by disorder depends on different features of the fluctuation spectrum from its thermal counterpart. In the notation of (3.6), the zero-point energy of quantum fluctuations out of the classical ground-state manifold generates an effective Hamiltonian

$$\mathcal{H}_{\mathrm{eff}}(\mathbf{x}) = \frac{1}{2}\sum_k \hbar\omega_k(\mathbf{x})\,. \tag{3.9}$$

Suppose $\mathcal{H}_{\mathrm{eff}}(\mathbf{x})$ has minima for preferred configurations $\mathbf{x} = \mathbf{x}_0$. There are quantum fluctuations of $\mathbf{x}$ about these configurations, because the set of ground-state coordinates includes canonically conjugate pairs of generalized positions and momenta. At large spin S, the amplitude of these fluctuations is small and one always expects order. Reducing S, we expect within this description that a quantum-disordered ground state may emerge via delocalization of the ground-state wavefunction in the landscape $\mathcal{H}_{\mathrm{eff}}(\mathbf{x})$.

Experimental demonstrations of fluctuation-induced order rely on there being a good characterization of interactions, so as to show that these do not drive ordering in a conventional way. For the garnet $Ca_3Fe_2Ge_3O_{12}$, a material with two interpenetrating magnetic lattices coupled via zero-point fluctuations, it has been demonstrated that a spin-wave gap in the Néel ordered state does indeed arise mainly in this way, by independent determination of the size of single-ion anisotropy (the other possible origin for the gap) [21], as well as via the characteristic temperature dependence of the gap [22]. In the context of highly frustrated systems, $Er_2Ti_2O_7$ has been thoroughly investigated as an example of a system with quantum order by disorder [23].

3.2 Classical spin liquids

The problem of finding a good description for the low-temperature states of classical frustrated magnets presents an obvious challenge. We would like to replace the high-energy spin degrees of freedom, which have strongly correlated fluctuations at low temperature, with a new set of low-energy degrees of freedom that are only weakly correlated. Remarkably, this turns out to be possible for some of the systems of most interest. Moreover, the emergent low-energy coordinates have simple and appealing interpretations, which we introduce in the following.

3.2.1 Simple approximations

As context for a discussion of low-temperature states in frustrated magnets, it is useful to examine what special features these systems present when they are treated using some of the standard approximations. In particular, it is worthwhile to see how in some cases the absence of ordering is signalled within mean-field theory, and how an alternative approach known as the self-consistent Gaussian or large-n approximation can often give a good description of low-temperature correlations.

Recall the essentials of mean-field theory: thermal averages $\langle \cdots \rangle$ with respect to the full Hamiltonian $\mathcal{H}$ are approximated by averages $\langle \cdots \rangle_0$ using a tractable Hamiltonian $\mathcal{H}_0$. This gives a variational bound

$$\mathcal{F} \leq \langle \mathcal{H} \rangle_0 - T\mathcal{S}_0 \tag{3.10}$$

on the free energy $\mathcal{F}$ of the system, in terms of the energy $\langle \mathcal{H} \rangle_0$ and entropy $\mathcal{S}_0$ computed from $\mathcal{H}_0$. Taking a single-site $\mathcal{H}_0$, these quantities are parametrized by the site magnetizations $\{m_i\}$, leading to an expansion of the form

$$\langle \mathcal{H} \rangle_0 - T\mathcal{S}_0 = \text{const} + \sum_{\langle ij \rangle} J_{ij} m_i m_j + \frac{1}{2} n k_{\mathrm{B}} T \sum_i m_i^2 + \mathcal{O}(m_i^4)\,, \tag{3.11}$$

with exchange interactions J_{ij}, where n is the number of spin components. Choosing the $\{m_i\}$ to minimize this estimate for F, one finds solutions of two types, depending on temperature: above the mean-field ordering temperature T_c, all $m_i = 0$, while for $T < T_c$ some $m_i \neq 0$. Within the mean-field approximation, the value of T_c and the ordering pattern below T_c are determined from the eigenvalues and eigenvectors of the matrix J_{ij}: denoting the minimum eigenvalue (which is negative) by ε_{min} and an associated eigenvector by φ_i, one has $\varepsilon_{\text{min}} + \frac{1}{2} n k_{\mathrm{B}} T_c = 0$ and $m_i \propto \varphi_i$ for $T \lesssim T_c$. The distinction that arises in this framework between unfrustrated and highly frustrated systems concerns the degeneracy of ε_{min}. In a conventional system, the minimum eigenvalues form a discrete set. For example, in a nearest-neighbour square-lattice antiferromagnet the eigenvalues of the interaction matrix, labelled by a wavevector $\mathbf{q}$, are $J(\mathbf{q}) = J(\cos q_x + \cos q_y)$, and so $\varepsilon_{\text{min}} = -2J$ at $\mathbf{q} = (\pi, \pi)$. By contrast, a number of important examples of highly frustrated magnets lead to minima of J_{ij} at all wavevectors, forming dispersionless or 'flat' bands across the Brillouin zone. Mean-field theory fails for these systems by wrongly predicting ordering at a temperature $T_c \sim |\theta_{\text{CW}}|$. A warning of this failure is provided by there being macroscopically many possible ordering patterns φ_i.

The appearance of flat bands with nearest-neighbour interactions J on lattices built from corner-sharing clusters of q sites can be understood via the same Maxwell-counting approach that we employed in (3.5) to discuss ground-state degeneracy of systems such as the classical Heisenberg model on the same lattices. We start from the fact that ε_{min} is the minimum of $\sum_{ij} J_{ij}\varphi_i\varphi_j$ subject to the constraint $\sum_i \varphi_i^2 = 1$. For corner-sharing clusters, using i, j as site labels and α as a cluster label, we have

$$\sum_{ij} J_{ij}\varphi_i\varphi_j = J\sum_{\alpha}\left|\sum_{i\in\alpha}\varphi_i\right|^2 + \text{const.} \tag{3.12}$$

Eigenvectors associated with ε_{min} therefore satisfy $\sum_{i\in\alpha}\varphi_i = 0$ for all α. Using the notation of (3.5), these conditions amount to $K = N_{\text{C}}$ constraints on the $F = N_{\text{S}} = \frac{1}{2}qN_{\text{C}}$ degrees of freedom $\{\varphi_i\}$. Omitting sub-extensive terms, the degeneracy of ε_{min} is therefore

$$D = F - K = \left(\frac{1}{2}q - 1\right)N_{\text{C}}\,. \tag{3.13}$$

For example, on the kagome lattice, one-third of eigenvectors have eigenvalue ε_{min}; as there are three sites in the unit cell, the matrix J_{ij} has three bands, of which the lowest is flat. Similarly, for the pyrochlore lattice, there are four bands, of which the lowest two are degenerate and flat.

A successful treatment of these systems at low temperature must involve an average over correlated low-energy states. The self-consistent Gaussian approach provides a simple way of approximating this average, and is widely applicable to systems in which all sites are symmetry-equivalent. The central idea is to replace an average over orientations of classical, fixed-length spins by independent Gaussian averages over the magnitudes of each component, with a variance chosen to maintain the correct spin length on average. These simplifications are exact in the limit that the number n of spin components is large, and in many instances they are remarkably accurate for $n = 3$ or even for $n = 1$ [24, 25]. Under this approximation, the trace over spin configurations is written

$$\prod_i \int \mathrm{d}\mathbf{S}_i \cdots \delta(|\mathbf{S}_i| - 1) \approx \prod_i \int \mathrm{d}\mathbf{S}_i \cdots \mathrm{e}^{-\frac{1}{2}\lambda|\mathbf{S}_i|^2}, \tag{3.14}$$

with the Lagrange multiplier λ determined by the condition $\langle|\mathbf{S}_i|^2\rangle = 1$. Denoting the partition function by $\mathcal{Z}$, a general thermal average then takes the form

$$\langle\cdots\rangle = \mathcal{Z}^{-1}\int \mathrm{d}\{\mathbf{S}_i\}\cdots \mathrm{e}^{-\frac{1}{2}\sum_{ij}\mathbf{S}_i\cdot(\beta J_{ij}+\lambda\delta_{ij})\mathbf{S}_j}\,. \tag{3.15}$$

In particular, the spin correlator is

$$\langle\mathbf{S}_i\cdot\mathbf{S}_j\rangle = n\left[(\beta\mathsf{J}+\lambda)^{-1}\right]_{ij} \tag{3.16}$$

and λ satisfies

$$n\left[(\beta\mathsf{J}+\lambda)^{-1}\right]_{ii} \equiv \frac{n}{N_\mathrm{S}}\mathrm{Tr}\,\frac{1}{\beta\mathsf{J}+\lambda} = 1\,. \tag{3.17}$$

If the minimum eigenvalues of J form a flat band, then, in the low-temperature limit, $(\beta\mathsf{J}+\lambda)^{-1}$ is proportional to the projector $\mathbb{P}$ onto this band, so that $\langle\mathbf{S}_i\cdot\mathbf{S}_j\rangle \propto \mathbb{P}_{ij}$. In many of the systems we are concerned with (those with a Coulomb phase: see Section 3.4.2), $\mathbb{P}_{ij}$ falls off at large separation r_{ij} as $\mathbb{P}_{ij}\sim r_{ij}^{-d}$, where d is the spatial dimension.

3.2.2 The triangular lattice Ising antiferromagnet and height models

Moving beyond these simple approximations, we would like to find a description of low-energy spin configurations that respects microscopic constraints imposed by the Hamiltonian but is amenable to coarse-graining. An early and illuminating example is provided by a mapping that we now describe, from the triangular lattice Ising antiferromagnet to a height model.

As background, we note that this Ising model (probably the first highly frustrated magnet to be studied in detail [26]) has a macroscopically degenerate ground state, and ground-state spin correlations that are known from an exact solution to decay with distance as $r^{-1/2}$. In ground states, every elementary triangle of the lattice has two spins parallel and one antiparallel. Degeneracy arises because many such configurations include some spins that are subject to zero net exchange field and can therefore be reversed at zero energy cost, as illustrated in Fig. 3.5(a). To avoid confusion, we should point out that the degeneracy on this lattice is specific to the Ising model, and not a consequence of flat bands.

Spin configurations can be mapped onto a new variable, termed a height field, in such a way that the ground-state condition in the spin model translates into a condition that the height field is single-valued. The mapping associates an integer height h_i with each site i [27, 28]. To describe it, we introduce a direction on each bond of the lattice in such a way that there is (say) anticlockwise circulation around 'up' triangles and clockwise circulation around 'down' triangles. With the height of an origin site chosen arbitrarily, the height change on traversing a bond in the positive direction is $+1$ if the bond links antiparallel (unfrustrated) spins and -2 if it links parallel (frustrated) spins, as illustrated in Fig. 3.5(b). A convenient further stage is to define heights $h(\mathbf{r})$ at the centres of triangles that are averages of the three values at corners; see Fig. 3.5(c).

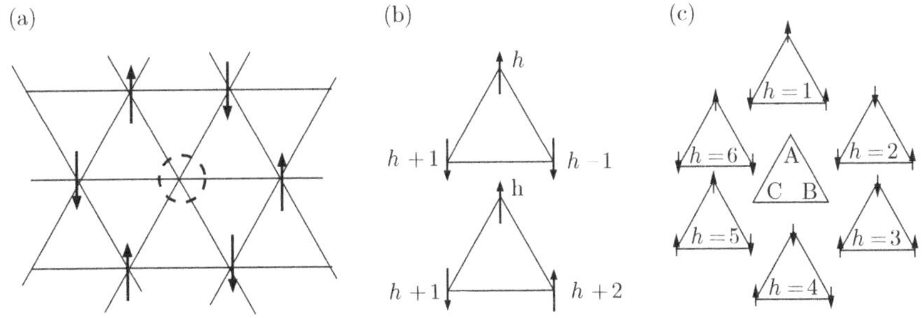

Fig. 3.5 Triangular-lattice Ising antiferromagnet spin configurations. (a) State with a flippable spin, marked with a circle. (b) Mapping from spins to heights at sites. (c) Mapping from spin orientations on the three sublattices (marked A, B, and C) to triangle heights in flat states.

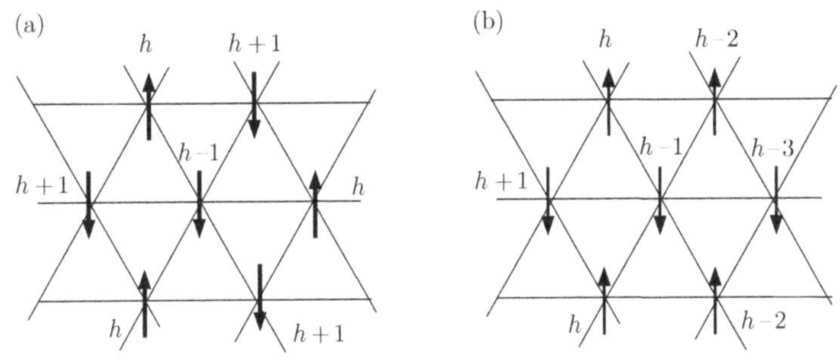

Fig. 3.6 Triangular lattice Ising antiferromagnet spin configurations and mappings to a height field. (a) State with flippable spin generating a flat height field (with the value h at all triangle centres). (b) Spin state without flippable spins that generates a height field with maximum gradient (heights at triangle centres decrease by 1 on going from any triangle to its right-hand neighbour).

As demonstrated with examples in Fig. 3.6, configurations with a flippable spin are locally flat, and those in which the gradient of the height field is maximal have no flippable spins. These facts motivate a coarse-grained theory in which $h(\mathbf{r})$ is taken to be a real-valued function of a continuous coordinate, with an entropic weight on configurations that (in the first approximation) has the form

$$P[h(\mathbf{r})] = \mathcal{Z}^{-1}\mathrm{e}^{-\mathcal{H}}\,, \tag{3.18}$$

where

$$\mathcal{H} = \frac{1}{2}K\int \mathrm{d}^2\mathbf{r}\,|\nabla h(\mathbf{r})|^2 \tag{3.19}$$

and $\mathcal{Z}$ is the usual normalization. The inverse mapping between spins and heights (modulo 6) is shown in Fig. 3.5(c) for the six distinct flat states. It has the algebraic form

$$S(\mathbf{r}) \simeq \cos\left[\frac{1}{3}\pi h(\mathbf{r}) + \varphi_\alpha\right], \tag{3.20}$$

where we have omitted higher Fourier components of $h(\mathbf{r})$, and the phase φ_α takes the values $0, \pm\frac{2}{3}\pi$, depending on the sublattice α.

The height field $h(\mathbf{r})$ has fluctuations that diverge logarithmically with separation, as we see from an explicit calculation. For a system of size $L \times L$, define the Fourier transform

$$h_{\mathbf{q}} = \frac{1}{L}\int \mathrm{d}^2\mathbf{r}\, \mathrm{e}^{\mathrm{i}\mathbf{q}\cdot\mathbf{r}} h(\mathbf{r}), \tag{3.21}$$

so that

$$\mathcal{H} = \frac{1}{2}K\sum_{\mathbf{q}} q^2 |h_{\mathbf{q}}|^2 . \tag{3.22}$$

Then, with short-distance cutoff a,

$$\begin{aligned} \langle [h(\mathbf{0}) - h(\mathbf{r})]^2 \rangle &= \frac{2}{L^2}\sum_{\mathbf{q}}(1 - \cos\mathbf{q}\cdot\mathbf{r})\langle |h_{\mathbf{q}}|^2\rangle \\ &= \frac{1}{2\pi^2}\int \mathrm{d}^2\mathbf{q}\, \frac{1-\cos\mathbf{q}\cdot\mathbf{r}}{Kq^2} \\ &\simeq \frac{1}{\pi K}\ln(r/a) . \end{aligned} \tag{3.23}$$

Putting these ingredients together, we can evaluate the spin correlator using the coarse-grained theory. For two sites on the same sublattice, we obtain

$$\langle \sigma(\mathbf{0})\sigma(\mathbf{r})\rangle \propto \langle \mathrm{e}^{\mathrm{i}\frac{1}{3}\pi[h(\mathbf{0})-h(\mathbf{r})]}\rangle = \mathrm{e}^{-\frac{1}{18}\pi^2\langle [h(\mathbf{0})-h(\mathbf{r})]^2\rangle} = (r/a)^{-\pi/18K} . \tag{3.24}$$

The power-law form illustrates the consequences of large but strongly correlated ground-state fluctuations, and a comparison with exact results for the Ising model fixes the value of the height model stiffness as $K = \frac{1}{9}\pi$.

The leading approximation that has been made in using the height model to represent the triangular-lattice Ising antiferromagnet is to treat $h(\mathbf{r})$ as a real, rather than integer-valued, field. To correct this, one can consider the replacement

$$\mathcal{H} \to \mathcal{H} + \mathcal{H}_1, \qquad \text{with} \quad \mathcal{H}_1 = -g\int \mathrm{d}^2\mathbf{r}\, \cos 2\pi h(\mathbf{r}), \tag{3.25}$$

so that values of $h(\mathbf{r})$ close to integers are preferred. One finds that $\mathcal{H}_1$ is an irrelevant perturbation at the renormalization group fixed point represented by $\mathcal{H}$ if $K < \frac{1}{2}\pi$, as is the case for the height model that represents the triangular-lattice Ising antiferromagnet. Changes to the Ising model (e.g. higher spin [28]) may increase K so that

$\mathcal{H}_1$ is relevant. The height field than locks to a particular integer value, representing long-range order of the Ising spins.

Excitations out of the ground state have an attractively simple description in height-model language. From the rules of Fig. 3.5, we see that if the three Ising spins of an elementary triangle have the same orientation, the height field is no longer everywhere single-valued: it changes by ± 6 on encircling the triangle that carries the excitation. A single spin flip can introduce such excitations into a ground-state configuration only as vortex–antivortex pairs, since a local move leaves the distant height field unchanged. A vortex and antivortex can be separated by additional spin flips without further increase in exchange energy, and constitute our first example of a fractionalized excitation.

Vortices are dilute at low temperature, because they have an energy cost $4J$. Their presence also changes the number of ground states available to the system, and so they have an entropy cost. This can be calculated within the height description. A vortex at the origin leads to an average height gradient at radius r of $|\nabla h(\mathbf{r})| = 6/(2\pi r)$. In a system of linear size L, this generates a contribution to $\mathcal{H}$ of

$$\frac{1}{2}K \int \mathrm{d}^2\mathbf{r}\,|\nabla h(\mathbf{r})|^2 = \frac{9K}{\pi}\ln(L/a)\,. \qquad (3.26)$$

Similarly, the presence of a vortex–antivortex pair at fixed positions with separation r has an entropy cost that depends on their separation. As a consequence, the pair is subject to an entropic attractive potential $V(r) \simeq (9K/\pi)\ln(r/a)$.

The entropy cost for vortices should be compared with the entropy gain $2\ln(L/a)$ arising from translations: vortex–antivortex pairs are unbound if $9K/\pi < 2$, as is the case here. In the setting of the Ising model, this means that the power-law correlations of the ground state are cut off at non-zero temperature by a finite correlation length, set by the vortex separation. This correlation length is much larger than the lattice spacing if $T \ll J/k_{\mathrm{B}}$, and it diverges as $T \to 0$.

In summary, the triangular-lattice Ising antiferromagnet provides an illustration of a system that, in its ground state, combines finite entropy with long-range correlations. The height model shows how these features can be captured in a long-wavelength description. The physics of the triangular-lattice Ising antiferromagnet at low temperature, including power-law correlations and fractionalized excitations, has important generalizations to other systems, including most notably spin ices. In addition, some of the main theoretical tools used in a long-wavelength description of these generalized problems are extensions of those underlying the height model. Many of these ideas are exemplified in classical dimer models, which we now introduce.

3.3 Classical dimer models

Classical dimers models [29–31] offer a setting in which to discuss some general features of the statistical physics of systems that are both highly degenerate and strongly constrained. They are important in their own right and also serve as the foundation for a treatment of quantum spin liquids using quantum dimer models.

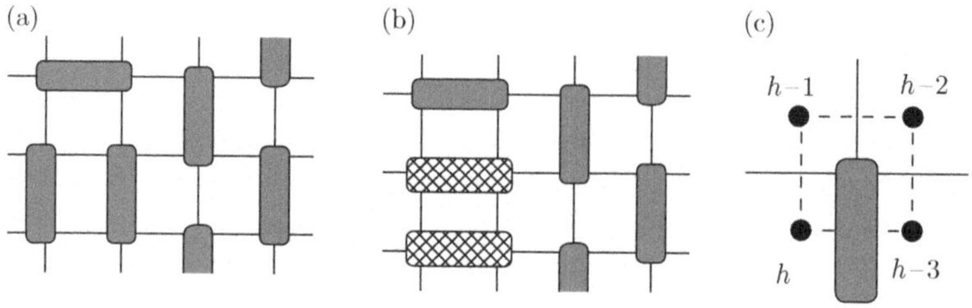

Fig. 3.7 States of square-lattice dimer model. (a) Close-packed configuration. (b) Local rearrangement. (c) Mapping to height model.

3.3.1 Introduction

The configurations of a dimer model are close-packed coverings of a lattice, with dimers arranged on bonds in such a way that there is, in the simplest case, exactly one dimer touching each lattice site. Examples are shown in Fig. 3.7. Entropy arises from rearrangements of dimers. Consider, in an initial covering, a closed loop of bonds that are alternately empty and occupied by dimers. The configuration on this loop can be flipped, exchanging empty and occupied bonds, independently of the configurations on other loops that do not intersect this one.

Close-packed dimer configurations on a planar lattice admit a height representation, provided that the lattice is bipartite. Let z be the coordination number of the lattice, and introduce the dual lattice, which has sites at the centres of the plaquettes of the original lattice and links intersecting the edges of these plaquettes. The height field is defined at sites of the dual lattice: traversing in (say) an anticlockwise direction the plaquettes of the dual lattice that enclose A-sublattice sites of the original lattice, we take the height difference to be $\Delta h = +1$ on crossing an empty bond and $\Delta h = 1 - z$ on crossing the occupied bond, as in Fig. 3.7(c). A simple generalization is to take dimer configurations in which exactly n dimers touch each site. Then $\Delta h = +n$ for empty bonds and $n - z$ for occupied ones. These choices ensure that the height field is single-valued.

There is in fact an exact correspondence between close-packed dimer configurations on the hexagonal lattice and ground states of the triangular-lattice Ising antiferromagnet. To establish this, note that the hexagonal lattice has as its dual the triangular lattice. Under the correspondence, dimers on the hexagonal lattice lie across frustrated bonds of the triangular lattice, as in Fig. 3.8. In a ground state of the Ising model, the exchange interaction on exactly one edge of every elementary triangle is frustrated, and so, in the corresponding dimer covering, every site of the hexagonal lattice is touched by exactly one dimer.

3.3.2 General formulation

While the mapping from dimer coverings to a height field is particular to two-dimensional, bipartite lattices, it can be reformulated in language that generalizes

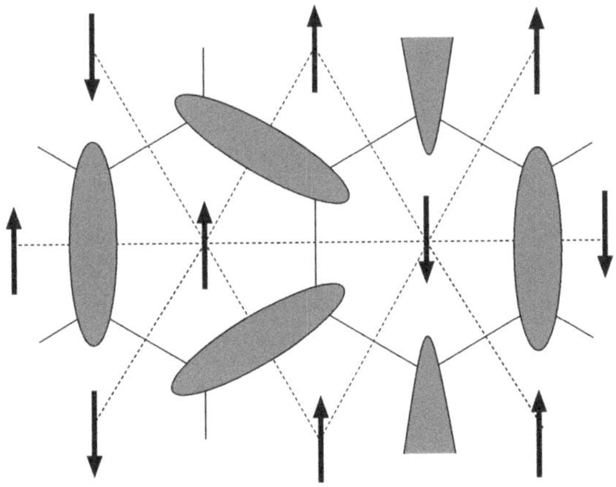

Fig. 3.8 Hexagonal-lattice dimer model and its correspondence with ground states of the triangular-lattice Ising antiferromagnet.

directly to higher dimensions [32]. To do so, we first make use of the bipartite nature of the lattice to define an orientation convention on nearest-neighbour bonds, taking the direction to be from (say) the A-sublattice to the B-sublattice. We then define for each dimer configuration a flux in this direction on each link, which (for a lattice with coordination number z) is $1 - z$ on links occupied by a dimer, and $+1$ on unoccupied links.

This flux is constrained by its construction to be divergenceless for a dimer covering that everywhere obeys the rule of exactly one dimer meeting each site. The constraint can be resolved in the usual way, by taking the flux to be the curl of a vector potential $\vec{A}(\mathbf{r})$. In the continuum, this is of course familiar for a three-dimensional system. It also applies to a two-dimensional system: one takes $\vec{A}(\mathbf{r}) = \hat{z}h(x, y)$, with $\hat{z}$ the normal to the plane of the system and $h(x, y)$ a scalar, which we will see is simply the height field.

The next step is to write a coarse-grained free energy for dimer configurations that generalizes the height model. In three dimensions as in two, coarse-grained states with high entropy arise from configurations in which there are many short loops of bonds alternately occupied and unoccupied by dimers, around which dimer occupations can be flipped. Those loops correspond roughly to closed flux lines: flux has a constant direction around each loop, although its magnitude alternates between occupied and unoccupied bonds. Configurations containing mainly small closed flux loops generate small values of coarse-grained flux, and conversely there is an entropy penalty attached to large flux. This motivates the conjecture [32]

$$\mathcal{H} = \frac{1}{2}K \int \mathrm{d}^3\mathbf{r}\, |\vec{\nabla} \times \vec{A}(\mathbf{r})|^2 . \tag{3.27}$$

as a generalization of (3.19), from which the height model is recovered in two dimensions via the substitution $\vec{A}(\mathbf{r}) = \hat{z}h(x, y)$.

Despite the equivalent forms of $\mathcal{H}$ in two and three dimensions, there are important differences. For the two-dimensional model, the value of the stiffness K determines the location of the theory on a line of fixed points and controls which possible perturbations are renormalization-group-relevant. By contrast, for the three-dimensional model there are no symmetry-allowed perturbations that are renormalization-group-relevant, and the value of K simply sets the amplitude of fluctuations.

Dimer correlations can be evaluated straightforwardly since the theory is Gaussian. We introduce a dimer number operator $\sigma_{\hat{r}}(\mathbf{r})$, which takes the value $\sigma_{\hat{r}}(\mathbf{r}) = +1$ if the bond in direction $\hat{r}$ centred at $\mathbf{r}$ is occupied by a dimer, and otherwise has the value $\sigma_{\hat{r}}(\mathbf{r}) = -1$. Defining the flux density $\vec{B}(\mathbf{r}) = \vec{\nabla} \times \vec{A}(\mathbf{r})$, the connected dimer density correlation function is

$$C_{kl}(\mathbf{r}_1, \mathbf{r}_2) \equiv \langle \sigma_{\hat{k}}(\mathbf{r}_1) \sigma_{\hat{l}}(\mathbf{r}_2) \rangle - \langle \sigma_{\hat{k}}(\mathbf{r}_1) \rangle \langle \sigma_{\hat{l}}(\mathbf{r}_2) \rangle \propto \langle B^k(\mathbf{r}_1) B^l(\mathbf{r}_2) \rangle . \tag{3.28}$$

In two dimensions, $B^i(\mathbf{r}) = \epsilon_{ij} \partial_j h(\mathbf{r})$ and we can compute $\langle B^i(\mathbf{0}) B^j(\mathbf{r}) \rangle$ by differentiating (3.23), with the result

$$\langle B^i(0) B^j(\mathbf{r}) \rangle = \frac{1}{2\pi K} \frac{2x_i x_j - \delta_{ij} r^2}{r^4} . \tag{3.29}$$

Similarly, in three dimensions, defining Fourier transforms in a system of linear size L via

$$\vec{B}_{\mathbf{q}} = L^{-3/2} \int \mathrm{d}^3\mathbf{r}\, \mathrm{e}^{\mathrm{i}\mathbf{q}\cdot\mathbf{r}} \vec{B}(\mathbf{r}), \qquad \vec{B}(\mathbf{r}) = L^{-3/2} \sum_{\mathbf{q}} \mathrm{e}^{-\mathrm{i}\mathbf{q}\cdot\mathbf{r}} \vec{B}_{\mathbf{q}} , \tag{3.30}$$

one finds

$$\langle B^i_{\mathbf{q}} B^j_{\mathbf{k}} \rangle = \frac{1}{K} \left(\delta_{ij} - \frac{q_i q_j}{q^2} \right) \delta_{\mathbf{q},-\mathbf{k}} , \tag{3.31}$$

and hence [32]

$$\langle B^i(0) B^j(\mathbf{r}) \rangle = \frac{1}{4\pi K} \frac{3r_i r_j - \delta_{ij} r^2}{r^5} . \tag{3.32}$$

We see that while dimer coverings of bipartite lattices described by this coarse-grained theory are disordered, with finite entropy density and no local symmetry breaking, the constraints of close packing and hard-core exclusion lead to power-law correlations with a characteristic form. States with these correlations are known as Coulomb phases [37].

3.3.3 Flux sectors, and $U(1)$ and $\mathbb{Z}_2$ theories

It is an important feature of dimer models on bipartite lattices that the configuration space divides into sectors that are not connected by any local dimer rearrangements. For a system with periodic boundary conditions, these sectors are distinguished by the values of total flux encircling the system in each direction. Dimer rearrangements around short loops leave these global fluxes unchanged, and the order characterized by the fluxes is referred to as topological.

This definition of global flux is illustrated in Fig. 3.9. Let $N_\uparrow$ be the number of dimers that cross the line A–B on upward-directed links and let $N_\downarrow$ be the number on downward-directed links. (In d dimensions, this line is replaced by a $(d-1)$-dimensional hypersurface.) Then the net flux across the line (or hypersurface) is $\Phi = -z(N_\uparrow - N_\downarrow)$, where the coordination number is $z = 4$ for the square lattice. The value of Φ is unchanged by flipping dimers on contractable loops (ones that do not wrap around the system), as shown in Fig. 3.9(b, c). Moreover, although microscopic values of the flux Φ are discrete, this restriction is unimportant after coarse-graining: for a three-dimensional system in the continuum limit, we can view $\vec{A}(\mathbf{r})$ as the vector potential for a continuous-valued flux density $\vec{B}(\mathbf{r}) = \vec{\nabla} \times \vec{A}(\mathbf{r})$. Then (3.27) is simply the action for a U(1) gauge theory.

Dimer models can equally be defined on lattices that are not bipartite, but the long-distance physics in these cases is very different. For these systems, it is not possible to define a local divergenceless flux, and sectors of configuration space are labelled by $\mathbb{Z}_2$ rather than U(1) quantum numbers. For a d-dimensional system, there are d of these quantum numbers, giving the parity of the number N of dimers intersecting a set of $(d-1)$-dimensional hypersurfaces. A change in the dimer configuration produced by flipping dimers on a contractable loop leaves these parities unchanged. To see this, note that the loop intersects the surface an even number of times, and consider the contribution to overall parity from two successive interactions. If the length of the loop between the crossings is even, then both intersections make the same contribution (both 0 or both 1) to N, and the combined contribution modulo 2 is unchanged when dimers on the loop are flipped; alternatively, if the length between crossings is odd, the two interactions make opposite contributions (one 0 and the other 1) to N, and their individual contributions swap when dimers are flipped.

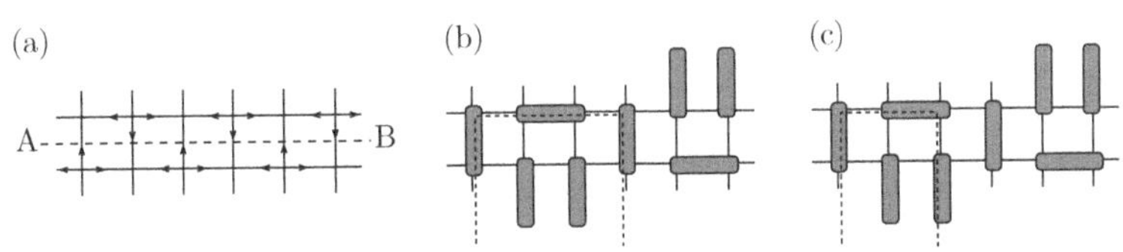

Fig. 3.9 Flux sectors in a bipartite lattice dimer model. (a) Orientation convention on edges of the lattice. (b) A dimer flip on the marked loop changes both $N_\uparrow$ and $N_\downarrow$ by 1. (c) A dimer flip on a different marked loop changes local contributions to $N_\uparrow$ by ± 1 but leaves its net value unaltered.

It is known from exact solutions for two-dimensional lattices using Pfaffians [34] and from Monte Carlo simulations in three dimensions [32] that dimer–dimer correlations generically decay exponentially with separation for close-packed dimer coverings of non-bipartite lattices.

3.3.4 Excitations

Vortices in the height representation can arise from defects of more than one type in the dimer covering. One of these is obvious from the mapping between triangular-lattice spin configurations and hexagonal-lattice dimer configurations: a triangle in which three spins have the same orientation maps to a site of the hexagonal lattice at which three dimers meet. This excitation acts as a source or sink of the flux we have introduced, depending on which sublattice the site belongs to.

An alternative type of height vortex, which is of interest for the dimer model, is one in which a dimer is removed, or, equivalently, replaced by two monomers, as in Fig. 3.10. The two monomers can be separated by subsequent dimer moves, one always remaining on the A-sublattice and the other on the B-sublattice. The two monomers are represented by a vortex–antivortex pair in the height model. From the arguments leading to (3.26), we see that such a pair will be subject to an attractive entropic potential that increases logarithmically with separation.

This entropic potential is a natural consequence of the fact that a monomer acts as a source for flux $\vec{B}(\mathbf{r})$ if it is on one sublattice, and as a sink if it is on the other sublattice, and the logarithmic dependence on distance is characteristic of the Coulomb interaction in two dimensions. Similarly, the entropic potential $V(r)$ in three dimensions between a pair of monomers on opposite sublattices at separation r can be evaluated straightforwardly within the continuum description of (3.27). The presence of the excitations results in an additional contribution to $\vec{B}(\mathbf{r})$. Let $\vec{B}_{\text{source}}$ be the field configuration that minimizes $\mathcal{H}$ in the presence of the pair, and write $\vec{B}(\mathbf{r}) = \vec{B}_{\text{source}} + \delta\vec{B}$. Since $\mathcal{H}$ is quadratic in $\vec{B}(\mathbf{r})$, integration over fluctuations $\delta\vec{B}$ yields a weight that is unaffected by the presence or separation of the pair. We can therefore determine $V(r)$ simply from $\vec{B}_{\text{source}}$, and by the usual arguments of electrostatics we find

$$V(r) = -\frac{K}{4\pi r} \,. \tag{3.33}$$

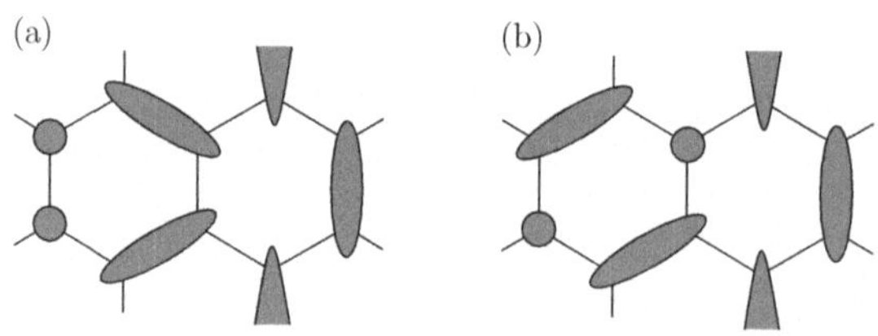

Fig. 3.10 Monomers in the hexagonal-lattice dimer model. (a) Dimer replaced with two monomers. (b) Separation of the monomers by dimer flips.

Since the entropic cost of separating the pair to infinity is bounded, monomer excitations in a Coulomb phase in three dimensions are deconfined.

In a dimer model that has long-range order, monomers are subject to a potential that grows much more rapidly with separation. For example, in two dimensions, the coarse-grained height field steps between different pinned values along a line joining the vortex–antivortex pair. This generates an interaction that is linear in separation, and the same result holds in three dimensions.

On non-bipartite lattices in both two and three dimensions, the entropic interaction potential between a pair of monomers approaches a finite limiting value exponentially fast with increasing separation, provided the dimer coverings are disordered. The deconfinement of monomers is an important property distinguishing $\mathbb{Z}_2$ from U(1) phases in two dimensions.

It is characteristic of topologically ordered systems that transitions between ground states of a system on a torus can be engineered by a sequence of steps, consisting of the generation of a pair of excitations, followed by transport of one excitation around the torus, and ending with recombination. For a dimer model, the first of these steps is the replacement of a dimer by a pair of monomers. The flux quantum number for the dimer covering is changed if one of these monomers is transported around the torus by flipping dimers, until the two monomers are again adjacent and can be replaced with a dimer.

3.4 Spin ices

The spin-ice materials $Ho_2Ti_2O_7$ and $Dy_2Ti_2O_7$ provide fascinating realizations of the Ising antiferromagnet on the pyrochlore lattice, in which both the nearest-neighbour and the long-range dipolar contributions to spin interactions make very distinctive contributions to the physical behaviour [35]. In this section, we give an overview of the resulting physics, making use of some of the general ideas developed in our discussion of Coulomb phases.

3.4.1 Materials

Isolated Ho^{3+} and Dy^{3+} ions have high angular momentum ($J = 8$ and $J = \frac{15}{2}$, respectively) and large magnetic moments ($10\mu_B$ in both cases). In spin-ice materials, the effect of the electrostatic environment of the rare-earth ions is to split the $2J + 1$ degenerate states of the free ions into crystal field levels.

Approximating the crystal field Hamiltonian by $-D(J_z)^2$, one has for positive D a ground-state doublet $M_J = \pm J$. Since excited crystal field levels are several hundred kelvin higher in energy and the scale for interactions between spins is only a few kelvin, the moments can be represented by Ising pseudospins S_i. The easy axis at given a site is the local $\langle 111 \rangle$ direction joining the centres of the two tetrahedra that share it, and so moments are directed either into or out of tetrahedra, as shown in Fig. 3.11. The centres of tetrahedra of the pyrochlore lattice lie on a diamond lattice, which is bipartite. We take the convention that $S_i = +1$ represents a spin directed out of a tetrahedron on the A-sublattice.

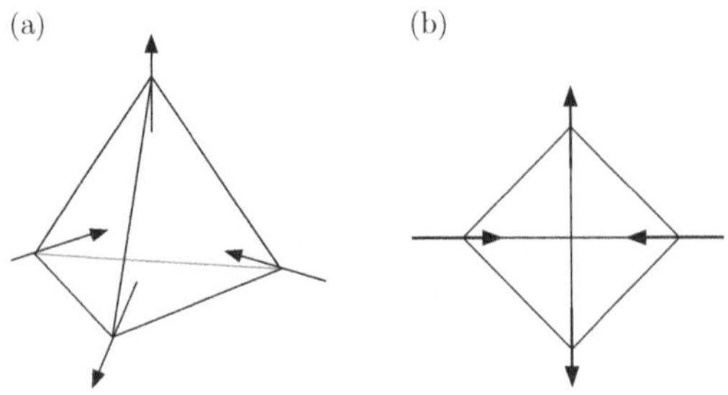

Fig. 3.11 (a) Magnetic moments at sites of a pyrochlore lattice, orientated along local $\langle 111 \rangle$ axes in a 'two-in two-out' state. In later figures, for ease of drawing, we represent the tetrahedron in the flattened way shown in (b).

Strikingly, frustration arises from the combination of this local easy-axis anisotropy with *ferromagnetic* nearest-neighbour coupling ($\theta_{\mathrm{CW}} \simeq +1.9\,\mathrm{K}$ and $+0.5\,\mathrm{K}$ for the Ho and Dy compounds, respectively), and energy is minimized for the two-in two-out states of the type illustrated in Fig. 3.11 [36]. The term *spin ice* is chosen because these spin arrangements mimic the proton positions in water ice. As the values of θ_{CW} are relatively small and the magnetic moments are large, long-range dipolar interactions are important in addition to the nearest-neighbour coupling; we discuss their consequences in Section 3.4.3, but first examine the physics of the nearest-neighbour model.

3.4.2 Coulomb phase correlations

We would like to develop a description of ground states of the nearest-neighbour model for spin ice that is analogous to the height representation for the triangular-lattice Ising antiferromagnet and amenable to coarse-graining. The approach [25, 37] parallels the one introduced for three-dimensional dimer models in Section 3.3

In order to describe in a general way the ideas that are involved, it is useful to introduce some terminology. For a given system of corner-sharing frustrated clusters, we will be concerned with two types of lattice. One is simply the magnetic lattice on which the moments reside, also known as the medial lattice, and we denote this by $\mathcal{L}$. The other is the cluster (or simplex) lattice, also known as the pre-medial or parent lattice, which we denote by $\mathcal{B}$. The sites of $\mathcal{L}$ lie at the midpoints of the links of $\mathcal{B}$, and, in the notation of graph theory, $\mathcal{L}$ is the line graph associated with the graph $\mathcal{B}$. For spin ices, $\mathcal{L}$ is the pyrochlore lattice and $\mathcal{B}$ is the diamond lattice. Alternatively, if we take $\mathcal{L}$ to be the kagome lattice, then $\mathcal{B}$ is the hexagonal lattice.

A key requirement in the following is that $\mathcal{B}$ should be a bipartite lattice. We can then orient the links of $\mathcal{B}$, say from sites of sublattice A to sites of sublattice B. Let $\hat{e}_i$ be the unit vector in this direction on link i, and note that i also labels a site of $\mathcal{L}$.

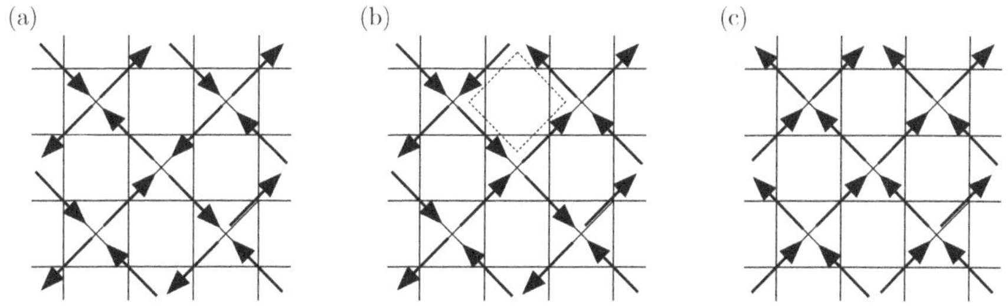

Fig. 3.12 Ground-state configurations on the two-dimensional pyrochlore lattice. (a) A state containing short flippable loops of spins. (b) The state obtained from this by flipping four spins on the marked loop. (c) A state with no flippable spins and maximal flux $\vec{B}$.

The central idea is to introduce a vector field $\vec{B}$, defined on the links of $\mathcal{B}$, that is a representation of a configuration of Ising spins $\{S_i\}$ on $\mathcal{L}$ and is given by the relation

$$\vec{B}_i = S_i \hat{e}_i \,. \tag{3.34}$$

The field $\vec{B}$ is a useful construction because the ground-state condition for a spin ice—that two spins are directed into each tetrahedron and two are directed out—translates into the condition that $\vec{B}$ has zero lattice divergence at each node of $\mathcal{B}$. The field $\vec{B}$ is therefore an emergent gauge field.

The next step is to conjecture a probability distribution for a coarse-grained version of $\vec{B}$. Some ground states contain short loops of flippable spins: closed loops on the $\mathcal{B}$ lattice, around which all spins are directed in the same sense. Further ground states with a similar coarse-grained $\vec{B}$ are obtained by reversing all the spins on one of these loops, and so entropy favours states with a high density of short loops, as illustrated in Fig. 3.12 using a two-dimensional version of the pyrochlore lattice. These considerations suggest that states in which $\vec{B}$ is large have lower entropy. This motivates for the probability distribution the form

$$P[\vec{B}(\mathbf{r})] = \mathcal{Z}^{-1} \mathrm{e}^{-\mathcal{H}} \,, \tag{3.35}$$

with

$$\mathcal{H} = \frac{1}{2} K \int \mathrm{d}^3\mathbf{r}\, |\vec{B}(\mathbf{r})|^2 \,, \tag{3.36}$$

where $\vec{B}(\mathbf{r})$ is a continuum field, subject to the ground-state constraint $\vec{\nabla} \cdot \vec{B} = 0$, just as in our earlier discussion of Coulomb phases in dimer models. Indeed, spin-ice ground states can be represented directly by dimer coverings on the diamond lattice with two dimers touching every site, simply by using dimers on $\mathcal{B}$ to represent spins $S_i = +1$ on $\mathcal{L}$. Ground-state spin correlations in a spin ice therefore have the dipolar form given in (3.32).

The angular dependence of this correlator means that a pair of well-separated spins on sites of the same sublattice of $\mathcal{L}$ (and hence with the same orientation for $\hat{e}$

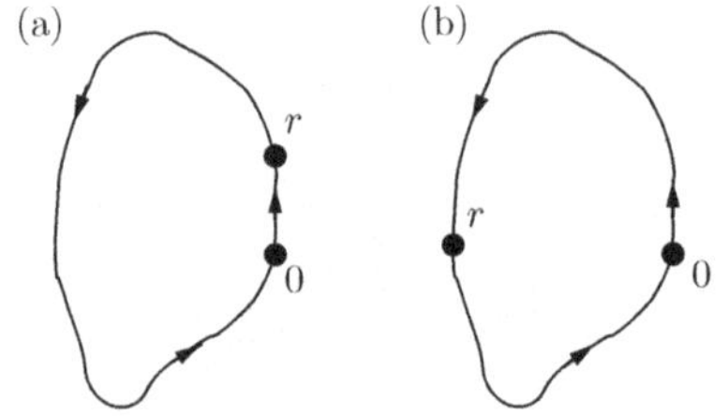

Fig. 3.13 Illustration of the origin of dipolar correlations in a spin ice. Spins at 0 and $\mathbf{r}$ are correlated if a flux line of the emergent field $\vec{B}(\mathbf{r})$ passes through both sites. The orientations of the flux line at the two sites depends on the direction of their spatial separation, so that for (a) $\langle S_0 S_\mathrm{r}\rangle > 0$ and for (b) $\langle S_0 S_\mathrm{r}\rangle < 0$.

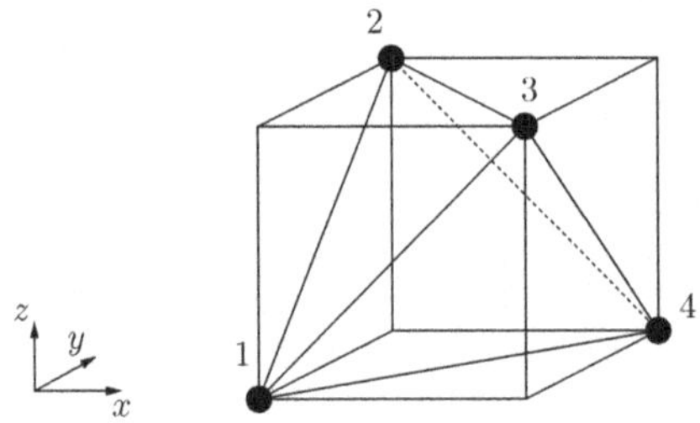

Fig. 3.14 Choice of axes and sublattice labels for the pyrochlore lattice.

at both sites) are positively correlated if their separation vector $\mathbf{r}$ is in the direction of $\hat{e}$, but are most likely to be anti-aligned if $\mathbf{r}$ is perpendicular to $\hat{e}$. Such behaviour can be understood by considering the geometry of flux lines of the emergent gauge field $\vec{B}(\mathbf{r})$. A spin configuration in which $\sigma_0 = +1$ is one with a flux line passing through the origin in the direction $\hat{e}$, and this flux line must close on itself since the field is divergenceless. The spin $\sigma_\mathbf{r}$ is correlated with σ_0 only in those configurations in which the same flux line passes through $\mathbf{r}$, and the most likely orientation of this flux line at $\mathbf{r}$ depends on the relative directions of $\mathbf{r}$ and $\hat{e}$, as shown in Fig. 3.13. The reciprocal space signature, (3.31), of these correlations consists of so-called pinch-point structures, sharp but without divergences, observed in elastic neutron diffraction [10].

It is interesting to connect the results obtained from a continuum treatment of the emergent gauge field to those arising from the self-consistent Gaussian approximation of Section 3.2.1. Adopting the sublattice labels and axis orientations shown in Fig. 3.14, the net Ising moment M and flux components B_x, B_y, B_z arising from a spin configuration S_1, S_2, S_3, S_4 on the four sublattices are [38]

$$\begin{pmatrix} M \\ B_x \\ B_y \\ B_z \end{pmatrix} = \frac{1}{2}\begin{pmatrix} 1 & 1 & 1 & 1 \\ 1 & 1 & -1 & -1 \\ 1 & -1 & 1 & -1 \\ 1 & -1 & -1 & 1 \end{pmatrix}\begin{pmatrix} S_1 \\ S_2 \\ S_3 \\ S_4 \end{pmatrix}. \tag{3.37}$$

The statistical weight of fluctuations is determined within the self-consistent Gaussian approximation by their energy and by a Lagrange multiplier λ. From (3.1), the energy per unit cell of a configuration is $\frac{1}{2}JM^2$. Moreover, the net Ising moment M is simply the lattice version of $\vec{\nabla} \cdot \vec{B}$. In continuum notation, (3.15) therefore amounts to a Boltzmann factor $\mathrm{e}^{-\mathcal{H}}$ with

$$\mathcal{H} = \int \mathrm{d}^3\mathbf{r} \left\{ \frac{1}{2}\lambda|\vec{B}(\mathbf{r})|^2 + \frac{1}{2}\beta J M^2 \right\} . \tag{3.38}$$

In this way, we see that λ sets the value of the stiffness K for fluctuations of $\vec{B}(\mathbf{r})$. We also recover the condition $\vec{\nabla} \cdot \vec{B}(\mathbf{r}) = 0$ in the low-temperature limit $\beta J \to \infty$.

3.4.3 Monopoles

Some beautiful physics becomes apparent when we examine spin-ice configurations that do not obey the two-in two-out rule for ground states of the model with nearest-neighbour interactions [39]. Consider the configuration obtained from a ground state by reversing a single spin. As illustrated in Fig. 3.15, two separate elementary excitations are obtained from it through further spin reversals. Because, like vortex excitations in the triangular-lattice Ising antiferromagnet, these elementary excitations are not produced singly by local spin flips, they are said to be fractionalized. Moreover, since one member of the pair is a source for the emergent gauge flux, and the other a sink, they form a monopole–antimonopole pair.

The energy of a monopole–antimonopole pair arising from exchange interactions is independent of their separation in a nearest-neighbour model. There is, however, an entropic interaction between the pair, since the number of ground states available to the background spins depends on the separation r of the excitations. The entropic potential $V(r)$, as for monomers in the Coulomb phase of a three-dimensional dimer model on a bipartite lattice, is given by (3.33). Likewise, since the entropic cost of separating the pair to infinity is bounded, monopole excitations are deconfined.

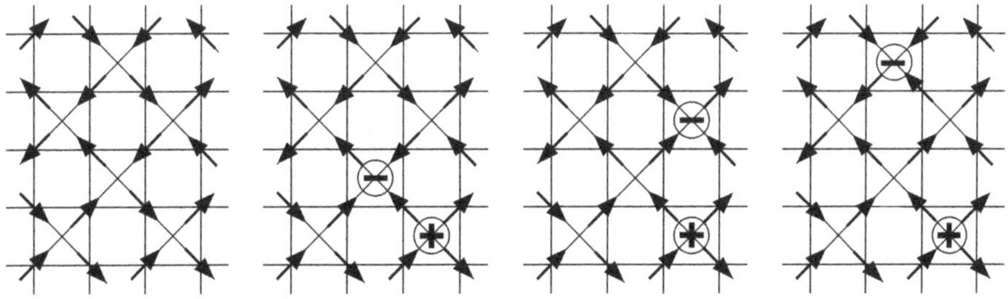

Fig. 3.15 Generation of a monopole–antimonopole pair from the ground state of a two-dimensional spin ice, and their separation, by successive spin flips.

3.4.4 Dipolar interactions

Our treatment of spin ices to this point has omitted the long-range part of dipolar interactions. Clearly, its inclusion will lift the high degeneracy of ground states of the nearest-neighbour model, and one might expect that it would simply set an unwelcome limit on the physics we have discussed so far. Rather than being just a bug, however, dipolar interactions turn out to add a spectacular feature to spin-ice physics [39].

A very convenient framework for thinking about dipolar effects is provided by an approximation known as the dumbbell model (Fig. 3.16). Here, in the first instance, magnetic dipoles μ of atomic size are replaced by 'dumbbells' of positive and negative magnetic charge $\frac{1}{2}Q$ at a separation a equal to the distance between the centres of adjacent tetrahedra of the lattice. Setting the dipole moment of the dumbbell equal to the microscopic moment, the leading contribution to long-range interactions is captured exactly. The power of this description stems from the fact that for all ground states of the nearest-neighbour model, the positive and negative magnetic charges at the centre of each tetrahedron cancel. In turn, this fact is a demonstration that the leading contribution to the energy from dipolar interactions is the same for all these states. Subleading terms follow from the multipole expansion and fall off with distance as r^{-5}. The estimated ordering temperature of spin-ice materials, $T_c \lesssim 0.2\,\theta_{\mathrm{CW}}$, is rather low for that reason [40]. Such ordering is not observed under ordinary experimental conditions because spin dynamics is very slow at low temperature.

Turning to monopole excitations, the dumbbell model serves to expose a striking consequence of dipolar couplings, since dumbbell charges fail to cancel in tetrahedra that contain these quasiparticles. As a result, a well-separated monopole–antimonopole pair is subject to a Coulomb interaction

$$U(r) = -\frac{\mu_0 Q^2}{4\pi r} \tag{3.39}$$

of *magnetic* origin. The charge Q is related to the atomic dipole moment and the lattice spacing by $Q = 2\mu/a$, and the monopole chemical potential is fixed by the nearest-neighbour contributions to spin interactions. Note that while the entropic and dipolar contributions to monopole interactions, (3.33) and (3.39), have the same dependence on separation, the entropic one makes a temperature-independent contribution to Boltzmann weights and so the dipolar one is dominant at low temperature. The magnetic Coulomb interaction between monopoles is a remarkable example of an emergent longer-range interaction ($1/r$) arising from shorter-range ($1/r^3$) microscopic interactions. It appears because of the interplay between these microscopic interactions and the correlations of the Coulomb phase, and it stands in contrast to the familiar situation (for example, in a plasma) in which correlations serve to screen long-range microscopic interactions, leaving only a short-range effective potential.

A second, and much more conventional, consequence of atomic dipole moments is that spins couple to an external magnetic field. To appreciate the form of this coupling, recall that moments are aligned along local crystal field axes, and that these are differently orientated on each of the four sublattices, as shown in Fig. 3.11. The

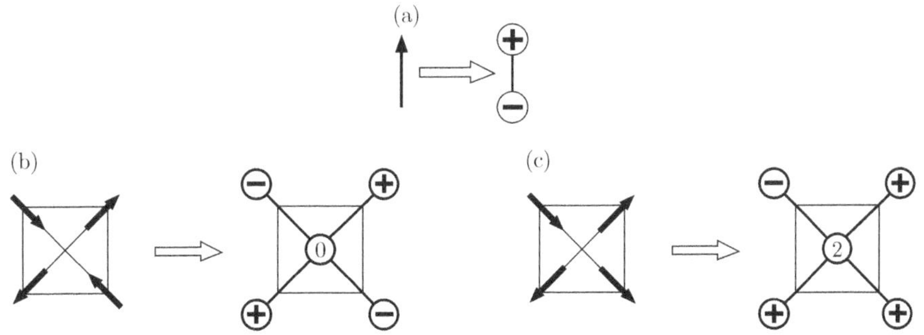

Fig. 3.16 The dumbbell approximation. Dipoles are replaced with a dumbbell of opposite charges at finite separation, as shown in (a). If this separation is chosen to be the distance between tetrahedron centres, then two-in two-out tetrahedra are charge-neutral (b), but one-in three-out tetrahedra have net charge (c).

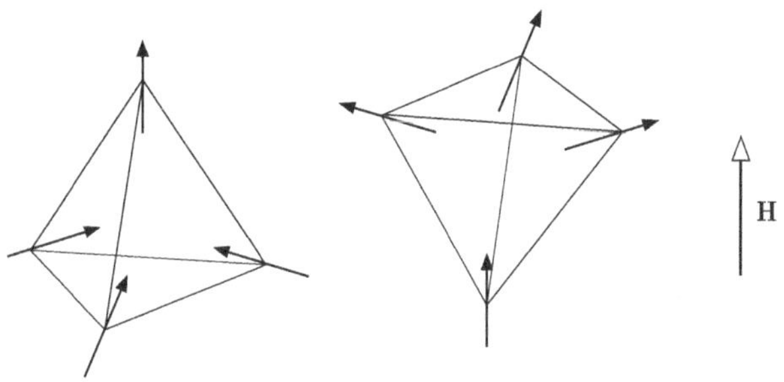

Fig. 3.17 Spin configurations favoured by a [111] field: magnetic charges $\pm Q$ are induced in tetrahedra on opposite sublattices.

Zeeman contribution to the Hamiltonian in the presence of a field $\vec{H}$ is therefore

$$\mathcal{H}_{\rm Z} = -\mu\mu_0 \sum_i (\vec{H} \cdot \hat{e}_i) S_i \,, \tag{3.40}$$

and so the strength of coupling to spins on different sublattices depends on the orientation of $\vec{H}$ relative to the crystal axes.

This sublattice-dependent Zeeman coupling can be exploited to control monopole density in a way that provides rather direct evidence for magnetic Coulomb interactions. Specifically, a field directed along the [111] axis acts as a staggered chemical potential for monopoles, favouring monopoles of charge Q in the tetrahedra on one sublattice of $\mathcal{B}$ and charge $-Q$ on the other sublattice, as indicated in Fig. 3.17. In this way, by varying field strength, one can drive a transition between low- and high-monopole-density phases. Experimentally, this is observed to be of first order. Theory for a charged system, and simulations including magnetic dipolar interactions, reproduce this first-order transition, in contrast to theory and simulations for models with only nearest-neighbour interactions, where the transition is continuous [39].

3.5 Quantum spin liquids

We now turn to the quantum physics of frustrated magnets, where frustration is interesting particularly because it provides a mechanism that suppresses Néel order and promotes alternative, quantum-disordered phases.

3.5.1 Introduction

To understand that destruction of conventional order is likely in frustrated magnets, we can examine the reduction of ordered moments in the Néel state by zero-point fluctuations. Within the framework of harmonic spin-wave theory, we start from a classical ground state for a spin model and choose axes at each site that have $\hat{z}$ oriented along the local ground-state spin direction. Using the Holstein–Primakoff transformation, spin operators are expressed in terms of bosonic ones via the relation $S^z_{\mathbf{r}} = S - a^\dagger_{\mathbf{r}} a_{\mathbf{r}}$. Fluctuations lower the ground-state moment $\langle S^z_{\mathbf{r}} \rangle = S - \Delta S$ by an amount $\Delta S \equiv \langle a^\dagger_{\mathbf{r}} a_{\mathbf{r}} \rangle$ that can be expressed as an average over contributions from each mode. The schematic form (the details depend on the system) in terms of spin-wave frequencies $\omega_{\mathbf{k}}$ and exchange interaction J is [41]

$$\Delta S \sim \frac{J}{\Omega} \int_{\mathrm{BZ}} \frac{\mathrm{d}^d \mathbf{k}}{\hbar \omega_{\mathbf{k}}}, \tag{3.41}$$

where Ω is the Brillouin zone (BZ) volume. As we have seen in Section 3.1.2, frustration promotes macroscopic classical ground-state degeneracy and branches of soft modes. Here we find that these modes make divergent contributions to ΔS, destabilizing long-range order.

Is the resulting state a quantum spin liquid? A necessary requirement is that *(i) the ground state leaves all symmetries of the Hamiltonian unbroken*, and the absence of Néel order is one aspect of this. To appreciate that we should demand more, consider as an example a bilayer, square-lattice spin-$\frac{1}{2}$ Heisenberg antiferromagnet, having nearest-neighbour exchange J within layers and J' between layers. This model has two phases: the ground state is Néel-ordered for $J' \ll J$, but consists of interlayer singlets for $J' \gg J$. Although the large J'/J state breaks no symmetries, it is 'ordinary' rather than 'exotic', in the sense that it is continuously connected to a band insulator. That is to say, there is a path in the space of Hamiltonians that connects this phase of the spin model to a tight-binding model without interactions that has one filled band (symmetric under layer interchange) and one empty band (antisymmetric under interchange).

To exclude such ordinary possibilities, we require in addition that a quantum spin liquid *(ii) has half-odd-integer spin per unit cell.* The combination of (i) and (ii) together implies for a large class of models that a system with a gapped ground state has topological order, as we now discuss.

3.5.2 Lieb–Schultz–Mattis theorem

Some strong constraints on the nature of ground states and excitations in spin models that have half-odd-integer spin per site and (at least) U(1) symmetry are revealed by the Lieb–Schultz–Mattis theorem [42]. This was originally proved for one-dimensional models, but has subsequently been applied to quasi-one-dimensional and higher-dimensional systems [43, 44]. The theorem shows for a chain of length L that the energy gap between the ground and first excited states vanishes as $L \to \infty$. We know of three distinct ways in which this can happen. Two of these are conventional. First, if the model spontaneously breaks a symmetry, the ground state belongs to a low-lying multiplet, and splittings within the multiplet vanish as L diverges. An example is a state in which the size of the unit cell is doubled spontaneously by spin dimerization. Second, if the model has a branch of gapless excitations, the lowest excitation energy decreases as a power of system size. This is the case for spinon excitations in the spin-$\frac{1}{2}$ Heisenberg chain. The third, unconventional, possibility concerns systems that do not show symmetry breaking, and in which excitations that can be created by local operators are gapped. For these, the implication of the theorem is that the ground state belongs to a low-lying multiplet that does not originate from symmetry breaking. Instead, it has its origin in topological order.

We will sketch the proof as it applies to the spin-$\frac{1}{2}$ XXZ chain, and then discuss more general implications. Consider the Hamiltonian

$$\mathcal{H} = J\sum_n \left[\frac{1}{2}(S_n^+ S_{n+1}^- + S_n^- S_{n+1}^+) + \Delta S_n^z S_{n+1}^z\right] \tag{3.42}$$

for a chain of L sites with periodic boundary conditions and L even. Suppose the ground state $|0\rangle$ is unique (if it is degenerate, there is nothing to prove, since the gap is zero) and denote its energy by E_0. We construct a second state $|\psi\rangle = U|0\rangle$ from it by acting with an operator

$$U = \exp\left(2\pi \mathrm{i} \sum_{n=1}^{L} \frac{n}{L} S_n^z\right) \tag{3.43}$$

that generates a long-wavelength twist of spin configurations about the z axis. We will show that $\langle\psi|0\rangle = 0$, and will use $\langle\psi|(\mathcal{H} - E_0)|\psi\rangle$ to obtain a variational bound on the separation in energy between the ground and first excited states of $\mathcal{H}$.

To show orthogonality of $|0\rangle$ and $|\psi\rangle$, consider the effect of translations on $|0\rangle$ and on $|\psi\rangle$. Let the operator T effect translation by one lattice spacing. We have $\mathrm{T}\,|0\rangle = |0\rangle$ since the ground state is unique. On the other hand,

$$\mathrm{T}\,U\,\mathrm{T}^{-1} = U \mathrm{e}^{-(2\pi\mathrm{i}/L)\sum_n S_n^z} \mathrm{e}^{2\pi \mathrm{i} S_1^z} . \tag{3.44}$$

The factor $\mathrm{e}^{-(2\pi\mathrm{i}/L)\sum_n S_n^z} = +1$ if $\sum_n S_n^z = 0$ (and there is ground-state degeneracy if $\sum_n S_n^z \neq 0$), but for half-odd-integer spins the factor $\mathrm{e}^{2\pi\mathrm{i}S_1^z} = -1$. Hence, $\mathrm{T}|\psi\rangle = -|\psi\rangle$, and therefore $\langle\psi|0\rangle = 0$.

In order to evaluate $\langle\psi|(\mathcal{H} - E_0)|\psi\rangle$, we first examine how U transforms a single spin operator. We have

$$\mathrm{e}^{-\mathrm{i}\theta S^z} S^+ \mathrm{e}^{\mathrm{i}\theta S^z} = \mathrm{e}^{-\mathrm{i}\theta} S^+, \quad \text{and so} \quad U^\dagger S_n^+ S_{n+1}^- U = \mathrm{e}^{2\pi\mathrm{i}/L} S_n^+ S_{n+1}^- \, . \tag{3.45}$$

From this, we find

$$\langle\psi|(\mathcal{H}-E_0)|\psi\rangle = -\frac{1}{2}J[1-\cos(2\pi/L)]\sum_n\langle S_n^+S_{n+1}^- + S_n^-S_{n+1}^+\rangle\,. \qquad (3.46)$$

Since the factor $[1-\cos(2\pi/L)]$ decreases as $1/L^2$, while $\sum_n |\langle S_n^+S_{n+1}^- + S_n^-S_{n+1}^+\rangle| \leq 2L$, the energy gap separating E_0 from the next eigenstate vanishes at least as fast as $1/L$, which is the result we seek.

These ideas can most simply be extended to higher dimensions by considering a system on a strip, with the width M chosen to be an odd integer so that the spin per unit cell of the strip remains half an odd integer [43]. The bound on the energy gap implied by (3.46) is then $\mathcal{O}(M/L)$, which again vanishes provided that we take the thermodynamic limit in an anisotropic fashion, a restriction that is not required in a more sophisticated approach [44].

The possibility of asymptotic degeneracy without symmetry breaking or gapless excitations is very striking. One route to understanding how it can arise is provided by quantum dimer models.

3.5.3 Quantum dimer models

3.5.3.1 RVB picture

In order to discuss spin-liquid states, we need a suitable language. It should provide an alternative to the picture we have of Néel order, which starts from a product wavefunction based on the classical ground state. Anderson's resonating valence bond (RVB) state [1, 45] offers this language: we describe the spin-liquid wavefunction using a basis of short-range singlets. In one such basis state, each spin is paired with another nearby spin to form a singlet, with different basis states arising from different pairings. This idea is depicted in Fig. 3.18.

Efforts to develop this picture directly face many difficult issues and questions [33]. Different basis states are not orthogonal, and it is not immediately apparent whether the basis is complete in the space of total singlets. Equally, one might ask what the prescription should be for choosing expansion coefficients, and how the Néel state can be written in this basis.

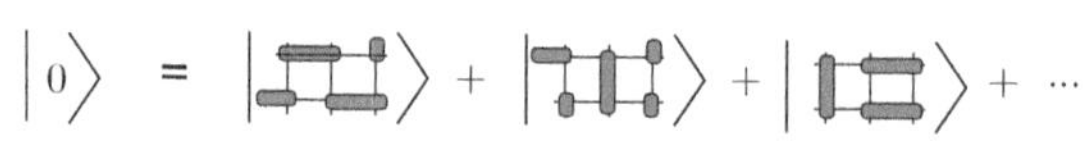

Fig. 3.18 Schematic illustration of the RVB state as a superposition of short-range singlets.

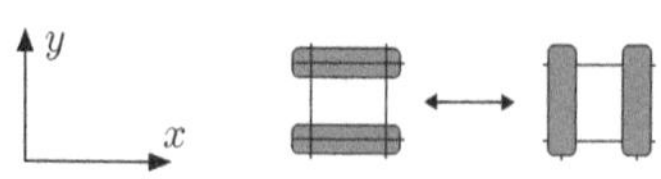

Fig. 3.19 Resonance for quantum dimer model on a square lattice.

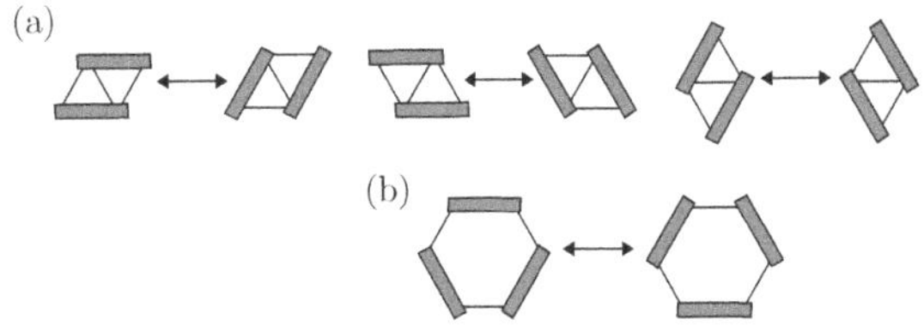

Fig. 3.20 Resonances for quantum dimer models on (a) triangular and (b) honeycomb lattices.

3.5.3.2 Quantum dimer models

Quantum dimer models [33] short-circuit many of these problems by defining a quantum-mechanical problem on a Hilbert space that has the correlations of a classical Coulomb phase built in from the start. The key idea is simply to define an orthonormal basis set $\{|\mathcal{C}\rangle\}$ to be the close-packed dimer coverings $\mathcal{C}$ of a given lattice, so that an arbitrary state in this space has the form $|\psi\rangle = \sum_{\mathcal{C}} A_{\mathcal{C}}|\mathcal{C}\rangle$. The other ingredient is a choice of Hamiltonian. In general, it will have 'potential' terms, which are diagonal in the dimer covering basis, and 'kinetic' terms, which are off-diagonal. The form proposed for the square lattice by Rokhsar and Kivelson [33] is

$$\mathcal{H} = \sum \{-t[\,|{=}\rangle\langle{|\!|}|+|{|\!|}\rangle\langle{=}|\,] + v[\,|{|\!|}\rangle\langle{|\!|}| + |{=}\rangle\langle{=}|\,]\}\ . \tag{3.47}$$

The notation used is intuitive although compact. Unpacking it: the sum runs over all elementary plaquettes of the lattice and the symbol $|{=}\rangle\langle{=}|$ denotes a projection operator onto states that have a horizontal pair of dimers in this plaquette. Similarly, the symbol $|{=}\rangle\langle{|\!|}|$ represents an operator that converts a horizontal pair of dimers in this plaquette to a vertical pair, and yields zero otherwise. It therefore produces the dimer resonances shown in Fig. 3.19.

Extensions to different lattices are straightforward. For example, on the triangular lattice, one allows resonances of pairs of dimers on four-site plaquettes of three types, as shown in Fig. 3.20(a), while on the honeycomb lattice, three-dimer resonances are required, as shown in Fig. 3.20(b). In general, one wants to include in the kinetic energy a set of resonances that is sufficient to connect all dimer configurations within a given U(1) or $\mathbb{Z}_2$ sector, but is otherwise as local as possible.

The nature of the ground state of the quantum dimer model Hamiltonian (3.47) depends on the values of the parameters v and t. We will discuss only $t > 0$, so that the kinetic energy favours a nodeless wavefunction. A special role is played by the Rokhsar–Kivelson (RK) point in parameter space, $v = t$, because here the ground-state wavefunction is given exactly by an equal-amplitude superposition $|\mathcal{G}\rangle$ of all dimer coverings within a given sector. To see this, note that the Hamiltonian at the RK point has the form

$$\mathcal{H}_{\rm RK} = t\sum (|{|\!|}\rangle - |{=}\rangle)(\langle{|\!|}| - \langle{=}|)\,, \tag{3.48}$$

which is a sum of projectors with a positive coefficient. Its eigenvalues are therefore non-negative. Moreover, $|\mathcal{G}\rangle$ is annihilated by the projection operators, and so is an eigenstate with energy zero. It is unique by the Perron–Frobenius theorem, provided that the kinetic term is ergodic within the sector.

3.5.3.3 Correlations

Knowledge of the ground-state wavefunction enables us to evaluate equal-time correlation functions. In particular, the ground-state expectation value of an observable that is diagonal in the dimer basis is given by an average over dimer coverings. From this, we can deduce at once that dimer correlations at the RK point of a quantum dimer model take the power-law form characteristic of a Coulomb phase (see (3.29) and (3.32) for two and three dimensions, respectively) on bipartite lattices, and are exponentially decaying on non-bipartite lattices.

We can also consider the quantum dimer model in the presence of a pair of static monomers. At the RK point, equal-amplitude superpositions of dimer coverings continue to define zero-energy eigenfunctions, and so the ground-state energy is independent of the separation between monomers. From this, one concludes that monomers are deconfined in the ground state. Consider for comparison the behaviour at infinite temperature. In this limit (as discussed in Section 3.3), there is an entropic contribution to the monomer–monomer potential. It is weakly (logarithmically) divergent at large separations in two dimensions on bipartite lattices, leading to confinement in this case. In other cases, monomers are deconfined: the potential has the Coulomb form on three-dimensional bipartite lattices, and approaches its limiting value exponentially fast with separation on non-bipartite lattices in both two and three dimensions.

3.5.3.4 Phase diagram

Moving away from the solvable RK point, we would like to understand the ground-state phase diagrams of quantum dimer models on various lattices as a function of the dimensionless coupling v/t [33, 46, 47]. The behaviour in some regimes and limits is clear from simple arguments.

First, for $v > t$, we can write $\mathcal{H}$ in terms of the Hamiltonian $\mathcal{H}_{\text{RK}}$ at the RK point and a non-negative remainder, as

$$\mathcal{H} = \mathcal{H}_{\text{RK}} + (v - t) \sum \left(|{\parallel}\rangle\langle{\parallel}| + |{=}\rangle\langle{=}| \right). \tag{3.49}$$

The so-called staggered state shown in Fig. 3.21(a) is annihilated by both terms in this Hamiltonian, and hence is the ground state for all $v > t$. In terms of the description of Coulomb phases on bipartite lattices, this is the state with maximal flux. It also has analogues on non-bipartite lattices; see Fig. 3.21(d).

In the opposite limit, $v \to -\infty$, the ground state is a columnar state, maximizing the number of flippable plaquettes, as illustrated in Fig. 3.21(b). Further possibilities at intermediate values of v/t include plaquette states, shown schematically for the square lattice in Fig. 3.21(c): in these states, dimers resonate independently on different plaquettes between horizontal and vertical pairs.

Note that all of the states shown in Fig. 3.21 break spatial symmetries and are degenerate for that reason. By contrast, the ground state at the RK point leaves spatial symmetries intact; its degeneracy arises topologically, from the existence of different sectors, labelled by U(1) or $\mathbb{Z}_2$ quantum numbers according to the lattice type. The fact that they support a set of distinct ground states labelled by topological fluxes is a crucial feature of quantum dimer models, inherited from their classical counterparts. It illustrates how the degeneracy demanded by the Lieb–Schultz–Mattis theorem can arise without local symmetry breaking.

A full determination of the phase diagram requires Monte Carlo calculations. Here, quantum dimer models have a tremendous advantage over most spin Hamiltonians for frustrated systems, because they avoid the so-called sign problem and can be simulated efficiently using worm algorithms. The resulting phase diagrams are shown schematically in Fig. 3.22.

Some aspects of the phase diagram are generic, but others depend on spatial dimension and on whether or not the lattice is bipartite. Properties precisely at the RK point are known in all cases by reference to the corresponding classical dimer problem: dimers are disordered, with correlations that decay exponentially on non-bipartite lattices and as a power law on bipartite lattices. Moreover, the RK point lies at a phase boundary, since the staggered state is the ground state for all $v > t$. Crucial differences in behaviour appear on the other side of the RK point ($v < t$). A simple first step to rationalizing these differences is to use classical, high-temperature properties as a basis for guessing the nature of the quantum ground state. Specifically,

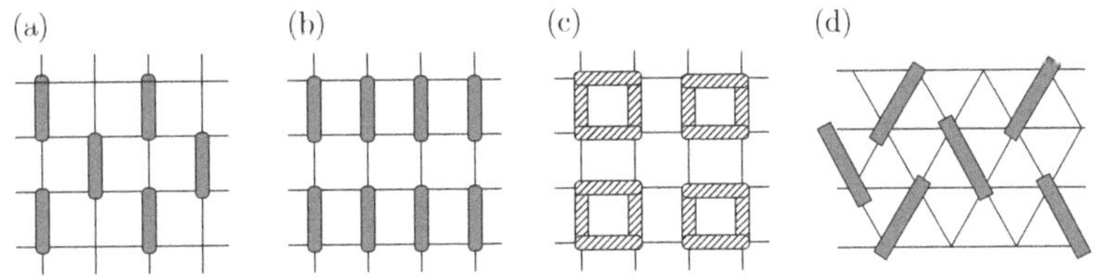

Fig. 3.21 Some possible ordered phases for quantum dimer models: (a) staggered, (b) columnar and (c) plaquette states on the square lattice; (d) staggered state on the triangular lattice.

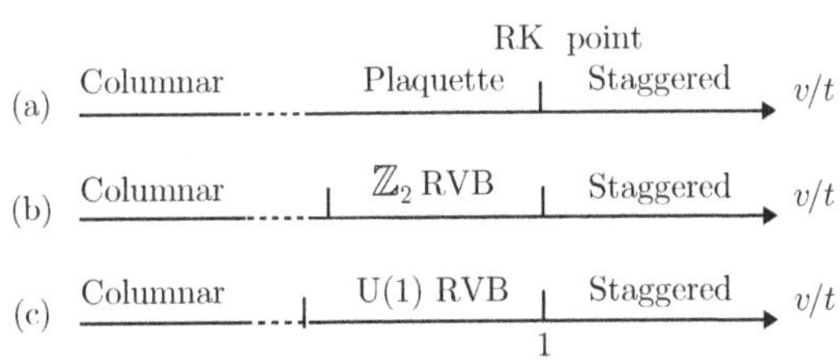

Fig. 3.22 Schematic phase diagrams for the quantum dimer model on (a) square, (b) triangular, and (c) diamond lattices.

we know that the high-temperature, entropic interaction between monomers yields confinement on bipartite lattices in two dimensions, but not in three dimensions or on non-bipartite lattices. Correspondingly, the ground state of quantum dimer models is generically ordered on bipartite lattices in two dimensions. Deconfinement of monomers at the RK point on the square lattice must therefore be seen as a special feature of the transition between one ordered phase (the plaquette phase for $v < t$) and another (the staggered phase for $v > t$). In contrast, deconfinement of monomers at the RK point on the triangular or diamond lattices is a reflection of behaviour throughout a dimer liquid phase that extends from the RK point to smaller values of v/t. As we have seen in Section 3.3, topological order in this liquid phase is characterized respectively by $\mathbb{Z}_2$ and U(1) quantum numbers, which distinguish different sectors of the dimer configuration space, and therefore different quantum ground states.

A summary of the main features of the phase diagrams for quantum dimers models shown in Fig. 3.22 is that deconfined phases are found: (i) in both two and three spatial dimensions for $\mathbb{Z}_2$ models, which arise on non-bipartite lattices, and (ii) only in three spatial dimensions in U(1) models, which arise on bipartite lattices [46, 47]. These are general properties of lattice gauge theories: whereas $\mathbb{Z}_2$ theories support confined and deconfined phases in $2+1$ and $3+1$ dimensions, compact U(1) theories are known always to be confining in $2+1$ dimensions, but to have both types of phase in $3+1$ dimensions [48].

3.5.3.5 Excitations

Three types of excitation are important in quantum dimer models. Of these, monomers (mentioned already) involve a relaxation of the dimer covering constraint, while the others—visons and emergent photons—are excited states of complete dimer coverings, which are respectively point-like and wave-like.

The energy cost of introducing monomers is a free parameter of the quantum dimer model, not fixed by the parameters v and t. It is natural, however, to regard them as gapped excitations, arising in pairs from breaking dimers. Viewing the dimer as a spin singlet, the monomer carries spin $\frac{1}{2}$ and so is also referred to as a spinon. An isolated monomer in a dimer covering of a bipartite lattice is either a source or sink of U(1) flux, depending on which sublattice it occupies, and in that sense is a monopole. As we have discussed in Section 3.3.4, transitions between ground states in different topological sectors can be produced by generation of a pair of quasiparticles, followed by transport of one quasiparticle around the torus, and ending with recombination. The relevant quasiparticles here are monomer excitations.

Variational wavefunctions offer a language in which to discuss excitations of $\mathcal{H}_{\rm RK}$ within the space of close-packed dimer coverings. At the RK point, we know that the ground-state wavefunction has equal amplitude for all coverings within a given sector. An excited-state wavefunction must have a phase that varies with dimer covering or it would not be orthogonal to this ground state.

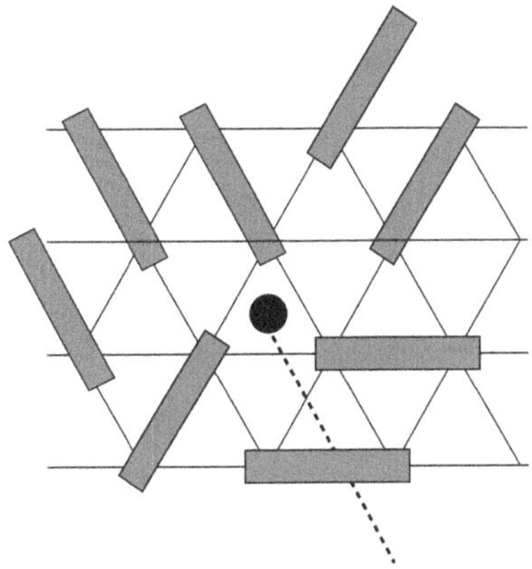

Fig. 3.23 Vison excitation.

The vison [49] is a vortex excitation of a two-dimensional system. It can be represented at the RK point by a variational wavefunction of the form

$$|\psi_{\text{vison}}\rangle = \sum_{\mathcal{C}} (-1)^{n_{\mathcal{C}}} |\mathcal{C}\rangle \,. \tag{3.50}$$

Here $n_{\mathcal{C}}$ is the number of dimers in the configuration $\mathcal{C}$ that cross a line on the dual lattice that extends from the centre of the plaquette on which the excitation is based to the system boundary (or, for a system on a torus, to the centre of another vison), as illustrated in Fig. 3.23. In order for a wavefunction of this form to be reasonable, its physical properties should depend only on the line's endpoints, and not on the line's path: it is a simple exercise to check that this is true. To understand the significance of the form of this wavefunction, consider two configurations, $\mathcal{C}$ and $\mathcal{C}'$, which are related by flipping dimers around a single loop. The dimer flip changes the sign of $(-1)^{n_{\mathcal{C}}}$ if the loop encloses the vison, but otherwise has no effect. This suggests several conclusions. First, far from the vison, the state $|\psi_{\text{vison}}\rangle$ is very similar to $|\mathcal{G}\rangle$, since many different components of the wavefunction, related by dimer flips around loops that do not enclose the vison, contribute all with the same phase to $|\psi_{\text{vison}}\rangle$, just as they do to $|\mathcal{G}\rangle$. Second, $|\psi_{\text{vison}}\rangle$ and $|\mathcal{G}\rangle$ are orthogonal, since the average of $(-1)^{n_{\mathcal{C}}}$ over configurations is zero. Third, close to the vison, the states $|\psi_{\text{vison}}\rangle$ and $|\mathcal{G}\rangle$ are quite different, and so we expect a finite energy gap for vison creation.

An emergent photon is an excitation involving density waves of the dimer orientations. It is a gapless excitation of a U(1) quantum dimer liquid, and so of interest at the RK point on bipartite lattices in $2+1$ and $3+1$ dimensions, and also away from the RK point in the U(1) phase in $3+1$ dimensions. At the RK point itself, discussion of a trial wavefunction is again a very useful approach [33]. The excitation is characterized by a polarization $\hat{\tau}$ and a wavevector $\mathbf{q}$. To specify the trial wavefunction, we use the dimer number operator $\sigma_{\hat{\tau}}(\mathbf{r})$. Its Fourier transform is

$$\sigma_{\hat{\tau}}(\mathbf{q}) = \sum_{\mathbf{r}} \mathrm{e}^{\mathrm{i}\mathbf{q}\cdot\mathbf{r}} \sigma_{\hat{\tau}}(\mathbf{r}) \tag{3.51}$$

and the trial wavefunction we consider is

$$|\psi_{\text{photon}}\rangle = \sigma_{\hat{\tau}}(\mathbf{q})|\mathcal{G}\rangle\,. \tag{3.52}$$

Note that $\langle\mathcal{G}|\psi_{\text{photon}}\rangle = 0$ for $|\mathbf{q}| \neq 0$ since $|\mathcal{G}\rangle$ and $|\psi_{\text{photon}}\rangle = 0$ are both eigenstates of translation, but with different eigenvalues. In addition, for small $|\mathbf{q}|$, the state $|\psi_{\text{photon}}\rangle$ is locally similar to $|\mathcal{G}\rangle$, in the sense that dimer flips around short loops induce only small changes in the phases of its expansion coefficients $A_{\mathcal{C}}$.

Excitation energies $\mathcal{E}(\mathbf{q})$ in the quantum dimer model can be determined variationally using this trial wavefunction, taking inspiration from Feynman's treatment of phonon modes in Bose condensates [50]. We start from

$$\mathcal{E}(\mathbf{q}) \leq \frac{\langle\mathcal{G}|\sigma^{\dagger}_{\hat{\tau}}(\mathbf{q})\mathcal{H}_{\text{RK}}\sigma_{\hat{\tau}}(\mathbf{q})|\mathcal{G}\rangle}{\langle\mathcal{G}|\sigma^{\dagger}_{\hat{\tau}}(\mathbf{q})\sigma_{\hat{\tau}}(\mathbf{q})|\mathcal{G}\rangle} - \frac{\langle\mathcal{G}|\mathcal{H}_{\text{RK}}|\mathcal{G}\rangle}{\langle\mathcal{G}|\mathcal{G}\rangle} = \frac{f(\mathbf{q})}{2s(\mathbf{q})}\,, \tag{3.53}$$

where

$$f(\mathbf{q}) = \langle\mathcal{G}|[\sigma^{\dagger}_{\hat{\tau}}(\mathbf{q}), [\mathcal{H}_{\text{RK}}, \sigma_{\hat{\tau}}(\mathbf{q})]]|\mathcal{G}\rangle\,, \qquad s(\mathbf{q}) = \langle\mathcal{G}|[\sigma^{\dagger}_{\hat{\tau}}(\mathbf{q})\sigma_{\hat{\tau}}(\mathbf{q})|\mathcal{G}\rangle\,. \tag{3.54}$$

From (3.53), we see that excitations are gapless if there are wavevectors $\mathbf{q}$ at which $f(\mathbf{q})$ vanishes and $s(\mathbf{q})$ remains finite or approaches zero more slowly. In turn, $f(\mathbf{q}) = 0$ if $[\mathcal{H}_{\text{RK}}, \sigma_{\hat{\tau}}(\mathbf{q})] = 0$, and for the latter to hold we require both of the two dimer configurations shown in Fig. 3.19 to make the same contribution to $\sigma_{\hat{\tau}}(\mathbf{q})$. Setting $\hat{\tau} = \hat{x}$ for definiteness, this is the case in two dimensions if $q_y = \pi$ or in three dimensions if $q_y = q_z = \pi$. Writing $\mathbf{q} = (\pi, \pi) + \mathbf{k}$ or $\mathbf{q} = (\pi, \pi, \pi) + \mathbf{k}$, one finds $f(\mathbf{q}) \propto k_y^2$ or $f(\mathbf{q}) \propto k_y^2 + k_z^2$, respectively. It remains to compute $s(\mathbf{q})$, which is a correlator for classical dimer coverings and can be evaluated using the treatment of Coulomb phases established in Section 3.3. As dimer occupancy is represented by flux, we require the flux correlator; and because the mapping between dimers and fluxes uses an alternating orientation convention on links, long-wavelength flux correlations reflect dimer correlations near the Brillouin zone corner. From the relation

$$\langle\mathcal{G}|[\sigma^{\dagger}_{\hat{x}}(\mathbf{q})\sigma_{\hat{x}}(\mathbf{q})|\mathcal{G}\rangle \propto \langle B^x(-\mathbf{k})B^x(\mathbf{k})\rangle\,, \tag{3.55}$$

one finds $s(\mathbf{q}) \propto k_y^2/k^2$ in two dimensions and $s(\mathbf{q}) \propto (k_y^2 + k_z^2)/k^2$ in three. This yields the important result

$$\mathcal{E}(\mathbf{q}) \leq ck^2\,, \tag{3.56}$$

with c a numerical constant, showing that the quantum dimer model has gapless excitations at the RK point.

We shall see that the quadratic dispersion shown in (3.56) reflects true behaviour, rather than simply providing an upper bound, but it is specific to the RK point and is replaced by a linear dispersion within the U(1) phase in $3+1$ dimensions. To discuss the relevant physics, we consider how to extend the continuum description of Coulomb phases from classical systems in D space dimensions to quantum systems in $D+1$ spacetime dimensions.

Taking $D=2$, we require an imaginary-time action for a height field $h(\mathbf{r},t)$ that is compatible with (3.56). The quadratic expression

$$S_{\mathrm{RK}} = \frac{1}{2}\int \{[\partial_t h(\mathbf{r},t)]^2 + K^2[\nabla^2 h(\mathbf{r},t)]^2\}\,\mathrm{d}^2\mathbf{r}\,\mathrm{d}t \tag{3.57}$$

is implied. It yields for the Fourier transform of the height field the correlator

$$\langle |h(\mathbf{k},\omega)|^2\rangle = \frac{1}{(Kk^2)^2+\omega^2} \tag{3.58}$$

and hence an equal-time correlator

$$\int \mathrm{d}\omega\,\langle |h(\mathbf{k},\omega)|^2\rangle \propto \frac{1}{Kk^2} \tag{3.59}$$

of the form required for the classical height model that describes dimer correlations in the ground-state wavefunction at the RK point. This action is, however fine-tuned; adding further symmetry-allowed terms, we arrive at

$$S_{2+1} = \frac{1}{2}\int \{[\partial_t h(\mathbf{r},t)]^2 + \rho_2|\vec{\nabla} h(\mathbf{r},t)|^2 + K^2[\nabla^2 h(\mathbf{r},t)]^2 - g\cos[2\pi h(\mathbf{r},t)]\}\,\mathrm{d}^2\mathbf{r}\,\mathrm{d}t\,. \tag{3.60}$$

We can identify the stiffness ρ_2 with the parameter combination $1-v/t$ in the quantum dimer model. At the RK point, the stiffness vanishes, g is irrelevant under renormalization, and we return to (3.57). On one side of the RK point, $v>t$, ρ_2 is negative, promoting a state with a large gradient $|\vec{\nabla} h(\mathbf{r},t)|$ in the height field: the staggered phase. On the other side, $v<t$, positive ρ_2 suppresses fluctuations of $h(\mathbf{r},t)$ and $g\cos[2\pi h(\mathbf{r},t)]$ is relevant, producing a pinned phase. In this way, we see that the RK point for a bipartite lattice in $2+1$ dimensions is an isolated critical point between two conventional ordered phases.

Contrastingly, quantum dimer models on bipartite lattices in $3+1$ dimensions generically support a dimer liquid phase. Again we require an imaginary-time action compatible with (3.56), now expressed in terms of the vector potential $\vec{A}(\mathbf{r},t)$. Picking the Coulomb gauge $\vec{\nabla}\cdot\vec{A}(\mathbf{r},t)=0$, an expansion in space and time derivatives gives

$$S_{3+1} = \frac{1}{2}\int \{|\partial_t \vec{A}(\mathbf{r},t)|^2 + \rho_2|\vec{\nabla}\times\vec{A}(\mathbf{r},t)|^2 + K^2|\vec{\nabla}\times[\vec{\nabla}\times\vec{A}(\mathbf{r},t)]|^2\}\,\mathrm{d}^3\mathbf{r}\,\mathrm{d}t\,. \tag{3.61}$$

Again, we expect $\rho_2 \propto 1-v/t$. At the RK point, with $\rho_2=0$, we recover a quadratic dispersion relation for excitations, and equal-time correlations of the classical Coulomb phase. On one side of the RK point, with $v>t$, negative ρ_2 drives the system into a staggered phase, as in $2+1$ dimensions. On the other side, however, with $v<t$, by introducing a scalar potential $\Phi(\mathbf{r},t)$ and writing $\vec{E}(\mathbf{r},t)=\partial\vec{A}(\mathbf{r},t)-\vec{\nabla}\Phi(\mathbf{r},t)$, the action can be expressed in the form [51, 52]

$$S_{3+1}=\frac{1}{2}\int\{|\vec{E}(\mathbf{r},t)|^2+\rho_2|\vec{B}(\mathbf{r},t)|^2+K^2|\vec{\nabla}\times\vec{B}(\mathbf{r},t)|^2\}\,\mathrm{d}^3\mathbf{r}\,\mathrm{d}t\,, \tag{3.62}$$

familiar from quantum electrodynamics. Weak perturbations to this action allowed by symmetry are irrelevant in the renormalization group sense. It describes a liquid phase of the quantum dimer model and has linearly dispersing emergent photon excitations. Monomers introduced into the dimer covering are monopole sources for the $\vec{B}$ field, and the three-dimensional analogue of visons are sources for the $\vec{E}$ field.

3.6 Concluding remarks

The presentation has necessarily been a very selective one, chosen particularly to bring out common strands in the treatment of frustration in classical models and in quantum systems. It is reassuring to find that the main ideas also emerge from quite different treatments.

3.6.1 Slave particles

A large and important class of theories follows from representing spins in terms of particles subject to a local constraint. The constraint can be imposed using a gauge field, which takes the same place in a description of a spin liquid as the gauge fields that appear in dimer models. One starting point is Schwinger's representation of spin-S operators using two species of bosons, as [41]

$$S^z=\frac{1}{2}(b_1^\dagger b_1-b_2^\dagger b_2),\qquad S^+=b_1^\dagger b_2,\qquad S^-=b_2^\dagger b_1\,, \tag{3.63}$$

with $[b_i b_j^\dagger]=\delta_{ij}$ and the constraint $b_1^\dagger b_1+b_2^\dagger b_2=2S$. Fluctuations can be controlled by generalizing from SU(2) to SU(N) [53] or, on non-bipartite lattices, to Sp(N) [54], and from two to N species of boson.

Alternatively, we can recall the origin of local moments in itinerant fermions [55], writing

$$\vec{S}_i=f_{i\alpha}^\dagger\vec{\sigma}_{\alpha\beta}f_{i\beta}\,,\qquad\text{with}\quad f_{i\uparrow}^\dagger f_{i\uparrow}+f_{i\downarrow}^\dagger f_{i\downarrow}=1\,, \tag{3.64}$$

where $\{f_{i\alpha}^\dagger,f_{j\beta}\}=\delta_{ij}\delta_{\alpha\beta}$. The Heisenberg exchange term $\vec{S}_i\cdot\vec{S}_i$ then corresponds to a four-fermion interaction, and a mean-field decoupling leads to a quadratic fermion Hamiltonian of the form

$$\mathcal{H}_{\mathrm{MF}}=\sum_{ij,\sigma}\left\{t_{ij}f_{i\sigma}^\dagger f_{j\sigma}+\left(\Delta_{ij}f_{i\uparrow}^\dagger,f_{j\downarrow}^\dagger+\mathrm{h.c.}\right)\right\}\,, \tag{3.65}$$

where the hopping coefficients t_{ij} and pairing amplitudes Δ_{ij} serve as variational coefficients. Denoting the ground state of $\mathcal{H}_{\text{MF}}$ by $|\text{Slater}\rangle$, components with double site occupancy can be removed by Gutzwiller projection to yield a spin-liquid wavefunction

$$|\mathcal{G}\rangle = \prod_i (1 - n_{i\uparrow} n_{i\downarrow}) \, |\text{Slater}\rangle \,. \tag{3.66}$$

Within this approach, depending on the phases of the hopping amplitudes, the fermions move in a $\mathbb{Z}_2$, U(1), or SU(2) gauge field, and have a gapped, Dirac or metallic spectrum—in decreasing order of stability. Spinons are represented by Bogoliubov quasiparticles, and visons by flux excitations encoded in the phases of the hopping amplitudes [56].

3.6.2 Numerics

Unbiased numerical methods play a key role in this research field, and often represent the only way to find out what phases are supported by a particular physical model; reviews can be found in [57] and [58]. In general, while exact diagonalization of the many-body Hamiltonian has the advantages of flexibility (for example, providing an early identification of a gapped spin liquid [59]), alternative techniques are important to avoid the difficulties stemming from the exponential growth with system size of the Hilbert space dimension. These include the design of Hamiltonians that avoid Monte Carlo sign problems [58], and the use of the density matrix renormalization group and related methods to study quasi-one-dimensional samples (see e.g. [61]). At the same time, new approaches to identifying exotic states are increasingly important, such as the evaluation of entanglement entropy to probe topological order [60].

3.6.3 Summary

Some common strands run through much of the physics that has been presented. Frustration gives rise to classical degeneracy, and the correlations that are built onto these classical degenerate states lead to the ideas of topological sectors and deconfined, fractionalized excitations. By adding quantum dynamics to dimer models, two important types of stable quantum liquid phase can be realised. One is a gapped $\mathbb{Z}_2$ phase, stable in both two and three spatial dimensions, which has point-like excitations of two types: spinons and visons. The other is the U(1) liquid, stable only in three spatial dimensions, and having gapped electric and magnetic monopole excitations as well as gapless emergent photon modes. Models realizing $\mathbb{Z}_2$ phases include, in two dimensions, the triangular-lattice quantum dimer model [46], the toric code [62], and the spin-$\frac{1}{2}$ kagome-lattice Heisenberg antiferromagnet [63]. Examples of U(1) liquids are provided by the diamond-lattice quantum dimer model [64] and by hard-core bosons on the pyrochlore lattice [65].

Acknowledgements

I am very grateful to all my collaborators in this field, and especially to Eugene Shender, Peter Holdsworth, and Roderich Moessner. I acknowledge financial support from EPSRC.

References

[1] P. W. Anderson, *Mater. Res. Bull.* **8**, 153 (1973).
[2] A. P. Ramirez, *Annu. Rev. Mater. Sci.* **24**, 453 (1994).
[3] P. A. Lee, *Science* **321**, 1306 (2008).
[4] L. Balents, *Nature* **464**, 199 (2010).
[5] C. Lacroix, P. Mendels, and F. Mila (eds.), *Introduction to Frustrated Magnetism*, Springer (2011).
[6] O. Tchernyshyov, R. Moessner, and S. L. Sondhi, *Phys. Rev. B* **66**, 064403 (2002).
[7] J. Villain, *Z. Phys. B* **33**, 31 (1979).
[8] B. Martinez, F. Sandiumenge, A. Rouco, A. Labarta, J. Rodriguezcarvajal, M. Tova, M. T. Causa, S. Gali, and X. Obradors, *Phys. Rev. B* **46**, 10786 (1992).
[9] S.-H. Lee, C. Broholm, G. Aeppli, A. P. Ramirez, T. G. Perring, C. J. Carlile, M. Adams, T. J. L. Jones, and B. Hessen, *Europhys. Lett.* **35**, 127 (1996).
[10] T. Fennell, P. P. Deen, A. R. Wildes, K. Schmalzl, D. Prabhakaran, A. T. Boothroyd, R. J. Aldus, D. F. McMorrow, and S. T. Bramwell, *Science* **326**, 415 (2009).
[11] T.-H. Han, J. S. Helton, S. Chu, D. G. Nocera, J. A. Rodriguez-Rivera, C. Broholm, and Y. S. Lee, *Nature* **492**, 406 (2012).
[12] S. Yamashita , Y. Nakazawa, M Oguni, Y. Oshimi, H. Nojiri, Y. Shimizu, K. Miyagawa, and K. Kanoda, *Nat. Phys.* **4**, 459 (2008).
[13] L. Pauling, *The Nature of the Chemical Bond*, Cornell University Press (1945).
[14] A. P. Ramirez, A. Hayashi, R. J. Cava, R. Siddharthan, and B. S. Shastry, *Nature* **399**, 333 (1999).
[15] J. C. Maxwell, *Philos. Mag.* **27**, 294 (1864).
[16] J. N. Reimers, A. J. Berlinsky, and A.-C. Shi, *Phys. Rev. B* **43**, 865 (1991).
[17] R. Moessner and J. T. Chalker, *Phys. Rev. Lett.* **80**, 2929 (1998); *Phys. Rev. B* **58**, 12049 (1998).
[18] J. Villain, R. Bidaux, J. P. Carton, and R. J. Conte, *J. Phys. Paris* **41**, 1263 (1980).
[19] E. F. Shender, *Sov. Phys. JETP* **56**, 178 (1982).
[20] J. T. Chalker, P. C. W. Holdsworth, and E. F. Shender, *Phys. Rev. Lett.* **68**, 855 (1992).
[21] A. G. Gukasov, T. Brückel, B. Dorner, V. P. Plakhty, W. Prandtl, E. F. Shender, and O. P. Smirnov, *Europhys. Lett.* **7**, 83 (1988).
[22] T. Brückel, B. Dorner, A. G. Gukasov, and V. P. Plakhty, *Phys. Lett. A* **162**, 357 (1992).
[23] M. E. Zhitomirsky, M. V. Gvozdikova, P. C. W. Holdsworth, and R. Moessner, *Phys. Rev. Lett.* **109**, 077204 (2012); L. Savary, K. A. Ross, B. D. Gaulin, J. P. C. Ruff, and L. Balents, *Phys. Rev. Lett.* **109**, 167201 (2012).

[24] D. A. Garanin and B. Canals, *Phys. Rev. B* **59**, 443 (1999); B. Canals and D.A. Garanin, *Can. J. Phys.* **79**, 1323 (2001).
[25] S. V. Isakov, K. Gregor, R. Moessner, and S. L. Sondhi, Phys. Rev. Lett. **93**, 167204 (2004).
[26] G. Wannier, *Phys. Rev.* **79**, 357 (1950); R. M. F. Houtappel, *Physica* **16**, 425 (1950).
[27] H. W. J. Blöte and H. J. Hilhorst, *J. Phys. A* **15**, L631 (1982); B. Nienhuis, H. J. Hilhorst, and H. W. Blöte, *J. Phys A* **17**, 3559 (1984); B. Nienhuis, *Phys. Rev. Lett.* **49**, 1062 (1982).
[28] C. Zeng and C. L. Henley, *Phys. Rev. B* **55**, 14 953 (1997).
[29] M. E. Fisher, *Phys. Rev.* **124**, 1664 (1961); P. W. Kastelyn, *Physica* **27**, 1209 (1961).
[30] R. Kenyon, An introduction to the dimer model, ICTP Lecture Notes, arXiv:math/0310326.
[31] C. L. Henley, *J. Stat. Phys.* **89**, 483 (1997).
[32] D. A. Huse, W. Krauth, R. Moessner, and S. L. Sondhi, *Phys. Rev. Lett.* **91**, 167004 (2003).
[33] D. S. Rokhsar and S. A. Kivelson, *Phys. Rev. Lett.* **61**, 2376 (1998).
[34] P. Fendley, R. Moessner, and S.L. Sondhi, *Phys. Rev. B* 66, 214513 (2002).
[35] S. T. Bramwell and M. J. P. Gingras, *Science* **294**, 1495 (2001).
[36] M. J. Harris, S. T. Bramwell, D. F. McMorrow, T. Zeiske, and K. W. Godfrey, *Phys. Rev. Lett.* **79**, 2554 (1997).
[37] C. L. Henley, *Annu. Rev. Condens. Matter Phys.* **1**, 179 (2010).
[38] P. H. Conlon and J. T. Chalker, *Phys. Rev. B* **81**, 224413 (2010).
[39] C. Castelnovo, R. Moessner, and S. L. Sondhi, *Nature* **451**, 42 (2008).
[40] R. G. Melko and M. J. P. Gingras, *J. Phys.: Condens. Matter* **16**, R1277 (2004).
[41] A. Auerbach, *Interacting Electrons and Quantum Magnetism*, Springer (1998).
[42] E. H. Lieb, T. D. Schultz, and D. C. Mattis, *Ann. Phys. (NY)* **16**, 407 (1961).
[43] I. Affleck, *Phys. Rev. B* **37**, 5186 (1988).
[44] M. B. Hastings, *Phys. Rev.* **69**, 104431 (2004).
[45] P. W. Anderson, *Science* **235**, 1196 (1987).
[46] R. Moessner and S. L. Sondhi, *Phys. Rev. Lett.* **86**, 1881 (2001).
[47] R. Moessner and K. Raman, in C. Lacroix, P. Mendels, and F. Mila (eds.), *Introduction to Frustrated Magnetism*, p. 437, Springer (2011).
[48] J. B. Kogut, *Rev. Mod. Phys.* **51**, 659 (1979).
[49] T. Senthil and M. P. A. Fisher, *Phys. Rev. B* **62**, 7850 (2000).
[50] R. P. Feynman, *Phys. Rev.* **94**, 262 (1954).
[51] R. Moessner and S. L. Sondhi, *Phys. Rev. B* **68**, 184512 (2003).
[52] M. Hermele, M. P. A. Fisher, and L. Balents, *Phys. Rev. B* **69**, 064404 (2004).
[53] N. Read and S. Sachdev, *Phys. Rev. Lett.* **62**, 1694 (1989).
[54] N. Read and S. Sachdev, *Phys. Rev. Lett.* **66**, 1773 (1991).
[55] J. B. Marston and I. Affleck, *Phys. Rev. B* **39**, 11539 (1989).
[56] X. G. Wen, Phys. *Rev. Phys. B* **65**, 165113 (2002); P. Lee, N. Nagaosa, and X.-G. Wen, *Rev. Mod. Phys.* **78**, 17 (2006).

[57] A. M. Läuchli, in C. Lacroix, P. Mendels, and F. Mila (eds.), *Introduction to Frustrated Magnetism*, p. 481, Springer (2011).
[58] A. W. Sandvik, *AIP Conf. Proc.* **1297**,135 (2010).
[59] G. Misguich, C. Lhuillier, B. Bernu, and C. Waldtmann, *Phys. Rev. B* **60**, 1064 (1999).
[60] S. V. Isakov, R. G. Melko, and M. B. Hastings, *Science* **335**, 193 (2012).
[61] M. S. Block, R. V. Mishmash, R. K. Kaul, D. N. Sheng, O. I. Motrunich, and M. P. A. Fisher, *Phys. Rev. Lett.* **106**, 046402 (2011).
[62] A. Yu. Kitaev, *Ann. Phys. (NY)* **303**, 2 (2003).
[63] S. Yan, D. A. Huse, and S. R. White, *Science* **332**, 1173 (2011); S. Depenbrock, I. P. McCulloch, and U. Schollwoeck, *Phys. Rev. Lett.* **109**, 067201 (2012); H.C. Jiang, Z. Wang, and L. Balents, *Nat. Phys.* **8**, 902 (2012).
[64] O. Sikora, F. Pollmann, N. Shannon, K. Penc, and P. Fulde, *Phys. Rev. Lett.* **103**, 247001 (2009).
[65] A. Banerjee, S. V. Isakov, K. Damle, and Y. B. Kim, *Phys. Rev. Lett.* **100**, 047208 (2008).

4
Entanglement spectroscopy and its application to the quantum Hall effects

Nicolas Regnault

Department of Physics, Princeton University
Princeton, NJ 08544, USA
Laboratoire Pierre Aigrain, Ecole Normale Supérieure-PSL Research University, CNRS
Université Pierre et Marie Curie-Sorbonne Universités, Université Paris Diderot-Sorbonne Paris Cité
24 rue Lhomond, 75231 Paris Cedex 05, France

Topological Aspects of Condensed Matter Physics. First Edition. C. Chamon et al.

Chapter Contents

Colour figures. For those figures in this chapter that use colour, please see the version of these lecture notes at arXiv:1510.07670 [cond-mat.str.el]. These figures are indicated by '[Colour online]' at the start of the caption.

Preface

Entanglement spectroscopy, initially introduced by Li and Haldane in the context of the fractional quantum Hall effect, has stimulated an extensive range of studies. The entanglement spectrum is the spectrum of the reduced density matrix when we partition the system into two. For many quantum systems, it unveils a unique feature: computed from the bulk ground-state wavefunction, the entanglement spectrum gives access to the physics of edge excitations. Using this property, entanglement spectroscopy has proved to be a highly valuable tool to diagnose topological ordering.

The aim of these lectures is to provide an overview of entanglement spectroscopy, mostly in the context of the fractional quantum Hall effect. We introduce the basic concepts through the example of quantum spin chains. We discuss the connection with entanglement entropy and the matrix product state representation. We show how the entanglement spectrum can be computed for non-interacting topological phases and how it reveals the edge excitation from the ground state. We then present an extensive review of entanglement spectra as applied to fractional quantum Hall phases, showing how much information is encoded within the ground state and how different partitions probe different type of excitations. Finally, we discuss the application of this tool to study fractional Chern insulators.

4.1 Introduction

In the past decade, it has become clear that Landau's theory of phase transitions, which involves the appearance of a broken-symmetry order parameter, does not apply to a series of phases of matter with so-called topological order. Topological phases exhibit the surprising property that their quantum ground state can be degenerate and no local measurement can distinguish these degenerate states. This feature is the key to topological quantum computing: this robustness is used to solve the problem of local decoherence (for example due to disorder) by construction instead of by quantum error correction (a hardware versus a software approach). This inherent robustness is also the source of a major issue: the absence of a local order parameter makes the identification of a topological order a difficult task.

Most intrinsic topologically ordered phases are strongly correlated systems. Thus, numerical simulations have been an important tool to understand the emergence of these phases from microscopic models. The absence of a local order parameter, coupled with the few finite sizes that can be reached through simulations, restrict our ability to characterize these systems. From this perspective, we would like to achieve the following two goals:

1. Convince ourselves that the phase is indeed emerging. The more signatures we have, the more confidently might we claim to have strong evidence of this emergence.
2. Minimize our effort—meaning that we compute only the ground state of our system. This is actually more a limitation of the algorithm (or technique) or of computational power than a consequence of being lazy. If we work with quantum

many-body systems, the Hilbert-space dimension grows exponentially with system size (think about a spin-$\frac{1}{2}$ system). This will be our bottleneck sooner or later (and most probably sooner).

So, if we only have access to the ground state, we can wonder how much information about the system can be extracted. If we are lucky, we might have a model ground state we can compare with. The simplest way to compare the ground state from a simulation and this model ground state would be to compute an overlap (meaning a scalar product). Unfortunately, this might be tricky to do (for example, the two states might be expressed in two different bases). Moreover, we end up with a number between 0 and 1, scaling with the system size. Thus, we have to decide what is a good overlap.

Other routes would be to use a global order parameter, looking for the ground-state degeneracy (playing with the genus of the surface we are working on, which might affect the ground-state degeneracy for a topological phase), play with the boundary conditions (twist the boundary, flux insertion, etc.)

A promising tool to extract topological information from the ground-state wavefunction is the entanglement entropy [12, 35, 40]. The key idea is to break the system ground state into pieces and look at the entanglement between these pieces. In the simplest case, we consider the bipartite entanglement between two parts A and B. As we will see, this technique will reveal lots of information about the phase itself, such as its excitations. In many examples, this provides an in-depth view of the information encoded within the wavefunction of a topological phase. As a corollary, it means that we can store a wavefunction in a more efficient way when we perform numerical simulations, just by keeping the relevant information.

In the simplest case, we consider the bipartite entanglement between two parts A and B of the system in its ground state $|\Psi\rangle$. This partition is characterized by the reduced density matrix $\rho_A = \mathrm{Tr}_B |\Psi\rangle \langle\Psi|$ of subsystem A, obtained by tracing out all the B degrees of freedom. Among the various entropies that have been considered as an entanglement measurement, the entanglement entropy is the most popular (see [5] for an extensive review). It is defined as the von Neumann entropy associated with ρ_A, that is, $\mathcal{S}_A = -\mathrm{Tr}_A [\rho_A \ln \rho_A]$. For a system in d dimensions with a finite correlation length l, the entanglement entropy satisfies the area law [68]

$$\mathcal{S}_A \simeq \alpha \mathcal{L}^{d-1}, \tag{4.1}$$

where $\mathcal{L} \gg l$ is the typical length that defines the size of the region A and α is a non-universal constant. The area law indicates that the dominant part of the entanglement entropy is controlled by the area ($\mathcal{L}^{d-1}$) that separates the two domains. Physically, it means that the entanglement between A and B is located at the interface of the two regions.

For two-dimensional topological phases, it was shown in [35] and [40] that the first correction to the area law is a topological term: $\mathcal{S}_A \sim \alpha\mathcal{L} - \gamma$. The subleading term γ is called the topological entanglement entropy: it is a constant for a given topologically ordered phase, $\gamma = \ln \mathcal{D}$. Here $\mathcal{D}$ is the total quantum dimension characterizing the topological field theory describing the phase and thus the nature of the system excitations. The topological entanglement entropy appears as a way to characterize the topological order of a phase. However, its practical calculation depends on scaling arguments, which might be hard to obtain to sufficient accuracy from numerical calculations [31, 36]. Moreover, it does not uniquely determine the topological order in the state.

While the topological entanglement entropy compresses the information contained in the reduced density matrix into a single number, the concept of entanglement spectroscopy have been shown to be a powerful tool to probe the topological order. Indeed, the aim of using the entanglement spectrum (ES) is to have a deeper look at ρ_A by analysing its full spectrum. The ES was initially introduced by Li and Haldane [41] in the context of the fractional quantum Hall effect (FQHE), stimulating an extensive range of studies [3, 6, 15, 19, 33, 37, 43, 60, 63, 64, 69, 70, 75, 76, 88]. It has also been studied and applied to a wide variety of topological and non-topological phases, including spin systems [13, 16, 34, 38, 46, 54–56, 65, 74, 86], as well as topological insulators [24, 57, 78], Bose–Hubbard models [45], and complex paired superfluids [18]. Moreover, the partition of the system has to be thought in a broad sense: it can be done in real space, in momentum or Fourier space, or in particle space. For many model states such as the Laughlin wavefunction [39] and the AKLT spin chain [1, 2], the counting (the number of non-zero eigenvalues) is exponentially lower than expected. This counting is related to the nature of the system excitations. The salient feature is that this information about the excitations is obtained only from the ground state. The ES is a way to extract this information, and each type of cut reveals different aspects of these excitations.

These lecture notes will try to give an overview of entanglement spectroscopy, but with an emphasis on its application to the FQHE and similar phases. This bias is motivated by the large number of studies of the ES that have been carried out for these phases and the detailed understanding of the ES that has thereby been obtained. These notes are organized as follows. Section 4.2 provides an introduction to the relevant notation and the concept of the ES. We exemplify these notions with simple spin systems and relate the ES to the matrix product state (MPS) representation. In Section 4.3, we show how the chiral edge mode can be observed from the ES of a non-interacting system, illustrating this feature with the integer quantum Hall effect (IQHE) and the Chern insulator (the simplest example of a topological insulator). In Section 4.4, we present an extensive overview of the ES for the FQHE. We show the different bipartite partitions that have been considered for these systems and the kind of information that has been revealed by determining the ES. Finally, in Section 4.5, we discuss how the ES has been used as a tool to probe the phases that emerge in fractional Chern insulators (FCIs).

4.2 Entanglement spectrum and entanglement entropy

As a first step, we discuss the concept of entanglement spectroscopy in some simple cases. We also briefly cover the definition and relevant properties of the entanglement entropy. We introduce the Li–Haldane conjecture in the case of the AKLT spin chain. We discuss the important situation where the number of non-zero eigenvalues of the reduced density matrix is massively reduced. In particular, we show the relation between the latter property and the MPS representation.

4.2.1 Definitions

Let us consider a generic n-body quantum state $|\Psi\rangle$ that can be decomposed on the orthonormal basis $\{|\lambda\rangle\}$. We now assume that this basis can be written as the tensor product of two orthonormal bases $\{|\mu_A\rangle\}$ and $\{|\mu_B\rangle\}$, that is, $\{|\lambda\rangle = |\mu_A\rangle \otimes |\mu_B\rangle\}$, providing a natural bipartition of the system into A and B. The decomposition of the state $|\Psi\rangle$ reads

$$|\Psi\rangle = \sum_{\mu_A,\mu_B} c_{\mu_A,\mu_B} |\mu_A\rangle \otimes |\mu_B\rangle. \tag{4.2}$$

The entanglement matrix M is defined such that its matrix elements are given by $M_{\mu_A,\mu_B} = c_{\mu_A,\mu_B}$. The size of M is given by the dimensions of the subspaces A and B, which we denote respectively by $\dim_A$ and $\dim_B$. Note that we do not assume that $\dim_A = \dim_B$, and thus M is generically a rectangular matrix and not necessarily square. One can perform a singular value decomposition (SVD) of M. The SVD allows one to write a rectangular matrix

$$M = UDV^\dagger, \tag{4.3}$$

where U is a $\dim_A \times \min(\dim_A, \dim_B)$ matrix that satisfies $U^\dagger U = 1$ (i.e. it has orthonormalized columns), V is a $\dim_B \times \min(\dim_A, \dim_B)$ matrix that satisfies $VV^\dagger = 1$ (i.e. it has orthonormalized rows). D is a diagonal square of dimension $\min(\dim_A, \dim_B)$, where all entries are non-negative and can be expressed as $\{e^{-\xi_i/2}\}$.

Using the SVD, one can derive the Schmidt decomposition of $|\Psi\rangle$:

$$|\Psi\rangle = \sum_{i=1}^{\min(\dim_A,\dim_B)} e^{-\xi_i/2} |A:i\rangle \otimes |B:i\rangle, \tag{4.4}$$

where

$$|A:i\rangle = \sum_{\mu_A} U^\dagger_{i,\mu_A} |\mu_A\rangle, \tag{4.5}$$

$$|B:i\rangle = \sum_{\mu_B} V^\dagger_{i,\mu_B} |\mu_B\rangle. \tag{4.6}$$

To be a Schmidt decomposition, the states $|A:i\rangle$ and $|B:i\rangle$ have to obey $\langle A:i|A:j\rangle = \langle B:i|B:j\rangle = \delta_{i,j}$. This property is trivially verified using the identities on U and V. The Schmidt decomposition provides a nice and numerically efficient way to compute the spectrum of the reduced density matrix. Consider the density matrix of the pure state $\rho = |\Psi\rangle\langle\Psi|$. We compute the reduced density matrix of A by tracing out the degree of freedom related to B, that is, $\rho_A = \mathrm{Tr}_B \rho$. Using (4.4), we deduce that

$$\rho_A = \sum_i e^{-\xi_i} |A:i\rangle\langle A:i|. \tag{4.7}$$

Thus, the spectrum of ρ_A can be obtained from the coefficient of the Schmidt decomposition or the SVD of the entanglement matrix and is given by the set $\{e^{-\xi_i}\}$. From a numerical perspective, getting the spectrum of ρ_A is more accurate using the SVD of M than a brute-force calculation of ρ_A in the $\{|\mu_A\rangle\}$ basis followed by its diagonalization. In a similar way, we can obtain the reduced density matrix of B:

$$\rho_B = \mathrm{Tr}_A\,\rho = \sum_i e^{-\xi_i}|B:i\rangle\langle B:i|. \tag{4.8}$$

Note that ρ_A and ρ_B have the same spectrum. While these two square matrices might have different dimensions ($\dim_A$ and $\dim_B$, respectively), they both have the same number of non-zero eigenvalues. This number has to be less than or equal to $\min(\dim_A, \dim_B)$. Thus, studying the properties of ρ_A for various partitions (i.e. choices of A and B) can be restricted to the cases where $\dim_A \leq \dim_B$.

With these tools and properties, we can now define the entanglement spectrum (ES). The latter corresponds to the set $\{\xi_i\}$, the logarithms of the reduced density matrix eigenvalues. The key idea of the original article of Li and Haldane [41] was to look not only at this whole spectrum, but at a specific subset of these values (or a block of ρ_A) with well-defined quantum numbers. Assume an operator $\mathcal{O}$ that can be decomposed as $\mathcal{O}_A + \mathcal{O}_B$, where $\mathcal{O}_A$ (respectively $\mathcal{O}_B$) acts only on the A (respectively B) subspace. One can think about $\mathcal{O}$ as the projection of the spin operator or the momentum. If $[\mathcal{O}, \rho] = 0$, we also have $0 = \mathrm{Tr}_B[\mathcal{O}_A, \rho] + \mathrm{Tr}_B[\mathcal{O}_B, \rho] = [\mathcal{O}_A, \mathrm{Tr}_B\rho] = [\mathcal{O}_A, \rho_A]$ as the trace over the B degrees of freedom of a commutator operator in the B part vanishes. If $|\Psi\rangle$ is an eigenstate of $\mathcal{O}$, then the latter commutes with ρ. We can simultaneously diagonalize ρ_A and $\mathcal{O}_A$, and label the $\{\xi_i\}$ according to the quantum number of $\mathcal{O}_A$.

4.2.2 A simple example: two spin-$\frac{1}{2}$

To exemplify the notation and concepts described above, we consider a system of two spin-$\frac{1}{2}$ as depicted in Fig. 4.1(a). Any state $|\Psi\rangle$ can be decomposed onto the four basis states:

$$|\Psi\rangle = c_{\uparrow\uparrow}|\uparrow\uparrow\rangle + c_{\uparrow\downarrow}|\uparrow\downarrow\rangle + c_{\downarrow\uparrow}|\downarrow\uparrow\rangle + c_{\downarrow\downarrow}|\downarrow\downarrow\rangle. \tag{4.9}$$

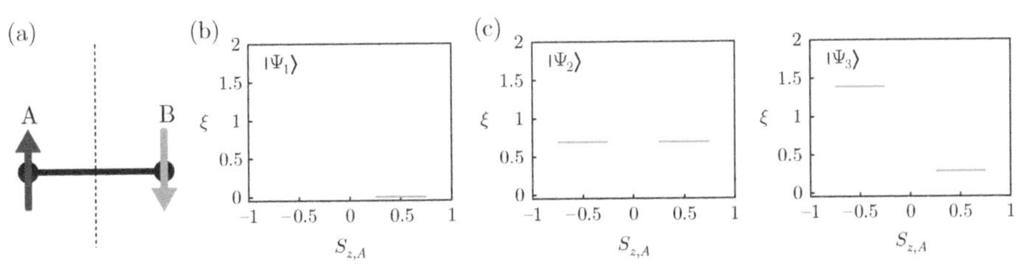

Fig. 4.1 [Colour online] (a) Schematic picture of the two-spin-$\frac{1}{2}$ system. (b) Entanglement spectrum (ES) for the state $|\Psi_1\rangle = |\uparrow\uparrow\rangle$. (c) ES for the state $|\Psi_2\rangle = \frac{1}{\sqrt{2}}(|\uparrow\downarrow\rangle - |\downarrow\uparrow\rangle)$. (d) ES for the state $|\Psi_3\rangle = \frac{1}{2}|\uparrow\downarrow\rangle + \frac{\sqrt{3}}{2}|\downarrow\uparrow\rangle$.

A natural way to cut this system into two parts consists of the A (respectively B) part being the left (respectively right) spin. The entanglement matrix is given by

$$M = \begin{pmatrix} c_{\uparrow\uparrow} & c_{\uparrow\downarrow} \\ c_{\downarrow\uparrow} & c_{\downarrow\downarrow} \end{pmatrix}, \qquad \text{such that} \quad |\Psi\rangle = \sum_{i,j=\uparrow,\downarrow} M_{i,j}|A:i\rangle \otimes |B:j\rangle. \tag{4.10}$$

We consider three examples: a product state $|\Psi_1\rangle = |\uparrow\uparrow\rangle$, a maximally entangled state $|\Psi_2\rangle = \frac{1}{\sqrt{2}}(|\uparrow\downarrow\rangle - |\downarrow\uparrow\rangle)$, and a generic entangled state $|\Psi_3\rangle = \frac{1}{2}|\uparrow\downarrow\rangle + \frac{\sqrt{3}}{2}|\downarrow\uparrow\rangle$. The entanglement matrices for these three states are

$$M_1 = (10\ 00)\,, \qquad M_2 = \begin{pmatrix} 0 & \frac{1}{\sqrt{2}} \\ -\frac{1}{\sqrt{2}} & 0 \end{pmatrix}, \qquad M_3 = \begin{pmatrix} 0 & \frac{1}{2} \\ \frac{\sqrt{3}}{2} & 0 \end{pmatrix}. \tag{4.11}$$

Performing the SVD on the first state $|\Psi_1\rangle$ is trivial: being a product state, it is already written as a Schmidt decomposition. For $|\Psi_2\rangle$, we can do the SVD

$$M_2 = \begin{pmatrix} 1 & 0 \\ 0 & 1 \end{pmatrix} \begin{pmatrix} \frac{1}{\sqrt{2}} & 0 \\ 0 & \frac{1}{\sqrt{2}} \end{pmatrix} \begin{pmatrix} 0 & -1 \\ 1 & 0 \end{pmatrix}, \tag{4.12}$$

such that the Schmidt decomposition is

$$|\Psi_2\rangle = \frac{1}{\sqrt{2}}(+|\uparrow\rangle) \otimes (+|\downarrow\rangle) + \frac{1}{\sqrt{2}}(+|\downarrow\rangle) \otimes (-|\uparrow\rangle)\,. \tag{4.13}$$

A similar calculation can be performed for $|\Psi_3\rangle$.

The projection of the total spin along the z axis, S_z, is the sum of individual components $S_{z,A}$ and $S_{z,B}$. Thus, when we perform the cut into the two parts A and B, $S_{z,A}$ is a good quantum number that can be used to label the eigenvalues of the ES according to the discussion in Section 4.2.1. The ET for the three states $|\Psi_1\rangle$, $|\Psi_2\rangle$, and $|\Psi_3\rangle$ are shown in Fig. 4.1(b–d). For the product state $|\Psi_1\rangle$, there is just a single level, since the reduced density matrix has a single non-zero eigenvalue. For the two other examples, there are two levels, each with a given $S_{z,A}$ value. The calculation of the entanglement entropy, which is a measure of the entanglement, indicates directly that $|\Psi_1\rangle$ is a product state. We can derive the same conclusion from the number of levels in the ES. While this example is rather a trivial result obtained from the ES, it stresses one of the strong points of this technique. Some properties of the states can be deduced just by counting the non-zero eigenvalues of the reduced density matrix.

4.2.3 Entanglement entropy

They are several ways to quantify the entanglement between two parts of a system and there is an extensive literature on this topic (see [5] for an extensive review). The goal of these lectures is not to give a detailed introduction to entanglement entropies. So we will restrict ourselves to a few useful examples in the context of topological phases. Perhaps the most common measure of entanglement is the von Neumann entanglement entropy

$$\mathcal{S}_A = -\mathrm{Tr}_A\left[\rho_A \ln \rho_A\right]. \tag{4.14}$$

From a practical point of view, the calculation of the Von Neumann entanglement entropy is easy once the Schmidt decomposition or the spectrum of the reduced density matrix has been obtained:

$$\mathcal{S}_A = -\sum_i \lambda_i \ln \lambda_i = \sum_i \xi_i e^{-\xi_i}. \tag{4.15}$$

Similarly, we can define the entanglement entropy for the B part of the system $S_B = -\mathrm{Tr}_B\left[\rho_B \ln \rho_B\right]$. Using 4.8, we immediately see that $S_A = S_B$. If A and B are not entangled (i.e. $|\Psi\rangle = |\Psi_A\rangle \otimes |\Psi_B\rangle$), we get $S_A = 0$. For the full system $A + B$, the entanglement entropy is also zero. As a consequence, we have in general that $S_A + S_B \neq S_{A+B}$ (the entanglement entropy is actually strongly subadditive).

We will now turn to the entanglement entropy of some specific systems. In many situations, it is useful to look at the case of a random state. Especially for people interested in numerical simulations, it is always a good idea to compare with what a random output would give. For example, consider the calculation of the overlap (the simplest way to compare two wavefunctions). Let us take two random states $|\Psi_1\rangle$ and $|\Psi_2\rangle$ defined in a Hilbert space of dimension $\mathcal{D}$. Then the average overlap $|\langle\Psi_1|\Psi_2\rangle|^2 \simeq 1/\mathcal{D}$. This result gives a simple bound for what is a bad overlap in finite systems (note that one should not cheat and define $\mathcal{D}$ as the dimension of the Hilbert space with all the symmetries the system has).

For the entanglement entropy, we recall the notation $\dim_A$ for the dimension of the Hilbert associated with the A part and $\dim_B$ for the dimension of the Hilbert space of the B part. In the limit $\dim_B \geq \dim_A \gg 1$, it has been shown [50] that

$$S_A \simeq \ln(\dim_A) - \frac{\dim_A}{2\dim_B}. \tag{4.16}$$

In particular, when $\dim_B \gg \dim_A \gg 1$, we find that $S_A \simeq \ln(\dim_A)$.

To get a more physical picture of this formula, we can consider the system to be made of spin-$\frac{1}{2}$, with V_A spin-$\frac{1}{2}$ for A and V_B spin-$\frac{1}{2}$ for B. The Hilbert-space dimensions are $\dim_A = 2^{V_A}$ and $\dim_B = 2^{V_B}$, leading to $S_A \simeq V_A \ln 2$. Thus, for a random state, the entanglement entropy is proportional to the volume of the subsystem A, meaning that the entanglement entropy obeys a volume law.

We can now move on to the case of gapped phases. We denote the correlation length by η. We consider a geometrical bipartition of the system into A and B as depicted in Fig. 4.2. For one-dimensional gapped systems, if the size of A is large enough compared with η, then the entanglement entropy does not depend on the length V_A, that is, S_A is constant. This statement can be proved and an upper bound on the constant can be found [32].

For higher-dimensional systems, it is conjectured that the entanglement entropy (see e.g. [20] for an extensive discussion) satisfies

$$\mathcal{S}_A \simeq \alpha\mathcal{L}, \tag{4.17}$$

where $\mathcal{L} \gg \eta$ denotes the area of the surface that separates A from B and α is a constant. Thus, the entanglement entropy for a gapped system satisfies an area law

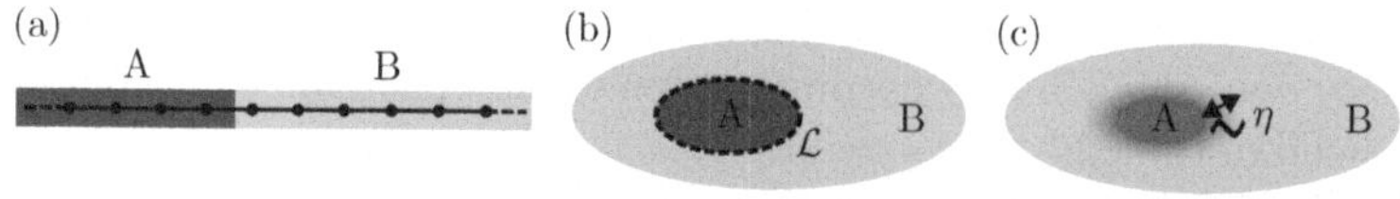

Fig. 4.2 [Colour online] Schematic description of the bipartite geometrical partition for a one-dimensional system (a) and for a two-dimensional system (b). (c) illustrates the small region around the boundary between A and B (with a thickness of the order of the correlation length η) that is relevant in the entanglement entropy when considering a gapped phase.

(as opposed to the volume law of the random state). In two dimensions, $\mathcal{L}$ is just the perimeter of the boundary between A and B (see Fig. 4.2(b)). Here we should make two remarks. First, one-dimensional gapped systems also obey the area law (simply set $\mathcal{L}$ to 1, the boundary being just a point). Second, this is a major difference from the case of a random state, where one gets a volume law for the entanglement entropy. Intuitively, if the correlation length is finite, then we expect that only the region around the boundary between A and B, whose thickness is of the order of a few η's (as shown in Fig. 4.2(c)) should matter in the entanglement between A and B.

For two-dimensional topological phases, we can go beyond the area-law contribution. It was shown in [35] and [40] that the first correction to this area law is a constant term γ:

$$\mathcal{S}_A \sim \alpha\mathcal{L} - \gamma. \tag{4.18}$$

While α is non-universal, this is not the case for the sub-leading term γ. This latter is called the topological entanglement entropy. It is a constant for a given topologically ordered phase:

$$\gamma = \ln\left(\frac{\mathcal{D}}{d_a}\right). \tag{4.19}$$

For a given type of excitation a, the quantum dimension d_a defines how the Hilbert-space dimension exponentially increases with the number of such excitations. Each type of excitation corresponds to a topological sector. Abelian excitations have a quantum dimension equal to 1, while non-Abelian ones have $d_a > 1$. The total quantum dimension is given by $\mathcal{D} = \sqrt{\sum_a d_a^2}$. These quantum dimensions characterize the topological field theory describing the phase and thus the nature of the system excitations. Note that in (4.19), the a of the d_a term corresponds to the topological sector of the wavefunction $|\Psi\rangle$ whose entanglement entropy is computed.

The topological entanglement entropy appears as a way to characterize the topological order of a phase. However, its practical calculation depends on scaling arguments, which might be hard to obtain to sufficient accuracy from numerical calculations [31, 36]. Moreover, it does not uniquely determine the topological order in the state. For that reason, it is interesting to look at the full spectrum of the reduced density matrix and not to reduce it to a single number.

4.2.4 The AKLT spin chain

We now move to a typical example of strongly correlated n-body quantum systems: the quantum spin chains. One of the simplest examples of a strongly correlated gapped system is the antiferromagnetic spin-1 chain. It is also one of the simplest examples of a symmetry-protected topological phase (here protected by spin rotation or time-reversal symmetries; see e.g. [66] for a short review) . More precisely, we focus on the Affleck–Kennedy–Lieb–Tasaki (AKLT) model [1, 2]. This system is the prototype of a gapped spin-1 chain [27]. The AKLT Hamiltonian of the one-dimensional spin-1 chain reads

$$H_{\text{AKLT}} = \sum_j \vec{S}_j \cdot \vec{S}_{j+1} + \frac{1}{3} \sum_j (\vec{S}_j \cdot \vec{S}_{j+1})^2. \tag{4.20}$$

The ground state of the AKLT Hamiltonian is a valence bond state. It can be understood within a simple picture sketched in Fig. 4.3. Each spin-1 can be written as two spin-$\frac{1}{2}$ combined in the triplet state. Between two neighbouring sites, two of the four spin-$\frac{1}{2}$ (one per site) are combined in the singlet sector. When an open chain is considered, the two extreme unpaired spin-$\frac{1}{2}$ (see Fig. 4.3) correspond to the edge excitations, leading to a fourfold-degenerate ground state (one singlet state and one triplet state).

To compute the ES of the AKLT ground state for an open chain, we first have to decide which of the four degenerate states we would like to analyse. In the sector of total spin $S_z = \pm 1$, there is only one state, so the choice is simple, while in the sector $S_z = 0$, there are two states. For the sake of simplicity, we focus on the $S_z = 1$ case. To cut the system into two parts, we can follow the same procedure as that described in Section 4.2.2: the A part will be made from the l_A consecutive leftmost sites and the B part from the remaining rightmost sites (see Fig. 4.3).

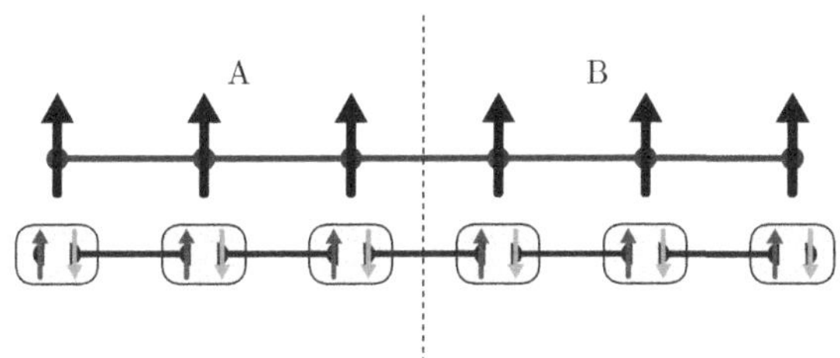

Fig. 4.3 [Colour online] Schematic description of the AKLT ground state. The upper chain shows the spin-1 AKLT chain and the lower chain its valence bond description. Each spin-1 is decomposed into two spin-$\frac{1}{2}$, one dark grey [red] and one light grey [blue], that are projected on the triplet state (depicted by a box). The AKLT ground state is obtained by projecting one dark grey [red] spin-$\frac{1}{2}$ of one site with one light grey [blue] spin-$\frac{1}{2}$ of the neighbouring site in the singlet state. We observed the two unpaired spin-$\frac{1}{2}$, one at each end of the spin chain. This figure also shows how the system is cut into two parts A and B during the ES calculation.

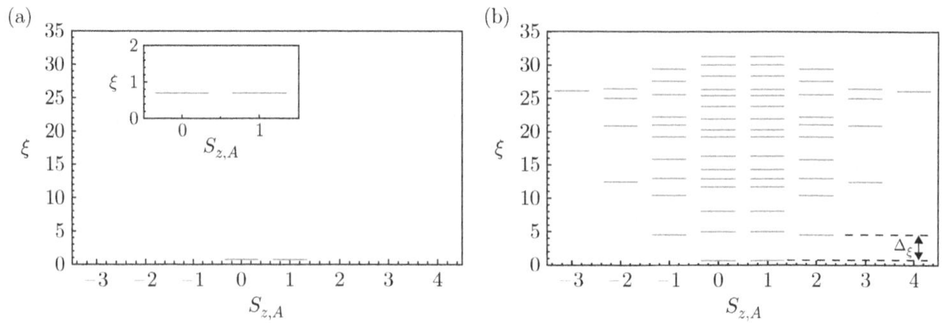

Fig. 4.4 (a) The ES for the AKLT ground state with 8 sites in the $S_z = 1$ sector. The system is cut into two equal parts of size $l_A = 4$. The ES only contains two levels, i.e. two non-zero eigenvalues in the reduced density matrix. This reflects the edge excitation (a spin-$\frac{1}{2}$) of the AKLT ground state. The inset is a zoom on these two levels. (b) The ES for the Heisenberg spin-1 chain with 8 sites in the $S_z = 1$ sector. The system is cut in a similar way to the AKLT case. The two states with the lowest entanglement energies (i.e. the two largest eigenvalues of the reduced density matrix) are similar to those of the AKLT ground state. We also show the entanglement gap Δ_ξ between the low-entanglement-energy structure similar to the AKLT ground state and the higher-entanglement-energy states.

Figure 4.4(a) displays the ES for a AKLT open chain with 8 sites and $l_A = 4$. The entanglement energies ξ are plotted versus $S_{z,A}$, the z projection of the A-part total spin. The reduced density matrix has only two non-zero eigenvalues, whereas the size of the reduced density matrix is 81×81. This dramatic reduction in the number of non-zero eigenvalues compared with a random state is a major characteristic that we will observe for many model states. If we think about the cut as an artificial edge that we have introduced in the system, then the physical interpretation becomes obvious: what we observe here is a spin-$\frac{1}{2}$ edge excitation of the AKLT chain. This is the first example where the Li–Haldane conjecture [41] can be observed: for this gapped phase, the ES is directly related to the spectrum of the edge excitation. Note that the true edge excitations of the system do not play any role here, since our choice of the AKLT ground state in the $S_z = 1$ sector freezes these excitations.

The AKLT Hamiltonian is the prototype of the gapped quantum spin-1 chain, and so it is interesting to look at the behaviour of the ES away from this specific case. The simplest case that one can consider is the Heisenberg spin-1 chain, where the Hamiltonian is simply given by $H = \sum_j \vec{S}_j \cdot \vec{S}_{j+1}$. In Fig. 4.4(b), we consider a similar situation to that for the AKLT model of Fig. 4.4(a). At the bottom of the ES, we recover two states, with the same quantum numbers as the AKLT case. In contrast to the latter, however, we also observe some levels with higher entanglement energy. Because the AKLT ground state and the Heisenberg spin-1 chain are adiabatically connected, we would like to argue that the low-entanglement-energy structure in the Heisenberg spin-1 ES will characterize the system. We define the entanglement gap Δ_ξ as the minimum difference in entanglement energy between the low-entanglement-energy structure similar to a model state (the AKLT model in this example) and the entanglement energy levels above this structure. The meaning of Δ_ξ is actually the Li–Haldane conjecture away from model states: if this entanglement gap stays finite in the thermodynamic limit, then the edge excitations of the system will be in the same universality class as the model state whose ES reduces to the same low-entanglement-energy structure. Indeed, numerical simulations based on the density matrix renormalization group [14] have shown such a property for the Heisenberg model.

To summarize, this example has already been able to show us several features of the ES. For some model states, the number of non-zero eigenvalues might be related to the edge excitation of the system. This number can be exponentially smaller than what we would expect from a random state, which is a non-trivial signature. Away from this ideal situation and as long as we stay in the same universality class, we would expect to observe a similar fingerprint to that of the model state in the low-entanglement-energy part of the spectrum. This structure should be protected from the higher-entanglement-energy levels by an entanglement gap.

4.2.5 Matrix product states and the entanglement spectrum

The understanding and simulation of quantum many-body states in one space dimension has experienced revolutionary progress with the advent of the density matrix renormalization group [81]. In modern language, this method can be viewed as a variational optimization over the set of matrix product states (MPS) [23, 53]. Let us consider a quantum state

$$|\Psi\rangle = \sum_{\{m_i\}} c_{\{m_i\}} |m_1, \ldots, m_{N_{\text{orb}}}\rangle, \tag{4.21}$$

where the $\{m_i\} = \{m_1, \ldots, m_{N_{\text{orb}}}\}$ are a set of physical indices such as a spin up or down or an occupied or empty orbital:

$$|\Psi\rangle = \sum_{\{m_i\}} (C^{[m_1]} \cdots C^{[m_{N_{\text{orb}}}]})_{\alpha_L, \alpha_R} |m_1, \ldots, m_{N_{\text{orb}}}\rangle, \tag{4.22}$$

where $\{C^{[m]}\}$ is a set of matrices (each orbital might require a different set of matrices) and α_L and α_R are boundary conditions that pick one matrix element of the matrix product (taking the trace being another option). The $C^{[m]}_{\alpha,\beta}$ matrices have two types of indices: $[m]$ is the physical index and (α, β) are the bond indices (or auxiliary space indices), with $\alpha, \beta, = 1, \ldots, \chi$, where χ is called the bond dimension. Such a rewriting of a state decomposition is always possible. When the bond dimension χ of the matrix $C^{[m]}$ is much smaller than the size of the n-body Hilbert space, this formulation provides a more economical representation of the state. The crucial question is how small can χ be for (4.22) to still be an exact statement. Generic one-dimensional gapped systems can be approximated by finite χ [79]. Critical systems, however, require an MPS with an infinite bond dimension [17, 49].

The AKLT ground state that we have discussed in the previous section can be expressed in a rather simple MPS form. In that case, N_{orb} is the number of spin-1, and the physical index m can take three different values $-1, 0, +1$ corresponding to the three values of S_z. The MPS representation requires three 2×2 matrices

$$C^{[0]} = \begin{pmatrix} -\sqrt{\frac{1}{3}} & 0 \\ 0 & \sqrt{\frac{1}{3}} \end{pmatrix}, \qquad C^{[+1]} = \begin{pmatrix} 0 & \sqrt{\frac{2}{3}} \\ 0 & 0 \end{pmatrix}, \qquad C^{[-1]} = \begin{pmatrix} 0 & 0 \\ -\sqrt{\frac{2}{3}} & 0 \end{pmatrix}. \tag{4.23}$$

As an exercise, we can check that we indeed reproduce the ground state of the AKLT model. We focus on the ground state $\Psi^N_{1,1}$ in the $S_z = 1$ sector. For $N = 2$ spins, it decomposes into

$$\Psi^{N=2}_{1,1} = \tfrac{\sqrt{2}}{2}|+1, 0\rangle - \tfrac{\sqrt{2}}{2}|0, +1\rangle, \tag{4.24}$$

where $|s_{z,1}, s_{z,2}\rangle = |s_{z,1}\rangle \otimes |s_{z,2}\rangle$ is the many-body basis where the first (respectively second) spin has its spin projection along z equal to $s_{z,1}$ (respectively $s_{z,2}$). Using the matrices of (4.23), we find that, picking the entry in the first row and second column, we recover (up to a normalization factor) the coefficients of the decomposition of (4.24). Choosing the entry in the matrix product (i.e. α_L and α_R) is akin to select the boundary conditions. Note that this matrix element is zero for products such as $C^{[+1]}C^{[+1]}$ or $C^{[0]}C^{[0]}$, as it should be since $\Psi^{N=2}_{1,1}$ has $S_z = 1$.

A similar calculation can be performed for $N = 3$ spins. There, the decomposition of $\Psi_{1,1}$ is

$$\Psi^{N=3}_{1,1} = \tfrac{1}{\sqrt{7}}\left(|+1, 0, 0\rangle - |0, +1, 0\rangle + |+1, 0, 0\rangle\right) - \tfrac{2}{\sqrt{7}}|+1, -1, +1\rangle. \tag{4.25}$$

Once again, we can explicitly check that the MPS description leads to the correct decomposition. Note that $\Psi^{N=3}_{1,1}$ has weight neither on $|-1, +1, +1\rangle$ nor on $|+1, +1, -1\rangle$, despite having $S_z = 1$. The MPS description also gives such a result.

In this example, the size of the $C^{[m]}$ matrices is equal to the number of non-zero eigenvalues observed in the ES. As we will now show, these two quantities are related. A way to create a bipartite partition of the system is to consider A being made of the indices $\{m_1, \ldots, m_{l_A}\}$ and B built from the indices $\{m_{l_A+1}, \ldots, m_{N_{\text{orb}}}\}$. Following the notation of (4.2), we have $\{|\mu_A\rangle = |m_1, \ldots, m_{l_A}\rangle\}$ and $\{|\mu_B\rangle = |m_{l_A+1}\rangle, \ldots, m_{N_{\text{orb}}}\}$. The MPS formulation of (4.22) can be rewritten to make this partition apparent:

$$|\Psi\rangle = \sum_{\alpha=1}^{\chi} \sum_{\{m_i\}} (C^{[m_1]} \cdots C^{[m_{l_A}]})_{\alpha_L,\alpha} \\ (C^{[m_{l_A+1}]} \cdots C^{[m_{N_{\text{orb}}}]})_{\alpha,\alpha_R} |m_1, \ldots, m_{N_{\text{orb}}}\rangle. \tag{4.26}$$

Thus, we obtain that

$$|\Psi\rangle = \sum_{\alpha=1}^{\chi} |A : \alpha\rangle \otimes |B : \alpha\rangle, \tag{4.27}$$

with

$$|A:\alpha\rangle = \sum_{\{m_i\}} (C^{[m_1]}\dots C^{[m_{l_A}]})_{\alpha_L,\alpha}|m_1,\dots,m_{l_A}\rangle, \tag{4.28}$$

$$|B:\alpha\rangle = \sum_{\{m_i\}} (C^{[m_{l_A+1}]}\dots C^{[m_{N_{\rm orb}}]})_{\alpha,\alpha_R}|m_{l_A+1}\rangle,\dots,m_{N_{\rm orb}}. \tag{4.29}$$

While this decomposition looks similar to the Schmidt decomposition of (4.4), the states $|A:\alpha\rangle$ and $|B:\alpha\rangle$ are neither orthonormal nor linearly independent. Two extra steps are actually required to obtain the true Schmidt decomposition: we need to extract an orthonormal complete basis from $\{|A:\alpha\rangle\}$ and $\{|B:\alpha\rangle\}$ and then perform an SVD on the entanglement matrix. But these extra steps can only (at worst) reduce the number of terms in the sum of (4.27). Indeed, denoting by $\{|A:\tilde{\alpha}\rangle\}$ (respectively $\{|B:\tilde{\beta}\rangle\}$) the orthonormalized basis extracted from $\{|A:\alpha\rangle\}$ (respectively $\{|B:\alpha\rangle\}$), we can introduce two transformation matrices U and V such that

$$|A:\alpha\rangle = \sum_{\tilde{\alpha}} U_{\alpha,\tilde{\alpha}}|A:\tilde{\alpha}\rangle, \qquad |B:\alpha\rangle = \sum_{\tilde{\beta}} V_{\alpha,\tilde{\beta}}|B:\tilde{\beta}\rangle\,. \tag{4.30}$$

These bases have dimensions lower than or equal to the bond dimension χ. We immediately find that

$$|\Psi\rangle = \sum_{\tilde{\alpha},\tilde{\beta}} \left(U^t V\right)_{\tilde{\alpha},\tilde{\beta}} |A:\tilde{\alpha}\rangle \otimes |B:\tilde{\beta}\rangle\,. \tag{4.31}$$

The entanglement matrix can be directly read out from the previous equation, leading to the ES once the SVD is performed on U^tV. As a consequence, if we want to write an exact MPS for $|\Psi\rangle$, the bond dimension χ cannot be lower than the number of non-zero eigenvalues of the reduced density matrix. The latter number gives the optimal size for the MPS representation of a state (as discussed in the case of the AKLT ground state). Thus, any massive reduction of the system ES should be the sign of an efficient MPS representation of a quantum state.

4.3 Observing an edge mode through the entanglement spectrum

In the previous section, we have shown how the ES is able to reveal the gapped edge physics of a system such as the AKLT model. We will now discuss the case of gapless edge excitation of a topological phase. Remarkably, even a non-interacting system such as the IQHE allows us to illustrate this unique property of the ES.

4.3.1 The integer quantum Hall effect

4.3.1.1 Overview and notation

We consider a completely filled single Landau level. For the sake of simplicity and without loss of generality, we focus on the lowest Landau level (LLL). Using the symmetric gauge $\vec{A} = (-yB, xB, 0)$, the one-body wavefunctions are given by

$$\phi_m(z) = \frac{1}{\sqrt{2\pi 2^m m!}} z^m e^{-|z|^2/4l_B^2}, \tag{4.32}$$

where $l_B = \sqrt{\hbar/eB}$ is the magnetic length, which we will set to 1, and $z = x + iy$ is the particle complex coordinate in the plane. $m \geq 0$ is an integer corresponding to the angular momentum. If we have a radial confining potential with a slow variation compared with the magnetic length, then the Landau levels are just bent and follow the confining potential as depicted in Fig. 4.5. The edge mode of the IQHE corresponds to a non-interacting chiral state, with a linear dispersion relation $E = (2\pi v/\mathcal{L})n$, where $n \geq 0$ is an integer, v is the edge mode velocity, and $\mathcal{L}$ is the length of the edge. From this one-particle spectrum, we can build the many-body spectrum of the chiral edge as depicted in Fig. 4.6. For simplicity, we drop the energy scale $2\pi v/\mathcal{L}$. We start by filling all the levels up to the Fermi level (Fig. 4.6(a)). We define the total energy as the energy reference (i.e. $E = 0$). The lowest-energy excitation $E = 1$ is unique and shown in Fig. 4.6(b): it gives to a single level in the many-body spectrum at momentum $n = 1$. At the energy $E = 2$, we have two possible excitations, Fig. 4.6(c) and (d), leading to two levels at momentum $n = 2$. Performing a similar reasoning for the other excitations, we end up with the counting per momentum $1, 1, 2, 3, 5, \ldots$ depicted in Fig. 4.6(e). This is the counting of the chiral $U(1)$ bosons.

The bulk of the IQHE is actually quite simple to describe. Since the system is a filled band, we are filling all the orbitals starting from the one with the lowest momentum. The number of occupied orbitals is given by the number of flux quanta N_Φ in the system. For N electrons, the many-body wavefunction simply reads

$$|\Psi_{\text{IQHE}}\rangle = \left(\prod_{m=0}^{N-1} c_m^\dagger\right)|0\rangle, \tag{4.33}$$

where $c_m^\dagger$ is the creation operator associated with the one-body wavefunction $\phi_m(z)$. In first-quantized notation, and up to a normalization factor, (4.33) simply becomes

$$\Psi_{\text{IQHE}}(z_1, \ldots, z_N) = \prod_{i<j}(z_i - z_j)\exp(-\tfrac{1}{4}\sum_i |z_i|^2). \tag{4.34}$$

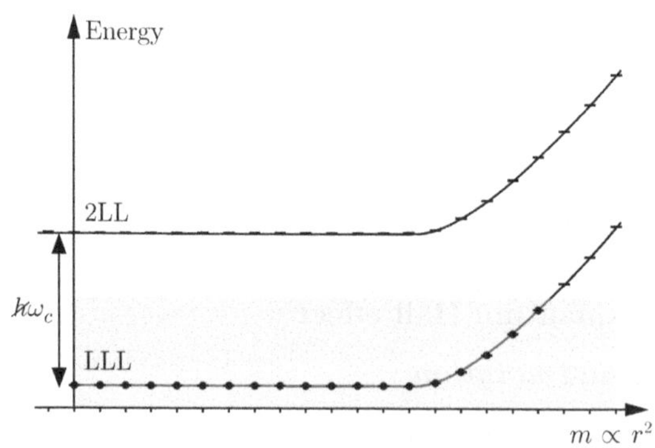

Fig. 4.5 Schematic description of the Landau levels in the presence of a slowly varying confining potential. Here only the lowest Landau level (LLL) and the second Landau level (2LL) are depicted. The shading denotes the region below the Fermi level.

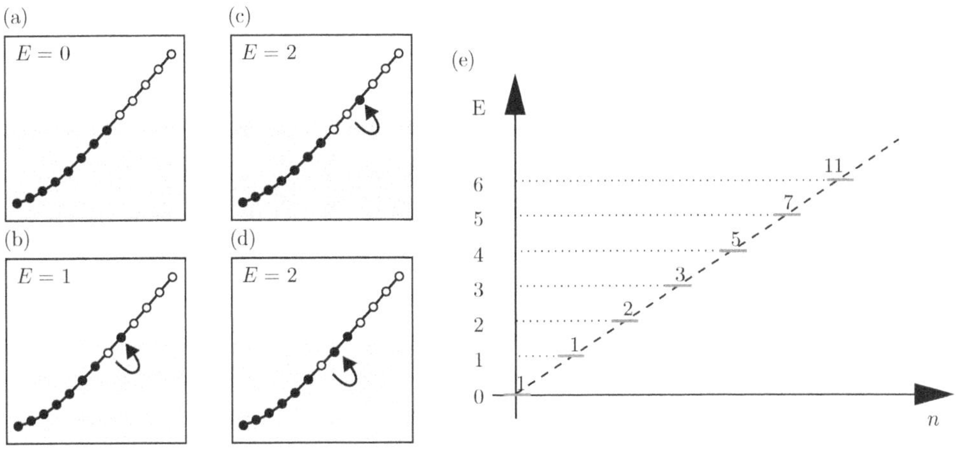

Fig. 4.6 Construction of the many-body energy spectrum of the IQHE chiral edge mode. (a–d) Starting with the filled situation (a) defining our reference energy, we can consider the lowest-energy excitation (b) by moving the topmost particle to the next empty level. (c) and (d) show the two possible excitations with energy $E = 2$. (e) The many-body energy E of the IQHE chiral edge mode as a function of the momentum m. The number above each level gives its degeneracy.

4.3.1.2 Real-space partition

We will now discuss how to perform a partition in real space of our system. Note that this discussion is generic and not specific to the IQHE. Our goal is to divide the space into two separate regions A and B. Consider the creation operator $c^\dagger$ for a given orbital. We want to split this operator into two parts A and B such that

$$c^\dagger = \alpha c_A^\dagger + \beta c_B^\dagger, \tag{4.35}$$

where $c_A^\dagger$ (respectively $c_B^\dagger$) creates an electron living only in A (respectively B) and α and β are constants. Since $\{c, c^\dagger\} = \{c_A, c_A^\dagger\} = \{c_B, c_B^\dagger\} = 1$, the two constants should satisfy $|\alpha|^2 + |\beta|^2 = 1$. Let $\Psi(\vec{r}) = \langle \vec{r}|c^\dagger|0\rangle$ denote the wavefunction in first-quantized notation. We can perform the following decomposition:

$$\begin{aligned}\Psi(\vec{r}) &= \Theta_A(\vec{r})\Psi(\vec{r}) + [1 - \Theta_A(\vec{r})]\,\Psi(\vec{r}),\\ &= \mathcal{N}_A\Psi_A(\vec{r}) + \mathcal{N}_B\Psi_B(\vec{r}),\end{aligned} \tag{4.36}$$

where $\Theta_A(\vec{r})$ is 1 if $\vec{r}$ belongs to A and 0 otherwise, and where the two normalized (and orthogonal) wavefunctions Ψ_A and Ψ_B are given by

$$\Psi_A(\vec{r}) = \frac{1}{\mathcal{N}_A}\Theta_A(\vec{r})\Psi(\vec{r}), \qquad \mathcal{N}_A = \int_A d\vec{r}\,|\Psi(\vec{r})|^2, \tag{4.37}$$

$$\Psi_B(r) = \frac{1}{\mathcal{N}_B}[1 - \Theta_A(\vec{r})]\,\Psi(\vec{r}), \qquad \mathcal{N}_B = \int_B d\vec{r}\,|\Psi(\vec{r})|^2. \tag{4.38}$$

Notice that $\mathcal{N}_A^2 + \mathcal{N}_B^2 = 1$. Comparing (4.35) and (4.36), we immediately deduce that $\alpha = \mathcal{N}_A$ and $\beta = \mathcal{N}_B$. The treatment that we have performed here is generic and does not depend on any specific property of Ψ.

4.3.1.3 Back to the IQHE

Equipped with the decomposition mentioned in Section 4.3.1.2, we can now perform the real-space ES of the IQHE. While there is in principle no constraint on the choice of A, we would like to preserve as many quantum numbers as possible. If we use the symmetric gauge, a natural choice is to preserve the rotational symmetry. Thus, we can decide to take for A a disc centred at the origin and with a radius R. Such a cut was described in [62] to compute the entanglement entropy of the IQHE. Each orbital $\Phi_m(z)$ can be decomposed, following (4.36), as

$$\Phi_m(z) = \alpha_m \Phi_{m,A}(z) + \beta_m \Psi_{m,B}(z), \tag{4.39}$$

with

$$\alpha_m^2 = \frac{1}{2^m m!} \int_0^R dr\, r^{2m+1} e^{-r^2/2}, \qquad \beta_m^2 = 1 - \alpha_m^2. \tag{4.40}$$

α_m^2 and β_m^2 are incomplete gamma functions and can easily be computed numerically. Since $\Phi_{m,A}(z)$ are eigenstates of the angular momentum, they are still orthogonal (similarly for $\Psi_{m,B}(z)$).

Performing the Schmidt decomposition for such a partition is rather easy. For pedagogical reasons, we first consider the IQHE with only two particles occupying the two first orbitals $m = 0$ and $m = 1$. The ground-state wavefunction $|\Psi_{\text{IQHE}}\rangle$ reads

$$|\Psi_{\text{IQHE}}\rangle = c_0^\dagger c_1^\dagger |0\rangle. \tag{4.41}$$

Substituting the expression (4.35) for each creation operator, we get

$$\begin{aligned} |\Psi_{\text{IQHE}}\rangle &= \alpha_0 \alpha_1 c_{0,A}^\dagger c_{1,A}^\dagger |0\rangle \\ &\quad + \beta_0 \beta_1 c_{0,B}^\dagger c_{1,B}^\dagger |0\rangle \\ &\quad + (\alpha_0 \beta_1 c_{0,A}^\dagger c_{1,B}^\dagger |0\rangle - \alpha_1 \beta_0 c_{1,A}^\dagger c_{0,B}^\dagger |0\rangle). \end{aligned} \tag{4.42}$$

The first term (respectively second term) on the right hand side of (4.42) corresponds to the case where we have two particles in A (respectively B). The third term corresponds to the case where we have one particle in A and one in B (notice the minus sign due to the fermionic statistics). When performing such a cut, we have two good quantum numbers:

- The total number of particles $N = N_A + N_B$, where N_A (respectively N_B) is the number of particles in the region A (respectively B).
- The total angular momentum $L_z = L_{z,A} + L_{z,B}$, where $L_{z,A}$ (respectively $L_{z,B}$) is the total angular momentum of the particles in A (respectively B).

The ES can be read out directly from the Schmidt decomposition of (4.42) and sorted per quantum numbers:

$N_A = 0$, $L_{z,A} = 0$: one level at $-\ln(|\beta_0 \beta_1|^2)$;

$N_A = 1,\ L_{z,A} = 0$: one level at $-\ln(|\alpha_0\beta_1|^2)$;
$N_A = 1,\ L_{z,A} = 1$: one level at $-\ln(|\alpha_1\beta_1|^2)$;
$N_A = 2,\ L_{z,A} = 1$: one level at $-\ln(|\alpha_0\alpha_1|^2)$.

As an exercise, we let the reader check that the reduced density matrix associated with this ES is properly normalized to 1. For the generic case (N electrons occupying the N first orbitals), a similar calculation of the Schmidt decomposition would give

$$|\Psi_{\text{IQHE}}\rangle = \sum_{N_A=0}^{N} \sum_{L_{z,A}} \sum_{\substack{\{m_1,\ldots,m_{N_A}\},\\ \sum_i m_i = L_{z,A}}} \prod_{i=1}^{N_A} \alpha_{m_i} \prod_{j=N_A+1}^{N} \beta_{m_j} |\{m_1,\ldots,m_{N_A}\}\rangle \otimes |\{m_{N_A+1},\ldots,m_N\}\rangle. \tag{4.43}$$

The third sum runs over all the possible ways to choose N_A particles among the N occupied orbitals with the constraint that these N_A particles have a total angular momentum $L_{z,A}$. Each valid configuration $\{m_1,\ldots,m_{N_A}\}$ where the m_i are ordered from the smallest to the largest integer leads to a state $|\{m_1,\ldots,m_{N_A}\}\rangle$ up to a sign (a consequence of the possible orbital reordering and the fermionic statistics, as was mentioned after (4.42)). Once the $\{m_1,\ldots,m_{N_A}\}$ have been selected, the occupied orbitals $\{m_{N_A+1},\ldots,m_N\}$ for the particles in B are automatically defined, giving the state $|\{m_{N_A+1},\ldots,m_N\}\rangle$. Once again, the ES follows directly from such a decomposition since it is a Schmidt decomposition: the orthogonality conditions are satisfied since two different sets $\{m_1,\ldots,m_{N_A}\}$ give two orthogonal (and normalized) states.

If we focus on a single N_A sector, we can count how many entanglement energies we have per angular momentum. For concreteness, let us take $N_A = 6$ and $N = 12$. The system has 12 orbitals with angular momentum going from 0 to 11. Each $\{m_1,\ldots,m_{N_A}\}$ can be represented with boxes corresponding to each orbitals as depicted in Fig. 4.7. The smallest angular momentum $L_{z,A} = 15$ is obtained by occupying the first six orbitals as shown in Fig. 4.7(a). Thus, there is a single level in the ES for $N_A = 6$ and $L_{z,A} = 15$. For $L_{z,A} = 16$, there is also a unique configuration (and thus a unique level), as shown in Fig. 4.7(b). Moving on to $L_{z,A} = 17$, we now have two options, shown in Fig. 4.7(c, d). We immediately see that this construction is identical to that of the many-body spectrum for the chiral edge mode. So this latest ES and the ES in a given particle number sector have the same counting of levels.

Do the entanglement energies also follow a linear dispersion relation? The ES for $N_A = N/2 = 6$ is shown in Fig. 4.8. The finite-size effects lead to two deviations in the physical spectrum of the edge mode:

- They spoil the edge-mode counting for large angular momentum. Indeed, for a given number of flux quanta, there is a maximum angular momentum that can be reached.
- As can be observed in Fig. 4.8, levels are not strictly degenerate for a given angular momentum, as they should be. This property is only recovered in the thermodynamic limit.

(a)

$m=0$	1	2	3	4	5	6	7	8	9	10	11
1	1	1	1	1	1	0	0	0	0	0	0

(b)

$m=0$	1	2	3	4	5	6	7	8	9	10	11
1	1	1	1	1	0	1	0	0	0	0	0

(c)

$m=0$	1	2	3	4	5	6	7	8	9	10	11
1	1	1	1	1	0	0	1	0	0	0	0

(d)

$m=0$	1	2	3	4	5	6	7	8	9	10	11
1	1	1	1	0	1	1	0	0	0	0	0

Fig. 4.7 Valid configurations that appear in the Schmidt decomposition for $N_A = N/2 = 6$. Each box represents an orbital with momentum m (the label on top) and can be occupied (1) or empty (0). (a) The configuration with the smallest angular momentum $L_{z,A} = 15$. (b) The only configuration with $L_{z,A} = 16$. For $L_{z,A} = 17$, we have the two options shown in (c) and (d).

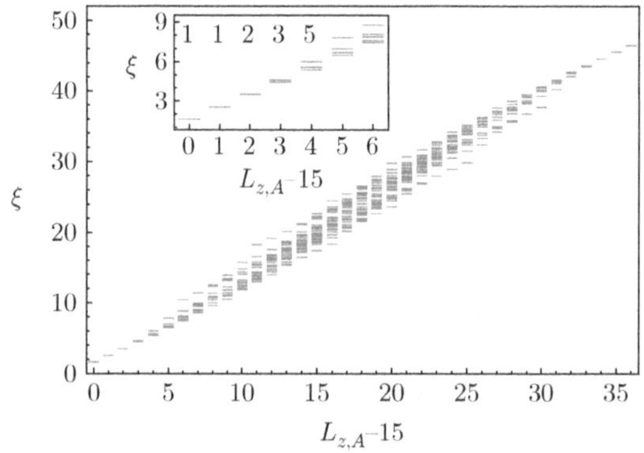

Fig. 4.8 Real-space ES of the $\nu = 1$ IQHE on the disc geometry for $N = 12$ fermions and $N_\Phi = 11$. We focus on the sector $N_A = 6$. The inset displays the counting in the thermodynamic part (the RSES has some exact degeneracies). It matches that of a $U(1)$ chiral boson.

We might wonder why we looked at a single sector of N_A. A key idea of Li and Haldane's approach when they introduced the ES was to look at some specific block of the reduced density matrix. When we cut the system into two, we want A to be a smaller droplet of the same quantum fluid up to edge excitation. In this picture, it is natural to look at a fixed number of particles. We can even be more quantitative. Consider the eigenstates of the reduced density matrix (i.e. the $|A : i\rangle$ of the Schmidt decomposition of (4.4)). The eigenstate related to the level with the smallest $L_{z,A}$ is a many-body state where all the first N_A lowest orbitals in the $\phi_{A,m}$ basis (i.e. each orbital from $m = 0$ to $m = N_A - 1$) are occupied. This is the densest state that one can create for $\nu = 1$ IQHE with N_A particles.

4.3.2 Chern insulators

The construction of the ES that we have done for the IQHE can be extended to any non-interacting topological insulators. For simplicity, we will focus on the first and simplest example of a topological insulator, the Chern insulator (CI) that was introduced in a theoretical work by F. D. M. Haldane in 1988 [28]. It is defined by

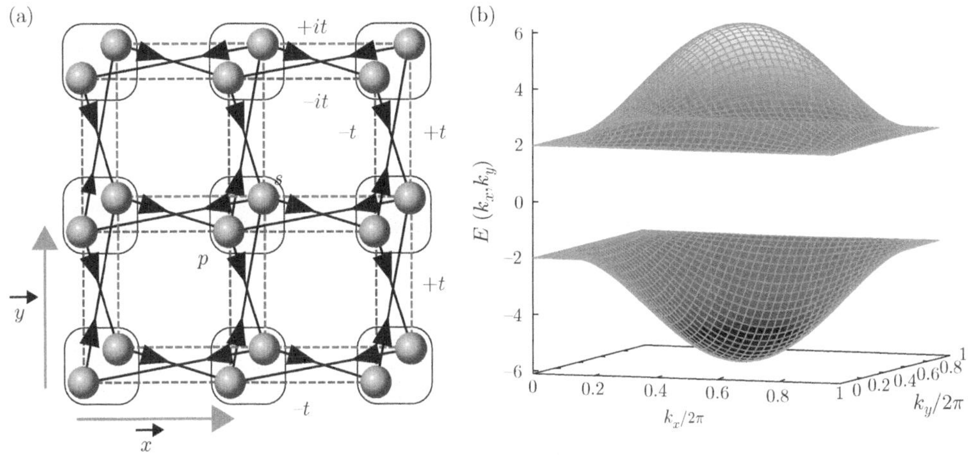

Fig. 4.9 [Colour online] (a) The two-orbital lattice model with one s and one p orbital per site. (b) The band structure for the two-orbital lattice model with a mass term set to $M = 1$, plotted as a function of the momenta k_x and k_y. Each band carries a Chern number $C = \pm 1$.

a non-zero Chern number C of the occupied bands. This (first) Chern number is a topological invariant, computed over the Brillouin zone, that characterizes a given band. A key feature is that a non-zero Chern number results [77] in a quantized Hall conductance $\sigma_{xy} = (e^2/h)C$, similar to the quantum Hall effect, but now without the requirement for an external magnetic field.

A typical example of a Chern insulator is shown in Fig. 4.9(a). It is based on a simple tight-binding model on a square lattice with two orbitals (one s and one p orbital) per site (see [82] for a more detailed description). With a suitable choice for the hopping amplitudes, the Bloch Hamiltonian for this model reads

$$\mathcal{H}(\mathbf{k}) = \sum_{i=x,y} [-\sin(k_i)\sigma_i] - \left[M - \sum_{i=x,y} \cos(k_i)\right] \sigma_z, \tag{4.44}$$

where $\sigma_x, \sigma_y, \sigma_z$ are the three Pauli matrices and M is a mass term (chemical potential between the two orbitals). The band structure is shown in Fig. 4.9(b), with two bands, each carrying a Chern number equal to ± 1 if $|M| < 2$ and separated by a gap. If the model is put on a cylinder when the system is in a topological phase (meaning that the Chern number is non-zero), the energy spectrum clearly exhibits one chiral gapless edge mode at each end of the cylinder, as shown in Fig. 4.10(b). Setting $|M| > 2$ to drive the system to a trivial phase will suppress these gapless edge modes as in Fig. 4.10(c).

4.3.3 Entanglement spectrum for a CI

By tuning the mass parameter M of the two-orbital model described previously, we can drive the system from a topological insulator to a trivial insulator. In these two

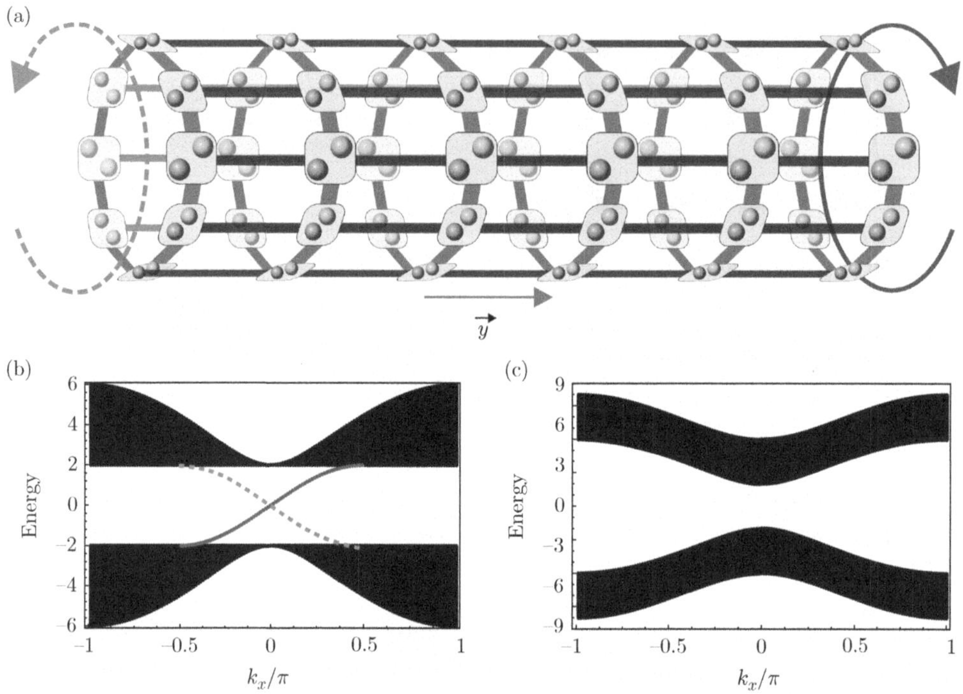

Fig. 4.10 [Colour online] (a) Two-orbital lattice model on a cylinder. At each end, there is one chiral gapless edge mode (the full [blue] and dashed [red] arrows). (b) Band structure for the two-orbital lattice model on a cylinder with a mass term set to $M = 1$, plotted as a function of the momentum k_x along the cylinder radius. In the gap region, we clearly observe the dispersion relation of the two gapless edge modes (full [blue] and dashed [red]), going from one band to another (black). (c) Band structure for the two-orbital lattice model on a cylinder with a mass term set to $M = 3$. The system is in trivial (non-topological) phase and no gapless edge modes are observed.

cases, the energy spectrum would look similar, exhibiting a bulk gap. Here we will show that the ES is able to distinguish between the trivial and the topological phases. For that purpose, we consider the case where we completely fill the lower band up to the Fermi energy ϵ_F located in the system bulk gap. The quantum state $|\Psi_{\rm CI}\rangle$ of the system is just a simple Slater determinant that can be written in second-quantized notation as

$$|\Psi_{\rm CI}\rangle = \prod_{k_x;\epsilon<\epsilon_F} c^\dagger_{k_x,\epsilon}|0\rangle, \tag{4.45}$$

where $c^\dagger_{k_x,\epsilon}$ is the creation operator related to the state with momentum k_x and energy ϵ. We fill all the states with an energy ϵ lower than the Fermi energy ϵ_F (see Fig. 4.11(a)). Any choice of ϵ_F is valid as long as it is located in the bulk gap. For the sake of simplicity, the calculation is done on the cylinder geometry, motivating

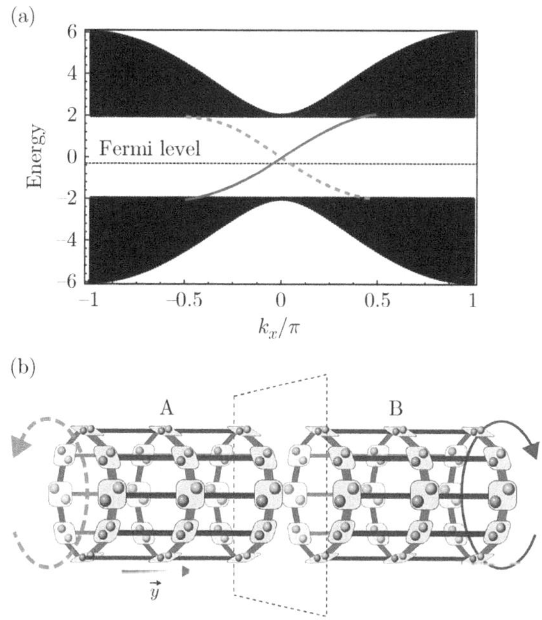

Fig. 4.11 [Colour online] (a) Band structure for the two-orbital lattice model on a cylinder with a mass term set to $M = 1$, where we set the Fermi level in the bulk gap. (b) Two-orbital lattice model on a cylinder cut into parts A (left) and B (right).

our choice to label the states using the momentum along the x direction, where we apply periodic boundary conditions. In this situation, the edge-mode excitations are completely frozen in a similar manner to the way in which choosing one of the ground state of the AKLT in Section 4.2.4 fixes the excitations at the edge.

We now cut the cylinder into two parts A (left part) and B (right part) as depicted in Fig. 4.11(b) (our partition is performed far from the edges). Performing the same decomposition as in (4.39), each creation operator can be written as a sum of two creation operators, one for each part of the system:

$$c^\dagger_{k_x,\epsilon} = \alpha^*_{k_x,\epsilon} c^\dagger_{k_x,\epsilon;A} + \beta^*_{k_x,\epsilon} c^\dagger_{k_x,\epsilon;B}, \tag{4.46}$$

where $\alpha_{k_x,\epsilon}$ (respectively $\beta_{k_x,\epsilon}$) is the weight of the state (k_x, ϵ) on the A (respectively B) part. These weights satisfy $|\alpha_{k_x,\epsilon}|^2 + |\beta_{k_x,\epsilon}|^2 = 1$. Using this decomposition, $|\Psi_{\rm CI}\rangle$ can be rewritten as

$$|\Psi_{\rm CI}\rangle = \sum_{N_A} \sum_{\{k_{x,A},\epsilon_A\}} \mathcal{N}_{\{k_{x,A},\epsilon_A\}} \mathcal{N}_{\{k_{x,B},\epsilon_B\}} |\{k_{x,A}, \epsilon_A\}\rangle \otimes |\{k_{x,B}, \epsilon_B\}\rangle. \tag{4.47}$$

Here N_A is the number of particles in part A, and the sum over $\{k_{x,A}, \epsilon_A\}$ corresponds to all the possible ways to choose N_A states among the original states used to build the Slater determinant of the full band. Note that once we have fixed the choice for the states we consider to be in A, the state for B is completely determined and unique, labelled here by $\{k_{x,B}, \epsilon_B\}$. This equation is the exact analogue of (4.43) for the IQHE. The other quantities in (4.47) are

$$\mathcal{N}_{\{k_{x,A},\epsilon_A\}} = \prod_{(k_x,\epsilon)\in\{k_{x,A},\epsilon_A\}} \alpha^*_{k_x,\epsilon}, \tag{4.48}$$

$$|\{k_{x,A,\epsilon_A}\}\rangle = \prod_{(k_x,\epsilon)\in\{k_{x,A},\epsilon_A\}} c^\dagger_{k_x,\epsilon;A}|0\rangle, \tag{4.49}$$

and similarly for the quantity with a B index.

By construction, we have just realized the Schmidt decomposition of $|\Psi_{\text{CI}}\rangle$. So the ES can be read out directly from (4.47). Indeed, the spectrum of the reduced density matrix is just given by $\{|\mathcal{N}_{\{k_{x,A},\epsilon_A\}}\mathcal{N}_{\{k_{x,B},\epsilon_B\}}|^2\}$. Unfortunately, this spectrum still requires a factorial effort to be computed owing to the combinatorial factor involved in choosing N_A states among the occupied ones, as discussed in Section 4.3.1.3. But, as pointed out in [24] and [62], this many-body ES can be deduced from a one-body ES similarly to the way in which one constructs the many-body energy spectrum of non-interacting particles from the one-body spectrum. For such a system, the reduced density matrix ρ_A can be rewritten as

$$\rho_A = \sum_\alpha e^{-\xi_\alpha}|A:\alpha\rangle\langle A:\alpha| = e^{-\hat{H}_{\text{ent}}}, \tag{4.50}$$

where the entanglement Hamiltonian $\hat{H}_{\text{ent}}$ is a one-body Hamiltonian

$$\hat{H}_{\text{ent}} = \sum_{k_x,i,j} h(k_x)_{i,j} c^\dagger_{k_x,i} c_{k_x,j}. \tag{4.51}$$

Here the indices i and j denote any site (or orbital) that belongs to the A part. Note that one can always define an entanglement Hamiltonian using (4.50). The possibility of expressing it as a one-body Hamiltonian is a specific feature of $|\Psi_{\text{CI}}\rangle$ being a product state (i.e. a single Slater determinant) or more generally a Gaussian state [11].

To compute the spectrum of $h(k_x)$, we use its relation to the propagator

$$G^A_{i,j}(k_x) = \langle\Psi_{\text{CI}}|c^\dagger_{k_x,i}c_{k_x,j}|\Psi_{\text{CI}}\rangle. \tag{4.52}$$

Indeed, theses two matrices $G^A(k_x)$ and $h(k_x)$ are related by the following equation (see [4] for a detailed derivation):

$$G^{A^t} = \frac{1}{1+e^{h(k_x)}}. \tag{4.53}$$

Note that this propagator can be written of a sum of projectors onto the the A part. As a consequence, its eigenvalues are between 0 and 1.

While the many-body ES exhibits the chiral edge mode of the CI (this ES is actually similar to the one that we will describe in Section 4.4.5 for interacting systems), the spectrum $G^A(k_x)$ (or $h(k_x)$) allows us to unveil the same information from a one-body

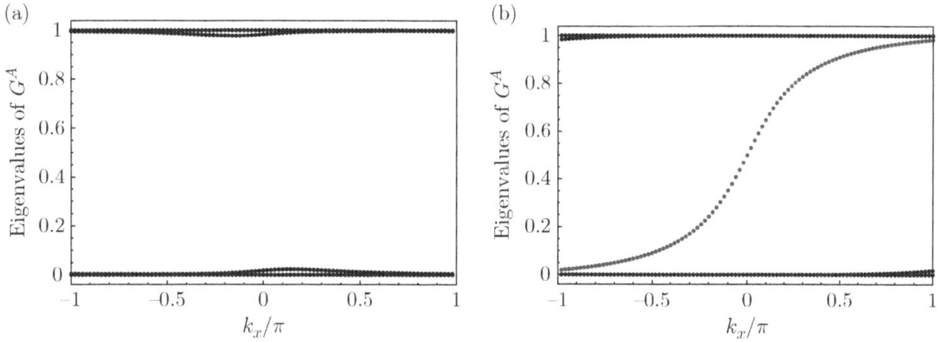

Fig. 4.12 [Colour online] (a) Eigenvalues of the propagator (4.52) for the two-orbital lattice with a mass term set to $M = 3$ (trivial phase). The partition is exactly half of the cylinder. There is a clear gap between eigenvalues localized around 0 and eigenvalues localized around 1. (b) Eigenvalues of the propagator (4.52) for the two-orbital lattice with a mass term set to $M = 1$ (topological phase). The partition is exactly half of the cylinder. The [blue] dots that mimic the chiral edge mode interpolate between the two bands. This is similar to the true edge mode shown in Fig. 4.10(b).

calculation. In Fig. 4.12, we show the spectrum of $G^A(k_x)$ for both the topological (Fig. 4.12(a)) and the trivial phases (Fig. 4.12(b)). Similar to its energy spectrum, the one-body ES of the trivial phases exhibits two bands (one located around 0, the other around 1) separated by a gap. On the other hand, in the topological case, there is clearly a chiral mode connecting the two bands. Like the AKLT model discussed in Section 4.2.4, the partition has introduced an artificial edge and the ES mimics the true edge spectrum of the system.

Compared with the spin chain discussed in Section 4.2.4, we see that plotting the entanglement energies as a function of conserved quantum number (here the momentum along the x direction) is really helpful to directly observe the edge mode. But it is more a matter of convenience than a requirement. Indeed, the ES can be used to diagnose topological order without this additional information. In the presence of disorder, k_x would not be a good quantum number. Still, by looking at the level statistics in the ES [57], we can differentiate a trivial phase from a topological phase. The CI case is one example of a two-dimensional system where the ES can be computed analytically. Actually, this derivation holds true for any of the non-interacting topological insulators. Unfortunately, such an analytical derivation is not feasible, in general, for strongly interacting systems, as we will now discuss in the case of the fractional quantum Hall system.

4.4 Fractional quantum Hall effect and entanglement spectra

In this section, we review the different aspects of entanglement spectra applied to the FQHE. We provide a short (and partial) introduction to this topic. We discuss the different partitions that have been proposed and their relation. In particular, we show

how much information about the excitations can be extracted from the ground state by using the entanglement spectra.

4.4.1 Fractional quantum Hall effect: overview and notation

In these lecture notes, we restrict attention to the case of spinless particles occupying the lowest Landau level. The natural geometry to consider is the plane (or disc). For technical reasons, other geometries having periodic boundary conditions (in one or two directions), such as the cylinder [61], the torus [30], and the sphere [26], are more convenient when it comes to finite-size (numerical) studies. In the following, we will mostly focus on genus-zero surfaces, in particular the disc and the sphere. We denote by N the number of particles in the system and by N_Φ the number of flux quanta. The filling factor is defined (in the thermodynamic limit) as $\nu = N/N_\Phi$. A convenient choice for the one-body basis on the plane (using the symmetric gauge as discussed in Section 4.3.1.1) and on the sphere leads to the following set of wavefunctions:

$$\phi_m(\mathbf{r}) = \begin{cases} \dfrac{1}{\sqrt{2\pi 2^m m!}} z^m e^{-|z|^2/4} & \text{(plane)}, \\ \sqrt{\dfrac{(N_\Phi+1)!}{4\pi m!(N_\Phi - m)!}}\, u^m v^{N_\Phi - m} & \text{(sphere)}. \end{cases} \tag{4.54}$$

On the plane (or disc), $z = x + iy$ is the particle coordinate and $L_z = m$ is the angular momentum (where $m \geq 0$ is an integer). On the sphere, $u = \cos(\frac{1}{2}\theta)\, e^{i\varphi/2}$ and $v = \sin(\frac{1}{2}\theta)\, e^{-i\varphi/2}$ are the spinor coordinates with the polar coordinates (θ, φ), and $L_z = \frac{1}{2}N_\Phi - m$ is the angular momentum along z, where $m = 0, 1, \ldots, N_\Phi$. N_Φ is the number of flux quanta that pierce through the sphere. On such a closed geometry, both the radius ($\propto \sqrt{N_\Phi}$) of the sphere and the number of orbitals ($N_\Phi + 1$) are fixed by the strength of the magnetic monopole at its centre. Figure 4.13(a, b) schematically describe these orbitals for the two geometries.

On the plane geometry, a general quantum Hall wavefunction for N particles in the lowest Landau level can be expressed as

$$\Psi(z_1, \ldots, z_N) = P(z_1, \ldots, z_N)\, e^{-\sum_i |z_i|^2/4}, \tag{4.55}$$

where P is a polynomial in the N complex variables associated with the particle positions $z_1, \ldots, z_N$. If we restrict attention to fermionic wavefunctions, this polynomial has to be antisymmetric. Note that any wavefunction written on the disc can also be obtained on the sphere using the stereographic projection by identifying $z \equiv u/v$ (up to some global factor). To simplify our equations, we will drop the Gaussian factor in any wavefunctions. So, when discussing model wavefunction on the plane or the sphere geometry, it is sufficient to provide P and we can drop all the other factors. We can decompose this wavefunction in the occupation basis, using the orbitals of (4.54):

$$\Psi(z_1, \ldots, z_N) = \sum_{\{\lambda\}} c_\lambda \mathcal{M}_\lambda(z_1, \ldots, z_N). \tag{4.56}$$

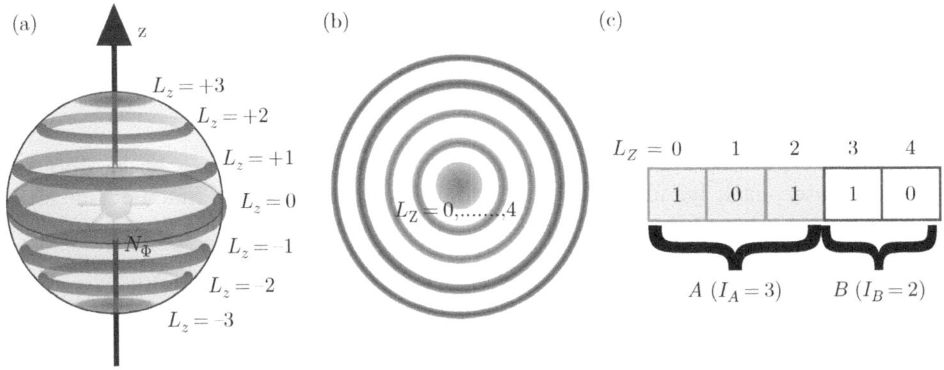

Fig. 4.13 [Colour online] Schematic representation of the orbital basis (a) on the sphere geometry for $N_\Phi = 6$ and (b) on the disc geometry. (c) Typical n-body state of the occupation basis having three particles in an orbital with angular momentum $L_z = 0, 2$ and 3. When we perform an orbital partition into the l_A leftmost orbitals (shaded boxes [in red]), the A part in real space is roughly the domain that contains these orbitals.

$\mathcal{M}_\lambda$ is the normalized Slater determinant that has its orbital occupation given by the configuration λ (such a configuration is shown in Fig. 4.13(c)). The functions $\mathcal{M}_\lambda$ form a set of orthonormal free many-body states. When the wavefunction is obtained through numerical simulation, such a decomposition is directly accessible: one diagonalizes a Hamiltonian by expressing it in a convenient basis, which is generally the occupation basis for the FQHE. For model wavefunctions such as the Laughlin [39] or Moore–Read [47] states, one can use an efficient recursive algorithm [8] that provides the corresponding decomposition.

The archetypal fractional quantum Hall (FQH) model wavefunction is the celebrated Laughlin state [39]

$$\Psi_{\mathrm{Lgh}}(z_1, \ldots, z_N) = \prod_{i<j} (z_i - z_j)^m. \tag{4.57}$$

m is the only variational parameter. Actually, m is related to the filling factor $\nu = 1/m$. On the sphere geometry, (4.57) implies the relation $N_\Phi = m(N-1)$. For a fermionic wavefunction, m has to be odd. $m = 1$ corresponds to the completely filled lowest Landau level as indicated in (4.34). It is a single Slater determinant (Vandermonde determinant) and thus is a product state in the occupation basis. At $\nu = \frac{1}{3}$, the Laughlin wavefunction is a very accurate approximation of the FQH ground state obtained through any realistic simulation. Being an intrinsic topological phase in its full glory (as opposed to the IQHE), the Laughlin wavefunction is degenerate when placed on a higher-genus surface. For example, it is m-fold-degenerate on the torus geometry.

Bulk excitations can be nucleated by removing (for quasielectron) or inserting (for quasihole) fluxes. Each excitation carries a fractional charge ($\pm e/m$) and obeys fractional statistics. For one quasihole located at the position η, we can write the corresponding wavefunction

$$\Psi_{\mathrm{Lgh,\ 1qh}}(z_1, \ldots, z_N; \eta) = \prod_i (z_i - \eta) \Psi_{\mathrm{Lgh}}(z_1, \ldots, z_N). \tag{4.58}$$

Changing the quasihole position η spans a subspace described by a basis of $N+1$ quasihole states, each having a well and uniquely defined angular momentum. More generally, the number of quasihole states for given values of N and N_Φ is a signature of the phase and acts as a fingerprint that can be tracked in numerical simulations. This counting of states can be obtained by Haldane's exclusion principle [29] (or Haldane statistics). For the $\nu = 1/m$ Laughlin wavefunction, this number is identical to the number of configurations with N particles and $N_\Phi + 1$ orbitals where there is no more than 1 particle in m consecutive orbitals. Figure 4.14 gives some simple examples of compatible configurations. Note that both the Laughlin wavefunction and its quasihole excitations are the only zero-energy states of a local two-body model interaction [26], (4.58) being the unique densest zero-energy state among them.

The edge excitations of the Laughlin state are described by a chiral $U(1)$ boson. For an edge of length $\mathcal{L}$, the dispersion relation is given by $E \simeq (2\pi v/\mathcal{L})n$, where n is an integer and v is the edge-mode velocity. The degeneracy of each energy level can be deduced from the picture described in Fig. 4.15. Using the Haldane statistics for the Laughlin $\nu = 1/m$ state, starting from the ground state, we obtain the sequence $1, 1, 2, 3, \ldots$ irrespective of m, identical to the counting of the IQHE discussed in Section 4.3.1.1. As in the case of the quasihole state, this counting is a fingerprint of the edge excitations.

Finally, we give another example of model wavefunctions: the Moore–Read state [47]. This model is considered to be the prototype model wavefunction to explain the appearance of a Hall conductance plateau at filling factor $\nu = \frac{5}{2}$ (i.e. in the second Landau level). It can be written as

$$\Psi_{\mathrm{MR}}(z_1, \ldots, z_N) = \mathrm{Pf}\left(\frac{1}{z_i - z_j}\right) \prod_{i<j} (z_i - z_j)^2. \tag{4.59}$$

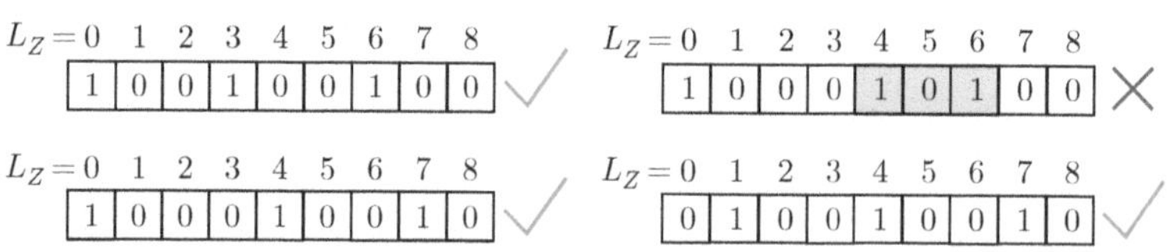

Fig. 4.14 An example of Haldane's exclusion principle. We consider a system with $N = 3$ particles in 9 orbitals on the disc with momenta going from $L_z = 0$ to $L_z = 8$. Among the four configurations described here, only three satisfy Haldane's exclusion principle for the $\nu = \frac{1}{3}$ Laughlin state: no more than one particle in three consecutive orbitals. The violation of this principle in the top right configuration is shown by the shaded boxes. For each compatible partition, one can easily compute the corresponding total L_z value (e.g. 9 for the left topmost configuration).

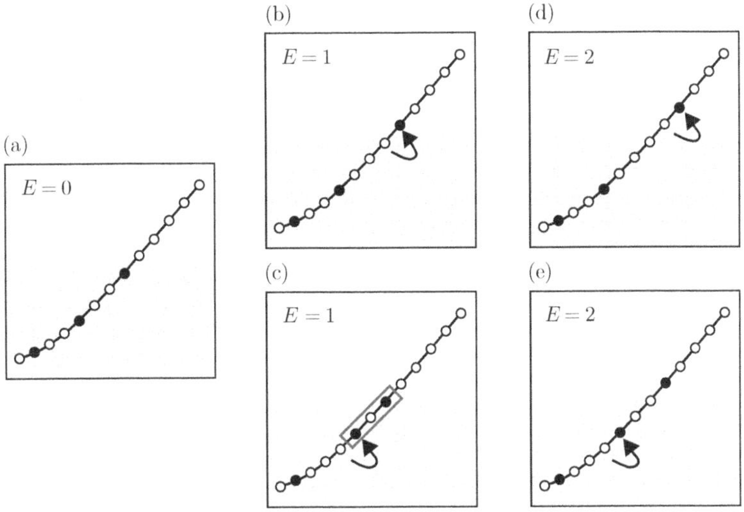

Fig. 4.15 A description of the chiral $U(1)$ edge mode counting at $\nu = \frac{1}{3}$. (a) The ground state with energy $E = 0$. It obeys Haldane statistics (no more than one particle in three consecutive orbitals). The shading denotes the region below the Fermi level. (b) The lowest-energy excitation ($E = 1$) that satisfies the Haldane statistics. (c) An example of an excitation at the same energy but violating (outlined by the rectangular box) the Haldane statistics. (d) and (e) The two possible excitations at $E = 2$.

The Moore–Read state possesses two kind of excitations: Abelian excitations with a charge $\pm e/2$ and non-Abelian excitations carrying a charge $\pm e/4$. In a similar way to the Laughlin case, the number of quasihole states can be derived from Haldane's exclusion principle (in that case, no more than two particles in four consecutive orbitals). The Moore–Read state has two edge modes: a charge edge mode similar to that of the Laughlin state and a neutral Majorana fermion edge mode. Note that a natural way to build the Moore–Read state is based on conformal field theory (CFT) [47], rewriting (4.59) as a correlator (using the CFT of the Ising model in the present case).

4.4.2 Orbital entanglement spectrum

Li and Haldane [41] proposed to compute the ES of a FQH state using a partition in the orbital basis. We call this type of ES an orbital entanglement spectrum (OES). As already pointed out in [31], where the authors tried to extract the topological entanglement entropy of the Laughlin state from the wavefunction, a cut in the orbital basis roughly mimics a cut in real space. The OES is defined by the number of consecutive orbitals that are considered. This number will be denoted by l_A, the number of orbitals for the B part being l_B and satisfying $l_A + l_B = N_\Phi + 1$ on the sphere geometry. When we compute the OES for a FQH state on the geometry such as the sphere or the disc, we can use two good quantum numbers to label the blocks of the reduced density matrix: N_A the number of particles in A and $L_{z,A}$ the projection of the total angular momentum of the particles in A. The OES is generally represented in terms

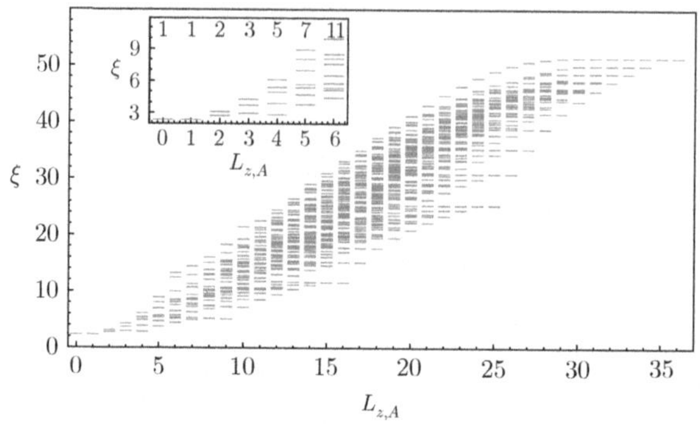

Fig. 4.16 OES for the $\nu = \frac{1}{3}$ Laughlin state with $N = 12$ fermions on the sphere geometry, keeping $l_A = 17$ orbitals and looking at the sector with a fixed number of particles, $N_A = 6$. The inset is a zoom on the ES related to the $U(1)$ edge mode counting of the Laughlin state. As expected, this counting is $1, 1, 2, 3, 5, 7, 11$. For this cut, the deviation of the OES counting to the edge mode counting (due to finite-size effects) starts at $L_{z,A} = 7$: the OES gives 13 levels, while the $U(1)$ counting is 15.

of the entanglement energies ξ as a function of $L_{z,A}$ for a fixed value of N_A. A typical example is shown in Fig. 4.16 for the $\nu = \frac{1}{3}$ Laughlin state. Note that for the sake of simplicity and in contrast to many of the original publications, the non-trivial part of the OES here is located at the left-hand side of the plot. We also shift the origin of $L_{z,A}$ such that the leftmost entanglement level state has $L_{z,A} = 0$.

The OES of the Laughlin state is highly specific: any random state with the same symmetry would produce many more entanglement energy levels, i.e. it would have many fewer zero eigenvalues in the reduced density matrix. Actually, not only would additional entanglement energy levels be present in the sector of $L_{z,A}$ where there is no level for the Laughlin state, but also the total number of levels would be exponentially larger. Thus, such a model state induces large constraints on the reduced density matrix. In Fig. 4.16, we observe that the counting of entanglement energies starting from the left matches the sequence $1, 1, 2, 3, 5, 7, 11$. This is the expected counting sequence for a chiral $U(1)$ boson edge mode as discussed in Section 4.4.1. Beyond a given $L_{z,A}$, the OES counting becomes lower than the $U(1)$ counting. Knowing we are dealing with a finite-size system (in both the numbers of orbitals and of particles), there is a maximum value of $L_{z,A}$ that can be reached and there is a single state with l_A orbitals and N_A fermions that can reach it. Thus, it is clear than both countings should differ at some point, since the $U(1)$ counting keeps growing. We call the thermodynamic region that part of the OES where there is no size effect on the edge-mode counting. This region increases with the system size. In a simplified picture, we can think of the unique state at $L_{z,A} = 0$ of the OES as a Laughlin liquid droplet for N_A particles. Slightly increasing $L_{z,A}$ corresponds to generating edge-mode excitations. A more rigorous derivation of this schematic point of view will be described in Section 4.4.4.

From this observation, Li and Haldane conjectured that in the thermodynamic limit, the OES should be identical to the energy spectrum of the edge mode of the model state. This statement goes beyond the counting argument, which is itself a signature of the edge physics. To corroborate this idea, one can look at the evolution of the entanglement energies when the system size is increased [76]. These energies should mimic the dispersion relation of the gapless edge mode, $(2\pi v/\mathcal{L})L_{z,A}$, where $\mathcal{L}$ is the cut perimeter. Despite some indication that this description is correct, the finite-size calculations are unable to lead to a definitive conclusion. A more accurate approach will be discussed in Section 4.4.5, and will provide more convincing evidence of this conjecture.

A similar calculation can be performed on other geometries. Figure 4.17 shows the OES of the $\nu = \frac{1}{3}$ Laughlin state for the disc, the cylinder, and the thin annulus (or conformal limit [76]). While the shape of the OES depends on the geometry, the counting remains identical as long as one considers genus-zero surfaces. The OES on the two different cylinders in Fig. 4.17(b, c) are a clear consequence of the area law. While the OES is an approximation of a real-space cut, its shape depends on the length of the cut. On the cylinder, this length is the cylinder circumference (or perimeter), and does not vary with the number of flux quanta, N_Φ (the usual hemisphere cut for the sphere would give a length proportional to $\sqrt{N_\Phi}$). A smaller perimeter, and thus a smaller entanglement entropy, results in an OES with a steeper slope.

Moving to a higher-genus surface like the torus leads to a slightly different picture [37]. The usual orbital basis on the torus is translationally invariant along one direction of this geometry. Performing the bipartite partition gives rise to two

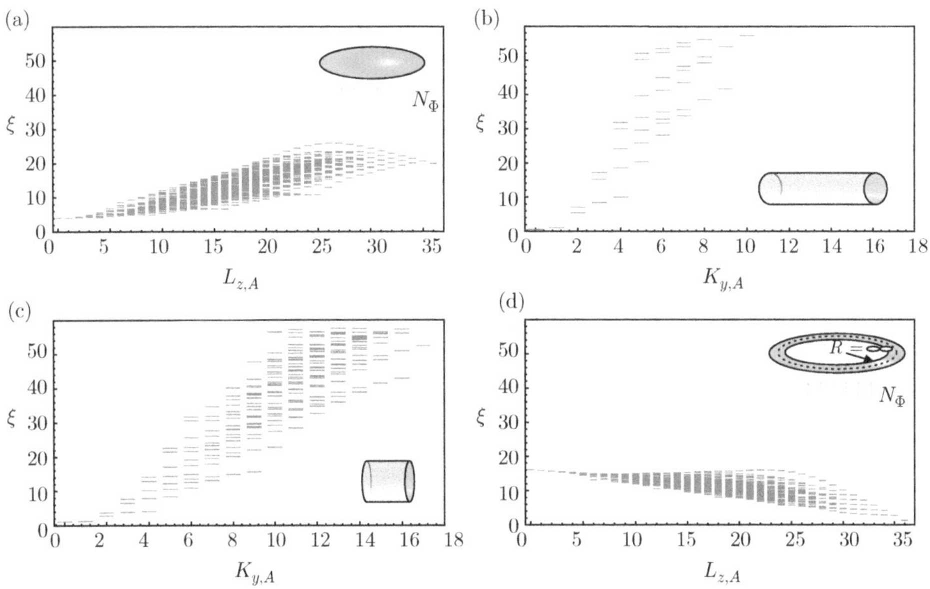

Fig. 4.17 OES for the $\nu = \frac{1}{3}$ Laughlin state with $N = 12$ fermions, keeping $l_A = 17$ orbitals and looking at the sector with a fixed number of particles, $N_A = 6$, on different geometries: (a) the disc geometry; (b) a thin cylinder with perimeter $L = 10l_B$; (c) a thicker cylinder with perimeter $L = 15l_B$; (d) the thin-annulus limit. Note that for the cylinder geometry, we use the momentum along the cylinder perimeter, $K_{y,A}$, instead of the angular momentum $L_{z,A}$.

artificial and spatially separated edges. The OES mimics the physics of two counter-propagating edge modes. A consequence of this interplay between these two modes is the absence of zero eigenvalue in the reduced density matrix. This is a major difference from the OES for the genus-zero geometries.

Beyond the thermodynamic region of the OES, finite-size effects start to appear. There, the spectrum also has a non-trivial structure compared with a generic wavefunction. For most of the model wavefunctions, there is no quantitative understanding of this non-thermodynamic part of the OES. In the case of the $\nu = 1/m$ Laughlin state, it has been shown [33] that the counting of this region can actually be deduced from a generalized exclusion principle that depends on m. Actually, it is a nice example where finite-size effects allow one to get more information than in the thermodynamic limit: while all $\nu = 1/m$ Laughlin states have the same edge theory, a chiral $U(1)$ boson, the compactification radius of the bosons depends on m ($\sqrt{m}$ in that case). The thermodynamic region gives access to the $U(1)$ counting, whereas the finite-size effects encode the value m.

More complex model wavefunctions exhibit a richer OES structure. We focus on the Moore–Read state. Figure 4.18 shows the OES for this state using the same partition (here $l_A = 14$) but looking at two different blocks of the reduced density matrix, namely $N_A = 8$ (a) and $N_A = 7$ (b). For these two cases, the counting in the thermodynamic region is different, reflecting the two sectors of the CFT (namely the identity and the ψ sectors) used to build this state. A surprising result here is that the state itself is built only using one of the two sectors, whereas the OES exhibits both.

For the time being, we have only looked at the OES for the ground state (i.e. in the absence of excitations) of model wavefunctions. But the OES in the presence of pinned excitations is also quite insightful [51]. Figure 4.19 shows the OES for the Moore–Read state in the presence of pinned quasihole excitations. In order to preserve the rotational symmetry along the z axis of the sphere, the excitations are located at

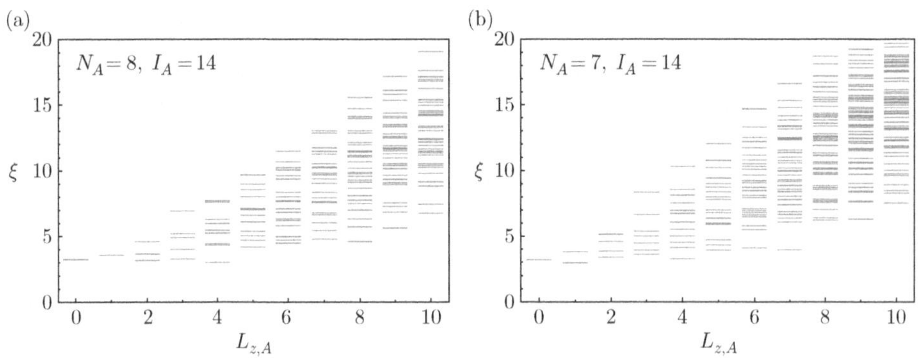

Fig. 4.18 OES for the $\nu = 2 + \frac{1}{2}$ Moore–Read state with $N = 16$ fermions keeping $l_A = 14$ orbitals. (a) Setting $N_A = 8$, we observe the counting $1, 1, 3, 5, \ldots$, which is related to the identity sector. (b) Setting $N_A = 7$, another counting emerges, $1, 2, 4, 7, \ldots$, corresponding to the ψ sector. In both cases and for this system size, the counting starts to deviate from the CFT one starting from $L_z = 4$.

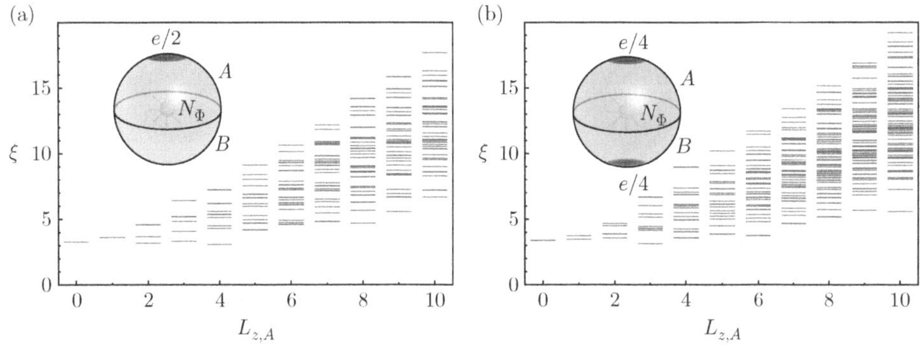

Fig. 4.19 OES for the $\nu = 2 + \frac{1}{2}$ Moore–Read state with quasiholes for $N = 16$ fermions and $N_\Phi = 30$ flux quanta, keeping $N_A = 8$ particles and $l_A = 15$ orbitals. We show two different situations: in (a), there is one Abelian excitation of charge $e/2$ located at the north pole in the hemisphere A; in (b), there are two non-Abelian excitations, each carrying a charge $e/4$, one at each pole. Both situations clearly exhibit a different counting, namely here $1, 1, 3, 5, \ldots$ for (a) and $1, 2, 4, 8, \ldots$ (non-Abelian sector).

the poles. We consider two situations: (a) an Abelian excitation of charge $e/2$ is located at the north pole in the hemisphere A; (b) there two non-Abelian excitations, each carrying a charge $e/4$, one at each pole. In both cases, the system has the same number of particles and flux quanta, and the OES is computed for the same parameters for l_A and N_A. Still, the OES clearly exhibits a different counting: in Fig. 4.19(a) we recover the counting of the vacuum sector (as in Fig. 4.18(a)), and in Fig. 4.19(b) we recover the counting of the non-Abelian sector sector (as in Fig. 4.18(b)). Thus, the OES can be used as a probe to check the parity of the number of non-Abelian excitations in a region of the system. Note that the OES of the Laughlin state is not modified by the presence of pinned quasihole excitations.

As a final note about the OES for model wavefunctions, we should stress once again that the rank of the reduced density matrix being exponentially smaller than any random state with the same symmetry is a major property. We have indicated in Section 4.2.5 that such a feature is a signal that an efficient MPS formula might exist. In the case of the FQHE, the edge modes are gapless. As the ES reflects the edge physics, we expect this MPS to be infinite, in contrast to the AKLT example. Indeed, recent developments [21, 22, 87] have shown that an (infinite) MPS formulation is available for a large class of model wavefunctions, with a well-controlled truncation parameter that allows numerical calculations.

4.4.3 OES beyond model wavefunctions

While the OES has already allowed us to gain some insight into the information encoded within the ground state of a topological phase, we would like to use it as a probe to detect topological order. For that purpose, we need to move away from model states. When dealing with more realistic descriptions of FQH systems, several assumptions are made. In general, we suppose that there is no Landau-level mixing, and in many cases we also assume that electrons are spin-polarized. For low filling factor (such as $\nu = \frac{1}{3}$), these hypothesis are quite accurate. Moreover, the disorder is neglected. In this scenario and for $\nu < 1$, the effective Hamiltonian reads

$$\mathcal{H} = \mathcal{P}_{\mathrm{LLL}} \sum_{i<j} V(\vec{r}_i - \vec{r}_j) \mathcal{P}_{\mathrm{LLL}}, \tag{4.60}$$

where $\mathcal{P}_{\mathrm{LLL}}$ is the projector onto the lowest Landau level. The two-particle interaction V has to be thought as an effective interaction, including effects such as screening and finite confinement of the electron gas. In a crude approach, it is generally assumed that this interaction is just the three-dimensional Coulomb interaction, $V(\vec{r}) = 1/r$. The ground state of this Hamiltonian can be computed for a small number of particles and flux quanta using exact diagonalization techniques such as the Lanćzos algorithm.

In Fig. 4.20, we have computed the OES for the ground state of the projected Coulomb interaction $|\Psi_{\mathrm{exact}}\rangle$, using exact diagonalization. The overlap between this state and the $\nu = \frac{1}{3}$ Laughlin state $|\Psi_{\mathrm{Lgh}}\rangle$ is $|\langle\Psi_{\mathrm{exact}}|\Psi_{\mathrm{Lgh}}\rangle|^2 = 0.9819$. In the low-entanglement-energy part of the spectrum, we clearly distinguish a structure similar to that of the Laughlin state in Fig. 4.16, which we have related to the edge-mode excitations. In contrast to the example of the spin-1 Heisenberg model discussed in Section 4.2.4, the entanglement gap does not extend along all momentum sectors. But the edge-mode counting is clearly separated from the higher entanglement-energy levels. The low-energy part related to the edge physics of the mode state is called the

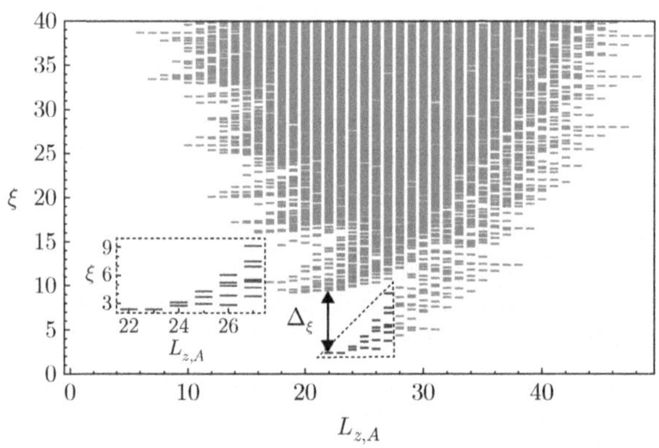

Fig. 4.20 [Colour online] OES for the ground state of the Coulomb interaction with $N = 12$ fermions and $N_\Phi = 33$ on the sphere geometry, keeping $l_A = 17$ orbitals and looking at the sector with a fixed number of particles, $N_A = 6$. We use the same system size and parameters as for the OES of the $\nu = \frac{1}{3}$ Laughlin state in Fig. 4.16. The levels enclosed by the dotted line [the levels shown in blue] are those related to the edge mode of the Laughlin state. Δ_ξ denotes the entanglement gap between the edge-mode counting and the non-universal part of the spectrum. The inset is a zoom on the ES related to the $U(1)$ edge-mode counting of the Laughlin state.

universal part of the ES. The higher-energy part is dubbed the non-universal part of the ES. In this example, the idea of looking at the ES per momentum sector is crucial: without resolving the OES as a function of $L_{z,A}$, the entanglement gap would not be visible.

The fact that the entanglement gap Δ_ξ does not spread over the full spectrum could appear as a failure of the OES to find the universality class. First, we should focus on the part of the spectrum that has reached the thermodynamic limit, i.e. in the Li and Haldane picture, the region that should match the edge physics. From that perspective, what should be relevant is the presence of the entanglement gap in this region that grows when we increase the system size. In the paper that introduced the ES [41], convincing numerical results were provided that Δ_ξ does not collapse when the system size is increased. Moreover, the extension of the region where there is an entanglement gap tightly depends on the geometry in finite size calculations. For example, performing the OES of the same state but on a thin annulus (also called the conformal limit [76]) leads to a modified picture as shown in Fig. 4.21. In some cases, one can clearly separate the full universal part of the ES (that of some model state) from the non-universal part. As discussed in [76], this can happen even when the exact state has a moderate overlap with the model state. In some examples, one can even adiabatically go from the model state to the exact state without closing the entanglement gap.

Up to now, we have mostly focused on the universal part of the ES. Looking at Fig. 4.22, we observe that the non-universal part exhibits several branches. Indeed, these branches can be related to the neutral excitations (the excitations that do not involve a change in the number of particles or in the number of flux quanta) of the system [69]. For the FQHE, these neutral excitations are quasihole–quasielectron excitons. Two approaches are available to test this idea. One can build an approximation of the

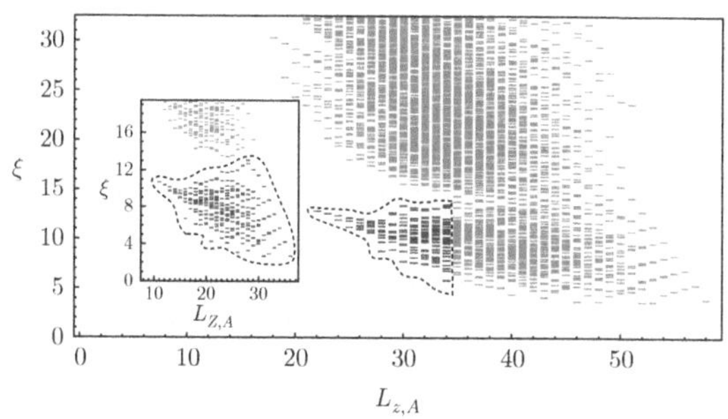

Fig. 4.21 [Colour online] OES for the ground state of the Coulomb interaction on the thin-annulus geometry, with the same system size and parameters as in Fig. 4.20. The levels enclosed by the dotted line [the levels shown in blue] have an identical counting to those of the Laughlin state and are separated from the non-universal part of the spectrum. The inset shows the OES for the same kind of system but with a lower number of particles ($N = 10$). In that case, the structure associated with the Laughlin state (enclosed by the dotted line [shown in blue]) clearly detaches from the non-universal part.

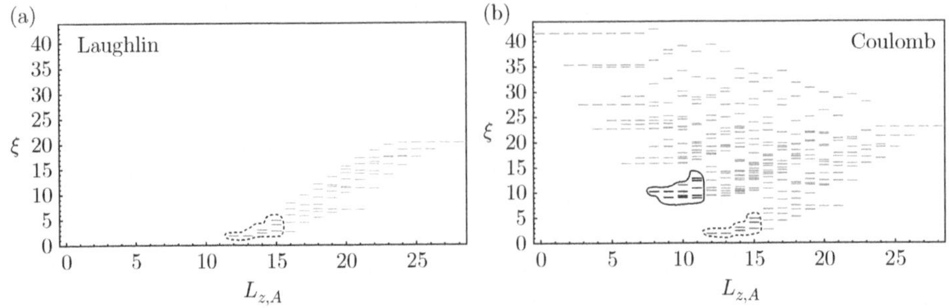

Fig. 4.22 [Colour online] OES for $N = 8$ fermions and $N_\Phi = 21$ flux quanta on the sphere geometry, setting $N_A = 4$ and $l_A = 11$: (a) OES of the $\nu = \frac{1}{3}$ Laughlin state; (b) OES of the ground state of the Coulomb ground state. We clearly observe three branches, the lowest (enclosed by the dotted line [shown in blue]) being related to the $\nu = \frac{1}{3}$ Laughlin state. The second lowest branch is enclosed by the full line [shown in green].

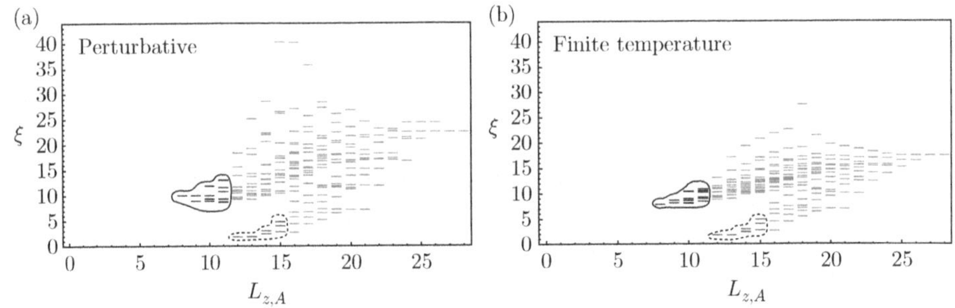

Fig. 4.23 [Colour online] (a) OES from the linear combination of the $\nu = \frac{1}{3}$ Laughlin state and the first neutral excitation having the same symmetry as the Laughlin state. For convenience, the linear combination is optimized to maximize the overlap with the Coulomb ground state. Fine tuning is not required to see that this technique reproduces the two lowest branches, indicating that the second branch (enclosed by the full line [shown in green]) is related to neutral excitations. (b) OES from the finite-temperature calculation as defined in (4.61). We truncate the energy spectrum to include only the lowest-energy neutral excitations (the magneto-roton mode). The temperature is set to $\beta = 7$ to mimic the OES of the Coulomb state. Once again, we clearly deduce that the second low-entanglement-energy branch is related to the lowest-energy neutral excitations. For both (a) and (b), we use the same system sizes as in Fig. 4.22.

exact ground state based on the model state and the lowest-energy neutral excitation that has the same symmetry as both the model and exact states; see Fig. 4.23(a). The other option is to consider the model state but at finite temperature, where the full density matrix is given by

$$\rho = \frac{1}{\sum_n e^{-\beta E_n}} \sum_n e^{-\beta E_n} |\Psi_n\rangle\langle\Psi_n|, \tag{4.61}$$

where $|\Psi_n\rangle$ and E_n are respectively the eigenstates and eigenvalues of the Hamiltonian that produces the model state; see Fig. 4.23(b). In both cases, we see that the resulting OES correctly captures the non-universal part. This exercise also appears to support the idea that the ES of the ground state of a realistic Hamiltonian contains information not only about the universality class of the ground state but also about its excitations.

4.4.4 Particle entanglement spectrum

The concepts of entanglement entropy and entanglement spectrum are not specifically related to a partition in real space. Indeed, the OES is, strictly speaking, a partition in momentum space, which in the specific case of the FQHE can be roughly related to a spatial cut. Partitioning a system in different ways can unveil different type of information, as has been shown in the case of quantum spin chains [74]. Among the possible partitions, a simple one is based on removing particles from the system, realizing a particle partition. In the context of the entanglement entropy for the FQHE, such a partition was introduced in [31] and [89]. The related ES, called the particle entanglement spectrum (PES),was introduced later in [71]. In contrast to the OES, the geometry (i.e. the number of orbitals) is preserved, and the particles are divided into two groups A and B, containing N_A and N_B particles, respectively. In first-quantized notation and for a generic wavefunction $\Psi(x_1,\ldots,x_N)$ for $N = N_A + N_B$ particles, the reduced density matrix is given by

$$\rho_A(x_1,\ldots,x_{N_A};x'_1,\ldots,x'_{N_A}) =$$
$$\int dx_{N_A+1}\cdots dx_N\,\Psi^*(x_1,\ldots,x_{N_A},x_{N_A+1},\ldots,x_N)\Psi(x'_1,\ldots,x'_{N_A},x_{N_A+1},\ldots,x_N). \tag{4.62}$$

As a first example, one can look at the completely filled lowest Landau level, i.e. the $\nu = 1$ IQHE. The ground state on the sphere geometry for $N = N_\Phi + 1$ fermions is given by

$$|\Psi_{\nu=1}\rangle = \left|-\tfrac{1}{2}N_\Phi,\ldots,\tfrac{1}{2}N_\Phi\right\rangle. \tag{4.63}$$

This state is a product state in the orbital basis, leading to a trivial OES with a single non-zero eigenvalue. For the PES, the picture is different: The counting is given by the number of ways one can choose N_A particles among the N particles of the system (it is the same counting that the real-space ES for the IQHE at finite size discussed in Section 4.3.1.3). This case clearly stresses that different partitions probe different properties of the same system.

We now turn to the cases of interacting states, focusing on the Laughlin $\nu = \frac{1}{3}$ state. Figure 4.24 shows the PES for the sphere and disc geometries. As in the case of the OES, the counting is non-trivial (i.e. the number of non-zero eigenvalues is much lower than the naive dimension of the reduced density matrix) and does not depend on the geometry. What was empirically found in [71] is that the counting matches (per momentum sector) the number of quasihole states of the same state with N_A and N_Φ flux quanta (the particle partition does not affect N_Φ). This statement was

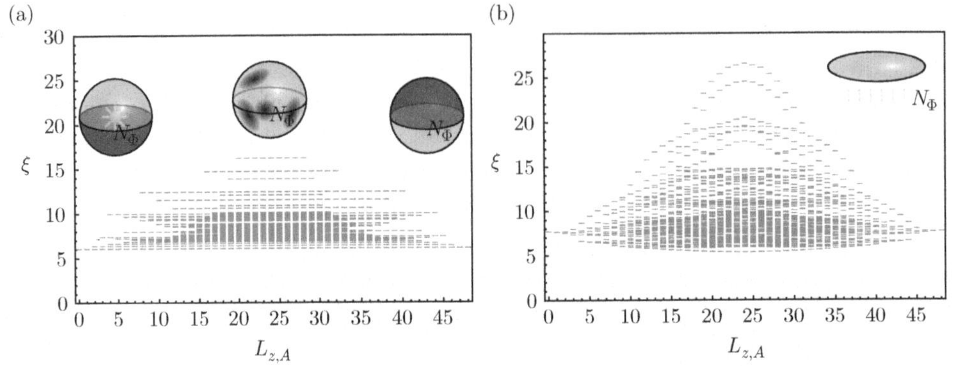

Fig. 4.24 PES for the $\nu = \frac{1}{3}$ Laughlin state with $N = 8$ fermions, $N_\Phi = 21$, and $N_A = 4$ on the sphere geometry (a) and on the disc geometry (b). In both cases, the counting per momentum is given by the number of quasihole states of the Laughlin state for N_A particles and $N_\Phi = 21$. The high degeneracy observed for the PES on the sphere is a consequence of the total angular momentum L_A^2 being a good quantum number when the PES is performed on a state with a total angular momentum equal to zero (such as the Laughlin ground state). For the sphere case in (a), we also present a schematic representation of the types of quasihole states that correspond to the leftmost, centre, and rightmost levels.

checked for a large series of model wavefunctions. When these model wavefunctions are unique zero-energy states of some local model Hamiltonian, one can prove that the counting is bounded by the number of quasihole states. Indeed, any eigenstate of ρ_A corresponding to a non-zero eigenvalue has to be a quasihole state (meaning a zero-energy state of the model Hamiltonian). So far, there is no mathematical proof in the generic case that this bound has to be saturated. Note that the PES for $\nu = 1$ that we have discussed above can also be understood as the quasihole excitations of the integer quantum Hall state.

If we admit that the conjecture about the bound saturation is valid, then we completely understand the counting of the PES, including any finite-size effect (as opposed to the OES). Both entanglement spectra, the OES and the PES, are actually related in the thermodynamic region [15]. In Fig. 4.24(a), we present a schematic representation of the quasihole states in each part of the PES. The leftmost angular momentum sector ($L_{z,A} = 0$) corresponds to the case where all quasiholes are located in the southern hemisphere which is then completely depleted. We are left with a Laughlin droplet occupying the northern hemisphere. Slightly moving away from $L_{z,A} = 0$ is equivalent to slight deformations of the droplet, i.e. the edge excitations. Indeed, the counting starting from $L_{z,A} = 0$ is $1, 1, 2, 3, \ldots$, as expected from the Laughlin edge mode. It was proved in [15] that the entanglement matrices (as defined in Section 4.2.1) associated with the thermodynamic region in both the PES and the OES must have the same rank. Using this bulk–edge (or PES–OES) correspondence, the proof of the Li–Haldane conjecture is reduced (at least for the class of model states that have been considered) to the proof of the bound saturation.

The PES can also be computed on the torus geometry. As the FQH phases are topological phases, the degeneracy of their ground state changes with the genus of the surface on which they live. For example, on the torus, the Laughlin $\nu = 1/m$ state is m-fold-degenerate and the Moore–Read state is 6-fold-degenerate. Thus, multiple choices for the density matrix are available. For the PES, we use the incoherent density matrix, where we sum all the sectors:

$$\rho = \frac{1}{d} \sum_{i=0}^{d} |\Psi_i\rangle\langle\Psi_i|, \tag{4.64}$$

where $\{|\Psi_i\rangle\}$ with $i = 1, \ldots, d$ forms an orthogonal basis of the degenerate ground-state manifold (d being the total degeneracy). As defined, this density matrix commutes with the magnetic translation operators and does not depend on a particular basis choice. The PES calculations are performed using the translational symmetry along one direction (here y), and the eigenvalues of ρ_A can be labelled by the corresponding $K_{y,A}$ momentum. Figure 4.25(a) shows the PES for the Laughlin state on the torus. The properties are identical to those of the PES on the sphere: the counting matches that of the quasihole states, and the corresponding eigenstates of ρ_A span the subspace of the quasihole states. This is a clear difference from the OES on the torus, where the counting is trivial, as discussed in Section 4.4.2. For the ground state of the Coulomb interaction at $\nu = \frac{1}{3}$, the PES is quite interesting: as observed in Fig. 4.25(b), there is a clear entanglement gap separating a low-entanglement-energy structure having the same counting as the PES of the Laughlin state and a higher-entanglement-energy part. From the different examples that have been studied, the PES behaves nicely on the torus geometry. This property will be used as a powerful tool to probe the physics of fractional Chern insulators in Section 4.5.

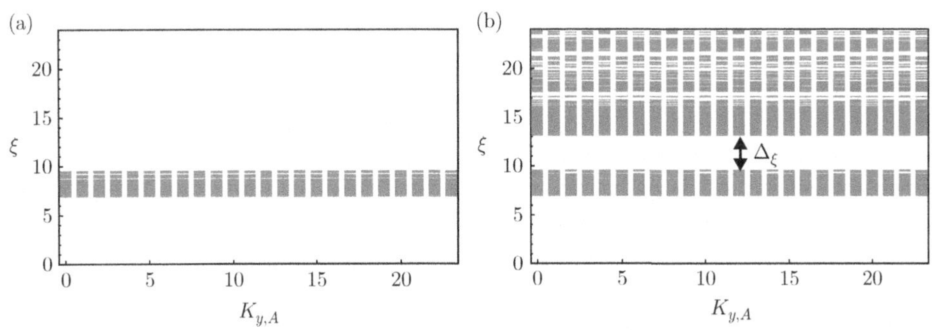

Fig. 4.25 PES on the torus geometry for $N = 8$ fermions and $N_\Phi = 24$, keeping $N_A = 4$ particles. (a) The $\nu = \frac{1}{3}$ Laughlin state. The counting per momentum sector is given exactly by the number of quasihole states with $N_A = 4$ fermions and $N_\Phi = 24$. (b) The Coulomb ground state. We observe a clear entanglement gap Δ_ξ between a low-entanglement-energy structure having the same counting as the PES of the Laughlin state and a higher-entanglement-energy part.

4.4.5 Real-space entanglement spectrum

When we described the OES in Section 4.4.2, we argued that this type of partition is an approximation of a partition in real space, thanks to the specific properties of the orbital basis. The OES appears as a fuzzy cut and not a sharp cut. Several articles [19, 63, 70] have addressed the question of the real-space entanglement spectrum (RSES) using a sharp real-space partition for FQH states. It is an natural extension from the non-interacting IQHE case described in Section 4.3.1. If one chooses a cut that preserves the rotation along z for the sphere or the disc, then $L_{z,A}$ is still a good quantum number. This makes the connection with the other entanglement spectra easier. As in the case of the OES, N_A is also a good quantum number. The generic principle of the RSES was described in Section 4.3.3. Following (4.46), we can split the creation operation associated with the orbital m as follows:

$$c_m^\dagger = \alpha_m c_{m;A}^\dagger + \beta_m c_{m;B}^\dagger, \tag{4.65}$$

where α_m^2 (respectively β_m^2) is the weight of the orbital m in the A (respectively B) part. Whereas α_m^2 and β_m^2 on the disc are related to incomplete gamma functions as shown in (4.40), these coefficients on the sphere can be expressed as incomplete beta functions. A key property of the RSES is that a block of its entanglement matrix $M_{N_A}^{\text{RSES}}$ with a fixed N_A can be related to the entanglement matrix $M_{N_A}^{\text{PES}}$ of the PES for N_A particles. Using (4.65), we can deduce the relation $M_{N_A}^{\text{RSES}} = SM_{N_A}^{\text{PES}}Q$, where S and Q are diagonal matrices with non-zero diagonal elements. These elements are purely one-body geometrical factors α_m and β_m coming from the space partition. As a consequence, the two matrices $M_{N_A}^{\text{PES}}$ and $M_{N_A}^{\text{RSES}}$ have the same rank and thus the two entanglement spectra have the same counting. It should be noted that if we take the weights $\alpha_m = \beta_m = 1/\sqrt{2}$, we recover the PES exactly. As discussed in Section 4.4.4, the OES for the $\nu = 1$ state (i.e. the IQHE) is trivial, where ρ_A has a single non-zero eigenvalue. Having the same counting as the PES, the RSES thus differs strongly from the OES, as shown in Fig. 4.7.

In Fig. 4.26(a), we show the RSES of the $\nu = \frac{1}{3}$ Laughlin state on the sphere when A consists of the northern hemisphere, with a sharp cut at the equator. As expected, the counting per momentum sector is identical to that of the PES. The shape of the spectrum itself is reminiscent of the OES, being due to the geometrical cut. Beyond the counting, one could ask whether the entanglement energies of RSES mimics the dispersion relation of the edge mode in a better way than the OES. In both cases, the spread between the smallest and largest entanglement energies in a given angular momentum sector seems to converge to zero. Figure 4.26(b) shows the extrapolation of the average entanglement energy per angular momentum sector to the limit of a large number of particles. If in this limit the RSES were equivalent to the edge-mode dispersion relation, we would expect these energies to be of the form $(2\pi v/\mathcal{L})n$, where n is an integer, v is the edge-mode velocity, and $\mathcal{L}$ is the cut perimeter. The finite-size calculation shows fairly good agreement with this picture. Most of these properties have been confirmed for much larger system sizes using the MPS description of the Laughlin state [87]. These results underline once again that the ES of the system

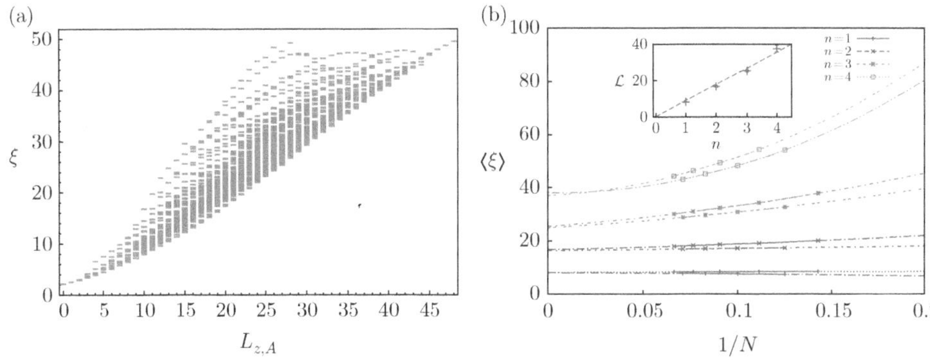

Fig. 4.26 [Colour online] (a) RSES of the $\nu = \frac{1}{3}$ Laughlin state on the sphere geometry for $N = 8$ fermions and $N_\Phi = 21$. We have used the (sharp) hemisphere cut and display the RSES in the sector $N_A = 4$. The counting is identical to that in Fig. 4.24(a). (b) Average entanglement energy $\langle\xi\rangle$ times the perimeter of the cut $\mathcal{L}$ per momentum sector (here $n = L_z^A$) extrapolated at the thermodynamic limit as a function of $1/N$. The even–odd effect is just a consequence of N_A being the integer part of $N/2$. The velocity of the edge mode is $v = 1.41(5)$ (see inset). Such a value would be compatible with a rescaling of the $\nu = 1$ edge mode velocity by a factor $1/\sqrt{3}$.

ground state contains a description of the edge excitations, reinforcing the bulk–edge correspondence.

4.5 Entanglement spectrum as a tool: probing the fractional Chern insulators

As a practical application of entanglement spectroscopy, we discuss the physics of Chern insulators in the strongly interacting regime. We emphasize that the ES can conveniently replace overlap calculations when those are not available. We show that entanglement spectroscopy can discriminate between two phases when simple energetic analysis fails.

4.5.1 From Chern insulators to fractional Chern insulators

In the context of the quantum Hall effects, strong interactions are known to give rise to the exotic physics of the FQHE. Current work suggests that, analogously to the FQHE, introducing strong interactions coupled with fractional filling of the topological insulator bands can give rise to novel and remarkable topological phases of matter. The first class of topological insulators that was studied in the strongly interacting regime were the Chern insulators described in Section 4.3.2. With the addition of strong interactions and fractionally filled bands, these systems are known as fractional Chern insulators (FCIs). At the beginning of 2011, several papers presented evidence from numerical simulations [48, 59, 67] that demonstrated that FCIs could be implemented in principle for model systems (see [7, 52] for reviews).

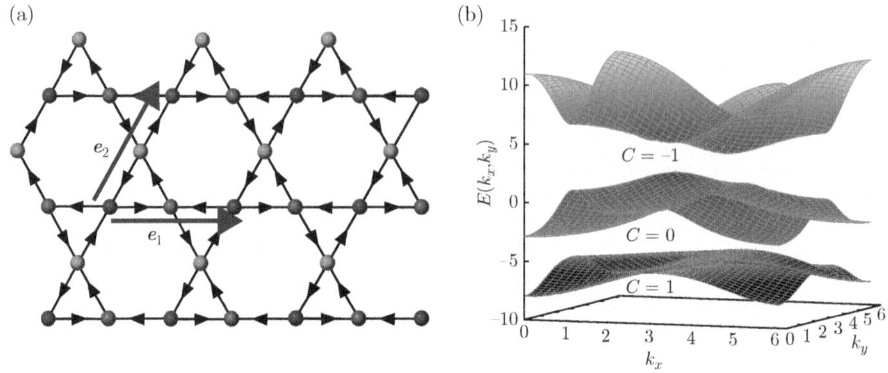

Fig. 4.27 [Colour online] (a) The kagome lattice model as discussed in [73] with three sites per unit cell. In the simplest case, this model has only a single complex nearest-neighbour hopping term. (b) Band structure for the Kagome lattice model with a hopping term of $e^{i\pi/4}$. The topmost and lowest bands have a Chern number $C = \pm 1$, while the middle band is trivial (i.e. with a Chern number $C = 0$).

The emergence of a FQH-like phase in FCIs depends strongly on the underlying one-body model [82]. In this subsection, we consider a slightly more complex model than the two-orbital model discussed in Section 4.3.2 (it was shown that this model does not exhibit any FQH state in the strongly interacting regime). It is based on the Kagome lattice [73] (see Fig. 4.27(a)), a triangular lattice with three sites per unit cell, with a complex hopping term $te^{i\varphi}$ between neighbouring sites. The Bloch Hamiltonian for this model is given by

$$\mathcal{H}(\mathbf{k}) = -\begin{pmatrix} 0 & e^{i\varphi}(1+e^{-ik_x}) & e^{-i\varphi}(1+e^{-ik_y}) \\ & 0 & e^{i\varphi}(1+e^{i(k_x-k_y)}) \\ \text{h.c.} & & 0 \end{pmatrix}, \tag{4.66}$$

where $k_x = \mathbf{k}\cdot\boldsymbol{e}_1$ and $k_y = \mathbf{k}\cdot\boldsymbol{e}_2$, $\boldsymbol{e}_1$ and $\boldsymbol{e}_2$ being the lattice translation vectors as shown in Fig. 4.27(a). The magnitude of the hopping term is set to 1. The dispersion relation is displayed in Fig. 4.27(b), showing the three bands, two of which carry a non-zero Chern number. This model with short-range repulsion was shown to host Laughlin-like phases both for fermions [82] and for bosons [42].

The simplest way to look at the FCI is to work in the flat-band limit [59]: we focus on the interaction and the topological properties of the band structure, discarding the effects of band dispersion and band mixing. This allows us to mimic the usual hypothesis of the FQHE calculations as described in Section 4.4.3. We start from the original Bloch Hamiltonian $\mathcal{H}(\mathbf{k}) = \sum_n E_n(\mathbf{k})P_n(\mathbf{k})$, where $E_n(\mathbf{k})$ and $P_n(\mathbf{k})$ are respectively the dispersion and the projector onto the nth band. Then we focus on the ith band (the lowest band in the case of the Kagome model). We can conveniently consider an equivalent system with the same one-body wavefunctions but with perfectly flat bands $\mathcal{H}_{\mathrm{FB}}(\mathbf{k}) = \sum_n nP_n(\mathbf{k})$. From the energy perspective, this is the same situation as a single Landau level.

For the interacting case, we consider N spinless fermions on a lattice made of N_x unit cells in the $\boldsymbol{e}_1$ direction and N_y in the $\boldsymbol{e}_2$ direction, with periodic boundary conditions. The filling factor is defined as $\nu = N/(N_x N_y)$. The simplest repulsive interaction that can be used for spinless fermions is just the nearest-neighbour repulsive interaction

$$H_{\mathrm{int}} = U \sum_{\langle i,j \rangle} : n_i n_j :, \tag{4.67}$$

where $\langle i,j \rangle$ denotes the sum over nearest-neighbouring sites. Projecting this interaction onto the lowest band and using the flat-band limit, the total effective Hamiltonian is just given by the projected interaction, similar to the FQHE case in (4.60). Exact diagonalizations can be performed to probe this system. A typical energy spectrum for the interacting Kagome lattice at filling factor $\nu = \frac{1}{3}$ is shown in Fig. 4.28(a). Similar to the FQHE on a torus (see Fig. 4.28(b)), we observe an (almost) threefold degenerate ground state clearly separated from the higher-energy excitations. Note that the ground state is not exactly degenerate, as expected for the FQHE phase on a torus such as the Laughlin state. This is a consequence of the absence of an exact magnetic translation symmetry [9, 25, 52], in contrast to the FQHE.

Since Chern insulators are equivalent to quantum Hall systems without an external magnetic field, one might imagine that FCIs should give rise to topological phases analogous to those exhibited by the FQHE. However, as stated previously and contrary to expectations, not all CI models [82] have been found to exhibit such 'fractional' phases. For the time being, the emergence of FQH-like phases for a given model can

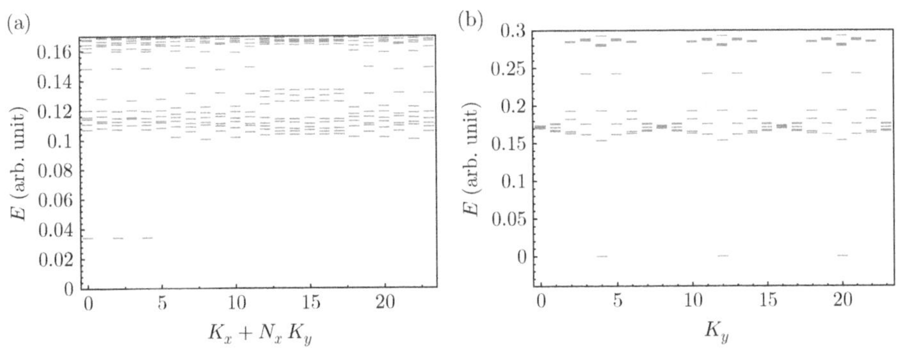

Fig. 4.28 (a) Low-energy spectrum for $N = 8$ fermions on a $N_x \times N_y = 6 \times 4$ unit-cell Kagome lattice with periodic boundary conditions. K_x and K_y denote the total momenta in the x and y directions. We clearly observe an almost threefold-degenerate ground state (the energy splitting between these three states is 3.1×10^{-5}). (b) Low-energy spectrum for the FQHE with $N = 8$ fermions and $N_\Phi = 24$ on a torus. The Hamiltonian that we have used is the hollow-core interaction, for which the Laughlin state is the exact zero-energy ground state.

only be probed through numerical simulations of that model. Moreover, many of the signatures obtained through the energy spectrum could also be obtained for charge density waves (CDWs) in finite-size calculations. As we will now discuss, the concept of the ES has proved to be a powerful tool for probing these systems.

4.5.2 Entanglement spectrum for fractional Chern insulators

FCIs are lattice models, and thus one might expect the real-space partition to be a rather trivial task to perform. However, the projection onto the flattened lowest band, which is done in momentum space, makes such a calculation rather non-trivial. Fortunately, the particle ES does not suffer from this problem and can be performed using a specific representation. Indeed, we can apply the same procedure as for the FQHE on the torus that we described in Section 4.4.4. We will use the same definition for the total density matrix as in (4.64), even though for an FCI the degeneracy of the ground state is not exact. In Fig. 4.29, we present the PES for the almost threefold-degenerate ground state of the kagome system in Fig. 4.28(a). This PES, which can be plotted as a function of the momenta in both the x and y directions, exhibits a clear and large entanglement gap. It is reminiscent of the PES in Fig. 4.25(b) for the Coulomb ground state of the FQHE on a torus. The counting below this gap is exactly that predicted for a Laughlin-like phase. One might wonder whether an overlap calculation could identify a Laughlin-like state. Writing the Laughlin state on an FCI in a suitable way for numerical simulations is a difficult task [58, 83], but the results that have been obtained confirm what was already concluded from the PES.

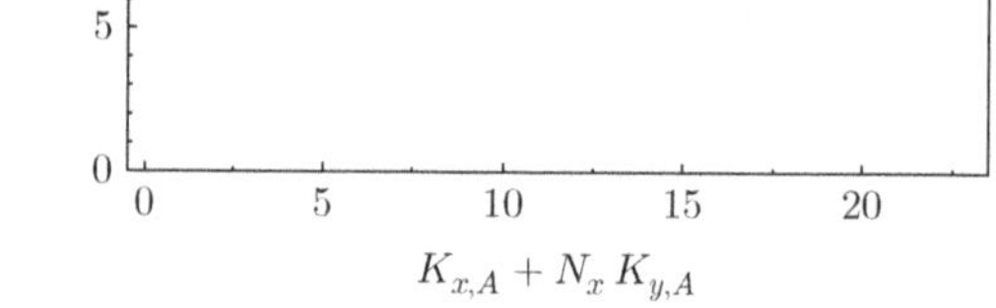

Fig. 4.29 [Colour online] PES for the three lowest-energy states of the kagome FCI model for $N_x \times N_y = 6 \times 4$, keeping $N_A = 4$ particles. $K_{x,A}$ and $K_{y,A}$ denote the total momenta in the x and y directions, and are good quantum numbers when the PES is determined. There is a clear entanglement gap below which the number of levels [in blue] exactly matches the counting of Laughlin quasihole excitations on a system with 4 fermions on a system with $6 \times 4 = 24$ flux quanta. The counting per momentum sector below the entanglement gap matches that predicted by the folding formula of [9].

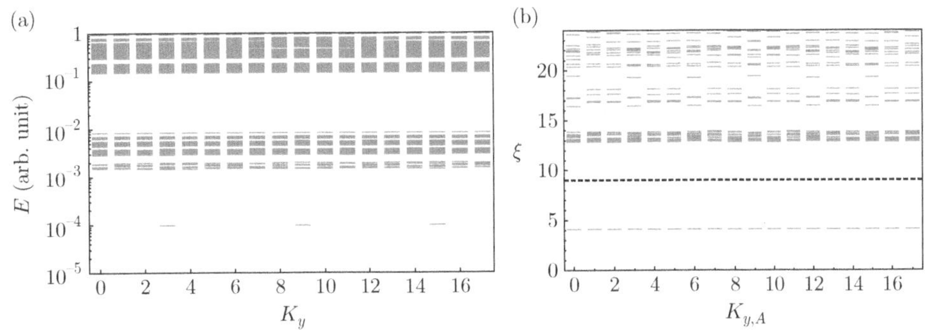

Fig. 4.30 (a) The kagome model in the one-dimensional limit (i.e. $N_x = 1$) for $N = 6$ and $N_y = 18$. While we use a logarithmic scale for the energy, we still observe an almost threefold-degenerate ground state. (b) PES for this almost threefold-degenerate ground state, keeping $N_A = 3$ particles. There are 59 states per momentum sector below the entanglement gap (depicted by the black dotted line), as expected for a CDW (a Laughlin state would give 2530 states per momentum sector).

Since these systems could also host CDW-like phases, one might also wonder whether such a phase would be detected by the PES. A simple way to force the system into a CDW phase involves considering the one-dimensional limit of an FCI [10], keeping only one unit cell in one direction (let us say x). For such a case, the signature from the energy spectra is actually quite similar to that of a regular, two-dimensional FCI. For example, we still observe a threefold-degenerate ground state at filling factor $\nu = \frac{1}{3}$; see Fig. 4.30(a). Obtaining the PES reveals a completely different perspective. As observed in Fig. 4.30(b), there is still a large entanglement gap, but the counting does not match that expected for a Laughlin-like state. Indeed, it has been shown that this counting is that of a CDW [10].

As a final remark about ES for FCIs, we point out that this technique has again been quite successful in probing unusual phases. While FCIs share many common features with the FQHE, some striking differences result in them hosting new physics. The most remarkable example is that a single band can have a Chern number C greater than 1. Indeed, whereas usually a single Landau level carries a Chern number equal to 1, and thus a completely filled Landau level has a Hall conductance equal to h/e^2, this restriction does not apply to Chern insulators. The physics of non-interacting $C > 1$ is actually similar to C copies of a Landau level. These systems have been investigated numerically [44, 72, 80, 85]. However, the studies using ES [72, 84] have revealed that the picture of a simple multicomponent FQH-like system breaks down when strong interactions are allowed.

4.6 Conclusions

In these lecture notes, we have discussed some basic concepts of entanglement spectroscopy and have illustrated some of its features. The most remarkable result of the ES is its ability to reveal how much information is encoded within many of the quantum ground states, even for finite-size systems. Of course, relations between the ground state (the bulk of the system) and the low-energy excitations (the edge modes) had already been pointed out before the introduction of the ES. For FQH model wave-functions built from a CFT, the equivalence between the CFT of the bulk and that associated with the edge was conjectured in [47]. In a similar manner, the use of the reduced density matrix in strongly correlated systems is at the heart of the density matrix renormalization group [81]. The fundamental step made by Li and Haldane was to look at the data stored in the reduced density matrix in the right way, guided by the idea that the ES should mimic the energy spectrum of the edge modes.

The FQHE is a nice sandbox where the concept of the ES has been developed and tested. We have seen that several types of bipartition allow the extraction of different types of information about the system excitations. Although some of these results are still empirical, several steps have been made to give them a more robust analytical basis. Maybe the most intriguing concept is still that of the entanglement gap. For the FQH phases, there is a good understanding of the universal (or low-entanglement-energy) part. On the other hand, the 'non-universal' part also has its own structure, related to neutral excitations. But a quantitative understanding of the entanglement gap is still missing. How large should it be for a phase to be driven by the universal part? Future studies should address this issue.

In the early days of the ES, most of the results were derived from situations where many properties were already known (e.g. the case of model states). Recent work on FCIs has proved that the ES can be used as a tool to probe new systems. It has helped to discriminate between different phases, especially when no expression for model states has been available. Since computing the ES is generally a relatively straightforward numerical calculation, it should now be part of the toolbox used to analyse quantum systems. By picking the right quantum numbers, the ES can be a powerful way to unveil the physics hidden in gigabytes of data.

Acknowledgements

I would like to acknowledge the organizers of the Les Houches 2014 Summer School 'Topological Aspects of Condensed Matter Physics'. These notes have also been used for the lectures that I gave in Florence at the school 'SFT 2015—Lectures on Statistical Field Theories'. Part of the text is based on my 'Habilitation à diriger des recherches' manuscript, and so I am grateful to N. Cooper, D. Poilblanc, B. Douçot, J. Dalibard, V. Pasquier, and P. Lecheminant for being part of my habilitation committee and for their comments. I thank A. Sterdyniak for his useful comments about an earlier version of these lecture notes. I was supported by the Princeton Global Scholarship.

References

[1] Affleck, I., Kennedy, T., Lieb, E. H., and Tasaki, H. (1987). Rigorous results on valence-bond ground states in antiferromagnets. *Phys. Rev. Lett.*, **59**, 799–802.

[2] Affleck, I., Kennedy, T., Lieb, E. H., and Tasaki, H. (1988). Valence bond ground states in isotropic quantum antiferromagnets. *Commun. Math. Phys.*, **115**, 477–528.

[3] Alba, V., Haque, M., and Läuchli, A. M. (2012). Boundary-locality and perturbative structure of entanglement spectra in gapped systems. *Phys. Rev. Lett.*, **108**, 227201.

[4] Alexandradinata, A., Hughes, T. L., and Bernevig, B. A. (2011). Trace index and spectral flow in the entanglement spectrum of topological insulators. *Phys. Rev. B*, **84**, 195103.

[5] Amico, L., Fazio, R., Osterloh, A., and Vedral, V. (2008). Entanglement in many-body systems. *Rev. Mod. Phys.*, **80**, 517–576.

[6] Ardonne, E. and Regnault, N. (2011). Structure of spinful quantum Hall states: a squeezing perspective. *Phys. Rev. B*, **84**, 205134.

[7] Bergholtz, E. J. and Liu, Z. (2013). Topological flat band models and fractional Chern insulators. *Int. J. Mod. Phys. B*, **27**, 1330017.

[8] Bernevig, B. A. and Regnault, N. (2009). The anatomy of Abelian and non-Abelian fractional quantum Hall states. *Phys. Rev. Lett.*, **103**, 206801.

[9] Bernevig, B. A. and Regnault, N. (2012). Emergent many-body translational symmetries of Abelian and non-Abelian fractionally filled topological insulators. *Phys. Rev. B*, **85**, 075128.

[10] Bernevig, B. A. and Regnault, N. (2012). Thin-torus limit of fractional topological insulators. arXiv:1204.5682 [cond-mat.str-el].

[11] Bravyi, Sergey (2005). Lagrangian representation for fermionic linear optics. *Quantum Inf. Comput.*, **5**, 216–238.

[12] Calabrese, P. and Cardy, J. (2004). Entanglement entropy and quantum field theory. *J. Stat. Mech.*, P06002.

[13] Calabrese, P. and Lefevre, A. (2008). Entanglement spectrum in one-dimensional systems. *Phys. Rev. A*, **78**, 032329.

[14] Capponi, S. and Lecheminant, P. (2013). Personal communication.

[15] Chandran, A., Hermanns, M., Regnault, N., and Bernevig, B. A. (2011). Bulk–edge correspondence in entanglement spectra. *Phys. Rev. B*, **84**, 205136.

[16] Cirac, J. I., Poilblanc, D., Schuch, N., and Verstraete, F. (2011). Entanglement spectrum and boundary theories with projected entangled-pair states. *Phys. Rev. B*, **83**, 245134.

[17] Cirac, J. I. and Sierra, G. (2010). Infinite matrix product states, conformal field theory, and the Haldane–Shastry model. *Phys. Rev. B*, **81**, 104431.

[18] Dubail, J. and Read, N. (2011, Oct). Entanglement spectra of complex paired superfluids. *Phys. Rev. Lett.*, **107**, 157001.

[19] Dubail, J., Read, N., and Rezayi, E. H. (2012). Real-space entanglement spectrum of quantum Hall systems. *Phys. Rev. B*, **85**, 115321.

[20] Eisert, J., Cramer, M., and Plenio, M. B. (2010). Colloquium: Area laws for the entanglement entropy. *Rev. Mod. Phys.*, **82**, 277–306.

[21] Estienne, B., Papić, Z., Regnault, N., and Bernevig, B. A. (2013). Matrix product states for trial quantum Hall states. *Phys. Rev. B*, **87**, 161112.

[22] Estienne, B., Regnault, N., and Bernevig, B. A. (2013). Fractional quantum Hall matrix product states for interacting conformal field theories. arXiv:1311.2936 [cond-mat.str-el].

[23] Fannes, M., Nachtergaele, B., and Werner, R.F. (1992). Finitely correlated states on quantum spin chains. *Commun. Math. Phys.*, **144**, 443–490.

[24] Fidkowski, L. (2010). Entanglement spectrum of topological insulators and superconductors. *Phys. Rev. Lett.*, **104**, 130502.

[25] Goerbig, M.O. (2012). From fractional Chern insulators to a fractional quantum spin Hall effect. *Eur. Phys. J. B*, **85**, 15.

[26] Haldane, F. D. M. (1983). Fractional quantization of the Hall effect: a hierarchy of incompressible quantum fluid states. *Phys. Rev. Lett.*, **51**, 605–608.

[27] Haldane, F. D. M. (1983). Nonlinear field theory of large-spin Heisenberg antiferromagnets: semiclassically quantized solitons of the one-dimensional easy-axis Néel state. *Phys. Rev. Lett.*, **50**, 1153–1156.

[28] Haldane, F. D. M. (1988, October). Model for a quantum Hall effect without Landau levels: condensed-matter realization of the 'parity anomaly'. *Phys. Rev. Lett.*, **61**, 2015–2018.

[29] Haldane, F. D. M. (1991). 'Fractional statistics' in arbitrary dimensions: a generalization of the Pauli principle. *Phys. Rev. Lett.*, **67**, 937–940.

[30] Haldane, F. D. M. and Rezayi, E. H. (1985). Periodic Laughlin–Jastrow wave functions for the fractional quantized Hall effect. *Phys. Rev. B*, **31**, 2529–2531.

[31] Haque, M., Zozulya, O., and Schoutens, K. (2007). Entanglement entropy in fermionic Laughlin states. *Phys. Rev. Lett.*, **98**, 060401.

[32] Hastings, M. B. (2007). An area law for one-dimensional quantum systems. *J. Stat. Mech.: Theory Exp.*, **2007**, P08024.

[33] Hermanns, M., Chandran, A., Regnault, N., and Bernevig, B. A. (2011). Haldane statistics in the finite-size entanglement spectra of $1/m$ fractional quantum Hall states. *Phys. Rev. B*, **84**, 121309.

[34] Huang, C.-Y. and Lin, F.-L. (2011). Topological order and degenerate singular value spectrum in two-dimensional dimerized quantum Heisenberg model. *Phys. Rev. B*, **84**, 125110.

[35] Kitaev, A. and Preskill, J. (2006). Topological entanglement entropy. *Phys. Rev. Lett.*, **96**, 110404.

[36] Läuchli, A. M., Bergholtz, E. J., and Haque, M. (2010). Entanglement scaling of fractional quantum Hall states through geometric deformations. *New. J. Phys.*, **12**, 075004.

[37] Läuchli, A. M., Bergholtz, E. J., Suorsa, J., and Haque, M. (2010). Disentangling entanglement spectra of fractional quantum Hall states on torus geometries. *Phys. Rev. Lett.*, **104**, 156404.

[38] Läuchli, A. M. and Schliemann, J. (2012). Entanglement spectra of coupled $s = \frac{1}{2}$ spin chains in a ladder geometry. *Phys. Rev. B*, **85**, 054403.

[39] Laughlin, R. B. (1983). Anomalous quantum Hall effect: an incompressible quantum fluid with fractionally charged excitations. *Phys. Rev. Lett.*, **50**, 1395–1398.

[40] Levin, M. and Wen, X.-G. (2006). Detecting topological order in a ground state wave function. *Phys. Rev. Lett.*, **96**, 110405.
[41] Li, H. and Haldane, F. D. M. (2008). Entanglement spectrum as a generalization of entanglement entropy: identification of topological order in non-Abelian fractional quantum Hall effect states. *Phys. Rev. Lett.*, **101**, 010504.
[42] Liu, T., Repellin, C., Bernevig, B. A., and Regnault, N. (2013). Fractional Chern insulators beyond Laughlin states. *Phys. Rev. B*, **87**, 205136.
[43] Liu, Z., Bergholtz, E. J., Fan, H., and Läuchli, A. M. (2012). Edge-mode combinations in the entanglement spectra of non-Abelian fractional quantum Hall states on the torus. *Phys. Rev. B*, **85**, 045119.
[44] Liu, Z, Bergholtz, E. J., Fan, H., and Läuchli, A. M. (2012). Fractional Chern insulators in topological flat bands with higher Chern number. *Phys. Rev. Lett.*, **109**, 186805.
[45] Liu, Z., Guo, H.-L., Vedral, V., and Fan, H. (2011). Entanglement spectrum: identification of the transition from vortex-liquid to vortex-lattice state in a weakly interacting rotating Bose–Einstein condensate. *Phys. Rev. A*, **83**, 013620.
[46] Lou, J., Tanaka, S., Katsura, H., and Kawashima, N. (2011). Entanglement spectra of the two-dimensional Affleck–Kennedy–Lieb–Tasaki model: correspondence between the valence-bond-solid state and conformal field theory. *Phys. Rev. B*, **84**, 245128.
[47] Moore, G. and Read, N. (1991). Nonabelions in the fractional quantum Hall effect. *Nucl. Phys. B*, **360**, 362–396.
[48] Neupert, T., Santos, L., Chamon, C., and Mudry, C. (2011). Fractional quantum Hall states at zero magnetic field. *Phys. Rev. Lett.*, **106**, 236804.
[49] Nielsen, A. E. B., Cirac, J. I., and Sierra, G. (2011). Quantum spin Hamiltonians for the $SU(2)_k$ WZW model. *J. Stat. Mech.: Theory Exp.*, **2011**, P11014.
[50] Page, D. N. (1993). Average entropy of a subsystem. *Phys. Rev. Lett.*, **71**, 1291–1294.
[51] Papić, Z., Bernevig, B. A., and Regnault, N. (2011). Topological entanglement in Abelian and non-Abelian excitation eigenstates. *Phys. Rev. Lett.*, **106**, 056801.
[52] Parameswaran, S. A., Roy, R., and Sondhi, S. L. (2012). Fractional Chern insulators and the W_∞ algebra. *Phys. Rev. B*, **85**, 241308.
[53] Perez-Garcia, D., Verstraete, F., Wolf, M. M., and Cirac, J. I. (2007). *Quantum Inf. Comput.*, **7**, 401.
[54] Peschel, I. and Chung, M.-C.(2011). On the relation between entanglement and subsystem Hamiltonians. *EPL (Europhys. Lett.)*, **96**, 50006.
[55] Pollmann, F. and Moore, J. E. (2010). Entanglement spectra of critical and near-critical systems in one dimension. *New J. Phys.*, **12**(2), 025006.
[56] Pollmann, F., Turner, A. M., Berg, E., and Oshikawa, M. (2010). Entanglement spectrum of a topological phase in one dimension. *Phys. Rev. B*, **81**, 064439.
[57] Prodan, E., Hughes, T. L., and Bernevig, B. A. (2010). Entanglement spectrum of a disordered topological Chern insulator. *Phys. Rev. Lett.*, **105**, 115501.
[58] Qi, X.-L. (2011). Generic wave-function description of fractional quantum anomalous Hall states and fractional topological insulators. *Phys. Rev. Lett.*, **107**, 126803.

[59] Regnault, N. and Bernevig, B. A. (2011). Fractional Chern insulator. *Phys. Rev. X*, **1**, 021014.
[60] Regnault, N., Bernevig, B. A., and Haldane, F. D. M. (2009, Jun). Topological entanglement and clustering of Jain hierarchy states. *Phys. Rev. Lett.*, **103**, 016801.
[61] Rezayi, E. H. and Haldane, F. D. M. (1994). Laughlin state on stretched and squeezed cylinders and edge excitations in the quantum Hall effect. *Phys. Rev. B*, **50**, 17199–17207.
[62] Rodríguez, I. D. and Sierra, G. (2009). Entanglement entropy of integer quantum Hall states. *Phys. Rev. B*, **80**, 153303.
[63] Rodríguez, I. D., Simon, S. H., and Slingerland, J. K. (2012). Evaluation of ranks of real space and particle entanglement spectra for large systems. *Phys. Rev. Lett.*, **108**, 256806.
[64] Schliemann, J. (2011). Entanglement spectrum and entanglement thermodynamics of quantum Hall bilayers at $\nu = 1$. *Phys. Rev. B*, **83**, 115322.
[65] Schliemann, J. and Läuchli, A. M (2012). Entanglement spectra of Heisenberg ladders of higher spin. *J. Stat. Mech.: Theory Exp.*, **2012**, P11021.
[66] Senthil, T. (2015). Symmetry-protected topological phases of quantum matter. *Annu. Rev. Condens. Matter Phys.*, **6**, 299–324.
[67] Sheng, D. N., Gu, Z.-C., Sun, K., and Sheng, L. (2011). Fractional quantum Hall effect in the absence of Landau levels. *Nat. Commun.*, **2**, 389.
[68] Srednicki, M. (1993). Entropy and area. *Phys. Rev. Lett.*, **71**, 666–669.
[69] Sterdyniak, A., Bernevig, B. A., Regnault, N., and Haldane, F. D. M. (2011). The hierarchical structure in the orbital entanglement spectrum of fractional quantum Hall systems. *New J. Phys.*, **13**, 105001.
[70] Sterdyniak, A., Chandran, A., Regnault, N., Bernevig, B. A., and Bonderson, Parsa (2012, Mar). Real-space entanglement spectrum of quantum Hall states. *Phys. Rev. B*, **85**, 125308.
[71] Sterdyniak, A., Regnault, N., and Bernevig, B. A. (2011). Extracting excitations from model state entanglement. *Phys. Rev. Lett.*, **106**, 100405.
[72] Sterdyniak, A., Repellin, C., Bernevig, B. A., and Regnault, N. (2013). Series of Abelian and non-Abelian states in $c > 1$ fractional Chern insulators. *Phys. Rev. B*, **87**, 205137.
[73] Tang, E., Mei, J.-W., and Wen, X.-G. (2011). High-temperature fractional quantum Hall states. *Phys. Rev. Lett.*, **106**, 236802.
[74] Thomale, R., Arovas, D. P., and Bernevig, B. A. (2010). Nonlocal order in gapless systems: entanglement spectrum in spin chains. *Phys. Rev. Lett.*, **105**, 116805.
[75] Thomale, R., Estienne, B., Regnault, N., and Bernevig, B. A. (2011). Decomposition of fractional quantum Hall model states: product rule symmetries and approximations. *Phys. Rev. B*, **84**, 045127.
[76] Thomale, R., Sterdyniak, A., Regnault, N., and Bernevig, B. A. (2010). Entanglement gap and a new principle of adiabatic continuity. *Phys. Rev. Lett.*, **104**, 180502.

[77] Thouless, D. J., Kohmoto, M., Nightingale, M. P., and den Nijs, M. (1982). Quantized Hall conductance in a two-dimensional periodic potential. *Phys. Rev. Lett.*, **49**, 405–408.

[78] Turner, A. M., Zhang, Y., and Vishwanath, A. (2010). Entanglement and inversion symmetry in topological insulators. *Phys. Rev. B*, **82**, 241102.

[79] Verstraete, F., Cirac, J. I., Latorre, J. I., Rico, E., and Wolf, M. M. (2005). *Phys. Rev. Lett.*, **94**, 140601.

[80] Wang, Y.-F., Yao, H., Gong, C.-D., and Sheng, D. N. (2012). Fractional quantum Hall effect in topological flat bands with Chern number two. *Phys. Rev. B*, **86**, 201101.

[81] White, S. R. (1992). Density matrix formulation for quantum renormalization groups. *Phys. Rev. Lett.*, **69**, 2863–2866.

[82] Wu, Y.-L., Bernevig, B. A., and Regnault, N. (2012). Zoology of fractional Chern insulators. *Phys. Rev. B*, **85**, 075116.

[83] Wu, Y.-L., Regnault, N., and Bernevig, B. A. (2012). Gauge-fixed Wannier wavefunctions for fractional topological insulators. *Phys. Rev. B*, **86**, 085129.

[84] Wu, Y.-L., Regnault, N., and Bernevig, B. A. (2013). Bloch model wave functions and pseudopotentials for all fractional Chern insulators. *Phys. Rev. Lett.*, **110**, 106802.

[85] Yang, S., Gu, Z.-C., Sun, K., and Das Sarma, S. (2012). Topological flat band models with arbitrary Chern numbers. *Phys. Rev. B*, **86**, 241112.

[86] Yao, H. and Qi, X.-L. (2010). Entanglement entropy and entanglement spectrum of the Kitaev model. *Phys. Rev. Lett.*, **105**, 080501.

[87] Zaletel, M. P. and Mong, R. S. K. (2012). Exact matrix product states for quantum Hall wavefunctions. *Phys. Rev. B*, **86**, 245305.

[88] Zozulya, O. S., Haque, M., and Regnault, N. (2009). Entanglement signatures of quantum Hall phase transitions. *Phys. Rev. B*, **79**, 045409.

[89] Zozulya, O. S., Haque, M., Schoutens, K., and Rezayi, E. H. (2007). Bipartite entanglement entropy in fractional quantum Hall states. *Phys. Rev. B*, **76**, 125310.

Part II

Topical lectures

5 Duality in generalized Ising models

Franz J. Wegner

Institut für Theoretische Physik
Ruprecht-Karls-Universität Heidelberg
Germany

Chapter Contents

Colour figures. For those figures in this chapter that use colour, please see the version of these lecture notes at http://topo-houches.pks.mpg.de and arXiv:1411.5815 [hep-lat]. These figures are indicated by '[Colour online]' at the start of the caption.

Preface

This chapter rests to a large extent on a paper I wrote some time ago on *Duality in generalized Ising models and phase transitions without local order parameter.* It deals with Ising models with interactions containing products of more than two spins. In contrast to this old paper, I will first give examples before I come to the general statements.

Of particular interest is a gauge-invariant Ising model in four dimensions. It has important properties in common with models for quantum chromodynamics as developed by Ken Wilson. One phase yields an area law for the Wilson loop, yielding an interaction increasing proportionally to the distance and thus corresponding to quark confinement. The other phase yields a perimeter law allowing for a quark–gluon plasma.

5.1 Introduction

In this contribution, I consider a number of Ising models, which arose out of the question of whether there is duality for Ising models in dimensions larger than two. Indeed, the idea of duality can be used to construct a whole class of such systems, which, however, differ from conventional Ising models in some properties. First, these models contain interactions with products of more than two Ising spins. Second, they no longer have local order parameters, although they can still have two phases. For a number of these systems, the order appears in the expectation value of the product of the spins along a loop, called a Wilson loop. This shows, in the limit of large loops, an area law at high temperatures and a perimeter law at low temperatures.

Such models, where the Ising spins are replaced by elements of groups, typically the groups U(1), SU(2), and SU(3), have become important as lattice gauge models in high-energy physics for the description of quarks and gluons.

In Section 5.2, I review the Kramers–Wannier duality for two-dimensional Ising models. In Section 5.3, I introduce the model dual to the conventional three-dimensional Ising model. In Section 5.4 I introduce the general concept of Ising models and duality. In Section 5.5, this is applied to general lattices and in Section 5.6 to models on hypercubic lattices. The correlation functions are considered in Section 5.7. The basic idea of lattice gauge theory is given in Section 5.8, and a useful lattice for the discretization of Maxwell's equations is mentioned in Section 5.9.

5.2 Kramers–Wannier duality

Kramers and Wannier [7, 10] predicted in 1941 the exact critical temperature of the two-dimensional Ising model on a square lattice. They did this by comparing the high- and low-temperature expansions for the partition function of this model. Consider a square lattice with $N_s = N_1 \times N_2$ lattice points and periodic boundary conditions. There is an Ising spin $S_{i,j} = \pm 1$ at each lattice site. The Hamiltonian reads

$$H = -J \sum_{i=1}^{N_1} \sum_{j=1}^{N_2} (S_{i,j}S_{i,j+1} + S_{i,j}S_{i+1,j}). \tag{5.1}$$

5.2.1 High-temperature expansion (HTE)

We may rewrite the Boltzmann factor

$$\begin{aligned} e^{-\beta H} &= \prod_{i,j} (\cosh K + S_{i,j}S_{i,j+1}\sinh K)(\cosh K + S_{i,j}S_{i+1,j}\sinh K) \\ &= (\cosh K)^{N_b} \prod_{i,j} (1 + S_{i,j}S_{i,j+1}\tanh K)(1 + S_{i,j}S_{i+1,j}\tanh K), \qquad (5.2) \end{aligned}$$

where $K = \beta J$ and N_b is the number of bonds. To determine the partition function, we may expand this expression in powers of $SS' \tanh K$ and sum over all spin configurations. This summation yields zero unless all spins appear with even powers. In this latter case, the sum is 2^{N_s}. This is the case when the interaction bonds form closed loops. That is, at each lattice site, there meet an even number of bonds, as shown in the upper row of Fig. 5.1.

The partition function can be expanded as

$$Z(K) = 2^{N_s}(\cosh K)^{N_b} f(\tanh K), \qquad (5.3)$$

$$f(a) = \sum_l c_l a^l. \qquad (5.4)$$

The coefficients c_l count the number of closed loops of length l, $c_0 = 1$, $c_2 = 0$, $c_4 = N_s$, $c_6 = 2N_s$, $c_8 = N_s(N_s + 9)/2$, etc., with $c_l = 0$ for odd l.

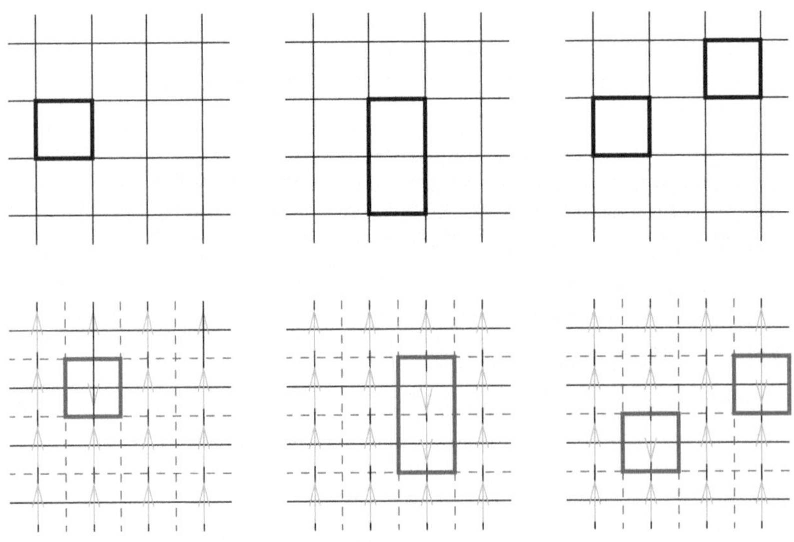

Fig. 5.1 [Colour online] Examples of closed loops in the HTE and Bloch walls in the LTE on the dual lattice.

5.2.2 Low-temperature expansion (LTE)

We now consider the low-temperature expansion on the dual lattice. The dual lattice is obtained by placing a spin $S^*(r^*)$ inside each of the squares (in general polygons) of the original lattice. We multiply spins S^* in polygons with a common edge and sum over these products, which in the case of the square lattice is written as

$$H^* = -J^* \sum_{i,j} (S^*_{i-1/2,j-1/2} S^*_{i-1/2,j+1/2} + S^*_{i-1/2,j-1/2} S^*_{i+1/2,j-1/2}). \tag{5.5}$$

Assuming positive J^*, the states lowest in energy are those where all S^* are equal. Their energy is

$$E^*_{\text{min}} = -N_{\text{b}} J^*, \tag{5.6}$$

where $N_{\text{b}} = 2N^*_{\text{s}}$ is the number of bonds.

Excited states are found by turning some spins, as in the bottom row of Fig. 5.1. Reversing one spin costs an excitation energy $2lJ$ if the spin interacts with l other spins. Quite generally, the excitation energy is given by $2lJ$ if the overturned spins are surrounded by Bloch walls with a total number of l edges. In the case of the square lattice, one obtains

$$Z^*(K^*) = 2\mathrm{e}^{N_{\text{b}} K^*} f(\mathrm{e}^{-2K^*}), \tag{5.7}$$

with f defined in (5.4).

5.2.3 Comparison

Kramers and Wannier argued that if the partition function, or equivalently the free energy, has a singularity at the critical point and no other singularity, then it must be determined by

$$\mathrm{e}^{-2K_{\text{c}}} = \tanh K_{\text{c}}, \tag{5.8}$$

which yields

$$K_{\text{c}} = \frac{1}{2} \ln(1 + \sqrt{2}) = 0.4407, \tag{5.9}$$

which indeed turned out to be correct from Onsager's exact solution [8]. Thus, there is a relationship between the partition function and similarly the free energy at high ($K < K_{\text{c}}$) and low ($K^* > K_{\text{c}}$) temperatures for

$$\tanh K = \mathrm{e}^{-2K^*} \iff \tanh K^* = \mathrm{e}^{-2K}$$
$$\implies \sinh(2K) \sinh(2K^*) = 1. \tag{5.10}$$

The square lattice is called self-dual, since the HTE and LTE are performed on the same lattice. This is in contrast to the case of the triangular lattice, for which the HTE is performed on the triangular lattice and the LTE on the honeycomb lattice; see Fig. 5.2. Then, however, the HTE of the triangular lattice and the LTE of the honeycomb lattice are given by the same sum $f(a)$:

$$Z_3^{\text{hte}}(K) = 2^{N_{s3}} (\cosh K)^{N_b} f_3(\tanh K), \quad Z_6^{\text{lte}}(K) = 2\mathrm{e}^{N_b K} f_3(\mathrm{e}^{-2K}), \tag{5.11}$$
$$Z_6^{\text{hte}}(K) = 2^{N_{s6}} (\cosh K)^{N_b} f_6(\tanh K), \quad Z_3^{\text{lte}}(K) = 2\mathrm{e}^{N_b K} f_6(\mathrm{e}^{-2K}), \tag{5.12}$$

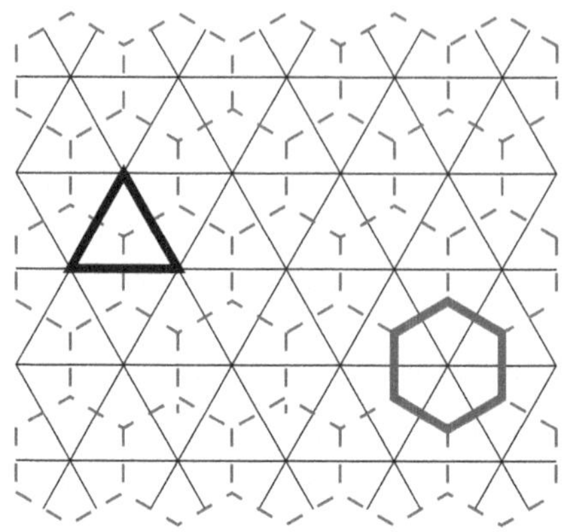

Fig. 5.2 [Colour online] Triangular and dual hexagonal lattices. The thick [black] triangle indicates a product of three interactions on the triangular lattice contributing to HTE and the Bloch wall for an overturned spin on the hexagonal lattice. Similarly, the thick [red] hexagon indicates a product of six interactions on the hexagonal lattice contributing to the HTE and the Bloch wall of an overturned spin on the triangular lattice.

where the numbers N_b of bonds are equal in both lattices and $N_{s3} = N_b/3$ and $N_{s6} = 2N_b/3$. The coefficients c_l in f_3 and f_6 count the number of closed loops on the triangular and honeycomb lattices, respectively.

As a consequence, the partition functions $Z_3(K)$ and $Z_6(K^*)$ are directly related for K and K^* given by (5.10). One cannot directly read off the critical values K_c for these lattices. However, the Ising model on the honeycomb lattice can be related to that on the triangular lattice by means of the star–triangle transformation [10]. To do this, one eliminates every other spin of the hexagonal lattice by summing $\sum_{S_0} e^{KS_0(S_1+S_2+S_3)}$. This gives $Ce^{K'(S_1S_2+S_1S_3+S_2S_3)}$, which yields the Boltzmann factor of the Ising model on the triangular lattice.

5.3 Duality in three dimensions

The basic question I asked myself when I started my paper [11] on duality in generalized Ising models was: *Does there exist a dual model to the three-dimensional Ising model?* It turned out that there is such a model, but of a different kind of interaction. (Compare also [1].)

In order to see this, I consider the low-temperature expansion of the three-dimensional Ising model on a cubic lattice. I start out from the ordered state and then change single spins. These single spins are surrounded by closed Bloch walls. The expansion of the partition function is again of the form (5.4), (5.7), but now with $c_2 = 0$, $c_4 = 0$, $c_6 = N_s$, $c_8 = 0$, $c_{10} = 3N_s$, $c_{12} = N_s(N_s - 7)/2$, etc.

The HTE of the dual model must be given by an interaction such that only closed surfaces yield a contribution. Thus, we locate a spin at each edge and introduce the interaction as a product of the spins surrounding an elementary square called a plaquette. Then the interaction of the dual model reads

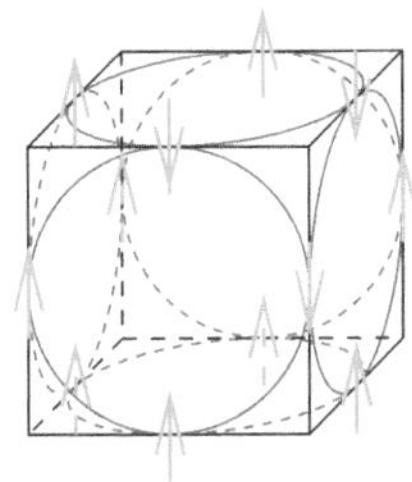

Fig. 5.3 [Colour online] Elementary cube with spins. The full and dashed [red] circles (ellipses) connect the four spins that are multiplied in the interaction.

$$
\begin{aligned}
H = -J \sum_{i,j,k} (& S_{i+1/2,j,k+1/2} S_{i+1/2,j+1/2,k} S_{i+1/2,j,k-1/2} S_{i+1/2,j-1/2,k} \\
& + S_{i+1/2,j+1/2,k} S_{i,j+1/2,k+1/2} S_{i-1/2,j+1/2,k} S_{i,j+1/2,k-1/2} \\
& + S_{i+1/2,j,k+1/2} S_{i,j+1/2,k+1/2} S_{i-1/2,j,k+1/2} S_{i,j-1/2,k+1/2}).
\end{aligned} \quad (5.13)
$$

It is a sum over three differently oriented plaquettes. They are shown in Fig. 5.3.

S-independent products of $R(b)$ The $R(b)$'s denote the products of four spins on a plaquette as they appear in the Hamiltonian (5.13). From Fig. 5.3, it is obvious that the product of the six $R(b)$s around the cube does not depend on the spin configuration, since each spin appears twice in the product.

Gauge invariance This model has a local gauge invariance. Turning all spins around the corner of a cube does not change the energy of the configuration. As an example, in Fig. 5.3, the three spins around the corner close to the centre are reversed from the state in which all spins are aligned upwards.

5.4 General Ising models and duality

5.4.1 General Ising models

We consider models with N_s Ising spins on lattice sites r described by a Hamiltonian

$$
\beta H = -\sum_b K(b)R(b), \qquad R(b) = \prod_r [S(r)]^{\theta(b,r)}, \qquad \theta(b,r) \in \{0,1\}, \qquad (5.14)
$$

where $\{0,1\}$ contains the two elements 0 and 1 of the ring modulo 2 with

$$
\begin{aligned}
0+0 = 1+1 = 0, \qquad 0+1 = 1+0 = 1, \\
0 \cdot 0 = 0 \cdot 1 = 1 \cdot 0 = 0, \qquad 1 \cdot 1 = 1.
\end{aligned} \qquad (5.15)
$$

We call the b bonds; there are N_b of them. The element $\theta(b,r)$ of the incidence matrix θ assumes the value 1 if r belongs to the bond b; otherwise it is 0. Thus, $R(b)$ is the product of the Ising spins $S(r)$ with $\theta(b,r) = 1$. We denote the rank of the matrix

θ modulo 2 by N_θ. Thus, at least one $N_\theta \times N_\theta$ subdeterminant equals 1 modulo 2, whereas all $(N_\theta + 1) \times (N_\theta + 1)$ subdeterminants equal 0 modulo 2. If we write

$$S(r) = (-)^{\sigma(r)}, \qquad R(b) = (-)^{\rho(b)}, \qquad \sigma(r), \rho(b) \in \{0, 1\}, \tag{5.16}$$

then

$$\rho(b) = \sum_r \theta(b, r)\sigma(r) \tag{5.17}$$

is the image of θ. If all $K(b)$ are positive, then one ground state is given by $S(r) = +1$. In general, there will be several ground states. They obey $\rho(b) \equiv 0 \mod 2$ for all b. These configurations $\sigma_0(r)$ constitute the kernel of θ:

$$\sum_r \theta(b, r)\sigma_0(r) \equiv 0 \mod 2. \tag{5.18}$$

There are $N_\mathrm{g} = N_\mathrm{s} - N_\theta$ linearly independent solutions $\{\sigma_0\}$, which yield 2^{N_g} ground-state configurations.

5.4.2 Duality

Besides the Ising model described by the Hamiltonian (5.14), we consider a second Hamiltonian

$$\beta^* H^* = -\sum_b K^*(b) R^*(b), \qquad R^*(b) = \prod_{r^*} [S^*(r^*)]^{\theta^*(b, r^*)}, \tag{5.19}$$

with N_s^* spins $S^*(r^*)$ on lattice sites r^*. The bonds b are common to both Hamiltonians. Similarly, we introduce the rank N_θ^* and obtain the ground-state degeneracy $2^{N_\mathrm{g}^*}$, with $N_\mathrm{g}^* = N_\mathrm{s}^* - N_\theta^*$.

The two models are called dual to each other if two conditions are fulfilled:

(i) the closure condition

$$\sum_b \theta(b, r)\theta^*(b, r^*) \equiv 0 \mod 2 \tag{5.20}$$

for all pairs r, r^*;

(ii) the completeness relation $N_\mathrm{m} = 0$, where

$$N_\mathrm{m} := N_\mathrm{b} - N_\theta - N_{\theta^*}. \tag{5.21}$$

If these two conditions are fulfilled, and $K(b)$ and $K^*(b)$ are connected by (5.10), then the partition functions of the two models are related by

$$Y\{K\} = Y^*\{K^*\}, \tag{5.22}$$

where

$$Y\{K\} = Z\{K\} 2^{-(N_\mathrm{s}+N_\mathrm{g})/2} \prod_b [\cosh 2K(b)]^{-1/2}, \tag{5.23}$$

and similarly for $Y^*\{K^*\}$.

5.4.2.1 *Derivation of* (5.22) *and* (5.23)

The partition function $Z\{K\}$ can be written in the HTE as

$$\begin{aligned} Z\{K\} &= \sum_{\{S(r)\}} \mathrm{e}^{-\beta H} \\ &= \sum_{\{S(r)\}} \prod_b \mathrm{e}^{K(b)R(b)} \\ &= \prod_b \cosh K(b) \sum_{\{\phi(b)\}} \prod_b [\tanh K(b)]^{\phi(b)} \sum_{\{S(r)\}} \prod_b [R(b)]^{\phi(b)}, \end{aligned} \tag{5.24}$$

with $\phi(b) \in \{0, 1\}$ independent for all b. Since

$$\prod_b [R(b)]^{\phi(b)} = \prod_r [S(r)]^{\sum_b \theta(b,r)\phi(b)}, \tag{5.25}$$

those $\phi(b) = \phi_0(b)$ contribute that satisfy the set of homogeneous equations

$$\sum_b \theta(b, r)\phi_0(b) = 0 \mod 2 \tag{5.26}$$

for all r. Thus, ϕ_0 is the kernel of the transpose θ^{t} of θ. Its dimension is $N_{\mathrm{b}} - N_\theta$. Thus, there are in total $2^{N_{\mathrm{b}}-N_\theta}$ solutions $\{\phi_0\}$. They contribute with a factor $2^{N_{\mathrm{s}}}$. Thus,

$$Z\{K\} = 2^{N_{\mathrm{s}}} \prod_b \cosh K(b) \sum_{\{\phi_0(b)\}} \prod_b [\tanh K(b)]^{\phi_0(b)}. \tag{5.27}$$

In the LTE, the partition function $Z^*\{K^*\}$ reads

$$\begin{aligned} Z^*\{K^*\} &= \sum_{\{S^*(r^*)\}} \mathrm{e}^{-\beta^* H^*} \\ &= 2^{N_{\mathrm{g}}^*} \sum_{\mathrm{closed}\{b\}} \prod_b \mathrm{e}^{K^*(b)R^*(b)} \\ &= 2^{N_{\mathrm{g}}^*} \prod_b \mathrm{e}^{K^*(b)} \sum_{\{\rho^*\}} \prod_b \mathrm{e}^{-2K^*(b)\rho^*(b)}, \end{aligned} \tag{5.28}$$

since $R^*(b) = 1 - 2\rho^*(b)$. ρ^* is the image of θ^*:

$$\rho^*(b) = \sum_{r^*} \theta^*(b, r^*)\sigma(r^*). \tag{5.29}$$

Its dimension is $\dim \mathrm{im}(\theta^*) = N_\theta^*$. Owing to the closure relation, ρ^* satisfies the homogeneous equations

$$\sum_b \theta(b, r)\rho^*(b) = \sum_b \sum_{r^*} \theta(b, r)\theta^*(b, r^*)\sigma(r^*) \equiv 0 \mod 2. \tag{5.30}$$

Thus, ρ^* belongs to the kernel of $\theta^{\rm t}$ with dimension $\dim\ker(\theta) = N_{\rm b} - N_\theta$. If both dimensions are equal, $\dim{\rm im}(\theta^*) = \dim\ker(\theta)$, then the completeness relation is fulfilled, $N_{\rm m} = 0$, and the partition function reads

$$Z^*\{K^*\} = 2^{N_{\rm g}^*} \prod_b {\rm e}^{K^*(b)} \sum_{\{\phi_0\}} \prod_b {\rm e}^{-2K^*(b)\phi_0(b)}, \tag{5.31}$$

with the kernel ϕ_0 of $\theta^{\rm t}$ as given in (5.26).

The sums over $\{\phi_0\}$ are the same for $Z\{K\}$ and $Z^*\{K^*\}$ in (5.27) and (5.31). Denoting

$$f\{a\} := \sum_{\{\phi_0\}} \prod_b a(b)^{\phi_0(b)}, \qquad C := \prod_b \frac{\cosh K(b)}{[\cosh 2K(b)]^{1/2}}, \tag{5.32}$$

we obtain

$$\prod_b \frac{{\rm e}^{K^*(b)}}{[\cosh 2K^*(b)]^{1/2}} = 2^{N_{\rm b}} C \tag{5.33}$$

and

$$Y\{K\} = 2^{(N_{\rm s}-N_{\rm g})/2} C f(\tanh K), \tag{5.34}$$

$$Y^*\{K^*\} = 2^{(N_{\rm b}+N_{\rm g}^*-N_{\rm s}^*)/2} C f({\rm e}^{-2K^*}) = 2^{(N_{\rm s}-N_{\rm g}+N_{\rm m})/2} C f({\rm e}^{-2K^*}). \tag{5.35}$$

This yields the duality relation (5.22) for $N_{\rm m} = 0$.

If $N_{\rm m} > 0$, then the summation in (5.31) does not extend over the full set $\{\phi_0\}$. Denoting the sum (5.31) by $f'\{a\}$ instead of $f\{a\}$, we then have

$$Y^*\{K^*\} = 2^{(N_{\rm s}-N_{\rm g}+N_{\rm m})/2} C f'({\rm e}^{-2K^*}). \tag{5.36}$$

Since all terms in the sum f are positive, we obtain $f' < f$ and thus the inequality

$$Y^*\{K^*\} < 2^{N_{\rm m}/2} Y\{K\}. \tag{5.37}$$

We have obtained this relation from the HTE of Z and the LTE of Z^*. If instead we consider the HTE of Z^* and the LTE of Z, then we obtain a similar second inequality, and, in total,

$$2^{-N_{\rm m}/2} Y\{K\} < Y^*\{K^*\} < 2^{N_{\rm m}/2} Y\{K\}. \tag{5.38}$$

The difference in the free energy per lattice site vanishes in the thermodynamic limit owing to the factors $N_{\rm m}^{\pm 1/2}$ in (5.38), if $N_{\rm m}$ does not increase in the thermodynamic limit.

5.4.2.2 Example: two-dimensional Ising model

The two-dimensional Ising model with N_s spins on the square lattice yields $N_s^* = N_s$, $N_b = 2N_s$, $N_g = N_g^* = 1$, and thus $N_m = 2$. The closed loops that show up in the HTE but not in the LTE are those where one loop runs around the torus in one or the other or both directions. This corresponds to antiperiodic boundary conditions. Denoting the partition function with boundary conditions $S_{i,j} = s_x S_{i+N_1,j} = s_y S_{i,j+N_2}$ by Z_{s_x,s_y}, we obtain the exact relation

$$Y\{K\} = \frac{1}{2}(Y_{++}\{K^*\} + Y_{+-}\{K^*\} + Y_{-+}\{K^*\} + Y_{--}\{K^*\}). \tag{5.39}$$

The difference in the free energy per lattice site due to the factors $N_m^{\pm 1/2}$ in (5.38) vanishes in the thermodynamic limit.

5.5 Lattices and Ising models

5.5.1 Lattices and their dual lattices

The models considered so far, will now be generalized to arbitrary dimension d. We call k-dimensional hypercells k-cells. We divide the d-dimensional hypervolume into C_d d-cells $B^{(d)}$. These are bounded by $(d-1)$-cells $B^{(d-1)}$. Generally, the k-cells $B^{(k)}$ are bounded by $(k-1)$-cells $B^{(k-1)}$ and the number of k-cells is denoted by C_k. The 0-cells are simply the C_0 corners $B^{(0)}$ of the d-cells.

We associate lattice points $r^{(k)}$ with the k-cells $B^{(k)}$. Their location will be specified more precisely below.

The dual lattice is obtained in the following way. The points $r^{(d)}$ are the corners of the dual lattice. Pairs of points $r^{(d)}$ are connected by 1-cells $B^{*(1)}$ if the corresponding two cells are separated by a common $B^{(d-1)}$. The 1-cells $B^{*(1)}$ crossing the cells $B^{(d-1)}$ around a given cell $B^{(d-2)}$ form the boundary of a 2-cell $B^{*(2)}$. Generally, the k-cells $B^{*(k)}$ crossing the cells $B^{(d-k)}$ around a cell $B^{(d-k-1)}$ form the boundary of a $(k+1)$-cell $B^{*(k+1)}$. It is reasonable to define the intersection of a cell $B^{*(k)}$ with its corresponding cell $B^{(d-k)}$ as the point $r^{*(k)} = r^{(d-k)}$. Thus, the number of cells $B^{*(k)}$ is equal to the number of points $r^{*(k)}$: $C_k^* = C_{d-k}$.

We define the incidence matrix

$$\theta(r^{(k+1)}, r^{(k)}) = \begin{cases} 1 & \text{if } B^{(k)} \text{ is on the boundary of } B^{(k+1)}, \\ 0 & \text{if } B^{(k)} \text{ is not on the boundary of } B^{(k+1)}. \end{cases} \tag{5.40}$$

Closure relation An important property of the lattices is the closure relation. Consider a pair $r^{(k+1)}$ and $r^{(k-1)}$. They lie in cells $B^{(k+1)}$ and $B^{(k-1)}$. Then

$$\sum_{r^{(k)}} \theta(r^{(k+1)}, r^{(k)})\theta(r^{(k)}, r^{(k-1)}) \equiv 0 \mod 2. \tag{5.41}$$

Proof. If $B^{(k-1)}$ is on the boundary of $B^{(k+1)}$, then two cells $B^{(k)}$ on the boundary of $B^{(k+1)}$ have $B^{(k-1)}$ as boundaries. If $B^{(k-1)}$ is not at the boundary of $B^{(k+1)}$, then none of the $B^{(k)}$ on the boundary of $B^{(k+1)}$ has $B^{(k-1)}$ as boundary. This proves (5.41).

5.5.2 Models on the lattice

The model M_{dn} has C_{n-1} spins on lattice sites $r^{(n-1)}$ with an interaction defined by the bonds:

$$R(b) = \prod_{r^{(n-1)}} [S(r^{(n-1)})]^{\theta(r^{(n)}(b),r^{(n-1)})}. \tag{5.42}$$

The dual model $M^*_{d,d-n}$ has $C^*_{d-n+1} = C_{n-1}$ spins at lattice sites $r^{(n+1)}$:

$$R^*(b) = \prod_{r^{(n+1)}} [S^*(r^{(n+1)})]^{\theta(r^{(n+1)},r^{(n)}(b))}. \tag{5.43}$$

Together, these define, with couplings K and K^*, the models (5.14) and (5.19). Since there is a one-to-one correspondence between the bonds b and the sites $r^{(n)}$, we shall use interchangeably $b(r^{(n)})$ and $r^{(n)}(b)$.

5.5.2.1 *Gauge invariance*

Changing all spins close to a point $r^{(n-2)}$,

$$S(r^{(n-1)}) \to (-)^{\theta(r^{(n-1)},r^{(n-2)})} S(r^{(n-1)}), \tag{5.44}$$

does not change the energy of the system, since any $R(b)$ is multiplied by

$$(-)^{\sum_{r^{(n-1)}} \theta(r^{(n-1)},r^{(n-2)}))\theta(r^{(n)}(b),r^{(n-1)})}, \tag{5.45}$$

which, owing to the closure relation (5.41), yields 1.

5.5.2.2 *Spin-independent products of $R(b)$*

The product over all $R(b)$ around a given $r^{(n+1)}$, i.e.

$$\prod_b R(b)^{\theta(r^{(n+1)},r^{(n)}(b))} = \prod_{r^{(n-1)}} S(r^{(n-1)})^{\sum_{r^{(n)}} \theta(r^{(n)},r^{(n-1)})\theta(r^{(n+1)},r^{(n)})} = 1, \tag{5.46}$$

does not depend on the spin configuration, since it yields 1 owing to the closure relation (5.41). Of course, products of these products are also spin-independent.

5.5.3 Euler characteristic and degeneracy

5.5.3.1 *Generalized Euler characteristic*

The well-known Euler characteristic in $d = 2$ dimensions,

$$\chi = C_0 - C_1 + C_2, \tag{5.47}$$

where C_0 is the number of vertices (corners), C_1 the number of edges, and C_2 the number of faces, depends only on the topology of the surface. For the plane, $\chi = 2$, if the outer face is also counted. For the torus, $\chi = 0$. This characteristic can be generalized to an arbitrary dimension d:

$$\chi = \sum_{m=0}^{d} (-)^m C_m. \tag{5.48}$$

Any lattice with the same boundaries (topology) can be created from any other by repeated application of the following action and its inverse:

- An m-cell is divided into two such cells by creating an $(m+1)$-cell between them. Then both C_m and C_{m+1} increase by 1 and χ is conserved.

For periodic boundary conditions, $\chi = 0$, since we may cut the lattice in one direction, double it, and glue the two parts together. Then all C_m have doubled, and $\chi = 2\chi$ and thus vanishes. It is presumed that it is not possible to introduce an additional 'wall' $B^{(d-1)}$ in any periodic direction that does not intersect any of the original cells $B^{(d-1)}$. See Fig. 5.4.

5.5.3.2 *Degeneracy*

We consider the change in N_g resulting from the application of the action defined below (5.48). For $m > n$, the Hamiltonian is unchanged. For $m = n$, one bond and thus one interaction is duplicated without change of degeneracy. For $m = n-1$, one spin is duplicated, but for the ground state both must be equal. For $m = n-2$, there is also one additional spin. Taking this spin aligned upwards, we again obtain the ground state. But, by changing the signs of all spins lying on bonds adjacent to one $B^{(n-1)}$ at the boundary of the new bond, we obtain another ground state. Then the system has twice the degeneracy of the original system. The Hamiltonian does not change for $m < n-2$. Therefore, we obtain

$$N_g = t_g + \sum_{m=0}^{n-2} (-)^{n-m} C_m, \tag{5.49}$$

where t_g depends only on the boundary condition. Similarly, we obtain

$$N_g^* = t_g^* + \sum_{m=n+2}^{d} (-)^{m-n} C_m. \tag{5.50}$$

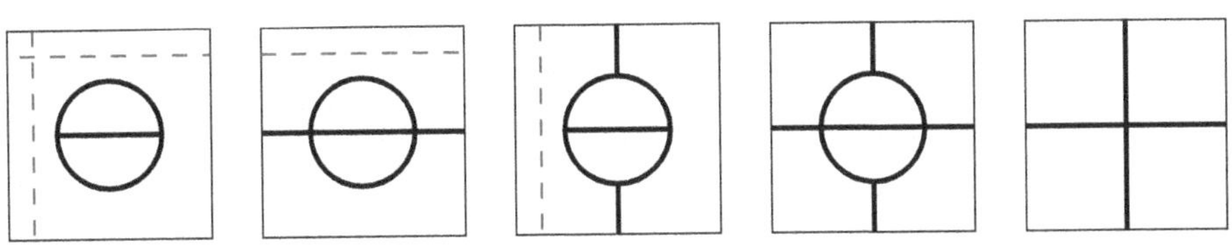

Fig. 5.4 [Colour online] Example of two-dimensional lattices in a periodicity square. The first three examples do not yield $\chi = 0$, since the walls indicated by the dashed [red] lines do not intersect any edges. The last two examples yield $\chi = 0$.

Thus,

$$\begin{aligned} N_{\mathrm{m}} &= N_{\mathrm{b}} - N_{\mathrm{s}} + N_{\mathrm{g}} - N_{\mathrm{s}}^* + N_{\mathrm{g}}^* \\ &= C_n - C_{n-1} - C_{n+1} + N_{\mathrm{g}} + N_{\mathrm{g}}^* \\ &= \sum_{m=0}^{d} (-)^{n-m} C_m + t_{\mathrm{g}} + t_{\mathrm{g}}^* \\ &= (-)^n \chi + t_{\mathrm{g}} + t_{\mathrm{g}}^*. \end{aligned} \tag{5.51}$$

We argue after (5.56) that

$$t_{\mathrm{g}} = \binom{d-1}{n-1}, \qquad t_{\mathrm{g}}^* = \binom{d-1}{d-n-1}, \qquad N_{\mathrm{m}} = \binom{d}{n}, \tag{5.52}$$

for periodic boundary conditions. Thus, N_{m} does not depend on the size of the model.

5.6 The models $M_{d,n}$ on hypercubic lattices

We consider now the models $M_{d,n}$ on hypercubic lattices. The k-cells are k-dimensional hypercubes with edges of unit length around $r^{(k)}$. The lattice points $r^{(k)}$ have k integer coordinates and $(d-k)$ half-integer coordinates, i.e. they are $\frac{1}{2}$ modulo 1. The k-cells are defined by

$$\begin{aligned} r_i^{(k)} - \frac{1}{2} < x_i < r_i^{(k)} + \frac{1}{2} \qquad &\text{for} \quad r_i^{(k)} \in \mathbb{Z}, \\ x_i = r_i^{(k)} \qquad &\text{for} \quad r_i^{(k)} \in \mathbb{Z} + \frac{1}{2}. \end{aligned} \tag{5.53}$$

The coordinates of the dual model are $r^{*(k)} = r^{(d-k)}$, and the corresponding k-dimensional hypercubes are given by

$$\begin{aligned} r_i^{*(k)} - \frac{1}{2} < x_i < r_i^{*(k)} + \frac{1}{2} \qquad &\text{for} \quad r_i^{*(k)} \in \mathbb{Z} + \frac{1}{2}, \\ x_i = r_i^{*(k)} \qquad &\text{for} \quad r_i^{*(k)} \in \mathbb{Z}. \end{aligned} \tag{5.54}$$

We assume periodic boundary conditions, and then

$$C_k = C_{d-k} = C_k^* = C_{d-k}^* = \binom{d}{k} C_d. \tag{5.55}$$

The model M_{dn} has spins on sites $r^{(n-1)}$. The dual model $M_{d,d-n}^*$ has spins on sites $r^{(n+1)} = r^{*(d-n-1)}$. Thus, the model $M_{d,d-n}^*$ is the model $M_{d,d-n}$ shifted by $\frac{1}{2}$ in all coordinates.

Because of the above conditions, $r^{(n+1)}$ and $r^{(n-1)}$ can only have bonds in common if they agree in $d-2$ coordinates and differ only in 2 coordinates. Let these different coordinates be (i,j) and $(i \pm \frac{1}{2}, j \pm \frac{1}{2})$. Then they have two bonds in common as claimed before: $(i, j \pm \frac{1}{2})$ and $(i \pm \frac{1}{2}, j)$ and fulfil the closure condition.

5.6.1 Gauge invariance and degeneracy

If $n > 1$, then one may reverse all spins closest to a given point $r^{(n-2)}$ without changing the energy of the system. Thus, these systems have a local gauge invariance. This leads to a high degeneracy of the ground state. From (5.49), we obtain

$$N_{\mathrm{g}} = t_{\mathrm{g}} + \sum_{m=0}^{n-2} (-)^{n-m} \binom{d}{m} C_d = t_{\mathrm{g}} + \binom{d-1}{n-2} C_d. \tag{5.56}$$

t_{g} is determined by considering only one hypercube $C_d = 1$ in the periodic lattice. We obtain $N_{\mathrm{g}} = N_{\mathrm{s}}$, since periodic boundary conditions require that the spins in the products $R(b)$ be pairwise equal, and we obtain t_{g} as given in (5.52). Thus,

$$N_{\mathrm{g}} = \binom{d-1}{n-1} + \binom{d-1}{n-2} C_d. \tag{5.57}$$

Similarly, we obtain

$$N_{\mathrm{g}}^* = \binom{d-1}{d-n-1} + \binom{d-1}{d-n-2} C_d, \tag{5.58}$$

and t_{g}^* and N_{m}, as given in (5.52).

5.6.2 Self-duality

The model M_{dn} on the hypercubic lattice is self-dual if $d = 2n$. This is the case for $M_{2,1}$, which is the two-dimensional Ising model on the square lattice. But the four-dimensional model $M_{4,2}$ with the plaquette interaction is also self-dual. Both have a phase transition at $K_{\mathrm{c}} = 0.4407$, (5.9). The Ising model $M_{2,1}$ shows a continuous transition. Creutz, Jacobs, and Rebbi [2] investigated the model $M_{4,2}$ using Monte Carlo techniques. They determined $\langle R(b) \rangle$ as a function of K and found a first-order transition with hysteresis. When K was decreased, the system showed superheating until about 0.48, and when K was increased, undercooling was found until about 0.40. Starting from a mixed phase, the phase transition was located between 0.43 and 0.45.

Duality can be generalized to Abelian groups $Z(N)$. Let $S(r) = \mathrm{e}^{2\pi \mathrm{i} p/N}$, with $p = 0, \ldots, N-1$, and let the energy assigned to the product of two spins in states p and p' be $E_{p-p'}$. The weights $\omega_{p-p'} = \mathrm{e}^{-\beta E_{p-p'}}$ and their duals are then related by the Fourier transform [12]

$$\omega_p^* = N^{-1/2} \sum_{p'} \mathrm{e}^{2\pi \mathrm{i} pp'/N} \omega_{p'}. \tag{5.59}$$

This can be generalized to the models M_{dn}. The models M_{42} are self-dual for Z_N with $N = 3, 4$ and the critical K_{c}'s have been determined [6, 16]. Monte Carlo calculations [3] confirm these transition temperatures for $N = 3, 4$. Corresponding calculations yield two phase transitions for $N \geq 5$. For more general aspects of duality for Abelian groups, see Section 6.1.4 in [4].

5.7 Correlations

Non-vanishing correlations are only obtained for gauge-invariant products. These are products of $R(b)$. In particular, we consider the product of spins on the boundary of an n-dimensional hypercube of M_{dn}. The HTE yields

$$\left\langle \prod_r S(r) \right\rangle = \begin{cases} [\tanh K + 2(d-n)(\tanh K)^{1+2n} + \dots]^v & \text{for } n > 1, \\ \frac{1}{2}[\tanh K + [2(d-1)]^{1/2}(\tanh K)^2 + \dots]^v & \\ \quad + \frac{1}{2}[\tanh K - [2(d-1)]^{1/2}(\tanh K)^2 + \dots]^v & \text{for } n = 1, \end{cases} \tag{5.60}$$

where v is the volume of the hypercube, which for $n = 1$ is the distance between the two spins and for $n = 2$ the area spanned by the spins. The LTE yields

$$\left\langle \prod_r S(r) \right\rangle = \begin{cases} (1 - e^{-4(d-n+1)K} + \dots)^f & \text{for } n < d, \\ (1 - 2e^{-2K} + \dots)^v & \text{for } n = d, \end{cases} \tag{5.61}$$

where f is the hyperarea of the boundary of the hypercube, which for $n = 1$ is the number $f = 2$ of ends of the line and for $n = 2$ the perimeter of the square. Thus, the behaviour of the correlation functions of large hypercubes differs in the high- and low-temperature phases, and we expect

$$\left\langle \prod_r S(r) \right\rangle \propto \begin{cases} e^{-v/v_0(T)} & \text{for } T > T_c, \ n < d, \\ e^{-f/f_0(T)} & \text{for } T < T_c, \ n < d. \end{cases} \tag{5.62}$$

We attribute the qualitatively different asymptotic behaviour in the two temperature regions to different states of the system above and below a critical temperature T_c.

5.7.1 The model M_{dd}

The only restriction on the $R(b)$ is that the product of all of them equals 1. Consequently, the partition function reads

$$Z(K) = 2^{N_s} \left[(\cosh K)^{N_b} + (\sinh K)^{N_b} \right]. \tag{5.63}$$

The expectation value of a product of v factors R yields

$$\left\langle \prod_b R(b) \right\rangle = \frac{(\tanh K)^v + (\tanh K)^{N_b - v}}{1 + (\tanh K)^{N_b}}. \tag{5.64}$$

The models M_{dd} do not show a phase transition. Among these models is M_{11}, a closed linear chain of Ising spins.

5.7.2 Dislocations

We consider systems with magnetic dislocations. Let the operator $M(b)$ change the sign of $K(b)$. We introduce $\phi^*(b) = 1$ for bonds with changed signs and $\phi^*(b) = 0$ for bonds with unchanged coupling. Then the expectation value of the product of the $M(b)$'s is

$$\left\langle \prod_b M(b)^{\phi^*(b)} \right\rangle = \left\langle \prod_b \mathrm{e}^{-2\phi^*(b)K(b)R(b)} \right\rangle$$
$$= \frac{Z\{(-)^{\phi^*}K\}}{Z\{K\}} = \frac{Y\{(-)^{\phi^*}K\}}{Y\{K\}}. \tag{5.65}$$

From (5.10), we obtain $\tanh[(-)^{\phi^*}K] = \mathrm{e}^{-2K^* - \mathrm{i}\pi\phi^*}$ and thus

$$\left\langle \prod_b M(b)^{\phi^*(b)} \right\rangle = \frac{Y\{K^* + \mathrm{i}\pi\phi^*/2\}}{Y\{K^*\}}$$
$$= \mathrm{i}^{-\sum_b \phi^*(b)} \left\langle \prod_b \mathrm{e}^{\mathrm{i}\pi\phi^*(b)R^*(b)/2} \right\rangle \{K^*\}$$
$$= \left\langle \prod_b R^*(b)^{\phi^*(b)} \right\rangle \{K^*\}. \tag{5.66}$$

5.7.2.1 Interpretation

Kadanoff and Ceva[5] introduced this concept for the two-dimensional Ising model $M_{2,1}$. Take a sequence of bonds b indicated by either the black [blue online] or the grey [green] bars between the two spins at the sites indicated in Fig. 5.5 by two black

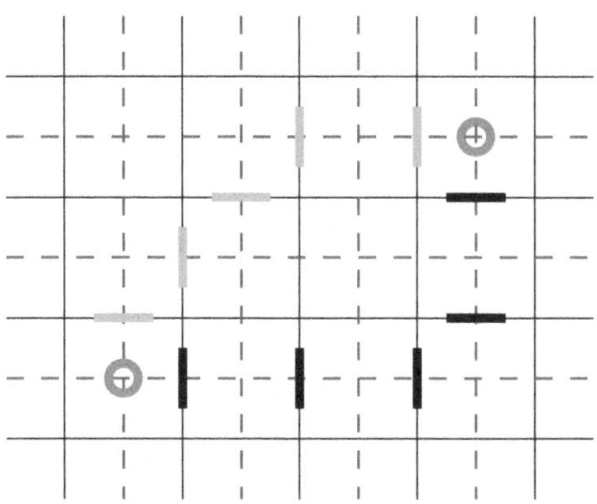

Fig. 5.5 [Colour online] Correlation between the two spins at the dots.

[red] circles. Then $\langle \prod_b R^*(b) \rangle$ is the product of these two Ising spins at K^*. It is equal to the ratio of the partition functions with the changed bonds and with the unchanged bonds, and thus to the exponential of the difference ΔF of the free energy without and with the changed bonds at K:

$$\langle S(r^*)S(r^{*\prime})\rangle(K^*) = \mathrm{e}^{-\beta\Delta F(K)}. \tag{5.67}$$

If K is in the paramagnetic region, then the disturbance of the bonds yields a contribution to ΔF only close to the points, where this line of bonds ends. Thus, for large separation of the two spins, it approaches a finite value, which corresponds to the square of the magnetization at K^*. On the other hand, if K is in the ferromagnetic region, then the disturbance will change the free energy proportionally to the distance between the two spins $S(r^*)$ and $S(r^{*\prime})$, which yields an exponential decay of the correlation function.

Let us now consider $M_{3,1}$ and $M_{3,2}$. Change the sign of the interaction $\sum_{ij} S_{ijk}S_{ijk+1}$ over a whole region (area) in the plane spanned by ij. Analogous to the two-dimensional Ising model, the change $\Delta F(K)$ will be proportional to the perimeter f for paramagnetic K and proportional to the area v for ferromagnetic K. The product $\prod_b R^*(b)$ is now the product of the Ising spins along the perimeter of the dislocations. Consequently, the expectation value decays proportionally to $\mathrm{e}^{-f/f_0(T^*)}$ at low temperatures T^* and proportionally to $\mathrm{e}^{-v/v_0(T^*)}$ at high temperatures T^*, in accordance with (5.62).

5.7.2.2 Local order parameter

If all states are taken into account, non-zero correlations are obtained only from products of R's. For $n = 1$, the product of two spins $S(0)S(r)$ can be written as a product of R's. For $n > 1$, products of spins $\prod_k S(a_k) \prod_l S(r + a_l)$ with a_k and a_l restricted to some finite region $|a_k| < c$, $|a_l| < c$ yield non-vanishing correlations for distances $r > 2c$ only if both $\prod_k S(a_k)$ and $\prod_l S(r + a_l)$ are separately gauge-invariant, i.e. if they are expressed as finite products of R's. However, with (5.65) and (5.66), expectations of products of R's in one phase can be expressed as correlations in the other phase:

$$\left\langle \prod_{\text{some } b} R(b) \right\rangle \{K\} = \left\langle \prod_{\text{same } b} [\cosh 2K^*(b) - R^*(b) \sinh 2K^*(b)] \right\rangle. \tag{5.68}$$

Thus, since there is no long-range order in the high-temperature phase, there can be none in the low-temperature phase:

$$\lim_{r\to\infty} \left[\left\langle \prod_k S(a_k) \prod_l S(r + a_l) \right\rangle - \left\langle \prod_k S(a_k) \right\rangle \left\langle \prod_l S(a_l) \right\rangle \right] = 0. \tag{5.69}$$

Thus, there is no local order parameter for models M_{dn} with $n > 1$. This argument does not apply for $n = 1$, since in this case the number of R's in the product increases with $|r|$.

5.8 Lattice gauge theories

We have seen that models M_{dn} with $n > 1$ show local gauge invariance. Such models are related to quantum chromodynamics (QCD). The basic idea first formulated by Wilson [14] is to start from the lattice that we introduced as M_{42}. (For a retrospect by Wilson, see [15]. Many reprints on this subject are compiled in Rebbi's book [9]). The degrees of freedom are now denoted by U in place of S. These U's are elements of a group. It may be a finite or a continuous group, it may be Abelian or non-Abelian. In the case of QCD, one considers the 'colour' group SU(3). Let us denote the U placed on the link between lattice sites i and j by U_{ij}, where one requires $U_{ji} = U_{ij}^{-1}$. The action is a sum of terms

$$g^{-2} \sum_{\text{plaquettes}} \left\{1 - \frac{1}{N}\,\mathrm{Re}[\mathrm{tr}(U_{ij}U_{jk}U_{kl}U_{li})]\right\}, \tag{5.70}$$

where N is the dimension of U. In addition, we introduce quarks (fermions) with interaction

$$g'^{-2} \sum_{\text{links}} {\psi_i}^\dagger U_{ij} \psi_j. \tag{5.71}$$

These interaction terms are invariant under local gauge transformations

$$\psi_j \to G_j \psi_j, \qquad {\psi^\dagger}_j \to {\psi^\dagger}_j {G^\dagger}_j, \qquad U_{ij} \to G_i U_{ij} {G^\dagger}_j. \tag{5.72}$$

The couplings depend on the temperature and pressure of the hadron system. At low temperature and pressure, the correlations fall off with an area law. Since the action is an integral over time, this behaviour corresponds to an increase in the effective potential between quarks proportional to the distance between them. The gradient of the potential is called the string tension and is given by $1/v_0(T)$ in (5.62). This potential binds three quarks, which constitute a hadron. Or one quark and one antiquark are bound, constituting a meson. Generally, the difference between the numbers of quarks and antiquarks has to be a multiple of three. At high temperature and high pressure, the system forms a quark–gluon plasma. This corresponds to the phase in which the correlation increases proportionally to the perimeter of the loop. Then the effective potential between the quarks stays finite at large distances and the quarks are free to move in this plasma.

5.9 Electromagnetic field

The electromagnetic field in quantum electrodynamics (QED) and its coupling to charged particles can be described similarly, with the group $U(1)$:

$$U_{ij} = \exp\left(\mathrm{i} \int_j^i A_\mu \, \mathrm{d}x^\mu\right). \tag{5.73}$$

Then

$$\mathrm{tr}(U_{r,r+a^\mu e_\mu} U_{r+a^\mu e_\mu, r+a^\mu e_\mu + a^\nu e_\nu} U_{r+a^\mu e_\mu + a^\nu e_\nu, r+a^\nu e_\nu} U_{r+a^\nu e_\nu, r})$$
$$\approx \exp\left[\mathrm{i} a^\mu a^\nu F_{\mu\nu}\left(r + \frac{1}{2}(a^\mu e_\mu + a^\nu e_\nu)\right)\right], \quad (5.74)$$

with the electromagnetic field tensor

$$F_{\mu\nu} = \partial_\mu A_\nu - \partial_\nu A_\mu. \quad (5.75)$$

Since only the real part of $\mathrm{tr}(\prod U)$ contributes, one obtains in leading order the well-known action of the electromagnetic field proportional to $F_{\mu\nu}F^{\mu\nu}$. If one takes the continuum limit ($a \to 0$), then only these terms survive.

The discretized Maxwell equations can be solved on such a lattice [13]. One places the components A_μ on sites $r^{(1)}$, the six electromagnetic field components $F_{\mu\nu}$ on sites $r^{(2)}$, and the components of the charge and current densities on sites $r^{(1)}$. Lorenz gauge and charge conservation can be put on sites $r^{(0)}$.

References

[1] R. Balian, J.-M. Drouffe, and C. Itzykson, Gauge fields on a lattice. II. Gauge-invariant Ising model, *Phys. Rev. D* **11** (1975) 2098.

[2] M. Creutz, L. Jacobs, and C. Rebbi, Experiments with a gauge-invariant Ising system, *Phys. Rev. Lett.* **42** (1979) 1390.

[3] M. Creutz, L. Jacobs, and C. Rebbi, Monte Carlo study of Abelian lattice gauge theories, *Phys. Rev. D* **20** (1973) 1915.

[4] C. Itzykson and J.-M. Drouffe, *Statistical Field Theory*, Vol. 1, Cambridge University Press (1989).

[5] L. P. Kadanoff and H. Ceva, Determination of an operator algebra for the two-dimensional Ising model, *Phys. Rev. B* **3** (1971) 3918.

[6] C. P. Korthals Altes, Duality for $Z(N)$ gauge theories, *Nucl. Phys. B* **142** (1978) 315.

[7] H. A. Kramers and G. H. Wannier, Statistics of the two-dimensional ferromagnet. Parts I and II, *Phys. Rev.* **60** (1941) 252 and 263.

[8] L. Onsager, Crystal statistics. I. A two-dimensional model with an order–disorder transition, *Phys. Rev.* **66** (1944) 117.

[9] C. Rebbi, *Lattice Gauge Theories and Monte Carlo Simulations*, World Scientific (1983).

[10] G. H. Wannier, The statistical problem in cooperative phenomena, *Rev. Mod. Phys.* **17** (1945) 50.

[11] F. J. Wegner, Duality in generalized Ising models and phase transitions without local order parameter, *J. Math. Phys.* **12** (1971) 2259.

[12] F. J. Wegner, A transformation including the weak-graph theorem and the duality transformation, *Physica* **68** (1973) 570.

[13] T. Weiland, A discretization model for the solution of Maxwell's equations for six-component fields, *Arch. Elektron. Übertragungstech.* **31** (1977) 116.
[14] K. G. Wilson, Confinement of quarks, *Phys. Rev. D* **10** (1974) 2445.
[15] K. G. Wilson, The origin of lattice gauge theory, *Nucl. Phys. Proc. Suppl.* **140** (2005) 3.
[16] T. Yoneya, $Z(N)$ topological excitations in Yang–Mills theories: duality and confinement, *Nucl. Phys. B* **144** (1978) 195.

6
Topological insulators and related phases with strong interactions

Ashvin VISHWANATH

University of California, Berkeley
USA

Topological Aspects of Condensed Matter Physics. First Edition. C. Chamon et al.

Chapter Contents

Colour figures. For those figures in this chapter that use colour, please see the version of these lecture notes at http://topo-houches.pks.mpg.de. These figures are indicated by '[Colour online]' at the start of the caption.

6.1 Overview

In these lectures, we will mainly be interested in how the concept of topological insulators generalizes when we include interactions. More generally, we discuss the interplay of symmetry and topology. Traditionally, phases of matter were distinguished on the basis of symmetry alone. On the other hand, fractional quantum Hall (FQH) phases are examples of topological states whose essential character does not require a discussion of symmetry. However, topological insulators are an example of a new phase of matter that combines both symmetry and topology.

To generalize the concept of topological insulators to strongly interacting systems, we will need some definitions to limit the set of states we study and hence allow us to make progress. Previously, the thinking was that phases like the integer quantum Hall (IQH) state can occur in free-fermion models, and the new physics that interactions bring are FQH phases. Now we understand that there is some space between these two—there are states that retain the essential physical character of IQH states, but *require* interactions. Partly, the advances occurred by sharply defining what we mean by 'IQH-like' states—by identifying short-range entanglement as an essential property.

Throughout, we will discuss phases with an energy gap in the bulk and will focus on zero temperature properties. Often, we will be interested in new phases of matter, but some of the most striking results will expose phenomena connected to well-known phases like topological insulators and superconductors that are obscured by the free-particle description.

To whet the reader's appetite, we begin by mentioning three striking theoretical results that emerge on including the effects of interactions. Establishing these will be the goal of these lectures.

- IQH states of electrons have long been believed to be characterized by, of course, an integer ($\mathbb{Z}$), which is the Hall conductance in units of e^2/h. We will see that this is modified in the presence of interactions—*actually, the classification is by two integers* ($\mathbb{Z} \times \mathbb{Z}$)—and this family of states retains the essential properties of IQH phases.
- It was believed that if the conducting surface of a three-dimensional (3D) topological insulator is made insulating, this must be because the time-reversal symmetry is broken, either spontaneously or by the application of external fields. It has recently been understood that you can have your cake and eat it too—that *there exist strongly interacting surface phases of a 3D topological insulator that are insulating but retain time-reversal symmetry.* The price to be paid is that this state must have fractional excitations at the surface, i.e. excitations with fractional charge and fractional (or anyonic) statistics. In fact, it must realize a particularly exotic version of fractional statistics, namely non-Abelian statistics. The simplest version of this state is closely related to the celebrated Read–Moore Pfaffian state, but with a twist.
- We are used to thinking of the surface states of topological phases (say of 3D phases) as being 'impossible' to realize in a purely 2D system with the same symmetries. Indeed, this is generally true—even the new interacting surface phase of a topological insulator cannot be realized in a purely 2D system. However, the

set of topological superconductors protected by time-reversal symmetry (class DIII) are labelled by an integer ν according to the free-fermion classification. Roughly, this counts the number of Majorana cones (which are like 'half' of a Dirac cone) present at the surface. From the free-fermion point of view, all these surface states are 'impossible' 2D states. However, with interactions, we will show that while $\nu = 1, 2, \ldots, 15$ are all indeed impossible in 2D states, *the surface of* $\nu = 16$ *can be realized in a purely 2D but interacting model.* This also means that the integer classification is broken down: $\mathbb{Z} \to \mathbb{Z}_{16}$.

All these are statements about adding interactions to many-electron states. But to make progress, we will need to take a diversion and study topological phases of *bosons* or, equivalently, spins. Results there are integrally connected to a deeper understanding of interacting electronic topological insulators and superconductors. Moreover, they might be realized in experiments on ultracold bosons or frustrated magnetic models. We will discuss some ideas along these lines, but it is fair to say that conceptual theory is well ahead of experiments and model building in this area. However, the spectacular success of the theory of topological insulators in connecting with experiments makes us optimistic. There are even strongly correlated materials that have been proposed to be in this phase.

6.2 Quantum phases of matter. Short-range versus long-range entanglement

How do we distinguish different phases of matter? We will be particularly interested in the zero-temperature state, i.e. the ground state of an interacting bunch of particles. Typically, the phases of matter are only sharply defined in the limit of an infinite number of particles. Then, two states belong to different phases if they are necessarily separated by a phase transition at which properties change in a singular fashion. For some time, people thought that they knew how to diagnose this. The answer, they believed, had to do with symmetry—at the fundamental level, breaking symmetry in different ways led to different phases. For example, in the quantum Ising model, with two-level systems arranged on a line, with Hamiltonian

$$H = -J \sum_i \left(\sigma_i^z \sigma_j^z + g \sigma_i^x \right), \tag{6.1}$$

there is a symmetry or flipping of the spin: $\sigma^z \to -\sigma^z$ (similarly for the y spin direction). This $\mathbb{Z}_2$ symmetry is spontaneously broken if g is small, while it is restored if g is sufficiently large. Thus, there are two phases, which can be distinguished by the order parameter $\langle \sigma_i^z \rangle$. Symmetry is key to having a sharp distinction—if it is broken by hand, for example by adding a field along σ^z, then the phase transition can be converted into a crossover. For a while, it was thought that all phases of matter (apart from a few well-characterized outliers like metals) could be identified by such a procedure.

However, Wegner came up a model that could be shown to have two phases that shared the same symmetry. Today, we understand that they differ in their topology.

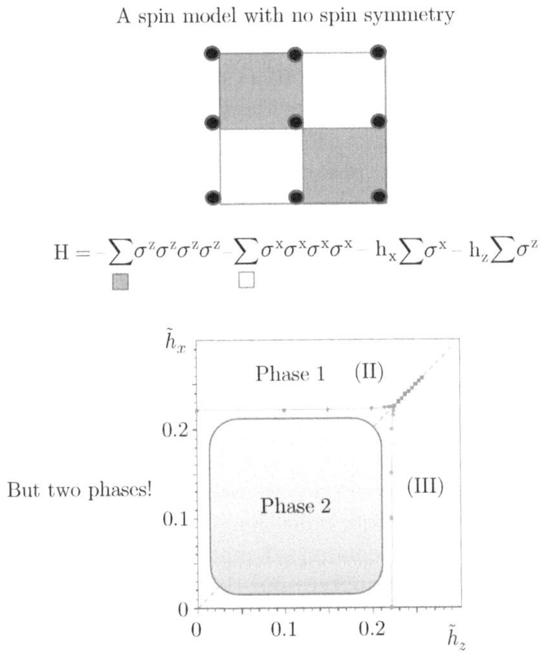

Fig. 6.1 [Colour online] The toric code model with generic perturbations, which has two phases although they have the same symmetry. Phase 2 is gapped but has long-range entanglement—as evidenced by the presence of ground-state degeneracy with periodic boundary conditions and by anyon excitations with non-trivial mutual statistics. (Phase diagram adapted with permission from [28]. Copyrighted by the American Physical Society.)

Here is the modern avatar of that model, the Kitaev toric code, which also has two-level systems on the vertices of a 2D square lattice [16]. The coupling takes the following form, and includes four spin couplings around the plaquettes (see Fig. 6.1):

$$H = -\sum_{\text{black}} \sigma_i^z \sigma_j^z \sigma_k^z \sigma_l^z - \sum_{\text{white}} \sigma_i^x \sigma_j^x \sigma_k^x \sigma_l^x - h_z \sum_i \sigma_i^z - h_x \sum_i \sigma_i^x. \tag{6.2}$$

One of the two phases in this model is a 'trivial' phase, which can be thought of as a product state of spins pointing along a certain direction. The other phase does not have any representation as a product state of a site or finite group of sites. It can be thought of as a condensate of closed loops, where the loops are formed by linking $\sigma^z = -1$ spins for example. There are two kinds of point excitations in this phase, which violate the individual plaquette terms. One is called a 'charge' and the other a 'vortex'. Despite ultimately being built out of bosons (the spins can equally be thought of as occupying sites with hard-core bosons), the excitations have unusual statistics. Taking one around another leads to a −1 sign, and hence they are mutual semions. This is an example of fractional statistics, in the generalized sense, which includes both exchange of identical particles as well as mutual statistics. It is an indication of long-range entanglement (LRE). Another signature is that when the system is defined with periodic boundary conditions (i.e. on a torus), there is a ground-state degeneracy. The degenerate states appear identical with respect to any local operator (the degeneracy itself can be understood since the Hamiltonian is a local operator). We define a short-range-entangled phase as one that does *not* have these properties:

- A *short-range-entangled (SRE) state* is a gapped phase with a unique ground state on any closed manifold. All excitations (particles with short-range interactions) have conventional statistics.

For example, if the phase is built of bosons, all excitations are bosonic with trivial mutual statistics.

We also allow for the possibility of a symmetry specified by a group G. We will restrict attention to *internal symmetries*; i.e. we will not consider symmetries that change spatial coordinates, such as inversion, reflection, or translation. Common internal symmetries that are encountered in condensed matter physics are charge conservation, various types of spin-rotation symmetry, and time-reversal symmetry. The advantage of working with internal symmetries is that we can consider disordered systems that respect the symmetry. Also, the symmetries can be defined at the edge, whereas for spatial symmetries, one may require a special edge configuration to preserve symmetry. Some spatial symmetries, such as inversion, are always broken at the edge.

Gapped SRE ground states that preserve their internal symmetries differ from the trivial phase only if they possess edge states. (For 1D systems, the edge states are always gapless excitations, and rigorous statements can be made using a matrix product state representation of gapped phases [7, 10, 23, 29].)

The fact that SRE topological phases differ only at the edge, not in the bulk (unlike LRE states), makes them much easier to study. The set of SRE topological phases in a given dimension with symmetry G actually has more structure than just a set. If we add the trivial phase as an 'identity' element, the set of phases is actually an Abelian group. The operations for the group are shown in Fig. 6.2. The addition operation is obvious: take two states and put them side by side. But it is not so obvious that a state has an inverse: how is it possible to cancel out edge states? Two copies of a topological insulator cancel one another, because the Dirac points can be coupled by a scattering term that makes a gap. The inverse of a phase is its mirror image (i.e. with one of the coordinates reversed). To see this, we must show that the state and its mirror image cancel; the argument is illustrated at the bottom of Fig. 6.2. In one dimension, for example, take a closed loop of the state and flatten it. The ends are really part of the bulk of the loop before it was squashed, so they are gapped. Therefore, this state has no edge states. Because topological SRE phases are classified by their edge states, it must be the trivial state. Therefore, a Hamiltonian describing the inverse of a particular state $H(x, y, z, \ldots)$ is, for example, $H(-x, y, z, \ldots)$. SRE phases protected by a symmetry are termed symmetry-protected topological (SPT) phases [5].

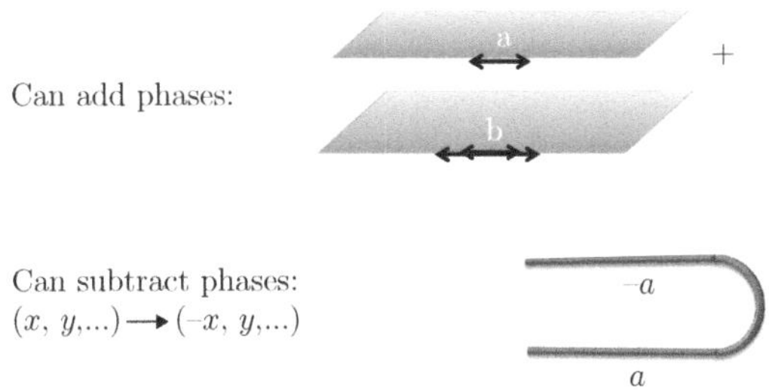

Fig. 6.2 SRE phases that preserve a symmetry must form an Abelian group. This is not true for LRE phases, which typically become more complicated when they are combined.

6.3 Examples of SRE topological phases

Let us give a couple of concrete examples of SRE topological phases of bosons/spins. These are necessarily interacting—unlike free fermions, there is no 'band' picture here.

6.3.1 Haldane phase of $S = 1$ antiferromagnet in $d = 1$

The following simple Hamiltonian actually leads to a gapped phase with SRE, but gapless edge states:

$$H = J \sum_i \vec{S}_i \cdot \vec{S}_{i+1}. \tag{6.3}$$

The edge realizes effectively a $S = \frac{1}{2}$ state, despite the chain being built of $S = 1$ spins. This phenomena has been observed experimentally in some nickel-based insulating magnets, such as Y_2BaNiO_5. The Ni atoms form $S = 1$ spins, organized into chain-like structures that are relatively well isolated from one another .

The symmetry that is crucial to protecting this phase is $SO(3)$ spin-rotation symmetry. However, it turns out that the full rotation symmetry is not required. It is sufficient to retain just the 180° rotations about the x, y, and z axes. This symmetry group $\{I, X, Y, Z\}$ contains the identity and the three rotation elements. This can be written as $\{I, X\} \times \{1, Y\}$, since $Z = X \times Y$, i.e. the combination of two rotations is the third rotation. Mathematically, this group is $\mathbb{Z}_2 \times \mathbb{Z}_2$. We will write down a model with this rotation symmetry that, although it does not quite reduce the $S = 1$ down to this symmetry, has the advantage of being exactly soluble—not just for the ground state but also for all the excited states. This model also has a nice interpretation, namely of arising from condensing domain walls bound to spin flips.

6.3.2 An exactly soluble topological phase in $d = 1$

Consider a spin model with $\mathbb{Z}_2 \times \mathbb{Z}_2$ symmetry. There is a $\mathbb{Z}_2$ set of topological phases with this symmetry in $d = 1$, and we will explicitly construct the non-trivial topological phase. We will implement this symmetry by means of a pair of Ising models (labelled σ and τ) that live on a zigzag lattice as shown in Fig. 6.3. Consider beginning in

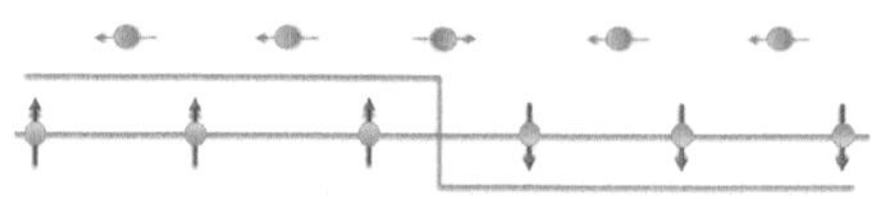

Fig. 6.3 [Colour online] An exactly soluble model of a 1D SRE phase with gapless edge states protected by $\mathbb{Z}_2 \times \mathbb{Z}_2$ symmetry. Terms in this Hamiltonian encourage binding of domain walls to spin flips. The topological phase emerges on condensing these 'decorated' domain walls.

the ordered state of the σ, but with the τ disordered and pointing along a transverse field τ^x. Now, we would like to restore the $\mathbb{Z}_2$ symmetry of the σ spins. We do this by condensing the domain walls of the σ spins. If we directly condense domain walls, we get the trivial symmetric state. However, we can choose to condense domain walls with a spin flip of τ attached. We will see that this gives the topological phase [8].

One way to do this is to write down a Hamiltonian that would lead to this binding. Note that the operator $\sigma_i^z \sigma_{i+1}^z$ detects a domain wall. Consider

$$H = -\sum_i \left(\sigma_{2i}^z \tau_{2i+1}^x \sigma_{2i+2}^z + \tau_{2i-1}^z \sigma_{2i}^x \tau_{2i+1}^z\right), \tag{6.4}$$

where we have placed the σ (τ) on the even (odd) sites of the lattice. In the absence of a domain wall, we have the usual transverse field term, whose sign changes when a domain wall is encountered. We will show that this is a gapped phase with SRE, but has gapless edge states. In fact this Hamiltonian was previously discussed in the context of generating 'cluster states' [27]

First, consider the system with periodic boundary conditions. We will leave it as an exercise to show that each of the terms in the Hamiltonian (6.4) commutes with all the others. Then, for a system with N sites, we have exactly N terms, which can be written as

$$H = -\sum_i \left(\tilde{\sigma}_{2i}^x + \tilde{\tau}_{2i+1}^x\right), \tag{6.5}$$

where the tilde indicates the three spin operators in the Hamiltonian (6.4). Hence, this simply looks as if each site has a modified transverse field, which implies a unique ground state and a gap, in this system with periodic boundary conditions.

Now consider open boundary conditions as shown. Let us focus on the left edge, where the end of the chain implies that we lose the $\tilde{\sigma}_x$ operator. This will result in a twofold degeneracy, as we will show. The first term in the Hamiltonian is now $-\sigma_0^z \tau_1^x \sigma_2^z$. We can easily show that the two operators $\Sigma^z = \sigma_0^z$ and $\Sigma^x = \sigma_0^x \tau_1^z$ commute with the Hamiltonian. However, they anticommute with one another. Hence, we can show that the ground state must be at least twofold-degenerate. Suppose that we have a unique ground state of the Hamiltonian, $|\psi\rangle$. This must be an eigenstate of Σ^z, since it commutes with the Hamiltonian. Let us say $\Sigma^z|\psi\rangle = \lambda|\psi\rangle$, where $\lambda = \pm 1$. However, we can find an independent state $|\psi'\rangle = \Sigma^x|\psi\rangle$. This is a degenerate state, since $[\Sigma^x, H] = 0$. It is also a distinct state, since it has a different eigenvalue $\Sigma^z|\psi\rangle = -\lambda|\psi\rangle$, owing to $\Sigma^z\Sigma^x = -\Sigma^x\Sigma^z$. Hence, there are at least two ground states, $(|\psi\rangle, |\psi'\rangle)$. They differ only by application of an edge operator, and hence this is an edge degeneracy.

Note that it is important that we preserve the symmetry—if we add Σ^a to the Hamiltonian, then we can gap the edge state, but at the expense of also breaking the $\mathbb{Z}_2 \times \mathbb{Z}_2$ symmetry. Hence, this is called a symmetry-protected topological (SPT) phase. This model has special properties that make it exactly soluble—but the addition of general perturbations that are local and preserve the symmetry leads to a more generic state. The presence of an energy gap implies that the state is stable against weak perturbations, which means that it will remain in the same phase.

Some of this material can be found in the review [30].

Exercise 6.1 Verify that the terms in (6.4) commute with one another, in a chain with periodic boundary conditions. With open boundaries, explicitly write out a Hamiltonian and check that the edge operators Σ^a commute with it.

Exercise 6.2 Use the Jordan–Wigner procedure to map (6.4) onto fermion operators. Recall that, for the 1D quantum Ising model, the transformation is $c_j^\dagger = \sigma_j^+ S_j$, where the string operator is $S_j = \prod_{i>j} \sigma_i^x$. Show that the same mapping leads to a topological phase of these non-local fermions. Note, however, that there are some important differences from a topological phase of electrons. Argue that in the latter case there must always be a residual symmetry that cannot be broken by any physical operator, unlike in the fermionized version of the problem here.

Let us give a couple of concrete examples of SRE topological phases of bosons/spins. These are necessarily interacting—unlike in the case of free fermions, there is no 'band' picture here. Later, we will see how they combine with the free-fermion topological phases, in particular the IQH effect, to extend the classification from a single integer to a pair of integers.

6.4 SRE phase of bosons in two dimensions

Let us consider a system of two species of bosons, 'A' and 'B' (e.g. two species of atoms in an optical lattice), whose numbers are individually conserved. By analogy with the 1D example, we want to find a disordered phase that respects these symmetries and can be obtained by condensing a vortex combined with a charge. In particular, say we begin in the superfluid state of the 'A' bosons. We want to exit from it by condensing vortices (and antivortices). However, in order to avoid reaching the regular Mott insulator, we will bind a +1 charge of a 'B' boson to this vortex (and a −1 charge on the antivortex) before condensing them. We will show that this gives rise to an SRE topological phase. The phase will have a gapless edge state protected by symmetry. We will implement this in two ways [17]: (i) by a coupled-wire construction and (ii) by writing down an effective field theory.

6.4.1 Coupled-wire construction

The field theory for a single chain of bosons takes the form

$$H = J(\partial_x \phi)^2 + U(\partial_x \theta)^2, \tag{6.6}$$

where ϕ is the boson phase and θ is defined as

$$\partial_x \theta(x) = 2\pi n(x), \tag{6.7}$$

with n being the particle density. Using the standard density–phase commutation relation, we have $[\partial_x \theta(x), \phi(x')] = -2\pi i \delta(x - x')$. This can be integrated to give

$$[\theta(x), \phi(x')] = -\pi i \, \mathrm{Sign}(x - x'), \tag{6.8}$$

where $\mathrm{Sign}(x) = +(-)1$ for $x > 0$ $(x < 0)$.

Exercise 6.3 Show that this commutation relation is symmetric on interchanging the fields $\theta \leftrightarrow \phi$.

Normally, when the bosons are at commensurate filling, one also has vortex tunnelling operators $H_v = -\sum_n \lambda_n \cos(n\theta)$. When these are relevant, a Mott insulator results—for example, when $-\cos\theta$ is large, the field is pinned at $\theta = 0$, which implies that the density is uniform. as in a Mott state. An excitation is a soliton, that is, $\theta(x \ll 0) \to 0$ while $\theta(x \gg 0) \to 2\pi$. This costs a finite energy, which is the gap to particle excitations in the Mott state.

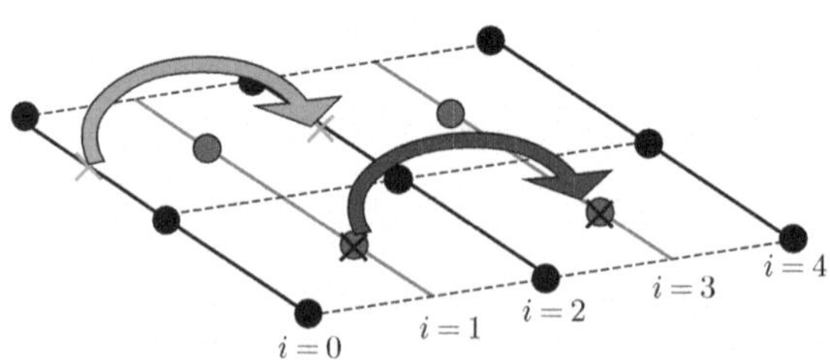

Fig. 6.4 [Colour online] Coupled-wire construction of bosonic topological phase. 'A' and 'B' bosons live on the even-numbered and odd-numbered tubes, respectively. Vortices of one species are bound to bosons of the opposite species, and their tunnelling is shown by the arrows. These processes are shown to commute and gap the bulk, but leave behind a topological edge state.

However, here we will be interested in a different type of vortex condensate, one that binds charge. To this end, consider building a 2D system by coupling 1D systems (tubes) together. We will assume that on alternate tubes we have bosons of the two different species—thus, on the even-numbered tubes ($2i$) we have 'A' and on the odd-numbered tubes ($2i+1$) we have 'B'. In the absence of interchain coupling, we have the decoupled Luttinger-liquid Hamiltonian

$$H_0 = \sum_i \left[U(\partial_x\theta_i)^2 + J(\partial_x\phi_i)^2\right] \tag{6.9}$$

and

$$[\theta_j(x), \phi_k(x')] = -i\pi\delta_{j,k}\delta \tag{6.10}$$

We ignore the effect of conventional vortices. Let us instead attempt to condense vortices of the 'A' bosons with 'B' charge, and vice versa. We will do this by allowing the composite objects to 'hop', which will lower their energy and make them eventually 'Bose-condense'. Of course, since this is a vortex condensate, we are led to an insulator, as elaborated previously. We will see that this is an exotic insulator.

Note that the vortices of the 'A' bosons (in Fig. 6.4) naturally live in the centres of plaquettes of the sites available to 'A'. This happens to be on the 'B' tubes as shown by the crosses in Fig. 6.4. Hence, if we hop a vortex from one of the crosses to the adjacent one, this can be represented as a space–time event on tube $i = 2$, represented as $e^{i\theta_2}$. However, we also simultaneously want to hop 'B' bosons, and we have arranged for their lattice sites to coincide with the locations of the 'B' vortices. This combined process is then written as $e^{i(-\phi_1+\phi_3+\theta_2)}$. The reverse process binds an antivortex to a 'hole', $e^{i(+\phi_1-\phi_3-\theta_2)}$, and, taken together with the process of hopping B vortices bound with A bosons, this leads to the following set of terms:

$$H_{\text{int}} = -\lambda\sum_i \cos(\phi_{i-1} - \theta_i - \phi_{i+1}). \tag{6.11}$$

First, we note that these terms all commute with one another; for example, if we denote $\tilde{\theta}_i = \theta_i + (\phi_{i+1} - \phi_{i-1})$, then any two terms commute:

$$[\tilde{\theta}_i(x), \tilde{\theta}_j(x')] = 0. \tag{6.12}$$

Exercise 6.4 Calculate the commutator of a pair of fields $\Phi_{l,m}$ and $\Phi_{l',m'}$ where $\Phi_{l,m} = \sum_i(l_i\phi_i + m_i\theta_i)$. Use this to prove the result (6.12).

Thus, all of these can be simultaneously satisfied. If we have periodic boundary conditions, then there is a unique ground state, rather like the way in which pinning the θ fields gives a unique Mott insulating state. The same count of variables leads to a unique state in this case. However, interesting edge states appear if we have an open slab as in Fig. 6.4. Let us focus on the left edge. Clearly, we are missing a cosine pinning field for $\tilde{\theta}_0$. This means that the first non-vanishing cosine terms are $\tilde{\theta}_1 = -\phi_0 + \theta_1 + \phi_2$ and $\tilde{\theta}_2 = -\phi_1 + \theta_2 + \phi_3$. A field that commutes with both of these is $\Phi = \phi_0$. However, there is also a conjugate field that we require to define the edge dynamics. In an isolated chain, this would have been θ_0. However, here this does not commute with one of the first cosines. It can be rectified by adding $\Theta = \theta_0 + \phi_1$, which commutes with all the cosines and has the standard Luttinger-liquid commutator with Φ: $[\Theta(x), \Phi(x')] = -i\pi(x - x')$. Hence, we have a gapless edge mode, which is described by the usual Luttinger-liquid theory. However, there is an important difference between this Luttinger liquid at the edge and one that can be realized in purely 1D. It is impossible to gap this edge without breaking one of the symmetries, which happens to be number conservation of the two boson species. This is an internal symmetry—and in a purely 1D system it is always possible to find a gapped state that preserves all internal symmetries: we just combine degrees of freedom until they transform in a trivial way under the symmetry and condense them. However, this is not possible at the edge: one cannot condense Φ, since it is charged under the $U(1)$ symmetry that protects 'A' particle conservation, and similarly we cannot introduce a cosine of the Θ field, since it is charged under the other $U(1)$. This is an indication that it is a topological phase—we will see that this also implies a quantized Hall conductance.

6.4.2 Effective field theory

Let us write down an effective theory to describe a fluid built out of boson–vortex composites [17]. The 'A' particles acquire a phase of 2π on circling vortices, and hence the effect of vortices can be modelled by a vector potential whose curl is centred at the vortex locations:

$$\partial_x a_y - \partial_y a_x = 2\pi \sum_j n_j^v \delta(r - r_j^v), \tag{6.13}$$

where n_j^v and r_j^v are respectively the strength and location of the vortices. This vector potential will couple minimally to the current $\mathcal{L} = \vec{j}_A \cdot \vec{a}$, where the vectors are 2-vectors. A rewriting of this formalism to include the motion of vortices results in a generalization to a 3-current $j^\mu = (\rho, j_x, j_y)$ and a gauge 3-potential $a_\mu = (a_0, a_x, a_y)$. Also, since we assume that the vortices are bound to the 'B' bosons, we can rewrite the equation for a as

$$\epsilon^{\mu\nu\lambda}\partial_\nu a_\lambda^B = 2\pi j_B^\mu, \tag{6.14}$$

where we have introduced a superscript B for the vector potential. We can also utilize the continuity equation for the current j_A, $\partial_\mu j_A^\mu = 0$, to write

$$\epsilon^{\mu\nu\lambda}\partial_\nu a_\lambda^A = 2\pi j_A^\mu. \tag{6.15}$$

To keep track of the charge densities of 'A' and 'B' bosons, it is useful to introduce external vector potentials $A^{(A,B)}$ that couple to the currents of these bosons. This leads to our final topological Lagrangian

$$\mathcal{L}_{\text{topo}} = \frac{\epsilon^{\mu\nu\lambda}}{2\pi}\left(a_\mu^A \partial_\nu a_\lambda^B + A_\mu^A \partial_\nu a_\lambda^A + A_\mu^B \partial_\nu a_\lambda^B\right) \tag{6.16}$$

and the corresponding partition function

$$\mathcal{Z}_{\text{topo}}[A^A, A^B] = e^{iS_{\text{topo}}} = \int \mathcal{D}a^A\, \mathcal{D}a^B\, e^{i\int dx\, dy\, dt\, \mathcal{L}_{\text{topo}}}. \tag{6.17}$$

First, we would like to establish that this describes a SRE phase. Note that the mutual phases involved are all 2π, implying the absence of fractional statistics. We can also compute the ground-state degeneracy on the torus—this turns out to be directly computable from this theory: if we write

$$\mathcal{L} = \frac{K_{IJ}}{4\pi}\epsilon^{\mu\nu\lambda} a_\mu^I \partial_\nu a_\lambda^J, \tag{6.18}$$

then the ground-state degeneracy is $|\det K|$. In this case, $K = \sigma^x$, and there is a unique ground state.

Given that we have an SRE phase, we can deduce two important consequences from this theory. The first concerns edge states, which can be shown to be equivalent to that derived before, and the quantized Hall conductivity. The latter is obtained by integrating out the a fields to obtain an action purely in terms of the external probe fields A. The current is then defined as $j_A = \delta S/\delta A^A$, where $\mathcal{Z} = e^{iS}$. A Gaussian integration of (6.16) yields

$$S_{\text{topo}} = -\int dx\, dy\, dt\, \frac{\epsilon^{\mu\nu\lambda}}{2\pi} A_\mu^A \partial_\nu A_\lambda^B, \tag{6.19}$$

and thus we have

$$j_A^\mu = -\frac{1}{2\pi}\epsilon^{\mu\nu\lambda}\partial_\nu A_\lambda^B. \tag{6.20}$$

If we consider the spatial components of this equation, we find: $j_A^x = (1/2\pi)E_B^y$, where E_B is the electric field applied to species 'B'. Thus, we have a crossed-response Hall conductivity $\sigma_{xy}^{AB} = 1/2\pi$, which, on replacing charge Q_a for the bosons and $\hbar$, gives $\sigma_{xy}^{AB} = Q_A Q_B/h$.

We would like to apply these insights to electronic systems, where one may combine pairs of electrons to form Cooper pairs with charge $Q = 2e$. However, in that case, there is a single conservation law. Topological phases with a single $U(1)$ can be described by the above field theory, (6.16), if we assume that the two species of bosons can tunnel into one another and collapse the combined $U(1) \times U(1)$ symmetry into a single common $U(1)$. This amounts to replacing the pair of external vector potentials by a single one, and the resulting topological response theory is

$$S_{\text{topo}} = -\int dx\, dy\, dt\, \frac{\epsilon^{\mu\nu\lambda}}{2\pi} A_\mu \partial_\nu A_\lambda. \tag{6.21}$$

Now, differentiating with respect to the vector potential, we get *two* contributions, and hence $j^\mu = (2/2\pi)\epsilon^{\mu\nu\lambda}\partial_\nu A_\lambda$, which implies a Hall conductivity, in units of the boson charge, of $\sigma_{xy} = 2Q^2/h$. This is the bosonic IQH (BIQH) phase. Somewhat surprisingly, its Hall conductance is always an *even* integer. Potential realizations of this phase in bilayer systems of bosons in the lowest Landau level with net filling $\nu = 2$ have been discussed in numerical work [11, 22, 37].

Note that we have assumed commensurate filling to admit an insulator. Also, these models are not exactly soluble in the same way that the previous models were—for other approaches to constructing models in this phase, see [12].

6.4.3 Implications for IQH state of electrons

It is well known that free-fermion IQH states have a quantized Hall conductance $\sigma_{xy} = ne^2/h$. At the same time, they have a quantized thermal Hall effect $\kappa_{xy}/T = c\pi^2 k_B^2/3h$, where $c = n$. The latter simply counts the difference between the numbers of right-moving and left-moving edge states. This equality is an expression of the Wiedemann–Franz law that relates thermal and electrical conductivities for weakly interacting electrons. This leads to the familiar integer classification of the IQH in terms of $\mathbb{Z}$. How is this modified in the presence of interactions? We will continue to assume SRE—so FQH states are excluded from our discussion. It has long been known that n must remain an integer if charge is to remain unfractionalized. However, the equality $n = c$ can be modified. In fact, if we assume that the electrons can combine into Cooper pairs, which form the BIQH state, then the latter has Hall conductance $\sigma_{xy} = 8e^2/h$, but $\kappa_{xy} = 0$. Thus, we can have $n - c = 8m$. Indeed, this implies that the classification of interacting quantum Hall states of electrons with SRE is in terms of $\mathbb{Z} \times \mathbb{Z}$ at least. Note, that this also predicts a phase where $n = 0$ but $c = 8$. This can be achieved by combining an $n = 8$ free-fermion quantum Hall state with a BIQH state of Cooper pairs to cancel the electrical Hall conductance. The remaining thermal Hall conductance is $c = 8$. It can be shown that a π flux inserted in this state has trivial statistics and can be condensed—which implies that all electrons are confined into bosonic particles without disturbing the topological response of this phase [14]. Alternatively, one can show that neutral bosons with short-range interaction can lead to a topological phase with chiral edge states, if they appear in multiples of eight. Indeed, one can write down a multicomponent Chern–Simons theory to describe this topological phase of neutral bosons, in terms of a K-matrix as described in detail below [17].

A phase without topological order is characterized by a symmetric K-matrix that satisfies $|\det K| = 1$. A chiral state in $(2+1)$ dimensions requires the signature (n_+, n_-) of its K-matrix to satisfy $n_+ \neq n_-$. We therefore seek a K-matrix with the following properties:

1. $|\det K| = 1$.
2. The diagonal elements $K_{I,I}$ are all even integers so that all excitations are bosons.
3. There is a maximally chiral phase, where all the edge states propagate in a single direction. Then, all eigenvalues of K must have the same sign (say positive), so K is a positive-definite symmetric unimodular matrix.

It is helpful to map the problem of finding such a K to the following crystallographic problem. By diagonalizing K and multiplying each normalized eigenvector by the square root of its eigenvalue, one obtains a set of primitive lattice vectors $\mathbf{e}_I$ such that $K_{IJ} = \mathbf{e}_I \cdot \mathbf{e}_J$. The inner product of a pair of vectors $l_I \mathbf{e}_I$ and $l'_I \mathbf{e}_I$ is given by $l'_I K_{IJ} l_J$, while the volume of the unit cell is given by $(\det K)^{1/2}$. The latter can be seen by writing the components of the vectors as a square matrix: $[k]_{aI} = [e_I]_a$. Then, $\det k$ is the volume of the unit cell. However, $K_{IJ} = \sum_a k_{aI} k_{aJ} = (k^T k)_{IJ}$. Thus, $\det K = (\det k)^2$.

Thus, for a phase without topological order, we require the volume of the lattice unit cell to be unity: $\det k = 1$ (unimodular lattice). Furthermore, for a bosonic state, we require that all lattice vectors have even length, $l_I K_{IJ} l_J =$ even integer, since the K-matrix has even diagonal entries (even lattice). It is known that the minimum dimension in which this can occur is eight. In fact, the root lattice of the exceptional Lie group E_8 is the smallest-dimensional unimodular even lattice.[1] Such lattices occur only in dimensions that are multiples of 8. See [3] for more details.

A specific form of the K-matrix is

$$K = \begin{pmatrix} 2 & -1 & 0 & 0 & 0 & 0 & 0 & 0 \\ -1 & 2 & -1 & 0 & 0 & 0 & 0 & 0 \\ 0 & -1 & 2 & -1 & 0 & 0 & 0 & -1 \\ 0 & 0 & -1 & 2 & -1 & 0 & 0 & 0 \\ 0 & 0 & 0 & -1 & 2 & -1 & 0 & 0 \\ 0 & 0 & 0 & 0 & -1 & 2 & -1 & 0 \\ 0 & 0 & 0 & 0 & 0 & -1 & 2 & 0 \\ 0 & 0 & -1 & 0 & 0 & 0 & 0 & 2 \end{pmatrix}. \tag{6.22}$$

This matrix has unit determinant and all of its eigenvalues are positive. It defines a topological phase of bosons without topological order, with eight chiral bosons at the edge. We will call this the E_8 state since it is related to the E_8 group.

6.5 SPT phases of bosons in three dimensions

We would like to understand how to describe the surface states of a 3D SRE topological phase of bosons. While for free fermions one has Dirac or Majorana surface cones, the bosonic analogue is less clear, particularly since the surface is 2D and one does not have access to bosonization and other powerful tools available for the previous problem of 1D edges.

Based on our previous experience with 1D edges, we will consider the surface directly and ask how symmetry can act in an anomalous way, to produce a topological

[1] See the Wikipedia entry for the E_8 root lattice (Gosset lattice): https://en.wikipedia.org/wiki/E8_lattice.

surface [31]. The simplest example to consider is a system with $U(1)$ and time-reversal ($\mathcal{T}$) symmetry, where the $U(1)$ may be considered as a conserved spin component (S_z) rather than charge. This corresponds to $U(1) \times \mathcal{T}$. One option is to break the symmetry at the surface—this is a valid surface state even for a topological bulk. Say we break the $U(1)$ to get an ordered surface (which we will call a 'superfluid' since it breaks a $U(1)$ symmetry). To restore this symmetry, we would like to proliferate vortices, which can revert us to the fully symmetric state. However, for the surface of a topological bulk, there should be an obstruction to proliferating vortices. Rolling is a potential mechanism—note that the vortices here preserve time-reversal symmetry, i.e. a vortex is mapped to a vortex under $\mathcal{T}$. This follows from the fact that our phase degree of freedom transforms like magnetic order with $\phi \to \phi + \pi$ under time reversal, so that $e^{i\phi} \to -e^{i\phi}$. The vorticity, which is defined via $\nabla \times \nabla\phi$, is invariant under this operation. Hence we can ask—how does a vortex transform under $\mathcal{T}$?

There are two physically distinct options: namely whether the vortex transforms as a regular or as a projective representation. In the former case, there is no obstruction to condensing vortices and restoring the symmetric phase—hence this cannot represent a topological surface state. However, the vortices can also transform as a projective representation, since they are non-local objects. On a closed surface, one must make a vortex–antivortex pair. Taken together, these must transform as $\mathcal{T}^2 = +1$. However, individually, they can transform as $T^2 = -1$, i.e. the vortex is a Kramers doublet. Denote by ψ_σ the two-component vortex field $\sigma = \uparrow, \downarrow$, which transforms as $\psi_\uparrow \to \psi_\downarrow$, $\psi_\downarrow \to -\psi_\uparrow$ under time reversal. The effective Lagrangian is

$$\mathcal{L} = |(\partial_\mu - ia_\mu)\psi_\sigma|^2 + (\partial_\mu a_\nu - \partial_\nu a_\mu)^2 + m|\psi_\sigma|^2 + \ldots, \tag{6.23}$$

where the gauge field is determined by the boson 3-current j^μ, which includes the boson charge density j^0 and current $j^{1,2}$:

$$\epsilon^{\mu\nu\lambda}\partial_\nu a_\lambda = 2\pi j^\mu. \tag{6.24}$$

The vortex–gauge field coupling is intuitively rationalized by the fact that a vortex moving around a boson acquires a 2π phase. Hence, the gauge potential a that implements this satisfies $\partial_x a_y - \partial_y a_x = 2\pi j^0$. This is one component of (6.24), while the other components follow from the continuity equation $\partial_\mu j^\mu = 0$.

In this dual language, when the vortices are gapped, the $U(1)$ symmetry is broken, while if they are condensed, the $U(1)$ symmetry is restored. The key difference between a single-component vortex field and the Kramers doublet vortex is that in the latter case the vortex condensate always breaks time-reversal symmetry. This can be seen by considering the operator $\psi^\dagger_\sigma \sigma^a_{\sigma\sigma'} \psi_\sigma = n^a$, where σ^a are Pauli matrices. Since it is a product of a vortex–antivortex pair, it is a local operator, unlike an operator that insets a vortex. In a vortex condensate, this operator will acquire a non-zero expectation value. Under time reversal, it is readily seen that $n^a \to -n^a$, indicating that time-reversal symmetry is broken. Thus, the $U(1)$ symmetry is restored at the expense of breaking $\mathcal{T}$. This is a candidate for a topological surface state.

Exercise 6.5 Establish the above result by introducing an external 'probe' electromagnet field that couples to the bosons, $\mathcal{L}_{\text{int}} = j^{\mu}A_{\mu} = A_{\mu}\epsilon^{\mu\nu\lambda}\partial_{\nu}a_{\lambda}/2\pi$, and integrate out the other fields to obtain an effective action in terms of A. Consider doing this in the two limits $m > 0$ and $m < 0$, where vortices are respectively gapped and condensed. Show that when the vortices are gapped, the effective Lagrangian is $\mathcal{L}_{\text{eff}} \sim A_{\perp}^2$, where $A_{\perp}$ is the transverse part, and this represents a $U(1)$-broken phase (a 'superfluid'). On the other hand, when the vortices are condensed, show that $\mathcal{L}_{\text{eff}} \sim (\partial_{\mu}A_{\nu} - \partial_{\nu}A_{\mu})^2$ (an 'insulator').

We now turn to two other possible surface states that can occur in this theory.

6.5.1 The $m = 0$ critical point

The first is the critical point at $m = 0$, where symmetries are unbroken but the surface is gapless. This is the bosonic analog of the gapless Dirac cone of fermionic topological insulators. However, since bosons are either gapped or condensed, this requires tuning of a parameter for its realization. This field theory (the non-compact $\mathbb{C}P^1$ model [21]) has appeared before in the theory of 'deconfined quantum critical points', describing a direct transition between Néel and valence bond solid order in spin models on a square lattice [25, 26]. However, there the vortices transformed projectively under spatial symmetries, such as translation and rotation. Here, an internal symmetry (time reversal) is involved—which can only occur on the surface of a 3D topological phase.

6.5.2 Surface topological order of 3D bosonic SRE phases

The second possibility is to consider condensing a pair of vortices $\Phi = \epsilon_{\sigma\sigma'}\psi_{\sigma}(r)\psi'_{\sigma}(r')$, which is a Kramers singlet. This leads to a restoration of the $U(1)$ symmetry (insulator), while preserving $\mathcal{T}$. However, this is an 'exotic' insulator with topological order (excitations that have fractional statistics). Note that the bulk 3D state is still SRE and the exotic excitations are confined to the surface. It is readily shown that the topological order is the same as that in the toric code. To show this, we need to identify an e and an m particle that are bosons but with π mutual statistics. The m particle is just the unpaired vortex, which remains as a gapped excitation in this phase. Additionally, we can discuss defects in the 2-vortex condensate. These are nothing but particles—however, the 2-condensate allows for a fractional particle. To see this, consider the the effective 2-vortex theory

$$\mathcal{L}_{2v} = |(\partial_{\mu} - 2ia_{\mu})\Phi|^2 + (\partial_{\mu}a_{\nu} - \partial_{\nu}a_{\mu})^2 + m_2|\Phi|^2 + \ldots, \tag{6.25}$$

which can be obtained from (6.23) by considering an interaction that pairs vortices and by ignoring the gapped single vortices. In the 2-vortex condensate, one can consider vortices, which are obtained from the flux-quantization condition $2(\partial_x a_y - \partial_y a_x) = 2\pi$; however, since the flux is related to particle density, this implies a particle with charge one-half that of the fundamental bosons. Clearly, taking a half charge around a vortex leads to a π phase. Hence, this is the m particle.

This surface topological order provides a powerful way to characterize a 3D topological phase [31]. The surfaces of SRE topological phases should be distinct from states that can be realized purely in the lower dimension. The way this works with surface topological order is that although the topological order itself can be realized in 2D, the way the excitations transform under symmetry cannot be realized in a purely 2D setup. For example, here the m particle is a Kramers doublet, while the e particle carries half the charge of the boson (and may or may not be a Kramers doublet).

While in this case it is not immediately apparent that this is forbidden in 2D, we can give another example that arises where this is obvious. Consider the situation where both e and m particles carry half charge—this is one of the surface topological orders associated with $U(1)$ charge and $\mathcal{T}$ symmetry. We can show that this state cannot be $\mathcal{T}$-symmetric if realized in 2D, where it can be described by a K-matrix Chern–Simons theory:

$$\mathcal{L}_{\text{CS}} = \frac{2}{2\pi}\vec{a}_1 \cdot \vec{\nabla} \times \vec{a}_2 - \frac{\vec{\nabla} \times \vec{A}}{2\pi} \cdot (\vec{a}_1 + \vec{a}_2). \tag{6.26}$$

Coupling to the external $\vec{A}$ ensures that we can keep track of the charge. Note that $K = 2\sigma_x$ ensures that we have toric-code-type $\mathbb{Z}_2$ topological order ($|\det K| = 4$). Now, by integrating out $\vec{a}$, we obtain

$$\mathcal{L}_{\text{eff}} = -\frac{1}{4\pi}\vec{A} \cdot \vec{\nabla} \times \vec{A}. \tag{6.27}$$

This implies that if this state is realized in 2D, then it will have a non-vanishing Hall conductance $\sigma_{xy} = Q^2/h$, contradicting the fact that it is $\mathcal{T}$-symmetric. However, it can be realized while still retaining $\mathcal{T}$ symmetry on the surface of a topological phase.

The simplest way to argue this is through the following coupled-layer construction [34], analogous to the 1D and 2D cases that we discussed before. Consider layers of 2D toric-code models where just the e particle carries half charge. Now, in parallel to the constructions in lower dimensions, we consider a set of three layers, and form a bound state of $e_i m_{i+1} e_{i+2}$. This is a boson, which commutes with other triplets. For example, $e_0 m_1 e_2$ and $e_1 m_2 e_3$ are mutual bosons. Also, it has integer charge and can be neutralized by a physical boson. Hence, condensing these triplets leads to an SRE 3D state, with all symmetries. However, it leaves behind an edge state: for example, e_0 is not confined. Similarly, $m_0 e_1$ also commutes with the condensate. This is the new m particle of the toric-code topological order, which is confined to the top layer. Note that it carries half charge, which is precisely what we wanted to construct. Here, time reversal is explicitly preserved. Realizing this state on the surface of a 3D system ensures that we never have to declare the edge physics (which would break time-reversal symmetry).

This state corresponds to the surface of a 3D bosonic topological insulator (3D BTI), and models a surface that is 'half' the 2D bosonic IQH phase, which has $\sigma_{xy} = Q^2/h$. Note that one can draw the following analogy to the free-fermion topological insulator. A time-reversal symmetry-breaking perturbation can render the surface of the 3D topological insulator insulating. However, a domain wall between two opposite $\mathcal{T}$-breaking domains on the surface necessarily has a single chiral mode along it; see Fig. 6.5. Therefore, the difference in Hall conductivity between the two domains is

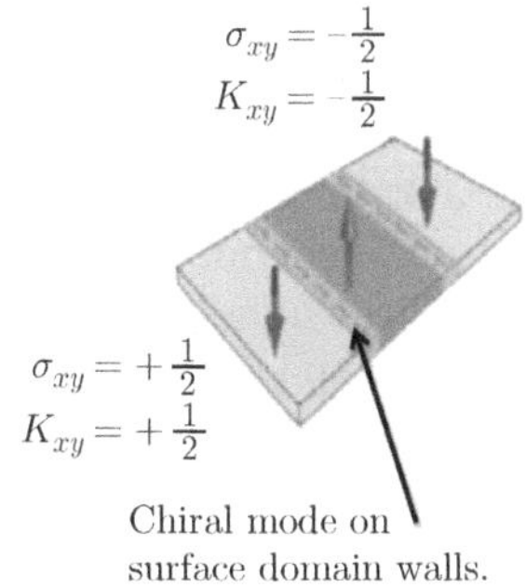

Fig. 6.5 [Colour online] Magnetic domains on the surface of a 3D free-electron topological insulator. The resulting insulating surface has $\sigma_{xy} = \pm\frac{1}{2}e^2/h$ and $\kappa_{xy} = \pm\frac{1}{2}$, since the domain wall carries a single chiral edge mode.

$\Delta\sigma_{xy} = 1(e^2/h)$. Also $\Delta\kappa_{xy} = 1$. By time-reversal symmetry, the two domains should have opposite Hall conductivities—hence, we are forced to assign $\sigma_{xy} = +\frac{1}{2}e^2/h$, $\kappa_{xy} = +\frac{1}{2}$, and the time-reversed version to the other domain. Since a purely 2D free-fermion system cannot have fractional Hall conductivities, it is not possible to screen this with a 2D layer. In a similar way, we can build a 3D topological phase from the 2D IQH state of bosons by including time-reversal symmetry. This is the 3D BTI, whose surface state is described above.

In a very similar fashion, one can model a state with a surface that is 'half' the chiral E_8 state, but time-reversal-symmetric when realized in 3D. This is the 3D bosonic topological superconductor (3D BTSc), and, although a symmetry-protected topological phase, it is not captured by the 'cohomology' approach [5, 6]. The surface topological order is the fermionic variant of the toric code—it has three non-trivial particles that have mutual π statistics, like the toric code, but all three particles have fermionic statistics. At first sight, it might appear that this state is time-reversal-symmetric, but in fact it must carry chiral edge modes if realized in 2D. An explicit Chern–Simons representation of this state is provided through the K-matrix

$$K = \begin{pmatrix} 2 & -1 & -1 & -1 \\ -1 & 2 & 0 & 0 \\ -1 & 0 & 2 & 0 \\ -1 & 0 & 0 & 2 \end{pmatrix}. \tag{6.28}$$

The eigenvalues of this matrix are all the same sign, implying that there are four chiral modes if this state is realized in 2D, and hence the state always breaks $\mathcal{T}$ symmetry. However, it may be realized on the surface of a 3D topological state with $\mathcal{T}$. Again, this result can be obtained via a coupled-layer construction. A different approach [2] to realizing this phase is via an exactly soluble model, based on the following observation. It is well known in the context of the 2D FQH effect that ground-state wavefunctions can be related to correlation functions of the edge conformal field theory. The two coordinates of particles in the wavefunction are traded for a single spatial coordinate at the edge together with time. Can a similar approach be taken for 3D topological phases? While the obvious generalization is to relate the wavefunction written in terms of particle coordinates, a useful generalization is obtained by representing the wavefunction in terms of loops. Now, the amplitude of a particular loop configuration $\mathcal{C}$ in 3D space, $\Psi(\mathcal{C})$, is related to the space–time amplitude for a process in which the loops are imagined as world lines of particles in the surface topological order. Hence, the loops come in different 'colours' corresponding to the non-trivial particles in the theory, and rules concerning how they fuse together etc. are determined by the topological data of the surface theory. For example, in the case of the three-fermion topological order, the amplitude is

$$\Psi(\mathcal{C}) = \int \mathcal{D}a \, \exp\Big(i \oint_{\mathcal{C}} j_I^{\mu} a_{\mu}^I\Big) \exp\Big(iK_{IJ} \int \epsilon^{\mu\nu\lambda} a_{\mu}^I \partial_{\nu} a_{\lambda}^J\Big), \tag{6.29}$$

where the j define the loop structure. This can be converted into a exactly soluble model (the Walker–Wang model) on a cubic lattice.

6.6 Surface topological order of fermionic topological insulators and superconductors

The well-known fermionic $\mathbb{Z}_2$ topological insulator is usually associated with a single Dirac-cone surface state. Breaking time-reversal symmetry at the surface (e.g. by introducing magnetic moments that order) can open a surface gap and render it insulating. However, this is not the only way to obtain an insulating surface—one can preserve all symmetries and obtain an insulator with topological order as for the bosonic SRE phases. Note that, by the same logic as for the 3D BTI and 3D BTSc, the topological order is such that when realized in purely 3D, it breaks $\mathcal{T}$ symmetry and has $\sigma_{xy} = \frac{1}{2}e^2/h$ and $\kappa_{xy} = \frac{1}{2}$. That is, it is a candidate for an FQH effect of electrons in a half-filled Landau level. The most famous such candidate is the Moore–Read Pfaffian state, which may be thought of as Ising $\times$ $U(1)_8$. More physically, one can imagine beginning with a superconductor of electrons, in a $p_x + ip_y$ state, where the Cooper pairs are effectively at $\nu_{\text{Cooper}} = \frac{1}{8}$ filling ($\nu_{\text{Cooper}} = \frac{1}{4}\nu_{\text{electron}}$, since there are half as many Cooper pairs as electrons, and the magnetic field measured in units of the new flux quantum $h/2e$ is twice as large). When the Cooper pairs form a bosonic Laughlin state, the Moore–Read state results. Unfortunately, while this state has the right σ_{xy}, it has $\kappa_{xy} = \frac{3}{2}$. Moreover, a quick glance at the topological spins of the quasiparticles reveals that it cannot be made time-reversal-symmetric even on the surface of a 3D topological insulator. Fortunately, a simple variant is much more promising [1, 4]: one considers a $p_x - ip_y$ superconductor in conjunction with the same Cooper-pair Laughlin state, i.e. Ising* $\times$ $U(1)_8$. This state, dubbed the T-Pfaffian [4], can be made time-reversal-symmetric on the surface of a 3D topological insulator, but of course breaks this symmetry in 2D since it has a finite Hall conductance. A different but equivalent solution [20, 32] features the Moore–Read state in conjunction with a neutral antisemion theory $U(1)_{-2}$. The surface topological order helps to understand how this classification is augmented in the presence of strong interactions wherein the electrons may pair to form bosons that exhibit a topological phase. The electrons could form Cooper pairs that then go into a 3D BTI phase. Or the electrons could combine into neutral bosons that then enter a 3D BTSc phase. Both of these extend the original $\mathbb{Z}_2$ classification by an additional factor of $\mathbb{Z}_2$. Wang, Potter, and Senthil [33] showed that this exhausts the set of 3D topological phases of interacting electrons with charge conservation and $\mathcal{T}$ symmetry.

Table 6.1 shows the topological phases with SRE that can exist in 3D, together with their physically most relevant symmetries.

Topological superconductors in 3D are protected by $\mathcal{T}$, and, for the physical case of $\mathcal{T}^2 = -1$ when acting on fermions, this gives rise to an integer set of topological phases. One might imagine that by combining pairs of electrons into neutral bosons, one could augment this classification by $\mathbb{Z}_2$, by including the 3D BTSc in this list. However, it turns out that this phase is already present in the free-fermion classification and corresponds to $\nu = 8$ of the $\mathbb{Z}$ classification [9]. Hence, there is no new phase. On the other hand, since this phase has a $\mathbb{Z}_2$ classification, this implies that the case with two copies of $\nu = 8$, i.e. $\nu = 16$, is trivial. Therefore, the interacting topological superconductor classification is reduced from the free-fermion one: $\mathbb{Z} \to \mathbb{Z}_{16}$ [9, 15]. This observation is interesting since it represents a non-perturbative result in $(2+1)$ dimensions and has been verified by other means [13, 18, 35, 36]. The surface of a topological superconductor with $\nu = 1$ has a Majorana cone with low-energy dispersion:

$$H = -i\chi^T(\sigma^x \partial_x + \sigma^z \partial_y)\chi, \tag{6.30}$$

where $\chi^T = (\chi_1, \chi_2)$. The surface with index ν then has ν flavours with the above dispersion. It is readily verified that weak interactions at the surface are irrelevant. From (6.30), requiring that the action corresponding to the kinetic term be dimensionless (and that time and space have the same dimensions $[t] = [x] \sim L$), the scaling dimensions of the χ fields are $[\chi] \sim 1/L$. The interaction term, written schematically as $\mathcal{S}_{\text{int}} \sim \int d^2\, dt\, (\chi_a^T \sigma_y \chi_a)(\chi_b^T \sigma_y \chi_b)$, then has scaling dimension $[\mathcal{S}_{\text{int}}] \sim 1/L$, which means that it is irrelevant at long scales. Therefore, the way $\nu = 16$ is connected to $\nu = 0$ is via strong interactions, as shown in Fig. 6.6. Establishing this therefore

Table 6.1 Topological phases in 3D with SRE. The physically most relevant symmetries, corresponding to the topological insulator and topological superconductor (TSc), are shown

SYMMETRIES	FREE FERMIONS	INTERACTING BOSONS	INTERACTING FERMIONS
$U(1)$ (charge) and $\mathcal{T}$ Topological insulator	$\mathbb{Z}_2$ Class AII	$\mathbb{Z}_2 \times \mathbb{Z}_2 \times \mathbb{Z}_2$	$\mathbb{Z}_2 \times \mathbb{Z}_2 \times \mathbb{Z}_2$
$\mathcal{T}$ $\mathcal{T}^2 = (-1)^{N_F}$ TSc	$\mathbb{Z}$ Class DIII	$\mathbb{Z}_2 \times \mathbb{Z}_2$	$\mathbb{Z}_{16}$

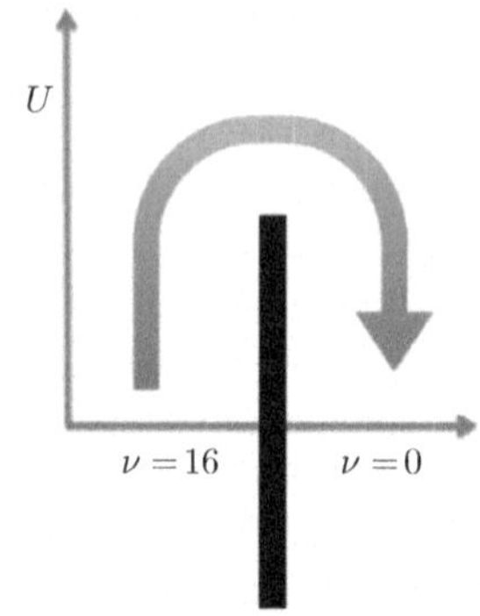

Fig. 6.6 Interaction effects on the surface of a 3D topological superconductor are irrelevant for weak interactions, given the linear dispersion of the Majorana cone surface states. However, strong interactions U can connect two surfaces without a phase transition—which implies that the bulk phases are equivalent in the presence of interactions. Establishing this requires non-perturbative techniques.

requires a non-perturbative analysis. While bosonization provides such a tool for a $(1 + 1)$-dimensional edge, establishing this for $(2 + 1)$ dimensions requires new non-perturbative tools, such as working with the surface topological order or with a dual vortex/monopole theory [19]. A nice review of these recent developments can be found in [24].

Acknowledgements

We thank Adrian P for comments on the manuscript.

References

[1] Bonderson, P., Nayak, C., and Qi, X.-L. (2013). A time-reversal invariant topological phase at the surface of a 3D topological insulator. *J. Stat. Mech.: Theory Exp.*, **2013**, P09016.

[2] Burnell, F. J., Chen, X., Fidkowski, L., and Vishwanath, A. (2014). Exactly soluble model of a three-dimensional symmetry-protected topological phase of bosons with surface topological order. *Phys. Rev. B*, **90**, 245122.

[3] Cano, J., Cheng, M., Mulligan, M., Nayak, C., Plamadeala, E., and Yard, J. (2014). Bulk–edge correspondence in $(2 + 1)$-dimensional Abelian topological phases. *Phys. Rev. B*, **89**, 115116.

[4] Chen, X., Burnell, F. J., Vishwanath, A., and Fidkowski, L. (2015). Anomalous symmetry fractionalization and surface topological order. *Phys. Rev. X*, **5**, 041013.

[5] Chen, X., Gu, Z.-C., Liu, Z.-X., and Wen, X.-G. (2012). Symmetry-protected topological orders in interacting bosonic systems. *Science*, **338**, 1604–1606.

[6] Chen, X., Gu, Z.-C., Liu, Z.-X., and Wen, X.-G. (2013). Symmetry protected topological orders and the group cohomology of their symmetry group. *Phys. Rev. B*, **87**, 155114.

[7] Chen, X., Gu, Z.-C., and Wen, X.-G. (2011). Classification of gapped symmetric phases in one-dimensional spin systems. *Phys. Rev. B*, **83**, 035107.

[8] Chen, X., Lu, Y.-M., and Vishwanath, A. (2014). Symmetry-protected topological phases from decorated domain walls. *Nature Commun.*, **5**, 3507.

[9] Fidkowski, L., Chen, X., and Vishwanath, A. (2013). Non-Abelian topological order on the surface of a 3D topological superconductor from an exactly solved model. *Phys. Rev. X*, **3**, 041016.

[10] Fidkowski, L. and Kitaev, A. (2011). Topological phases of fermions in one dimension. *Phys. Rev. B*, **83**, 075103.

[11] Furukawa, S. and Ueda, M. (2013). Integer quantum Hall state in two-component Bose gases in a synthetic magnetic field. *Phys. Rev. Lett.*, **111**, 090401.

[12] Geraedts, S. D. and Motrunich, O. I. (2013). Exact realization of integer and fractional quantum Hall phases in $U(1) \times U(1)$ models in $(2+1)$d. *Ann. Phys. (NY)*, **334**, 288–315.

[13] Kapustin, A., Thorngren, R., Turzillo, A., and Wang, Z. (2015). Fermionic symmetry protected topological phases and cobordisms. *JHEP*, **2015**, 052.

[14] Kitaev, A. (2006). Anyons in an exactly solved model and beyond. *Ann. Phys. (NY)*, **321**, 2 – 111.

[15] Kitaev, A. (2015). Unpublished talk at IPAM Conference.

[16] Kitaev, A. Yu. (2003). Fault-tolerant quantum computation by anyons. *Ann. Phys. (NY)*, **303**, 2–30.

[17] Lu, Y.-M. and Vishwanath, A. (2012). Theory and classification of interacting integer topological phases in two dimensions: a Chern–Simons approach. *Phys. Rev. B*, **86**, 125119.

[18] Metlitski, M. A., Fidkowski, L., Chen, X., and Vishwanath, A. (2014). Interaction effects on 3D topological superconductors: surface topological order from vortex condensation, the 16 fold way and fermionic Kramers doublets. arXiv:1406.3032 [cond-mat.str-el].

[19] Metlitski, M. A., Kane, C. L., and Fisher, M. P. A. (2013). Bosonic topological insulator in three dimensions and the statistical Witten effect. *Phys. Rev. B*, **88**, 035131.

[20] Metlitski, M. A., Kane, C. L., and Fisher, M. P. A. (2015). Symmetry-respecting topologically ordered surface phase of three-dimensional electron topological insulators. *Phys. Rev. B*, **92**, 125111.

[21] Motrunich, O. I. and Vishwanath, A. (2004). Emergent photons and transitions in the $O(3)$ sigma model with hedgehog suppression. *Phys. Rev. B*, **70**, 075104.

[22] Regnault, N. and Senthil, T. (2013). Microscopic model for the boson integer quantum Hall effect. *Phys. Rev. B*, **88**, 161106.

[23] Schuch, Norbert, Pérez-García, D, and Cirac, I. (2011). Classifying quantum phases using matrix product states and projected entangled pair states. *Phys. Rev. B*, **84**, 165139.

[24] Senthil, T. (2015). Symmetry-protected topological phases of quantum matter. *Ann. Rev. Condens. Matter Phys.*, **6**, 299–324.

[25] Senthil, T., Balents, L., Sachdev, S., Vishwanath, A., and Fisher, M. P. A. (2004). Quantum criticality beyond the Landau–Ginzburg–Wilson paradigm. *Phys. Rev. B*, **70**, 144407.

[26] Senthil, T., Vishwanath, A., Balents, L., Sachdev, S., and Fisher, M. P. A. (2004). Deconfined quantum critical points. *Science*, **303**, 1490–1494.

[27] Smacchia, P., Amico, L., Facchi, P., Fazio, R., Florio, G., Pascazio, S., and Vedral, V. (2011). Statistical mechanics of the cluster Ising model. *Phys. Rev. A*, **84**, 022304.

[28] Tupitsyn, I. S., Kitaev, A., Prokof'ev, N. V., and Stamp, P. C. E. (2010). Topological multicritical point in the phase diagram of the toric code model and three-dimensional lattice gauge Higgs model. *Phys. Rev. B*, **82**, 085114.

[29] Turner, A. M., Pollmann, F., and Berg, E. (2011). Topological phases of one-dimensional fermions: an entanglement point of view. *Phys. Rev. B*, **83**, 075102.

[30] Turner, A. M. and Vishwanath, A. (2013). Beyond band insulators: topology of semimetals and interacting phases. In *Topological Insulators* (ed. M. Franz and L. Molenkamp), pp. 293–324. Elsevier.

[31] Vishwanath, A. and Senthil, T. (2013). Physics of three-dimensional bosonic topological insulators: surface-deconfined criticality and quantized magnetoelectric effect. *Phys. Rev. X*, **3**, 011016.

[32] Wang, C., Potter, A. C., and Senthil, T. (2013). Gapped symmetry preserving surface state for the electron topological insulator. *Phys. Rev. B*, **88**, 115137.

[33] Wang, C., Potter, A. C., and Senthil, T. (2014). Classification of interacting electronic topological insulators in three dimensions. *Science*, **343**, 629–631.

[34] Wang, C. and Senthil, T. (2013). Boson topological insulators: a window into highly entangled quantum phases. *Phys. Rev. B*, **87**, 235122.

[35] Wang, C. and Senthil, T. (2014). Interacting fermionic topological insulators/superconductors in three dimensions. *Phys. Rev. B*, **89**, 195124.

[36] Witten, E. (2015). Fermion path integrals and topological phases. arXiv:1508.04715 [cond-mat.mes-hall].

[37] Wu, Y.-H. and Jain, J. K. (2013). Quantum Hall effect of two-component bosons at fractional and integral fillings. *Phys. Rev. B*, **87**, 245123.

7

Fractional Abelian topological phases of matter for fermions in two-dimensional space

Christopher MUDRY

Condensed Matter Theory Group
Paul Scherrer Institute
CH-5232 Villigen PSI
Switzerland

Topological Aspects of Condensed Matter Physics. First Edition. C. Chamon et al.

Chapter Contents

Colour figures. For those figures in this chapter that use colour, please see the version of these lecture notes at http://topo-houches.pks.mpg.de. These figures are indicated by '[Colour online]' at the start of the caption.

7.1 Introduction

In these lectures, I will focus exclusively on two-dimensional realizations of fractional topological insulators. However, before doing so, I need to revisit the definition of non-interacting topological phases of matter for fermions and, for this matter, I would like to attempt to place some of the recurrent concepts that have been used during this school on a time line that starts in 1931.

Topology in physics entered the scene in 1931 when Dirac [23] showed that the existence of magnetic monopoles in quantum mechanics implies the quantization of electric and magnetic charge.

In the same decade, Tamm [118] and Shockley [113] surmised from the band theory of Bloch that surface states can appear at the boundaries of band insulators (see Fig. 7.1).

The dramatic importance of static and local disorder for electronic quantum transport was overlooked until 1957, when Anderson [3] showed that sufficiently strong disorder 'generically' localizes a bulk electron. That there can be exceptions to this rule follows from reinterpreting the demonstration in 1953 by Dyson [24] that disordered phonons in a linear chain can acquire a diverging density of states at zero energy with the help of bosonization tools in one-dimensional space (see Fig. 7.2).

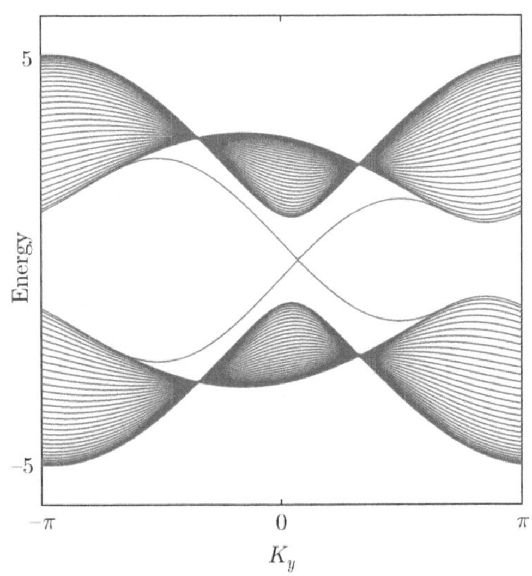

Fig. 7.1 [Colour online] Single-particle spectrum of a Bogoliubov–de Gennes superconductor in a cylindrical geometry that is the direct sum of a $p_x + ip_y$ and of a $p_x - ip_y$ superconductor. A twofold-degenerate dispersion of two chiral edge states is seen to cross the mean-field superconducting gap. There is a single pair of Kramers-degenerate edge states that disperses along one edge of the cylinder. (Adapted from [15].)

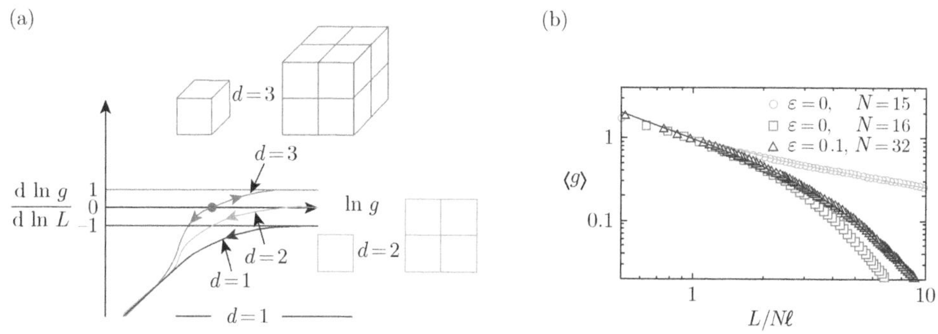

Fig. 7.2 [Colour online] (a) The beta function of the dimensionless conductance g is plotted (qualitatively) as a function of the linear system size L in the orthogonal symmetry class ($\beta = 1$) for space dimensions $d = 1$, 2, and 3. (b) The dependence of the mean Landauer conductance $\langle g \rangle$ for a quasi-one-dimensional wire as a function of the length of the wire L in symmetry class BD1. The number N of channels is varied, as well as the chemical potential ε of the leads. (Part (b) taken from [11].)

Following the proposal by Wigner to model nuclear interactions with the help of random matrix theory, Dyson [25] in 1962 introduced the threefold way, i.e. the study of the joint probability distribution

$$P(\theta_1, \ldots, \theta_N) \propto \prod_{1 \le j < k \le N} \left| e^{i\theta_j} - e^{i\theta_k} \right|^{\beta}, \qquad \beta = 1, 2, 4, \tag{7.1}$$

for the eigenvalues of unitary matrices of rank N generated by random Hamiltonians without any symmetry ($\beta = 2$), by random Hamiltonians with time-reversal symmetry corresponding to spin-0 particles ($\beta = 1$), and by random Hamiltonians with time-reversal symmetry corresponding to spin-$\frac{1}{2}$ particles ($\beta = 4$).

Topology acquired a mainstream status in physics in 1973 with the discovery by Berezinskii [7] and by Kosterlitz and Thouless [65, 66] that topological defects in magnetic classical textures can drive a phase transition. In turn, there is an intimate connection between topological defects of classical background fields in the presence of which electrons propagate and fermionic zero modes, as demonstrated by Jackiw and Rebbi [51] in 1976 (see Fig. 7.3); see also the work by Su, Schrieffer, and Heeger [117].

The 1970s witnessed the birth of lattice gauge theory as a means to regularize quantum chromodynamics (QCD_4). Regularizing the standard model on the lattice proved to be more difficult because of the Nielsen–Ninomiya no-go theorem that prohibits defining a theory of chiral fermions on a lattice in odd-dimensional space without violating locality or time-reversal symmetry [92–94]. This is known as the fermion-doubling problem when regularizing the Dirac equation in d-dimensional space on a d-dimensional lattice.

The 1980s opened with a big bang. The integer quantum Hall effect (IQHE) was discovered in 1980 by von Klitzing, Dorda, and Pepper [63] (see Fig. 7.4), while the fractional quantum Hall effect (FQHE) was discovered in 1982 by Tsui, Stormer, and

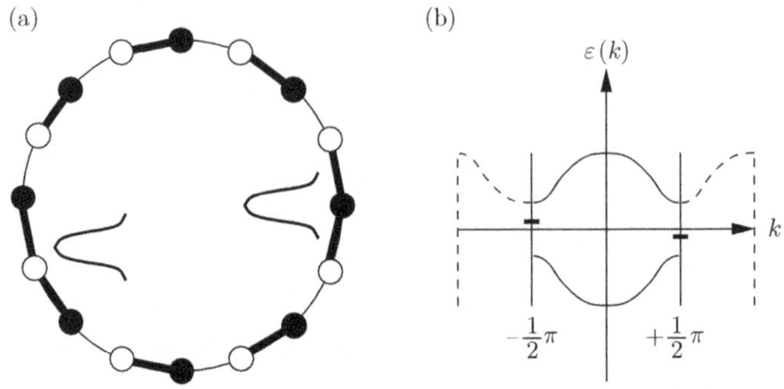

Fig. 7.3 (a) Nearest-neighbour hopping of a spinless fermion along a ring with a real-valued hopping amplitude that is larger on the thick bonds than on the thin bonds. There are two defective sites, each of which is shared by two thick bonds. (b) The single-particle spectrum is gapped at half-filling. There are two bound states within this gap, each exponentially localized around one of the defective sites, whose energy is split from the band centre by an energy that decreases exponentially fast with the separation of the two defects. (Reprinted with permission from [104]. Copyright 2012 by the American Physical Society.)

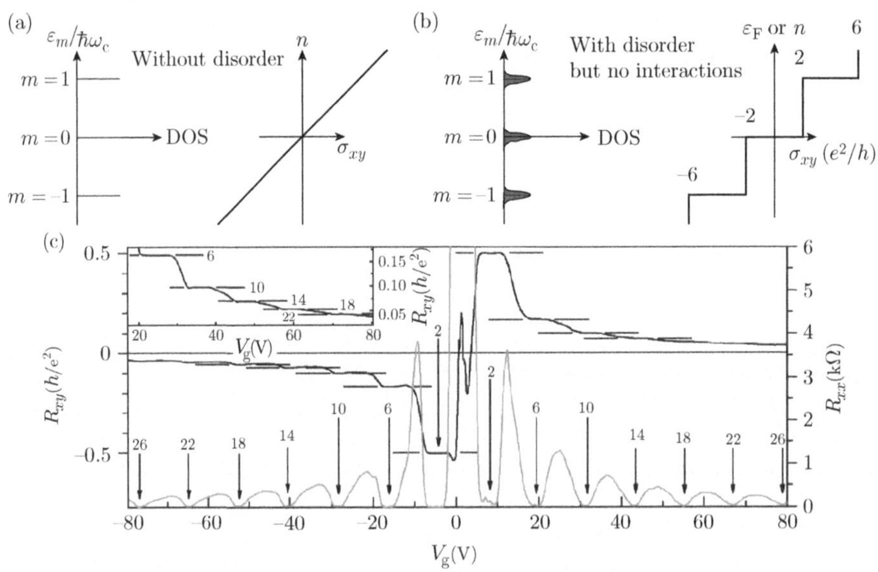

Fig. 7.4 [Colour online] (a) The Hall conductivity is a linear function of the electron density if Galilean invariance holds. (b) Galilean invariance is broken in the presence of disorder, so that plateaus become evident at integral filling fractions of the Landau levels. (c) Graphene deposited on SiO_2/Si at $T = 1.6$K and $B = 9$T (the inset is for $T = 30$mK) supports the integer quantum Hall effect at filling fractions $\nu = \pm 2, \pm 6, \pm 10, \ldots = \pm 2(2n+1)$, $n \in \mathbb{N}$. (Part (c) taken from [137]. Reprinted by permission from Macmillan Publishers Ltd: *Nature*, copyright 2005.)

Gossard [122]. At integer fillings of the Landau levels, the non-interacting ground state is unique and the screened Coulomb interaction V_{int} can be treated perturbatively, as long as transitions between Landau levels or outside the confining potential V_{conf} along the magnetic field are suppressed by the single-particle gaps $\hbar\omega_{\mathrm{c}}$ and V_{conf}, respectively:

$$V_{\mathrm{int}} \ll \hbar\omega_{\mathrm{c}} \ll V_{\mathrm{conf}}, \qquad \omega_{\mathrm{c}} = \frac{eB}{mc}. \tag{7.2}$$

When Galilean invariance is not broken, the conductivity tensor is then given by the classical Drude formula

$$\lim_{\tau\to\infty} \boldsymbol{j} = \begin{pmatrix} 0 & +\left(BR_{\mathrm{H}}\right)^{-1} \\ -\left(BR_{\mathrm{H}}\right)^{-1} & 0 \end{pmatrix} \boldsymbol{E}, \qquad R_{\mathrm{H}}^{-1} := -nec, \tag{7.3}$$

that relates the (expectation value of the) electronic current density $\boldsymbol{j} \in \mathbb{R}^2$ to an applied electric field $\boldsymbol{E} \in \mathbb{R}^2$ within the plane perpendicular to the applied static and uniform magnetic field $\boldsymbol{B}$ in the ballistic regime ($\tau \to \infty$ is the scattering time). The electronic density, the electronic charge, and the speed of light are denoted by n, e, and c, respectively. Moderate disorder is an essential ingredient to observe the IQHE, since it allows the Hall conductivity to develop plateaus at sufficiently low temperatures that are readily visible experimentally (see Fig. 7.4). These plateaus are a consequence of the fact that most single-particle states in a Landau level are localized by disorder, according to Anderson's insight that any quantum interference induced by a static and local disorder almost always leads to localization in one- and two-dimensional space. The caveat 'almost' is crucial here, since the very observation of transitions between Landau plateaus implies that not all single-particle Landau levels are localized.

The explanation for the IQHE followed rapidly after its discovery owing to a very general argument of Laughlin [68] based on gauge invariance that implies that the Hall conductivity must take a fractional value if the longitudinal conductivity vanishes (mobility gap). This argument was complemented by an argument of Halperin [47] stressing the crucial role played by edge states when electrons in the quantum Hall effect are confined to a strip geometry (see Fig. 7.5), while works from Thouless, Kohmoto, Nightingale, den Nijs, and Niu [95, 96, 114, 120] demonstrated that the Hall response is, within linear response theory, proportional to the topological invariant

$$C := -\frac{\mathrm{i}}{2\pi} \int_0^{2\pi} \mathrm{d}\phi \int_0^{2\pi} \mathrm{d}\varphi \left[\left\langle \frac{\partial \Psi}{\partial \phi} \middle| \frac{\partial \Psi}{\partial \varphi} \right\rangle - \left\langle \frac{\partial \Psi}{\partial \varphi} \middle| \frac{\partial \Psi}{\partial \phi} \right\rangle \right] \tag{7.4}$$

that characterizes the many-body ground state $|\Psi\rangle$ obeying twisted boundary conditions in the quantum Hall effect. Together, these arguments constitute the first example of the bulk–edge correspondence with observable consequences, namely the distinctive signatures of both the integer and fractional quantum Hall effects.

The transitions between plateaus in the quantum Hall effect are the manifestations at finite temperature and for a system of finite size of a continuous quantum phase transition, i.e. of a singular dependence of the conductivity tensor on the magnetic field (filling fraction) that is rounded by a non-vanishing temperature or by the finite linear size of a sample. In the non-interacting limit, as was the case for the Dyson singularity

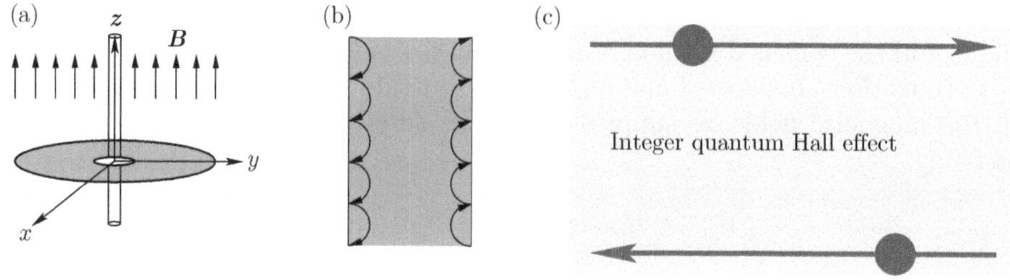

Fig. 7.5 [Colour online] (a) A gas of electrons confined to the geometry of a punctured disc is subjected to a uniform magnetic field $\boldsymbol{B}$ perpendicular to the disc. (b) The classical skipping orbits of electrons confined to the geometry of a Hall bar. The Hall-bar geometry of (b) is topologically equivalent to the Corbino geometry of (a). (c) Chiral edges are immune to backscattering within each traffic lane.

at the band centre, an isolated bulk single-particle state must become critical in the presence of not-too-strong disorder. The one-parameter scaling theory of Anderson localization that had been initiated by Wegner and was encoded by a class of nonlinear sigma models (NLSMs) has to be incomplete [1, 50, 127]. Khmelnitskii [57], on the one hand, and Levine, Libby, and Pruisken [72, 99], on the other, introduced in 1983 a two-parameter scaling theory for the IQHE on phenomenological grounds. They also argued that the NLSM for the IQHE, when augmented by a topological θ term, would reproduce the two-parameter flow diagram (see Fig. 7.6). This remarkable development took place simultaneously with the work of Haldane [42, 44] on encoding the difference between half-integer and integer spin chains (Haldane's conjecture) by the presence of a

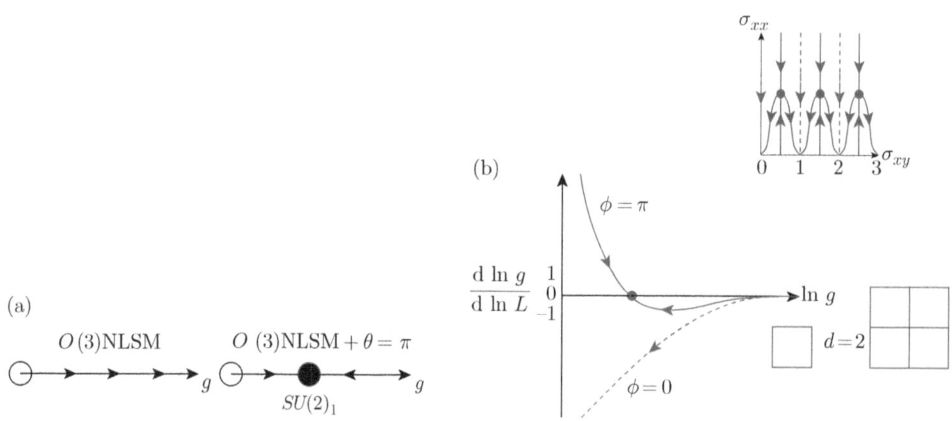

Fig. 7.6 [Colour online] (a) A topological $\theta = \pi$ term modifies the renormalization group flow to strong coupling in the two-dimensional $O(3)$ nonlinear sigma model (NLSM). There exists a stable critical point at intermediate coupling that realizes the conformal field theory $SU(2)_1$. (b) Pruisken argued that the phenomenological two-parameter flow diagram of Khmelnitskii is a consequence of augmenting the NLSM in the unitary symmetry class by a topological term.

$\theta = \pi$ topological term in the $O(3)$ NLSM and by the work of Witten [134] on principal chiral models augmented by a Wess–Zumino–Novikov–Witten (WZNW) term.

Deciphering the critical theory for the plateau transition is perhaps the most tantalizing challenge in the theory of Anderson localization. Among the many interesting avenues that have been proposed to reach this goal (which so far remains elusive), Ludwig, Fisher, Shankar, and Grinstein [76] studied random Dirac fermions in two-dimensional space in 1994 (see Fig. 7.7), motivated as they were by the fact that a massive Dirac fermion in two-dimensional space carries the fractional value

$$\sigma_{\mathrm{H}}^{\mathrm{Dirac}} = \pm\frac{1}{2}\frac{e^2}{h} \tag{7.5}$$

according to Deser, Jackiw, and Templeton [22] and that it is possible to regularize two such massive Dirac fermions on a two-band lattice model realizing a Chern insulator according to Haldane [45].

The early 1990s were also the golden age of mesoscopic physics, the application of random matrix theory to condensed matter physics. The threefold way had been

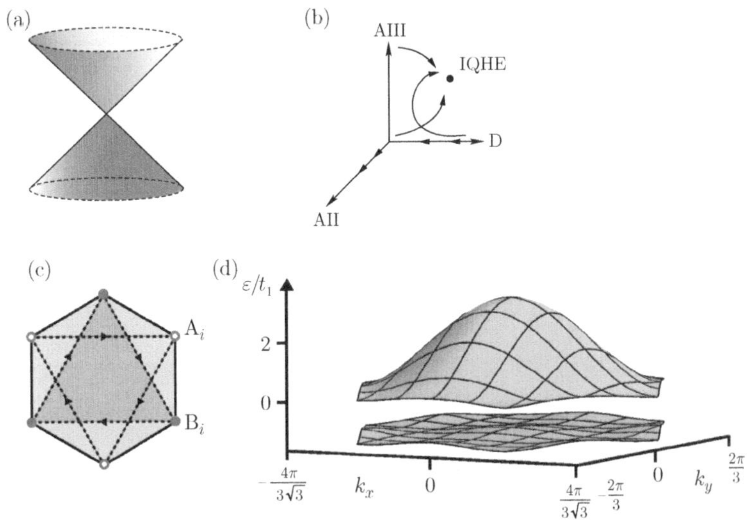

Fig. 7.7 [Colour online] (a) A single (non-degenerate) cone of Dirac fermions in two-dimensional space realizes a critical point between two massive phases of Dirac fermions, each of which carries the Hall conductance $\sigma_{\mathrm{H}}^{\mathrm{Dirac}} = \pm\frac{1}{2}$ in units of e^2/h. (b) A generic static and local random perturbation of a single Dirac cone is encoded by three channels. There is a random vector potential that realizes the symmetry class AIII if it is the only one present. There is a random scalar potential that realizes the symmetry class AII if it is the only one present. There is a random mass that realizes the symmetry class D if it is the only one present. It is conjectured in [76] that in the presence of all three channels, the renormalization group flow to strong coupling (the variance of the disorder in each channel) is to the plateau transition in the universality class of the IQHE. (c) Unit cell of the honeycomb lattice with the pattern of nearest- and next-nearest-neighbour hopping amplitude that realizes a Chern insulator with two bands, shown in (d), each of which carries Chern number ± 1.

applied successfully to quantum dots and quantum transport in quasi-one-dimensional geometries. Zirnbauer [138] in 1996 and Altland and Zirnbauer [2] in 1997 extended the threefold way of Dyson to the tenfold way by including three symmetry classes of relevance to quantum chromodynamics called the chiral classes, and four symmetry classes of relevance to superconducting quantum dots (see Table 7.1) [49]. Quantum

Table 7.1 The ten Altland–Zirnbauer (AZ) symmetry classes of single-particle Hamiltonians $\mathcal{H}$, classified according to their behaviour under time-reversal symmetry $\mathcal{T}$, charge conjugation (particle–hole) symmetry $\mathcal{C}$, and 'sublattice' ('chiral') symmetry $\mathcal{S}$. The entries in the columns headed T, C, and S indicate the presence/absence of these respective symmetries as well as their types. These operations square to + or − times the unit operator when they are symmetries. An entry 0 indicates that the operation is not a symmetry. The column 'Hamiltonian' lists, for each AZ symmetry class, the symmetric space of which the quantum mechanical time-evolution operator $\exp(\mathrm{i}t\mathcal{H})$ is an element. The column 'Cartan label' shows the name given to the corresponding symmetric space listed in the column 'Hamiltonian' in Élie Cartan's classification scheme (dating back to the year 1926). The column 'G/H (fermionic NLSM)' lists the (compact sectors of the) target space of the NLSM describing the physics of Anderson localization at long wavelength in the given symmetry class.

CARTAN LABEL	T	C	S	HAMILTONIAN	G/H (FERMIONIC NLSM)
A (unitary)	0	0	0	$U(N)$	$\frac{U(2n)}{U(n)\times U(n)}$
AI (orthogonal)	+1	0	0	$\frac{U(N)}{O(N)}$	$\frac{Sp(2n)}{Sp(n)\times Sp(n)}$
AII (symplectic)	−1	0	0	$\frac{U(2N)}{Sp(2N)}$	$\frac{O(2n)}{O(n)\times O(n)}$
AIII (chiral unitary)	0	0	+1	$\frac{U(N+M)}{U(N)\times U(M)}$	$U(n)$
BDI (chiral orthogonal)	+1	+1	+1	$\frac{O(N+M)}{O(N)\times O(M)}$	$\frac{U(2n)}{Sp(2n)}$
CII (chiral symplectic)	−1	−1	+1	$\frac{Sp(N+M)}{Sp(N)\times Sp(M)}$	$\frac{U(2n)}{O(2n)}$
D (BdG)	0	+1	0	$SO(2N)$	$\frac{O(2n)}{U(n)}$
C (BdG)	0	−1	0	$Sp(2N)$	$\frac{Sp(2n)}{U(n)}$
DIII (BdG)	−1	+1	−1	$\frac{SO(2N)}{U(N)}$	$O(2n)$
CI (BdG)	+1	−1	−1	$\frac{Sp(2N)}{U(N)}$	$Sp(2n)$

Table 7.2 AZ symmetry classes for disordered quantum wires. Symmetry classes are defined by the presence or absence of time-reversal symmetry and spin-rotation symmetry (SRS) and by the single-particle spectral symmetries of sublattice symmetry (SLS) (random hopping model at the band centre) also known as chiral symmetries, and particle–hole symmetry (zero-energy quasiparticles in superconductors). For historical reasons, the first three rows are referred to as the orthogonal (O), unitary (U), and symplectic (S) symmetry classes when the disorder is generic. A prefix 'ch' ('chiral') is added when the disorder respects a SLS, as in the next three rows. The last four rows correspond to dirty superconductors and are named after the symmetric spaces associated with their Hamiltonians. $m_{o\pm}$ and m_l are the multiplicities of the ordinary and long roots of the symmetric spaces associated with the transfer matrix. For the three chiral classes, $m_{o+} = 0$, $m_{o-} = m_o$; otherwise, $m_{o+} = m_{o-} = m_o$. D is the degeneracy of the transfer matrix eigenvalues. The next two columns show the symbols for the symmetric spaces associated with the transfer matrix $\mathcal{M}$ and the Hamiltonian $\mathcal{H}$. With g denoting the dimensionless Landauer conductance and $\rho(\varepsilon)$ the (self-averaging) density of states (DOS) per unit energy and per unit length, the last three columns list theoretical results for the weak-localization correction δg for $\ell \ll L \ll N\ell$, the disorder average $\overline{\ln g}$ of $\ln g$ for $L \gg N\ell$, and the DOS near $\varepsilon = 0$. The results for $\overline{\ln g}$ and $\rho(\varepsilon)$ in the chiral classes are for even N. For odd N, $\overline{\ln g}$ and $\rho(\varepsilon)$ are the same as in class D.

SYMMETRY CLASS	m_o	m_l	D	$\mathcal{M}$	$\mathcal{H}$	δg	$-\overline{\ln g}$	$\rho(\varepsilon)$ for $0 < \varepsilon\tau_c \ll 1$
AI	1	1	2	CI	AI	$-\frac{2}{3}$	$2L/(\gamma\ell)$	ρ_0
A	2	1	2(1)	AIII	A	0	$2L/(\gamma\ell)$	ρ_0
AII	4	1	2	DIII	AII	$+\frac{1}{3}$	$2L/(\gamma\ell)$	ρ_0
BDI	1	0	2	AI	BDI	0	$2m_oL/(\gamma\ell)$	$\rho_0 \lvert\ln \lvert\varepsilon\tau_c\rvert\rvert$
AIII	2	0	2(1)	A	AIII	0	$2m_oL/(\gamma\ell)$	$\pi\rho_0 \lvert\varepsilon\tau_c \ln \lvert\varepsilon\tau_c\rvert\rvert$
CII	4	0	2	AII	CII	0	$2m_oL/(\gamma\ell)$	$\frac{1}{3}\pi\rho_0\lvert(\varepsilon\tau_c)^3 \ln \lvert\varepsilon\tau_c\rvert\rvert$
CI	2	2	4	C	CI	$-\frac{4}{3}$	$2m_lL/(\gamma\ell)$	$\frac{1}{2}\pi\rho_0\lvert\varepsilon\tau_c\rvert$
C	4	3	4	CII	C	$-\frac{2}{3}$	$2m_lL/(\gamma\ell)$	$\rho_0\lvert\varepsilon\tau_c\rvert^2$
DIII	2	0	2	D	DIII	$+\frac{2}{3}$	$4\sqrt{L/(2\pi\gamma\ell)}$	$\pi\rho_0/\lvert\varepsilon\tau_c \ln^3 \lvert\varepsilon\tau_c\rvert\rvert$
D	1	0	1	BDI	D	$+\frac{1}{3}$	$4\sqrt{L/(2\pi\gamma\ell)}$	$\pi\rho_0/\lvert\varepsilon\tau_c \ln^3 \lvert\varepsilon\tau_c\rvert\rvert$

transport in quasi-one-dimensional wires belonging to the chiral and superconducting classes was studied by Brouwer, Mudry, Simons, and Altland and by Brouwer, Furusaki, Gruzberg, and Mudry, respectively [10–13] (see Fig. 7.8 in Section 7.2.3). In contrast to the threefold way, the three chiral symmetry classes and two of the four superconducting classes (symmetry classes D and DIII) were shown to realize a quantum critical point separating localized phases in quasi-one-dimensional arrays of wires. The diverging nature of the density of states at the band centre (the disorder is of vanishing mean) for five of the ten symmetry classes in Table 7.2 is a signature of topologically protected zero modes bound to point defects. These point defects are vanishing values of an order parameter (domain walls). The order parameter, if translation symmetry holds, is here also responsible for a spectral gap.

Lattice realizations of $\mathbb{Z}_2$ topological band insulators in two-dimensional space were proposed by Kane and Mele [54, 55] and in three-dimensional space by Moore and Balents [83], Roy [102], and Fu, Kane, and Mele [34]. This theoretical discovery initiated a search by Ryu, Mudry, Ludwig, Obuse, and Furusaki [104, 105] for Dirac Hamiltonians belonging to the two-dimensional symmetry classes AII and CII from Table 7.1, for which the corresponding NLSMs encoding the effects of static and local disorder were augmented by a topological term so as to evade Anderson localization on the boundary of a $d = 3$-dimensional topological insulator. Following this route for all symmetry

Table 7.3 Classification of topological insulators and superconductors as a function of spatial dimension d and AZ symmetry class, indicated by the Cartan label (first column). The definition of the ten AZ symmetry classes of single-particle Hamiltonians is given in Table 7.1. The symmetry classes are grouped in two separate lists of complex and real cases, respectively, depending on whether the Hamiltonian is complex or whether one (or more) reality conditions (arising from time-reversal or charge conjugation symmetries) are imposed on it; the AZ symmetry classes are ordered in such a way that a periodic pattern in dimensionality becomes visible [58]. Entries $\mathbb{Z}$, $2\mathbb{Z}$, and $\mathbb{Z}_2$ indicate that the topologically distinct phases within a given symmetry class of topological insulators (superconductors) are characterized by an integer invariant, an even-integer invariant, or a $\mathbb{Z}_2$ quantity, respectively. An entry '0' indicates that there exists no topological insulator (superconductor), i.e. a case where all quantum ground states are topologically equivalent to the trivial state.

CARTAN LABEL	$d=0$	1	2	3	4	5	6	7	8	9	10	11	...
Complex case													
A	$\mathbb{Z}$	0	$\mathbb{Z}$	0	$\mathbb{Z}$	0	$\mathbb{Z}$	0	$\mathbb{Z}$	0	$\mathbb{Z}$	0	...
AIII	0	$\mathbb{Z}$	0	$\mathbb{Z}$	0	$\mathbb{Z}$	0	$\mathbb{Z}$	0	$\mathbb{Z}$	0	$\mathbb{Z}$	...
Real case													
AI	$\mathbb{Z}$	0	0	0	$2\mathbb{Z}$	0	$\mathbb{Z}_2$	$\mathbb{Z}_2$	$\mathbb{Z}$	0	0	0	...
BDI	$\mathbb{Z}_2$	$\mathbb{Z}$	0	0	0	$2\mathbb{Z}$	0	$\mathbb{Z}_2$	$\mathbb{Z}_2$	$\mathbb{Z}$	0	0	...
D	$\mathbb{Z}_2$	$\mathbb{Z}_2$	$\mathbb{Z}$	0	0	0	$2\mathbb{Z}$	0	$\mathbb{Z}_2$	$\mathbb{Z}_2$	$\mathbb{Z}$	0	...
DIII	0	$\mathbb{Z}_2$	$\mathbb{Z}_2$	$\mathbb{Z}$	0	0	0	$2\mathbb{Z}$	0	$\mathbb{Z}_2$	$\mathbb{Z}_2$	$\mathbb{Z}$	...
AII	$2\mathbb{Z}$	0	$\mathbb{Z}_2$	$\mathbb{Z}_2$	$\mathbb{Z}$	0	0	0	$2\mathbb{Z}$	0	$\mathbb{Z}_2$	$\mathbb{Z}_2$	...
CII	0	$2\mathbb{Z}$	0	$\mathbb{Z}_2$	$\mathbb{Z}_2$	$\mathbb{Z}$	0	0	0	$2\mathbb{Z}$	0	$\mathbb{Z}_2$	...
C	0	0	$2\mathbb{Z}$	0	$\mathbb{Z}_2$	$\mathbb{Z}_2$	$\mathbb{Z}$	0	0	0	$2\mathbb{Z}$	0	...
CI	0	0	0	$2\mathbb{Z}$	0	$\mathbb{Z}_2$	$\mathbb{Z}_2$	$\mathbb{Z}$	0	0	0	$2\mathbb{Z}$	...

classes and for all dimensions, Ryu, Schnyder, Furusaki, and Ludwig [106, 109, 110] arrived at the periodic table shown here as Table 7.3. The same table was derived independently by Kitaev [58] using a mathematical construction known as K-theory, which he applied to gapped Hamiltonians in the bulk (upon the imposition of, for example, periodic boundary conditions) in the clean limit. This table specifies, in any given dimension d of space, the symmetry classes for which it is possible to realize a many-body ground state for non-interacting fermions subject to a static and local disorder such that all bulk states are localized but there exist a certain (topological) number of boundary states that remain delocalized.

The goals of these lectures are the following: first, to rederive the tenfold way for non-interacting fermions in the presence of local interactions and static local disorder, and, second, to determine whether interactions between fermions can produce topological phases of matter with protected boundary states that are not captured by the tenfold way. This programme will be applied in two-dimensional space.

Section 7.2 motivates the tenfold way by deriving it explicitly in quasi-one-dimensional space. Section 7.3 reviews Abelian chiral bosonization, a technical tool allowing one to go beyond the tenfold way to incorporate the effects of many-body interactions. Abelian chiral bosonization is applied in Section 7.4 to demonstrate the stability of the gapless helical edge states in symmetry class AII in the presence of disorder and many-body interactions and then in Section 7.5 to construct microscopically long-ranged entangled phases of two-dimensional quantum matter. Sections 7.4 and 7.5 follow the presentations made in [91] and [89], respectively.

7.2 The tenfold way in quasi-one-dimensional space

This section is dedicated to a non-vanishing density of non-interacting fermions hopping between the sites of quasi-one-dimensional lattices or between the sites defining the one-dimensional boundary of a two-dimensional lattice. According to the Pauli exclusion principle, the non-interacting ground state is obtained by filling all the single-particle energy eigenstates up to the Fermi energy fixed by the fermion density. The fate of this single-particle energy eigenstate when a static and local random potential is present is known as the problem of Anderson localization. The effect of disorder on a single-particle extended energy eigenstate state can be threefold:

- The extended nature of the single-particle energy eigenstate is robust to disorder.
- The extended single-particle energy eigenstate is turned into a critical state.
- The extended single-particle energy eigenstate is turned into a localized state.

There are several methods allowing one to decide which of these three outcomes takes place. Irrespective of the dimensionality d of space, the symmetries obeyed by the static and local random potential matter for the outcome in a dramatic fashion. To illustrate this point, let us consider the problem of Anderson localization in quasi-one-dimensional space.

7.2.1 Symmetries for the case of one one-dimensional channel

For simplicity, consider first the case of an infinitely long one-dimensional chain with lattice spacing $\mathrm{a} \equiv 1$ along which a non-vanishing but finite density of spinless fermions hop with the uniform nearest-neighbour hopping amplitude t. If periodic boundary conditions are imposed, then the single-particle Hamiltonian is the direct sum over all momenta $-\pi \leq k \leq +\pi$ within the first Brillouin zone of

$$\mathcal{H}(k) := -2t\cos k. \tag{7.6}$$

The Fermi energy ε_{F} intersects the dispersion (7.6) at the two Fermi points $\pm k_{\mathrm{F}}$. Linearization of the dispersion (7.6) about these two Fermi points delivers the Dirac Hamiltonian

$$\mathcal{H}_{\mathrm{D}} := -\tau_3 \mathrm{i}\partial_x, \tag{7.7a}$$

in units defined by

$$\hbar \equiv 1, \qquad v_{\mathrm{F}} = 2t|\sin k_{\mathrm{F}}| \equiv 1. \tag{7.7b}$$

Here, τ_3 is the third Pauli matrix among the four 2×2 matrices

$$\tau_0 := \begin{pmatrix} 1 & 0 \\ 0 & 1 \end{pmatrix}, \qquad \tau_1 := \begin{pmatrix} 0 & 1 \\ 1 & 0 \end{pmatrix}, \qquad \tau_2 := \begin{pmatrix} 0 & -\mathrm{i} \\ +\mathrm{i} & 0 \end{pmatrix}, \qquad \tau_3 := \begin{pmatrix} +1 & 0 \\ 0 & -1 \end{pmatrix}. \tag{7.7c}$$

The momentum eigenstate

$$\Psi_{\mathrm{R},p}(x) := \mathrm{e}^{+\mathrm{i}px} \begin{pmatrix} 1 \\ 0 \end{pmatrix} \tag{7.8a}$$

is an eigenstate with single-particle energy $\varepsilon_{\mathrm{R}}(p) = +p$. The momentum eigenstate

$$\Psi_{\mathrm{L},p}(x) := \mathrm{e}^{+\mathrm{i}px} \begin{pmatrix} 0 \\ 1 \end{pmatrix} \tag{7.8b}$$

is an eigenstate with single-particle energy $\varepsilon_{\mathrm{L}}(p) = -p$. The plane waves

$$\Psi_{\mathrm{R},p}(x,t) := \mathrm{e}^{+\mathrm{i}p(x-t)} \begin{pmatrix} 1 \\ 0 \end{pmatrix} \tag{7.9a}$$

and

$$\Psi_{\mathrm{L},p}(x,t) := \mathrm{e}^{+\mathrm{i}p(x+t)} \begin{pmatrix} 0 \\ 1 \end{pmatrix} \tag{7.9b}$$

are respectively right-moving and left-moving solutions to the massless Dirac equation

$$\mathrm{i}\partial_t \Psi = \mathcal{H}_{\mathrm{D}} \Psi. \tag{7.9c}$$

We perturb the massless Dirac Hamiltonian (7.7) with the most generic static and local one-body potential

$$\mathcal{V}(x) := a_0(x)\tau_0 + m_1(x)\tau_1 + m_2(x)\tau_2 + a_1(x)\tau_3. \tag{7.10}$$

The real-valued function a_0 is a space-dependent chemical potential. It couples to the spinless fermions as the scalar part of the electromagnetic gauge potential does. The real-valued function a_1 is a space-dependent modulation of the Fermi point. It couples to the spinless fermions as the vector part of the electromagnetic gauge potential does. Both a_0 and a_1 multiply Pauli matrices such that each commutes with the massless Dirac Hamiltonian (7.7). Neither channel is confining (localizing). The real-valued functions m_1 and m_2 are space-dependent mass terms, since they multiply Pauli matrices such that each anticommutes with the massless Dirac Hamiltonian and with each other. Each channel is confining (localizing).

7.2.1.1 Symmetry class A

The only symmetry preserved by

$$\mathcal{H} := \mathcal{H}_{\mathrm{D}} + \mathcal{V}(x) \tag{7.11}$$

with $\mathcal{V}$ defined in (7.10) is the global symmetry under multiplication of all states in the single-particle Hilbert space over which $\mathcal{H}$ acts by the same $U(1)$ phase. Correspondingly, the local two-current

$$J^{\mu}(x,t) := (\Psi^{\dagger}\Psi, \Psi^{\dagger}\tau_3\Psi)(x,t), \tag{7.12}$$

obeys the continuity equation

$$\partial_{\mu} J^{\mu} = 0, \qquad \text{with} \quad \partial_0 := \partial_t, \qquad \partial_1 := \partial_x. \tag{7.13}$$

The family of Hamiltonians (7.11) labelled by the potential $\mathcal{V}$ of the form (7.10) is said to belong to symmetry class A because of the conservation law (7.13).

One would like to reverse time in the Dirac equation

$$(\mathrm{i}\partial_t \Psi)(x,t) = (\mathcal{H}\Psi)(x,t), \tag{7.14}$$

where $\mathcal{H}$ is defined by (7.11). Under time reversal,

$$t = -t', \tag{7.15}$$

the Dirac equation (7.14) becomes

$$(-\mathrm{i}\partial_{t'} \Psi)(x,-t') = (\mathcal{H}\Psi)(x,-t'). \tag{7.16}$$

Complex conjugation removes the minus sign on the left-hand side:

$$(\mathrm{i}\partial_{t'} \Psi^{*})(x,-t') = (\mathcal{H}\Psi)^{*}(x,-t'). \tag{7.17}$$

Form invariance of the Dirac equation under time reversal then follows if one postulates the existence of a unitary 2×2 matrix $\mathcal{U}_{\mathcal{T}}$ and of a phase $0 \leq \phi_{\mathcal{T}} < 2\pi$ such that (with complex conjugation denoted by K)

$$(\mathcal{U}_{\mathcal{T}}\mathsf{K})^2 = \mathrm{e}^{\mathrm{i}\phi_{\mathcal{T}}}\tau_0, \qquad \Psi^{*}(x,-t) =: \mathcal{U}_{\mathcal{T}}\Psi_{\mathcal{T}}(x,t), \qquad \mathcal{H}_{\mathcal{T}} := \mathcal{U}_{\mathcal{T}}^{-1}\mathcal{H}^{*}\mathcal{U}_{\mathcal{T}}, \tag{7.18a}$$

in which case

$$(\mathrm{i}\partial_t \Psi_{\mathcal{T}})(x,t) = (\mathcal{H}_{\mathcal{T}}\Psi_{\mathcal{T}})(x,t). \tag{7.18b}$$

Time-reversal symmetry then holds if and only if

$$\mathcal{U}_{\mathcal{T}}^{-1}\mathcal{H}^{*}\mathcal{U}_{\mathcal{T}} = \mathcal{H}. \tag{7.19}$$

Time-reversal symmetry must hold for the massless Dirac equation. By inspection of the right- and left-moving solutions (7.9), one deduces that $\mathcal{U}_{\mathcal{T}}$ must interchange right- and left-movers. There are two possibilities: either

$$\mathcal{U}_{\mathcal{T}} = \tau_2, \qquad \phi_{\mathcal{T}} = \pi, \tag{7.20}$$

or

$$\mathcal{U}_{\mathcal{T}} = \tau_1, \qquad \phi_{\mathcal{T}} = 0. \tag{7.21}$$

7.2.1.2 *Symmetry class AII*

Imposing time-reversal symmetry using the definition (7.20) restricts the family of Dirac Hamiltonians (7.11) to

$$\mathcal{H}(x) := -\mathrm{i}\tau_3\partial_t + a_0(x)\tau_0. \tag{7.22}$$

The family of Hamiltonians (7.11) labelled by the potential $\mathcal{V}$ of the form (7.22) is said to belong to symmetry class AII because of the conservation law (7.13) and of the time-reversal symmetry (7.19) with the representation (7.20).

7.2.1.3 *Symmetry class AI*

Imposing time-reversal symmetry using the definition (7.21) restricts the family of Dirac Hamiltonians (7.11) to

$$\mathcal{H}(x) := -\mathrm{i}\tau_3\partial_t + a_0(x)\tau_0 + m_1(x)\tau_1 + m_2(x)\tau_2. \tag{7.23}$$

The family of Hamiltonians (7.11) labelled by the potential $\mathcal{V}$ of the form (7.23) is said to belong to symmetry class AI because of the conservation law (7.13) and of the time-reversal symmetry (7.19) with the representation (7.21).

Let us take advantage of the fact that the dispersion relation (7.6) obeys the symmetry

$$\mathcal{H}(k) = -\mathcal{H}(k+\pi). \tag{7.24}$$

This spectral symmetry is a consequence of the fact that the lattice Hamiltonian anticommutes with the local gauge transformation that maps the basis of single-particle localized wavefunctions

$$\psi_i : \mathbb{Z} \to \mathbb{C}, j \mapsto \psi_i(j) := \delta_{ij} \tag{7.25}$$

into the basis

$$\psi'_i : \mathbb{Z} \to \mathbb{C}, j \mapsto \psi'_i(j) := (-1)^j \delta_{ij}. \tag{7.26}$$

Such a spectral symmetry is an example of a sublattice symmetry in condensed matter physics. So far, the chemical potential

$$\varepsilon_{\mathrm{F}} \equiv -2t\cos k_{\mathrm{F}} \tag{7.27}$$

defined in (7.6) has been arbitrary. However, in view of the spectral symmetry (7.24), the single-particle energy eigenvalue

$$0 = \varepsilon_{\mathrm{F}} \equiv -2t\cos k_{\mathrm{F}}, \qquad k_{\mathrm{F}} = \frac{1}{2}\pi, \tag{7.28}$$

is special. It is the centre of symmetry of the single-particle spectrum (7.6). The spectral symmetry (7.24) is also known as a chiral symmetry of the Dirac equation (7.7a) by which

$$\mathcal{H}_{\mathrm{D}} = -\tau_1\mathcal{H}_{\mathrm{D}}\tau_1, \tag{7.29}$$

after an expansion to leading order in powers of the deviation of the momenta away from the two Fermi points $\pm\frac{1}{2}\pi$.

7.2.1.4 Symmetry class AIII

If charge conservation holds together with the chiral symmetry

$$\mathcal{H} = -\tau_1 \mathcal{H} \tau_1, \tag{7.30a}$$

then

$$\mathcal{H} = -\tau_3 \mathrm{i}\partial_x + a_1(x)\tau_3 + m_2(x)\tau_2 \tag{7.30b}$$

is said to belong to symmetry class AIII.

7.2.1.5 Symmetry class CII

It is not possible to write down a 2×2 Dirac equation in symmetry class CII. For example, if charge conservation holds together with the chiral and time-reversal symmetries

$$\mathcal{H} = -\tau_1 \mathcal{H} \tau_1, \qquad \mathcal{H} = +\tau_2 \mathcal{H}^* \tau_2, \tag{7.31a}$$

respectively, then

$$\mathcal{H} = -\tau_3 \mathrm{i}\partial_x \tag{7.31b}$$

does not belong to symmetry class CII, since the composition of the chiral transformation with time reversal squares to unity instead of minus unity.

7.2.1.6 Symmetry class BDI

If charge conservation holds together with the chiral and time-reversal symmetries

$$\mathcal{H} = -\tau_1 \mathcal{H} \tau_1, \qquad \mathcal{H} = +\tau_1 \mathcal{H}^* \tau_1, \tag{7.32a}$$

then

$$\mathcal{H} = -\tau_3 \mathrm{i}\partial_x + m_2(x)\tau_2 \tag{7.32b}$$

is said to belong to symmetry class BDI.

The global $U(1)$ gauge symmetry responsible for the continuity equation (7.13) demands that one treats the two components of the Dirac spinors as independent. This is not desirable if the global $U(1)$ gauge symmetry is to be restricted to a global $\mathbb{Z}_2$ gauge symmetry, as occurs in a mean-field treatment of superconductivity. If the possibility of restricting the global $U(1)$ to a global $\mathbb{Z}_2$ gauge symmetry is to be accounted for, then four more symmetry classes are permissible.

7.2.1.7 Symmetry class D

If we impose a particle–hole symmetry through

$$\mathcal{H} = -\mathcal{H}^{*}, \tag{7.33a}$$

then

$$\mathcal{H} = -\tau_{3}\mathrm{i}\partial_{x} + m_{2}(x)\tau_{2} \tag{7.33b}$$

is said to belong to symmetry class D.

7.2.1.8 Symmetry class DIII

If we impose particle–hole and time-reversal symmetries through

$$\mathcal{H} = -\mathcal{H}^{*}, \qquad \mathcal{H} = +\tau_{2}\mathcal{H}^{*}\tau_{2}, \tag{7.34a}$$

respectively, then

$$\mathcal{H} = -\tau_{3}\mathrm{i}\partial_{x} \tag{7.34b}$$

is said to belong to symmetry class DIII.

7.2.1.9 Symmetry class C

If we impose a particle–hole symmetry through

$$\mathcal{H} = -\tau_{2}\mathcal{H}^{*}\tau_{2}, \tag{7.35a}$$

then

$$\mathcal{H} = a_{1}(x)\tau_{3} + m_{2}(x)\tau_{2} + m_{1}(x)\tau_{1} \tag{7.35b}$$

is said to belong to symmetry class C. The Dirac kinetic energy is prohibited for a 2×2 Dirac Hamiltonian from symmetry class C.

7.2.1.10 Symmetry class CI

If we impose particle–hole and time-reversal symmetries through

$$\mathcal{H} = -\tau_{2}\mathcal{H}^{*}\tau_{2}, \qquad \mathcal{H} = +\tau_{1}\mathcal{H}^{*}\tau_{1}, \tag{7.36a}$$

respectively, then

$$\mathcal{H} = m_{2}(x)\tau_{2} + m_{1}(x)\tau_{1} \tag{7.36b}$$

is said to belong to symmetry class CI. The Dirac kinetic energy is prohibited for a 2×2 Dirac Hamiltonian from symmetry class CI.

7.2.2 Symmetries for the case of two one-dimensional channels

Imagine two coupled linear chains along which non-interacting spinless fermions are allowed to hop. If the two chains are decoupled and the hopping is a uniform nearest-neighbour hopping along any one of the two chains, then the low-energy and long-wavelength effective single-particle Hamiltonian in the vicinity of the chemical potential $\varepsilon_{\mathrm{F}} = 0$ is the tensor product of the massless Dirac Hamiltonian (7.7a) with the 2×2 unit matrix σ_0. Let the three Pauli matrices $\boldsymbol{\sigma}$ act on the same vector space as σ_0 does. For convenience, introduce the 16 Hermitian 4×4 matrices

$$X_{\mu\nu} := \tau_\mu \otimes \sigma_\nu, \qquad \mu, \nu = 0, 1, 2, 3. \tag{7.37}$$

7.2.2.1 *Symmetry class A*

The generic Dirac Hamiltonian of rank $r = 4$ is

$$\mathcal{H} := -X_{30}\mathrm{i}\partial_x + a_{1,\nu}(x)X_{3\nu} + m_{2,\nu}(x)X_{2\nu} + m_{1,\nu}(x)X_{1\nu} + a_{0,\nu}(x)X_{0\nu}. \tag{7.38}$$

The summation convention over the repeated index $\nu = 0, 1, 2, 3$ is implied. There are four real-valued parameters for the components $a_{1,\nu}$ with $\nu = 0, 1, 2, 3$ of a $U(2)$ vector potential, eight for the components $m_{1,\nu}$ and $m_{2,\nu}$ with $\nu = 0, 1, 2, 3$ of two independent $U(2)$ masses, and four for the components $a_{0,\nu}$ with $\nu = 0, 1, 2, 3$ of a $U(2)$ scalar potential. As it should be, there are 16 real-valued free parameters (functions, if one opts to break translation invariance). If all components of the spinors solving the eigenvalue problem

$$\mathcal{H}\Psi(x) = \varepsilon\Psi(x) \tag{7.39}$$

are independent, then the Dirac Hamiltonian (7.38) belongs to symmetry class A.

In addition to conservation of the fermion number, one can impose time-reversal symmetry on the Dirac Hamiltonian (7.38) There are two possibilities to do this.

7.2.2.2 *Symmetry class AII*

If charge conservation holds together with time-reversal symmetry through

$$\mathcal{H} = +X_{12}\mathcal{H}^{*}X_{12}, \tag{7.40a}$$

then

$$\mathcal{H} = -X_{30}\mathrm{i}\partial_x + \sum_{\nu=1,2,3} a_{1,\nu}(x)X_{3\nu} + m_{2,0}(x)X_{20} + m_{1,0}(x)X_{10} + a_{0,0}(x)X_{00} \tag{7.40b}$$

is said to belong to symmetry class AII.

7.2.2.3 Symmetry class AI

If charge conservation holds together with time-reversal symmetry through

$$\mathcal{H} = +X_{10}\mathcal{H}^{*}X_{10}, \tag{7.41a}$$

then

$$\mathcal{H} = -X_{30}\mathrm{i}\partial_{x} + a_{1,2}(x)X_{32} + \sum_{\nu=0,1,3}\left[m_{2,\nu}(x)X_{2\nu} + m_{1,\nu}(x)X_{1\nu} + a_{0,\nu}(x)X_{0\nu}\right] \tag{7.41b}$$

is said to belong to symmetry class AI.

The standard symmetry classes A, AII, and AI can be further constrained by imposing chiral symmetry. This gives the following three possibilities.

7.2.2.4 Symmetry class AIII

If charge conservation holds together with the chiral symmetry

$$\mathcal{H} = -X_{10}\mathcal{H}X_{01}, \tag{7.42a}$$

then

$$\mathcal{H} = -X_{30}\mathrm{i}\partial_{x} + a_{1,\nu}(x)X_{3\nu} + m_{2,\nu}(x)X_{2\nu} \tag{7.42b}$$

is said to belong to symmetry class AIII.

7.2.2.5 Symmetry class CII

If charge conservation holds together with the chiral and time-reversal symmetries

$$\mathcal{H} = -X_{10}\mathcal{H}X_{10}, \qquad \mathcal{H} = +X_{12}\mathcal{H}^{*}X_{12}, \tag{7.43a}$$

respectively, then

$$\mathcal{H} = -X_{30}\mathrm{i}\partial_{x} + \sum_{\nu=1,2,3} a_{1,\nu}(x)X_{3\nu} + m_{2,0}(x)X_{20} \tag{7.43b}$$

is said to belong to symmetry class CII.

7.2.2.6 Symmetry class BDI

If charge conservation holds together with the chiral and time-reversal symmetries

$$\mathcal{H} = -X_{10}\mathcal{H}X_{10}, \qquad \mathcal{H} = +X_{10}\mathcal{H}^{*}X_{10}, \tag{7.44a}$$

respectively, then

$$\mathcal{H} = -X_{30}\mathrm{i}\partial_{x} + a_{1,2}(x)X_{32} + \sum_{\nu=0,1,3} m_{2,\nu}(x)X_{2\nu} \tag{7.44b}$$

is said to belong to symmetry class BDI.

Now, we move to the four Bogoliubov–de Gennes (BdG) symmetry classes by relaxing the condition that all components of a spinor on which the Hamiltonian acts be independent. This means that changing each component of a spinor by a multiplicative global phase factor is no longer legitimate. However, changing each component of a spinor by a global sign remains legitimate. The constraints among the components of a spinor come about by imposing a particle–hole symmetry.

7.2.2.7 Symmetry class D

If we impose particle–hole symmetry through

$$\mathcal{H} = -\mathcal{H}^{*}, \tag{7.45a}$$

then

$$\mathcal{H} = -X_{30}\mathrm{i}\partial_x + a_{1,2}(x)X_{32} + \sum_{\nu=0,1,3} m_{2,\nu}(x)X_{2\nu} + m_{1,2}(x)X_{12} + a_{0,2}(x)X_{02} \tag{7.45b}$$

is said to belong to symmetry class D.

7.2.2.8 Symmetry class DIII

If we impose particle–hole and time-reversal symmetries through

$$\mathcal{H} = -\mathcal{H}^{*}, \qquad \mathcal{H} = +X_{20}\mathcal{H}^{*}X_{20}, \tag{7.46a}$$

respectively, then

$$\mathcal{H} = -X_{30}\mathrm{i}\partial_x + a_{1,2}(x)X_{32} + m_{1,2}(x)X_{12} \tag{7.46b}$$

is said to belong to symmetry class DIII.

7.2.2.9 Symmetry class C

If we impose particle–hole symmetry through

$$\mathcal{H} = -X_{02}\mathcal{H}^{*}X_{02}, \tag{7.47a}$$

respectively, then

$$\mathcal{H} = -X_{30}\mathrm{i}\partial_x + \sum_{\nu=1,2,3} a_{1,\nu}(x)X_{3\nu} + m_{2,0}(x)X_{20} + \sum_{\nu=1,2,3} \left[m_{1,\nu}(x)X_{1\nu} + a_{0,\nu}(x)X_{0\nu}\right] \tag{7.47b}$$

is said to belong to symmetry class C.

7.2.2.10 Symmetry class CI

If we impose particle–hole and time-reversal symmetries through

$$\mathcal{H} = -X_{02}\mathcal{H}^{*}X_{02}, \qquad \mathcal{H} = +X_{10}\mathcal{H}^{*}X_{10}, \tag{7.48a}$$

respectively, then

$$\mathcal{H} = -X_{30}\mathrm{i}\partial_{x} + a_{1,2}(x)X_{32} + m_{2,0}(x)X_{20} + m_{1,1}(x)X_{11} + m_{1,3}(x)X_{13} + a_{0,3}(x)X_{03} \tag{7.48b}$$

is said to belong to symmetry class CI.

7.2.3 Definition of the minimum rank

In Sections 7.2.1 and 7.2.2, we have imposed ten symmetry restrictions corresponding to the tenfold way introduced by Altland and Zirnbauer on Dirac Hamiltonians with Dirac matrices of rank $r = 2$ and 4, respectively. These Dirac Hamiltonians describe the propagation of single-particle states in one-dimensional space. These ten symmetry classes will be called the Altland–Zirnbauer (AZ) symmetry classes.

Observe that some of the AZ symmetries can be very restrictive for those Dirac Hamiltonians with Dirac matrices of small rank r. For example, it is not possible to write down a Dirac Hamiltonian of rank $r = 2$ in symmetry class CII, symmetry classes C and CI do not admit a Dirac kinetic energy of rank $r = 2$, and symmetry classes AII and DIII do not admit Dirac masses in their Dirac Hamiltonians of rank $r = 2$.

This observation suggests the definition of a minimum rank r_{min} for which the Dirac Hamiltonian describing propagation in d-dimensional space for a given AZ symmetry class admits a Dirac mass. Hence, r_{min} depends implicitly on the dimensionality of space and on the AZ symmetry class. In one-dimensional space, we have found that

$$\begin{aligned} &r^{\mathrm{A}}_{\mathrm{min}} = 2, \quad r^{\mathrm{AII}}_{\mathrm{min}} = 4, \quad r^{\mathrm{AI}}_{\mathrm{min}} = 2, \\ &r^{\mathrm{AIII}}_{\mathrm{min}} = 2, \quad r^{\mathrm{CII}}_{\mathrm{min}} = 4, \quad r^{\mathrm{BDI}}_{\mathrm{min}} = 2, \\ &r^{\mathrm{D}}_{\mathrm{min}} = 2, \quad r^{\mathrm{DIII}}_{\mathrm{min}} = 4, \quad r^{\mathrm{C}}_{\mathrm{min}} = 4, \quad r^{\mathrm{CI}}_{\mathrm{min}} = 4. \end{aligned} \tag{7.49}$$

This definition is useful for a number of reasons.

First, Anderson localization in a given AZ symmetry class is impossible for any random Dirac Hamiltonian with Dirac matrices of rank r smaller than r_{min}. This is the case for symmetry classes AII and DIII for a Dirac Hamiltonian of rank $r = 2$ in one-dimensional space. The lattice realization of these Dirac Hamiltonians is along the boundary of a two-dimensional insulator in symmetry classes AII and DIII when the bulk realizes a topologically non-trivial insulating phase owing to the fermion-doubling problem. This is why an odd number of helical pairs of edge states in symmetry class AII and an odd number of helical pairs of Majorana edge states in symmetry class DIII can evade Anderson localization. The $r = 2$ limit for the Dirac Hamiltonians encoding one-dimensional propagation in symmetry classes AII and DIII are the signatures for the topologically non-trivial entries of the group $\mathbb{Z}_2$ in the $d = 2$ column of Table 7.3. For symmetry classes A and D, we can consider the $r = 1$ limit as a special limit that shares with a Dirac Hamiltonian the property that it is a first-order differential operator in space, but, unlike a Dirac Hamiltonian, this limit does no treat right- and left-movers on equal footing (and thus breaks time-reversal symmetry explicitly). Such a first-order differential operator encodes the propagation of right-movers on the inner boundary of a two-dimensional ring (the Corbino geometry of Fig. 7.5), while its complex conjugate encodes the propagation of left-movers on the outer boundary of a two-dimensional ring or vice versa. For symmetry class C, one must consider two copies of opposite spins of the $r = 1$ limit of symmetry class D. The $r = 1$ limits for Dirac Hamiltonians encoding one-dimensional propagation in symmetry classes A, D, and C are realized on the boundaries of two-dimensional insulating phases supporting the IQHE, the thermal IQHE, and the spin-resolved thermal IQHE, respectively. These limits are the signatures for the non-trivial entries ± 1 and ± 2 of the groups $\mathbb{Z}$ and $2\mathbb{Z}$, respectively, in the $d = 2$ column of Table 7.3.

Second, one can always define the quasi-d-dimensional Dirac Hamiltonian

$$\mathcal{H}(\boldsymbol{x}) := -\mathrm{i}(\boldsymbol{\alpha} \otimes I) \cdot \frac{\partial}{\partial \boldsymbol{x}} + \mathcal{V}(\boldsymbol{x}), \tag{7.50a}$$

where $\boldsymbol{\alpha}$ and β are a set of matrices that anticommute pairwise and square to the $r_{\min} \times r_{\min}$ unit matrix, I is the $N \times N$ unit matrix, and

$$\mathcal{V}(\boldsymbol{x}) = m(\boldsymbol{x})\beta \otimes I + \dots, \tag{7.50b}$$

with '...' representing all other masses, vector potentials, and scalar potentials allowed by the AZ symmetry class. For one-dimensional space, the stationary eigenvalue problem

$$\mathcal{H}(x)\Psi(x;\varepsilon) = \varepsilon\Psi(x;\varepsilon) \tag{7.51}$$

with the given 'initial value' $\Psi(y;\varepsilon)$ is solved through the transfer matrix

$$\Psi(x;\varepsilon) = \mathcal{M}(x|y;\varepsilon)\Psi(y;\varepsilon) \tag{7.52a}$$

where

$$\mathcal{M}(x|y;\varepsilon) := \mathcal{P}_{x'} \exp\left(\int_y^x \mathrm{d}x'\, \mathrm{i}(\alpha \otimes I)[\varepsilon - \mathcal{V}(x')]\right). \tag{7.52b}$$

The symbol $\mathcal{P}_{x'}$ represents path ordering. The limit $N \to \infty$ with all entries of $\mathcal{V}$ independently and identically distributed (iid) up to the AZ symmetry constraints (averaging over the disorder is denoted by an overline),

$$\overline{\mathcal{V}_{ij}(x)} \propto \mathsf{v}_{ij}, \tag{7.53a}$$

$$\overline{[\mathcal{V}_{ij}(x) - \mathsf{v}_{ij}]\,[\mathcal{V}_{kl}(y) - \mathsf{v}_{kl}]} \propto \mathrm{g}^2 \mathrm{e}^{-|x-y|/\xi_{\mathrm{dis}}}, \tag{7.53b}$$

for $i, j, k, l = 1, \dots, r_{\min}N$, defines the thick-quantum-wire limit.

The consequences of (7.52) are as follows. First, the local symmetries defining symmetry classes A, AII, and AI obeyed by $\varepsilon - \mathcal{V}(x')$ carry through to the transfer matrix at any single-particle energy ε. The local unitary spectral symmetries defining symmetry classes AIII, CII, and BDI and the local antiunitary spectral symmetries defining symmetry classes D, DIII, C, and CI carry through to the transfer matrix at the single-particle energy $\varepsilon = 0$. Second, the diagonal matrix entering the polar decomposition of the transfer matrix at the band centre $\varepsilon = 0$ is related to the non-compact symmetric spaces from the $\mathcal{M}$ column of Table 7.2. Third, the composition law obeyed by the transfer matrix that encodes enlarging the length of a disordered wire coupled to perfect leads is matrix multiplication. It is then possible to derive a Fokker–Planck equation for the joint probability obeyed by the radial coordinates on the non-compact symmetric spaces from the $\mathcal{M}$ column of Table 7.2 as the length L of of a disordered wire coupled to perfect leads is increased. In this way, the moments of the dimensionless Landauer conductance g in the δg and $-\overline{\ln g}$ columns of Table 7.2 can be computed An infinitesimal increase in the length of the disordered region for one of the ten symmetry classes induces an infinitesimal Brownian motion (see Fig. 7.8) of the Lyapunov exponents that is solely controlled by the multiplicities of the ordinary, long, and short roots of the corresponding classical semisimple Lie algebra under suitable assumptions on the disorder (locality, weakness, and isotropy between all channels). When the transfer matrix describes the stability of the metallic phase in the thick-quantum-wire limit of non-interacting fermions perturbed by static one-body random potentials with local correlations and of vanishing means in the bulk of a quasi-one-dimensional lattice model, the multiplicities of the short root entering the Brownian motion of the Lyapunov exponents always vanish. However, when the transfer matrix describes the quasi-one-dimensional boundary of a two-dimensional topological band insulator moderately perturbed by static one-body random potentials with local correlations, the multiplicities of the short root are non-vanishing in the Brownian motion of the Lyapunov exponents in the five AZ symmetry classes A, AII, D, DIII, and C. Correspondingly, the conductance is of order unity along the infinitely

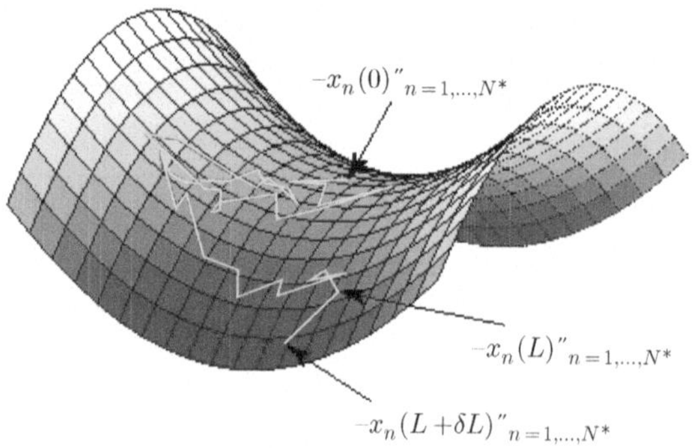

Fig. 7.8 [Colour online] The ‘radial coordinate’ of the transfer matrix $\mathcal{M}$ from Table 7.2 is responsible for Brownian motion on an associated non-compact symmetric space. (Taken from [11].)

long boundary, i.e. the insulating bulk supports extended edge states. These extended edge states can be thought of as realizing a quasi-one-dimensional ballistic phase of quantum matter robust to disorder.

7.2.4 Topological spaces for the normalized Dirac masses

To study systematically the effects of static and local disorder on the insulating phases of quasi-d-dimensional phases, it is very useful to explore the topological properties of the normalized Dirac masses entering a generic random Dirac Hamiltonian of the form (7.50) within any given AZ symmetry class. This approach allows us to construct the periodic table in Table 7.3 [85]. The $d = 1, \ldots, 8$ columns of Table 7.3 can be derived by brute force if one constructs the generic Dirac Hamiltonian with Dirac matrices of rank r_{min} belonging to any one of the ten AZ symmetry classes and repeats this construction with Dirac matrices of rank $2r_{\text{min}}$, $3r_{\text{min}}$, and so on. It then becomes apparent that for any dimension d:

1. Five of the ten AZ symmetry classes accommodate one normalized Dirac matrix up to sign when the Dirac matrices have rank $r = r_{\text{min}}$. These are the symmetry classes that realize topologically distinct localized phases of d-dimensional quantum matter.
   (a) Three of these symmetry classes are characterized by having one normalized Dirac mass matrix that commutes with all other Dirac matrices when $r = N r_{\text{min}}$, with $N = 2, 3, \ldots$. These are the entries with the group $\mathbb{Z}$ (or $2\mathbb{Z}$ when there is a degeneracy of 2) in Table 7.3. Mathematically, the group $\mathbb{Z}$ is the zeroth homotopy group of the normalized Dirac masses in the limit $N \to \infty$.
   (b) Two of these symmetry classes are characterized by the fact that the sum of all mass terms can be associated with a $2N \times 2N$ Hermitian and antisymmetric matrix for any $r = r_{\text{min}} N$, with $N = 1, 2, \ldots$. The sign ambiguity of the Pfaffian of this matrix indexes the two group elements in the entries with the group $\mathbb{Z}_2$ in Table 7.3. Mathematically, the group $\mathbb{Z}_2$ is the zeroth homotopy group of the normalized Dirac masses for N sufficiently large.
2. The topological space of normalized Dirac masses is compact and path-connected for the remaining five symmetry classes, i.e. its zeroth homotopy group is trivial. No Dirac mass is singled out. The localized phase of matter is topologically trivial.

The observed periodicity of two for the complex classes and of eight for the real classes in Table 7.3 follows from Bott periodicity in K-theory.

7.3 Fractionalization from Abelian bosonization

7.3.1 Introduction

Abelian bosonization is attributed to Coleman [21], Mandelstam [79], and Luther and Peschel [77]. However, Abelian bosonization can be traced to the much older works of

Kronig [67], Tomonaga [121], and Luttinger [78], in which it is observed that gapless fermions in one-dimensional space are equivalent to gapless phonons. Here, we follow the more general formulation of Abelian bosonization given by Haldane [46], since it lends itself to a description of one-dimensional quantum effective field theories arising in the low-energy sector along the boundary in space of $(2+1)$-dimensional topological quantum field theories.

7.3.2 Definition

Let us define the quantum Hamiltonian (in units with the electric charge e, the characteristic speed, and $\hbar$ set equal to 1)

$$\begin{aligned}\widehat{H} &:= \int_0^L \mathrm{d}x \left[\frac{1}{4\pi} V_{ij}\,(\mathrm{D}_x\hat{u}_i)\,(\mathrm{D}_x\hat{u}_j) + A_0\Big(\frac{q_i}{2\pi}K^{-1}_{ij}\,(\mathrm{D}_x\hat{u}_j)\Big)\right](t,x),\\ \mathrm{D}_x\hat{u}_i(t,x) &:= (\partial_x\hat{u}_i + q_i A_1)(t,x).\end{aligned} \tag{7.54a}$$

The indices $i,j=1,\dots,N$ label the bosonic modes. Summation over repeated indices is implied. The N real-valued quantum fields $\hat{u}_i(t,x)$ obey the equal-time commutation relations

$$[\hat{u}_i(t,x),\hat{u}_j(t,y)] := \mathrm{i}\pi[K_{ij}\,\mathrm{sgn}(x-y) + L_{ij}] \tag{7.54b}$$

for any pair $i,j=1,\dots,N$. The function $\mathrm{sgn}(x)=-\mathrm{sgn}(-x)$ gives the sign of the real variable x and will be assumed to be periodic with periodicity L. The $N\times N$ matrix K is symmetric, invertible, and integer-valued. Given the pair $i,j=1,\dots,N$, any of its matrix elements thus obey

$$K_{ij}=K_{ji}\in\mathbb{Z},\qquad K^{-1}_{ij}=K^{-1}_{ji}\in\mathbb{Q}. \tag{7.54c}$$

The $N\times N$ matrix L is antisymmetric:

$$L_{ij}=-L_{ji}=\begin{cases}0 & \text{if } i=j,\\ \mathrm{sgn}(i-j)(K_{ij}+q_iq_j) & \text{otherwise,}\end{cases} \tag{7.54d}$$

for $i,j=1,\dots,N$. The sign function $\mathrm{sgn}(i)$ of any integer i is here not made periodic and is taken to vanish at the origin of $\mathbb{Z}$. The external scalar gauge potential $A_0(t,x)$ and vector gauge potential $A_1(t,x)$ are real-valued functions of the time t and space x coordinates. They are also chosen to be periodic under $x\mapsto x+L$. The $N\times N$ matrix V is symmetric and positive-definite:

$$V_{ij}=V_{ji}\in\mathbb{R},\qquad v_iV_{ij}v_j>0,\qquad i,j=1,\dots,N, \tag{7.54e}$$

for any non-vanishing vector $v=(v_i)\in\mathbb{R}^N$. The charges q_i are integer-valued and satisfy

$$(-1)^{K_{ii}}=(-1)^{q_i},\qquad i=1,\dots,N. \tag{7.54f}$$

Finally, we shall impose the boundary conditions

$$\hat{u}_i(t, x+L) = \hat{u}_i(t, x) + 2\pi n_i, \qquad n_i \in \mathbb{Z}, \tag{7.54g}$$

and

$$(\partial_x \hat{u}_i)(t, x+L) = (\partial_x \hat{u}_i)(t, x), \tag{7.54h}$$

for any $i = 1, \ldots, N$.

7.3.3 Chiral equations of motion

For any $i, j = 1, \ldots, N$, one verifies with the help of the equal-time commutation relations

$$[\hat{u}_i(t, x), \mathrm{D}_y \hat{u}_j(t, y)] = -2\pi \mathrm{i} K_{ij} \delta(x - y) \tag{7.55}$$

that the equations of motions are

$$\begin{aligned} \mathrm{i}(\partial_t \hat{u}_i)(t, x) &:= [\hat{u}_i(t, x), \widehat{H}] \\ &= -\mathrm{i} K_{ij} V_{jk} (\partial_x \hat{u}_k + q_k A_1)(t, x) - \mathrm{i} q_i A_0(t, x). \end{aligned} \tag{7.56}$$

We introduce the covariant derivatives

$$\mathrm{D}_\mu \hat{u}_k := \partial_\mu \hat{u}_k + q_k A_\mu, \qquad \text{with} \quad \partial_0 \equiv \partial_t, \qquad \partial_1 \equiv \partial_x, \tag{7.57}$$

for $\mu = 0, 1$ and $k = 1, \ldots, N$. The equations of motion

$$0 = \delta_{ik} \mathrm{D}_0 \hat{u}_k + K_{ij} V_{jk} \mathrm{D}_1 \hat{u}_k \tag{7.58}$$

are chiral. Making the substitutions $\hat{u}_i \mapsto \hat{v}_i$ and $K \mapsto -K$ everywhere in (7.54) delivers the chiral equations of motion

$$0 = \delta_{ik} \mathrm{D}_0 \hat{v}_k - K_{ij} V_{jk} \mathrm{D}_1 \hat{v}_k, \tag{7.59}$$

with the opposite chirality. Evidently, the chiral equations of motion (7.58) and (7.59) are first-order differential equations, in contrast to the Klein–Gordon equations of motion obeyed by a relativistic quantum scalar field.

7.3.4 Gauge invariance

The chiral equations of motion (7.58) and (7.59) are invariant under the local $U(1)$ gauge symmetry

$$\begin{aligned} \hat{u}_i(t,x) &=: \hat{u}'_i(t,x) + q_i\chi(t,x), \\ A_0(t,x) &=: A'_0(t,x) - (\partial_t\chi)(t,x), \\ A_1(t,x) &=: A'_1(t,x) - (\partial_x\chi)(t,x), \end{aligned} \tag{7.60a}$$

for any real-valued function χ that satisfies the periodic boundary conditions

$$\chi(t,x+L) = \chi(t,x). \tag{7.60b}$$

Functional differentiation of the Hamiltonian (7.54a) with respect to the gauge potentials allows us to define the 2-current with components

$$\hat{J}^0(t,x) := \frac{\delta\widehat{H}}{\delta A_0(t,x)} = \frac{1}{2\pi} q_i K^{-1}_{ij} (\mathrm{D}_1\hat{u}_j)(t,x), \tag{7.61a}$$

$$\hat{J}^1(t,x) := \frac{\delta\widehat{H}}{\delta A_1(t,x)} = \frac{1}{2\pi} q_i V_{ij} (\mathrm{D}_1\hat{u}_j)(t,x) + \frac{1}{2\pi}(q_i K^{-1}_{ij} q_j) A_0(t,x). \tag{7.61b}$$

We introduce the shorthand notation

$$\sigma_{\mathrm{H}} := \frac{1}{2\pi}(q_i K^{-1}_{ij} q_j) \in \frac{1}{2\pi}\mathbb{Q} \tag{7.62}$$

for the second term on the right-hand side of (7.61b). The subscript stands for 'Hall' as we shall shortly interpret σ_{H} as a dimensionless Hall conductance.

The transformation law of the 2-current (7.61) under the local gauge transformation (7.60) is

$$\hat{J}^0(t,x) = \hat{J}^{0\prime}(t,x), \tag{7.63a}$$

$$\hat{J}^1(t,x) = \hat{J}^{1\prime}(t,x) - \sigma_{\mathrm{H}}(\partial_t\chi)(t,x). \tag{7.63b}$$

The 2-current (7.61) is only invariant under gauge transformations (7.60) that are static when $\sigma_{\mathrm{H}} \neq 0$.

With the help of

$$[\mathrm{D}_x\hat{u}_i(t,x), \mathrm{D}_y\hat{u}_j(t,y)] = -2\pi\mathrm{i}K_{ij}\delta'(x-y) \tag{7.64}$$

for $i,j = 1,\ldots,N$, one verifies that the time derivative of $\hat{J}^0(t,x)$ is

$$\begin{aligned} \partial_t\hat{J}^0 &= -\mathrm{i}[\hat{J}^0, \widehat{H}] + \sigma_{\mathrm{H}}\partial_t A_1 \\ &= -\partial_x\hat{J}^1 + \sigma_{\mathrm{H}}\partial_t A_1. \end{aligned} \tag{7.65}$$

The continuity equation

$$\partial_\mu \hat{J}^\mu = 0 \tag{7.66}$$

then follows, provided that A_1 is time-independent or $\sigma_{\rm H} = 0$. This continuity equation delivers a conserved total charge if and only if A_0 and A_1 are both static for arbitrary $\sigma_{\rm H} \neq 0$.

For any non-vanishing $\sigma_{\rm H}$, the continuity equation

$$\partial_\mu \hat{J}^\mu = \sigma_{\rm H}\partial_t A_1 \tag{7.67}$$

is anomalous as soon as the vector gauge potential A_1 is time-dependent. The edge theory (7.54) is said to be chiral when $\sigma_{\rm H} \neq 0$, in which case the continuity equation (7.67) is anomalous. The anomalous continuity equation (7.67) is form-covariant under any smooth gauge transformation (7.60). The choice of gauge may be fixed by the condition

$$\partial_x A_0 = 0, \tag{7.68a}$$

for which the anomalous continuity equation (7.67) then becomes

$$(\partial_\mu \hat{J}^\mu)(t,x) = +\sigma_{\rm H} E(t,x), \tag{7.68b}$$

where

$$E(t,x) := +(\partial_t A_1)(t,x) \equiv -(\partial_t A^1)(t,x) \tag{7.68c}$$

represents the electric field in this gauge.

To interpret the anomalous continuity equation (7.68) of the bosonic chiral edge theory (7.54), we recall that x is a compact coordinate because of the periodic boundary conditions (7.54g), (7.54f), and (7.60). For simplicity, we assume

$$E(t,x) = E(t). \tag{7.69}$$

The interval $0 \leq x \leq L$ is thought of as a circle of perimeter L centred at the origin of the three-dimensional Euclidean space as shown in Fig. 7.9. The vector potential $A^1(t)$ and the electric field $E(t) = -(\partial_t A^1)(t)$ along the circle of radius $R \equiv L/2\pi$ are then the polar components of a three-dimensional gauge field $A^\mu(t,\boldsymbol{r}) = (A^0, \boldsymbol{A})(t,\boldsymbol{r})$ in a cylindrical geometry with the electromagnetic fields

$$\boldsymbol{E}(t,\boldsymbol{r}) = -(\boldsymbol{\nabla} A^0)(t,\boldsymbol{r}) - (\partial_t \boldsymbol{A})(t,\boldsymbol{r}), \qquad \boldsymbol{B}(t,\boldsymbol{r}) = (\boldsymbol{\nabla}\times\boldsymbol{A})(t,\boldsymbol{r}). \tag{7.70}$$

The dimensionless Hall conductance $\sigma_{\rm H}$ encodes the linear response of spin-polarized electrons confined to move along this circle in the presence of a uniform and static magnetic field normal to the plane that contains this circle. The time-dependent anomalous term on the right-hand side of the anomalous continuity equation (7.68b) is caused by a solenoid of radius $r_{\rm sln} \ll r \ll R$ in a puncture of the plane that contains the circle of radius $r_{\rm sln}$ supporting a time-dependent flux. The combination of this

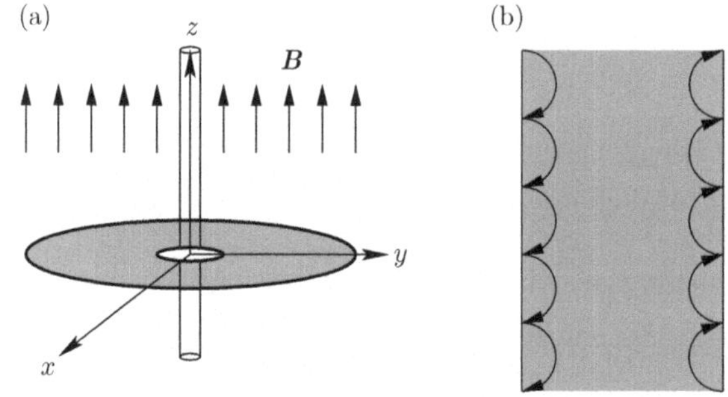

Fig. 7.9 (a) A ring of outer radius $R \equiv L/2\pi$ and inner radius r in which electrons are confined. A uniform and static magnetic field $\boldsymbol{B}$ normal to the ring is present. The hierarchy $\ell_B \ll r \ll R$ of lengthscales is assumed, where $\ell_B \equiv \hbar c/|eB|$ is the magnetic length. A time-dependent vector potential $\boldsymbol{A}(t, \boldsymbol{r})$ is induced by a time-dependent flux supported within a solenoid of radius $r_{\rm sln} \ll r$. This Corbino geometry has cylindrical symmetry. (b) The classical motion of electrons confined to a plane normal to a uniform static magnetic field is circular. In the limit $R \to \infty$ with R/r held fixed, the Corbino disc turns into a Hall bar. An electron within a magnetic length of the boundary undergoes a classical skipping orbit. On quantization, a classical electron undergoing a skipping orbit turns into a chiral electron. On bosonization, a chiral electron turns into a chiral boson.

time-dependent flux with the uniform static magnetic field exerts a Lorentz force on spin-polarized electrons moving along circles in the ring with inner edge of radius r and outer edge of radius R. This Lorentz force causes a net transfer of charge between the inner and outer edges,

$$\frac{1}{L}\int_0^T \mathrm{d}t\, Q(T) := \int_0^T \mathrm{d}t\, \langle \hat{J}^0(t)\rangle = \sigma_{\rm H} \int_0^T \mathrm{d}t\, E(t), \tag{7.71}$$

during the adiabatic evolution with period T of the normalized many-body ground state of the outer edge, provided that we may identify the anomalous continuity equation (7.68b) with that of chiral spin-polarized electrons propagating along the outer edge in Fig. 7.9. Here, for the many-body ground state at the outer edge to be separated from all spin-polarized electrons supported between the inner and outer edges, it is necessary for there to exist an energy scale separating the ground state from the many-body states in which these bulk spin-polarized electrons participate and for the inverse of this energy scale, a lengthscale, to be much smaller than $R - r$. This energy scale is brought about by the uniform and static magnetic field $\boldsymbol{B}$ in Fig. 7.9. That none of this pumped charge is lost in the shaded region of the ring follows if it is assumed that the spin-polarized electrons are unable to transport (dissipatively) a charge current across any circle of radius less than R and greater than r. The Hall conductance in the Corbino geometry of Fig. 7.9 is then a rank-two antisymmetric tensor proportional to the rank-two Levi–Civita antisymmetric tensor with $\sigma_{\rm H}$ the proportionality constant in units of e^2/h. The charge density and current density for the ring obey a continuity equation as full gauge invariance is restored in the ring.

The chiral bosonic theory (7.54) is nothing but a theory for chiral electrons at the outer edge of the Corbino disc, as we still have to demonstrate. Chiral fermions are a fraction of the original fermion (a spin-polarized electron). More precisely, low-energy fermions have been split into one half that propagate on the outer edge and another half that propagate on the inner edge of the Corbino disc. The price for this fractionalization is an apparent breakdown of gauge invariance and charge conservation, with each chiral edge being treated independently from the other. Manifest charge conservation and gauge invariance are only restored if all low-energy degrees of freedom from the Corbino disc are treated on an equal footing.

7.3.5 Conserved topological charges

Let us turn off the external gauge potentials:

$$A_0(t,x) = A_1(t,x) = 0. \tag{7.72}$$

For any $i = 1,\ldots,N$, we define the operator

$$\begin{aligned}\widehat{\mathcal{N}}_i(t) &:= \frac{1}{2\pi}\int_0^L \mathrm{d}x\,(\partial_x\hat{u}_i)(t,x)\\ &= \frac{1}{2\pi}[\hat{u}_i(t,L) - \hat{u}_i(t,0)].\end{aligned} \tag{7.73}$$

This operator is conserved if and only if

$$(\partial_x\hat{u}_i)(t,x) = (\partial_x\hat{u}_i)(t,x+L), \qquad 0 \le x \le L, \tag{7.74}$$

for

$$\mathrm{i}(\partial_t\widehat{\mathcal{N}}_i)(t) = -\frac{\mathrm{i}}{2\pi}K_{ik}V_{kl}[(\partial_x\hat{u}_l)(t,L) - (\partial_x\hat{u}_l)(t,0)]. \tag{7.75}$$

Furthermore, if we demand that there exists an $n_i \in \mathbb{Z}$ such that

$$\hat{u}_i(t,x+L) = \hat{u}_i(t,x) - 2\pi n_i, \tag{7.76}$$

then it follows that

$$\widehat{\mathcal{N}}_i = n_i. \tag{7.77}$$

A corollary to (7.77) is that the N conserved topological charges $\mathcal{N}_i$ with $i = 1,\ldots,N$ commute pairwise. The same conclusion follows from the brute force manipulations

$$\begin{aligned}[\widehat{\mathcal{N}}_i,\widehat{\mathcal{N}}_j] &= \frac{1}{2\pi}\int_0^L \mathrm{d}y\,[\widehat{\mathcal{N}}_i,(\partial_y\hat{u}_j)(y)]\\ &= \frac{1}{2\pi}\int_0^L \mathrm{d}y\,\partial_y[\widehat{\mathcal{N}}_i,\hat{u}_j(y)],\end{aligned} \tag{7.78}$$

where $j=1,\ldots,N$ and

$$[\widehat{\mathcal{N}}_i,\hat{u}_j(y)]=\mathrm{i}K_{ij} \tag{7.79}$$

is independent of y.

The local counterpart to the global conservation of the topological charge is

$$\partial_t\hat{\rho}^{\mathrm{top}}_i+\partial_x\hat{\jmath}^{\mathrm{top}}_i=0, \tag{7.80a}$$

where the local topological density operator is defined by

$$\hat{\rho}^{\mathrm{top}}_i(t,x):=\frac{1}{2\pi}(\partial_x\hat{u}_i)(t,x) \tag{7.80b}$$

and the local topological current operator is defined by

$$\hat{\jmath}^{\mathrm{top}}_i(t,x):=\frac{1}{2\pi}K_{ik}V_{kl}(\partial_x\hat{u}_l)(t,x) \tag{7.80c}$$

for $i=1,\ldots,N$. The local topological density operator obeys the equal-time algebra

$$[\hat{\rho}^{\mathrm{top}}_i(t,x),\hat{\rho}^{\mathrm{top}}_j(t,y)]=-\frac{\mathrm{i}}{2\pi}K_{ij}\partial_x\delta(x-y) \tag{7.81a}$$

for any $i,j=1,\ldots,N$. The local topological current operator obeys the equal-time algebra

$$[\hat{\jmath}^{\mathrm{top}}_i(t,x),\hat{\jmath}^{\mathrm{top}}_j(t,y)]=-\frac{\mathrm{i}}{2\pi}K_{ik}V_{kl}K_{jk'}V_{k'l'}K_{ll'}\partial_x\delta(x-y) \tag{7.81b}$$

for any $i,j=1,\ldots,N$. Finally,

$$[\hat{\rho}^{\mathrm{top}}_i(t,x),\hat{\jmath}^{\mathrm{top}}_j(t,y)]=-\frac{\mathrm{i}}{2\pi}K_{jk}V_{kl}K_{il}\partial_x\delta(x-y) \tag{7.81c}$$

for any $i,j=1,\ldots,N$.

Let us introduce the local charges and currents

$$\hat{\rho}_i(t,x):=K^{-1}_{ij}\,\hat{\rho}^{\mathrm{top}}_j(t,x) \tag{7.82a}$$

and

$$\hat{\jmath}_i(t,x):=K^{-1}_{ij}\hat{\jmath}^{\mathrm{top}}_j(t,x), \tag{7.82b}$$

respectively, for any $i=1,\ldots,N$. The continuity equation (7.80a) is unchanged under this linear transformation:

$$\partial_t\hat{\rho}_i+\partial_x\hat{\jmath}_i=0 \tag{7.82c}$$

for any $i = 1, \dots, N$. The topological current algebra (7.81) transforms into

$$[\hat{\rho}_i(t,x), \hat{\rho}_j(t,y)] = -\frac{\mathrm{i}}{2\pi} K^{-1}_{ij} \partial_x \delta(x-y), \tag{7.83a}$$

$$[\hat{\jmath}_i(t,x), \hat{\jmath}_j(t,y)] = -\frac{\mathrm{i}}{2\pi} V_{ik} V_{jl} K_{kl} \partial_x \delta(x-y), \tag{7.83b}$$

$$[\hat{\rho}_i(t,x), \hat{\jmath}_j(t,y)] = -\frac{\mathrm{i}}{2\pi} V_{ij} \partial_x \delta(x-y) \tag{7.83c}$$

for any $i, j = 1, \dots, N$.

Finally, if we contract the continuity equation (7.82c) with the integer-valued charge vector, we obtain the flavour-global continuity equation

$$\partial_t \hat{\rho} + \partial_x \hat{\jmath} = 0, \tag{7.84a}$$

where the local flavour-global charge operator is

$$\hat{\rho}(t,x) := q_i K^{-1}_{ij} \hat{\rho}^{\mathrm{top}}_j(t,x) \tag{7.84b}$$

and the local flavour-global current operator is

$$\hat{\jmath}(t,x) := q_i K^{-1}_{ij} \hat{\jmath}^{\mathrm{top}}_j(t,x). \tag{7.84c}$$

The flavour-resolved current algebra (7.83) turns into the flavour-global current algebra

$$[\hat{\rho}(t,x), \hat{\rho}(t,y)] = -\frac{\mathrm{i}}{2\pi} (q_i K^{-1}_{ij} q_j) \partial_x \delta(x-y), \tag{7.85a}$$

$$[\hat{\jmath}(t,x), \hat{\jmath}(t,y)] = -\frac{\mathrm{i}}{2\pi} (q_i V_{ik} K_{kl} V_{lj} q_j) \partial_x \delta(x-y), \tag{7.85b}$$

$$[\hat{\rho}(t,x), \hat{\jmath}(t,y)] = -\frac{\mathrm{i}}{2\pi} (q_i V_{ij} q_j) \partial_x \delta(x-y). \tag{7.85c}$$

7.3.6 Quasiparticle and particle excitations

When (7.72) holds, there exist N conserved global topological (i.e. integer-valued) charges $\widehat{\mathcal{N}}_i$, with $i = 1, \dots, N$, defined in (7.73) that commute pairwise. We define the N global charges

$$\widehat{Q}_i := \int_0^L \mathrm{d}x\, \hat{\rho}_i(t,x) = K^{-1}_{ij} \widehat{\mathcal{N}}_j, \qquad i = 1, \dots, N. \tag{7.86}$$

These charges will shortly be interpreted as the elementary Fermi–Bose charges.

For any $i = 1, \dots, N$, we define the following pair of vertex operators:

$$\widehat{\Psi}^{\dagger}_{\text{q-p},i}(t,x) := \mathrm{e}^{-\mathrm{i} K^{-1}_{ij} \hat{u}_j(t,x)}, \tag{7.87a}$$

$$\widehat{\Psi}^{\dagger}_{\text{f-b},i}(t,x) := \mathrm{e}^{-\mathrm{i} \delta_{ij} \hat{u}_j(t,x)}. \tag{7.87b}$$

The quasiparticle vertex operator $\widehat{\Psi}^{\dagger}_{\text{q-p},i}(t,x)$ is multivalued under a shift by 2π of all $\hat{u}_j(t,x)$, where $j=1,\ldots,N$. The Fermi–Bose vertex operator $\widehat{\Psi}^{\dagger}_{\text{f-b},i}(t,x)$ is single-valued under a shift by 2π of all $\hat{u}_j(t,x)$, where $j=1,\ldots,N$.

For any pair $i,j=1,\ldots,N$, the commutator (7.79) delivers the identities

$$[\widehat{\mathcal{N}}_i,\widehat{\Psi}^{\dagger}_{\text{q-p},j}(t,x)]=\delta_{ij}\widehat{\Psi}^{\dagger}_{\text{q-p},j}(t,x),\qquad [\widehat{\mathcal{N}}_i,\widehat{\Psi}^{\dagger}_{\text{f-b},j}(t,x)]=K_{ij}\widehat{\Psi}^{\dagger}_{\text{f-b},j}(t,x),\tag{7.88}$$

$$[\widehat{Q}_i,\widehat{\Psi}^{\dagger}_{\text{q-p},j}(t,x)]=K^{-1}_{ij}\widehat{\Psi}^{\dagger}_{\text{q-p},j}(t,x),\qquad [\widehat{Q}_i,\widehat{\Psi}^{\dagger}_{\text{f-b},j}(t,x)]=\delta_{ij}\widehat{\Psi}^{\dagger}_{\text{f-b},j}(t,x).\tag{7.89}$$

The quasiparticle vertex operator $\widehat{\Psi}^{\dagger}_{\text{q-p},i}(t,x)$ is an eigenstate of the topological number operator $\widehat{\mathcal{N}}_i$ with eigenvalue 1. The Fermi–Bose vertex operator $\widehat{\Psi}^{\dagger}_{\text{f-b},i}(t,x)$ is an eigenstate of the charge number operator $\widehat{Q}_i$ with eigenvalue 1.

The Baker–Campbell–Hausdorff formula implies that

$$e^{\hat{A}}e^{\hat{B}}=e^{\hat{A}+\hat{B}}e^{+\frac{1}{2}[\hat{A},\hat{B}]}=e^{\hat{B}}e^{\hat{A}}e^{[\hat{A},\hat{B}]}\tag{7.90}$$

whenever two operators $\hat{A}$ and $\hat{B}$ have a $\mathbb{C}$-number as their commutator. A first application of the Baker–Campbell–Hausdorff formula to any pair of quasiparticle vertex operators at equal time t but two distinct space coordinates $x\neq y$ gives

$$\widehat{\Psi}^{\dagger}_{\text{q-p},i}(t,x)\widehat{\Psi}^{\dagger}_{\text{q-p},j}(t,y)=e^{-i\pi\Theta^{\text{q-p}}_{ij}}\,\widehat{\Psi}^{\dagger}_{\text{q-p},j}(t,y)\widehat{\Psi}^{\dagger}_{\text{q-p},i}(t,x),\tag{7.91a}$$

where

$$\Theta^{\text{q-p}}_{ij}:=K^{-1}_{ji}\operatorname{sgn}(x-y)+(K^{-1}_{ik}K^{-1}_{jl}K_{kl}+q_kK^{-1}_{ik}K^{-1}_{jl}q_l)\operatorname{sgn}(k-l).\tag{7.91b}$$

Here and below, it is understood that

$$\operatorname{sgn}(k-l)=0\tag{7.92}$$

when $k=l=1,\ldots,N$. Hence, the quasiparticle vertex operators obey neither bosonic nor fermionic statistics, since $K^{-1}_{ij}\in\mathbb{Q}$.

The same exercise applied to the Fermi–Bose vertex operators yields

$$\widehat{\Psi}^{\dagger}_{\text{f-b},i}(t,x)\widehat{\Psi}^{\dagger}_{\text{f-b},j}(t,y)=\begin{cases}(-1)^{K_{ii}}\widehat{\Psi}^{\dagger}_{\text{f-b},i}(t,y)\widehat{\Psi}^{\dagger}_{\text{f-b},i}(t,x) & \text{if } i=j,\\[2mm] (-1)^{q_iq_j}\widehat{\Psi}^{\dagger}_{\text{f-b},j}(t,y)\widehat{\Psi}^{\dagger}_{\text{f-b},i}(t,x) & \text{if } i\neq j,\end{cases}\tag{7.93}$$

when $x\neq y$. The self-statistics of the Fermi–Bose vertex operators is carried by the diagonal matrix elements $K_{ii}\in\mathbb{Z}$. The mutual statistics of any pair of Fermi–Bose vertex operators labelled by $i\neq j$ is carried by the product $q_iq_j\in\mathbb{Z}$ of the integer-valued charges q_i and q_j. Had we not assumed that K_{ij} with $i\neq j$ are integers, the mutual statistics would not be Fermi–Bose, because of the non-local term $K_{ij}\operatorname{sgn}(x-y)$.

A third application of the Baker–Campbell–Hausdorff formula allows us to determine the boundary conditions obeyed by the quasiparticle and Fermi–Bose vertex operators:

$$\widehat{\Psi}^{\dagger}_{\text{q-p},i}(t,x+L)=\widehat{\Psi}^{\dagger}_{\text{q-p},i}(t,x)\mathrm{e}^{-2\pi\mathrm{i}K^{-1}_{ij}\widehat{\mathcal{N}}_j}\mathrm{e}^{-\pi\mathrm{i}K^{-1}_{ii}}, \tag{7.94}$$

$$\widehat{\Psi}^{\dagger}_{\text{f-b},i}(t,x+L)=\widehat{\Psi}^{\dagger}_{\text{f-b},i}(t,x)\mathrm{e}^{-2\pi\mathrm{i}\widehat{\mathcal{N}}_i}\mathrm{e}^{-\pi\mathrm{i}K_{ii}}. \tag{7.95}$$

The quasiparticle vertex operators obey twisted boundary conditions, the Fermi vertex operators obey antiperiodic boundary conditions, and the Bose vertex operators obey periodic boundary conditions.

We close this discussion with the following definitions. We introduce the operators

$$\widehat{Q}:=q_i\widehat{Q}_i, \tag{7.96}$$

$$\widehat{\Psi}^{\dagger}_{\text{q-p},\boldsymbol{m}}:=\mathrm{e}^{-\mathrm{i}m_iK^{-1}_{ij}\hat{u}_j(t,x)}, \tag{7.97}$$

$$\widehat{\Psi}^{\dagger}_{\text{f-b},\boldsymbol{m}}:=\mathrm{e}^{-\mathrm{i}m_i\delta_{ij}\hat{u}_j(t,x)}, \tag{7.98}$$

where $\boldsymbol{m}\in\mathbb{Z}^N$ is the vector with integer-valued components m_i for any $i=1,\ldots,N$. The N charges q_i with $i=1,\ldots,N$ that enter the Hamiltonian (7.54a) can also be viewed as the components of the vector $\boldsymbol{q}\in\mathbb{Z}^N$. Let us define the functions

$$\begin{aligned}q:\mathbb{Z}^N&\to\mathbb{Z},\\ \boldsymbol{m}&\mapsto q(\boldsymbol{m}):=q_im_i\equiv\boldsymbol{q}\cdot\boldsymbol{m},\end{aligned} \tag{7.99a}$$

and

$$\begin{aligned}K:\mathbb{Z}^N&\to\mathbb{Z},\\ \boldsymbol{m}&\mapsto K(\boldsymbol{m}):=m_iK_{ij}m_j.\end{aligned} \tag{7.99b}$$

On the one hand, for any distinct pair of space coordinates $x\neq y$, we deduce from (7.89), (7.91), and (7.94), respectively, that

$$[\widehat{Q},\widehat{\Psi}^{\dagger}_{\text{q-p},\boldsymbol{m}}(t,x)]=(q_iK^{-1}_{ij}m_j)\widehat{\Psi}^{\dagger}_{\text{q-p},\boldsymbol{m}}(t,x), \tag{7.100a}$$

$$\widehat{\Psi}^{\dagger}_{\text{q-p},\boldsymbol{m}}(t,x)\widehat{\Psi}^{\dagger}_{\text{q-p},\boldsymbol{n}}(t,y)=\mathrm{e}^{-\mathrm{i}\pi m_i\Theta^{\text{q-p}}_{ij}n_j}\widehat{\Psi}^{\dagger}_{\text{q-p},\boldsymbol{n}}(t,y)\widehat{\Psi}^{\dagger}_{\text{q-p},\boldsymbol{m}}(t,x), \tag{7.100b}$$

$$\widehat{\Psi}^{\dagger}_{\text{q-p},\boldsymbol{m}}(t,x+L)=\widehat{\Psi}^{\dagger}_{\text{q-p},\boldsymbol{m}}(t,x)\mathrm{e}^{-2\pi\mathrm{i}m_iK^{-1}_{ij}\widehat{\mathcal{N}}_j}\mathrm{e}^{-\pi\mathrm{i}m_iK^{-1}_{ij}m_j}. \tag{7.100c}$$

On the other hand, for any distinct pair of space coordinates $x\neq y$, we deduce from (7.89), (7.93), and (7.95), respectively, that

$$[\widehat{Q},\widehat{\Psi}^{\dagger}_{\text{f-b},\boldsymbol{m}}(t,x)]=q(\boldsymbol{m})\widehat{\Psi}^{\dagger}_{\text{f-b},\boldsymbol{m}}(t,x), \tag{7.101a}$$

$$\widehat{\Psi}^{\dagger}_{\text{f-b},\boldsymbol{m}}(t,x)\widehat{\Psi}^{\dagger}_{\text{f-b},\boldsymbol{n}}(t,y)=\mathrm{e}^{-\mathrm{i}\pi m_i\Theta^{\text{f-b}}_{ij}n_j}\widehat{\Psi}^{\dagger}_{\text{f-b},\boldsymbol{n}}(t,y)\widehat{\Psi}^{\dagger}_{\text{f-b},\boldsymbol{m}}(t,x), \tag{7.101b}$$

$$\widehat{\Psi}^{\dagger}_{\text{f-b},\boldsymbol{m}}(t,x+L)=\widehat{\Psi}^{\dagger}_{\text{f-b},\boldsymbol{m}}(t,x)\mathrm{e}^{-2\pi\mathrm{i}m_i\widehat{\mathcal{N}}_i}\mathrm{e}^{-\pi\mathrm{i}m_iK_{ij}m_j}, \tag{7.101c}$$

where

$$\Theta^{\text{f-b}}_{ij} := K_{ij}\,\text{sgn}(x-y) + (K_{ij} + q_i q_j)\,\text{sgn}(i-j). \tag{7.101d}$$

The integer quadratic form $K(\boldsymbol{m})$ thus dictates whether the vertex operator $\widehat{\Psi}^{\dagger}_{\text{f-b},\boldsymbol{m}}(t,x)$ realizes a fermion or a boson: $\widehat{\Psi}^{\dagger}_{\text{f-b},\boldsymbol{m}}(t,x)$ realizes

- a fermion if and only if $K(\boldsymbol{m})$ is an odd integer;
- a boson if and only if $K(\boldsymbol{m})$ is an even integer.

Because of the assumption (7.54f),

$$(-1)^{K(\boldsymbol{m})} = (-1)^{q(\boldsymbol{m})}. \tag{7.102}$$

Hence, the vertex operator $\widehat{\Psi}^{\dagger}_{\text{f-b},\boldsymbol{m}}(t,x)$ realizes

- a fermion if and only if $q(\boldsymbol{m})$ is an odd integer;
- a boson if and only if $q(\boldsymbol{m})$ is an even integer.

7.3.7 Bosonization rules

We are going to relate the theory of chiral bosons (7.54) without external gauge fields to a massless Dirac Hamiltonian. To this end, we proceed in three steps.

Step 1 Make the following choices in (7.54):

$$N = 2, \qquad i,j = 1,2 \equiv -,+, \tag{7.103a}$$

$$K := \begin{pmatrix} +1 & 0 \\ 0 & -1 \end{pmatrix}, \qquad V := \begin{pmatrix} +1 & 0 \\ 0 & +1 \end{pmatrix}, \qquad \boldsymbol{q} = \begin{pmatrix} 1 \\ 1 \end{pmatrix}. \tag{7.103b}$$

With these choices, the free-bosonic Hamiltonian on the real line is

$$\widehat{H}_{\text{B}} = \int_{\mathbb{R}} \mathrm{d}x\, \frac{1}{4\pi}[(\partial_x \hat{u}_-)^2 + (\partial_x \hat{u}_+)^2], \tag{7.104a}$$

where

$$[\hat{u}_-(t,x), \hat{u}_-(t,y)] = +\mathrm{i}\pi\,\text{sgn}(x-y), \tag{7.104b}$$

$$[\hat{u}_+(t,x), \hat{u}_+(t,y)] = -\mathrm{i}\pi\,\text{sgn}(x-y), \tag{7.104c}$$

$$[\hat{u}_-(t,x), \hat{u}_+(t,y)] = +\mathrm{i}\pi. \tag{7.104d}$$

From this, there follow the chiral equations of motion (recall (7.56))

$$\partial_t \hat{u}_- = -\partial_x \hat{u}_-, \qquad \partial_t \hat{u}_+ = +\partial_x \hat{u}_+ \tag{7.105}$$

obeyed by the right-mover $\hat{u}_-$ and the left-mover $\hat{u}_+$, the current algebra

$$[\hat{\rho}(t,x),\hat{\rho}(t,y)] = 0, \tag{7.106a}$$

$$[\hat{\jmath}(t,x),\hat{\jmath}(t,y)] = 0, \tag{7.106b}$$

$$[\hat{\rho}(t,x),\hat{\jmath}(t,y)] = -\frac{\mathrm{i}}{\pi}\partial_x\delta(x-y) \tag{7.106c}$$

obeyed by the density and current density[1]

$$\hat{\rho} = +\frac{1}{2\pi}(\partial_x\hat{u}_- - \partial_x\hat{u}_+) \equiv \hat{\jmath}_- + \hat{\jmath}_+, \tag{7.106d}$$

$$\hat{\jmath} = +\frac{1}{2\pi}(\partial_x\hat{u}_- + \partial_x\hat{u}_+) \equiv \hat{\jmath}_- - \hat{\jmath}_+, \tag{7.106e}$$

respectively, and the identification of the pair of vertex operators (recall (7.87b))

$$\hat{\psi}^\dagger_- := \sqrt{\frac{1}{4\pi\mathrm{a}}}\,\mathrm{e}^{-\mathrm{i}\hat{u}_-}, \qquad \hat{\psi}^\dagger_+ := \sqrt{\frac{1}{4\pi\mathrm{a}}}\,\mathrm{e}^{+\mathrm{i}\hat{u}_+} \tag{7.107}$$

with a pair of creation operators for fermions. The multiplicative prefactor $1/\sqrt{4\pi}$ is a matter of convention and the constant a carries the dimension of length, i.e. the fermion fields carry the dimension of $1/\sqrt{\text{length}}$. By construction, the chiral currents

$$\hat{\jmath}_- := +\frac{1}{2\pi}\partial_x\hat{u}_-, \qquad \hat{\jmath}_+ := -\frac{1}{2\pi}\partial_x\hat{u}_+ \tag{7.108}$$

obey the chiral equations of motion

$$\partial_t\hat{\jmath}_- = -\partial_x\hat{\jmath}_-, \qquad \partial_t\hat{\jmath}_+ = +\partial_x\hat{\jmath}_+, \tag{7.109}$$

i.e. they depend solely on $(t-x)$ and $(t+x)$, respectively. As with the chiral fields $\hat{u}_-$ and $\hat{u}_+$, the chiral currents $\hat{\jmath}_-$ and $\hat{\jmath}_+$ are right-moving and left-moving solutions, respectively, of the Klein–Gordon equation

$$(\partial_t^2 - \partial_x^2)f(t,x) = (\partial_t - \partial_x)(\partial_t + \partial_x)f(t,x) = 0. \tag{7.110}$$

Step 2 Let us define the free Dirac Hamiltonian

$$\widehat{H}_{\mathrm{D}} := -\int_{\mathbb{R}} \mathrm{d}x\,\hat{\psi}^\dagger\gamma^0\gamma^1\mathrm{i}\partial_x\hat{\psi} \equiv \int_{\mathbb{R}} \mathrm{d}x\,\hat{\bar{\psi}}\gamma^1\mathrm{i}\partial_x\hat{\psi}, \tag{7.111a}$$

where

$$\{\psi_\alpha(t,x),\hat{\psi}^\dagger_\beta(t,y)\} = \delta_{\alpha\beta}\delta(x-y) \tag{7.111b}$$

[1] Notice that the chiral equations of motion imply that $\hat{\jmath} = -\frac{1}{2\pi}(\partial_t\hat{u}_- - \partial_t\hat{u}_+)$.

delivers the only non-vanishing equal-time anticommutators. If we define the chiral projections ($\gamma_5 \equiv -\gamma^5 \equiv -\gamma^0\gamma^1$)

$$\hat{\psi}^{\dagger}_{\mp} := \hat{\psi}^{\dagger}\frac{1}{2}(1 \mp \gamma_5), \qquad \hat{\psi}_{\mp} := \frac{1}{2}(1 \mp \gamma_5)\hat{\psi}, \tag{7.112a}$$

we then have the chiral equations of motion

$$\partial_t \hat{\psi}_{-} = -\partial_x \hat{\psi}_{-}, \qquad \partial_t \hat{\psi}_{+} = +\partial_x \hat{\psi}_{+}. \tag{7.112b}$$

The annihilation operator $\hat{\psi}_{-}$ removes a right-moving fermion and the annihilation operator $\hat{\psi}_{+}$ a left-moving fermion. Moreover, for the Lagrangian density

$$\widehat{\mathcal{L}}_{\mathrm{D}} := \hat{\psi}^{\dagger}\gamma^0 \mathrm{i}\gamma^{\mu}\partial_{\mu}\hat{\psi}, \tag{7.113}$$

there is an additive decomposition

$$\widehat{\mathcal{L}}_{\mathrm{D}} = \hat{\psi}^{\dagger}_{-}\mathrm{i}(\partial_0 + \partial_1)\hat{\psi}_{-} + \hat{\psi}^{\dagger}_{+}\mathrm{i}(\partial_0 - \partial_1)\hat{\psi}_{+}, \tag{7.114}$$

with two independent chiral currents

$$\hat{\jmath}_{\mathrm{D}-} := 2\hat{\psi}^{\dagger}_{-}\hat{\psi}_{-}, \qquad \hat{\jmath}_{\mathrm{D}+} := 2\hat{\psi}^{\dagger}_{+}\hat{\psi}_{+} \tag{7.115a}$$

obeying the independent conservation laws

$$\partial_t \hat{\jmath}_{\mathrm{D}-} = -\partial_x \hat{\jmath}_{\mathrm{D}-}, \qquad \partial_t \hat{\jmath}_{\mathrm{D}+} = +\partial_x \hat{\jmath}_{\mathrm{D}+}. \tag{7.115b}$$

Finally, it can be shown that if the chiral currents are normal-ordered with respect to the filled Fermi sea with a vanishing chemical potential, then the only non-vanishing equal-time commutators are

$$[\hat{\jmath}_{\mathrm{D}-}(t,x), \hat{\jmath}_{\mathrm{D}-}(t,y)] = -\frac{\mathrm{i}}{2\pi}\partial_x \delta(x-y), \tag{7.116a}$$

$$[\hat{\jmath}_{\mathrm{D}+}(t,x), \hat{\jmath}_{\mathrm{D}+}(t,y)] = +\frac{\mathrm{i}}{2\pi}\partial_x \delta(x-y). \tag{7.116b}$$

Step 3 The Dirac chiral current algebra (7.116) is equivalent to the bosonic chiral current algebra (7.106). This equivalence can be interpreted as (i) the bosonic theory (7.104) being equivalent to the Dirac theory (7.111) and (ii) the existence of a one-to-one correspondence between the following operators acting on their respective Fock spaces. To establish this one-to-one correspondence, we introduce the pair of bosonic fields

$$\hat{\phi}(x^0, x^1) := \hat{u}_{-}(x^0 - x^1) + \hat{u}_{+}(x^0 + x^1), \tag{7.117a}$$

$$\hat{\theta}(x^0, x^1) := \hat{u}_{-}(x^0 - x^1) - \hat{u}_{+}(x^0 + x^1). \tag{7.117b}$$

Now, the relevant one-to-one correspondence between operators in the Dirac theory for fermions and operators in the chiral bosonic theory is given in Table 7.4.

Table 7.4 Abelian bosonization rules in two-dimensional Minkowski space. The conventions with regard to the scalar mass $\hat{\bar{\psi}}\hat{\psi}$ and the pseudoscalar mass $\hat{\bar{\psi}}\gamma_5\hat{\psi}$ are $\hat{\bar{\psi}} = \hat{\psi}^\dagger\gamma^0$ with $\hat{\psi}^\dagger = (\hat{\psi}^\dagger_-, \hat{\psi}^\dagger_+)$, whereby $\gamma^0 = \tau_1$ and $\gamma^1 = \mathrm{i}\tau_2$ so that $\gamma^5 = -\gamma_5 = +\gamma^0\gamma^1 = -\tau_3$.

	FERMIONS	BOSONS
Kinetic energy	$\hat{\bar{\psi}}\mathrm{i}\gamma^\mu\partial_\mu\hat{\psi}$	$\frac{1}{8\pi}(\partial^\mu\hat{\phi})(\partial_\mu\hat{\phi})$
Current	$\hat{\bar{\psi}}\gamma^\mu\hat{\psi}$	$\frac{1}{2\pi}\epsilon^{\mu\nu}\partial_\nu\hat{\phi}$
Chiral currents	$2\hat{\psi}^\dagger_\mp\hat{\psi}_\mp$	$\pm\frac{1}{2\pi}\partial_x\hat{u}_\mp$
Right- and left-movers	$\hat{\psi}^\dagger_\mp$	$\sqrt{\frac{1}{4\pi\mathrm{a}}}\,\mathrm{e}^{\mp\mathrm{i}\hat{u}_\mp}$
Backward scattering	$\hat{\psi}^\dagger_-\hat{\psi}_+$	$\frac{1}{4\pi\mathrm{a}}\mathrm{e}^{-\mathrm{i}\hat{\phi}}$
Cooper pairing	$\hat{\psi}^\dagger_-\hat{\psi}^\dagger_+$	$\frac{1}{4\pi\mathrm{a}}\mathrm{e}^{-\mathrm{i}\hat{\theta}}$
Scalar mass	$\hat{\psi}^\dagger_-\hat{\psi}_+ + \hat{\psi}^\dagger_+\hat{\psi}_-$	$\frac{1}{2\pi\mathrm{a}}\cos\hat{\phi}$
Pseudoscalar mass	$\hat{\psi}^\dagger_-\hat{\psi}_+ - \hat{\psi}^\dagger_+\hat{\psi}_-$	$\frac{-\mathrm{i}}{2\pi\mathrm{a}}\sin\hat{\phi}$

7.3.8 From the Hamiltonian to the Lagrangian formalism

What is the Minkowski path integral that is equivalent to the quantum theory defined by (7.54)? In other words, we seek the path integrals

$$Z^{(\pm)} := \int \mathcal{D}[u]\,\mathrm{e}^{\mathrm{i}S^{(\pm)}[u]} \tag{7.118a}$$

with the Minkowski actions

$$S^{(\pm)}[u] := \int_{-\infty}^{+\infty} \mathrm{d}t\, L^{(\pm)}[u] \equiv \int_{-\infty}^{+\infty} \mathrm{d}t \int_0^L \mathrm{d}x\, \mathcal{L}^{(\pm)}[u](t,x) \tag{7.118b}$$

such that one of the two Hamiltonians

$$H^{(\pm)} := \int_0^L \mathrm{d}x\,\{\Pi^{(\pm)}_i(\partial_t u_i) - \mathcal{L}^{(\pm)}[u]\} \tag{7.119}$$

can be identified with $\widehat{H}$ in (7.54a) after elevating the classical fields

$$u_i(t,x) \qquad \text{and} \qquad \Pi^{(\pm)}_i(t,x) := \frac{\delta\mathcal{L}^{(\pm)}}{\delta(\partial_t u_i)(t,x)} \tag{7.120a}$$

entering $\mathcal{L}^{(\pm)}[u]$ to the status of quantum fields $\hat{u}_i(t,x)$ and $\hat{\Pi}_j^{(\pm)}(t,y)$ upon imposing the equal-time commutation relations

$$[\hat{u}_i(t,x),\hat{\Pi}_j^{(\pm)}(t,y)] = \pm\tfrac{1}{2}\mathrm{i}\delta_{ij}\delta(x-y) \tag{7.120b}$$

for any $i,j=1,\ldots,N$. The unusual factor $\pm\frac{1}{2}$ (instead of 1) on the right-hand side of the commutator between pairs of canonically conjugate fields arises because each scalar field u_i with $i=1,\ldots,N$ is chiral, i.e. it represents 'one-half' of a canonical scalar field.

Without loss of generality, we set $A_0=A_1=0$ in (7.54a). We try

$$\mathcal{L}^{(\pm)} := \frac{1}{4\pi}[\mp(\partial_x u_i)K_{ij}^{-1}(\partial_t u_j) - (\partial_x u_i)V_{ij}(\partial_x u_j)] \tag{7.121a}$$

with the chiral equations of motion

$$\begin{aligned}
0 &= \partial_\mu \frac{\delta\mathcal{L}^{(\pm)}}{\delta\partial_\mu u_i} - \frac{\delta\mathcal{L}^{(\pm)}}{\delta u_i}\\
&= \partial_t \frac{\delta\mathcal{L}^{(\pm)}}{\delta\partial_t u_i} + \partial_x \frac{\delta\mathcal{L}^{(\pm)}}{\delta\partial_x u_i} - \frac{\delta\mathcal{L}^{(\pm)}}{\delta u_i}\\
&= \frac{1}{4\pi}(\mp K_{ji}^{-1}\partial_t\partial_x \mp K_{ij}^{-1}\partial_x\partial_t - 2V_{ij}\partial_x\partial_x)u_j\\
&= \mp\frac{K_{ij}^{-1}}{2\pi}\partial_x(\delta_{jl}\partial_t \pm K_{jk}V_{kl}\partial_x)u_l
\end{aligned} \tag{7.121b}$$

for any $i=1,\ldots,N$. Observe that the term that mixes time t and space x derivatives becomes imaginary only in Euclidean time $\tau=\mathrm{i}t$.

Proof The canonical momentum for the field u_i is

$$\begin{aligned}
\Pi_i^{(\pm)}(t,x) &:= \frac{\delta\mathcal{L}^{(\pm)}}{\delta(\partial_t u_i)(t,x)}\\
&= \mp\frac{1}{4\pi}K_{ij}^{-1}(\partial_x u_j)(t,x)
\end{aligned} \tag{7.122}$$

for any $i=1,\ldots,N$ owing to the symmetry of the matrix K. Evidently, the Legendre transform

$$\mathcal{H}^{(\pm)} := \Pi_i^{(\pm)}(\partial_t u_i) - \mathcal{L}^{(\pm)} \tag{7.123}$$

gives

$$\mathcal{H}^{(\pm)} = \frac{1}{4\pi}(\partial_x u_i)V_{ij}(\partial_x u_j). \tag{7.124}$$

The right-hand side does not depend on the chiral index $\pm$. We now quantize the theory by elevating the classical fields u_i to the status of operators $\hat{u}_i$ obeying the algebra (7.54b). This gives a quantum theory that meets all the demands of the quantum chiral edge theory (7.54) and is compatible with the canonical quantization rules (7.120b), since

$$\begin{aligned}
[\hat{u}_i(t,x), \hat{\Pi}^{(\pm)}_j(t,y)] &= \mp\frac{1}{4\pi}K^{-1}_{jk}\partial_y\,[\hat{u}_i(t,x), \hat{u}_k(t,y)] \\
&= \mp\frac{1}{4\pi}K^{-1}_{jk}(\pi\mathrm{i})K_{ik}(-2)\delta(x-y) && \text{[from (7.54}b\text{)]} \\
&= \pm\frac{1}{2}\mathrm{i}K^{-1}_{jk}K_{ki}\delta(x-y) && \text{[since } K_{ik}=K_{ki}\text{]} \\
&= \pm\frac{1}{2}\mathrm{i}\delta_{ij}\delta(x-y), && (7.125)
\end{aligned}$$

where $i,j=1,\ldots,N$. ■

Finally, analytical continuation to Euclidean time

$$\tau = \mathrm{i}t \tag{7.126a}$$

allows us to define the finite-temperature quantum chiral theory through the path integral

$$Z^{(\pm)}_\beta := \int\mathcal{D}[u]\exp\left(-\int_0^\beta \mathrm{d}\tau\int_0^L \mathrm{d}x\,\mathcal{L}^{(\pm)}\right), \tag{7.126b}$$

$$\begin{aligned}
\mathcal{L}^{(\pm)} &:= \frac{1}{4\pi}[(\pm)\mathrm{i}(\partial_x u_i)K^{-1}_{ij}(\partial_\tau u_j) + (\partial_x u_i)V_{ij}(\partial_x u_j)] \\
&\quad + J\left(\frac{q_i}{2\pi}K^{-1}_{ij}(\partial_x u_j)\right),
\end{aligned} \tag{7.126c}$$

in the presence of an external source field J that couples to the charges q_i like a scalar potential would do.

7.3.9 Applications to polyacetylene

Consider the Dirac Hamiltonian

$$\widehat{H}_{\mathrm{D}} := \widehat{H}_{\mathrm{D}0} + \widehat{H}_{\mathrm{D}1}, \tag{7.127a}$$

where the free-field and massless contribution is

$$\widehat{H}_{\mathrm{D}0} := \int_{\mathbb{R}} \mathrm{d}x(\hat{\psi}^\dagger_+\mathrm{i}\partial_x\hat{\psi}_+ - \hat{\psi}^\dagger_-\mathrm{i}\partial_x\hat{\psi}_-), \tag{7.127b}$$

while

$$\widehat{H}_{\mathrm{D}1} := \int_{\mathbb{R}} \mathrm{d}x\,[\phi_1(\hat{\psi}^\dagger_-\hat{\psi}_+ + \hat{\psi}^\dagger_+\hat{\psi}_-) + \mathrm{i}\phi_2(\hat{\psi}^\dagger_-\hat{\psi}_+ - \hat{\psi}^\dagger_+\hat{\psi}_-)] \tag{7.127c}$$

couples the Dirac field to two real-valued and classical scalar fields ϕ_1 and ϕ_2. The only non-vanishing equal-time anticommutators are given by (7.111b). This Hamiltonian was considered by Goldstone and Wilczek [36] in their study of charge fractionalization for polyacetylene.

According to the bosonization rules from Table 7.4 and with the help of the polar decomposition

$$\phi_1(t,x) = |\phi(t,x)| \cos\varphi(t,x), \quad \phi_2(t,x) = |\phi(t,x)| \sin\varphi(t,x), \tag{7.128}$$

the many-body bosonic Hamiltonian that is equivalent to the Dirac Hamiltonian (7.127) is

$$\widehat{H}_{\mathrm{B}} := \widehat{H}_{\mathrm{B}0} + \widehat{H}_{\mathrm{B}1}, \tag{7.129a}$$

where

$$\widehat{H}_{\mathrm{B}0} := \int_{\mathbb{R}} \mathrm{d}x \, \frac{1}{8\pi} [\widehat{\Pi}^2 + (\partial_x \hat{\phi})^2], \tag{7.129b}$$

while

$$\widehat{H}_{\mathrm{B}1} := \int_{\mathbb{R}} \mathrm{d}x \, \frac{1}{2\pi \mathrm{a}} |\phi| \cos(\hat{\phi} - \varphi). \tag{7.129c}$$

Here, the canonical momentum

$$\widehat{\Pi}(t,x) := (\partial_t \hat{\phi})(t,x) \tag{7.130a}$$

shares with $\hat{\phi}(t,x)$ the only non-vanishing equal-time commutator

$$[\hat{\phi}(t,x), \widehat{\Pi}(t,y)] = \mathrm{i}\delta(x-y). \tag{7.130b}$$

The Hamiltonian (7.129) is interacting, and its interaction (7.129c) can be traced to the mass contributions in the non-interacting Dirac Hamiltonian (7.127). The interaction (7.129c) is minimized when the operator identity

$$\hat{\phi}(t,x) = \varphi(t,x) + \pi \tag{7.131}$$

holds. This identity can only be met in the limit

$$|\phi(t,x)| \to \infty \tag{7.132}$$

for all times t and positions x in view of the algebra (7.130) and the competition between the contributions (7.129b) and (7.129c).

Close to the limit (7.132), the bosonization formula for the conserved current

$$\hat{\bar{\psi}} \gamma^\mu \hat{\psi} \to \frac{1}{2\pi} \epsilon^{\mu\nu} \partial_\nu \hat{\phi} \tag{7.133}$$

simplifies to

$$\frac{1}{2\pi} \epsilon^{\mu\nu} \partial_\nu \hat{\phi} \approx \frac{1}{2\pi} \epsilon^{\mu\nu} \partial_\nu \varphi. \tag{7.134}$$

On the one hand, the conserved charge

$$\widehat{Q} := \int_{\mathbb{R}} \mathrm{d}x\, (\hat{\bar{\psi}}\gamma^0\hat{\psi})(t,x) \to \frac{\epsilon^{01}}{2\pi}[\hat{\phi}(t,x=+\infty) - \hat{\phi}(t,x=-\infty)] \tag{7.135}$$

for the static profile $\varphi(x)$ is given approximately by

$$\widehat{Q} \approx \frac{\epsilon^{01}}{2\pi}\left[\varphi(x=+\infty) - \varphi(x=-\infty)\right]. \tag{7.136}$$

On the other hand, the number of electrons per period $T = 2\pi/\omega$ that flow across a point x,

$$\hat{I} := \int_0^T \mathrm{d}t\, (\hat{\bar{\psi}}\gamma^1\hat{\psi})(t,x) \to \frac{\epsilon^{10}}{2\pi}[\hat{\phi}(T,x) - \hat{\phi}(0,x)], \tag{7.137}$$

for the uniform profile $\varphi(t) = \omega t$, is given approximately by

$$\hat{I} \approx \frac{\epsilon^{10}}{2\pi}\omega T = \epsilon^{10}. \tag{7.138}$$

The results (7.136) and (7.138) are sharp operator identities in the limit (7.132). The small parameter in both expansions is $1/m$ where $m := \lim_{x\to\infty} |\boldsymbol{\phi}(t,x)|$.

7.4 Stability analysis for the edge theory in symmetry class AII

7.4.1 Introduction

The hallmark of the IQHE in an open geometry is the localized nature of all two-dimensional (bulk) states while an integer number of chiral edge states freely propagate along the one-dimensional boundaries [47, 63, 68]. These chiral edge states are immune to the physics of Anderson localization as long as backward scattering between edge states of opposite chiralities is negligible [47, 68].

Many-body interactions among electrons can be treated perturbatively in the IQHE provided that the characteristic many-body energy scale is less than the single-particle gap between Landau levels. This is no longer true if the chemical potential lies within a Landau level, since the non-interacting many-body ground state is then macroscopically degenerate. The lifting of this extensive degeneracy by the many-body interactions is a non-perturbative effect. At some 'magic' filling fractions that deliver the FQHE [43, 69, 116, 122], a screened Coulomb interaction selects a finitely degenerate family of ground states, each of which describes a featureless liquid separated from excitations by an energy gap in a closed geometry. Such a ground state is called an incompressible fractional Hall liquid. The FQHE is an example of topological order [128, 129, 133]. In an open geometry, there are branches of excitations that disperse across the spectral gap of the two-dimensional bulk, but these excitations are localized along the direction normal to the boundary, while they propagate freely along the boundary [130–133]. In contrast to the IQHE, these excitations need not all share the same chirality. However, they are nevertheless immune to the physics of Anderson localization, provided that scattering induced by the disorder between distinct edges in an open geometry is negligible.

The IQHE is the archetype of a two-dimensional topological band insulator. The two-dimensional $\mathbb{Z}_2$ topological band insulator is a close relative of the IQHE that occurs in semiconductors with sufficiently large spin–orbit coupling but no breaking of time-reversal symmetry [8, 9, 54, 55, 64]. As with the IQHE, the smoking gun for the $\mathbb{Z}_2$ topological band insulator is the existence of gapless Kramers-degenerate pairs of edge states that are delocalized along the boundaries of an open geometry as long as disorder-induced scattering between distinct boundaries is negligible. In contrast to the IQHE, it is the odd parity in the number of Kramers pairs of edge states that is robust to the physics of Anderson localization.

A simple example of a two-dimensional $\mathbb{Z}_2$ topological band insulator can be obtained by putting together two copies of an IQHE system with opposite chiralities for up and down spins. For instance, one could take two copies of Haldane's model [45] each of which realizes an IQHE on the honeycomb lattice but with Hall conductance differing in sign. In this case, the spin current is conserved, a consequence of the independent conservation of the up and down currents, and the spin Hall conductance inherits its quantization from the IQHE of each spin species. This example thus realizes an integer quantum spin Hall effect (IQSHE). However, although simple, this example is not generic. The $\mathbb{Z}_2$ topological band insulator does not necessarily have conserved spin currents, let alone quantized responses.

Along the same line of reasoning, two copies of an FQHE system put together, again with opposite chiralities for up and down particles, would realize a fractional quantum spin Hall effect (FQSHE), as proposed by Bernevig and Zhang [9] (see also [32] and [48]). Levin and Stern [71] proposed to characterize two-dimensional fractional topological liquids supporting the FQSHE by the criterion that their edge states are stable against disorder provided that they do not break time-reversal symmetry spontaneously.

In the discussion here, the condition that projection about some quantization axis of the electron spin from the underlying microscopic model is a good quantum number will *not* be imposed. Only time-reversal symmetry is assumed to hold. The generic cases of fractional topological liquids with time-reversal symmetry from the special cases of fractional topological liquids with time-reversal symmetry *and* with residual spin-$\frac{1}{2}$ $U(1)$ rotation symmetry will thus be distinguished. In the former cases, the electronic spin is not a good quantum number. In the latter cases, conservation of spin allows the FQSHE.

The subclass of incompressible time-reversal-symmetric liquids that we construct here is closely related to Abelian Chern–Simons theories. Other possibilities, which will not be discussed here, may include non-Abelian Chern–Simons theories [33, 82] or theories that include, additionally, conventional local order parameters (Higgs fields) [103].

The relevant effective action for the Abelian Chern–Simons theory is of the form [128, 130, 131, 133]

$$S := S_0 + S_e + S_s, \tag{7.139a}$$

where

$$S_0 := -\int \mathrm{d}t\, \mathrm{d}^2\boldsymbol{x}\, \epsilon^{\mu\nu\rho} \frac{1}{4\pi} K_{ij} a^i_\mu \partial_\nu a^j_\rho, \tag{7.139b}$$

$$S_e := \int \mathrm{d}t\, \mathrm{d}^2\boldsymbol{x}\, \epsilon^{\mu\nu\rho} \frac{e}{2\pi} Q_i A_\mu \partial_\nu a^i_\rho, \tag{7.139c}$$

$$S_s := \int \mathrm{d}t\, \mathrm{d}^2\boldsymbol{x}\, \epsilon^{\mu\nu\rho} \frac{s}{2\pi} S_i B_\mu \partial_\nu a^i_\rho. \tag{7.139d}$$

The indices i and j run from 1 to $2N$ and any pair thereof labels an integer-valued matrix element K_{ij} of the symmetric and invertible $2N \times 2N$ matrix K. The indices μ, ν, and ρ run from 0 to 2. They label the component x_μ of the coordinates $(t, \boldsymbol{x})$ in $(2+1)$-dimensional spacetime or the component $A_\mu(t, \boldsymbol{x})$ of an external electromagnetic gauge potential, or the component $B_\mu(t, \boldsymbol{x})$ of an external gauge potential that couples to the spin-$\frac{1}{2}$ degrees of freedom along some quantization axis, or the components of $2N$ flavours of dynamical Chern–Simons fields $a^i_\mu(t, \boldsymbol{x})$. The integer-valued component Q_i of the $2N$-dimensional vector Q represents the ith electric charge in units of the electronic charge e and obeys the compatibility condition

$$(-1)^{Q_i} = (-1)^{K_{ii}} \tag{7.139e}$$

for any $i = 1, \ldots, 2N$ in order for bulk quasiparticles or, in an open geometry, quasiparticles on edges to obey a consistent statistics. The integer-valued component S_i of the $2N$-dimensional vector S represents the ith spin charge in units of the spin charge s along some conserved quantization axis. The operation of time reversal is the map

$$A_\mu(t, \boldsymbol{x}) \mapsto +g^{\mu\nu} A_\nu(-t, \boldsymbol{x}), \tag{7.140a}$$

$$B_\mu(t, \boldsymbol{x}) \mapsto -g^{\mu\nu} b_\nu(-t, \boldsymbol{x}), \tag{7.140b}$$

$$a^i_\mu(t, \boldsymbol{x}) \mapsto -g^{\mu\nu} a^{i+N}_\nu(-t, \boldsymbol{x}) \tag{7.140c}$$

for $i = 1, \ldots, N$. Here, $g_{\mu\nu} = \mathrm{diag}(+1, -1, -1)$ is the Lorentz metric in $(2+1)$-dimensional spacetime. It will be shown that time-reversal symmetry imposes that the matrix K be of the block form

$$K = \begin{pmatrix} \kappa & \Delta \\ \Delta^{\mathsf{T}} & -\kappa \end{pmatrix}, \tag{7.141a}$$

$$\kappa^{\mathsf{T}} = \kappa, \qquad \Delta^{\mathsf{T}} = -\Delta, \tag{7.141b}$$

where κ and Δ are $N \times N$ matrices, while the integer-charge vectors Q and S are of the block forms

$$Q = \begin{pmatrix} \varrho \\ \varrho \end{pmatrix}, \qquad S = \begin{pmatrix} \varrho \\ -\varrho \end{pmatrix}. \tag{7.141c}$$

The matrix K together with the charge vector Q and spin vector S that characterize the topological field theory with the action (7.139a) define the charge filling fraction, a rational number,

$$\nu_e := Q^{\mathsf{T}} K^{-1} Q, \tag{7.142a}$$

and the spin filling fraction, another rational number,

$$\nu_s := \tfrac{1}{2} Q^{\mathsf{T}} K^{-1} S. \tag{7.142b}$$

The block forms of K and Q in (7.141) imply that

$$\nu_e = 0. \tag{7.142c}$$

The 'zero charge filling fraction' (7.142c) states nothing but the fact that there is no charge Hall conductance when time-reversal symmetry holds. On the other hand, time-reversal symmetry of the action (7.139a) is compatible with a non-vanishing FQSHE as measured by the non-vanishing quantized spin Hall conductance

$$\sigma_{\mathrm{sH}} := \frac{e}{2\pi} \times \nu_s. \tag{7.142d}$$

The origin of the FQSHE in the action (7.139a) is the $U(1) \times U(1)$ gauge symmetry when $(2+1)$-dimensional spacetime has the same topology as a manifold without boundary. It is always assumed in these lectures that the $U(1)$ symmetry associated with charge conservation holds. However, we shall not make the same assumption regarding the $U(1)$ symmetry responsible for the conservation of the 'spin' quantum number.

The special cases of the FQSHE treated in [9] and [71] correspond to imposing the condition

$$\Delta = 0 \tag{7.143}$$

on the matrix K in (7.141a). This restriction is, however, not necessary to treat either the FQSHE or the generic case when there is no residual spin-$\frac{1}{2}$ $U(1)$ symmetry in the underlying microscopic model.

The effective topological field theory (7.139) with the condition for time-reversal symmetry (7.141) is made of $2N$ Abelian Chern–Simons fields. As is the case with the FQHE, when two-dimensional space is a manifold without boundary of genus 1, i.e. when two-dimensional space is topologically equivalent to a torus, this theory is characterized by distinct topological sectors [128, 129, 133]. All topological sectors are in one-to-one correspondence with a finite number $\mathcal{N}_{\mathrm{GS}}$ of topologically degenerate ground states of the underlying microscopic theory [128, 129, 133]. This degeneracy is nothing but the magnitude of the determinant of K in (7.139a), which is, because of the block structure (7.141a), given in turn by

$$\begin{aligned}
\mathcal{N}_{\mathrm{GS}} &= \left|\det\begin{pmatrix}\kappa & \Delta\\ \Delta^{\mathsf{T}} & -\kappa\end{pmatrix}\right| \\
&= \left|\det\begin{pmatrix}\Delta^{\mathsf{T}} & -\kappa\\ \kappa & \Delta\end{pmatrix}\right| \\
&= \left|\mathrm{Pf}\begin{pmatrix}\Delta^{\mathsf{T}} & -\kappa\\ \kappa & \Delta\end{pmatrix}\right|^2 \\
&= (\text{integer})^2.
\end{aligned} \tag{7.144}$$

To reach the last line, the fact that the matrix K is integer-valued was used. It is thus predicted that the class of two-dimensional time-reversal-symmetric fractional topological liquids, whose universal properties are captured by (7.139) and (7.141), are characterized by a topological ground-state degeneracy that is always the square of an integer, even if $\Delta \neq 0$, when space is topologically equivalent to a torus. (Notice that the condition that Δ is antisymmetric implies that non-vanishing Δ can occur only for $N > 1$.)

The stability of the edge states associated with the bulk Chern–Simons action (7.139) obeying the condition for time-reversal symmetry (7.141) will now be discussed in detail. A single one-dimensional edge is considered and an interacting quantum field theory for $1 \leq N_{\mathrm{K}} \leq N$ pairs of Kramers-degenerate electrons subject to strong disorder that preserves time-reversal symmetry is constructed, where the integer $2N_{\mathrm{K}}$ is the number of odd charges entering the charge vector Q.[2] The conditions under which at least one Kramers-degenerate pair of electrons remains gapless in spite of the interactions and disorder are identified. The approach here is inspired by the stability analysis of the edge states performed by Haldane [46] for the single-layer FQHE (see also [53, 84]), by that of Naud et al. [87, 88] for the bilayer FQHE, and especially by that of Levin and Stern [71] for the FQSHE and that of Neupert et al. [91]. As for the FQSHE, our analysis departs from the analysis by Haldane in that we impose time-reversal symmetry. We also depart here from [71] by considering explicitly the effects of the off-diagonal elements Δ in the matrix K. Such terms are generically present for any realistic underlying microscopic model independently of whether or not this model supports the FQSHE. When considering the stability of the edge theory, we allow the residual spin-$\frac{1}{2}$ $U(1)$ symmetry responsible for the FQSHE to be broken by interactions among the edge modes or by a disorder potential. Hence, we seek a criterion for the stability of the edge theory that does not rely on the existence of a quantized spin Hall conductance in the bulk as was done in [71].

The stability of the edge states against disorder hinges on whether the integer

$$R := r\varrho^{\mathsf{T}}(\kappa - \Delta)^{-1}\varrho \tag{7.145}$$

[2] More precisely, to guarantee that there are N_{K} Kramers-degenerate pairs of electrons in the theory, we demand that there exists a space- and time-independent transformation $O = +\Sigma_1 O \Sigma_1 \in SL(2N, \mathbb{Z})$ such that $K \mapsto O^{\mathsf{T}} K O$, $V \mapsto O^{\mathsf{T}} V O$, and $Q \mapsto O^{\mathsf{T}} Q$, with the transformed charge vector containing $2N_{\mathrm{K}}$ odd integers.

is odd (stable) or even (unstable). The vector ϱ together with the matrices κ and Δ were defined in (7.141). The integer r is the smallest integer such that all N components of the vector $r(\kappa - \Delta)^{-1}\varrho$ are integers. We can quickly check a few simple examples. First, observe that, in the limit $\Delta = 0$, we recover the criterion derived in [71]. Second, when we impose a residual spin-$\frac{1}{2}$ $U(1)$ symmetry by appropriately restricting the interactions between edge channels, $\nu_{\uparrow} = -\nu_{\downarrow} = \varrho^{\mathsf{T}}(\kappa - \Delta)^{-1}\varrho$ can be interpreted as the Hall conductivity σ_{xy} in units of e^2/h for each of the separately conserved spin components along the spin quantization axis. The integer r has the interpretation of the number of fluxes needed to pump a unit of charge, or the inverse of the 'minimum charge' of [71]. Further restricting to the case where $\kappa = \mathbf{1}_N$ (where $\mathbf{1}_N$ is the $N \times N$ unit matrix) gives $R = N$, i.e. we have recovered the same criterion as for the two-dimensional non-interacting $\mathbb{Z}_2$ topological band insulator.

When there is no residual spin-$\frac{1}{2}$ $U(1)$ symmetry, one can no longer relate the index R to a physical spin Hall conductance. Nevertheless, the index R defined in (7.145) discriminates in all cases between whether there is or is not a remaining branch of gapless modes dispersing along the edge.

7.4.2 Definitions

Let us consider an interacting model for electrons in a two-dimensional cylindrical geometry as depicted in Fig. 7.10. We demand that (i) charge conservation and time-reversal symmetry are the only intrinsic symmetries of the microscopic quantum Hamiltonian, (ii) neither is broken spontaneously by the many-body ground state, and (iii) if periodic boundary conditions are assumed along the y coordinate in Fig. 7.10, then there is at most a finite number of degenerate many-body ground states and each many-body ground state is separated from its tower of many-body excited states by an energy gap. Had the condition of time-reversal symmetry been relaxed, the remaining assumptions would be realized for the FQHE.

In the open geometry of Fig. 7.10, the only possible excitations with an energy smaller than the bulk gap in the closed geometry of a torus must be localized along the y coordinate in the vicinities of the edges at $\pm\frac{1}{2}L_y$. If L_y is much larger than the characteristic linear extension into the bulk of edge states, then the two edges decouple from each other. It is then meaningful to define a low-energy and long-wavelength quantum field theory for the edge states propagating along either of the two boundaries in Fig. 7.10, which we take to be of length L each.

The low-energy and long-wavelength effective quantum field theory for the edge that we are going to construct is inspired by the construction by Wen of the chiral Luttinger edge theory for the FQHE [128, 130, 131]. As for the FQHE, this time-reversal-symmetric boundary quantum field theory has a correspondence to the effective time-reversal symmetric bulk topological quantum field theory built out of $2N$ Abelian Chern–Simons fields.

The simplest class of quantum Hamiltonians that fulfils requirements (i)–(iii) can be represented in terms of $2N$ real-valued chiral scalar quantum fields $\widehat{\Phi}_i(t,x)$ with $i = 1,\ldots,2N$ that form the components of the quantum vector field $\widehat{\Phi}(t,x)$. After setting the electronic charge e, the characteristic speed, and $\hbar$ all equal to unity, the

Hamiltonian for the system is given by[3]

$$\widehat{H} := \widehat{H}_0 + \widehat{H}_{\text{int}}, \tag{7.146a}$$

where

$$\widehat{H}_0 := \int_0^L \mathrm{d}x \, \frac{1}{4\pi} (\partial_x \widehat{\Phi}^{\mathsf{T}})(t,x) V(x) (\partial_x \widehat{\Phi})(t,x), \tag{7.146b}$$

with $V(x)$ a $2N \times 2N$ symmetric and positive-definite matrix that accounts, in this bosonic representation, for the screened density–density interactions between electrons. The theory is quantized according to the equal-time commutators

$$[\widehat{\Phi}_i(t,x), \widehat{\Phi}_j(t,x')] = -\mathrm{i}\pi [K^{-1}_{ij} \operatorname{sgn}(x-x') + \Theta_{ij}], \tag{7.146c}$$

where K is a $2N \times 2N$ symmetric and invertible matrix with integer-valued matrix elements, and the matrix Θ accounts for Klein factors that ensure that charged excitations in the theory (vertex operators) satisfy the proper commutation relations. Fermionic or bosonic charged excitations are represented by the normal-ordered vertex operators

$$\widehat{\Psi}^{\dagger}_T(t,x) := \; : \mathrm{e}^{-\mathrm{i}T_i K_{ij} \widehat{\Phi}_j(t,x)} : , \tag{7.146d}$$

where the integer-valued $2N$-dimensional vector T determines the charge (and statistics) of the operator. The operator that measures the total charge density is

$$\hat{\rho}(t,x) = \frac{1}{2\pi} Q_i (\partial_x \widehat{\Phi}_i)(t,x), \tag{7.146e}$$

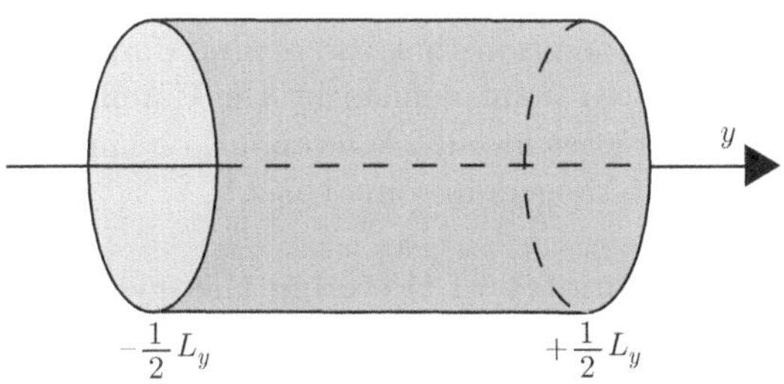

Fig. 7.10 Cylindrical geometry for a two-dimensional band insulator. The cylinder axis is labelled by the coordinate y. Periodic boundary conditions are imposed in the transverse direction labelled by the coordinate x. There is an edge at $y = -\frac{1}{2}L_y$ and another at $y = +\frac{1}{2}L_y$. Bulk states have support on the shaded surface of the cylinder. Edge states are confined in the y direction to the vicinity of the edges $y = \pm\frac{1}{2}L_y$. Topological band insulators have the property that, even in the presence of disorder, there are edge states freely propagating in the x direction with mean free path ℓ, provided that the limit $\ell/L_y \ll 1$ holds.

[3] Perform the linear transformation $\hat{u}_i \equiv K_{ij}\widehat{\Phi}_j$ in (7.54), recalling that the matrix K and its inverse K^{-1} are symmetric, so that we can write $\Theta_{ij} \equiv K^{-1}_{ii'} L_{i'j'} K^{-1}_{j'j}$.

where the integer-valued $2N$-dimensional charge vector Q, together with the matrix K, specify the universal properties of the edge theory. The charge q_T of the vertex operator in (7.146d) follows from its commutation with the charge density operator in (7.146e), yielding $q_T = T^{\mathsf{T}}Q$.

Tunnelling of electronic charge among the different edge branches is accounted for by

$$\widehat{H}_{\text{int}} := -\int_0^L \mathrm{d}x \sum_{T\in\mathbb{L}} h_T(x) : \cos[T^{\mathsf{T}}K\widehat{\Phi}(t,x) + \alpha_T(x)] : . \tag{7.146f}$$

The real functions $h_T(x) \geq 0$ and $0 \leq \alpha_T(x) \leq 2\pi$ encode information about the disorder along the edge when position-dependent. The set

$$\mathbb{L} := \left\{T \in \mathbb{Z}^{2N} \mid T^{\mathsf{T}}Q = 0\right\}, \tag{7.146g}$$

encodes all the possible charge-neutral tunnelling processes (i.e. those that just rearrange charge among the branches). This charge-neutrality condition implies that the operator $\hat{\Psi}^{\dagger}_{T}(t,x)$ that makes up (7.146f) is bosonic, since it has even charge. Observe that set $\mathbb{L}$ forms a lattice. Consequently, if T belongs to $\mathbb{L}$ so does $-T$. In turn, relabelling T to $-T$ in $\widehat{H}_{\text{int}}$ implies that $h_T(x) = +h_{-T}(x)$, whereas $\alpha_T(x) = -\alpha_{-T}(x)$.

The theory (7.146) inherently encodes interactions. The terms $\widehat{H}_0$ and $\widehat{H}_{\text{int}}$ encode single-particle *as well as* many-body interactions with matrix elements that preserve and break translation symmetry, respectively. Recovering the single-particle kinetic energy of N Kramers-degenerate pairs of electrons from (7.146b) corresponds to choosing the matrix V to be proportional to the $2N \times 2N$ unit matrix, with the proportionality constant being fixed by the condition that the scaling dimension of each electron is $\frac{1}{2}$ at the bosonic free-field fixed point defined by the Hamiltonian $\widehat{H}_0$. Of course, to implement the fermionic statistics for all $2N$ fermions, one must also demand that all diagonal entries of K be odd integers in some basis.[4]

7.4.3 Time-reversal symmetry of the edge theory

The operation of time reversal on the $\widehat{\Phi}$ fields is defined by

$$\mathcal{T}\widehat{\Phi}(t,x)\mathcal{T}^{-1} := \Sigma_1\widehat{\Phi}(-t,x) + \pi K^{-1}\Sigma_{\downarrow}Q, \tag{7.147a}$$

where

$$\Sigma_1 = \begin{pmatrix} 0 & 1 \\ 1 & 0 \end{pmatrix}, \qquad \Sigma_{\downarrow} = \begin{pmatrix} 0 & 0 \\ 0 & 1 \end{pmatrix}. \tag{7.147b}$$

This definition ensures that the fermionic and bosonic vertex operators defined in (7.146d) are properly transformed under time reversal. More precisely, one can then

[4] See footnote 2.

construct a pair of fermionic operators $\hat{\Psi}_1^\dagger$ and $\hat{\Psi}_2^\dagger$ of the form (7.146d) by suitably choosing a pair of vectors T_1 and T_2, respectively, in such a way that the time-reversal operation maps $\hat{\Psi}_1^\dagger$ into $+\hat{\Psi}_2^\dagger$ whereas it maps $\hat{\Psi}_2^\dagger$ into $-\hat{\Psi}_1^\dagger$. Thus, it is meaningful to interpret the block structure displayed in (7.147b) as arising from the upper or lower projection along some spin-$\frac{1}{2}$ quantization axis.

Time-reversal symmetry of the chiral edge theory (7.146) demands that

$$V = +\Sigma_1 V \Sigma_1, \tag{7.148a}$$

$$K = -\Sigma_1 K \Sigma_1, \tag{7.148b}$$

$$Q = \Sigma_1 Q, \tag{7.148c}$$

$$h_T(x) = h_{\Sigma_1 T}(x), \tag{7.148d}$$

$$\alpha_T(x) = -\alpha_{\Sigma_1 T}(x) + \pi T^{\mathsf{T}} \Sigma_{\downarrow} Q \mod 2\pi. \tag{7.148e}$$

Proof The first two conditions, (7.148a) and (7.148b), follow from the requirement that $\widehat{H}_0$ be time-reversal-invariant. In particular, the decomposition

$$K = \begin{pmatrix} \kappa & \Delta \\ \Delta^{\mathsf{T}} & -\kappa \end{pmatrix}, \qquad \kappa^{\mathsf{T}} = \kappa, \qquad \Delta^{\mathsf{T}} = -\Delta, \tag{7.149}$$

where κ and Δ are $N \times N$ matrices, follows from (7.148b) and $K = K^{\mathsf{T}}$.

The third condition, (7.148c), states that the charge density is invariant under time reversal. In particular, the decomposition

$$Q = \begin{pmatrix} \varrho \\ \varrho \end{pmatrix} \tag{7.150}$$

follows.

Finally, $\mathcal{T} \widehat{H}_{\mathrm{int}} \mathcal{T}^{-1} = \widehat{H}_{\mathrm{int}}$ requires that

$$\begin{aligned}
\sum_{T \in \mathbb{L}} & h_T(x) \cos[T^{\mathsf{T}} K \widehat{\Phi}(t,x) + \alpha_T(x)] \\
&= \sum_{T \in \mathbb{L}} \mathcal{T} \{ h_T(x) \cos[T^{\mathsf{T}} K \widehat{\Phi}(t,x) + \alpha_T(x)] \} \mathcal{T}^{-1} \\
&= \sum_{T \in \mathbb{L}} h_T(x) \cos[-(\Sigma_1 T)^{\mathsf{T}} K \widehat{\Phi}(-t,x) + \alpha_T(x) - \pi T^{\mathsf{T}} \Sigma_{\downarrow} Q] \\
&= \sum_{T \in \mathbb{L}} h_{\Sigma_1 T}(x) \cos[-T^{\mathsf{T}} K \widehat{\Phi}(-t,x) + \alpha_{\Sigma_1 T}(x) - \pi (\Sigma_1 T)^{\mathsf{T}} \Sigma_{\downarrow} Q] \\
&= \sum_{T \in \mathbb{L}} h_{\Sigma_1 T}(x) \cos[T^{\mathsf{T}} K \widehat{\Phi}(-t,x) - \alpha_{\Sigma_1 T}(x) + \pi (\Sigma_1 T)^{\mathsf{T}} \Sigma_{\downarrow} Q],
\end{aligned} \tag{7.151}$$

as the conditions needed to match the two trigonometric expansions. This leads to the last two relations, (7.148d) and (7.148e). ■

Disorder parametrized by $h_T(x) = +h_{-T}(x)$ and $\alpha_T(x) = -\alpha_{-T}(x)$ and for which the matrix T obeys

$$\Sigma_1 T = -T, \tag{7.152a}$$

and

$$T^{\mathsf{T}} \Sigma_{\downarrow} Q \text{ is an odd integer} \tag{7.152b}$$

cannot satisfy the condition (7.148e) for time-reversal symmetry. Such disorder is thus prohibited from entering $\widehat{H}_{\text{int}}$ in (7.146f), since it would break time-reversal symmetry explicitly. Moreover, we also prohibit any ground state that provides $\exp[\mathrm{i}T^{\mathsf{T}} K \widehat{\Phi}(t,x)]$ with an expectation value when T satisfies (7.152), since it would spontaneously break time-reversal symmetry.

7.4.4 Pinning the edge fields with disorder potentials: the Haldane criterion

Solving the interacting theory (7.146) is beyond the scope of these lectures. What can be done, however, is to identify those fixed points of the interacting theory (7.146) that are pertinent to the question of whether or not some edge modes remain extended along the edge in the limit of strong disorder $h_T(x) \to \infty$ for all tunnelling matrices $T \in \mathbb{L}$ entering the interaction (7.146f).

This question is related to one posed and answered by Haldane [46] for Abelian FQH states and which, in the context of these lectures, can be formulated as follows. Given an interaction potential caused by *weak* disorder on the edges as defined by the Hamiltonian (7.146f), what are the tunnelling vectors $T \in \mathbb{L}$ that can, in principle, describe relevant perturbations that will cause the system to flow to a strong-coupling fixed point characterized by $h_T \to \infty$ away from the fixed point $\widehat{H}_0$? (See [135] for an answer to this weak-coupling question in the context of the IQSHE and $\mathbb{Z}_2$ topological band insulators.) By focusing on the strong-coupling limit from the outset, we avoid the issue of following the renormalization group flow from weak to strong coupling. Evidently, this point of view presumes that the strong-coupling fixed point is stable and that no intermediate fixed point prevents it from being reached.

To identify the fixed points of the interacting theory (7.146) in the strong-coupling limit (strong-disorder limit) $h_T \to \infty$, we ignore the contribution $\widehat{H}_0$ and restrict the sum over the tunnelling matrices in $\widehat{H}_{\text{int}}$ to a subset $\mathbb{H}$ of $\mathbb{L}$ ($\mathbb{H} \subset \mathbb{L}$) with a precise definition of $\mathbb{H}$ that will follow in (7.159). For any choice of $\mathbb{H}$, there follows the strong-coupling fixed-point Hamiltonian

$$\widehat{H}_{\mathbb{H}} := -\int_0^L \mathrm{d}x \sum_{T \in \mathbb{H}} h_T(x) : \cos[T^{\mathsf{T}} K \widehat{\Phi}(x) + \alpha_T(x)] : . \tag{7.153}$$

Let us assume that the fixed-point Hamiltonian (7.153) is stable if and only if the set $\mathbb{H}$ is 'maximal'. The study of the renormalization group flows relating the weak, moderate (if any), and strong fixed points in the infinite-dimensional parameter space spanned by the non-universal data V, $h_T(x)$, and $\alpha_T(x)$ is again beyond the scope of these lectures.

One might wonder why we cannot simply choose $\mathbb{H} = \mathbb{L}$. As emphasized by Haldane [46], this is a consequence of the chiral equal-time commutation relations (7.146c), which prevent simultaneous locking of the phases of all the cosines through the condition, for some time-independent and real-valued function $C_T(x)$,

$$\partial_x[T^{\mathsf{T}} K \widehat{\Phi}(t,x) + \alpha_T(x)] = C_T(x) \tag{7.154}$$

on the canonical momentum

$$(4\pi)^{-1} K (\partial_x \widehat{\Phi})(t,x) \tag{7.155}$$

conjugate to $\widehat{\Phi}(t,x)$, when applied to the ground state. The locking condition (7.154) removes a pair of chiral bosonic modes with opposite chiralities from the gapless degrees of freedom of the theory. However, even in the strong-coupling limit, there are quantum fluctuations as a consequence of the chiral equal-time commutation relations (7.146c) that prevent minimizing the interaction $\widehat{H}_{\text{int}}$ by minimizing separately each contribution to the trigonometric expansion (7.146f). Finding the ground state in the strong-coupling limit is a strongly frustrated optimization problem.

To construct a maximal set $\mathbb{H}$, we demand that any $T \in \mathbb{H}$ must satisfy the locking condition (7.154). Furthermore, we require that the phases of the cosines entering the fixed-point Hamiltonian (7.153) be constants of motion:

$$[\partial_x(T^{\mathsf{T}} K \widehat{\Phi}(t,x)), \widehat{H}_{\mathbb{H}}] = 0. \tag{7.156}$$

To find the tunnelling vectors $T \in \mathbb{H}$, we thus need to consider the commutator

$$\begin{aligned}\int_0^L \mathrm{d}x' \, [\partial_x(T^{\mathsf{T}} K \widehat{\Phi}(t,x)),\, h_{T'}(x') \cos(T'^{\mathsf{T}} K \widehat{\Phi}(t,x') + \alpha_{T'}(x'))] \\ = -\mathrm{i}2\pi T^{\mathsf{T}} K T' h_{T'}(x) \sin(T'^{\mathsf{T}} K \widehat{\Phi}(t,x) + \alpha_{T'}(x)),\end{aligned} \tag{7.157}$$

and demand that it vanishes. This is achieved if $T^{\mathsf{T}} K T' = 0$. Equation (7.157) implies that any set $\mathbb{H}$ is composed of the charge-neutral vectors satisfying

$$T^{\mathsf{T}} K T' = 0. \tag{7.158}$$

It is by choosing a set $\mathbb{H}$ to be 'maximal' that we shall obtain the desired Haldane criterion for stability.

7.4.5 Stability criterion for edge modes

Section 7.4.1 presented and briefly discussed the criteria for at least one branch of edge excitations to remain delocalized even in the presence of strong disorder. Here, we prove these criteria.

The idea is to count the maximum possible number of edge modes that can be pinned (localized) along the edge by tunnelling processes. The set of pinning processes must satisfy

$$T^{\mathsf{T}}Q = 0, \qquad T^{\mathsf{T}}KT' = 0, \tag{7.159}$$

which defines the set $\mathbb{H}$ introduced in Section 7.4.4. (Note, however, that $\mathbb{H}$ is not uniquely determined by this condition.) Let us define the real extension $\mathbb{V}$ of a set $\mathbb{H}$ by allowing the tunnelling vectors T that satisfy (7.159) to take real values instead of integer values. Notice that $\mathbb{V}$ is a vector space over the real numbers. We demand that $\mathbb{H}$ form a lattice that is as dense as the lattice $\mathbb{L}$ by imposing

$$\mathbb{V} \cap \mathbb{L} = \mathbb{H}. \tag{7.160}$$

For any vector $T \in \mathbb{V}$, we consider the vector KT. It follows from (7.159) that $KT \perp T' \; \forall T' \in \mathbb{V}$. So K maps the space $\mathbb{V}$ into an orthogonal space $\mathbb{V}^{\perp}$. Since K is invertible, we have $\mathbb{V}^{\perp} = K\mathbb{V}$, as well as $\mathbb{V} = K^{-1}\mathbb{V}^{\perp}$, and thus $\dim \mathbb{V} = \dim \mathbb{V}^{\perp}$. Since $\dim \mathbb{V} + \dim \mathbb{V}^{\perp} \leq 2N$, it follows that $\dim \mathbb{V} \leq N$. Therefore (as could be anticipated physically) the maximum number of Kramers pairs of edge modes that can be pinned is N. If that happens, the edge has no gapless delocalized mode.

Next, we look at the conditions for which the maximum dimension N is achieved in order to establish a contradiction.

Let us assume that $\dim \mathbb{V} = \dim \mathbb{V}^{\perp} = N$. It follows that $\mathbb{V} \oplus \mathbb{V}^{\perp} = \mathbb{R}^{2N}$, exhausting the space of available vectors. In this case, the charge vector $Q \in \mathbb{V}^{\perp}$ because of (7.159). Consequently, $K^{-1}Q \in \mathbb{V}$. We can then construct an integer vector $\bar{T} \parallel K^{-1}Q$ by scaling $K^{-1}Q$ with the minimum integer r that accomplishes this. (This is always possible, because K^{-1} is a matrix with rational entries and Q is a vector of integers.) Because the inverse of K is not known, it seems hopeless to write $K^{-1}Q$ in closed form. However, K^{-1} must anticommute with Σ_1, given that K anticommutes with Σ_1, while Σ_1 squares to the unit matrix. Hence, $K^{-1}Q$ is an eigenstate of Σ_1 with eigenvalue -1. Now,

$$\bar{T} := r \begin{pmatrix} +(\kappa - \Delta)^{-1}\varrho \\ -(\kappa - \Delta)^{-1}\varrho \end{pmatrix} \tag{7.161}$$

is also an eigenstate of Σ_1 with eigenvalue -1. This suggests that we may use $\bar{T}$ instead of $K^{-1}Q$. Indeed, the existence of $(\kappa - \Delta)^{-1}$ follows from $\det K \neq 0$ and

$$\det K = (-1)^N [\det(\kappa - \Delta)]^2. \tag{7.162}$$

Moreover, we verify that $\bar{T}$ is orthogonal to the charge vector Q and that $K\bar{T}$ is orthogonal to $\bar{T}$.

Equipped with $\bar{T}$, we construct the integer

$$R := -\bar{T}^{\mathsf{T}}\Sigma_{\downarrow}Q. \tag{7.163}$$

It is the parity of this integer that will allow us to establish a contradiction, i.e. it is the parity of R that determines whether it is possible to localize all the modes with the

N tunnelling operators. To establish the contradiction, we employ (7.148e) together with the fact that $\Sigma_1\bar{T} = -\bar{T}$. In other words,

$$\begin{aligned}\pi R &= -\pi\bar{T}^{\mathsf{T}}\Sigma_{\downarrow}Q \\ &= -\alpha_{\bar{T}}(x) - \alpha_{\Sigma_1\bar{T}}(x) \mod 2\pi \\ &= -\alpha_{\bar{T}}(x) - \alpha_{-\bar{T}}(x) \mod 2\pi \\ &= 0 \mod 2\pi,\end{aligned} \tag{7.164}$$

where in the last line $\alpha_T(x) = -\alpha_{-T}(x)$ for all $T \in \mathbb{L}$ has been used. If $\bar{T}$ satisfies (7.164), then R must be an even integer. If (7.164) is violated (i.e. if R is an odd integer), then $\bar{T}$ is not allowed to enter $\widehat{H}_{\mathrm{int}}$ since it would then break time-reversal symmetry (thus, $h_{\bar{T}}(x) = 0$ must always hold in this case to prevent $\bar{T}$ from entering $\widehat{H}_{\mathrm{int}}$). One therefore arrives at the condition that

- if the maximum number of edge modes are localized or gaped, then R must be even.

A corollary is that

- if R is odd, then at least one edge branch is gapless and delocalized.

It remains to prove that if R is even, then one can indeed reach the maximum dimension N for the space of pinning vectors. This is done by construction. Take all eigenvectors of Σ_1 with $+1$ eigenvalue. Choose $N-1$ such vectors, namely all those orthogonal to Q. For the last vector, choose $\bar{T}$. One can check that these N vectors satisfy (7.159) with the help of $\Sigma_1 K \Sigma_1 = -K$ (given in (7.148b)) and of $\bar{T} \parallel K^{-1}Q$. Now, the $N-1$ vectors $\Sigma_1 T = +T$ are of the form $T^{\mathsf{T}} = (t^{\mathsf{T}}, t^{\mathsf{T}})$, where we need to satisfy $T^{\mathsf{T}}Q = 2t^{\mathsf{T}}\varrho = 0$. This leads to $T^{\mathsf{T}}\Sigma_{\downarrow}Q$ being even, and then (7.148e) brings no further conditions whatsoever. So we can take all these $N-1$ tunnelling vectors. Finally, we take $\bar{T}$ as constructed above, which is a legitimate choice since R is assumed to be even and thus consistent with (7.164). Hence, we have constructed the N tunnelling vectors that gap or localize all edge modes, and we can state that

- if R is even, then the maximum number of edge modes are localized or gaped.

As a by-product, we see that it is always possible to localize along the boundary at least all but one Kramers-degenerate pair of edge states via the $N-1$ tunnelling vectors that satisfy $\Sigma_1 T = +T$. Thus, either one or no Kramers-degenerate pair of edge state remains delocalized along the boundary when translation invariance is strongly broken along the boundary.

We close this subsection by demonstrating[5] that the spin Hall conductance (7.142d) is related to the pair of integers r and R defined by (7.161) and (7.163), respectively, through

$$\sigma_{\mathrm{sH}} := \frac{e}{2\pi} \times \frac{R}{r} \quad \Longleftrightarrow \quad \nu_{\mathrm{s}} = \frac{R}{r}. \tag{7.165}$$

[5] Private communication from Jyong-Hao Chen.

If we interpret

$$e^{\star} := \frac{e}{r} \tag{7.166}$$

as the minimal quasiparticle charge, we can state the stability criterion for the helical edge states in symmetry class AII to interactions or static disorder as the condition that the spin Hall conductance divided by $e^{\star}/2\pi$ must be an odd integer. This criterion was first established by Levin and Stern [71] for the FQSHE when the spin-$\frac{1}{2}$ rotation symmetry is broken to its $U(1)$ subgroup. We have shown that the same criterion holds even in the absence of any $U(1)$ residual symmetry of the spin-$SU(2)$ symmetry group.

Proof By the definitions of the charge vector Q, the spin vector S, and the matrix $\bar{T}$ (recall (7.141c) and (7.161)),

$$\bar{T}^{\mathsf{T}} S = r Q^{\mathsf{T}} K^{-1} S. \tag{7.167}$$

The dimensionless spin Hall conductance defined by (7.142b) is thus nothing but

$$\begin{aligned} \nu_{\mathrm{s}} &= \frac{1}{2r} S^{\mathsf{T}} \bar{T} \\ &= \rho^{\mathsf{T}} (\kappa - \Delta)^{-1} \rho. \end{aligned} \tag{7.168}$$

From the definition of R in (7.163),

$$\begin{aligned} R &= -\begin{pmatrix} 0 \\ \rho \end{pmatrix}^{\mathsf{T}} \bar{T} \\ &= +r \rho^{\mathsf{T}} (\kappa - \Delta)^{-1} \rho \\ &= +r \nu_{\mathrm{s}}. \end{aligned} \tag{7.169}$$

■

7.4.6 The stability criterion for edge modes in the FQSHE

What is the fate of the stability criterion when we impose the residual spin-$\frac{1}{2}$ $U(1)$ symmetry in the model so as to describe an underlying microscopic model that supports the FQSHE? The residual spin-$\frac{1}{2}$ $U(1)$ symmetry is imposed on the interacting theory (7.146) by positing the existence of a spin vector $S = -\Sigma_1 S \in \mathbb{Z}^{2N}$ associated with a conserved $U(1)$ spin current. This spin vector is the counterpart of the charge vector $Q = +\Sigma_1 Q \in \mathbb{Z}^{2N}$. The condition

$$S = -\Sigma_1 S \tag{7.170a}$$

is required for compatibility with time-reversal symmetry and is the counterpart of (7.148c). Compatibility of Q and S with time-reversal symmetry thus implies that they are orthogonal: $Q^{\mathsf{T}}S = 0$. If we restrict the interaction (7.146f) by demanding that the tunnelling matrices obey

$$T^{\mathsf{T}}S = 0, \tag{7.170b}$$

we probe the stability of the FQSHE described by $\hat{H}_0$ when perturbed by $\hat{H}_{\text{int}}$.[6]

To answer this question, we supplement the condition $T^{\mathsf{T}}Q = 0$ on tunnelling vectors that belong to $\mathbb{L}$ and $\mathbb{H}$ by $T^{\mathsf{T}}S = 0$. By construction, S is orthogonal to Q. Hence, it remains true that $\mathbb{H}$ is made of at most N linearly independent tunnelling vectors.

The strategy for establishing the condition for the strong-coupling limit of $\hat{H}_{\text{int}}$ to open a mobility gap for all the extended modes of $\hat{H}_0$ thus remains that of constructing the largest set $\mathbb{H}$ out of as few tunnelling vectors with $T = -\Sigma_1 T$ as possible, since these tunnelling vectors might spontaneously break time-reversal symmetry.

As before, there are $N - 1$ linearly independent tunnelling vectors with $T = +\Sigma_1 T$, while the tunnelling matrix $\bar{T}$ from (7.161) must belong to any $\mathbb{H}$ with N linearly independent tunnelling vectors.

At this stage, we need to distinguish the case

$$\bar{T}^{\mathsf{T}}S = 0 \tag{7.171a}$$

from the case

$$\bar{T}^{\mathsf{T}}S \neq 0. \tag{7.171b}$$

In the former case, the spin-neutrality condition (7.170b) holds for $\bar{T}$, and thus the stability criterion is unchanged for the FQSHE. In the latter case, (7.170b) is violated, so that $\hat{H}_{\text{int}}$ is independent of any tunnelling matrix proportional to $\bar{T}$. Thus, when (7.171b) holds, as could be the case when $\kappa \propto \mathbf{1}_N$ and $\Delta = 0$ say, the FQSHE carried by at least one Kramers pair of edge states of $\hat{H}_0$ is robust to the strong-coupling limit of the time-reversal-symmetric and residual spin-$\frac{1}{2}$ $U(1)$-symmetric perturbation $\hat{H}_{\text{int}}$.

[6] It is important to observe that the quadratic Hamiltonian (7.146b) has a much larger symmetry group than the interacting Hamiltonian (7.146f). For example, $\widehat{H}_0$ commutes with the transformation $\widehat{\Phi}(t,x) \longrightarrow \widehat{\Phi}(t,x) + \pi K^{-1}\Sigma_{\downarrow}S$. One verifies that the transformation law of a Kramers doublet of fermions under this transformation is that expected from a rotation about the quantization axis of the residual spin-$\frac{1}{2}$ $U(1)$ symmetry, provided that the parities of the components of S are the same as those of Q. Hence, $\widehat{H}_0$ has, by construction, a residual spin-$\frac{1}{2}$ $U(1)$ symmetry, even though a generic microscopic model with time-reversal symmetry does not. This residual spin-$\frac{1}{2}$ $U(1)$ symmetry of $\widehat{H}_0$ is broken by $\widehat{H}_{\text{int}}$, unless one imposes the additional constraint (7.170b) on the tunnelling matrices $T \in \mathbb{L}$ allowed to enter the interacting theory defined in (7.146).

7.5 Construction of two-dimensional topological phases from coupled wires

7.5.1 Introduction

One accomplishment in the study of topological phases of matter has been the theoretical prediction and experimental discovery of two-dimensional topological insulators [8, 9, 54, 55, 64]. The IQHE is an early example of how states can be classified into distinct topological classes using an integer, the Chern number, to express the quantized Hall conductivity [63, 68, 120]. In the IQHE, the number of delocalized edge channels is proportional to the quantized Hall conductivity through the Chern number. More recently, it has been found that symmetry under time reversal protects the parity in the number of edge modes in (bulk) insulators with strong spin–orbit interactions in two and three dimensions [34, 54]. Correspondingly, these systems are characterized by a $\mathbb{Z}_2$ topological invariant.

The discovery of $\mathbb{Z}_2$ topological insulators initiated a search for a classification of phases of fermionic matter that are distinguished by some topological attribute. For non-interacting electrons, a complete classification, the tenfold way, has been accomplished in arbitrary dimensions [58, 106, 109, 110]. In this scheme, three discrete symmetries that act locally in position space—time-reversal symmetry (TRS), particle–hole symmetry (PHS), and chiral or sublattice symmetry (SLS)—play a central role when defining the quantum numbers that identify the topological insulating fermionic phases of matter within one of the ten symmetry classes (see the first three columns in Table 7.5).

The tenfold way is believed to be robust to a perturbative treatment of short-ranged electron–electron interactions for the following reasons. First, the unperturbed ground state in the clean limit and in a closed geometry is non-degenerate. It is given by the filled bands of a band insulator. The band gap provides a small expansion parameter, namely the ratio of the characteristic interacting energy scale to the band gap. Second, the quantized topological invariant that characterizes the filled bands, provided that its definition and topological character survive the presence of electron–electron interactions as is the case for symmetry class A in two spatial dimensions, cannot change in a perturbative treatment of short-range electron–electron interactions [41].

On the other hand, the fate of the tenfold way when electron–electron interactions are strong is rather subtle [27, 41, 80, 126]. For example, short-range interactions can drive the system through a topological phase transition at which the energy gap closes [14, 100]. They may also break spontaneously a defining symmetry of the topological phase. Even when short-range interactions neither spontaneously break the symmetries nor close the gap, it may be that two phases from the non-interacting tenfold way cease to be distinguishable in the presence of interactions. Indeed, it was shown for symmetry class BDI in one dimension by Fidkowski and Kitaev that the non-interacting $\mathbb{Z}$ classification is too fine in that it must be replaced by a $\mathbb{Z}_8$ classification when generic short-range interactions are allowed. How to construct a counterpart to the tenfold way for interacting fermion (and boson) systems has thus attracted a lot of interest [16–18, 28, 39, 40, 74, 75, 98, 111, 123].

The FQHE is the paradigm for a situation by which interactions select topologically ordered ground states of a very different kind from the non-degenerate ground states of the tenfold way. On a closed two-dimensional manifold of genus g, interactions can stabilize incompressible many-body ground states with a g-dependent degeneracy. Excited states in the bulk must then carry fractional quantum numbers (see [132] and references therein). Such phases of matter, that follow the FQHE paradigm appear in the literature under different names: fractional topological insulators, long-range entangled phases, topologically ordered phases, or symmetry enriched topological phases. In this section, the terminology 'long-range entangled (LRE)' phase is used for all phases with non-trivial g-dependent ground-state degeneracy. All other phases, i.e. those that follow the IQHE paradigm, are called 'short-range entangled (SRE)' phases. (In doing so, we are borrowing the terminology of [74]. It differs slightly from that used in [17], which counts all chiral phases irrespective of their ground-state degeneracy as LRE.)

While there are non-trivial SRE and LRE phases in the absence of any symmetry constraint, many SRE and LRE phases are defined by some symmetry that they obey. If this symmetry is broken, then the topological attribute of the phase is no longer well defined. However, there is a sense in which LRE phases are more robust than SRE phases against a weak breaking of the defining symmetry. The topological attributes of LRE phases are not confined to the boundary in space between two distinct topological realizations of these phases, as they are for SRE phases. They also characterize intrinsic bulk properties such as the existence of gapped deconfined fractionalized excitations. Hence, whereas gapless edge states are gapped by any breaking of the defining symmetry, topological bulk properties are robust to a weak breaking of the defining symmetry as long as the characteristic energy scale for this symmetry breaking is small compared with the bulk gap in the LRE phase, since a small breaking of the protecting symmetry does not wipe out the gapped deconfined fractionalized bulk excitations.

The purpose of this section is to implement a classification scheme for interacting electronic systems in two spatial dimensions that treats SRE and LRE phases on an equal footing. To this end, a coupled-wire construction for each of the symmetry classes from the tenfold way is used. This approach was pioneered in [136] and [70] for the IQHE and in [56] and [119] for the FQHE (see also related work in [60, 62, 81, 112, 115, 125]).

The main idea here is the following. To begin with, non-chiral Luttinger liquids are placed in a periodic array of coupled wires. In doing so, forward-scattering two-body interactions are naturally accounted for within each wire. Backscatterings (i.e. tunnelling) within a given wire or between neighbouring wires are assumed to be the dominant energy scales. Imposing symmetries constrains these allowed tunnellings. Whether a given arrangement of tunnellings truly gaps out all bulk modes, except for some non-gapped edge states on the first and last wires, is verified with the help of a condition that applies to the limit of strong tunnelling. This condition is nothing but the Haldane criterion of Section 7.4.4 [46]. It will be shown that, for a proper choice of the tunnellings, all bulk modes are gapped. Moreover, in five out of the ten symmetry classes of the tenfold way, there remain gapless edge states, in agreement with the tenfold way. It is the character of the tunnellings that determines whether this wire construction selects an SRE or an LRE phase. Hence, this construction, predicated as it is on the strong-tunnelling limit, generalizes the tenfold way for SRE phases to LRE phases. Evidently, this edge-centred classification scheme does not distinguish between LRE phases of matter that do not carry protected gapless edge modes at their interfaces. For example, some fractional, time-reversal-symmetric, incompressible, and topological phases of matter can have fractionalized excitations in the bulk, while not supporting protected gapless modes at their boundaries [86, 107, 130].

The discussion in this section is inspired by [89]. It is organized as follows. The array of Luttinger liquids is defined in Section 7.5.2. The Haldane criterion, which plays an essential role for the stability analysis of the edge theory, is reviewed in Section 7.5.3.3. The five SRE entries and five LRE entries in Table 7.5/Fig. 7.11 are derived in Sections 7.5.4 and 7.5.5, respectively.

The main results of this section are presented in Table 7.5 and Fig. 7.11. For each of the symmetry classes A, AII, D, DIII, and C shown there, the ground state supports propagating gapless edge modes localized on the first and last wire that are immune to local and symmetry-preserving perturbations. The first column of the table labels the symmetry classes according to the Cartan classification of symmetric spaces. The second column dictates if the operations for reversal of time ($\widehat{\Theta}$ with the single-particle representation Θ), exchange of particles and holes ($\widehat{\Pi}$ with the single-particle representation Π), and reversal of chirality ($\widehat{C}$ with the single-particle representation C) are the generators of symmetries, with their single-particle representations squaring to $+1$ or -1, or are not present, in which case this is indicated by an entry 0.[7] The third

Table 7.5 Realization of a two-dimensional array of quantum wires in each symmetry class of the tenfold way.

CARTAN LABEL	Θ^2	Π^2	C^2		SRE TOPOLOGICAL PHASE	LRE TOPOLOGICAL PHASE
A	0	0	0	$\mathbb{Z}$	Fig. 7.11(a)	Fig. 7.11(b)
AIII	0	0	+		None	
AII	−	0	0	$\mathbb{Z}_2$	Fig. 7.11(c)	Fig. 7.11(d)
DIII	−	+	−	$\mathbb{Z}_2$	Fig. 7.11(e)	Fig. 7.11(f)
D	0	+	0	$\mathbb{Z}$	Fig. 7.11(g)	Fig. 7.11(h)
BDI	+	+	+		None	
AI	+	0	0		None	
CI	+	−	−		None	
C	0	−	0	$\mathbb{Z}$	Fig. 7.11(i)	Fig. 7.11(j)
CII	−	−	+		None	

[7] A chiral symmetry is present if there exists a chiral operator $\widehat{C}$ that is antiunitary and commutes with the Hamiltonian. The single-particle representation C of $\widehat{C}$ is a unitary operator that anticommutes with the single-particle Hamiltonian. In a basis in which C is strictly block off-diagonal, C reverses the chirality. This chirality is unrelated to the direction of propagation of left- and right-movers, which is also called chirality in these lectures.

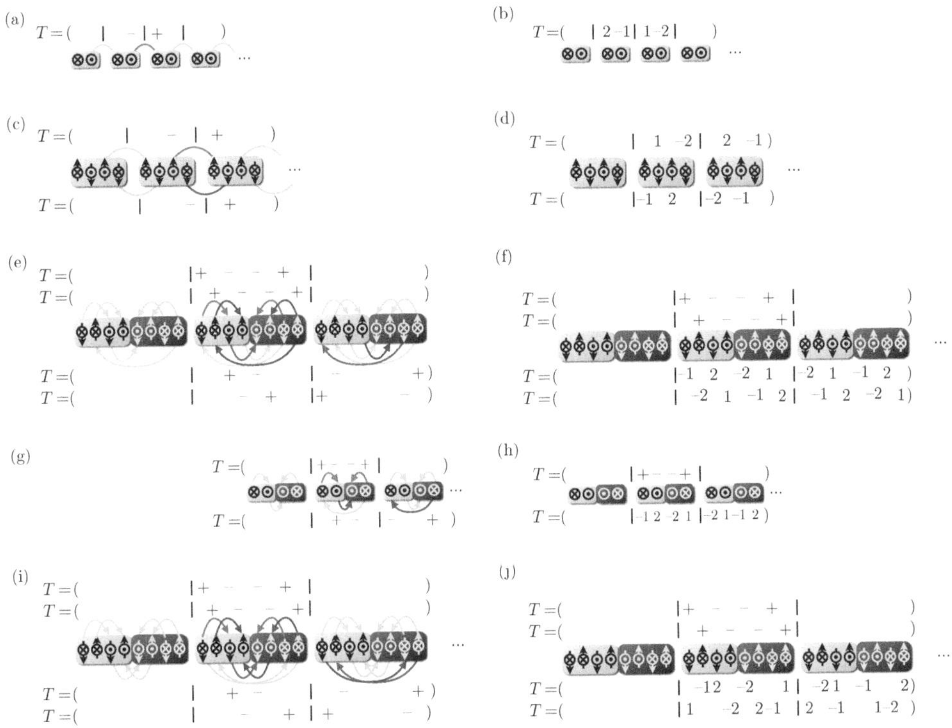

Fig. 7.11 [Colour online] Figures for Table 7.5. (Reprinted with permission from [89]. Copyright 2014 by the American Physical Society.)

column shows the set to which the topological index from the tenfold way, defined as it is in the non-interacting limit, belongs. The fourth column indicates the corresponding pictorial representation of the interactions (a set of tunnelling vectors T) for the two-dimensional array of quantum wires that delivers short-range entangled (SRE) gapless edge states. In these pictorial representations, shown in Fig. 7.11(a, c, e, g, i), a wire is represented by a shaded [coloured online] box with the minimum number of channels compatible with the symmetry class. Each channel in a wire is either a right-mover ($\otimes$) or a left-mover ($\odot$) that may or may not carry a spin quantum number ($\uparrow, \downarrow$) or a particle (light [yellow] shading) or hole (dark shading) attribute. The lines describe tunnelling processes within a wire or between consecutive wires in the array that are of one-body type when they do not carry an arrow or of strictly many-body type when they carry an arrow. Arrows point towards the sites on which creation operators act and away from the sites on which annihilation operators act. For example in symmetry class A, the single line connecting two consecutive wires in the SRE column represents a one-body backward scattering by which left and right movers belonging to consecutive wires are coupled. The lines have been omitted from the pictorial representations for the LRE case (fifth column of Table 7.5 and Fig. 7.11(b, d, f, h, j)), where only the tunnelling vectors are specified.

7.5.2 Definitions

Let us consider an array of N parallel wires that stretch along the x direction of the two-dimensional embedding Euclidean space (see Fig. 7.12). We label a wire by the Latin letter $i = 1, \ldots, N$. Each wire supports fermions that carry an even integer number M of internal degrees of freedom that discriminate between left- and right-movers, the projection along the spin-$\frac{1}{2}$ quantization axis, and particle–hole quantum numbers, among others (e.g. flavours). We label these internal degrees of freedom by the Greek letter $\gamma = 1, \ldots, M$. We combine those two indices in a collective index $\mathsf{a} \equiv (i, \gamma)$. Correspondingly, we introduce the $M \times N$ pairs of creation $\hat{\psi}^{\dagger}_{\mathsf{a}}(x)$ and annihilation $\hat{\psi}_{\mathsf{a}}(x)$ field operators obeying the fermionic equal-time algebra

$$\{\hat{\psi}_{\mathsf{a}}(x), \hat{\psi}^{\dagger}_{\mathsf{a}'}(x')\} = \delta_{\mathsf{a},\mathsf{a}'}\delta(x - x'), \tag{7.172a}$$

with all other anticommutators vanishing and the collective labels $\mathsf{a}, \mathsf{a}' = 1, \ldots, M \times N$. The notation

$$\widehat{\Psi}^{\dagger}(x) \equiv \begin{pmatrix} \hat{\psi}^{\dagger}_{1}(x) & \ldots & \hat{\psi}^{\dagger}_{MN}(x) \end{pmatrix}, \qquad \widehat{\Psi}(x) \equiv \begin{pmatrix} \hat{\psi}_{1}(x) \\ \vdots \\ \hat{\psi}_{MN}(x) \end{pmatrix} \tag{7.172b}$$

is used for the operator-valued row ($\widehat{\Psi}^{\dagger}$) and column ($\widehat{\Psi}$) vector fields. We assume that the many-body quantum dynamics of the fermions supported by this array of wires is governed by the Hamiltonian $\hat{H}$, whereby interactions within each wire are dominant over interactions between wires so that $\hat{H}$ may be represented as N coupled Luttinger liquids, each one of which is composed of M interacting fermionic channels.

By assumption, the $M \times N$ fermionic channels making up the array may thus be bosonized as was explained in Sections 7.3 and 7.4. Within Abelian bosonization [90],

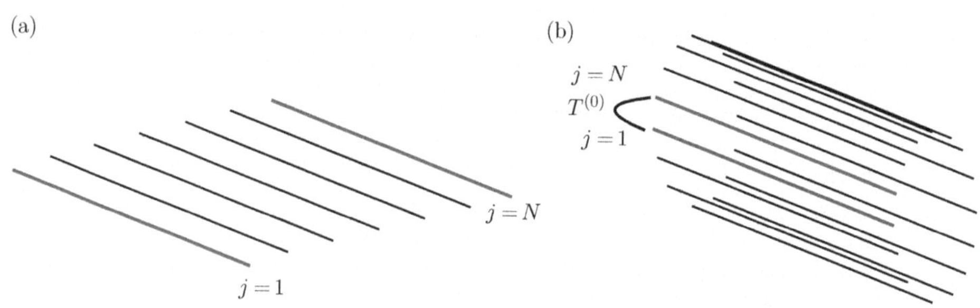

Fig. 7.12 [Colour online] The boundary conditions determine whether a topological phase has protected gapless modes or not. (a) With open boundary conditions, gapless modes exist near the wires $j = 1$ and $j = N$. Scattering between them is forbidden by imposing locality in the limit $N \to \infty$. (b) Periodic boundary conditions allow the scattering vector $\mathcal{T}^{(0)}$, which gaps modes that were protected by locality before. (Reprinted with permission from [89]. Copyright 2014 by the American Physical Society.)

this is done by first postulating that the $MN \times MN$ matrix

$$\mathcal{K} \equiv (\mathcal{K}_{\mathsf{aa}'}) \tag{7.173a}$$

is symmetric with integer-valued entries. Because an array of identical wires—each of which has its quantum dynamics governed by that of a Luttinger liquid—is assumed, it is natural to choose $\mathcal{K}$ to be reducible:

$$\mathcal{K}_{\mathsf{aa}'} = \delta_{ii'} K_{\gamma\gamma'}, \qquad \gamma, \gamma' = 1, \ldots, M, \qquad i, i' = 1, \ldots, N. \tag{7.173b}$$

A second $MN \times MN$ matrix is then defined by

$$\mathcal{L} \equiv (\mathcal{L}_{\mathsf{aa}'}) \tag{7.174a}$$

where

$$\mathcal{L}_{\mathsf{aa}'} := \mathrm{sgn}(\mathsf{a} - \mathsf{a}')\left(\mathcal{K}_{\mathsf{aa}'} + \mathcal{Q}_{\mathsf{a}}\mathcal{Q}_{\mathsf{a}'}\right), \qquad \mathsf{a}, \mathsf{a}' = 1, \ldots, MN, \tag{7.174b}$$

depends on the integer-valued charge vector $\mathcal{Q} \equiv (\mathcal{Q}_{\mathsf{a}})$ in addition to the matrix $\mathcal{K} \equiv (\mathcal{L}_{\mathsf{aa}'})$. The MN compatibility conditions

$$(-1)^{\mathcal{K}_{\mathsf{aa}}} = (-1)^{\mathcal{Q}_{\mathsf{a}}}, \qquad \mathsf{a} = 1, \ldots, MN, \tag{7.174c}$$

must hold. As we are after the effects of interactions between electrons, we choose $\mathcal{Q}_{\mathsf{a}} = 1$, so that

$$\mathcal{L}_{\mathsf{aa}'} := \mathrm{sgn}(\mathsf{a} - \mathsf{a}')\left(\mathcal{K}_{\mathsf{aa}'} + 1\right), \qquad \mathsf{a}, \mathsf{a}' = 1, \ldots, MN. \tag{7.174d}$$

Third, we verify that for any pair $\mathsf{a}, \mathsf{a}' = 1, \ldots, MN$, the Hermitian fields $\hat{\phi}_{\mathsf{a}}$ and $\hat{\phi}_{\mathsf{a}'}$, defined as they are by the Mandelstam formula

$$\hat{\psi}_{\mathsf{a}}(x) \equiv \ :\exp[+\mathrm{i}\mathcal{K}_{\mathsf{aa}'}\hat{\phi}_{\mathsf{a}'}(x)]:, \tag{7.175a}$$

obey the bosonic equal-time algebra

$$[\hat{\phi}_{\mathsf{a}}(x), \hat{\phi}_{\mathsf{a}'}(x')] = -\mathrm{i}\pi[\mathcal{K}^{-1}_{\mathsf{aa}'}\,\mathrm{sgn}(x - x') + \mathcal{K}^{-1}_{\mathsf{ab}}\mathcal{L}_{\mathsf{bc}}\mathcal{K}^{-1}_{\mathsf{ca}'}]. \tag{7.175b}$$

Here, the notation $:(\cdots):$ stands for normal ordering of the argument $(\cdots)$ and the summation convention over repeated indices is implied. In line with (7.172b), the notation

$$\widehat{\Phi}^{\mathsf{T}}(x) \equiv \begin{pmatrix}\hat{\phi}_1(x) & \ldots & \hat{\phi}_{MN}(x)\end{pmatrix}, \qquad \widehat{\Phi}(x) \equiv \begin{pmatrix}\hat{\phi}_1(x)\\ \vdots \\ \hat{\phi}_{MN}(x)\end{pmatrix} \tag{7.175c}$$

is used for the operator-valued row $\widehat{\Phi}^{\mathsf{T}}$ and column $\widehat{\Phi}$ vector fields. Periodic boundary conditions along the x direction parallel to the wires are imposed by demanding that

$$\mathcal{K}\widehat{\Phi}(x + L) = \mathcal{K}\widehat{\Phi}(x) + 2\pi\mathcal{N}, \qquad \mathcal{N} \in \mathbb{Z}^{MN}. \tag{7.175d}$$

Equipped with (7.173)–(7.175), the many-body Hamiltonian $\widehat{H}$ for the MN interacting fermions all carrying the same electric charge e and propagating on the array of wires is decomposed additively into

$$\widehat{H} = \widehat{H}_{\mathcal{V}} + \widehat{H}_{\{\mathcal{T}\}} + \widehat{H}_{\{\mathcal{Q}\}}. \tag{7.176a}$$

The Hamiltonian

$$\widehat{H}_{\mathcal{V}} := \int \mathrm{d}x\, (\partial_x \widehat{\Phi}^{\mathsf{T}})(t,x) \mathcal{V}(x) (\partial_x \widehat{\Phi})(t,x), \tag{7.176b}$$

even though quadratic in the bosonic field, encodes both local one-body terms as well as contact many-body interactions between the M fermionic channels in any given wire from the array through the block-diagonal, real-valued, and symmetric $MN \times MN$ matrix

$$\mathcal{V}(x) := (\mathcal{V}_{\mathsf{a}\mathsf{a}'}(x)) \equiv (\mathcal{V}_{(i,\gamma)(i',\gamma')}(x)) = \mathbb{1}_N \otimes (V_{\gamma\gamma'}(x)). \tag{7.176c}$$

The Hamiltonian

$$\begin{aligned}\widehat{H}_{\{\mathcal{T}\}} &:= \int \mathrm{d}x \sum_{\mathcal{T}} \frac{h_{\mathcal{T}}(x)}{2} \left[\mathrm{e}^{+\mathrm{i}\alpha_{\mathcal{T}}(x)} \prod_{\mathsf{a}=1}^{MN} \hat{\psi}_{\mathsf{a}}^{\mathcal{T}_{\mathsf{a}}}(t,x) + \mathrm{h.c.} \right] \\ &= \int \mathrm{d}x \sum_{\mathcal{T}} h_{\mathcal{T}}(x) : \cos[\mathcal{T}^{\mathsf{T}} \mathcal{K} \widehat{\Phi}(t,x) + \alpha_{\mathcal{T}}(x)] : \end{aligned} \tag{7.176d}$$

is not quadratic in the bosonic fields. With the understanding that the operator multiplication of identical fermion fields at the same point x along the wire requires point splitting, and with the shorthand notation $\hat{\psi}_{\mathsf{a}}^{-1}(x) \equiv \hat{\psi}_{\mathsf{a}}^{\dagger}(x)$, $\widehat{H}_{\{\mathcal{T}\}}$ is interpreted as a sum of all (possibly many-body) tunnellings between the fermionic channels. The set $\{\mathcal{T}\}$ comprises here *all* integer-valued tunnelling vectors

$$\mathcal{T} \equiv (\mathcal{T}_{\mathsf{a}}) \tag{7.176e}$$

obeying the condition

$$\sum_{\mathsf{a}=1}^{MN} \mathcal{T}_{\mathsf{a}} = \begin{cases} 0 \bmod 2 & \text{for D, DIII, C, and CI,} \\ 0 & \text{otherwise.} \end{cases} \tag{7.176f}$$

Moreover, each $\mathcal{T}$ from the set $\{\mathcal{T}\}$ is assigned the real-valued functions

$$h_{\mathcal{T}}(x) = h_{\mathcal{T}}^{*}(x) \geq 0 \tag{7.176g}$$

$$\alpha_{\mathcal{T}}(x) = \alpha_{\mathcal{T}}^{*}(x). \tag{7.176h}$$

The condition (7.176f) ensures that these tunnelling events preserve the parity of the total fermion number for the superconducting symmetry classes (symmetry classes D, DIII, C, and CI in Table 7.5), while they preserve the total fermion number for the non-superconducting symmetry classes (symmetry classes A, AIII, AI, AII, BDI, and CII in Table 7.5). The integer

$$q := \sum_{\mathsf{a}=1}^{MN} \frac{1}{2} |\mathcal{T}_{\mathsf{a}}| \tag{7.176i}$$

dictates that $\mathcal{T}$ encodes a q-body interaction in the fermion representation. The Hamiltonian

$$\widehat{H}_{\{\mathcal{Q}\}} := \int \mathrm{d}x \, \frac{1}{2\pi} A_0(x) \mathcal{Q}^{\mathsf{T}} (\partial_x \widehat{\Phi})(t,x) \tag{7.176j}$$

encodes the response to a static scalar potential A_0 through the charge vector $\mathcal{Q}$ chosen to be

$$\mathcal{Q} = \begin{pmatrix} 1 & \dots & 1 \end{pmatrix}^{\mathsf{T}} \tag{7.176k}$$

in units of the electron charge e.

The Hamiltonian (7.176) and the commutators (7.175b) are form-invariant under the transformation

$$\begin{aligned} \widehat{\Phi}(t,x) &=: \mathcal{W} \widetilde{\Phi}(t,x), && (7.177\text{a}) \\ \widetilde{\mathcal{V}}(x) &:= \mathcal{W}^{\mathsf{T}} \mathcal{V}(x) \mathcal{W}, && (7.177\text{b}) \\ \widetilde{\mathcal{K}} &:= \mathcal{W}^{\mathsf{T}} \mathcal{K} \mathcal{W}, && (7.177\text{c}) \\ \widetilde{\mathcal{T}} &:= \mathcal{W}^{-1} \mathcal{T}, && (7.177\text{d}) \\ \widetilde{\mathcal{Q}} &:= \mathcal{W}^{\mathsf{T}} \mathcal{Q}, && (7.177\text{e}) \end{aligned}$$

where the $MN \times MN$ integer-valued matrix $\mathcal{W}$ is assumed to be invertible, but not necessarily orthogonal! Observe that the tunnelling and charge vectors transform differently whenever $\mathcal{W}^{-1} \neq \mathcal{W}^{\mathsf{T}}$, since they enter the Hamiltonian (7.176) with and without the matrix $\mathcal{K}$, respectively.[8]

Even if the deviation of the matrix $\mathcal{W}$ from the $MN \times MN$ unit matrix is small, the relationship between the vertex operators

$$\begin{aligned} \widetilde{\psi}_{\tilde{\mathsf{a}}}(t,x) &\equiv \, : \exp[\, + \mathrm{i} (\widetilde{\mathcal{K}} \widetilde{\Phi})_{\tilde{\mathsf{a}}}(t,x)] : \\ &= \, : \exp[\, + \mathrm{i} (\mathcal{W}^{\mathsf{T}} \mathcal{K} \widehat{\Phi})_{\tilde{\mathsf{a}}}(t,x)] : , \qquad \tilde{\mathsf{a}} = 1, \dots, MN, \end{aligned} \tag{7.178}$$

and the vertex operators (7.175a) is non-perturbative. Performing a transformation of the form (7.177) to interpret a specific choice of interactions encoded by the tunnelling matrices $\{\mathcal{T}\}$ will play an essential role below.

[8] Alternatively, $\mathcal{K}$ and $\mathcal{Q}\mathcal{Q}^{\mathsf{T}}$ must transform in the same way because of the Klein factors (7.174b).

Because of the transformation laws (7.177c) and (7.177e), the dimensionless Hall conductivity is invariant under the (not necessarily orthogonal) transformation (7.177). Indeed,

$$\begin{aligned}\sigma_{\mathrm{H}} :=& \frac{1}{2\pi}(\mathcal{Q}^{\mathsf{T}}\mathcal{K}^{-1}\mathcal{Q}) \\ =& \frac{1}{2\pi}[\widetilde{\mathcal{Q}}^{\mathsf{T}}(\mathcal{W}^{-1}\mathcal{W})\widetilde{\mathcal{K}}^{-1}(\mathcal{W}^{-1}\mathcal{W})^{\mathsf{T}}\widetilde{\mathcal{Q}}]\end{aligned} \tag{7.179a}$$

equals

$$\widetilde{\sigma}_{\mathrm{H}} := \frac{1}{2\pi}\left(\widetilde{\mathcal{Q}}^{\mathsf{T}}\widetilde{\mathcal{K}}^{-1}\widetilde{\mathcal{Q}}\right). \tag{7.179b}$$

In the sequel, we shall choose a non-orthogonal integer-valued $\mathcal{W}$ with $|\det\mathcal{W}| = 1$ when studying SRE phases of matter outside the tenfold way, while we shall choose a non-orthogonal integer-valued $\mathcal{W}$ with $|\det\mathcal{W}| \neq 1$ in order to construct LRE phases of matter.

7.5.3 Strategy for constructing topological phases

The many-body Hamiltonian $\widehat{H}_{\mathcal{V}} + \widehat{H}_{\{\mathcal{T}\}}$ defined in (7.176) is to be chosen so that (i) it belongs to any one of the ten symmetry classes from the tenfold way (with the action of symmetries defined in Section 7.5.3.1) and (ii) all excitations in the bulk are gapped by a *specific* choice of the tunnelling vectors $\{\mathcal{T}\}$ entering $\widehat{H}_{\{\mathcal{T}\}}$ (with the condition for a spectral gap given in Section 7.5.3.3). The energy scales in $\widehat{H}_{\{\mathcal{T}\}}$ are assumed to be sufficiently large compared with those in $\widehat{H}_{\mathcal{V}}$ so that it is $\widehat{H}_{\mathcal{V}}$ that may be thought of as a perturbation of $\widehat{H}_{\{\mathcal{T}\}}$ and not the converse.

It will be shown that for five of the ten symmetry classes, there can be protected gapless edge states because of locality and symmetry. Step (ii) for each of the five symmetry classes supporting gapless edge states is represented pictorially as shown in Fig. 7.11. In each symmetry class, topologically trivial states that do not support protected gapless edge states in the tenfold classification can be constructed by gapping all states in each individual wire from the array.

7.5.3.1 Representation of symmetries

The classification is based on the presence or the absence of the TRS and the PHS that are represented by the antiunitary many-body operator $\widehat{\Theta}$ and the unitary many-body operator $\widehat{\Pi}$, respectively. Each of $\widehat{\Theta}$ and $\widehat{\Pi}$ can exist in two varieties such that their single-particle representations Θ and Π square to the identity operator up to a multiplicative factor of ± 1:

$$\Theta^2 = \pm 1, \qquad \Pi^2 = \pm 1. \tag{7.180}$$

By assumption, the set of all degrees of freedom in each given wire is invariant under the actions of $\widehat{\Theta}$ and $\widehat{\Pi}$. If so, then the actions of $\widehat{\Theta}$ and $\widehat{\Pi}$ on the fermionic fields can be represented in two steps. First, we introduce two $(M \times M)$-dimensional matrix representations P_Θ and P_Π of the permutation group of M elements, which are combined into the block-diagonal $MN \times MN$ real-valued and orthogonal matrices

$$\mathcal{P}_\Theta := \mathbf{1}_N \otimes P_\Theta, \qquad \mathcal{P}_\Pi := \mathbf{1}_N \otimes P_\Pi, \tag{7.181a}$$

where P_Θ and P_Π represent products of transpositions so that

$$P_\Theta = P_\Theta^{-1} = P_\Theta^{\mathsf{T}}, \qquad P_\Pi = P_\Pi^{-1} = P_\Pi^{\mathsf{T}}. \tag{7.181b}$$

Second, we introduce two column vectors $I_\Theta \in \mathbb{Z}^M$ and $I_\Pi \in \mathbb{Z}^M$, which are combined into two column vectors

$$\mathcal{I}_\Theta := \begin{pmatrix} I_\Theta \\ \vdots \\ I_\Theta \end{pmatrix}, \qquad \mathcal{I}_\Pi := \begin{pmatrix} I_\Pi \\ \vdots \\ I_\Pi \end{pmatrix} \tag{7.181c}$$

and $MN \times MN$ diagonal matrices

$$\mathcal{D}_\Theta := \operatorname{diag}(\mathcal{I}_\Theta), \qquad \mathcal{D}_\Pi := \operatorname{diag}(\mathcal{I}_\Pi), \tag{7.181d}$$

with the components of the vectors $\mathcal{I}_\Theta$ and $\mathcal{I}_\Pi$ as diagonal matrix elements. The vectors I_Θ and I_Π are not chosen arbitrarily. We demand that the vectors $(1+\mathcal{P}_\Theta)\mathcal{I}_\Theta$ and $(1+\mathcal{P}_\Pi)\mathcal{I}_\Pi$ contain only even (for the $+1$ in (7.180)) or only odd (for the -1 in (7.180)) integer entries, while

$$\mathrm{e}^{+\mathrm{i}\pi\mathcal{D}_\Theta}\mathcal{P}_\Theta = \pm\mathcal{P}_\Theta\mathrm{e}^{+\mathrm{i}\pi\mathcal{D}_\Theta} \tag{7.181e}$$

and

$$\mathrm{e}^{+\mathrm{i}\pi\mathcal{D}_\Pi}\mathcal{P}_\Pi = \pm\mathcal{P}_\Pi\mathrm{e}^{+\mathrm{i}\pi\mathcal{D}_\Pi}, \tag{7.181f}$$

in order to satisfy $\Theta^2 = \pm 1$ and $\Pi^2 = \pm 1$, respectively. The operations of time reversal and interchange of particles and holes are then represented for the fermions by

$$\widehat{\Theta}\widehat{\Psi}\widehat{\Theta}^{-1} = \mathrm{e}^{+\mathrm{i}\pi\mathcal{D}_\Theta}\mathcal{P}_\Theta\widehat{\Psi}, \tag{7.181g}$$

$$\widehat{\Pi}\widehat{\Psi}\widehat{\Pi}^{-1} = \mathrm{e}^{+\mathrm{i}\pi\mathcal{D}_\Pi}\mathcal{P}_\Pi\widehat{\Psi}, \tag{7.181h}$$

and for the bosons by

$$\widehat{\Theta}\widehat{\Phi}\widehat{\Theta}^{-1} = \mathcal{P}_\Theta\widehat{\Phi} + \pi\mathcal{K}^{-1}\mathcal{I}_\Theta, \tag{7.181i}$$

$$\widehat{\Pi}\widehat{\Phi}\widehat{\Pi}^{-1} = \mathcal{P}_\Pi\widehat{\Phi} + \pi\mathcal{K}^{-1}\mathcal{I}_\Pi, \tag{7.181j}$$

We can verify that (7.180) is fulfiled.

The Hamiltonian (7.176) is TRS if

$$\widehat{\Theta}\widehat{H}\widehat{\Theta}^{-1} = +\widehat{H}. \tag{7.182a}$$

This condition is met if

$$P_\Theta V P_\Theta^{-1} = +V, \tag{7.182b}$$

$$P_\Theta K P_\Theta^{-1} = -K, \tag{7.182c}$$

$$h_{\mathcal{T}}(x) = h_{-\mathcal{P}_\Theta \mathcal{T}}(x), \tag{7.182d}$$

$$\alpha_{\mathcal{T}}(x) = \alpha_{-\mathcal{P}_\Theta \mathcal{T}}(x) - \pi \mathcal{T}^{\mathsf{T}} \mathcal{P}_\Theta \mathcal{I}_\Theta. \tag{7.182e}$$

The Hamiltonian (7.176) is PHS if

$$\widehat{\Pi}\widehat{H}\widehat{\Pi}^{-1} = +\widehat{H}. \tag{7.183a}$$

This condition is met if

$$P_\Pi V P_\Pi^{-1} = +V, \tag{7.183b}$$

$$P_\Pi K P_\Pi^{-1} = +K, \tag{7.183c}$$

$$h_{\mathcal{T}}(x) = h_{+\mathcal{P}_\Pi \mathcal{T}}(x), \tag{7.183d}$$

$$\alpha_{\mathcal{T}}(x) = \alpha_{\mathcal{P}_\Pi \mathcal{T}}(x) + \pi \mathcal{T}^{\mathsf{T}} \mathcal{P}_\Pi \mathcal{I}_\Pi. \tag{7.183e}$$

7.5.3.2 Particle–hole symmetry in interacting superconductors

The total number of fermions is a good quantum number in any metallic or insulating phase of fermionic matter. This is no longer true in the mean-field treatment of superconductivity. In a superconductor, within a mean-field approximation, charge is conserved modulo 2 because Cooper pairs can be created and annihilated. The existence of superconductors and the phenomenological success of the mean-field approximation suggest that the conservation of the total fermion number operator should be relaxed down to its parity in a superconducting phase of matter. If one demands only that the parity of the total fermion number be conserved, then one may decompose any fermionic creation operator in the position basis into its real and imaginary parts, thereby obtaining two Hermitian operators called Majorana operators. Any Hermitian Hamiltonian that is build out of even powers of Majorana operators necessarily preserves the parity of the total fermion number operator, but it might break the conservation of the total fermion number. By definition, any such Hamiltonian belongs to symmetry class D.

The tool of Abelian bosonization allows a fermion operator to be represented as a single exponential of a Bose field. In Abelian bosonization, a Majorana operator is the sum of two exponentials, and this fact makes it cumbersome to apply Abelian bosonization for Majorana operators. It is possible to circumvent this difficulty by representing any Hamiltonian from symmetry class D in terms of the components of

Nambu–Gorkov spinors obeying a reality condition. Indeed, one may double the dimensionality of the single-particle Hilbert space by introducing Nambu–Gorkov spinors with the understanding that (i) a reality condition on these spinors must hold within the physical subspace of the enlarged single-particle Hilbert space and (ii) the dynamics dictated by the many-body Hamiltonian must be compatible with this reality condition. The reality condition keeps track of the fact that there are many ways to express an even polynomial of Majorana operators in terms of the components of a Nambu–Gorkov spinor. The complication brought about by this redundancy is compensated by the fact that it is straightforward to implement Abelian bosonization in the Nambu–Gorkov representation.

To implement this particle–hole doubling, we assign to every pair of fermionic operators $\hat{\psi}$ and $\hat{\psi}^{\dagger}$ (whose indices have been omitted for simplicity) related to each other by the reality condition

$$\widehat{\Pi}\hat{\psi}\widehat{\Pi}^{\dagger} = \hat{\psi}^{\dagger} \tag{7.184a}$$

a pair of bosonic field operators $\hat{\phi}$ and $\hat{\phi}'$ related by the reality condition

$$\widehat{\Pi}\hat{\phi}\widehat{\Pi}^{\dagger} = -\hat{\phi}'. \tag{7.184b}$$

Invariance under this transformation has to be imposed on the (interacting) Hamiltonian in the doubled (Nambu–Gorkov) representation. In addition to the PHS, when describing the superconducting symmetry classes, we also require that the parity of the total fermion number be conserved. This discrete global symmetry, the symmetry of the Hamiltonian under reversal of sign of all fermion operators, becomes a continuous $U(1)$ global symmetry that is responsible for the conservation of the electric charge in all non-superconducting symmetry classes. In this way, all nine symmetry classes from the tenfold way descend from symmetry class D by the imposition of a composition of TRS, $U(1)$ charge conservation, and the chiral (sublattice) symmetry.

The combined effect of disorder and interactions in superconductors was studied in [29–31, 52], starting from the Nambu–Gorkov formalism to derive a nonlinear sigma model for the Goldstone modes relevant to the interplay between the physics of Anderson localization and that of interactions. The stability of Majorana zero modes to interactions preserving PHS was studied in [35].

7.5.3.3 Conditions for a spectral gap

The Hamiltonian $\widehat{H}_{\mathcal{V}}$ in the decomposition (7.176) has MN gapless modes. However, $\widehat{H}_{\mathcal{V}}$ does not commute with $\widehat{H}_{\{\mathcal{T}\}}$ and the competition between $\widehat{H}_{\mathcal{V}}$ and $\widehat{H}_{\{\mathcal{T}\}}$ can gap some, if not all, the gapless modes of $\widehat{H}_{\mathcal{V}}$. For example, a tunnelling amplitude that scatters the right-mover into the left-mover of each flavour in each wire will gap out the spectrum of $\widehat{H}_{\mathcal{V}}$.

A term in $\widehat{H}_{\{\mathcal{T}\}}$ has the potential to gap out a gapless mode of $\widehat{H}_{\mathcal{V}}$ if, for some time-independent real-valued function $C_{\mathcal{T}}(x)$, the condition (in the Heisenberg representation) [45, 90]

$$\partial_x[\mathcal{T}^{\mathsf{T}}\mathcal{K}\widehat{\Phi}(t,x) + \alpha_{\mathcal{T}}(x)] = C_{\mathcal{T}}(x) \tag{7.185}$$

holds on the canonical momentum

$$(4\pi)^{-1}\mathcal{K}(\partial_x\widehat{\Phi})(t,x) \tag{7.186}$$

that is conjugate to $\widehat{\Phi}(t,x)$, when applied to the ground state. The locking condition (7.185) removes a pair of chiral bosonic modes with opposite chiralities from the gapless degrees of freedom of the theory. However, not all scattering vectors $\mathcal{T}$ can simultaneously lead to such a locking, owing to quantum fluctuations. The set of linear combinations $\{\mathcal{T}^{\mathsf{T}}\mathcal{K}\widehat{\Phi}(t,x)\}$ that can satisfy the locking condition (7.185) simultaneously is labelled by the subset $\{\mathcal{T}\}_{\text{locking}}$ of all tunnelling matrices $\{\mathcal{T}\}$ defined by (7.176e) and (7.176f) obeying the Haldane criterion [45, 90]

$$\mathcal{T}^{\mathsf{T}}\mathcal{K}\mathcal{T}=0 \tag{7.187a}$$

for any $\mathcal{T}\in\{\mathcal{T}\}_{\text{locking}}$ and

$$\mathcal{T}^{\mathsf{T}}\mathcal{K}\mathcal{T}'=0 \tag{7.187b}$$

pairwise for any $\mathcal{T}\neq\mathcal{T}'\in\{\mathcal{T}\}_{\text{locking}}$.

7.5.4 Reproducing the tenfold way

The first goal is to apply the wire construction in order to reproduce the classification of non-interacting topological insulators (symmetry classes A, AIII, AI, AII, BDI, and CII in Table 7.5) and superconductors (symmetry classes D, DIII, C,and CI in Table 7.5) in $(2+1)$ dimension [58, 106, 109, 110]. In this subsection, the classification scheme is carried out within the bosonized description of quantum wires. Here, the classification is restricted to one-body tunnelling terms, i.e. $q=1$ in (7.176i), for the non-superconducting symmetry classes, and to two-body tunnelling terms, i.e. $q=2$ in (7.176i), for the superconducting symmetry classes. In Section 7.5.5, this construction is generalized to the cases $q>1$ and $q>2$ of multiparticle tunnellings in the non-superconducting and superconducting symmetry classes, respectively. The topological stability of edge modes will be an immediate consequence of the observation that no symmetry-respecting local terms can be added to the models to be constructed below.

Within the classification of non-interacting Hamiltonians, superconductors are nothing but fermionic bilinears with a PHS. The physical interpretation of the degrees of freedom as Bogoliubov quasiparticles is of no consequence to the analysis. In particular, they still carry an effective conserved $U(1)$ charge in the non-interacting description.

7.5.4.1 Symmetry class A

SRE phases in the tenfold way. Topological insulators in symmetry class A can be realized without any symmetry aside from the $U(1)$ charge conservation. The wire construction starts from wires supporting spinless fermions, so that the minimal choice $M=2$ only counts left- and right-moving degrees of freedom. The K-matrix reads

$$K:=\operatorname{diag}(+1,-1). \tag{7.188a}$$

The entry $+1$ of the K-matrix corresponds to a right-mover. It is depicted by the symbol $\otimes$ in Fig. 7.11(a). The entry -1 corresponds to a left-mover. It is depicted by the symbol $\odot$ in Fig. 7.11(a). The operation for reversal of time in any one of the N wires is represented by (one can verify that (7.181e) holds)

$$P_{\Theta} := \begin{pmatrix} 0 & 1 \\ 1 & 0 \end{pmatrix}, \qquad I_{\Theta} := \begin{pmatrix} 0 \\ 0 \end{pmatrix}. \tag{7.188b}$$

We define $\widehat{H}_{\{\mathcal{T}\}}$ by choosing $N-1$ scattering vectors, whereby, for any $j = 1, \ldots, N-1$,

$$\mathcal{T}^{(j)}_{(i,\gamma)} := \delta_{i,j}\delta_{\gamma,2} - \delta_{i-1,j}\delta_{\gamma,1}, \tag{7.189a}$$

with $i = 1, \ldots, N$ and $\gamma = 1, 2$. In other words,

$$\mathcal{T}^{(j)} := (0,0|\ldots|0,+1|-1,0|\ldots|0,0)^{\mathsf{T}} \tag{7.189b}$$

for $j = 1, \ldots, N-1$. To clarify the interpretation of the tunnelling vectors, the $|$'s in (7.189b) are used to compartmentalize the elements within a given wire. Henceforth, there are $M = 2$ vector components within each pair of $|$'s that encode the $M = 2$ degrees of freedom within a given wire. The jth scattering vector (7.189b) labels a one-body interaction in the fermion representation that fulfils (7.176f) and breaks TRS, since the scattering vector $(0,+1)^{\mathsf{T}}$ is mapped into the scattering vector $(+1,0)^{\mathsf{T}}$ by the permutation P_{Θ} that represents time reversal in a wire by exchanging right- with left-movers. For any $j = 1, \ldots, N-1$, we introduce the amplitude

$$h_{\mathcal{T}^{(j)}}(x) \geq 0 \tag{7.189c}$$

and the phase

$$\alpha_{\mathcal{T}^{(j)}}(x) \in \mathbb{R} \tag{7.189d}$$

according to (7.182d) and (7.182e), respectively. The choices for the amplitude (7.189c) and the phase (7.189d) are arbitrary. In particular the amplitude (7.189c) can be chosen to be sufficiently large that it is $\widehat{H}_{\mathcal{V}}$ that may be thought of as a perturbation of $\widehat{H}_{\{\mathcal{T}\}}$ and not the converse.

We can verify that all $N-1$ scattering vectors (7.189a) satisfy the Haldane criterion (7.187), i.e.

$$\mathcal{T}^{(i)\mathsf{T}}\mathcal{K}\mathcal{T}^{(j)} = 0, \qquad i,j = 1, \ldots, N-1. \tag{7.190}$$

Correspondingly, the term $\widehat{H}_{\{\mathcal{T}\}}$ gaps out $2(N-1)$ of the $2N$ gapless modes of $\widehat{H}_{\mathcal{V}}$. Two modes of opposite chirality that propagate along the first and last wires, respectively, remain in the low-energy sector of the theory. These edge states are localized on wires $i = 1$ and N, respectively, since their overlaps with the gapped states from the bulk decay exponentially fast as a function of the distance away from the first and end wires. The energy splitting between the edge state localized on wire $i = 1$

and that localized on wire $i = N$ that is brought about by the bulk states vanishes exponentially fast with increasing N. Two gapless edge states with opposite chiralities emerge in the two-dimensional limit $N \to \infty$.

At energies much lower than the bulk gap, the effective $\mathcal{K}$-matrix for the edge modes is

$$\mathcal{K}_{\text{eff}} := \text{diag}(+1, 0|0, 0|\ldots|0, 0|0, -1). \tag{7.191}$$

Here, $\mathcal{K}_{\text{eff}}$ follows from replacing the entries in the $2N \times 2N$ $\mathcal{K}$-matrix for all gapped modes by 0. The pictorial representation of the topological phase in symmetry class A with one chiral edge state per end wire through the wire construction is shown in Fig. 7.11(a). The generalization to an arbitrary number n of gapless edge states sharing a given chirality on the first wire that is opposite to that of the last wire is as follows. We enlarge $M = 2$ to $M = 2n$ by making n identical copies of the model depicted in Fig. 7.11(a). The stability of the n chiral gapless edge states in wires 1 and N is guaranteed because backscattering among these gapless edges states is not allowed kinematically within wire 1 or wire N, while backscattering across the bulk is exponentially suppressed for large N by locality and the gap in the bulk. The number of robust gapless edge states of a given chirality is thus an integer. This is why the entry in the third column of the first row of Table 7.5 is $\mathbb{Z}$.

SRE phases beyond the tenfold way. It is imperative to ask whether the phases constructed so far exhaust all possible SRE phases in symmetry class A. By demanding that one-body interactions are dominant over many-body interactions, all phases from the (exhaustive) classification for non-interacting fermions in class A—and only those—were constructed. In these phases, the same topological invariant controls the Hall and thermal conductivities. However, it was observed that interacting fermion systems can host additional SRE phases in symmetry class A where this connection is lost [74]. These phases are characterized by an edge that includes charge-neutral chiral modes. While such modes contribute to the quantized energy transport (i.e. the thermal Hall conductivity), they do not contribute to the quantized charge transport (i.e. the charge Hall conductivity). Considering the thermal and charge Hall conductivities as two independent quantized topological responses enlarges the classification of SPT phases in symmetry class A to $\mathbb{Z} \times \mathbb{Z}$.

Starting from identical fermions of charge e, an explicit model for an array of wires will be constructed that stabilizes an SRE phase of matter in symmetry class A carrying a non-vanishing Hall conductivity but a vanishing thermal Hall conductivity. In order to build a wire construction of such a strongly interacting SRE phase in symmetry class A, three spinless electronic wires are grouped into one unit cell, i.e.

$$K := \text{diag}(+1, -1, +1, -1, +1, -1). \tag{7.192a}$$

It will be useful to arrange the charges $Q_\gamma = 1$ measured in units of the electron charge e for each of the modes $\hat{\phi}_\gamma$, $\gamma = 1, \ldots, M$, into a vector

$$Q = (1, 1, 1, 1, 1, 1)^{\mathsf{T}}. \tag{7.192b}$$

The physical meaning of the tunnelling vectors (interactions) to be defined below is most transparent if we employ the following linear transformation on the bosonic field variables:

$$\widehat{\Phi}(x) =: \mathcal{W}\widetilde{\Phi}(x), \tag{7.193a}$$

$$\widetilde{\mathcal{K}} := \mathcal{W}^{\mathsf{T}}\mathcal{K}\mathcal{W}, \tag{7.193b}$$

$$\widetilde{\mathcal{T}} := \mathcal{W}^{-1}\mathcal{T}, \tag{7.193c}$$

$$\widetilde{\mathcal{Q}} := \mathcal{W}^{\mathsf{T}}\mathcal{Q}, \tag{7.193d}$$

where $\mathcal{W}$ is a $MN \times MN$ block-diagonal matrix with the non-orthogonal block W having integer entries and the determinant being of unit magnitude. The transformation W and its inverse W^{-1} are given by

$$W := \begin{pmatrix} 0 & -1 & -1 & 0 & 0 & 0 \\ +1 & -1 & -1 & 0 & 0 & 0 \\ +1 & 0 & -1 & 0 & 0 & 0 \\ 0 & 0 & 0 & -1 & 0 & +1 \\ 0 & 0 & 0 & -1 & -1 & +1 \\ 0 & 0 & 0 & -1 & -1 & 0 \end{pmatrix}, \qquad W^{-1} := \begin{pmatrix} -1 & +1 & 0 & 0 & 0 & 0 \\ 0 & -1 & +1 & 0 & 0 & 0 \\ -1 & +1 & -1 & 0 & 0 & 0 \\ 0 & 0 & 0 & -1 & +1 & -1 \\ 0 & 0 & 0 & +1 & -1 & 0 \\ 0 & 0 & 0 & 0 & +1 & -1 \end{pmatrix}. \tag{7.194}$$

This brings K to the form

$$\widetilde{K} := \begin{pmatrix} 0 & +1 & 0 & 0 & 0 & 0 \\ +1 & 0 & 0 & 0 & 0 & 0 \\ 0 & 0 & +1 & 0 & 0 & 0 \\ 0 & 0 & 0 & -1 & 0 & 0 \\ 0 & 0 & 0 & 0 & 0 & -1 \\ 0 & 0 & 0 & 0 & -1 & 0 \end{pmatrix}. \tag{7.195}$$

As can be read off from (7.175b), the parity of $K_{\gamma\gamma}$ determines the self-statistics of particles of type $\gamma = 1, \dots, N$. As (7.175b) is form-invariant under the transformation (7.193), we conclude that, with the choice (7.194), the transformed modes $\gamma = 1, 2$ as well as the modes $\gamma = 5, 6$ are pairs of bosonic degrees of freedom, while the third and fourth modes remain fermionic. Furthermore, the charges transported by the transformed modes $\widetilde{\phi}_\gamma$ are given by

$$\widetilde{Q} = W^{\mathsf{T}}Q = (+2, -2, -3, -3, -2, +2)^{\mathsf{T}}. \tag{7.196}$$

Let us define the charge-conserving tunnelling vectors ($j = 1, \dots, N-1$)

$$\begin{aligned} \widetilde{\mathcal{T}}_1^{(j)} &:= (0,0,0,0,0,0|\dots|0,0,+1,-1,0,0|\dots|0,0,0,0,0,0)^{\mathsf{T}}, \\ \widetilde{\mathcal{T}}_2^{(j)} &:= (0,0,0,0,0,0|\dots|0,0,0,0,+1,0|0,-1,0,0,0,0|\dots|0,0,0,0,0,0)^{\mathsf{T}}, \\ \widetilde{\mathcal{T}}_3^{(j)} &:= (0,0,0,0,0,0|\dots|0,0,0,0,0,+1|-1,0,0,0,0,0|\dots|0,0,0,0,0,0)^{\mathsf{T}}. \end{aligned} \tag{7.197}$$

In the original basis, these three families of tunnelling vectors are of order 3, 2, and 2, respectively. They are given explicitly by

$$\begin{aligned}
\mathcal{T}_1^{(j)} &:= (0,0,0,0,0,0|\dots|-1,-1,-1,+1,+1,+1|\dots|0,0,0,0,0,0)^{\mathsf{T}},\\
\mathcal{T}_2^{(j)} &:= (0,0,0,0,0,0|\dots|0,0,0,0,-1,-1|+1,+1,0,0,0,0|\dots|0,0,0,0,0,0)^{\mathsf{T}},\\
\mathcal{T}_3^{(j)} &:= (0,0,0,0,0,0|\dots|0,0,0,+1,+1,0|0,-1,-1,0,0,0|\dots|0,0,0,0,0,0)^{\mathsf{T}}.
\end{aligned} \tag{7.198}$$

The tunnelling vectors (7.197) gap all modes in the bulk and the remaining gapless edge modes on the left edge are

$$\widetilde{K}_{\text{eff,left}} = \begin{pmatrix} 0 & 1 \\ 1 & 0 \end{pmatrix}, \qquad \widetilde{Q}_{\text{eff,left}} = \begin{pmatrix} +2 \\ -2 \end{pmatrix}. \tag{7.199}$$

The only charge-conserving tunnelling vector that could gap out this effective edge theory, $\widetilde{T} = (1,1)^{\mathsf{T}}$, is not compatible with Haldane's criterion (7.187). Thus, the edge theory (7.199) is stable against charge-conserving perturbations. The Hall conductivity supported by this edge theory is given by

$$\widetilde{Q}^{\mathsf{T}}_{\text{eff,left}} \widetilde{K}^{-1}_{\text{eff,left}} \widetilde{Q}_{\text{eff,left}} = -8, \tag{7.200}$$

in units of e^2/h. This is the minimal Hall conductivity of an SRE phase of bosons, if each boson is interpreted as a pair of electrons carrying electronic charge $2e$ [74]. On the other hand, the edge theory (7.199) supports two modes with opposite chiralities, since the symmetric matrix $\widetilde{K}_{\text{eff,left}}$ has the pair of eigenvalues ± 1. Thus, the net energy transported along the left edge, and with it the thermal Hall conductivity, vanishes.

7.5.4.2 Symmetry class AII

Topological insulators in symmetry class AII can be realized by demanding that $U(1)$ charge conservation holds and that TRS with $\Theta^2 = -1$ holds. The wire construction starts from wires supporting spin-$\frac{1}{2}$ fermions because $\Theta^2 = -1$, so that the minimal choice $M = 4$ counts two pairs of Kramers-degenerate left- and right-moving degrees of freedom carrying opposite spin projections on the spin quantization axis, i.e. two pairs of Kramers-degenerate helical modes. The K-matrix reads

$$K := \operatorname{diag}(+1,-1,-1,+1). \tag{7.201a}$$

The entries in the K-matrix represent, from left to right, a right-moving particle with spin up, a left-moving particle with spin down, a left-moving particle with spin up, and a right-moving particle with spin down. The operation of time reversal in any one of the N wires is represented by (we can verify that (7.181e) holds)

$$P_{\Theta} := \begin{pmatrix} 0 & 1 & 0 & 0 \\ 1 & 0 & 0 & 0 \\ 0 & 0 & 0 & 1 \\ 0 & 0 & 1 & 0 \end{pmatrix}, \qquad I_{\Theta} := \begin{pmatrix} 0 \\ 1 \\ 0 \\ 1 \end{pmatrix}. \tag{7.201b}$$

We define $\widehat{H}_{\mathcal{V}}$ by choosing any symmetric 4×4 matrix V that obeys

$$V = P_{\Theta} V P_{\Theta}^{-1}. \tag{7.201c}$$

We define $\widehat{H}_{\{\mathcal{T}_{\mathrm{SO}}\}}$ by choosing $2(N-1)$ scattering vectors as follows. For any $j = 1, \ldots, N-1$, we introduce a pair of scattering vectors

$$\mathcal{T}^{(j)}_{\mathrm{SO}} := (0,0,0,0|\ldots|0,0,+1,0|-1,0,0,0|\ldots|0,0,0,0)^{\mathsf{T}}, \tag{7.202a}$$

$$\overline{\mathcal{T}}^{(j)}_{\mathrm{SO}} := -\mathcal{P}_{\Theta}\mathcal{T}^{(j)}_{\mathrm{SO}}. \tag{7.202b}$$

The scattering vector (7.202a) labels a one-body interaction in the fermion representation that fulfils (7.176f). It scatters a left-mover with spin up from wire j into a right-mover with spin up on wire $j+1$. For any $j = 1, \ldots, N-1$, we introduce the pair of amplitudes

$$h_{\mathcal{T}^{(j)}_{\mathrm{SO}}}(x) = h_{\overline{\mathcal{T}}^{(j)}_{\mathrm{SO}}}(x) \geq 0 \tag{7.202c}$$

and the pair of phases

$$\alpha_{\mathcal{T}^{(j)}_{\mathrm{SO}}}(x) = \alpha_{\overline{\mathcal{T}}^{(j)}_{\mathrm{SO}}}(x) \in \mathbb{R} \tag{7.202d}$$

according to (7.182d) and (7.182e), respectively. The choices for the amplitude (7.202c) and the phase (7.202d) are arbitrary. The subscript SO refers to the intrinsic spin–orbit coupling. The rationale for using it will be explained shortly.

We verify that all $2(N-1)$ scattering vectors (7.201c) and (7.202a) satisfy the Haldane criterion (7.187), i.e.

$$\mathcal{T}^{(i)\mathsf{T}}_{\mathrm{SO}} \mathcal{K} \mathcal{T}^{(j)}_{\mathrm{SO}} = \overline{\mathcal{T}}^{(i)\mathsf{T}}_{\mathrm{SO}} \mathcal{K} \overline{\mathcal{T}}^{(j)}_{\mathrm{SO}} = \mathcal{T}^{(i)\mathsf{T}}_{\mathrm{SO}} \mathcal{K} \overline{\mathcal{T}}^{(j)}_{\mathrm{SO}} = 0 \tag{7.203}$$

for $i, j = 1, \ldots, N-1$. Correspondingly, the term $\widehat{H}_{\{\mathcal{T}_{\mathrm{SO}}\}}$ gaps out $4(N-1)$ of the $4N$ gapless modes of $\widehat{H}_{\mathcal{V}}$. Two pairs of Kramers-degenerate helical edge states that propagate along the first and last wires, respectively, remain in the low-energy sector of the theory. These edge states are localized on wires $i = 1$ and N, respectively, since their overlaps with the gapped states from the bulk decay exponentially fast as a function of the distance away from the first and end wires. The energy splitting between the edge state localized on wires $i = 1$ and $i = N$ brought about by the bulk states vanishes exponentially fast with increasing N. Two pairs of gapless Kramers-degenerate helical edge states emerge in the two-dimensional limit $N \to \infty$.

At energies much lower than the bulk gap, the effective $\mathcal{K}$-matrix for the two pairs of helical edge modes is

$$\mathcal{K}_{\mathrm{eff}} := \mathrm{diag}(+1,-1,0,0|0,0,0,0|\ldots|0,0,0,0|0,0,-1,+1). \tag{7.204}$$

Here, $\mathcal{K}_{\mathrm{eff}}$ follows from replacing the entries in the $4N \times 4N$ $\mathcal{K}$ matrix for all gapped modes by 0. It will be shown that the effective scattering vector

$$\mathcal{T}_{\mathrm{eff}} := (+1,-1,0,0|0,0,0,0|\ldots)^{\mathsf{T}}, \tag{7.205}$$

with the potential to gap out the pair of Kramers-degenerate helical edge modes on wire $i = 1$ since it fulfils the Haldane criterion (7.187), is not allowed by TRS.[9] On the one hand, $\mathcal{T}_{\mathrm{eff}}$ maps to itself under time reversal:

$$\mathcal{T}_{\mathrm{eff}} = -\mathcal{P}_{\Theta}\mathcal{T}_{\mathrm{eff}}. \tag{7.206}$$

On the other hand,

$$\mathcal{T}_{\mathrm{eff}}^{\mathsf{T}}\mathcal{P}_{\Theta}\mathcal{I}_{\Theta} = -1. \tag{7.207}$$

Therefore, the condition (7.182e) for $\mathcal{T}_{\mathrm{eff}}$ to be a TRS perturbation cannot be met, since the phase $\alpha_{\mathcal{T}_{\mathrm{eff}}}(x)$ associated with $\mathcal{T}_{\mathrm{eff}}$ would then obey

$$\alpha_{\mathcal{T}_{\mathrm{eff}}}(x) = \alpha_{\mathcal{T}_{\mathrm{eff}}}(x) - \pi, \tag{7.208}$$

a condition that cannot be satisfied.

Had a TRS with $\Theta = +1$ been imposed instead of $\Theta = -1$ as is appropriate for symmetry class AI, which describes spinless fermions with TRS, then we would only need to replace I_{Θ} in (7.201b) by the null vector. The scattering vector (7.205) would then be compatible with TRS, since the condition (7.182e) for TRS would then become

$$\alpha_{\mathcal{T}_{\mathrm{eff}}}(x) = \alpha_{\mathcal{T}_{\mathrm{eff}}}(x) \tag{7.209}$$

instead of (7.208). This is why symmetry class AI is always topologically trivial in two-dimensional space from the point of view of the wire construction.

Note also that had we not insisted on the condition of charge neutrality (7.176f), the tunnelling vector

$$\mathcal{T}'_{\mathrm{eff}} := (+1,+1,0,0|0,0,0,0|\ldots)^{\mathsf{T}}, \tag{7.210}$$

which satisfies the Haldane criterion and is compatible with TRS, could gap out the Kramers-degenerate pair of helical edge states.

[9] Even integer multiples of $\mathcal{T}_{\mathrm{eff}}$ would gap the edge states, but they must also be discarded as explained in [90].

To address the question of what happens if $M = 4$ is changed to $M = 4n$, with n any strictly positive integer, in each wire from the array, consider, without loss of generality, the case $n = 2$. To this end, it suffices to repeat all the steps that led to (7.205), except for the change

$$\begin{aligned}\mathcal{K}_{\text{eff}} := \operatorname{diag}(&+1,-1,0,0;+1,-1,0,0|0,0,0,0;0,0,0,0|\\ &\ldots|0,0,0,0;0,0,0,0|0,0,-1,+1;0,0,-1,+1).\end{aligned} \tag{7.211}$$

We can verify that the scattering vectors

$$\mathcal{T}'_{\text{eff}} := (+1,0,0,0;0,-1,0,0|0,0,0,0;0,0,0,0|\cdots)^{\mathsf{T}}, \tag{7.212}$$

$$\mathcal{T}''_{\text{eff}} := (0,-1,0,0;+1,0,0,0|0,0,0,0;0,0,0,0|\cdots)^{\mathsf{T}} \tag{7.213}$$

are compatible with the condition that TRS holds, in that the pair form a closed set under time reversal:

$$\mathcal{T}'_{\text{eff}} = -\mathcal{P}_{\Theta}\mathcal{T}''_{\text{eff}}. \tag{7.214}$$

We can verify that these scattering vectors fulfil the Haldane criterion (7.187). Consequently, inclusion in $\widehat{H}_{\{\mathcal{T}_{\text{SO}}\}}$ of the two cosine potentials with $\mathcal{T}'_{\text{eff}}$ and $\mathcal{T}''_{\text{eff}}$ entering in their arguments, respectively, gaps out the pair of Kramers-degenerate helical modes on wire $i = 1$. The same treatment of wire $i = N$ leads to the conclusion that TRS does not protect the gapless pairs of Kramers-egenerate edge states from perturbations when $n = 2$. The generalization to $M = 4n$ channels is that it is only when n is odd that a pair of Kramers-degenerate helical edge modes is robust to the most generic $\widehat{H}_{\{\mathcal{T}_{\text{SO}}\}}$ of the form depicted in Fig. 7.11(c). Since it is the parity of n in the number $M = 4n$ of channels per wire that matters for the stability of the Kramers-degenerate helical edge states, the group of two integers $\mathbb{Z}_2$ under addition modulo 2 appears in the third column of the third row of Table 7.5.

If conservation of the projection of the spin-$\frac{1}{2}$ quantum number on the quantization axis were imposed, then processes by which a spin is flipped would be precluded for all scattering vectors. In particular, the scattering vectors (7.212) and (7.213) would no longer be admissible. By imposing the $U(1)$ residual symmetry of the full $SU(2)$ symmetry group for a spin-$\frac{1}{2}$ degree of freedom, the group of integers $\mathbb{Z}$ under addition that encodes the topological stability in the quantum spin Hall effect (QSHE) is recovered.

This discussion of symmetry class AII is closed by justifying the interpretation of the index SO as an abbreviation for the intrinsic spin–orbit coupling. To this end, we introduce a set of $N - 1$ pairs of scattering vectors

$$\mathcal{T}^{(j)}_{\text{R}} := (0,0,0,0|\ldots|0,+1,0,0|-1,0,0,0|\ldots|0,0,0,0)^{\mathsf{T}} \tag{7.215a}$$

and

$$\overline{\mathcal{T}}^{(j)}_{\text{R}} := -\mathcal{P}_{\Theta}\mathcal{T}^{(j)}_{\text{R}} \tag{7.215b}$$

for $j = 1, \ldots, N-1$. The scattering vector (7.215a) labels a one-body interaction in the fermion representation that fulfils (7.176f). The index R stands for 'Rashba', since it describes a backward-scattering process by which a left-mover with spin down from wire j is scattered into a right-mover with spin up on wire $j+1$ and conversely. For any $j = 1, \ldots, N-1$, we introduce the pair of amplitudes

$$h_{\mathcal{T}_{\mathrm{R}}^{(j)}}(x) = h_{\overline{\mathcal{T}}_{\mathrm{R}}^{(j)}}(x) \geq 0 \tag{7.215c}$$

and the pair of phases

$$\alpha_{\mathcal{T}_{\mathrm{R}}^{(j)}}(x) = \alpha_{\overline{\mathcal{T}}_{\mathrm{R}}^{(j)}}(x) + \pi \in \mathbb{R} \tag{7.215d}$$

according to (7.182d) and (7.182e), respectively. In contrast to the intrinsic spin–orbit scattering vectors, the Rashba scattering vectors (7.215a) fail to meet the Haldane criterion (7.187) since

$$\mathcal{T}_{\mathrm{R}}^{(j)\mathsf{T}} \mathcal{K} \overline{\mathcal{T}}_{\mathrm{R}}^{(j+1)} = -1, \qquad j = 1, \ldots, N-1. \tag{7.216}$$

Hence, the Rashba scattering processes fail to open a gap in the bulk, as is expected of a Rashba coupling in a two-dimensional electron gas. On the other hand, the intrinsic spin–orbit coupling can lead to a phase with a gap in the bulk that supports the QSHE in a two-dimensional electron gas.

7.5.4.3 Symmetry class D

The simplest example among the topological superconductors can be found in symmetry class D, which is defined by the presence of a PHS with $\Pi^2 = +1$ and the absence of TRS.

With the understanding of PHS as discussed in Section 7.5.3.2, a representative phase in symmetry D is constructed from identical wires supporting right- and left-moving spinless fermions, each of which carries a particle or a hole label, i.e. $M = 4$. The K-matrix reads

$$K := \mathrm{diag}(+1, -1, -1, +1). \tag{7.217a}$$

The entries in the K-matrix represent, from left to right, a right-moving particle, a left-moving particle, a left-moving hole, and a right-moving hole. The operation of the exchange of particles and holes in any one of the N wires is represented by (we can verify that (7.181f) holds)

$$P_{\Pi} := \begin{pmatrix} 0 & 0 & 0 & 1 \\ 0 & 0 & 1 & 0 \\ 0 & 1 & 0 & 0 \\ 1 & 0 & 0 & 0 \end{pmatrix}, \qquad I_{\Pi} := \begin{pmatrix} 0 \\ 0 \\ 0 \\ 0 \end{pmatrix}. \tag{7.217b}$$

We define $\widehat{H}_{\mathcal{V}}$ by choosing any symmetric 4×4 matrix V that obeys

$$V = +P_{\Pi} V P_{\Pi}^{-1}. \tag{7.217c}$$

We define $\widehat{H}_{\{\mathcal{T}\}}$ by choosing $2N-1$ scattering vectors as follows. For any wire $j=1,\ldots,N$, we introduce the scattering vector

$$\mathcal{T}^{(j)} := (0,0,0,0|\ldots|+1,-1,-1,+1|\ldots|0,0,0,0)^{\mathsf{T}}. \tag{7.218a}$$

Between any pair of neighbouring wires, we introduce the scattering vector

$$\overline{\mathcal{T}}^{(j)} := (0,0,0,0|\ldots|0,+1,-1,0|-1,0,0,+1|\ldots|0,0,0,0)^{\mathsf{T}} \tag{7.218b}$$

for $j=1,\ldots,N-1$. Observe that both $\mathcal{T}^{(j)}$ and $\overline{\mathcal{T}}^{(j)}$ are eigenvectors of the particle–hole transformation in that

$$\mathcal{P}_{\Pi}\mathcal{T}^{(j)} = +\mathcal{T}^{(j)}, \qquad \mathcal{P}_{\Pi}\overline{\mathcal{T}}^{(j)} = -\overline{\mathcal{T}}^{(j)}. \tag{7.218c}$$

Thus, to comply with PHS, we demand that the phases

$$\alpha_{\overline{\mathcal{T}}^{(j)}}(x) = 0, \tag{7.218d}$$

while $\alpha_{\mathcal{T}^{(j)}}(x)$ are unrestricted. Similarly, the amplitudes $h_{\mathcal{T}^{(j)}}(x)$ and $h_{\overline{\mathcal{T}}^{(j)}}(x)$ can take arbitrary real values.

We verify that the set of scattering vectors defined by (7.218a) and (7.218b) satisfies the Haldane criterion. Correspondingly, the term $\widehat{H}_{\{\mathcal{T}\}}$ gaps out $4N-2$ of the $4N$ gapless modes of $\widehat{H}_{\mathcal{V}}$. Furthermore, we identify with

$$\overline{\mathcal{T}}^{(0)} = (-1,0,0,+1|0,0,0,0|\ldots|0,0,0,0|0,+1,-1,0)^{\mathsf{T}} \tag{7.219}$$

a unique (up to an integer multiplicative factor) scattering vector that satisfies the Haldane criterion with all existing scattering vectors (7.218a) and (7.218b) and could thus potentially gap out the remaining pair of modes. However, the tunnelling $\overline{\mathcal{T}}^{(0)}$ is non-local since it connects the two edges of the system when open boundary conditions are chosen. We thus conclude that the two remaining modes are exponentially localized near wires $i=1$ and N, respectively, and propagate with opposite chirality.

To give a physical interpretation of the resulting topological (edge) theory in this wire construction, we must keep in mind that the degrees of freedom were artificially doubled. We find, in this doubled theory, a single chiral boson (with chiral central charge $c=1$). To interpret it as the edge of a chiral $(p_x+\mathrm{i}p_y)$ superconductor, the reality condition is imposed to obtain a single chiral Majorana mode with chiral central charge $c=\frac{1}{2}$.

The pictorial representation of the topological phase in symmetry class D through the wire construction is shown in Fig. 7.11(g). The generalization to an arbitrary number n of gapless chiral edge modes is analogous to the case discussed for symmetry class A. The number of robust gapless chiral edge states of a given chirality is thus an integer. This is why the group of integers $\mathbb{Z}$ is found in the third column of the fifth row of Table 7.5.

7.5.4.4 Symmetry classes DIII and C

The remaining two topologically non-trivial superconducting classes DIII (TRS with $\Theta^2 = -1$ and PHS with $\Pi^2 = +1$) and C (PHS with $\Pi^2 = -1$) involve spin-$\frac{1}{2}$ fermions. Each wire thus features no fewer than $M = 8$ internal degrees of freedom corresponding to the spin-$\frac{1}{2}$, chirality and particle/hole indices. The construction is very similar to the cases already presented. Details can be found in [89].

The scattering vectors that are needed to gap out the bulk for each symmetry class DIII and C are represented pictorially in Fig. 7.11(e, i).

7.5.4.5 Summary

An explicit construction has been provided by way of an array of wires supporting fermions that realizes all five insulating and superconducting topological phases of matter with a non-degenerate ground state in two-dimensional space according to the tenfold classification of band insulators and superconductors. The topological protection of edge modes in the bosonic formulation follows from imposing the Haldane criterion (7.187) along with the appropriate symmetry constraints. It remains for us to extend the wire construction to allow many-body tunnelling processes that deliver fractionalized phases with degenerate ground states.

7.5.5 Fractionalized phases

The power of the wire construction goes way beyond what has been used in Section 7.5.4 to reproduce the classification of the SRE phases. It is possible to construct models for interacting phases of matter with intrinsic topological order and fractionalized excitations by relaxing the condition on the tunnellings between wires that they be of the one-body type. While these phases are more complex, the principles for constructing the models and proving the stability of edge modes remain the same: all allowed tunnelling vectors have to obey the Haldane criterion (7.187) and the respective symmetries.

7.5.5.1 Symmetry class A: fractional quantum Hall states

First, the models of quantum wires that are topologically equivalent to the Laughlin state in the FQHE are reviewed [69] following the construction in [56] for Abelian fractional quantum Hall states. Here, the choice of scattering vectors is determined by the Haldane criterion (7.187) and at the same time prepares the ground for the construction of fractional topological insulators with TRS in Section 7.5.5.2.

We need the fermionic Laughlin states indexed by a positive odd integer m [69] (using the same method, other fractional quantum Hall phases from the Abelian hierarchy could be constructed [56]). The elementary degrees of freedom in each wire are spinless right- and left-moving fermions with K-matrix

$$K = \mathrm{diag}(+1, -1), \tag{7.220a}$$

as was used in (7.188a). Time reversal is defined through P_Θ and I_Θ given in (7.188b). Instead of (7.189), the scattering vectors that describe the interactions between the wires are now defined by

$$\mathcal{T}^{(j)} := (0,0\,|\dots|m_+,-m_-\,|m_-,-m_+\,|\dots|\,0,0)^{\mathsf{T}} \tag{7.220b}$$

for any $j = 1,\dots,N-1$, where $m_\pm = (m\pm 1)/2$ (see Fig. 7.11 for an illustration of the scattering process).

For any $j = 1,\dots,N-1$, the scattering (tunnelling) vectors (7.220b) preserve the conservation of the total fermion number in that they obey (7.176f), and they encode a tunnelling interaction of order $q = m$, with q defined in (7.176i). As a set, all tunnelling interactions satisfy the Haldane criterion (7.187), since

$$\mathcal{T}^{(i)\mathsf{T}}\mathcal{K}\mathcal{T}^{(j)} = 0, \qquad i,j = 1,\dots,N-1. \tag{7.221}$$

Note that the choice of tunnelling vector in (7.220b) is unique (up to an integer multiplicative factor) if one insists on charge conservation, compliance with the Haldane criterion (7.187), and only includes scattering between neighbouring wires.

The bare counting of tunnelling vectors shows that the wire model gaps out all but two modes. However, one still needs to show that the remaining two modes (i) live on the edge, (ii) cannot be gapped out by other (local) scattering vectors, and (iii) are made out of fractionalized quasiparticles.

To address (i) and (ii), we note that the remaining two modes can be gapped out by a unique (up to an integer multiplicative factor) charge-conserving scattering vector that satisfies the Haldane criterion (7.187) with all existing scatterings, namely

$$\mathcal{T}^{(0)} := (m_-,-m_+|0,0|\dots|0,0|m_+,-m_-)^{\mathsf{T}}. \tag{7.222}$$

Connecting the opposite ends of the array of wires through the tunnelling $\mathcal{T}^{(0)}$ is not an admissible perturbation, since it violates locality in the two-dimensional thermodynamic limit $N\to\infty$. Had periodic boundary conditions corresponding to a cylinder geometry (i.e. a tube as in Fig. 7.12) by which the first and last wires are nearest neighbours been chosen, then $\mathcal{T}^{(0)}$ would have been admissible. Hence, the gapless nature of the remaining modes when open boundary conditions are chosen depends on the boundary conditions. These gapless modes have support near the boundary only and are topologically protected.

Applying the transformation (7.177) with

$$W := \begin{pmatrix} -m_- & m_+ \\ m_+ & -m_- \end{pmatrix}, \qquad \det W = -m, \qquad W^{-1} = \frac{1}{m}\begin{pmatrix} m_- & m_+ \\ m_+ & m_- \end{pmatrix} \tag{7.223a}$$

transforms the matrix K into

$$\widetilde{K} = W^{\mathsf{T}}KW = \begin{pmatrix} -m_- & m_+ \\ m_+ & -m_- \end{pmatrix}\begin{pmatrix} +1 & 0 \\ 0 & -1 \end{pmatrix}\begin{pmatrix} -m_- & m_+ \\ m_+ & -m_- \end{pmatrix} = \begin{pmatrix} -m & 0 \\ 0 & +m \end{pmatrix}. \tag{7.223b}$$

As its determinant is not unity, the linear transformation (7.223a) changes the compactification radius of the new field $\widetilde{\Phi}(x)$ relative to that of the old field $\widehat{\Phi}(x)$ accordingly. Finally, the transformed tunnelling and charge vectors are given by

$$\widetilde{\mathcal{T}}^{(j)} = \mathcal{W}^{-1}\mathcal{T}^{(j)} = (0,0|\dots|0,0|0,+1|-1,0|0,0|\dots|0,0)^{\mathsf{T}} \neq \mathcal{T}^{(j)}, \tag{7.223c}$$

$$\widetilde{\mathcal{Q}} = \mathcal{W}^{\mathsf{T}}\mathcal{Q} = (1,1|\dots|1,1|1,1|1,1|1,1|\dots|1,1)^{\mathsf{T}} = \mathcal{Q}, \tag{7.223d}$$

respectively, where $\mathcal{W} := \mathbf{1}_N \otimes W$ and $j = 1,\dots,N-1$. In contrast to to the tunnelling vectors, the charge vector is invariant under the non-orthogonal linear transformation (7.223a).

In view of (7.223c), the remaining effective edge theory is described by

$$\widetilde{\mathcal{K}}_{\text{eff}} = \text{diag}(-m,0|0,0|\dots|0,0|0,+m). \tag{7.224}$$

This is a chiral theory at each edge that cannot be gapped by local perturbations. In combination with (7.223d), (7.224) is precisely the edge theory for anyons with statistical angle $1/m$ and charge e/m [132], where e is the charge of the original fermions.

7.5.5.2 Symmetry class AII: fractional topological insulators

Having understood how fractionalized quasiparticles emerge out of a wire construction, it is imperative to ask what other phases can be obtained when symmetries are imposed on the topologically ordered phase. Such symmetry-enriched topological phases have been classified by methods of group cohomology [16]. Here, the case of TRS with $\Theta^2 = -1$ will provide an example of how the wire construction can be used to build up an intuition for these phases and to study the stability of their edge theory.

The elementary degrees of freedom in each wire are spin-$\frac{1}{2}$ right- and left-moving fermions with K-matrix

$$K := \text{diag}(+1,-1,-1,+1), \tag{7.225a}$$

as in (7.201a). Time reversal is defined through P_Θ and I_Θ given in (7.201b). Instead of (7.202a), the scattering vectors that describe the interactions between the wires are now defined by

$$\mathcal{T}^{(j)} := (0,0,0,0|\dots|-m_-,0,+m_+,0|-m_+,0,+m_-,0|\dots|0,0,0,0)^{\mathsf{T}} \tag{7.225b}$$

$$\overline{\mathcal{T}}^{(j)} := -\mathcal{P}_\Theta\mathcal{T}^{(j)}, \tag{7.225c}$$

for any $j = 1,\dots,N-1$, m a positive odd integer, and $m_\pm = (m \pm 1)/2$.

For any $j = 1,\dots,N-1$, the scattering (tunnelling) vectors (7.225b) preserve conservation of the total fermion number in that they obey (7.176f), and they encode a tunnelling interaction of order $q = m$, with q defined in (7.176i). They also satisfy the Haldane criterion (7.187) as a set (see Fig. 7.11 for an illustration of the scattering process).

Applying the transformation (7.177) with

$$W := \begin{pmatrix} -m_- & 0 & m_+ & 0 \\ 0 & -m_- & 0 & m_+ \\ m_+ & 0 & -m_- & 0 \\ 0 & m_+ & 0 & -m_- \end{pmatrix} \tag{7.226}$$

to the bosonic fields leaves the representation of time-reversal invariant,

$$W^{-1} P_\Theta W = P_\Theta, \tag{7.227}$$

while casting the theory in a new form with the transformed matrix $\widetilde{K}$ given by

$$\widetilde{K} = \mathrm{diag}(-m, +m, +m, -m) \tag{7.228}$$

and, for any $j = 1, \ldots, N-1$, with the transformed pair of scattering vectors $(\widetilde{\mathcal{T}}^{(j)}, \overline{\widetilde{\mathcal{T}}}^{(j)})$ given by

$$\widetilde{\mathcal{T}}^{(j)} = (0,0,0,0|\ldots|+1,0,0,0|0,0,-1,0|\ldots|0,0,0,0)^{\mathsf{T}} \tag{7.229}$$

$$\overline{\widetilde{\mathcal{T}}}^{(j)} = (0,0,0,0|\ldots|0,-1,0,0|0,0,0,+1|\ldots|0,0,0,0)^{\mathsf{T}}. \tag{7.230}$$

When these scattering vectors have gapped out all modes in the bulk, the effective edge theory is described by

$$\widetilde{\mathcal{K}}_{\text{eff}} = \mathrm{diag}(0,0,+m,-m|0,0,0,0|\cdots|0,0,0,0|-m,+m,0,0). \tag{7.231}$$

This effective K-matrix describes a single Kramers-degenerate pair of $1/m$ anyons propagating along the first wire and another single Kramers-degenerate pair of $1/m$ anyons propagating along the last wire. Their robustness to local perturbations is guaranteed by TRS.

In contrast to the tenfold way, the correspondence between the bulk topological phase and the edge theories of LRE phases is not one-to-one. For example, while a bulk topological LRE phase supports fractionalized topological excitations in the bulk, its edge modes may be gapped out by symmetry-allowed perturbations. For the phases discussed in this section, namely the Abelian and TRS fractional topological insulators, it was shown in [90] and [71] that the edge, consisting of Kramers-degenerate pairs of edge modes, supports at most one stable Kramers-degenerate pair of delocalized quasiparticles that are stable against disorder. (Note that this does not preclude the richer edge physics of non-Abelian TRS fractional topological insulators [108].)

It turns out that the wire constructions with edge modes given by (7.231) exhaust all stable edge theories of Abelian topological phases that are protected by TRS with $\Theta^2 = -1$ alone.

We suppose that the single protected Kramers-degenerate pair is characterized by the linear combination of bosonic fields

$$\hat{\varphi}(x) := \mathcal{T}^{\mathsf{T}} \mathcal{K}' \widehat{\Phi}(x) \tag{7.232}$$

and its time-reversed partner

$$\hat{\bar{\varphi}}(x) := \overline{\mathcal{T}}^{\mathsf{T}}\mathcal{K}'\widehat{\Phi}(x), \tag{7.233}$$

where the tunnelling vector $\mathcal{T}$ was constructed from the microscopic information from the theory in [90] and $\mathcal{K}'$ is the K-matrix of a TRS bulk Chern–Simons theory from the theory in [90]. (In other words, the theory encoded by $\mathcal{K}'$ has nothing to do a priori with the array of quantum wires defined by (7.225).) The Kramers-degenerate pair of modes $(\hat{\varphi}, \hat{\bar{\varphi}})$ is stable against TRS perturbations supported on a single edge if and only if

$$\frac{1}{2}|\mathcal{T}^{\mathsf{T}}\mathcal{Q}| \text{ is an odd number.} \tag{7.234}$$

Here, $\mathcal{Q}$ is the charge vector with integer entries that determines the coupling of the different modes to the electromagnetic field. Provided $(\hat{\varphi}, \hat{\bar{\varphi}})$ is stable, its equal-time commutation relations follow from (7.175b) as

$$[\hat{\varphi}(x), \hat{\varphi}(x')] = -\,\mathrm{i}\pi(\mathcal{T}^{\mathsf{T}}\mathcal{K}'\mathcal{T}\,\mathrm{sgn}(x-x') + \mathcal{T}^{\mathsf{T}}\mathcal{L}\mathcal{T}), \tag{7.235a}$$

$$[\hat{\bar{\varphi}}(x), \hat{\bar{\varphi}}(x')] = -\,\mathrm{i}\pi(\,-\,\mathcal{T}^{\mathsf{T}}\mathcal{K}'\mathcal{T}\,\mathrm{sgn}(x-x') + \overline{\mathcal{T}}^{\mathsf{T}}\mathcal{L}\overline{\mathcal{T}}), \tag{7.235b}$$

where the fact that $\mathcal{K}'$ anticommutes with $\mathcal{P}_{\Theta}$ according to (7.182c) has been used. By the same token, one can show that the fields $\hat{\varphi}$ and $\hat{\bar{\varphi}}$ commute, since

$$\mathcal{T}^{\mathsf{T}}\mathcal{K}'\overline{\mathcal{T}} = \mathcal{T}^{\mathsf{T}}\mathcal{P}_{\Theta}\mathcal{K}'\mathcal{T} = -\overline{\mathcal{T}}^{\mathsf{T}}\mathcal{K}'\mathcal{T} = 0. \tag{7.236}$$

We conclude that the effective edge theory for *any* Abelian TRS fractional topological insulator build from fermions has the effective form of one Kramers-degenerate pair

$$\mathcal{K}_{\text{eff}} = \begin{pmatrix} \mathcal{T}^{\mathsf{T}}\mathcal{K}'\mathcal{T} & 0 \\ 0 & -\mathcal{T}^{\mathsf{T}}\mathcal{K}'\mathcal{T} \end{pmatrix}, \tag{7.237}$$

and is thus entirely defined by the single integer

$$m := \mathcal{T}^{\mathsf{T}}\mathcal{K}'\mathcal{T}. \tag{7.238}$$

With the scattering vectors (7.225c) An explicit wire construction for each of these cases has been given, thus exhausting all possible stable edge theories for Abelian fractional topological insulators.

For each positive odd integer m, the fractionalized mode has a $\mathbb{Z}_2$ character. It can have either one stable Kramers-degenerate pair of m quasiparticles or none.

7.5.5.3 Symmetry class D: fractional superconductors

In Section 7.5.5.2.2, TRS was imposed on the wire construction of fractional quantum Hall states, from which the fractional topological insulators in symmetry class AII followed. In complete analogy, one may impose PHS with $\Pi^2 = +1$ on the wire construction of a fractional quantum Hall state, thereby promoting it to symmetry class D. Physically, there follows a model for a superconductor with 'fractionalized Majorana fermions or Bogoliubov quasiparticles.

Lately, interest in this direction has been revived by the investigation of exotic quantum dimensions of twist defects embedded in an Abelian fractional quantum Hall liquid [4–6], along with heterostructures of superconductors combined with the FQHE [20, 73, 124] or with fractional topological insulators [19]. Furthermore, the Kitaev quantum wire has been generalized to $\mathbb{Z}_n$ clock models hosting parafermionic edge modes [26, 59], along with efforts to extend the Read–Rezayi quantum Hall state [101] to spin liquids [37, 38] and superconductors [81], all of which exhibit parafermionic quasiparticles.

As in the classification of non-interacting insulators, the Bogoliubov quasiparticles are treated with Abelian bosonization as if they were Dirac fermions. The fractional phase is driven by interactions among the Bogoliubov quasiparticles.

The elementary degrees of freedom in each wire are spinless right- and left-moving fermions and holes as defined for symmetry class D in (7.217a)–(7.217c). We construct the fractional topological insulator using the set of PHS scattering vectors $\mathcal{T}^{(j)}$ for $j = 1, \ldots, N$ with $\mathcal{T}^{(j)}$ as defined in (7.218a) in each wire and the PHS as defined in (7.217b). We complement them with the set of PHS scattering vectors $\overline{\mathcal{T}}^{(j)}$ for $j = 1, \ldots, N-1$ defined by

$$\overline{\mathcal{T}}^{(j)} = (0,0,0,0|\ldots|-m_-, m_+, -m_+, m_-|-m_+, m_-, -m_-, m_+|\ldots|0,0,0,0)^\mathsf{T}, \tag{7.239}$$

where $m_\pm = (m \pm 1)/2$, with m an odd positive integer. Notice that $\overline{\mathcal{T}}^{(j)} := -\mathcal{P}_\Pi \overline{\mathcal{T}}^{(j)}$, so that we must have $\alpha_{\overline{\mathcal{T}}^{(j)}} = 0$ in order to comply with PHS. Thus, together, $\mathcal{T}^{(j)}$ and $\overline{\mathcal{T}}^{(j)}$ gap out $4N - 2$ of the $4N$ chiral modes in the wire. We identify the unique (up to an integer multiplicative factor) scattering vector ($m_\pm = (m \pm 1)/2$)

$$\overline{\mathcal{T}}^{(0)} = (-m_+, m_-, -m_-, m_+|0,0,0,0|\ldots|0,0,0,0|-m_-, m_+, -m_+, m_-)^\mathsf{T}, \tag{7.240}$$

with m the same odd positive integer as in (7.239) that satisfies the Haldane criterion with all $\mathcal{T}^{(j)}$ and $\overline{\mathcal{T}}^{(j)}$ and thus can potentially gap out the two remaining modes. However, it is physically forbidden, since it represents a non-local scattering from one edge to the other. It is concluded that each boundary supports a single remaining chiral mode that is an eigenstate of PHS.

To understand the nature of the single remaining chiral mode on each boundary, we use the following local linear transformation of the bosonic fields:

$$W = \begin{pmatrix} -m_- & +m_+ & 0 & 0 \\ +m_+ & -m_- & 0 & 0 \\ 0 & 0 & -m_- & +m_+ \\ 0 & 0 & +m_+ & -m_- \end{pmatrix}, \qquad m_\pm = \frac{m \pm 1}{2}, \tag{7.241}$$

with determinant $\det W = m^4$. When applied to the non-local scattering vector $\mathcal{T}^{(0)}$ that connects the two remaining chiral edge modes, this gives

$$\begin{aligned} \widetilde{\mathcal{T}}^{(0)} &= \mathcal{W}^{-1}\mathcal{T}^{(0)} \\ &= (0,-1,+1,0|0,0,0,0|\dots|0,0,0,0|+1,0,0,-1), \end{aligned} \tag{7.242}$$

while the matrix K changes under this transformation to

$$\widetilde{K} = \mathrm{diag}(-m, m, m, -m). \tag{7.243}$$

Noting that the representation of PHS is unchanged,

$$W^{-1} P_{\Pi} W = P_{\Pi}, \tag{7.244}$$

we may interpret the remaining chiral edge mode as a PHS superposition of a Laughlin quasiparticle and a Laughlin quasihole. It thus describes a fractional chiral edge mode on either side of the two-dimensional array of quantum wires. The definite chirality is an important difference from the case of the fractional $\mathbb{Z}_2$ topological insulator discussed in Section 7.5.5.2. It guarantees that any integer number $n \in \mathbb{Z}$ layers of this theory is stable, since no tunnelling vector that acts locally on one edge can satisfy the Haldane criterion (7.187). For each m, we may thus say that the parafermion mode has a $\mathbb{Z}$ character, as does the SRE phase in symmetry class D.

7.5.5.4 Symmetry classes DIII and C: more fractional superconductors

The construction for classes DIII and C is very similar to that for the cases already presented. The details can be found in [89]. For class DIII, the edge excitations (and bulk quasiparticles) of the phase are TRS fractionalized Bogoliubov quasiparticles that have also been discussed for one-dimensional realizations. (In the latter context, these TRS fractionalized Bogoliubov quasiparticles are rather susceptible to perturbations [61, 97].)

7.5.6 Summary

In this section, a wire construction was developed to construct models of SRE and LRE topological phases of two-dimensional quantum matter, so as to yield immediate information about the topological stability of their edge modes. In doing so, the periodic table for integer topological phases was promoted to its fractional counterpart. The following paradigms were applied.

1. Each Luttinger liquid wire describes (spinfull or spinless) electrons. Abelian bosonization is used.
2. Backscattering and short-range interactions within and between wires are added. Modes are gapped out if these terms acquire a finite expectation value.
3. A mutual compatibility condition, the Haldane criterion, is imposed among the terms that acquire an expectation value. It is an incarnation of the statement that the operators have to commute if they are to be replaced simultaneously by their expectation values.
4. A set of discrete and local symmetries are imposed on all terms in the Hamiltonian. When modes become massive, they may not break these symmetries.
5. The model is analysed in a strong-coupling limit, instead of the weak-coupling limit in which one derives the renormalization group flows for interactions.

It has become fashionable to write papers in condensed matter physics that take Majorana fermions as the building blocks of lattice models. Elegant mathematical results have been obtained in this way, some of which have the added merit of conceptual clarity. However, the elementary building blocks of condensed matter are ions and electrons, whose interactions are governed by quantum electrodynamics. Majorana fermions in condensed matter physics can only emerge in a non-perturbative way (i) through the interactions between the electrons from the valence bands of a material, or (ii) as the low-energy excitations of exotic quantum magnets. For Majorana fermions to be observable in condensed matter physics, a deconfining transition must have taken place—a notoriously non-perturbative phenomenon. One of the challenges that was confronted in this section was the need to find interacting models for itinerant electrons with local interactions that support Majorana fermions at low energies and long wavelengths. This goal was achieved, starting from non-interacting itinerant electrons, by constructing local many-body interactions that conserve the electron charge and that stabilize two-dimensional bulk superconductors supporting gapless Majorana fermions along their two-dimensional boundaries. This is why strictly many-body interactions are needed in symmetry classes D, DIII, and C to realize the SRE topological phases in Fig. 7.11(a, c, e, g, i).

Acknowledgements

I am grateful to Claudio Chamon, Titus Neupert, Luiz Santos, Shinsei Ryu, and Ronnie Thomale, collaborations with whom shaped Sections 7.4 and 7.5. I am also grateful to Maurizio Storni for his help with Section 7.3. Finally, I would like to thank my student Jyong-Hao Chen for observing that (7.165) holds.

References

[1] Abrahams, E., Anderson, P. W., Licciardello, D. C., and Ramakrishnan, T. V. (1979). Scaling theory of localization: absence of quantum diffusion in two dimensions. *Phys. Rev. Lett.*, **42**, 673–676.

[2] Altland, A. and Zirnbauer, M. R. (1997). Nonstandard symmetry classes in mesoscopic normal–superconducting hybrid structures. *Phys. Rev. B*, **55**, 1142–1161.

[3] Anderson, P. W. (1958). Absence of diffusion in certain random lattices. *Phys. Rev.*, **109**, 1492–1505.

[4] Barkeshli, M., Jian, C.-M., and Qi, X.-L. (2013). Theory of defects in abelian topological states. *Phys. Rev. B*, **88**, 235103.

[5] Barkeshli, M., Jian, C.-M., and Qi, X.-L. (2013). Twist defects and projective non-Abelian braiding statistics. *Phys. Rev. B*, **87**, 045130.

[6] Barkeshli, M. and Qi, X.-L. (2012). Topological nematic states and non-Abelian lattice dislocations. *Phys. Rev. X*, **2**, 031013.

[7] Berezinskiĭ, V. L. (1971). Destruction of long-range order in one-dimensional and two-dimensional systems having a continuous symmetry group I. Classical systems. *Sov. Phys. JETP*, **32**, 493.

[8] Bernevig, B. A., Hughes, T. L., and Zhang, S.-C. (2006). Quantum spin Hall effect and topological phase transition in HgTe quantum wells. *Science*, **314**, 1757–1761.

[9] Bernevig, B. A. and Zhang, S.-C. (2006). Quantum spin Hall effect. *Phys. Rev. Lett.*, **96**, 106802.

[10] Brouwer, P. W., Furusaki, A., Gruzberg, I. A., and Mudry, C. (2000). Localization and delocalization in dirty superconducting wires. *Phys. Rev. Lett.*, **85**, 1064–1067.

[11] Brouwer, P. W., Furusaki, A., Mudry, C., and Ryu, S. (2005). Disorder-induced critical phenomena—new universality classes in Anderson localization. *BUTSURI*, **60**, 935 (2005) [in Japanese, English version arXiv:cond-mat/0511622].

[12] Brouwer, P. W., Mudry, C., and Furusaki, A. (2000). Density of states in coupled chains with off-diagonal disorder. *Phys. Rev. Lett.*, **84**, 2913–2916.

[13] Brouwer, P. W., Mudry, C., Simons, B. D., and Altland, A. (1998). Delocalization in coupled one-dimensional chains. *Phys. Rev. Lett.*, **81**, 862–865.

[14] Budich, J. C., Thomale, R., Li, G., Laubach, M., and Zhang, S.-C. (2012). Fluctuation-induced topological quantum phase transitions in quantum spin-Hall and anomalous-Hall insulators. *Phys. Rev. B*, **86**, 201407.

[15] Chang, P.-Y., Mudry, C., and Ryu, S. (2014). Symmetry-protected entangling boundary zero modes in crystalline topological insulators. *J. Stat. Mech.: Theory Exp.*, **2014**, P09014.

[16] Chen, X., Gu, Z.-C., Liu, Z.-X., and Wen, X.-G. (2013). Symmetry protected topological orders and the group cohomology of their symmetry group. *Phys. Rev. B*, **87**, 155114.

[17] Chen, X., Gu, Z.-C., and Wen, X.-G. (2011). Classification of gapped symmetric phases in one-dimensional spin systems. *Phys. Rev. B*, **83**, 035107.

[18] Chen, X., Gu, Z.-C., and Wen, X.-G. (2011). Complete classification of one-dimensional gapped quantum phases in interacting spin systems. *Phys. Rev. B*, **84**, 235128.

[19] Cheng, M. (2012). Superconducting proximity effect on the edge of fractional topological insulators. *Phys. Rev. B*, **86**, 195126.

[20] Clarke, D. J., Alicea, J., and Shtengel, K. (2013). *Nature Commun.*, **4**, 1348.

[21] Coleman, S. (1975). Quantum sine–Gordon equation as the massive Thirring model. *Phys. Rev. D*, **11**, 2088–2097.

[22] Deser, S., Jackiw, R., and Templeton, S. (1982). Three-dimensional massive gauge theories. *Phys. Rev. Lett.*, **48**, 975–978.

[23] Dirac, P. A. M. (1931). Quantised singularities in the electromagnetic field. *Proc. R. Soc. Lond., Ser. A*, **133**, 60–72.

[24] Dyson, F. J. (1953). The dynamics of a disordered linear chain. *Phys. Rev.*, **92**, 1331–1338.

[25] Dyson, F. J. (1962). The threefold way. Algebraic structure of symmetry groups and ensembles in quantum mechanics. *J. Math. Phys.*, **3**, 1199–1215.

[26] Fendley, P. (2012). *J. Stat. Mech.: Theory Exp.*, **2012**, P11020.

[27] Fidkowski, L. and Kitaev, A. (2010). Effects of interactions on the topological classification of free fermion systems. *Phys. Rev. B*, **81**, 134509.

[28] Fidkowski, Lukasz and Kitaev, Alexei (2011). Topological phases of fermions in one dimension. *Phys. Rev. B*, **83**, 075103.

[29] Foster, M. S. and Ludwig, A. W. W. (2006). Interaction effects on two-dimensional fermions with random hopping. *Phys. Rev. B*, **73**, 155104.

[30] Foster, M. S. and Ludwig, A. W. W. (2008). Metal–insulator transition from combined disorder and interaction effects in Hubbard-like electronic lattice models with random hopping. *Phys. Rev. B*, **77**, 165108.

[31] Foster, M. S., Xie, H.-Yi, and Chou, Y.-Z. (2014). Topological protection, disorder, and interactions: survival at the surface of three-dimensional topological superconductors. *Phys. Rev. B*, **89**, 155140.

[32] Freedman, M., Nayak, C., Shtengel, K., Walker, K., and Wang, Z. (2004). A class of P,T-invariant topological phases of interacting electrons. *Ann. Phys. (NY)*, **310**, 428–492.

[33] Fröhlich, J. and Gabbiani, F. (1990). Braid statistics in local quantum theory. *Rev. Math. Phys.*, **2**, 251–353.

[34] Fu, L., Kane, C. L., and Mele, E. J. (2007). Topological insulators in three dimensions. *Phys. Rev. Lett.*, **98**, 106803.

[35] Goldstein, G. and Chamon, C. (2012). Exact zero modes in closed systems of interacting fermions. *Phys. Rev. B*, **86**, 115122.

[36] Goldstone, J. and Wilczek, F. (1981). Fractional quantum numbers on solitons. *Phys. Rev. Lett.*, **47**, 986–989.

[37] Greiter, M., Schroeter, D. F., and Thomale, R. (2014). Parent Hamiltonian for the non-Abelian chiral spin liquid. *Phys. Rev. B*, **89**, 165125.

[38] Greiter, M. and Thomale, R. (2009). Non-Abelian statistics in a quantum antiferromagnet. *Phys. Rev. Lett.*, **102**, 207203.

[39] Gu, Z.-C. and Wen, X.-G. (2009). Tensor-entanglement-filtering renormalization approach and symmetry-protected topological order. *Phys. Rev. B*, **80**, 155131.

[40] Gu, Z.-C. and Wen, X.-G. (2014). Symmetry-protected topological orders for interacting fermions: fermionic topological nonlinear σ models and a special group super-cohomology theory. *Phys. Rev. B*, **90**, 115141.

[41] Gurarie, V. (2011). Single-particle Green's functions and interacting topological insulators. *Phys. Rev. B*, **83**, 085426.

[42] Haldane, F. D. M. (1983). Continuum dynamics of the 1-D Heisenberg antiferromagnet: Identification with the $O(3)$ nonlinear sigma model. *Phys. Lett. A*, **93**, 464–468.

[43] Haldane, F. D. M. (1983). Fractional quantization of the Hall effect: a hierarchy of incompressible quantum fluid states. *Phys. Rev. Lett.*, **51**, 605–608.

[44] Haldane, F. D. M. (1983). Nonlinear field theory of large-spin Heisenberg antiferromagnets: semiclassically quantized solitons of the one-dimensional easy-axis Néel state. *Phys. Rev. Lett.*, **50**, 1153–1156.

[45] Haldane, F. D. M. (1988). Model for a quantum Hall effect without Landau levels: condensed-matter realization of the 'parity anomaly'. *Phys. Rev. Lett.*, **61**, 2015–2018.

[46] Haldane, F. D. M. (1995). Stability of chiral Luttinger liquids and Abelian quantum Hall states. *Phys. Rev. Lett.*, **74**, 2090–2093.

[47] Halperin, B. I. (1982). Quantized Hall conductance, current-carrying edge states, and the existence of extended states in a two-dimensional disordered potential. *Phys. Rev. B*, **25**, 2185–2190.

[48] Hansson, T. H., Oganesyan, V., and Sondhi, S. L. (2004). Superconductors are topologically ordered. *Ann. Phys. (NY)*, **313**, 497–538.

[49] Heinzner, P., Huckleberry, A., and Zirnbauer, M. R. (2005). Symmetry classes of disordered fermions. *Commun. Math. Phys.*, **257**, 725–771.

[50] Hikami, S. (1980). Renormalization group functions of orthogonal and symplectic non-linear σ models. *Prog. Theor. Phys.*, **64**, 1466–1469.

[51] Jackiw, R. and Rebbi, C. (1976). Solitons with fermion number 1/2. *Phys. Rev. D*, **13**, 3398–3409.

[52] Jeng, M., Ludwig, A. W. W., Senthil, T., and Chamon, C. (2001). Interaction effects on quasiparticle localization in dirty superconductors. arXiv:cond-mat/0112044.

[53] Kane, C. L., Fisher, M. P. A., and Polchinski, J. (1994). Randomness at the edge: theory of quantum Hall transport at filling $\nu = 2/3$. *Phys. Rev. Lett.*, **72**, 4129–4132.

[54] Kane, C. L. and Mele, E. J. (2005). Quantum spin Hall effect in graphene. *Phys. Rev. Lett.*, **95**(22), 226801.

[55] Kane, C.. L. and Mele, E. J. (2005). Z_2 topological order and the quantum spin Hall effect. *Phys. Rev. Lett.*, **95**, 146802.

[56] Kane, C. L., Mukhopadhyay, R., and Lubensky, T. C. (2002). Fractional quantum Hall effect in an array of quantum wires. *Phys. Rev. Lett.*, **88**, 036401.

[57] Khmelnitskii, D. E. (1983). Quantization of Hall conductivity. *ZhETF Pisma*, **38**, 454–458.

[58] Kitaev, A. (2009). Periodic table for topological insulators and superconductors. *AIP Conf. Proc.*, **1134**, 22.

[59] Kitaev, A. Y. (2001). *Phys. Usp.*, **44**, 131.

[60] Klinovaja, J. and Loss, D. (2013). Topological edge states and fractional quantum Hall effect from umklapp scattering. *Phys. Rev. Lett.*, **111**, 196401.

[61] Klinovaja, J. and Loss, D. (2014). Time-reversal invariant parafermions in interacting Rashba nanowires. *Phys. Rev. B*, **90**, 045118.

[62] Klinovaja, J. and Loss, D. (2014). Integer and fractional quantum Hall effect in a strip of stripes. *Eur. Phys. J. B*, **87**, 171.

[63] Klitzing, K. v., Dorda, G., and Pepper, M. (1980). New method for high-accuracy determination of the fine-structure constant based on quantized Hall resistance. *Phys. Rev. Lett.*, **45**, 494–497.

[64] König, M., Wiedmann, S., Brüne, C., Roth, A., Buhmann, H., Molenkamp, L. W., Qi, X.-L., and Zhang, S.-C. (2007). Quantum spin Hall insulator state in HgTe quantum wells. *Science*, **318**, 766–770.

[65] Kosterlitz, J. M. (1974). The critical properties of the two-dimensional xy model. *J. Phys. C: Solid State Phys.*, **7**, 1046–1060.

[66] Kosterlitz, J. M. and Thouless, D. J. (1973). Ordering, metastability and phase transitions in two-dimensional systems. *J. Phys. C: Solid State Phys.*, **6**, 1181–1203.

[67] Kronig, R. DeL. (1935). Zur neutrinotheorie des lichtes III. *Physica*, **2**, 968–980.

[68] Laughlin, R. B. (1981). Quantized Hall conductivity in two dimensions. *Phys. Rev. B*, **23**, 5632–5633.

[69] Laughlin, R. B. (1983). Anomalous quantum Hall effect: an incompressible quantum fluid with fractionally charged excitations. *Phys. Rev. Lett.*, **50**, 1395–1398.

[70] Lee, D.-H. (1994). Network models of quantum percolation and their field-theory representations. *Phys. Rev. B*, **50**, 10788–10791.

[71] Levin, M. and Stern, A. (2009). Fractional topological insulators. *Phys. Rev. Lett.*, **103**, 196803.

[72] Levine, H., Pruisken, A. M. M., and Libby, S. B. (1983). Electron delocalization by a magnetic field in two dimensions. *Phys. Rev. Lett.*, **51**, 1915–1918.

[73] Lindner, N. H., Berg, E., Refael, G., and Stern, A. (2012). Fractionalizing Majorana fermions: non-Abelian statistics on the edges of Abelian quantum Hall states. *Phys. Rev. X*, **2**, 041002.

[74] Lu, Y.-M. and Vishwanath, A. (2012). Theory and classification of interacting integer topological phases in two dimensions: a Chern–Simons approach. *Phys. Rev. B*, **86**, 125119.

[75] Lu, Y.-M. and Vishwanath, A. (2016). Classification and properties of symmetry enriched topological phases: Chern–Simons approach with applications to Z_2 spin liquids. *Phys. Rev. B*, **93**, 155121.

[76] Ludwig, A. W. W., Fisher, M. P. A., Shankar, R., and Grinstein, G. (1994). Integer quantum Hall transition: an alternative approach and exact results. *Phys. Rev. B*, **50**, 7526–7552.

[77] Luther, A. and Peschel, I. (1975). Calculation of critical exponents in two dimensions from quantum field theory in one dimension. *Phys. Rev. B*, **12**, 3908–3917.

[78] Luttinger, J. M. (1963). An exactly soluble model of a many-fermion system. *J. Math. Phys.*, **4**, 1154–1162.

[79] Mandelstam, S. (1975). Soliton operators for the quantized sine–Gordon equation. *Phys. Rev. D*, **11**, 3026–3030.

[80] Manmana, S. R., Essin, A. M., Noack, R. M., and Gurarie, V. (2012). Topological invariants and interacting one-dimensional fermionic systems. *Phys. Rev. B*, **86**, 205119.

[81] Mong, R. S. K., Clarke, D. J., Alicea, J., Lindner, N. H., Fendley, P., Nayak, C., Oreg, Y., Stern, A., Berg, E., Shtengel, K., and Fisher, M. P. A. (2014). Universal topological quantum computation from a superconductor–Abelian quantum Hall heterostructure. *Phys. Rev. X*, **4**, 011036.

[82] Moore, G. and Read, N. (1991). Nonabelions in the fractional quantum Hall effect. *Nucl. Phys. B*, **360**, 362–396.

[83] Moore, J. E. and Balents, L. (2007). Topological invariants of time-reversal-invariant band structures. *Phys. Rev. B*, **75**, 121306.

[84] Moore, J. E. and Wen, X.-G. (1998). Classification of disordered phases of quantum Hall edge states. *Phys. Rev. B*, **57**, 10138–10156.

[85] Morimoto, T., Furusaki, A., and Mudry, C. (2015). Anderson localization and the topology of classifying spaces. *Phys. Rev. B*, **91**, 235111.

[86] Mudry, C. and Fradkin, E. (1994). Separation of spin and charge quantum numbers in strongly correlated systems. *Phys. Rev. B*, **49**, 5200–5219.

[87] Naud, J. D., Pryadko, L. P., and Sondhi, S. L. (2000). Quantum Hall bilayers and the chiral sine–Gordon equation. *Nucl. Phys. B*, **565**, 572–610.

[88] Naud, J. D., Pryadko, L. P., and Sondhi, S. L. (2001). Edge dynamics in quantum Hall bilayers: exact results with disorder and parallel fields. *Phys. Rev. B*, **63**, 115301.

[89] Neupert, T., Chamon, C., Mudry, C., and Thomale, R. (2014). Wire deconstructionism of two-dimensional topological phases. *Phys. Rev. B*, **90**, 205101.

[90] Neupert, T., Santos, L., Chamon, C., and Mudry, C. (2011). Fractional quantum Hall states at zero magnetic field. *Phys. Rev. Lett.*, **106**, 236804.

[91] Neupert, T., Santos, L., Ryu, S., Chamon, C., and Mudry, C. (2011). Fractional topological liquids with time-reversal symmetry and their lattice realization. *Phys. Rev. B*, **84**(16), 165107.

[92] Nielsen, H. B. and Ninomiya, M. (1981). A no-go theorem for regularizing chiral fermions. *Phys. Lett. B*, **105**, 219–223.

[93] Nielsen, H. B. and Ninomiya, M. (1981). Absence of neutrinos on a lattice (I). Proof by homotopy theory. *Nucl. Phys. B*, **185**, 20–40.

[94] Nielsen, H. B. and Ninomiya, M. (1981). Absence of neutrinos on a lattice (II). Intuitive topological proof. *Nucl. Phys. B*, **193**, 173–194.

[95] Niu, Q. and Thouless, D. J. (1984). Quantised adiabatic charge transport in the presence of substrate disorder and many-body interaction. *J. Phys. A: Math. Gen.*, **17**, 2453–2462.

[96] Niu, Q., Thouless, D. J., and Wu, Y.-S. (1985). Quantized Hall conductance as a topological invariant. *Phys. Rev. B*, **31**, 3372–3377.

[97] Oreg, Y., Sela, E., and Stern, A. (2014). Fractional helical liquids in quantum wires. *Phys. Rev. B*, **89**, 115402.

[98] Pollmann, F., Turner, A. M., Berg, E., and Oshikawa, M. (2010). Entanglement spectrum of a topological phase in one dimension. *Phys. Rev. B*, **81**, 064439.

[99] Pruisken, A. M. M. (1984). On localization in the theory of the quantized Hall effect: a two-dimensional realization of the θ-vacuum. *Nucl. Phys. B*, **235**, 277–298.

[100] Raghu, S., Qi, X.-L., Honerkamp, C., and Zhang, S.-C. (2008). Topological Mott insulators. *Phys. Rev. Lett.*, **100**, 156401.

[101] Read, N. and Rezayi, E. (1999). Beyond paired quantum Hall states: parafermions and incompressible states in the first excited Landau level. *Phys. Rev. B*, **59**, 8084–8092.

[102] Roy, R. (2009). Topological phases and the quantum spin Hall effect in three dimensions. *Phys. Rev. B*, **79**, 195322.

[103] Ryu, S., Mudry, C., Hou, C.-Y., and Chamon, C. (2009). Masses in graphene-like two-dimensional electronic systems: topological defects in order parameters and their fractional exchange statistics. *Phys. Rev. B*, **80**, 205319.

[104] Ryu, S., Mudry, C., Ludwig, A. W. W., and Furusaki, A. (2012). Global phase diagram of two-dimensional Dirac fermions in random potentials. *Phys. Rev. B*, **85**, 235115.

[105] Ryu, S., Mudry, C., Obuse, H., and Furusaki, A. (2007). Z_2 topological term, the global anomaly, and the two-dimensional symplectic symmetry class of Anderson localization. *Phys. Rev. Lett.*, **99**, 116601.

[106] Ryu, S., Schnyder, A. P., Furusaki, A., and Ludwig, A. W. W. (2010). Topological insulators and superconductors: tenfold way and dimensional hierarchy. *New J. Phys.*, **12**, 065010.

[107] Sachdev, S. and Read, N. (1991). Large N expansion for frustrated and doped quantum antiferromagnets. *Int. J. Mod. Phys. B*, **05**, 219–249.

[108] Scharfenberger, B. Thomale, R., and Greiter, M. (2011). Non-Abelian statistics and a hierarchy of fractional spin liquids in spin-S antiferromagnets. *Phys. Rev. B*, **84**, 140404.

[109] Schnyder, A. P., Ryu, S., Furusaki, A., and Ludwig, A. W. W. (2008). Classification of topological insulators and superconductors in three spatial dimensions. *Phys. Rev. B*, **78**, 195125.

[110] Schnyder, A. P., Ryu, S., Furusaki, A., and Ludwig, A. W. W. (2009). Classification of topological insulators and superconductors. *AIP Conf. Proc.*, **1134**, 10.

[111] Schuch, N., Perez-Garcia, D., and Cirac, I. (2011). Classifying quantum phases using matrix product states and projected entangled pair states. *Phys. Rev. B*, **84**, 165139.

[112] Seroussi, I., Berg, E., and Oreg, Y. (2014). Topological superconducting phases of weakly coupled quantum wires. *Phys. Rev. B*, **89**, 104523.

[113] Shockley, W. (1939). On the surface states associated with a periodic potential. *Phys. Rev.*, **56**, 317–323.

[114] Simon, B. (1983). Holonomy, the quantum adiabatic theorem, and Berry's phase. *Phys. Rev. Lett.*, **51**, 2167–2170.

[115] Sondhi, S. L. and Yang, K. (2001). Sliding phases via magnetic fields. *Phys. Rev. B*, **63**, 054430.

[116] Stormer, H. L., Chang, A., Tsui, D. C., Hwang, J. C. M., Gossard, A. C., and Wiegmann, W. (1983). Fractional quantization of the Hall effect. *Phys. Rev. Lett.*, **50**, 1953–1956.

[117] Su, W. P., Schrieffer, J. R., and Heeger, A. J. (1980). Soliton excitations in polyacetylene. *Phys. Rev. B*, **22**, 2099–2111.

[118] Tamm, I. (1932). On the possible bound states of electrons on a crystal surface. *Phys. Z. Soviet Union*, **1**, 733.

[119] Teo, J. C. Y. and Kane, C. L. (2014). From Luttinger liquid to non-Abelian quantum Hall states. *Phys. Rev. B*, **89**, 085101.

[120] Thouless, D. J., Kohmoto, M., Nightingale, M. P., and den Nijs, M. (1982). Quantized Hall conductance in a two-dimensional periodic potential. *Phys. Rev. Lett.*, **49**, 405–408.

[121] Tomonaga, S. (1950). Remarks on Bloch?s method of sound waves applied to many-fermion problems. *Prog. Theor. Phys.*, **5**, 544–569.

[122] Tsui, D. C., Stormer, H. L., and Gossard, A. C. (1982). Two-dimensional magnetotransport in the extreme quantum limit. *Phys. Rev. Lett.*, **48**, 1559–1562.

[123] Turner, A. M., Pollmann, F., and Berg, E. (2011). Topological phases of one-dimensional fermions: an entanglement point of view. *Phys. Rev. B*, **83**, 075102.

[124] Vaezi, A. (2013). Fractional topological superconductor with fractionalized Majorana fermions. *Phys. Rev. B*, **87**, 035132.

[125] Vaezi, A.(2014). Superconducting analogue of the parafermion fractional quantum hall states. *Phys. Rev. X*, **4**, 031009.

[126] Wang, Z. and Zhang, S.-C. (2012). Simplified topological invariants for interacting insulators. *Phys. Rev. X*, **2**, 031008.

[127] Wegner, F. J. (1976). Electrons in disordered systems. scaling near the mobility edge. *Z. Phys. B: Condens. Matter*, **25**, 327–337.

[128] Wen, X. G. (1990). Electrodynamical properties of gapless edge excitations in the fractional quantum Hall states. *Phys. Rev. Lett.*, **64**, 2206–2209.

[129] Wen, X. G. (1990). Topological orders in rigid states. *Int. J. Mod. Phys. B*, **4**, 239–271.

[130] Wen, X. G. (1991). Edge excitations in the fractional quantum Hall states at general filling fractions. *Mod. Phys. Lett. B*, **5**, 39–46.

[131] Wen, X.-G. (1991). Edge transport properties of the fractional quantum Hall states and weak-impurity scattering of a one-dimensional charge-density wave. *Phys. Rev. B*, **44**, 5708–5719.

[132] Wen, X.-G. (1991). Topological orders and Chern–Simons theory in strongly correlated quantum liquid. *Int. J. Mod. Phys. B*, **5**, 1641–1648.

[133] Wen, X. G. and Niu, Q. (1990). Ground-state degeneracy of the fractional quantum Hall states in the presence of a random potential and on high-genus Riemann surfaces. *Phys. Rev. B*, **41**, 9377–9396.

[134] Witten, E. (1984). Non-Abelian bosonization in two dimensions. *Commun. Math. Phys.*, **92**, 455–472.

[135] Xu, C. and Moore, J. E. (2006). Stability of the quantum spin Hall effect: effects of interactions, disorder, and Z_2 topology. *Phys. Rev. B*, **73**, 045322.

[136] Yakovenko, V. M. (1991). Quantum Hall effect in quasi-one-dimensional conductors. *Phys. Rev. B*, **43**, 11353–11366.

[137] Zhang, Y., Tan, Y.-W., Stormer, H. L., and Kim, P. (2005). Experimental observation of the quantum Hall effect and Berry's phase in graphene. *Nature*, **438**, 201–204.

[138] Zirnbauer, M. R. (1996). Riemannian symmetric superspaces and their origin in random-matrix theory. *J. Math. Phys.*, **37**, 4986–5018.

8
Symmetry-protected topological phases in one-dimensional systems

Frank Pollmann

Max-Planck-Institut für Physik komplexer Systeme
01187 Dresden, Germany

Topological Aspects of Condensed Matter Physics. First Edition. C. Chamon et al.

Chapter Contents

8.1 Introduction

Classifying and understanding different phases of matter is an important task of condensed matter physics. The class of 'conventional' *symmetry-broken* phases is well understood in terms of Landau's theory [25]. A paradigmatic example is the $\mathbb{Z}_2$ symmetric Ising model with a symmetric (paramagnetic) and a symmetry-broken (ferromagnet) phase. The two phases can be distinguished by measuring the magnetization as a *local order parameter.* In contrast, *topological phases* of matter [51] are less understood and no complete classification is known so far. In these lecture notes, we are interested in finding schemes that allow us to understand and characterize certain topological phases of matter. Throughout, we consider systems that are described by *local Hamiltonians* (i.e. $H = \sum_n h_n$, with h_n acting on sites near n).

Gapped quantum phases of matter (i.e. phases in which the ground state is separated from the excitation continuum by a finite energy gap) can be defined very generally in terms of local unitary (LU) transformations [8, 48]. These LU transformation correspond to the application of a finite number of unitary operators that act only locally on the wavefunction. We say that two gapped ground states are in the same phase if and only if they can be transformed into each other by LU transformations. Alternatively, we can use the definition that two ground states are in the same phase if they are connected adiabatically to each other by a continuous parameter in the Hamiltonian. Using this definition, all states that differ from 'trivial' product states only by local fluctuations, i.e. short-ranged entanglement (SRE) states, are in the same phase. States that contain non-local quantum correlation, so-called *topologically ordered* states like quantum Hall states or gapped spin liquids, exhibit long-range (LRE) entanglement that cannot be removed by LU transformations. Using this definition, there exists one trivial (SRE) phase and various different topologically ordered phases that differ in terms of their LRE from each other. In one-dimensional (1D) bosonic systems, all gapped ground states have only SRE [8].

Once symmetries are imposed, a much richer variety of phases emerges. In terms of the LU transformations, this means that when respecting the imposed symmetry, not all SRE states are in the same phase. One class of states falling into this category are the symmetry-breaking states discussed above (e.g. the Ising ferromagnet). Another recently discovered class are *symmetry-protected topological* (SPT) phases [7, 9, 10, 16, 38, 43]. The defining property of SPT phases is that they do not break a particular symmetry; however, given a certain symmetry constraint, they cannot be adiabatically connected to a trivial product state. Examples of SPT phases include topological insulators [20], which can be protected by time-reversal symmetry. For free fermions, SPT phases are classified in the periodic table for topological insulators and superconductors [22, 40]. Another example is the Haldane phase [1, 17] in 1D, which is protected by either time reversal, bond-centred inversion, or the dihedral group of the spin rotations [38].

The main focus of these lecture notes lies on SPT phases in 1D bosonic systems. Based on the entanglement properties of 1D systems, we will motivate the matrix product state (MPS) representation of ground states. Using the MPS framework, we will then demonstrate how SPT phases can be classified using projective representations of the symmetries. As a concrete example, we will consider a spin-1 chain described by the Hamiltonian

$$H = J\sum_j \vec{S}_j \cdot \vec{S}_{j+1} + D\sum_j (S_j^z)^2. \tag{8.1}$$

The first term is the standard spin-1 Heisenberg model with antiferromagnetic exchange interactions. The spin-1 Heisenberg model has a gapped ground state that does not break any symmetries [17, 18] (i.e. it is in the Haldane phase). The second term represents a uniaxial single-ion anisotropy. As the parameter $D \geq 0$ is tuned, the system undergoes a phase transition between two gapped phases at $D \approx 1$ [6, 11, 45]. In both phases, the ground state has the full symmetry of the Hamiltonian! Thus, the phase transition cannot be understood in terms of spontaneous symmetry breaking. With the framework developed in these lectures notes, we will be able to distinguish the two phases in terms of a topological invariant and identify the Haldane phase as an SPT phase. Furthermore, we will discuss non-local order parameters that will allow us two detect the two symmetric phases in numerical simulations.

These notes are structured as follows: In Section 8.2, we start by deriving some basic concepts of entanglement, including its definition and the area law. We then introduce MPS and show that these describe efficiently gapped ground states in 1D. Based on symmetry transformations of MPS, we introduce in Section 8.3 the concept of SPT phases. Using an intuitive approach by studying the symmetry transformations of a segment of consecutive sites, we demonstrate the stability of SPT phases. In Section 8.4, we propose non-local order parameters to detect SPT phases in numerical simulations. We conclude by summarizing the main result and present a short outlook in Section 8.5.

8.2 Entanglement and matrix product states

Entanglement is one of the fundamental phenomena in quantum mechanics and implies that different degrees of freedom of a quantum system cannot be described independently. Over the past decades, it has been realized that the entanglement in quantum many-body systems can give access to a lot of useful information about quantum states. First, entanglement-related quantities provide powerful tools to extract universal properties of quantum states. For example, scaling properties of the entanglement entropy help to characterize critical systems [4, 5, 36, 44], and entanglement is the basis for the classification of topological orders [23, 26]. Second, the understanding of entanglement has helped to develop new numerical methods to efficiently simulate quantum many-body systems [41, 47]. In the following, we give a short introduction to entanglement in 1D systems and then focus on the MPS representation.

8.2.1 Schmidt decomposition and entanglement

Let us consider the bipartition of the Hilbert space $\mathcal{H} = \mathcal{H}_L \otimes \mathcal{H}_R$ of a 1D system as illustrated in Fig. 8.1(a), where $\mathcal{H}_L$ ($\mathcal{H}_R$) describes all the states defined on the left (right) of a given bond. In the so-called *Schmidt decomposition*, a state $|\Psi\rangle \in \mathcal{H}$ is decomposed as

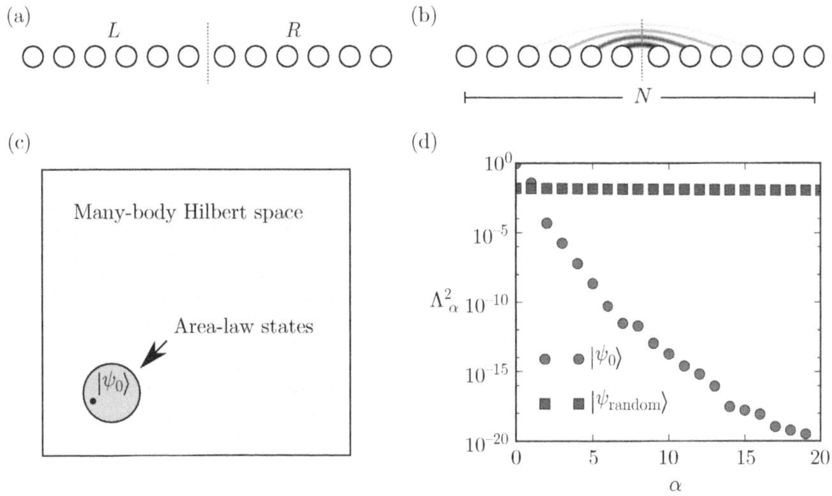

Fig. 8.1 (a) Bipartition of a 1D system into two half-chains. (b) Significant quantum fluctuations in gapped ground states occur only on short lengthscales. (c) 1D area-law states make up a very small fraction of the many-body Hilbert space but contain all gapped ground states. (d) Comparison of the entanglement spectrum of the ground state of the transverse field Ising model ($g = 1.5$) and a random state for a system consisting of $N = 16$ spins.

$$|\Psi\rangle = \sum_\alpha \Lambda_\alpha |\alpha\rangle_L \otimes |\alpha\rangle_R, \qquad |\alpha\rangle_{L(R)} \in \mathcal{H}_{L(R)}, \tag{8.2}$$

where the states $\{|\alpha\rangle_{L(R)}\}$ form an orthogonal basis of $\mathcal{H}_L$ ($\mathcal{H}_R$) and $\Lambda_\alpha \geq 0$. The Schmidt decomposition is unique up to degeneracies and for a normalized state $|\Psi\rangle$ we find that $\sum_\alpha \Lambda_\alpha^2 = 1$.

An important aspect is that the Schmidt decomposition gives direct insight into the *bipartite entanglement* (i.e. the entanglement between degrees of freedom in $\mathcal{H}_L$ and $\mathcal{H}_R$) of a state. In particular, only one term contributes to the Schmidt decomposition if and only if L and R are not entangled. If more than one term is required in the Schmidt decomposition to express the state, then the state is necessarily entangled. The relation between the Schmidt decomposition and the entanglement can be made more concrete. The *reduced density matrix*

$$\rho^R = \mathrm{Tr}_L(|\psi\rangle\langle\psi|) \tag{8.3}$$

has the Schmidt states $|\alpha\rangle_R$ as eigenstates and the Schmidt coefficients are the square roots of the corresponding eigenvalues, i.e. $\rho^R = \sum_\alpha \Lambda_\alpha^2 |\alpha\rangle_R \langle\alpha|_R$ (equivalently for ρ^L). The reduced density matrix of an entangled (pure) quantum state is the density matrix of a mixed state defined on the subsystem. Thus, the *entanglement entropy*, which is defined as the von Neumann entropy of the reduced density matrix, measures the amount of entanglement. In terms of the Schmidt values, it is given by

$$S_E = -\sum_\alpha \Lambda_\alpha^2 \log \Lambda_\alpha^2. \tag{8.4}$$

The entanglement entropy S is a very useful measure to quantify the amount of entanglement in a system for a given bipartition. Finally, the *entanglement spectrum* $\{\epsilon_\alpha\}$ [27] is defined in terms of the spectrum $\{\Lambda_\alpha^2\}$ of the reduced density matrix by $\Lambda_\alpha^2 = \exp(-\epsilon_\alpha)$ for each α.

8.2.2 Area law

A 'typical' state in the Hilbert space shows a *volume law*, i.e. the entanglement entropy grows proportionally with the volume of the partitions. In particular, it has been shown in [31] that a randomly drawn state $|\psi_{\text{random}}\rangle$ from the Hilbert space of a system of N sites with on-site Hilbert space dimension d has an entanglement entropy of $S \approx \frac{1}{2}N \log d - \frac{1}{2}$ for a bipartition into two parts of $\frac{1}{2}N$ sites.

Ground states $|\psi_0\rangle$ of gapped and local Hamiltonians follow instead an *area law*, i.e. the entanglement entropy grows proportionally with the area of the cut [12]. For a cut of an N-site chain as shown in Fig. 8.1(a), this implies that $S(N)$ is constant for $N \gtrsim \xi$ (with ξ being the correlation length). This can be intuitively understood from the fact that a gapped ground state contains only fluctuations within the correlation length ξ, and thus only degrees of freedom near the cut are entangled, as illustrated schematically in Fig. 8.1(b). A rigorous proof of the area law in 1D is given in [19]. In this respect, ground states are very special states and can be found within a very small corner of the Hilbert space, as illustrated in Fig. 8.1(c).

In slightly entangled states, only a relatively small number of Schmidt states contribute significantly. This is demonstrated in Fig. 8.1(d) by comparing the largest 20 Schmidt values of an area-law state and a volume-law state for a bipartition of an $N = 16$ chain into two half-chains.

As an example of an area-law state, we consider here the ground state of the transverse field Ising model

$$H = -\sum_n \sigma_n^z \sigma_{n+1}^z + g\sigma_n^x, \tag{8.5}$$

with σ_n^x and σ_n^z being the Pauli operators and $g > 0$. The $\mathbb{Z}_2$ symmetric model with a phase transition at $g = 1$ has two very simple limits. For $g = 0$, the ground state is twofold-degenerate and given by the ferromagnetic product state (symmetry-broken), while at $g \to \infty$, the ground state is a product state in which all spins are polarized (symmetric). For intermediate values of g, the ground states are area-law-type entangled states (except at the critical point). As shown in Fig. 8.1(d) for a representative example of $g = 1.5$, the ground state has essentially the entire weight contained in a few Schmidt states. Generic states fulfilling the area law show similar behaviour, and thus the above observation provides an extremely useful approach to compress quantum states by truncating the Schmidt decomposition. In particular, we can always truncate the Schmidt decomposition at some fixed finite χ such that

$$\left\| |\psi\rangle - \sum_{\alpha=1}^{\chi} \Lambda_\alpha |\alpha\rangle_L \otimes |\alpha\rangle_R \right\| < \epsilon, \quad \forall \epsilon > 0, \quad \forall N \tag{8.6}$$

This particular property of area-law states is intimately related to the MPS representation of 1D quantum states, as we will demonstrate in Section 8.2.3.

The situation is very different for a highly entangled (volume-law) random state: all the Schmidt values are roughly constant for all $2^{N/2}$ states and thus only little weight is contained in the 20 dominant states (assuming an equal weight, we find $\sim 2^{-N/2}$ per Schmidt state).

8.2.3 Matrix product states

A generic quantum state $|\Psi\rangle$ on a chain with N sites can be written in the following MPS form [13, 30, 39]:

$$|\Psi\rangle = \sum_{j_1,\ldots,j_N} A^{[1]j_1} A^{[2]j_2} \ldots A^{[N]j_N} |j_1, \ldots, j_N\rangle. \tag{8.7}$$

Here, $A^{[n]j_n}$ is a $\chi_{n-1} \times \chi_n$ matrix and $|j_n\rangle$ with $j_n = 1, \ldots, d$ is a basis of local states at site n. We call the indices of the matrices 'bond' indices. The matrices at the boundary, i.e. $n = 1$ and $n = N$, are vectors, i.e. $\chi_0 = \chi_N = 1$, such that the matrix product in (8.7) produces a complex number. The superscript $[n]$ denotes the fact that for a generic state, each site is represented by a different set of matrices.

In order to provide some intuition for the structure of MPS, we demonstrate how to transform a generic quantum state

$$|\psi\rangle = \sum_{j_1,j_2,\ldots,j_N} \psi_{j_1,j_2,\ldots,j_N} |j_1, j_2, \ldots j_N\rangle \tag{8.8}$$

into an MPS. This can be done exactly by performing successively Schmidt decompositions as shown diagrammatically in Fig. 8.2. This diagrammatic representation, in which a rank-N tensor is represented by a symbol with N legs, is very useful for representing tensor networks and related algorithms. Connecting the legs among tensors symbolizes a tensor contraction, i.e. summing over the relevant indices.

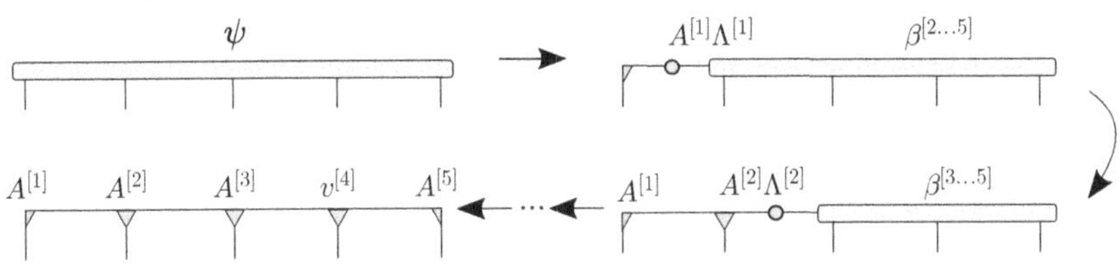

Fig. 8.2 Iterative conversion of a state $|\psi\rangle$ given by a rank-N tensor $\psi_{i_1,\ldots,i_N}$ using successive Schmidt decompositions in a diagrammatic representations. The horizontal lines represent the bond (Schmidt indices) $\alpha, \beta, \gamma, \ldots$ and the vertical lines the physical indices $j_n \in \{1, \ldots, d\}$. Connected lines between tensors denote summation over the corresponding indices (see the text for details).

We start by performing a Schmidt decomposition (8.2) of the state $|\psi\rangle$ into the first site and the rest such that

$$|\psi\rangle = \sum_{\alpha_1=1}^{d} \Lambda^{[2]}_{\alpha_1} |\alpha_1\rangle_{[1]} |\alpha_1\rangle_{[2,\ldots,N]}. \tag{8.9}$$

The states $|\alpha_1\rangle_{[1]}$ and $|\alpha_1\rangle_{[2,\ldots,N]}$ form orthogonal bases for the left and right parts, respectively. The first matrix, $A^{[i]j_1}_{\alpha_1}$, in the MPS is the matrix relating the left Schmidt states $|\alpha_1\rangle_{[1]}$ with the local states $|j_1\rangle$ (describing the local states on the first site) and is given by $A^{[1]j_1}_{\alpha_1} = \langle j_1 | \alpha_1 \rangle_{[1]}$. The resulting mixed representation of the state reads

$$|\psi\rangle = \sum_{j_1=1}^{d} \sum_{\alpha_1=1}^{d} A^{[1]j_1}_{\alpha_1} \Lambda^{[2]}_{\alpha_1} |j_1\rangle |\alpha_1\rangle_{[2,\ldots N]}. \tag{8.10}$$

We now proceed to the next bond and perform a Schmidt decomposition of the state such that

$$|\psi\rangle = \sum_{\alpha_2=1}^{d^2} \Lambda^{[3]}_{\alpha_2} |\alpha_2\rangle_{[1,2]} |\alpha_2\rangle_{[3,\ldots,N]}. \tag{8.11}$$

The second matrix, $A^{[2]j_2}_{\alpha_1\alpha_2}$, then relates the mixed basis states $|\alpha_1\rangle_{[1]}|j_2\rangle$ with the left Schmidt states $|\alpha_2\rangle_{[1,2]}$ and is given by $A^{[2]j_2}_{\alpha_1\alpha_2} = \left[\langle \alpha_1|_{[1]} \langle j_2|\right] |\alpha_2\rangle_{[1,2]}$. The resulting mixed representation of the state reads

$$|\psi\rangle = \sum_{\alpha_1=1}^{d} \sum_{\alpha_2=1}^{d^2} \sum_{j_1,j_2=1}^{d} A^{[1]j_1}_{\alpha_1} A^{[2]j_2}_{\alpha_1\alpha_2} \Lambda^{[3]}_{\alpha_2} |j_1, j_2\rangle |\alpha_2\rangle_{[3,\ldots,N]}. \tag{8.12}$$

This procedure can now be continued until the right end of the chain is reached. We choose the last matrix $A^{[N]j_n}$ to relate the states $\Lambda_{\alpha_N} |\alpha_n\rangle_{[N]}$ to the local basis $|j_n\rangle$. Then it is easy to see that we finally arrive at a representation of the state that has exactly the form (8.7).

The caveat is that the matrix dimension increases exponentially as we proceed towards the centre of the chain. However, we can make an approximation by neglecting the Schmidt states that have very small Schmidt values. For the ground state of the Ising model discussed above, we can find a very good approximation of the ground state as an MPS by keeping only a maximal bond dimension of ~20 with a truncation error that is of the order of the machine precision (independent of the system size). The same picture can be generalized to all states that obey an area law. On more general grounds, it has been proved that ground states of 1D gapped systems can be efficiently approximated by an MPS [15, 42].

8.2.3.1 Canonical form

The representation (8.7) is not unique, since an MPS with the transformed matrices

$$\tilde{A}^{[n]i_n} = X_n A^{[n]i_n} X_{n+1}^{-1} \tag{8.13}$$

represents the same state. In the following, we will show how to fix this degree of freedom by introducing a convenient *canonical* form of the MPS in which the bond index corresponds to the Schmidt decomposition.

Without loss of generality, we write the matrices $A^{[n]j_n}$ as products of $\chi_{j-1} \times \chi_j$ complex matrices $\Gamma^{[n]j_n}$ and positive, real, square diagonal matrices $\Lambda^{[n]}$,

$$|\Psi\rangle = \sum_{j_1,\ldots,j_N} \Gamma^{[1]j_1} \Lambda^{[2]} \Gamma^{[2]j_2} \Lambda^{[2]} \cdots \Lambda^{[N]} \Gamma^{[N]j_N} |j_1, \ldots, j_N\rangle, \tag{8.14}$$

as illustrated pictorially in Fig. 8.3(a, b). Let us now motivate the particular choice (8.14) for the MPS form. The freedom in choosing the MPS can be used to define a 'canonical form' of the MPS, following [49, 50]. As we will see later, the canonical form has several very useful features. Any bond n defines a bipartition of the system into sites $L = \{1, \ldots, n\}$ and $R = \{n+1, \ldots, N\}$ to the left and right of the bond.

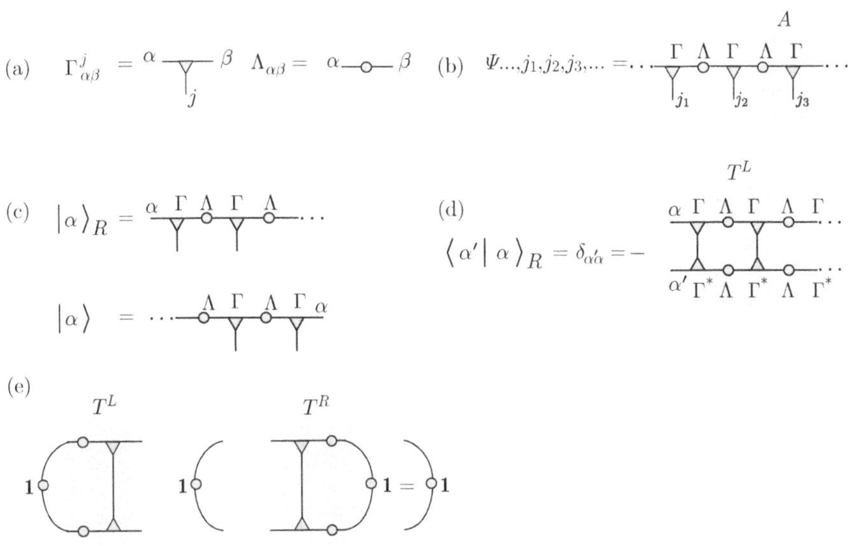

Fig. 8.3 (a) Diagrammatic representation of the tensors Γ and Λ. (b) MPS formed by the tensors Γ and Λ. (c) Definition of the right Schmidt basis states with respect to a partition on a bond with index α. (d) Condition for the MPS to be in the canonical form. The transfer matrix T^L of (8.17) has been shaded. The upward-pointing triangles are the complex conjugates of the Γ tensors. (e) If the state is in canonical form, then the dominant left eigenvector of T^L is the 'identity matrix' with eigenvalue equal to 1. A similar condition applies for the right transfer matrix T^R.

From the form of the MPS, we can define a set of χ_n wavefunctions $|\alpha\rangle_{[1,\dots,n]}$ and $|\alpha\rangle_{[n+1,\dots,N]}$ to the left and right of the bond (see Fig. 8.3(c)) such that the state takes the form

$$|\psi\rangle = \sum_{\alpha=1}^{\chi} \Lambda_{\alpha}^{[n+1]} |\alpha\rangle_{[1,\dots,n]} \otimes |\alpha\rangle_{[n+1,\dots,N]}. \tag{8.15}$$

The wavefunctions $|\alpha\rangle_{L/R}$ are formed by multiplying all matrices to the left and right, respectively. The MPS representation $\{\Gamma^{[1]}, \Lambda^{[2]}, \dots, \Gamma^{[N]}\}$ is in canonical form if *For every bond, the set of Schmidt states along with* $\Lambda^{[n]}$ *form a Schmidt decomposition of* Ψ. In other words, we must have $\langle\alpha'|\alpha\rangle_{[1,\dots,n]} = \delta_{\alpha'\alpha}$ and $\langle\alpha'|\alpha\rangle_{[n+1,\dots N]} = \delta_{\alpha'\alpha}$, along with $\sum (\Lambda_{\alpha}^{[n]})^2 = 1$ on every bond. For finite systems, a generic MPS can be transformed into canonical form by successively orthogonalizing the bonds starting from either the left or right end of the chain [41]. A great advantage of the canonical form is that local expectation values can be evaluated by only contracting the tensors locally by using the orthogonality. Note that the MPS form we obtained above by applying successive Schmidt decompositions provides naturally the canonical form with $A^{[n]j_n} = \Lambda^{[n]}\Gamma^{[n]j_n}$.

8.2.3.2 Infinite MPS

For infinite ($N \to \infty$) and translationally invariant systems, the set of matrices on any given site becomes the same, i.e. $\Gamma^{[n]j} = \Gamma^j$ and $\Lambda^{[n]} = \Lambda$ for all integers n. Computing the overlaps $\langle\alpha'|\alpha\rangle_R$ would appear to require an infinite tensor contraction. For an infinite MPS, the orthogonality condition can be conveniently expressed in terms of the *transfer matrix* T^R (illustrated in Fig. 8.3(d)) defined as

$$T^R_{\alpha\alpha';\beta\beta'} = \sum_j \Gamma^j_{\alpha\beta} (\Gamma^j_{\alpha'\beta'})^* \Lambda_\beta \Lambda_{\beta'}, \tag{8.16}$$

where * denotes complex conjugation [49]. The transfer matrix T^R relates the overlaps defined on bond n with the overlaps defined on bond $n+1$. Given that the right basis states $|\beta\rangle_R^{[n+1]}$ on bond $n+1$ are orthonormal, the states $|\alpha\rangle_R^{[n]}$ on bond n will also be orthonormal if T has a dominant *right* eigenvector $\delta_{\beta\beta'} (= \mathbf{1})$ with eigenvalue $\eta = 1$, as illustrated in Fig. 8.3(e). For the left set of states, we define an analogous transfer matrix T^L,

$$T^L_{\alpha\alpha';\beta\beta'} = \sum_j \Lambda_\alpha \Lambda_{\alpha'} \Gamma^j_{\alpha\beta} (\Gamma^j_{\alpha'\beta'})^* \tag{8.17}$$

which must have a *left* eigenvector $\delta_{\alpha\alpha'}$ with $\eta = 1$. These eigenvector criteria are clearly necessary conditions for all bonds to be canonical; in fact, assuming in addition that $\eta = 1$ is the dominant eigenvalue, they are sufficient.

A state is called *pure* if the dominant eigenvalue is unique and *mixed* if it is degenerate. In the following discussions, we will always assume that the state is *pure* (in fact, every mixed state can be uniquely decomposed into pure states). An algorithm to explicitly transform an arbitrary infinite MPS to the canonical form involves diagonalizing the two transfer matrices T^R and T^L and is given in [29]. If the infinite MPS is not translationally invariant with respect to a one-site unit cell, all the above can be simply generalized by considering a unit cell of L sites that repeats itself; for example, in the case of a two-site unit cell, the tensors are given by

$$\begin{aligned} \Gamma^{[2n]} &= \Gamma^A, & \Lambda^{[2n]} &= \Lambda^A, \\ \Gamma^{[2n+1]} &= \Gamma^B, & \Lambda^{[2n+1]} &= \Lambda^B, \end{aligned} \tag{8.18}$$

for $n \in \mathbb{Z}$. Reviews of MPS as well as the canonical form can be found in [29, 32, 49].

The infinite MPS representation in the canonical form has a number of important advantages. First, using the properties of the transfer matrices (Fig. 8.3(e)), it is very convenient to evaluate local expectation values as well as correlation functions. Second, with the help of efficient algorithms such as the *infinite time evolving block decimation* (iTEBD) [49] or the *infinite density matrix renormalization group method* (iDMRG) [28], the ground state of a given Hamiltonian can be found in the thermodynamic limit. A discussion of the two algorithms using the same notation as used in these lecture notes can be found in [24].

8.2.3.3 *Examples of infinite MPS*

To become more familiar with the infinite MPS representation, it is instructive to consider a few concrete examples.

1. *Néel state.* The state $|\ldots\uparrow\downarrow\uparrow\downarrow\ldots\rangle$ is a product state with a bond dimension $\chi = 1$ and a local Hilbert space of $d = 2$. The infinite MPS representation is given by

$$\begin{aligned} \Gamma^{[2n],\uparrow} &= \Gamma^{[2n+1],\downarrow} = 1, \\ \Gamma^{[2n],\downarrow} &= \Gamma^{[2n+1],\uparrow} = 0, \\ \Lambda^{[2n]} &= \Lambda^{[2n+1]} = 1. \end{aligned}$$

   Note that since the state is a simple product state, the matrices are actually simply complex numbers. It is easy to see that a contraction of the infinite MPS yields the desired Néel state. Furthermore, the corresponding transfer matrices trivially obey the conditions for the canonical form.
2. *Spin-1 AKLT state.* Affleck, Kennedy, Lieb, and Tasaki (AKLT) constructed an $S = 1$ Hamiltonian for which the ground state has valence bonds between all neighbouring sites (see Fig. 8.4) [2]. The AKLT Hamiltonian consists of a sum of projectors and reads

$$H = \sum_j \vec{S}_j \vec{S}_{j+1} + \frac{1}{3}(\vec{S}_j \vec{S}_{j+1})^2, \tag{8.19}$$

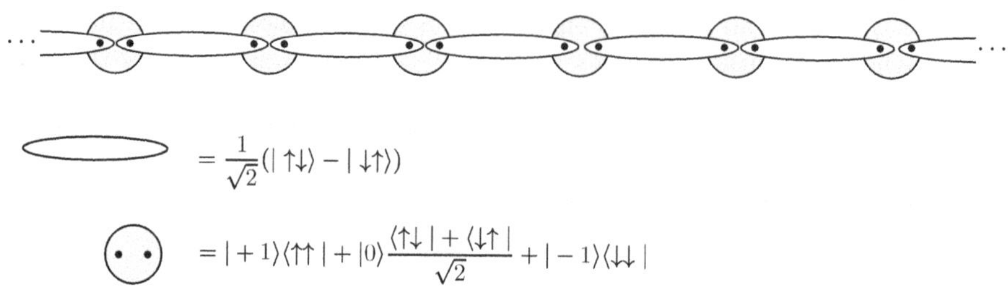

Fig. 8.4 Diagrammatic representation of the AKLT states. The $S = 1$ sites (grey circles) are decomposed into two $S = \frac{1}{2}$ that form a singlet with the neighbouring site (ellipses).

where $\vec{S}$ are the spin-1 operators. The ground state in the thermodynamic limit is unique and has a simple ($\chi = 2$) infinite MPS representation

$$\Gamma^{[n],-1} = \sqrt{\frac{4}{3}}\sigma^{+}, \qquad \Gamma^{[n],0} = -\sqrt{\frac{2}{3}}\sigma^{z}, \qquad \Gamma^{[n],1} = -\sqrt{\frac{4}{3}}\sigma^{-}, \qquad \Lambda^{[n]} = \sqrt{\frac{1}{2}}\begin{pmatrix}1 & 0\\ 0 & 1\end{pmatrix}. \tag{8.20}$$

The state can be shown to be in the canonical form by diagonalizing the corresponding left and right transfer matrices.

8.3 Symmetry-protected topological phases

8.3.1 Symmetry transformations of MPS

For the study of SPT phases, it will be essential to understand how symmetry operations act on MPS. Let us consider an on-site symmetry operation that is applied to all sites, i.e.

$$|\tilde{\psi}\rangle = \left[\bigotimes_n u_n(g)\right]|\psi\rangle, \tag{8.21}$$

where $u_n(g)$ is acting on site n with g being an element of the symmetry group G under which the state $|\psi\rangle$ is invariant. An example of such symmetry is the $\mathbb{Z}_2$ symmetry $\bigotimes_n \sigma_n^x$ of the transverse field Ising model (8.5). In the MPS formulation, the transformation corresponds to contracting the symmetry operation to all physical legs as shown in Fig. 8.5(a). In order for a state to be invariant, the overlap of the original state with the transformed state has to be of unit modulus, i.e. $|\langle\tilde{\psi}|\psi\rangle| = 1$. Thus, the mixed transfer matrices of the original and transformed MPS must have a dominant eigenvector X with eigenvalues $|\eta| = 1$. The right mixed transfer matrix has the form

$$T^{R}_{\alpha\alpha';\beta\beta'}(g) = \sum_{n}\left(\sum_{n'} u_{nn'}(g)\Gamma^{n'}_{\alpha\beta}\right)\left(\Gamma^{j}_{\alpha'\beta'}\right)^{*}\Lambda_{\beta}\Lambda_{\beta'} \tag{8.22}$$

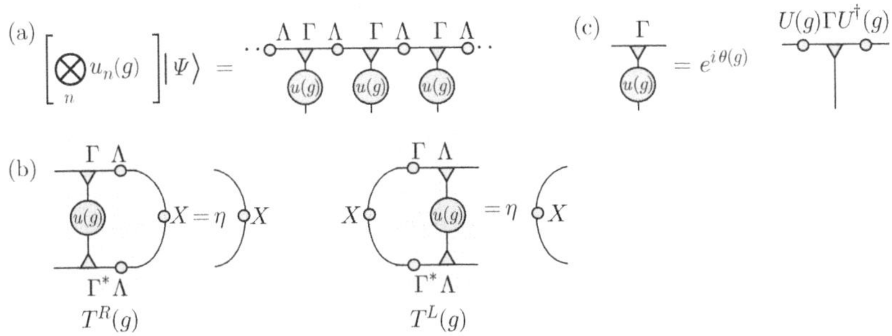

Fig. 8.5 (a) Transformation of an MPS under an on-site symmetry g applied to all sites. (b) Representation of a symmetry operation in terms of the MPS. (c) Mixed transfer matrices of the original and transformed states.

and fulfils

$$\sum_{\beta,\beta'} T^R_{\alpha\alpha';\beta\beta'}(g)X_{\beta\beta'} = \eta X_{\alpha\alpha'}. \tag{8.23}$$

Analogously, we find a similar relation for the left mixed transfer matrix $T^L(g)$. See also the diagrammatic representation in Fig. 8.5(b). If $|\eta| < 1$, the overlap between the original and transformed wavefunctions decays exponentially with the length of the chain and $|\psi\rangle$ is thus not invariant.

In [33], it was shown that for an MPS in canonical form, the matrices Γ^j transform under symmetry operations g as

$$\sum_{j'} u_{jj'}(g)\Gamma^{j'} = e^{i\theta_g} U^\dagger(g)\Gamma^j U(g), \tag{8.24}$$

with a diagrammatic representation as shown in Fig. 8.5(c). Here $U(g)$ is a unitary matrix that commutes with the Λ matrices, and $e^{i\theta(g)}$ is a phase.[1] It is clear that this is a sufficient condition for the mixed transfer matrices to have a dominant eigenvalue of unit modulus. To show that it is a necessary condition, one has to apply the Schwarz inequality and use the canonical form of the transfer matrix for the original state [33]. The matrices $U(g)$ form a χ-dimensional *projective* representation of the symmetry group of the wavefunction and $e^{i\theta(g)}$ is a linear (1D) representation [38]. The term 'projective' means that the $U(g)$ are a representation of the symmetry modulo a phase. As discussed in the following subsections, the fact that the $U(g)$ can be projective representations of the symmetries is key to understanding SPT phases. Note that the matrices $U(g)$ are actually a representation of the symmetry operations in the basis of Schmidt states (this can be seen by going back to the definition of the canonical form).

[1] As $U(g)$ commutes with Λ, it also commutes with the reduced density matrices ρ^L and ρ^R.

Similar relations can be derived for symmetries that are not on-site operations. Under a time-reversal transformation, Γ^j is transformed to $(\Gamma^j)^*$ (complex conjugate) on the left-hand side of (8.24) (including possible spin rotations). In the case of inversion symmetry, Γ^j is transformed to $(\Gamma^j)^T$ (transpose) on the left-hand side of (8.24). We refer to [38] for further details.

8.3.2 Classification of projective representations

Let us assume a group G with group elements $g_i \in G$. Then the matrices $U(g_i)$ form a *projective* representation of G if

$$U(g_i)U(g_j) = \omega(g_i, g_j)U(g_i g_j), \tag{8.25}$$

where $\omega(g_i, g_j) \in U(1)$ represent the so-called *factor set*. Thus, a projective representation is a linear representation modulo a $U(1)$ phase factor. In the case that all phase factors are unity, then the representation is a linear representation of the group. Because of the associativity of the group (i.e. the elements of G fulfil $g_i(g_j g_k) = (g_i g_j)g_k$), the factor set must satisfy

$$\omega(g_j, g_k)\omega(g_i, g_j g_k) = \omega(g_i g_j)\omega(g_i g_j, g_k). \tag{8.26}$$

Transforming the matrices as $\tilde{U}(g_i) = \beta(g_i)U(g_i)$, $\beta(g_i) \in U(1)$ yields a new factor set

$$\tilde{\omega}(g_i, g_j) = \frac{\beta(g_i g_j)}{\beta(g_i)\beta(g_j)}\omega(g_i, g_j). \tag{8.27}$$

Two projective representations $\tilde{U}(g)$ and $U(g)$ that are related by such a transformation are considered to be equivalent and belong to the same class.

It was Isaac Schur who derived in 1904 a classification of different types of projective representation using so called 'Schur multipliers' to label different classes. These correspond to the second cohomology group $H_2(G, U(1))$ of a group G. Instead of discussing the details of the proof, we refer for a general introduction to [21] and consider some simple examples.

1. *Group* $\mathbb{Z}_N$. The generators of the group are $\exp(i\pi/N)$ rotations and the group elements are $\{1, R, R^2, \ldots, R^N\}$. For a projective representation of the group, we can assign an arbitrary phase such that $U^N(R) = \exp(i\phi)$. However, a simple rescaling $U(R)$ by $\exp(i\phi/N)$ can always transform the projective representation to a linear one. Thus, this group has only one class and all projective representation can be transformed into a linear one.
2. *Group* D_2. This group is generated by π rotations R_x and R_z about two orthogonal axes. Clearly, $R_x^2 = R_z^2 = 1$ and $R_z R_x = R_x R_z$, and thus the group elements are $\{1, R_x, R_z, R_x R_z\}$. The group D_2 has two different classes of projective representations, which can be distinguished by the gauge-invariant phase factor

$$U(R_x)U(R_z)U^{-1}(R_x)U^{-1}(R_z) = \exp(i\phi),$$

with $\phi = 0, \pi$. Clearly, as each element occurs with its inverse, the phase of the commutator cannot be changed by rephasing the operators.

Both cases can be realized using a representation of the rotations in terms of spin operators by $U(R_x) = \exp(i\pi S^x)$ and $U(R_z) = \exp(i\pi S^z)$. The $S = 1$ representation with

$$S^x = \frac{1}{\sqrt{2}} \begin{pmatrix} 0 & 1 & 0 \\ 1 & 0 & 1 \\ 0 & 1 & 0 \end{pmatrix}, \qquad S^z = \begin{pmatrix} -1 & 0 & 0 \\ 0 & 0 & 0 \\ 0 & 0 & 1 \end{pmatrix} \tag{8.28}$$

is a linear ($\phi = 0$) representation. The $S = \frac{1}{2}$ spin matrices

$$S^x = \frac{1}{2} \begin{pmatrix} 0 & 1 \\ 1 & 0 \end{pmatrix}, \qquad S^z = \frac{1}{2} \begin{pmatrix} 1 & 0 \\ 0 & -1 \end{pmatrix} \tag{8.29}$$

form a projective ($\phi = \pi$) representation. This can be seen easily since $U(R_x) = \sigma_x$ and $U(R_z) = \sigma_z$ anticommute (σ_x and σ_z are the Pauli matrices).

8.3.3 Symmetry fractionalization

We now come to the core of these lecture notes and define SPT phases in 1D bosonic systems that are protected by an on-site symmetry group G. The classification scheme is based on the classification of projective representations in terms of the second cohomology classes $H_2(G, U(1))$ [9, 10, 38, 43]. While the general proof of the existence of SPT phases is given in terms of a classification of fixed-point wavefunctions of LU transformations [9, 10, 43], we follow here a more intuitive approach.

We consider systems in which the on-site representation is linear (e.g. integer spin systems). We show that a characteristic *symmetry fractionalization* occurs when a symmetry operation $g \in G$ is applied to the dominant Schmidt states of a segment S of length $\ell \gg \xi$ that we cut out of the ground state as shown in Fig. 8.6(a, b). Using the locality of the ground states, we show that the symmetry operations act non-trivially only on the boundary of the segment. We then argue that the representation of the symmetry actions onto the boundaries provides exactly the projective representations $U(g)$ as defined in the preceding subsections. As the two boundaries can be arbitrarily far away from each other, the class of the projective representation cannot be changed unless a phase transition occurs. Thus, we can use it to define a phase! *We define an SPT phase as a phase in which the boundary of the segment transforms projectively under a symmetry operation g while the bulk is in a linear representation.* Clearly, if the necessary symmetries are broken, the phase is no longer well defined.

We now discuss the details of the argument and relate it to the MPS formulation. Starting from a ground state $|\psi_0\rangle$ of a gapped Hamiltonian, we partition the system into regions ESE as shown in Fig. 8.6(a). The Schmidt decomposition for this bipartition is given by

$$|\psi_0\rangle = \sum_\gamma \Lambda_\gamma |\gamma\rangle_E |\gamma\rangle_S. \tag{8.30}$$

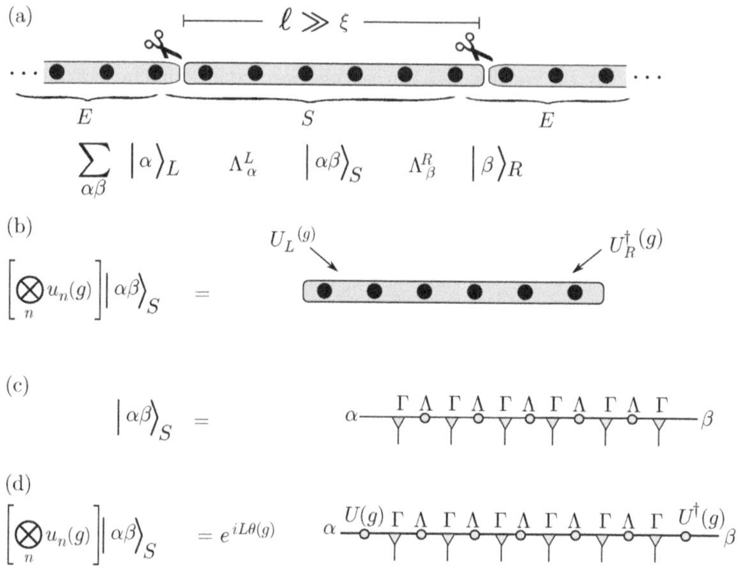

Fig. 8.6 (a) Partition of an infinite quasi-1D system into a segment S and an environment E. The segment S is a finite region of the system chosen to be large compared with the correlation length ξ. The Schmidt states on the segment of a ground state $|\psi_0\rangle$ are decomposed into tensor products of the left (α) and right (β) parts. (b) Symmetry operations acting on the dominant Schmidt states of the segment S can be represented in terms of operators acting on the boundaries. (c) Given that $|\psi_0\rangle$ is an infinite MPS formed by the χ dimensional matrices $\{\Gamma^j, \Lambda\}$, the contraction of the matrices yields the χ^2 dominant Schmidt states on the segment. (d) The Schmidt states on the segment transform under symmetry operations by local unitary transformations acting on the boundaries. These are the (projective) representations of the symmetries as defined in (8.24).

Here $|\gamma\rangle_{E/S}$ are the Schmidt states. A crucial point is that if the width ℓ of the segment is large compared with ξ, then the fluctuations across the left cut should be *independent* of the fluctuations across the right cut. The resulting Schmidt decomposition then has a tensor product structure for the important (dominant) Schmidt states. Now we label the left and right fluctuations by α and β and replace the Schmidt index γ by the pair $\gamma = (\alpha, \beta)$. Using $\Lambda_\gamma = \Lambda^L_\alpha \Lambda^R_\beta$ and $|\gamma\rangle_E = |\alpha\rangle_L \otimes |\beta\rangle_R$, we have

$$|\psi\rangle = \sum_{\alpha,\beta} \Lambda^L_\alpha \Lambda^R_\beta |\alpha\rangle_L |\alpha\beta\rangle_S |\beta\rangle_R, \tag{8.31}$$

as illustrated in Fig. 8.6(a). In this decomposition, we can think of the indices α and β as labelling the local fluctuations at the left and right of the segment. In each Schmidt state $|\alpha\beta\rangle_S$, the expectation values of any local operator have some particular spatial dependence near the ends of the chain, depending on α and β, but this decays exponentially to the ground state away from the ends. Therefore, it is possible to transform between these states by using operators defined on just the ends. A special case is the effective representations of symmetries in terms of operators $U(g)$ at the ends of the segments as illustrated in Fig. 8.6(b).

The above argument becomes more transparent when the states are represented as MPS. With the definition of the canonical form, the states $|\alpha\beta\rangle_S$ of a segment are given by

$$|\alpha\beta\rangle_S = \sum_{\{j_i\}} \left(\Gamma^{j_1}\Lambda\Gamma^{j_2}\cdots\Lambda\Gamma^{j_\ell}\right)_{\alpha\beta} |j_1, j_2, \ldots, j_\ell\rangle \tag{8.32}$$

as shown in Fig. 8.6(c). Here we assume that the state $|\psi_0\rangle$ is given as an infinite and translationally invariant MPS. When ℓ is large compared with the correlation length ξ, these states are nearly orthonormal, i.e. $\langle\alpha'\beta'|_S\alpha\beta\rangle_S \sim \delta_{\alpha'\alpha}\delta_{\beta'\beta}$. In this limit, $|\alpha\beta\rangle_S$ are the Schmidt eigenstates of the segment. If we transform the state using a symmetry operation g, then the matrices transform according to (8.24). As all unitaries $U(g)$ cancel each other with the bulk, only those at the two boundaries remain and we obtain the form shown in Fig. 8.6(d). Now, the two matrices $U(g)$ and $U^\dagger(g)$ are the representations of the symmetry action in terms of the boundaries and thus exactly those we introduced above. The phase factors $e^{i\theta(g)}$ sum up to an overall phase (which in fact characterizes different phases as long as translational invariance is preserved). Assuming the continuous change of the wave function, the class of the projective representation $\{U(g)\}$ of $g \in G$ cannot be changed locally, and thus it is 'stable' as long as the correlation length remains finite. This is why SPT phases are stable as long as the symmetry is unbroken!

In a closely related way, SPT phases can be defined in the presence of symmetries that do not have a simple on-site representation. Examples of symmetries are inversion and time-reversal symmetries and combinations thereof. We refer to [9, 10, 35, 38, 43] for details.

Note that the stability relies on the linear on-site representation of the symmetry. If a model allows for local fluctuations of the representations (e.g, by mixing integer and half-integer representations of the spin-rotation symmetry), then the SPT phase can be adiabatically connected to a trivial phase [3].

8.3.4 Spin-1 chain and the Haldane phase

We will now illustrate the main ideas of the classification of SPT phases by discussing a specific example, namely the spin-1 chain described by the Hamiltonian

$$H = J\sum_j \vec{S}_j \cdot \vec{S}_{j+1} + D\sum_j (S_j^z)^2. \tag{8.33}$$

As already discussed in the introduction, the first term is the standard spin-1 Heisenberg model with antiferromagnetic exchange interactions that stabilizes the Haldane phase [17, 18]. When the single-ion anisotropy $D \geq 0$ is tuned, the system undergoes a phase transition between two gapped *symmetric* phases at $D \approx 1$ [6, 11, 45]. The 'Haldane' phase at small D is an SPT phase and cannot be adiabatically connected to any product state as long as we preserve time reversal, bond-centred inversion, or the dihedral group of the spin-rotation symmetry [16, 38]. It has been shown numerically that the Haldane phase is adiabatically connected to the AKLT state $|\psi_{\mathrm{AKLT}}\rangle$. The so-called 'large-$D$' phase is adiabatically connected to a simple product state

$$|\psi_{\mathrm{large}\ D}\rangle = \cdots|0\rangle|0\rangle|0\rangle\cdots ,$$

which is the ground state of the Hamiltonian for $D \to \infty$. As the phase is adiabatically connected to a simple product state, we refer to the large-D phase as the trivial phase.

We now use the two simple representatives $|\psi_{\mathrm{AKLT}}\rangle$ and $|\psi_{\mathrm{large\ D}}\rangle$ to characterize the two phases. Besides many other symmetries, the Hamiltonian (8.33) has a D_2 on-site symmetry. The D_2 symmetry is a subgroup of the $U(1) \rtimes \mathbb{Z}_2$ spin-rotation symmetry, namely the Hamiltonian is invariant under rotation of all spins continuously about the z axis and π rotations about the x axis. Clearly, the on-site representation of the D_2 in terms of the spin-1 degrees of freedom is a linear one. Let us now analyse how the MPS representation (8.20) of $|\psi_{\mathrm{AKLT}}\rangle$ transforms under the D_2 symmetry. Using (8.24) with g_k being the π rotation about the x and z axis, we find that

$$U(R_x) = \sigma_x, \qquad U(R_z) = \sigma_z$$

and $\theta = \pi$ (To arrive at this result, one can simply apply the on-site symmetry operations to the MPS). The representation of D_2 is a projective one with the gauge-invariant phase factor $U(R_x)U(R_z)U^\dagger(R_x)U^\dagger(R_z) = -1$. The MPS representation of $|\psi_{\mathrm{large\ D}}\rangle$ is a simple product state of matrices in the $|0\rangle$ state. Thus, the MPS transforms trivially under the D_2 rotations:

$$U(R_x) = 1, \qquad U(R_z) = 1,$$

with $U(R_x)U(R_z)U^\dagger(R_x)U^\dagger(R_z) = 1$ and $\theta = 0$. As argued above, these phase factors characterize the two phases, since they cannot be changed unless the symmetry is broken or the system undergoes a phase transition.

8.4 Detection

The definitions in the previous section tell us exactly what kind of topological phases exist in 1D bosonic systems and how to classify them. They do not, however, give us a direct method to detect different phases. Here we discuss some practical ideas about how to detect SPT phases in numerical simulations.

8.4.1 Degeneracies in the entanglement spectrum

In [38], it is pointed out that topologically non-trivial phases must have degeneracies in the entanglement spectrum. This is, *all* eigenvalues of the reduced density matrices ρ^L and ρ^R for the bipartition of the system into two half-chains are degenerate. To see this, let us assume that the ground state is represented as an MPS and is symmetric under a symmetry group G. Using (8.24), we find the symmetry representation $U(g)$ in terms of the auxiliary indices that commutes with the reduced density matrices. If the $U(g)$ for $g \in G$ form a projective representation of the symmetry group, we can find a set of non-commuting elements such that, for example, $U(g_i)U(g_j)U(g_i)^\dagger U(g_j)^\dagger = \exp(i\phi)$. The non-trivial commutation relations require that the irreducible representations have dimensions larger than 1, which yields degeneracies in the spectrum of ρ^L and (ρ^R). For example, if $\phi = \pi$ (which is the case in the Haldane phase), then the spectrum is doubly degenerate, since ρ^L and ρ^R commute with the two unitary matrices U_x and U_z, which anticommute among themselves.

8.4.2 Extraction of projective representations from the mixed transfer matrix

The existence of degeneracies in the entanglement spectrum is a necessary condition for non-trivial cohomology classes. However, this does not distinguish among various non-trivial topological states (when there is more than one).

Let us now show how to directly obtain the projective representations $U(g)$ with $g \in G$ for a given infinite MPS state.

As we learned in Section 8.3, the matrices $U(g)$ tell us how the Schmidt states transform under the symmetry operation g. Thus, we can obtain all (projective) representations from the overlap of the Schmidt states with its symmetry transformed partners by

$$U_{\beta\beta'}(g) = \langle\beta|_R \left(\bigotimes_j u_j(g)|\beta'\rangle_R\right),$$

where $u_j(g)$ are the linear on-site representations and $|\beta\rangle_R$ are the Schmidt states describing the right half-chain. This overlap, represented in terms of an MPS in the canonical form, is shown in Fig. 8.7. In particular, it corresponds to multiplying infinitely many mixed transfer matrices $T^R(g)$ and hence only the dominant eigenvector $X_{\beta\beta'}$ remains. Clearly, as the state is symmetric under g, the dominant eigenvalue η

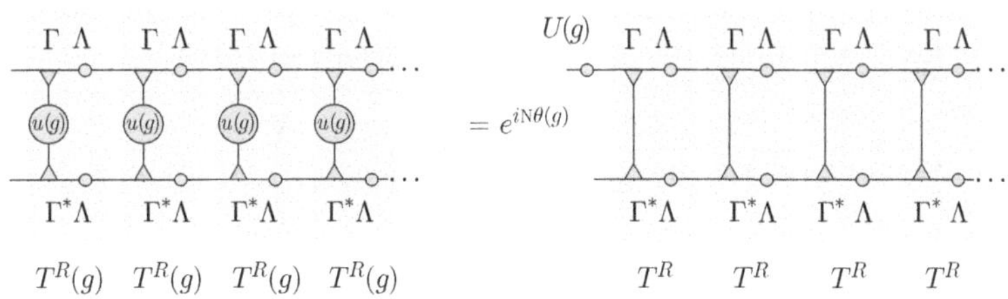

Fig. 8.7 Overlap of Schmidt states $|\alpha\rangle_R$ with its symmetry-transformed partners. If the chain is assumed to be very long, then the overlap can be expressed in terms of the eigenvector X corresponding to the largest-magnitude eigenvector $|\eta| = 1$ of the mixed transfer matrix (filled grey circles). The right boundary yields an overall phase factor $\exp iN\theta(g)$, which we ignore here (see the text for details).

of the generalized transfer matrix $T^R(g)$ is of unit modulus. On the other hand, we can apply the transformation (8.24) to each transformed matrix and see that only the $U(g)$ at the left end remains. Thus, we can read off that

$$U_{\beta\beta'}(g) = X_{\beta\beta'}. \tag{8.34}$$

Here we normalize X such that $XX^\dagger = \mathbf{1}$ and ignore a constant phase factor that results from the right end. Thus, the $U(g)$ matrices can be obtained by simply finding the dominant eigenvector of the mixed transfer matrix $T^R(g)$. Note that if the infinite MPS is not obtained in the canonical form, we need to multiply the right-hand side by the inverse of the eigenstate of the transfer matrix. Once we have obtained the $U(g)$ of each symmetry operation $g \in G$, we can read off the factor set and hence determine in which SPT phase the state is. In a similar way, SPT phases stabilized by time-reversal or inversion symmetry can be determined (see [37] for more details).

A useful procedure to detect SPT phases numerically from a microscopic Hamiltonian is to first use the iTEBD or iDMRG algorithm to find the infinite MPS representation of the ground state. The projective representations of the symmetries and thus the characterizing factor set can be obtained by diagonalizing the mixed transfer matrices. This procedure is demonstrated by a simple Python program [34].

8.4.3 String order parameters

The method demonstrated above is very useful to detect SPT phases using MPS-based algorithms. It is, however, not practical for Monte Carlo simulations or in any experimental setup. For this, we will show in this subsection how to derive non-local order parameters that detect SPT phases.

Pérez-García et al. [33] showed that the string order parameter that was originally defined for $Z_2 \times Z_2$ symmetric spin chains [11],

$$\mathcal{S}^\alpha_{\text{str}} \equiv \lim_{|j-k|\to\infty} \langle\psi_0|\, S^\alpha_j \exp\left(i\pi \sum_{j\le l<k} S^\alpha_l\right) S^\alpha_k \,|\psi_0\rangle,$$

can be generalized to systems with other symmetry groups. The generalized form for a state that is invariant under the symmetry operations $u(g)$ reads

$$\mathcal{S}(g, O^A, O^B) = \lim_{\ell\to\infty} \langle\psi_0|\, O^A(1) \left(\prod_{j=2}^{\ell-1} u_j(g)\right) O^B(\ell)\,|\psi_0\rangle. \tag{8.35}$$

The non-vanishing of this expression for generic operators only means that the state is symmetric, but does not distinguish among topologically distinct states. For example, the Z_2 symmetric Ising paramagnet (8.5) has a non-vanishing string order with $O^A = O^B = \mathbf{1}$ and $u(g) = \sigma_x$, even though it does not represent any non-trivial SPT phase! Nevertheless, we will now show that if the operators $O^A(1)$ and $O^B(n)$ are chosen appropriately, this order parameter can distinguish certain topological states. Besides the necessary condition of the state being symmetric under g, there is a second more

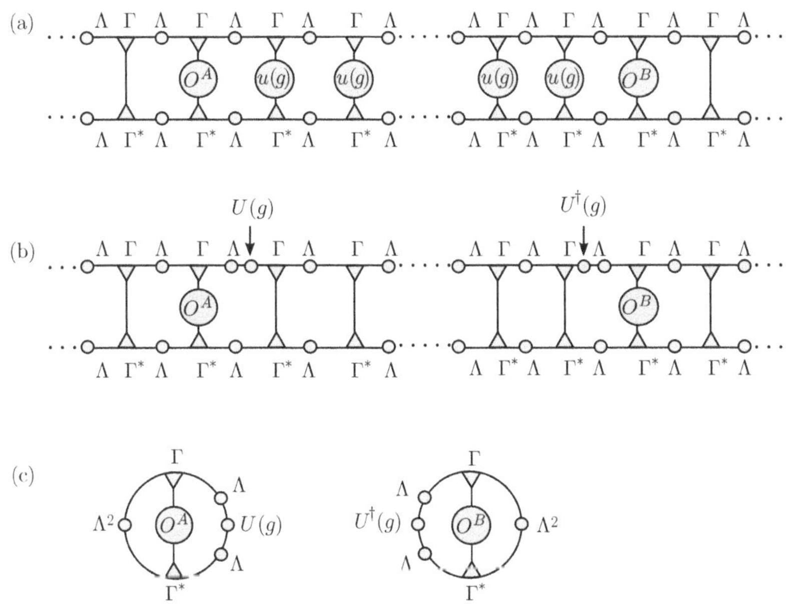

Fig. 8.8 Diagrammatic derivation of the string order $\mathcal{S}$ for a wavefunction that is symmetric under an internal transformation g and represented by an MPS in canonical form. (a) String order involving a segment of transformed sites terminated by operators O^A and O^B. (b) The matrices Γ_j transform according to (8.24) and all matrices U and $U^\dagger$ vanish except those at the edges. (c) Using the properties of the transfer matrices (defined in the text), the expectation value in (8.35) can be simplified for long segments.

refined condition for when the string order is non-zero, which *can* distinguish them. For example, the string order defined by $O^A = O^B = S^z$ vanishes in the large-D phase. Why does this occur even though the state is symmetric?

We will now closely follow a discussion given in [37] and show that (8.35) can distinguish certain SPT phases because of a *selection rule*. Intuitively, the string order corresponds to calculating the overlap between the wavefunction with g applied to ℓ consecutive sites and the wavefunction itself. Since g is a symmetry of the wavefunction, it does not change anything in the bulk of this segment and the overlap should not vanish. A diagrammatic representation of the string order is given in Fig. 8.8(a). If we represent the symmetry that is sandwiched in the middle using (8.24) and ignore the overall phase factor $e^{i\ell\theta}$, we obtain the expression that is represented diagrammatically in Fig. 8.8(b). If ℓ is large, then the part in between the $U(g)$ and the $U^\dagger(g)$ is a product of orthogonal Schmidt states of the segment, yielding a product of delta functions $\delta_{\alpha\alpha'}$ on the left and $\delta_{\beta\beta'}$ on the right (compare Fig. 8.8(c)). That is, the string order is equal to the product

$$\mathcal{S}(g, O^A, O^B) = \left\{\mathrm{tr}\left[\Lambda\bar{O}^A\Lambda U(g)\right]\right\}\left\{\mathrm{tr}\left[\Lambda\bar{O}^B\Lambda U^*(g)\right]\right\}, \tag{8.36}$$

where

$$\bar{O}^A_{\beta\beta'} = \sum_\alpha T^L_{\alpha\alpha;\beta\beta'}(O^A), \qquad \bar{O}^B_{\alpha'\alpha} = \sum_\beta T^R_{\alpha\alpha';\beta\beta'}(O^B), \tag{8.37}$$

with the generalized transfer matrices as defined in (8.22). The product (8.36) is non-zero unless at least one of the two factors is equal to zero. Whether the factors vanish depends on the symmetry of the operators $O^A(1)$ and $O^B(n)$ and can be seen as a selection rule for string order. Such selection rules exist only in the presence of additional symmetry. Thus, suppose that there are two symmetry operations g_j and g_k that commute but $U(g_j)U(g_k)U^\dagger(g_j)U^\dagger(g_j) = e^{i\phi}$. We consider the string correlator $\mathcal{S}(g_j, O^A, O^B)$ and focus on the left end of the interval. The operator O^A can be chosen to have a particular quantum number under g_k, i.e. $u(g_k)O^A u^\dagger(g_k) = e^{i\sigma}O^A$. Then a short calculation shows that $\bar{O}^A$ transforms in the same way under $U(g_j)$, i.e. $U(g_j)\bar{O}^A U^\dagger(g_j) = e^{i\sigma}\bar{O}^A$. It follows that

$$\begin{aligned} \mathrm{tr}\left[\Lambda\bar{O}^A\Lambda U^\dagger(g_j)\right] &= \mathrm{tr}\left[U(g_k)\Lambda\bar{O}^A\Lambda U^\dagger(g_j)U^\dagger(g_k)\right] \\ &= e^{i(\sigma-\phi)}\,\mathrm{tr}\left[\Lambda\bar{O}^A\Lambda U^\dagger(g_j)\right]. \end{aligned} \tag{8.38}$$

Thus, we obtain a string order selection rule: the string order parameter vanishes if $\sigma \neq \phi$. Without the second symmetry Σ^b, the string order would not vanish. Hence, a non-zero string order in a state (though intuitively surprising) is actually not so unusual; it is the *vanishing* of a string order that is the signature of a topological phase. Note that the string order might accidentally vanish (or become very small) at some points even in a phase where $\sigma = \phi$. In that case, one would have to find different operators O^A and O^B to distinguish the phases.

Note that the string order is not general, in that there are SPT phases that cannot be detected by this approach. First, there are SPT phases protected by non-on-site symmetries, for which this approach obviously fails. Second, more complex on-site symmetries exist that cannot be detected using simple string order parameters. Non-local order parameters that can detect every possible SPT phase can be found in [37].

The string order for the spin-1 Heisenberg chain can, for example, be derived simply in this way. Consider the Heisenberg chain with the symmetries $\mathcal{R}^x = \exp(i\pi S^x)$ and $\mathcal{R}^z = \exp(i\pi S^z)$. Then the selection rule implies that the string order vanishes in the trivial phase if one of the operators O^A or O^B is odd under π rotations about the x axis. The string order vanishes in the non-trivial phase if one of these operators is *even* (since U^z is odd under flips about the x axis in this phase). Thus, the string order $\mathcal{S}(R_z, \mathbf{1}, \mathbf{1})$ vanishes in the non-trivial ($\phi = \pi$) phase and $\mathcal{S}(R_z, S_z, S_z)$ does not, while the situation is reversed in the trivial ($\phi = 0$) phase. It is a nice exercise to derive the value of the string order for the AKLT state, which is known to be $\mathcal{S}(R_z, S_z, S_z) = 0.44444$.

8.5 Summary

In these lecture notes, we first introduced some basic concepts of entanglement and discussed the area law for ground states in 1D. We then derived the MPS representation and showed that these describe 1D ground states efficiently. Starting from the MPS transformation operations, we introduced the concept of SPT phases. Using an intuitive approach by studying the symmetry transformations of a segment of consecutive site, we demonstrated the stability of SPT phases. Finally, we proposed methods to detect SPT phases in numerical simulations.

While we discussed here the basics of 1D bosonic SPT, there exist a large number of generalizations. For example, based on fractionalization of symmetry operators, the classification can be extended to fermionic models. Using the concept of symmetry fractionalization, it can for example be shown that the $\mathbb{Z}$ classification of 1D topological superconductors breaks down to $\mathbb{Z}_8$ in the presence of interactions [14, 46]. Recently, there have been significant advances in understanding SPT phases in higher dimensions. In particular, it has been shown that SPT phases in higher dimensions can be classified by higher-order cocycles $H^n(G, U(1))$ [7].

Acknowledgement

I would like to thank the organizers of this school, Claudio Chamon, Mark Oliver Goerbig, Roderich Moessner, for giving me the opportunity to teach on this topic. I am grateful to my collaborators, Ari Turner, Erez Berg, and Masaki Oshikawa whose insights and work are reflected throughout my lectures.

References

[1] Affleck, I., Kennedy, T., Lieb, E. H., and Tasaki, H. (1987). Rigorous results on valence-bond ground states in antiferromagnets. *Phys. Rev. Lett.*, **59**, 799–802.

[2] Affleck, I., Kennedy, T., Lieh, E., and Tasaki, H. (1988). *Phys. Rev. Lett.*, **59**, 799.

[3] Anfuso, F. and Rosch, A. (2007). Fragility of string orders. *Phys. Rev. B*, **76**, 085124.

[4] Calabrese, P. and Cardy, J. (2004). Entanglement entropy and quantum field theory. *J. Stat. Mech.: Theory Exp.*, **2004**, P06002.

[5] Calabrese, P. and Lefevre, A. (2008). Entanglement spectrum in one-dimensional systems. *Phys. Rev. A*, **78**(032329), 032329.

[6] Chen, W., Hida, K., and Sanctuary, B. C. (2003). Ground-state phase diagram of $s = 1$ XXZ chains with uniaxial single-ion-type anisotropy. *Phys. Rev. B*, **67**, 104401.

[7] Chen, X., Gu, Z.-C., Liu, Z.-X., and Wen, X.-G. (2013). Symmetry protected topological orders and the group cohomology of their symmetry group. *Phys. Rev. B*, **87**, 155114.

[8] Chen, X., Gu, Z.-C., and Wen, X.-G. (2010). Local unitary transformation, long-range quantum entanglement, wave function renormalization, and topological order. *Phys. Rev. B*, **82**, 155138.

[9] Chen, X., Gu, Z.-C., and Wen, X.-G. (2011). Classification of gapped symmetric phases in one-dimensional spin systems. *Phys. Rev. B*, **83**, 035107.

[10] Chen, X., Gu, Z.-C., and Wen, X.-G. (2011). Complete classification of one-dimensional gapped quantum phases in interacting spin systems. *Phys. Rev. B*, **84**, 235128.

[11] den Nijs, M. and Rommelse, K. (1989). Preroughening transitions in crystal surfaces and valence-bond phases in quantum spin chains. *Phys. Rev. B*, **40**, 4709–4734.

[12] Eisert, J., Cramer, M., and Plenio, M. B. (2010). Colloquium: Area laws for the entanglement entropy. *Rev. Mod. Phys.*, **82**, 277–306.

[13] Fannes, M., Nachtergaele, B., and Werner, R. F. (1992). Finitely correlated states on quantum spin chains. *Commun. Math. Phys.*, **144**, 443–490.

[14] Fidkowski, L. and Kitaev, A. (2010). Effects of interactions on the topological classification of free fermion systems. *Phys. Rev. B*, **81**, 134509.

[15] Gottesman, D. and Hastings, M. B. (2010). Entanglement versus gap for one-dimensional spin systems. *New J. Phys.*, **12**, 025002.

[16] Gu, Z.-C. and Wen, X.-G. (2009). Tensor-entanglement-filtering renormalization approach and symmetry protected topological order. *Phys. Rev. B*, **80**, 155131.

[17] Haldane, F. D. M. (1983). Continuum dynamics of the 1-D Heisenberg antiferromagnet: identification with the $O(3)$ nonlinear sigma model. *Phys. Lett. A*, **93**, 464–468.

[18] Haldane, F. D. M. (1983). Nonlinear field theory of large-spin Heisenberg antiferromagnets: semiclassically quantized solitons of the one-dimensional easy-axis Néel state. *Phys. Rev. Lett.*, **50**, 1153.

[19] Hastings, M. B. (2007). An area law for one-dimensional quantum systems. *J. Stat. Mech.: Theory Exp.*, **2007**, P08024.

[20] Kane, C. L. and Mele, E. J. (2005). Z_2 topological order and the quantum spin Hall effect. *Phys. Rev. Lett.*, **95**, 146802.

[21] Karpilovsky, G. (1987). *The Schur Multiplier.* Clarendon Press, Oxford.

[22] Kitaev, A. (2009). Periodic table for topological insulators and superconductors. *AIP Conf. Proc.*, **1134**, 22.

[23] Kitaev, A. and Preskill, J. (2006). Topological entanglement entropy. *Phys. Rev. Lett.*, **96**, 110404.

[24] Kjäll, J. A., Zaletel, M. P., Mong, R. S. K., Bardarson, J. H., and Pollmann, F. (2013). Phase diagram of the anisotropic spin-2 XXZ model: infinite-system density matrix renormalization group study. *Phys. Rev. B*, **87**, 235106.

[25] Landau, L. D. (1937). *Phys. Z. Sowjetunion*, **11**, 26.

[26] Levin, M. and Wen, X.-G. (2006). Detecting topological order in a ground state wavefunction. *Phys. Rev. Lett.*, **96**, 110405.

[27] Li, H. and Haldane, F. D. M. (2008). Entanglement spectrum as a generalization of entanglement entropy: identification of topological order in non-abelian fractional quantum hall effect states. *Phys. Rev. Lett.*, **101**, 010504.

[28] McCulloch, I. P. (2007). From density-matrix renormalization group to matrix product states. *J. Stat. Mech.: Theory Exp.*, **2007**, P10014.

[29] Orús, R. and Vidal, G. (2008). Infinite time-evolving block decimation algorithm beyond unitary evolution. *Phys. Rev. B*, **78**, 155117.

[30] Östlund, S. and Rommer, S. (1995). Thermodynamic limit of density matrix renormalization. *Phys. Rev. Lett.*, **75**, 3537–3540.

[31] Page, D. (1993). Average entropy of a subsystem. *Phys. Rev. Lett.*, **71**, 1291–1294.

[32] Pérez-García, D., Verstraete, F., Wolf, M. M., and Cirac, J. I. (2007). Matrix product state representations. *Quantum Inf. Comput.*, **7**, 401.

[33] Pérez-García, D., Wolf, M. M., Sanz, M., Verstraete, F., and Cirac, J. I. (2008). String order and symmetries in quantum spin lattices. *Phys. Rev. Lett.*, **100**, 167202.

[34] Pollmann, F. Python code of an MPS-based simulation, demonstrating how to detect SPT phases. http://www.pks.mpg.de/ frankp/code/spt.py.

[35] Pollmann, F., Berg, E., Turner, A. M., and Oshikawa, M. (2012). Symmetry protection of topological phases in one-dimensional quantum spin systems. *Phys. Rev. B*, **85**, 075125.

[36] Pollmann, F., Mukerjee, S., Turner, A. A. M., and Moore, J. E. (2009). Theory of finite-entanglement scaling at one-dimensional quantum critical points. *Phys. Rev. Lett.*, **102**, 255701.

[37] Pollmann, F. and Turner, A. M. (2012). Detection of symmetry-protected topological phases in one dimension. *Phys. Rev. B*, **86**, 125441.

[38] Pollmann, F., Turner, A. M., Berg, E., and Oshikawa, M. (2010). Entanglement spectrum of a topological phase in one dimension. *Phys. Rev. B*, **81**, 064439.

[39] Rommer, S. and Östlund, S. (1997). Class of ansatz wavefunctions for one-dimensional spin systems and their relation to the density matrix renormalization group. *Phys. Rev. B*, **55**, 2164–2181.

[40] Schnyder, A. P., Ryu, S., Furusaki, A., and Ludwig, A. W. W. (2008). Classification of topological insulators and superconductors in three spatial dimensions. *Phys. Rev. B*, **78**, 195125.

[41] Schollwöck, U. (2011). The density-matrix renormalization group in the age of matrix product states. *Ann. Phys. (NY)*, **326**, 96–192.

[42] Schuch, N., Wolf, M. M., Verstraete, F., and Cirac, J. I. (2008). Entropy scaling and simulability by matrix product states. *Phys. Rev. Lett.*, **100**, 030504.

[43] Schuch, N. and Pérez-García, D. and Cirac, J. I. (2011). Classifying quantum phases using matrix product states and projected entangled pair states. *Phys. Rev. B*, **84**, 165139.

[44] Tagliacozzo, L., de Oliviera, T. R., Iblisdir, S., and Latorre, J. I. (2008). Scaling of entanglement support for matrix product states. *Phys. Rev. B*, **78**, 024410.

[45] Tasaki, H. (1991). Quantum liquid in antiferromagnetic chains: a stochastic geometric approach to the Haldane gap. *Phys. Rev. Lett.*, **66**, 798–801.

[46] Turner, A. M., Pollmann, F., and Berg, E. (2011). Topological phases of one-dimensional fermions: an entanglement point of view. *Phys. Rev. B*, **83**, 075102.

[47] Verstraete, F., Cirac, J. I., and Murg, V. (2008). Matrix product states, projected entangled pair states, and variational renormalization group methods for quantum spin systems. *Adv. Phys.*, **57**, 143.

[48] Verstraete, F., Cirac, J. I., Latorre, J. I., Rico, E., and Wolf, M. M. (2005). Renormalization-group transformations on quantum states. *Phys. Rev. Lett.*, **94**, 140601.

[49] Vidal, G. (2007). Classical simulation of infinite-size quantum lattice systems in one spatial dimension. *Phys. Rev. Lett.*, **98**, 070201.

[50] Vidal, G., Latorre, J. I., Rico, E., and Kitaev, A. (2003). Entanglement in quantum critical phenomena. *Phys. Rev. Lett.*, **90**, 227902.

[51] Wen, X.-G. (1989). Vacuum degeneracy of chiral spin states in compactified space. *Phys. Rev. B*, **40**, 7387.

9
Topological superconducting phases in one dimension

Felix von Oppen, Yang Peng, and Falko Pientka

Dahlem Center for Complex Quantum Systems and Fachbereich Physik
Freie Universität Berlin
Arnimallee 14, 14195 Berlin
Germany

Topological Aspects of Condensed Matter Physics. First Edition. C. Chamon et al.

Chapter Contents

Colour figures. For those figures in this chapter that use colour, please see the version of these lecture notes at http://topo-houches.pks.mpg.de. These figures are indicated by '[Colour online]' at the start of the caption.

9.1 Introduction

These lecture notes attempt to give a pedagogical account of the basic physics of Majorana bound states and the topological superconductors that host them. They introduce the basic concepts as well as possible experimental realizations, emphasizing one-dimensional systems, upon which much of the current experimental activity is focused. In writing these notes, we have preferred simple arguments and explicit derivations that illustrate the main points rather than the theoretically most pleasing or general approach. This is emphatically not a review, and the referencing is not meant to be complete or to accurately reflect the development of the field. Much more complete lists of references can be found in the original literature [1–4].

9.1.1 Motivation

Bosons are frequently their own antiparticles. Quite distinctly, however, no elementary fermion is known to have this property. A long time ago, Majorana developed a theoretical description of fermions that are their own antiparticles [5]. Ever since, there have been attempts to find such particles in nature, with neutrinos being the most likely candidate [6]. These experiments typically search for neutrinoless double-β decays (Fig. 9.1). In a β decay, a neutron decays into a proton, an electron, and a neutrino. In a neutrinoless double-β decay, two neutrons would be decaying into two protons and two electrons without emitting any neutrinos. A neutrinoless double-β decay is only possible when the neutrino is its own antiparticle and hence a Majorana fermion. Indeed, when a particle is its own antiparticle, creating and annihilating this particle is in some sense the same process. Then, the neutrino virtually emitted by one neutron can be absorbed in the β decay of the other neutron, with no real neutrino being created in the process. To date, no experiment has convincingly detected such a neutrinoless double-β decay.

In these notes, we are concerned with Majorana bound states. Unlike the particles that Majorana envisioned, these do not have any dynamics of their own in that they do not possess a dispersion as a function of a momentum quantum number. But they do share the property that annihilating and creating these excitations is described by the same operator:

$$\gamma = \gamma^\dagger, \tag{9.1}$$

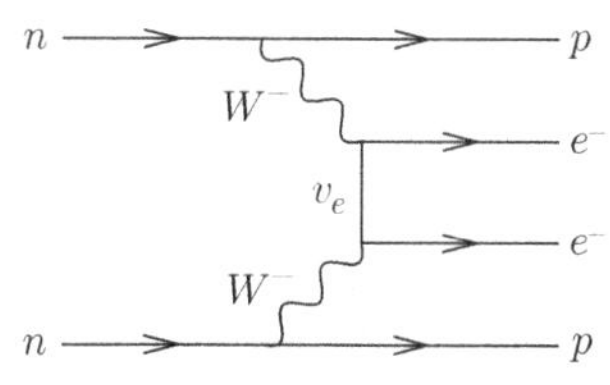

Fig. 9.1 Double-β decay is only possible when neutrinos are their own antiparticles. Then, the two neutrinos that are virtually emitted by the two neutrons can mutually annihilate.

i.e. γ is a Hermitian operator. As these bound states do not have any dynamics of their own, we can simply label them as γ_j, with $j = 1, 2, 3, \ldots$, where j enumerates, say, their locations. In a way, these condensed matter Majoranas are even more exotic and interesting than their high-energy counterparts. This is because of two of their essential properties. First, spatially isolated Majorana bound states have zero excitation energy, while non-Majorana excitations are separated by a finite energy gap. Thus, the existence of isolated Majorana bound states necessarily implies that the many-body ground state of the system is multiply degenerate. These degeneracies are highly non-local in that each twofold degeneracy of the ground state is associated with the existence of *two* Majorana bound states. Second, their quantum statistics turns out to be neither bosonic nor fermionic but rather of a new type called non-Abelian quantum statistics. When exchanging (also termed braiding) two Majorana bound states, the many-body state of the system neither remains unchanged (as for bosons) nor is it multiplied by a minus sign (as for fermions) nor even multiplied by a general phase factor (as for Abelian anyons). Instead, it undergoes a unitary rotation in the degenerate ground-state subspace. As is typical for rotations in higher dimensions, these unitary transformations within the ground-state manifold generally do not commute—hence the name non-Abelian statistics.

It is these two properties that make Majoranas attractive building blocks for topological quantum information processing [7, 8]. In his seminal work [7], Kitaev envisioned encoding quantum information within the degenerate ground-state manifold associated with the Majoranas and processing this information by means of braiding operations. Recall that the ground-state degeneracy associated with the Majoranas is robust as long as the Majoranas remain spatially isolated. This implies in particular that this degeneracy remains unaffected by local perturbations of the system, such as electric or magnetic fields, and that there are no relative dynamical phases that spoil the phase relations between different components of the many-body wavefunction within the ground-state manifold—at least as long as the fields vary adiabatically in time on the scale of the gap. This robustness to perturbing electric and magnetic fields can be thought of intuitively as a consequence of the Majoranas being their own antiparticles. Such particles can carry neither charge nor spin, since both would reverse sign for the antiparticle, and are thus unaffected by external fields.

The insensitivity to fields leads to a high degree of (topological) protection of the encoded quantum transformation against decoherence. But it also means that these fields cannot be used to manipulate this information. Instead, one hopes to use braiding operations of the Majorana bound states to manipulate the quantum information. Because of their non-Abelian statistics, such braiding operations effect unitary operations within the degenerate ground-state manifold and thus manipulation of the encoded quantum information. Just as the information storage is topologically protected, so is this strategy of information processing. Indeed, the effect of the braiding operation is insensitive to the specific geometry of the exchange path but depends only on its topology, i.e. the fact that we exchange two of the Majoranas.

Let us briefly address two concerns that one might have about this scheme. The first is simple: How does one braid the Majoranas given that they are insensitive to magnetic and electric fields? While the energy of the Majorana bound state is indeed insensitive to electric and magnetic fields, the position of the bound state is not! This is because the bound state is usually localized near a defect of the underlying phase, such as a domain wall or a superconducting vortex, and these defects can be moved by the application of electric or magnetic fields. The second concern turns out to be more serious: For a quantum computer to be universal, braiding operations must be able to implement any possible unitary transformation within the ground-state manifold. This is actually not the case for Majorana bound states. Thus, building a universal quantum computer based on Majoranas requires one to perform some operations in ways that are not topologically protected. Nevertheless, it turns out that one may still gain significantly from performing only a subset of operations in a topologically protected manner.

Above, we introduced the creation and annihilation operators associated with Majorana bound states through (9.1). In fact, any ordinary fermionic system can be discussed in terms of such operators. To see this, just note that we can always decompose conventional fermionic operators c_j (satisfying the anticommutation relations $\{c_i, c_j\} = \{c_i^\dagger, c_j^\dagger\} = 0$ and $\{c_i, c_j^\dagger\} = \delta_{ij}$) into their Hermitian and anti-Hermitian parts,

$$c_j = \frac{1}{2}(\gamma_{2j-1} + i\gamma_{2j}), \tag{9.2}$$

just as complex numbers can be decomposed into their real and imaginary parts (the factor of $\frac{1}{2}$ is a convention). Here, all γ_j are Hermitian operators, satisfying $\gamma_j = \gamma_j^\dagger$, and we have written the anti-Hermitian part of c_j as $i\gamma_{2j}$. Expressing the γ_j in terms of the original fermion operators by inverting (9.2), one readily finds that the γ_j satisfy the Majorana anticommutation relations

$$\{\gamma_i, \gamma_j\} = 2\delta_{ij}. \tag{9.3}$$

If we can discuss any fermionic system in terms of Majorana operators, what is so special about them? The answer is twofold. First, we will be concerned with situations in which the Majorana operators are eigenoperators of the system. In a generic fermionic system, the Majorana operators introduced through (9.2) are not eigenoperators. Second, the Majoranas introduced in this fashion do not correspond to spatially isolated bound states of the system. In contrast, the spatial isolation is absolutely essential for all the special properties of the Majorana bound states that we discuss in these lecture notes.

The Majorana anticommutation relation can be used to amplify the difference between Majoranas on the one hand and fermions and bosons on the other. Equation (9.3) implies that $(\gamma_j^\dagger)^2 = 1$. Thus, creating two Majoranas of the same kind brings the system back to the state it started from. This is of course eminently reasonable for a particle that is its own antiparticle—the second Majorana simply annihilates the first. On the other hand, this is very different from bosons or fermions. Fermions satisfy the Pauli principle, which implies that conventional fermionic operators square to zero. Adding two bosons to some (Fock) state by multiplying it by the square of a boson creation operator takes the system into a state that is orthogonal to the original one.

There have recently been numerous experiments that have provided possible evidence for Majorana bound states [9–17]. Many of these experiments are based on one-dimensional electron systems coupled to conventional s-wave superconductors. These notes focus on this class of systems. While there is substantial reason to be optimistic, the interpretation of many of these experiments is currently still under debate. For this reason, we will refrain from a detailed discussion of these experiments. Instead, we aim at providing the background knowledge required for understanding the experimental systems and the Majorana signatures that they are based on, allowing readers to make their own judgement.

These lecture notes are organized as follows. The remainder of this introduction will be concerned with heuristic considerations where one might reasonably look for Majorana bound states. Section 9.2 is concerned with simple model systems that exhibit Majorana bound states. We will focus on one-dimensional systems where Majoranas are associated with domain walls between topological and non-topological superconducting phases. While these models seem rather removed from experiment at first, it is now clear that they can be effectively realized in experimentally relevant systems. This is discussed at length in Sections 9.3, 9.4, and 9.5, which are concerned with proximity-coupled topological insulator edges, semiconductor quantum wires, and chains of magnetic adatoms, respectively. Section 9.6 discusses how to manipulate Majorana bound states and derives their non-Abelian statistics explicitly in a particularly simple setting. Section 9.7 discusses some popular techniques for detecting Majorana bound states experimentally. We conclude in Section 9.8.

9.1.2 Heuristic arguments

Let us start with a heuristic argument where we might be looking for Majorana excitations that are their own antiparticle, i.e. whose creation and annihilation operators satisfy $\gamma = \gamma^\dagger$. The building blocks at our disposal in a conventional metal or semiconductor material are electrons and holes, which can be viewed as particles and antiparticles. An excitation that is its own antiparticle should therefore consist in equal parts of electrons and holes. One such excitation in the solid state is the exciton, a bound state of an electron in the conduction band and a hole in the valence band. However, excitons are bound states of two fermions. These are created by products of two fermionic operators, and can thus be approximately described as bosonic excitations.

In order to realize Majorana excitations, we therefore need to consider operators satisfying $\gamma = \gamma^\dagger$ that are linear in the original fermionic operators. Such an operator consisting in equal parts of electrons and holes is[1]

$$\gamma = c + c^\dagger. \tag{9.4}$$

Written in this form, it becomes clear that we should be looking for Majorana excitations in BCS superconductors. As is familiar from the BCS theory of superconductivity,

[1] This is obviously not unique. We could also consider operators such as $i(c - c^\dagger)$ or $e^{i\varphi}c + e^{-i\varphi}c^\dagger$. This is not essential for the heuristic argument in this section.

these have fermionic quasiparticle excitations described by linear combinations of creation and annihilation operators,

$$\gamma = uc + vc^\dagger. \tag{9.5}$$

The prefactors in this linear combination depend on the energy of the (Bogoliubov) quasiparticle excitation.

An excitation far above the superconducting gap will be only weakly affected by the superconducting correlations and will consequently behave, to a good approximation, like an electron. We expect that annihilating such an excitation is essentially equivalent to annihilating an electron, and thus $u \approx 1$ and $v \approx 0$. Similarly, an excitation in the Fermi sea far below the superconducting gap will essentially look like a hole. To a good approximation, annihilating such an excitation just corresponds to filling the hole, i.e. we have $u \approx 0$ and $v \approx 1$. A Majorana excitation has equal amplitudes of c and $c^\dagger$, i.e. we are looking for an excitation with $u = v$.[2] In view of the energy dependence of the prefactors u and v, it is natural to expect $u = v$ halfway between the electron-like and hole-like excitations, i.e. for midgap excitations with excitation energy $E = 0$. We should thus be looking for Majoranas as zero-energy excitations in superconductors.

An attentive reader might object that these arguments do not make sense. The argument neglects the fact that electrons have spin and so do the Bogoliubov quasiparticles. In standard BCS theory, the Bogoliubov quasiparticles have the form

$$\gamma_\uparrow = uc_\uparrow + vc_\downarrow^\dagger, \tag{9.6}$$

which differs from (9.5) by the spin labels. Clearly, the spin indices spoil the Majorana property, i.e. these spinful Bogoliubov operators are no longer equal to their adjoint, $\gamma_\uparrow \neq \gamma_\uparrow^\dagger$, even when $u = v$.

However, there is an emergency exit that allows us to save the argument. We simply assume that we are considering BCS pairing of spinless fermions. Then the fermionic operators c and $c^\dagger$ do not have spin indices and (9.5) is the appropriate operator for the Bogoliubov excitations. Thus, we should be looking for Majoranas as zero-energy excitations in superconductors made of spinless fermions.

Finally, the assumption of spinless fermions immediately has one more consequence. In superconductors, the fermions pair into Cooper pairs. Because of the Pauli principle, the Cooper-pair wavefunction must be antisymmetric. In conventional (s-wave) superconductors, this is satisfied because the electrons are in an antisymmetric spin-singlet configuration while their orbital wavefunction is a symmetric s-state. For spinless fermions, there is no spin part of the Cooper-pair wavefunction and the antisymmetry must be in the orbital part. Then, the pairing symmetry can no longer be s-wave and the simplest antisymmetric option is p-wave pairing.

Thus, we can finally state where we should be looking for Majoranas: as *zero-energy excitations in spinless p-wave superconductors*!

[2] Or, in view of the previous footnote, more accurately $u = v^*$.

9.2 Spinless p-wave superconductors

9.2.1 Continuum model and phase diagram

Of course, the heuristic arguments of the previous section do not imply that spinless p-wave superconductors host Majoranas. To confirm that this can indeed be the case, we will now study a one-dimensional model of a spinless p-wave superconductors. To be specific, we will look at a continuum mean-field model with many-particle Hamiltonian[3]

$$\mathcal{H} = \int dx \left\{ \psi^\dagger(x) \left(\frac{p^2}{2m} - \mu \right) \psi(x) + \Delta' [\psi^\dagger(x) \partial_x \psi^\dagger(x) + \text{h.c.}] \right\}. \tag{9.7}$$

Here, $\psi^\dagger(x)$ creates a spinless fermion at position x and $\xi_p = p^2/2m - \mu$ is their normal-state dispersion. The pairing of strength Δ' (assumed real for definiteness) is of p-wave nature, as reflected in the presence of the derivative ∂_x in the pairing terms. Note that Δ' has units of velocity. The Bogoliubov-de Gennes (BdG) Hamiltonian associated with the many-body Hamiltonian is (cf. Appendix 9.A)

$$H = \begin{pmatrix} \xi_p & -i\Delta' p \\ i\Delta' p & -\xi_p \end{pmatrix} = \xi_p \tau_z + \Delta' p \tau_y, \tag{9.8}$$

where $\boldsymbol{\tau}$ is the vector of Pauli matrices τ_i in particle–hole space.

We can straightforwardly derive the excitation spectrum of the model from the BdG Hamiltonian. For an infinite system (or a system with periodic boundary conditions), momentum is a good quantum number and we obtain

$$E_k = \pm(\xi_k^2 + \Delta'^2 k^2)^{1/2} \tag{9.9}$$

by diagonalizing the 2×2 BdG Hamiltonian. This spectrum is gapped almost everywhere, except when $\xi_k = 0$ for $k = 0$, i.e. when $\mu = 0$.[4] Of course, the model also becomes gapless in the absence of pairing and any positive chemical potential, i.e. for $\Delta' = 0$ and $\mu > 0$.

The lines $\mu = 0$ as well as $\Delta' = 0$ for $\mu > 0$ delineate topological quantum phase transitions (see Fig. 9.2). To make this explicit, we rewrite the 2×2 BdG Hamiltonian in (9.8) as a spin Hamiltonian,

$$H_k = \mathbf{b}_k \cdot \boldsymbol{\tau}, \tag{9.10}$$

where $\mathbf{b}_k$ can be viewed as an effective Zeeman field acting in particle–hole space. According to the BdG Hamiltonian (9.8), we have

$$(b_k)_x = 0, \qquad (b_k)_y = \Delta' k, \qquad (b_k)_z = \xi_k. \tag{9.11}$$

[3] Often, one first discusses a lattice version of this model, the so-called Kitaev chain [18]. This is briefly discussed later in these notes in Section 9.5.3. This section can also be read at this point.

[4] It should be noted that this is distinctly different from s-wave pairing, for which the excitation spectrum $E_k = \pm(\xi_k^2 + \Delta^2)^{1/2}$ is always gapped for non-zero pairing, regardless of the chemical potential.

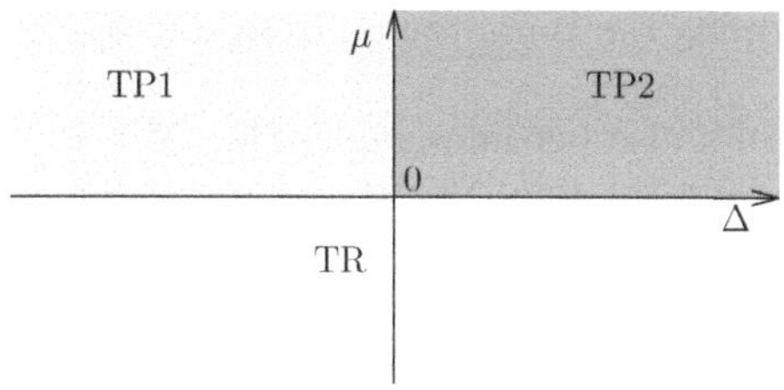

Fig. 9.2 [Colour online] Phase diagram of a one-dimensional p-wave superconductor as a function of p-wave pairing strength Δ' and chemical potential μ. There are topological superconducting phases for $\mu > 0$, while the system is topologically trivial for $\mu < 0$. The topological phases at positive μ differ in their winding numbers, depending on the sign of Δ'.

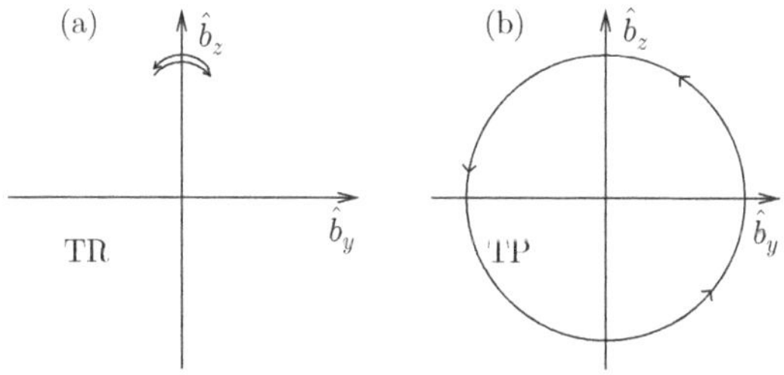

Fig. 9.3 Illustration of the mapping $k \to \hat{\mathbf{b}}_k$ in symmetry class BDI. (a) Trivial or non-topological phase. (b) Topological phase.

We can now consider the mapping from reciprocal space $k \in \mathbb{R}$ to the unit vector $\hat{\mathbf{b}}_k = \mathbf{b}_k/|\mathbf{b}_k|$. As the vector $\mathbf{b}_k$ lies in the yz plane and the corresponding unit vector on a circle, there is a topological winding number associated with this mapping, which counts the number of times the image winds around this circle. This mapping is illustrated in Fig. 9.3.

Consider first the case $\mu < 0$. Then, we have $\xi_k > 0$ for all k and the unit vector $\hat{\mathbf{b}}_k$ remains on the upper half-circle for all $k \in \mathbb{R}$, pointing in the positive z direction for $k = \pm\infty$. Thus, for $\mu < 0$, the unit vector has winding number zero.

Now consider $\mu > 0$. In this case, ξ_k changes sign from positive values at large $|k|$ to negative values near $k = 0$. Similarly, $(b_k)_y$ changes sign as k changes from negative to positive. As a result, the unit vector $\hat{\mathbf{b}}_k$ winds once around as k varies from $-\infty$ to $+\infty$. It is also evident that the direction of winding depends on the sign of Δ'. As a result, we have a winding number ± 1 depending on the sign of Δ'.

Thus, we find that the system has one phase—referred to as the non-topological or trivial phase—with zero winding number and a topological phase (or, more accurately topological phases—see below) with a non-zero winding number. The trivial phase occurs when the chemical potential is below the bottom of the normal-state band. We can change parameters in (a.k.a. deform) the Hamiltonian to the vacuum limit $\mu \to -\infty$ without ever closing the gap. This is a characteristic feature of a trivial gapped phase. In contrast, the topological phase occurs for $\mu > 0$ and there is always a gap closing when deforming the Hamiltonian from the topological phase towards the vacuum limit.

The excitation spectrum also becomes gapless for $\Delta' = 0$. This allows the winding number to change sign along the line $\Delta' = 0$ and $\mu > 0$. Thus, there are two distinct topological phases with opposite signs of the winding number. In fact, more generally, the winding number can take on any value in $\mathbb{Z}$, as long as the Hamiltonian satisfies a chiral symmetry. In the present case, this chiral symmetry is reflected in the fact that the Hamiltonian (9.10) involves only two of the three Pauli matrices, so that

$$\{\tau_x, H_k\} = 0, \tag{9.12}$$

which places this Hamiltonian into symmetry class BDI.[5] Clearly, it is this absence of the third Pauli matrix that makes the unit vector $\hat{\mathbf{b}}_k$ lie in a plane, which in turn allows the definition of a winding number.

We could also consider more general Hamiltonians that involve all three Pauli matrices, which fall into symmetry class D. Even in this case, we can define a topological index, which is now a $\mathbb{Z}_2$ index taking on only two distinct values corresponding to the trivial and topological phases. To understand this, it is perhaps easier to consider a lattice system with Brillouin zone $k \in [-\pi/a, \pi/a]$. The essential observation is that in a spinless system, there can be no pairing of the $k = 0$ and $k = \pm\pi/a$ states. The reason is that pairing is between states with opposite momenta. For $k = 0$, the opposite-momentum state would be the state itself. For $k = \pi/a$, this is actually also the case, since it differs from its opposite-momentum partner $k = -\pi/a$ by a reciprocal lattice vector, so that the two need to be identified. The absence of pairing at $k = 0$ and $k = \pm\pi/a$ implies that the unit vector $\hat{\mathbf{b}}_k$ necessarily points along the z direction at these points of the Brillouin zone, in either the positive or the negative z direction.

Now the mapping from the Brillouin zone to the unit vector $\hat{\mathbf{b}}_k$ is a mapping into the surface of a sphere and there can be two topologically distinct band structures (see Fig. 9.4): either $\hat{\mathbf{b}}_k$ has the same sign at $k = 0$ and $k = \pm\pi/a$, which corresponds

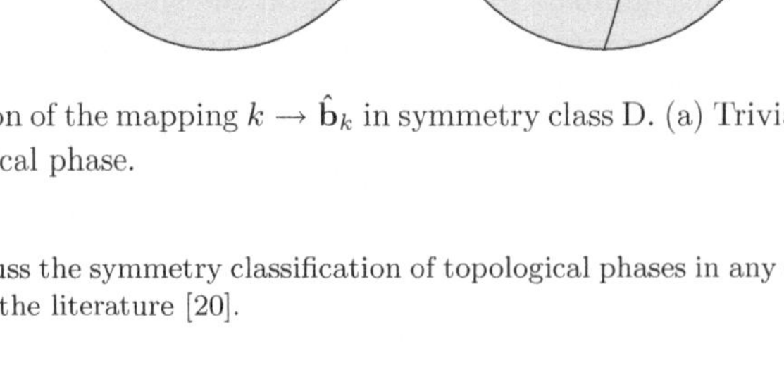

Fig. 9.4 Illustration of the mapping $k \to \hat{\mathbf{b}}_k$ in symmetry class D. (a) Trivial or non-topological phase. (b) Topological phase.

[5] We will not discuss the symmetry classification of topological phases in any detail. The interested reader is referred to the literature [20].

to the trivial phase, or it has opposite signs, which happens in the topological phase. Physically, this index measures whether the chemical potential falls within the band or not. Indeed, $k = 0$ and $k = \pm\pi/a$ correspond to the minimum and maximum of the normal-state band and the sign of the z component of $\hat{\mathbf{b}}_k$ is determined by whether the normal-state energy ϵ_k is below or above the chemical potential.

9.2.2 Domain walls and Majorana excitations

We can induce a domain wall between the topological and trivial phases by a spatially varying chemical potential. At the domain wall, the chemical potential changes from negative to positive values. Let us assume that μ changes linearly in x in the vicinity of the domain wall:[6]

$$\mu(x) = \alpha x. \tag{9.13}$$

Clearly, the gap in the excitation spectrum vanishes right at the domain wall and increases linearly away from it. Thus, it is natural to suspect that there are bound states trapped at the domain wall.

Quite generally, domain walls between topologically distinct phases are associated with gapless excitations. Examples are the chiral edge states of quantum Hall states or the helical edge states of topological insulators. We will now see explicitly that in the present case, the domain wall is also associated with a gapless excitation and that this gapless excitation is a zero-energy Majorana bound state.

While this is a general property of domain walls in this model, we will only consider a limit in which the calculation becomes particularly simple. Indeed, for a sufficiently smooth domain wall, the relevant momenta in the vicinity of the domain wall are small and we can neglect $p^2/2m$ in the BdG Hamiltonian (9.8).[7] Then, the BdG Hamiltonian of the domain wall takes on the form

$$H = -\alpha x \tau_z + \Delta' p \tau_y. \tag{9.14}$$

This has the form of a Dirac Hamiltonian with a spatially varying mass. In fact, the mass changes sign at the position of the domain wall. Following the seminal work of Jackiw and Rebbi [19] in the context of high-energy physics, this implies under rather general conditions that there is a zero-energy bound state localized at the domain wall.

Indeed, the spectrum of this Dirac Hamiltonian is readily obtained by squaring the Hamiltonian. As with any BdG Hamiltonian, the spectrum of H is symmetric about zero energy, i.e. for any eigenstate with energy E, there is another eigenstate with energy $-E$ (see Appendix 9.A). Thus, we do not lose any information on the

[6] It is also straightforward to study domain walls at which the chemical potential jumps abruptly from negative to positive values. It is left as an exercise for the reader to derive the subgap spectrum in this case.

[7] To see this, compare the chemical potential and the pairing term. This implies that there is a characteristic length $\sqrt{\Delta'/\alpha}$ and thus a characteristic energy $\sqrt{\Delta'\alpha}$. Then, the quadratic term is of order $\alpha/m\Delta'$, which is small compared with the characteristic energy as long as $\alpha \ll m^2\Delta'^3$.

spectrum when squaring the Hamiltonian. Using the fact that the Pauli matrices square to unity and anticommute,

$$\{\tau_i, \tau_j\} = 2\delta_{ij}, \tag{9.15}$$

we obtain

$$H^2 = (\alpha x)^2 + (\Delta' p)^2 - \Delta'\alpha[x, p]\tau_z\tau_y. \tag{9.16}$$

With the commutator $[x, p] = i$ and $\tau_z\tau_y = -i\tau_x$, this simplifies to

$$H^2 = (\alpha x)^2 + (\Delta' p)^2 - \Delta'\alpha\tau_x. \tag{9.17}$$

In the eigenbasis of τ_x (labelled by $\pm$), this is a harmonic-oscillator Hamiltonian up to a shift in energy, and the bound-state spectrum associated with the domain wall becomes

$$(E_n^{\pm})^2 = 2\Delta'\alpha\left(n + \tfrac{1}{2}\right) \mp \Delta'\alpha. \tag{9.18}$$

In line with general expectations for domain walls between topologically distinct phases, there is an eigenstate with exactly zero energy, $E_0^+ = 0$.

To better understand the quasiparticle excitation that is described by this zero-energy state, let us consider the associated Bogoliubov operator. The zero-energy eigenspinor of the BdG Hamiltonian is

$$\langle x|n = 0, +\rangle = u_0(x)\begin{pmatrix} 1 \\ 1 \end{pmatrix}, \tag{9.19}$$

where $u_0(x)$ is the Gaussian ground-state eigenfunction of the harmonic oscillator, centred at the domain wall. Then, the Bogoliubov operator follows in the usual way (cf. Appendix 9.A) by 'dotting' the bra $\langle n = 0, +|$ into the Nambu spinor $(\psi(x),\ \psi^\dagger(x))^T$. This gives

$$\gamma = \int dx\, u_0(x)[\psi(x) + \psi^\dagger(x)]. \tag{9.20}$$

This quasiparticle operator does indeed obey the Majorana property $\gamma = \gamma^\dagger$.[8]

9.2.3 Topological protection and many-body ground states

We have derived the zero-energy Majorana mode only for a special limit in which the calculation becomes particularly simple. It is thus natural to ask how general the result is. As mentioned above, such a zero-energy Majorana state is generically found at any domain wall between the topological and trivial phases. A simple argument that shows that this must be the case is the following. Consider a semi-infinite 'wire' in the topological phase. The end of the system is a domain wall between the topological phase and the trivial phase as represented by the 'vacuum' outside the system. Thus,

[8] Note that we can take the ground-state wavefunction $u_0(x)$ of the harmonic oscillator to be real.

there is one zero-energy Majorana bound state localized at the end of the wire. Its BdG spectrum consists of the zero-energy Majorana state and the symmetric quasiparticle continua outside the gap at positive and negative energies. Imagine that we are now deforming the Hamiltonian by changing its parameters. As long as the gap does not close, the symmetry of the BdG spectrum between positive and negative energies implies that the zero-energy state must stay put at zero energy!

Next consider a system of finite length. Then there are two Majorana bound states, one at each end of the wire. As long as we can neglect the overlap between these two localized Majorana bound states, both Majorana bound states have exactly zero energy. Overlap between the Majorana end states introduces a coupling between them, and the two Majorana zero modes can split into two states whose energies are non-zero and symmetric about zero energy. As the Majoranas are exponentially localized, the energy splitting decreases exponentially with the length of the wire.

Now consider a wire that is sufficiently long that we can neglect the overlap between the Majoranas at its ends. Then, the excitation spectrum has two zero-energy Majorana bound states with their corresponding quasiparticle operators, say, γ_1 and γ_2. We can combine these two Majorana operators into one conventional fermion operator

$$c = \frac{1}{2}(\gamma_1 + i\gamma_2). \tag{9.21}$$

As usual, this conventional fermion can be either empty or occupied. Since the Majorana bound states have zero excitation energy, both states have exactly the same many-body energy, and we find that there are two degenerate ground states.

Interestingly, these two states differ by fermion number parity. The mean-field Hamiltonian of superconductors breaks particle number conservation. Fermion number parity, however, remains a good quantum number, since the pairing terms add or remove particles only in pairs. In conventional superconductors, we expect the ground state to have even fermion parity. Any state with an odd number of fermions would necessarily have one unpaired electron, which is less favourable than a fully paired state. In contrast, the two ground states of our wire in a topologically non-trivial superconducting phase differ by the occupation of a single fermion state, so that we have one ground state with even and one ground state with odd fermion number parity. The fermion parity operator can be written as $P = 2c^\dagger c - 1$ with eigenvalues ± 1, or as

$$P = i\gamma_1\gamma_2 \tag{9.22}$$

when expressed in terms of the Majorana operators.

When there are $2N$ Majorana bound states, we can use the same strategy and group them into N pairs. Each pair of Majorana fermion operators γ_{2j-1} and γ_{2j} can be combined into a conventional fermion operator c_j. Each of these conventional fermion states can now be empty or occupied, leading to an overall ground-state degeneracy of 2^N. These states can again be grouped according to fermion parity. The fermion parity operator is just the product over the fermion parity operators $2c_j^\dagger c_j - 1$ for each pair, so that

$$P = i^N \gamma_1\gamma_2 \cdots \gamma_{2N}. \tag{9.23}$$

Thus, there are 2^{N-1} ground states of either parity.

9.2.4 Experimentally accessible systems

While it is perhaps theoretically pleasing that spinless p-wave superconductors host Majorana excitations, this result may seem rather unphysical. First and foremost, electrons do have spin. Second, the vast majority of superconductors in nature are s-wave, and there are only very few p-wave superconductors. Moreover, we will be looking for Majoranas in one-dimensional systems for which, strictly speaking, the mean-field BCS theory underlying these arguments is not appropriate owing to strong order-parameter fluctuations.

Nevertheless, starting with the seminal work of Fu and Kane [21, 22], it has become abundantly clear that this scenario can be realized experimentally in a variety of systems. The basic physical ingredients are the same in all of these platforms:

- proximity coupling to a conventional s-wave superconductor;
- spin polarization;
- spin–orbit coupling.

Employing proximity-induced superconductivity makes it appropriate to discuss the one-dimensional systems within mean-field theory, since the superconducting correlations are inherited from a bulk superconductor. Spin-polarized electron systems are a close relative of spinless fermion systems. Of course, there is a conflict in that it is impossible to proximity-induce s-wave pairing in a spin-polarized system. The reason is that, in order to satisfy the Pauli principle, the Cooper pairs are spin singlets. Such spin-singlet Cooper pairs obviously cannot enter into a spin-polarized system by spin-conserving processes.

This conflict is really an opportunity when involving spin–orbit coupling. To understand this, it might be simplest to locate the spin–orbit coupling in the superconductor rather than in the one-dimensional system. Then, orbital angular momentum is no longer a good quantum number in the superconductor and there can be a small p-wave admixture to the s-wave pairing. Unlike the s-wave correlations, the p-wave correlations can transfer to the spin-polarized system. As a result, the one-dimensional system effectively develops p-wave superconducting correlations by proximity.

In fact, the conditions for realizing topological superconductivity are less stringent than this argument may make it appear. For instance, the spin–orbit coupling can be in the proximitizing superconductor or in the one-dimensional system, and the assumption of full spin polarization can be relaxed. In the following three section, we will discuss some of the platforms that are most actively being pursued in experiments.

9.3 Topological insulator edges

9.3.1 Model and phases

We start illustrating the physics outlined at the end of the last section by two-dimensional topological insulators, proximity-coupled to an s-wave superconductor [22]. We assume that the Fermi energy is in the gap of the topological insulator, so that the only relevant electronic degrees of freedom are the helical edge states. Thus, there is just a single spin channel propagating in each direction at the Fermi energy, and there is perfect spin–orbit coupling as the propagation direction is directly tied to the spin polarization. We can gap out these edge states in two different ways, namely by proximity coupling to an s-wave superconductor and by applying a Zeeman field in a direction perpendicular to the spin quantization direction of the edge states.

The corresponding BdG Hamiltonian of the proximity-coupled topological insulator edge takes the form (see Appendix 9.A)

$$H = v_F p \sigma_x \tau_z - B\sigma_z + \Delta\tau_x. \tag{9.24}$$

Here, σ_i are the Pauli matrices in spin space. We assume that the helical edge states are polarized along the x direction, while the Zeeman field is applied in the z direction. The τ_i still denote Pauli matrices in particle–hole (Nambu) space. This way of writing the Hamiltonian assumes that we write the Nambu spinors as $(\psi_\uparrow, \psi_\downarrow, \psi_\downarrow^\dagger, -\psi_\uparrow^\dagger)^T$; see Appendix 9.A. For simplicity, we choose the chemical potential to be at $\mu = 0$ and thus right at the Dirac point associated with the edge states.

In the Hamiltonian (9.24), we have accounted for the proximity coupling to the superconductor through the induced s-wave gap Δ. In a more microscopic theory, we would describe both the topological insulator and the superconductor, including the coupling between the two. It turns out that, with certain caveats, this can then be reduced to the form of (9.24). For the most part, we will introduce the proximity-induced pairing correlations in the simplified manner of (9.24). The more microscopic approach is sketched in Appendix 9.B. We also need to rely on the microscopic approach in Section 9.5 when discussing chains of magnetic adatoms.

We can again obtain the spectrum of Hamiltonian (9.24) by squaring it. This yields

$$H^2 = (v_F p)^2 + B^2 + \Delta^2 - 2B\Delta\sigma_z\tau_x \tag{9.25}$$

and thus

$$E_k = \pm\sqrt{(v_F k)^2 + (\Delta \pm B)^2}, \tag{9.26}$$

where all combinations of signs are possible. Note that the gap closes for $B = \pm\Delta$, showing that the gaps due to Δ and B compete. This gap closing signals a topological phase transition.

At first sight, it may appear that the Hamiltonian (9.24) involves only s-wave pairing. However, in many ways, the Hamiltonian rather describes a p-wave superconductor owing to the anomalous kinetic energy. For instance, in the vicinity of the critical lines $B = \pm\Delta$, the Hamiltonian reduces to the same Dirac Hamiltonian as the spinless p-wave superconductor. To see this, we expand the Hamiltonian about the critical point $B = \Delta$. According to (9.25), the low-energy subspace is spanned by the eigenstates of $\sigma_z\tau_x$ with eigenvalue $+1$, i.e. by

$$|+\rangle = \frac{1}{\sqrt{2}}\begin{pmatrix}1\\0\\1\\0\end{pmatrix}, \qquad |-\rangle = \frac{1}{\sqrt{2}}\begin{pmatrix}0\\1\\0\\-1\end{pmatrix}. \tag{9.27}$$

Evaluating the matrix elements of H in this basis, we readily obtain the effective low-energy Hamiltonian

$$H \simeq \begin{pmatrix} \Delta - B & v_F p \\ v_F p & -(\Delta - B) \end{pmatrix}. \tag{9.28}$$

Indeed, this has the same structure as the domain-wall Hamiltonian in (9.14) for the spinless p-wave superconductor.[9] The Dirac mass is given by $\Delta - B$, which changes sign at the critical line $B = \Delta$.

For the spinless p-wave superconductor, we clearly identified one of the phases as topological while the other was topologically trivial. In the present case, such an identification is less obvious. In many ways, it turns out that the Δ-dominated phase is topological. However, this is not the full story. To start with, the underlying model of the topological insulator edge has a linear spectrum, and thus no well-defined vacuum (or atomic) limit that is obviously trivial. Moreover, the two phases of the proximity-coupled topological insulator edge are related by a superconductor–magnetism duality. To see this, let us rotate the Hamiltonian (9.24) about the y axis in spin space, such that $\sigma_x \to \sigma_z$ and $\sigma_z \to -\sigma_x$. Then, the Hamiltonian becomes

$$H = v_F p \sigma_z \tau_z + B\sigma_x + \Delta\tau_x. \tag{9.29}$$

Clearly, this Hamiltonian is invariant under the duality transformation $\tau_i \leftrightarrow \sigma_i$ and $B \leftrightarrow \Delta$, which just interchanges magnetic and superconducting quantities. This duality obviously maps the two phases into one another, since the interchange $B \leftrightarrow \Delta$ changes the sign of the Dirac mass. Strictly speaking, it is thus difficult to identify one of the phases as topological.

In fact, the duality of the model has physical consequences. As shown in [22], a Josephson junction between two Δ-dominated regions with a B-dominated junction region exhibits an anomalous 4π-periodic Josephson effect (this is discussed further in Section 9.7). The magnetism–superconductivity duality implies that there is also a 4π-periodic spin Josephson effect in the inverse B–Δ–B junction arrangement when the Zeeman field points in different directions (perpendicular to the spin–orbit field) on the two sides of the junction while the superconducting phase is uniform across the junction. Note that the direction of the magnetic field in the plane perpendicular to the spin–orbit direction maps onto the superconducting phase under the duality transformation. Incidentally, this magneto-Josephson effect may be easier to observe than it might seem, in that, owing to the spin–orbit coupling, the spin Josephson current is accompanied by a much more easily measurable charge current [23].

9.3.2 Zero-energy states and Majorana operators

Above, we explicitly constructed the Bogoliubov quasiparticle operator associated with the zero-energy domain wall state in spinless p-wave superconductors and showed that

[9] Strictly speaking, this Hamiltonian involves $v_F p\tau_x$, while the corresponding Hamiltonian for the spinless p-wave superconductor involved $\Delta' p\tau_y$. These two Hamiltonians can obviously be mapped onto each other by a trivial rotation about the z axis of particle–hole space.

it is a Majorana operator satisfying $\gamma = \gamma^\dagger$. Now that we are considering more physical spinful models, it may be useful to exhibit this connection more generally.

With the convention that the Nambu spinor is ordered as $(\psi_\uparrow, \psi_\downarrow, \psi_\downarrow^\dagger, -\psi_\uparrow^\dagger)^T$, the BdG Hamiltonian anticommutes with the product of time reversal $T = i\sigma_y k$ and charge conjugation $C = -i\tau_y$

$$\{H, CT\} = 0. \tag{9.30}$$

and the spectrum of H is symmetric about $E = 0$, i.e. for every eigenstate $|\psi\rangle$ with energy E, there is an eigenstate $CT|\psi\rangle$ with energy $-E$; see Appendix 9.A.

Now, let us assume that H has a zero-energy eigenstate $|\gamma\rangle$ that is spatially isolated from any other zero-energy solution. This is exactly the situation associated with a domain wall. Since T and C are local operations, we must conclude that

$$|\gamma\rangle = CT|\gamma\rangle. \tag{9.31}$$

To see what this implies, we write the corresponding BdG spinor as $|\gamma\rangle = (\chi_e, \chi_h)^T$, where χ_e and χ_h are themselves 2-component Pauli spinors with spin-up and spin-down components. Then, the relation (9.31) implies that the electron and hole spinors are related through

$$\chi_h = T\chi_e \tag{9.32}$$

as well as $\chi_e = -T\chi_h$ (note that $T^2 = -1$). Thus, we can write the BdG spinor $|\gamma\rangle$ explicitly as

$$\langle x|\gamma\rangle = (\chi_\uparrow, \chi_\downarrow, \chi_\downarrow^*, -\chi_\uparrow^*)^T, \tag{9.33}$$

and the corresponding Bogoliubov quasiparticle operator becomes

$$\begin{aligned} \gamma &= \int dx\, (\chi_\uparrow, \chi_\downarrow, \chi_\downarrow^*, -\chi_\uparrow^*) \cdot (\psi_\uparrow, \psi_\downarrow, \psi_\downarrow^\dagger, -\psi_\uparrow^\dagger)^T \\ &= \int dx\, (\chi_\uparrow\psi_\uparrow + \chi_\downarrow\psi_\downarrow + \chi_\downarrow^*\psi_\downarrow^\dagger + \chi_\uparrow^*\psi_\uparrow^\dagger). \end{aligned} \tag{9.34}$$

This operator γ clearly satisfies the Majorana relation $\gamma = \gamma^\dagger$.

9.4 Quantum wires

The minimal physics that turns the proximity-coupled topological insulator edge into a topological superconductor actually does not include the topological insulator properties, but merely the fact that there is only one left-moving and one right-moving channel each. This is sufficient to emulate the spinless-fermion situation discussed in the introduction.

This point is made explicit by the quantum-wire proposal [24, 25] to realize a topological superconducting phase and Majorana bound states. Let us consider a single-channel (i.e. strictly one-dimensional) quantum wire with Rashba spin–orbit coupling and applied Zeeman field B, proximity-coupled to an s-wave superconductor with induced pairing Δ:

$$H = \left(\frac{p^2}{2m} + up\sigma_x - \mu\right)\tau_z - B\sigma_z + \Delta\tau_x. \tag{9.35}$$

Here, u denotes the strength of the Rashba spin–orbit coupling and we include a chemical potential μ. Importantly, the spin–orbit field, taken along the x direction, is perpendicular to the Zeeman field, taken along the z direction. Note that this is just the topological insulator Hamiltonian (9.35) except for the kinetic-energy term. We will see that this term still allows topological superconducting phases, but also leads to important differences in the physics.

It is not very difficult to diagonalize and study the Hamiltonian (9.35) in full generality. However, it is perhaps more enlightening to restrict attention to limiting cases in which the physics becomes more transparent and which can be related to the models of topological superconducting phases that we have already discussed. Specifically, we consider two limits, depending on the strength of the Zeeman field relative to the spin–orbit coupling as measured by $\epsilon_{\mathrm{so}} = mu^2$; see Fig. 9.5. We will always assume that $\Delta \ll \max\{B, \epsilon_{\mathrm{so}}\}$. Then, we can consider the following limits:

- *Kitaev limit:* $B \gg \epsilon_{\mathrm{so}}$. First neglecting the spin–orbit coupling, the normal-state dispersion

  $$\epsilon_p = \frac{p^2}{2m} \pm B \tag{9.36}$$

  consists of two vertically shifted parabolas for the spin-up and spin-down electrons. The main effect of the spin–orbit coupling is that the spin polarizations of the parabolas are slightly tilted away from the Zeeman direction, with the tilt angle being proportional to p and thus having opposite signs for positive and negative momenta. Now, let us imagine that the chemical potential is placed below the band bottom of the spin-down band. Then, there is only a single left-moving and a single right-moving channel (or none at all), and we will see below that this limit maps to the spinless p-wave superconductor discussed above.

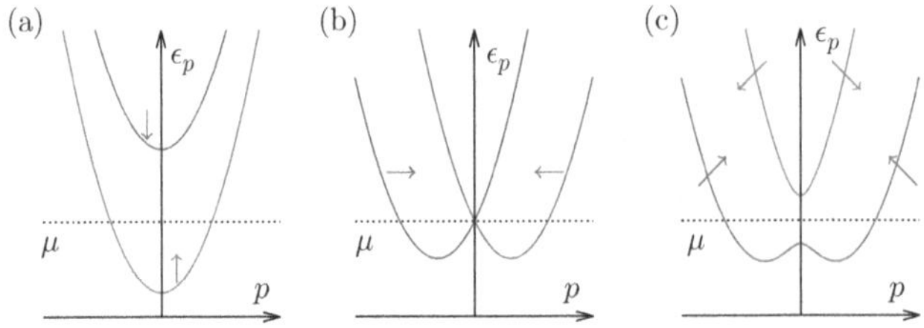

Fig. 9.5 [Colour online] Normal-state dispersions of the quantum wire in (a) the Kitaev limit, (b) the topological insulator limit without Zeeman field, and (c) the topological insulator limit with Zeeman field.

- *Topological insulator limit:* $B \ll \epsilon_{\rm so}$. First neglecting the Zeeman field, the normal-state dispersion

$$\epsilon_p = \frac{p^2}{2m} \pm up = \frac{1}{2m}(p \pm mu)^2 - \frac{1}{2}mu^2 \tag{9.37}$$

consists of two parabolas shifted relative to each other along the momentum axis owing to the Rashba spin–orbit coupling. The two parabolas correspond to spin-up and spin-down electrons with respect to the direction of the spin–orbit field (i.e. the x direction for the Hamiltonian in (9.35)) and cross at $p = 0$. The Zeeman field applied in a direction perpendicular to the spin–orbit field (the z direction for the Hamiltonian in (9.35)) mixes the two states at $p = 0$ and this opens a gap of size $2B$ in the spectrum, which now becomes[10]

$$\epsilon_p = \frac{p^2}{2m} \pm \sqrt{(up)^2 + B^2}\,. \tag{9.38}$$

When we adjust the Fermi energy to lie within this gap, we again have a situation in which there are only a single right-moving mode and a single left-moving mode at the Fermi energy. We will see that this limit is closely related to the topological insulator model discussed in the previous section.

9.4.1 Kitaev limit

First we consider the limit of strong Zeeman field with the Fermi energy lying far below the bottom of the spin-down parabola. In that case, we can project out the high-energy states associated with the spin-down parabola and derive an effective low-energy Hamiltonian. To do so, we first neglect the spin–orbit coupling and measure the Fermi energy from the bottom of the spin-down band, i.e. shift

$$H = \left(\frac{p^2}{2m} - (B + \mu)\right)\tau_z + \text{pairing terms} \longrightarrow \left(\frac{p^2}{2m} - \mu\right)\tau_z + \text{pairing terms}. \tag{9.39}$$

We now consider the pairing terms. In the absence of spin–orbit coupling, the low-energy space of the BdG equation is spanned by the spin-up electron

$$|\mathrm{e}\rangle = (1,\ 0,\ 0,\ 0)^T \tag{9.40}$$

and the spin-up hole

$$|\mathrm{h}\rangle = (0,\ 0,\ 0,\ 1)^T. \tag{9.41}$$

We can now readily see that within this subspace, there are no pairing terms. Indeed, we find that $\langle \mathrm{e}|\Delta\tau_x|\mathrm{e}\rangle = \langle \mathrm{h}|\Delta\tau_x|\mathrm{e}\rangle = \langle \mathrm{e}|\Delta\tau_x|\mathrm{h}\rangle = \langle \mathrm{h}|\Delta\tau_x|\mathrm{h}\rangle = 0$. This reflects that

[10] Note that the 'effective Zeeman field' acting on the electron spin now has orthogonal components up from spin–orbit and B from Zeeman, i.e. the overall strength of the effective Zeeman field is $\sqrt{(up)^2 + B^2}$.

spin-singlet Cooper pairing cannot induce proximity superconductivity in a perfectly spin-polarized system.

To find finite pairing terms, we need to include the spin–orbit coupling. Since the latter is weak, it can be included perturbatively. Using first-order perturbation theory, spin–orbit coupling modifies the low-energy spinors into

$$|\mathrm{e}\rangle = (1,\ -up/2B,\ 0,\ 0)^T, \tag{9.42}$$

$$|\mathrm{h}\rangle = (0,\ 0,\ -up/2B,\ 1)^T. \tag{9.43}$$

We can now repeat the calculation of the matrix elements of the pairing term within the low-energy subspace and obtain

$$\langle \mathrm{h}|\Delta\tau_x|\mathrm{e}\rangle = \langle \mathrm{e}|\Delta\tau_x|\mathrm{h}\rangle = -\frac{up}{B}\Delta, \tag{9.44}$$

as well as $\langle \mathrm{e}|\Delta\tau_x|\mathrm{e}\rangle = \langle \mathrm{h}|\Delta\tau_x|\mathrm{h}\rangle = 0$. Thus, the resulting projected Hamiltonian becomes

$$H \simeq \left(\frac{p^2}{2m} - \mu\right)\tau_z - \frac{up}{B}\Delta\tau_x. \tag{9.45}$$

This is just the BdG Hamiltonian of a spinless p-wave superconductor given in (9.8). By comparing with this Hamiltonian, we see that in the Kitaev limit, the effective p-wave pairing strength of the proximity-coupled quantum wire is $\Delta'_{\mathrm{eff}} = u\Delta/B$. The p-wave pairing is non-zero only as a result of the spin–orbit coupling and weakens as the Zeeman field increases and the spins become increasingly polarized. But, importantly, this implies that the spin–orbit-coupled quantum wire realizes a topological superconducting phase in the limit of strong Zeeman field.

9.4.2 Topological insulator limit

Now consider the opposite limit of strong spin–orbit coupling. For definiteness, let us place the chemical potential in the middle of the Zeeman-induced gap, i.e. we choose $\mu = 0$. The proximity coupling to an s-wave superconductor induces a gap Δ in the 'wings' of the spectrum, i.e. at momenta $p = \pm mu$. In contrast, there are now two mechanisms gapping out the system at $p = 0$, namely the Zeeman field and the proximity coupling. To understand the interplay of these two gapping mechanisms, we focus on small momenta, where we can neglect the kinetic energy in the Hamiltonian since it is quadratic in p. Then, the quantum wire Hamiltonian (9.35) reduces to

$$H \simeq up\sigma_x\tau_z - B\sigma_z + \Delta\tau_x. \tag{9.46}$$

which is just the topological insulator model (9.24) discussed in the previous section. Thus, the spectrum of the proximity-coupled quantum wire at small p becomes

$$E_p = \pm\sqrt{(up)^2 + (B \pm \Delta)^2}\,, \tag{9.47}$$

with all possible combinations of signs. As in the topological insulator case, the gap closes for $B = \pm\Delta$, indicating a topological phase transition.

The present model has a well-defined vacuum limit, so that we can clearly identify topological and trivial phases. As we already know that the large-B limit can be mapped to a spinless p-wave superconductor, it is natural to identify the topological phase with the high-B phase with $B > \Delta$. Indeed, it is easy to check that the gap does not close when the Zeeman field is increased from the topological insulator limit $\epsilon_{\text{so}} > B$ with $B > \Delta$ to the Kitaev limit $B > \epsilon_{\text{so}}$, say at fixed chemical potential $\mu = 0$, for the Hamiltonian (9.35).

The existence of two topologically distinct phases implies the existence of Majorana end states associated with domain walls. Both in the topological insulator limit and in the Kitaev limit, their wavefunctions and Bogoliubov operators can be obtained from the same calculations that we discussed above for spinless p-wave superconductors and proximity-coupled topological insulator edges.

The closing of the gap at $B = \Delta$ implies that the small-momentum region $p \ll mu$ dominates the low-energy physics. In fact, the gap $|B - \Delta|$ at $p = 0$ is much smaller than the gap of order Δ in the wings of the spectrum at $p = \pm mu$. At first sight, one may thus be tempted to assume that for these parameters, the physics is identical to the topological insulator case. Actually, this is not quite correct. In many ways, the phases of the present model are exactly reversed with respect to those of the topological insulator model! There, the 4π-periodic Josephson effect occurs for a Δ–B–Δ arrangement. In contrast, for the quantum wire, there is a 4π-periodic Josephson effect in the B–Δ–B arrangement.[11] This reversal of phases can be understood most easily in the limit of large spin–orbit energy ϵ_{so} with $\mu = 0$. In this limit, the Fermi points in the wings of the spectrum are far out and essentially decoupled from the $p \simeq 0$ physics. If we now consider the two Fermi points in the wings by themselves, there is only a single right-moving and a single left-moving channel, but with a proximity-induced superconducting gap. These are just the ingredients of a spinless p-wave superconductor in the topological phase! Thus, for large ϵ_{so} in the topological insulator limit, we can think of the system as a combination of a proximity-coupled topological insulator edge and a spinless p-wave superconductor. As the latter is always topological, the overall topological Z_2 index of the quantum-wire model is just the reverse of that of the proximity-coupled topological insulator.

9.5 Chains of magnetic adatoms on superconductors

Another proposal to realize Majorana bound states relies on a chain of magnetic impurities placed on atomically clean surfaces of conventional superconductors [26]. This system is a candidate for Majorana physics, since it combines the three essential ingredients: Zeeman coupling, superconductivity, and spin–orbit coupling. The Zeeman coupling is contributed by the magnetic adatoms and the substrate provides both superconductivity and spin–orbit coupling, provided the superconductor is made of a relatively heavy element.

[11] As above, we denote the phase by B if $B > \Delta$, and vice versa.

9.5.1 Shiba states

To understand this platform for topological superconductivity in more detail, we first consider the physics of individual magnetic adatoms. This is a classic problem in the theory of superconductors and was first studied in the late 1960s [27–30]. It will be useful for our discussion of adatom chains to explore the physics of individual adatoms at two levels. We will see that in both descriptions, the adatom induces localized subgap states in the superconductor, referred to as Yu–Shiba–Rusinov states or Shiba states for brevity. Such subgap states in superconductors can be readily probed using scanning tunnelling microscopy (STM) [31, 32].

9.5.1.1 Classical magnetic moment

The local magnetic moment of the adatoms is associated with their spin-split d-levels which will typically be far in energy from the Fermi level of the substrate superconductor. Then, the low-energy physics of the adatoms can be described in terms of their magnetic moments, while their electronic degrees of freedom are effectively frozen out. The large adatom spin $\mathbf{S}$ is exchange-coupled to the electrons of the superconductor and can be approximated as classical.

The BdG Hamiltonian for a local magnetic moment in a host superconductor is given by

$$H = \left(\frac{p^2}{2m} - \mu\right)\tau_z + (V\tau_z - J\mathbf{S}\cdot\boldsymbol{\sigma})\delta(\mathbf{r}) + \Delta\tau_x, \tag{9.48}$$

where J denotes the strength of the exchange coupling between the impurity spin located at the origin and the electrons in the superconductor. In addition to the exchange coupling, the impurity also induces potential scattering, which we parametrize through its strength V. This Hamiltonian has a pair of subgap bound states localized at the impurity site. The calculation is presented in Appendix 9.C (Section 9.C.1) and yields the symmetric bound-state energies

$$E = \pm\Delta\frac{1-\alpha^2+\beta^2}{\sqrt{(1-\alpha^2+\beta^2)^2+4\alpha^2}}, \tag{9.49}$$

where $\alpha = \pi\nu_0 SJ$ and $\beta = \pi\nu_0 V$ are dimensionless measures of the strengths of the exchange coupling and potential scattering, respectively, with ν_0 denoting the normal-state density of states of the superconductor.

These Shiba bound states possess two essential properties. First, they are *spin-polarized*, with the spin pointing parallel to the direction of $\mathbf{S}$. Second, their wavefunction is localized around the impurity, decaying as $1/r$ for distances r smaller than the (energy-dependent) coherence length

$$\xi_E = \frac{\hbar v_F}{\sqrt{\Delta^2 - E^2}}, \tag{9.50}$$

and exponentially beyond this length.

9.5.1.2 Anderson impurities

At a somewhat more refined level, we can include the electronic degrees of freedom of the adatom [33]. This description is required when the adatom d-levels are close in energy to the Fermi level of the substrate superconductor. We specifically consider a simplified model in which the magnetic adatom is a (spin-$\frac{1}{2}$) Anderson impurity, hybridized with the substrate superconductor. While this does not do full justice to the actual d-band nature of the magnetic adatoms, it captures much of the essential physics. We can follow Anderson's classic paper [34] and treat the local-moment formation within mean-field approximation.

The corresponding model Hamiltonian

$$\mathcal{H} = \mathcal{H}_d + \mathcal{H}_s + \mathcal{H}_T \tag{9.51}$$

contains a standard BCS Hamiltonian $\mathcal{H}_s$ for the host superconductor, the adatom's impurity level,

$$\mathcal{H}_d = \sum_\sigma (\epsilon_d - \mu) d_\sigma^\dagger d_\sigma + U n_\uparrow n_\downarrow, \tag{9.52}$$

and its hybridization with the superconductor,

$$\mathcal{H}_T = -t \sum_\sigma [\psi_\sigma^\dagger(\mathbf{0}) d_\sigma + d_\sigma^\dagger \psi_\sigma(\mathbf{0})]. \tag{9.53}$$

Here, d_σ annihilates a spin-σ electron in the Anderson impurity level, $n_\sigma = d_\sigma^\dagger d_\sigma$, and $\psi_\sigma(\mathbf{r})$ annihilates electrons at position $\mathbf{r}$ in the superconductor.

To analyse this Anderson model, we simplify the Hubbard term through a mean-field treatment,

$$U n_\uparrow n_\downarrow \to \frac{1}{2} U \sum_\sigma [\langle n \rangle n_\sigma - \langle m \rangle \sigma n_\sigma], \tag{9.54}$$

where we have defined the occupation $n = \sum_\sigma n_\sigma$ and the site polarization $m = n_\uparrow - n_\downarrow$. The first term merely renormalizes ϵ_d. The second term introduces a local exchange coupling in the adatom orbital.[12] If we assume that the adatom is singly occupied and develops a local moment, we have $\langle n \rangle = 1$ and $\langle m \rangle = 1$. Then, the two spin-split levels of the adatom have mean-field energies

$$E_{d\uparrow} = \epsilon_d - \mu, \qquad E_{d\downarrow} = \epsilon_d - \mu + U, \tag{9.55}$$

where we measure these energies relative to the chemical potential μ.

[12] In principle, we could have also included pairing terms localized on the adatom, whose strength would be determined self-consistently in the presence of the coupling to the substrate superconductor. Here, we assume that the Hubbard repulsion U strongly suppresses on-site pairing effects on the adatom so that these can be neglected.

Within mean-field theory, the Hamiltonian reduces to a BdG problem that is readily solved for subgap excitations. Details of this calculation are presented in Appendix 9.C (Section 9.C.2). As for a classical magnetic moment, one finds that there is a pair of subgap excitations at energies

$$E = \pm\Delta \frac{\Gamma^2 + E_{d\uparrow}E_{d\downarrow}}{\sqrt{(\Gamma^2 + E_{d\uparrow}E_{d\downarrow})^2 + \Gamma^2(E_{d\downarrow} - E_{d\uparrow})^2}}. \tag{9.56}$$

This expression is valid for $\langle n \rangle = 1$ and $\langle m \rangle = 1$, i.e. when, to a good approximation, the spin-up impurity level is occupied and the spin-down level is empty. In this limit, we can also eliminate the impurity levels by a Schrieffer–Wolf approximation and recover the description in terms of a local spin. Indeed, (9.49) and (9.56) for the Shiba-state energies connect when we identify

$$\alpha = -\frac{\frac{1}{2}\Gamma U}{\left(\frac{1}{2}U\right)^2 - \left(\epsilon_d - \mu + \frac{1}{2}U\right)^2} = -\frac{\Gamma(E_{d\uparrow} - E_{d\downarrow})}{2E_{d\uparrow}E_{d\downarrow}}, \tag{9.57}$$

$$\beta = \frac{\Gamma\left(\epsilon_d - \mu + \frac{1}{2}U\right)}{\left(\frac{1}{2}U\right)^2 - \left(\epsilon_d - \mu + \frac{1}{2}U\right)^2} = -\frac{\Gamma(E_{d\downarrow} + E_{d\uparrow})}{2E_{d\uparrow}E_{d\downarrow}} \tag{9.58}$$

as the dimensionless exchange and potential scattering amplitudes. More generally, the description in terms of an Anderson impurity can be made fully self-consistent, describing the local-moment formation. It can also be used to calculate the relative fractions of the spectral weight of the Shiba states that are located on the impurity and in the host superconductor.

9.5.2 Adatom chains

Armed with this understanding of individual adatoms, we now consider chains of adatoms. For an isolated impurity, the d-levels are typically far from the Fermi level of the substrate superconductor. In a chain of adatoms, there is direct hopping between neighbouring adatom orbitals, and the d-levels form one-dimensional bands of adatom states. If the adatom chain is dilute, then the hopping amplitude is small and the bandwidth remains negligible compared with the distance of the atomic d-levels from the Fermi energy. In this limit, illustrated in Fig. 9.6(a, b), we can discuss the physics of Shiba chains at subgap energies starting from the Shiba states associated with the individual impurities. As mentioned above, the Shiba wavefunctions decay away from the impurity only as a weak power law, so we need to account for their hybridization and the formation of Shiba bands at subgap energies of the host superconductor. The formation of Majorana bound states depends on the physics of these Shiba bands.

If, on the other hand, the adatoms are densely packed (as is presumably the case in the experiments [15, 17]), then the bandwidth of the adatom bands can exceed the energetic distance of the atomic d-levels from the Fermi level of the host superconductor. One or several adatom bands cross the superconductor's Fermi level and can no longer be treated as electronically inert. In this limit, illustrated in Fig. 9.6(c, d), it is more appropriate to extend the Anderson impurity approach for a single adatom.

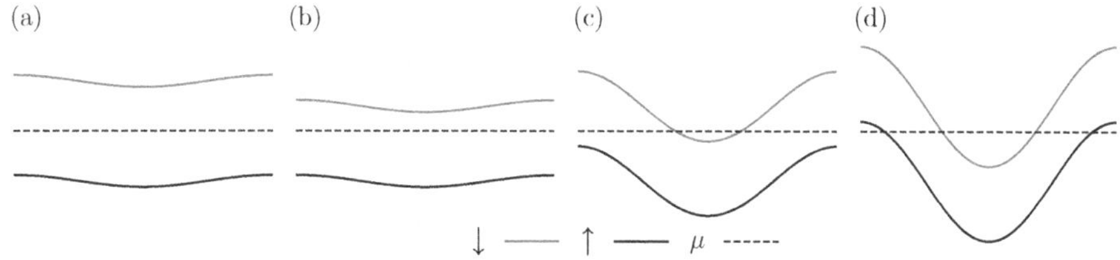

Fig. 9.6 [Colour online] Behaviour of adatom bands for various hopping strengths w between the adatoms. As the adatoms with their d-bands are modelled as Anderson impurities, there are two spin-split bands. (a) Weak hopping between Anderson impurity states that are symmetric about the Fermi energy. In this case, the electronic degrees of freedom of the adatoms are essentially frozen out and the system can be modelled in terms of bands of Shiba states. Owing to the symmetry, there is no potential scattering associated with the individual Anderson impurities. (b) Weak hopping between generic Anderson impurity states that are asymmetric about the Fermi energy. This case can be modelled as in (a), except that there is non-zero potential scattering associated with the individual Anderson impurities. (c) Strong hopping between Anderson impurity states such that the spin-down band crosses the Fermi energy while the spin-up band is entirely below the Fermi level. This requires that the spin-split Anderson impurity bands be asymmetric about the Fermi energy, as is generically the case. This situation realizes proximity-coupled spinless bands and is prone to develop topological superconductivity in the presence of spin–orbit coupling in the superconductor. (d) Very strong hopping between Anderson impurity states such that both spin-down and spin-up bands cross the Fermi level. As there are an even number of channels, this situation will no longer be topological.

In fact, it is quite natural to expect topological superconductivity in this regime. Let us assume that the adatoms can be described as spin-$\frac{1}{2}$ Anderson impurities and that the adatoms spin-polarize (say, into a ferromagnetic state or a spin helix). When the spin-up and spin-down levels of an individual adatom are sufficiently asymmetric about the host Fermi level, there will be a wide range of parameters for which only the spin-down band crosses the Fermi level while the spin-up band is completely occupied. Notice that this is a very close realization of a spin-polarized system, and hence prone to developing effectively spinless p-wave superconductivity by proximity when there is sufficient spin–orbit coupling in the host superconductor.[13]

We now discuss both of these scenarios in more detail.

9.5.2.1 Dilute adatom chains

In the dilute limit, the low-energy physics is governed by the Shiba states of the individual adatoms. In a chain of spin-polarized Shiba states, the neighbouring Shiba states will couple and broaden out into subgap Shiba bands. If the Shiba states of

[13] Spin–orbit coupling in the adatom chain would also place the system into a topological phase. However, this system has a very large Zeeman (exchange) splitting, comparable to atomic energy scales. This is presumably much larger than the spin–orbit coupling in the chain, so the induced p-wave pairing strength would be quite small; cf. (9.45). There is no such suppression when the spin–orbit coupling is provided by the host superconductor, as follows from the calculations below.

energy $\pm E_0$ are sufficiently deep and the bandwidth sufficiently small, we can focus attention on only the Shiba states and project out the quasiparticle continua above the gap. In this limit, it is tempting to describe the Shiba chain by a Hamiltonian of the form

$$\mathcal{H} = E_0 \sum_j c_j^\dagger c_j - t \sum_j (c_{j+1}^\dagger c_j + c_j^\dagger c_{j+1}) + \Delta \sum_j (c_{j+1} c_i + c_j^\dagger c_{j+1}^\dagger). \tag{9.59}$$

Here, we denote the fermionic annihilation (and creation) operators of the spin-polarized Shiba state at site j by c_j (and $c_j^\dagger$).[14] The Shiba states hybridize between neighbouring sites with amplitude t. Moreover, the Shiba chain is embedded in the host superconductor and thus it is natural to include a pairing term of strength Δ in this Hamiltonian. Importantly, the pairing term necessarily involves pairing correlations between different sites due to the perfect spin polarization. The Shiba states have energy $\pm E_0$ (measured from the Fermi energy).

The Hamiltonian (9.59) assumes that both hopping and pairing are dominated by nearest-neighbour terms. This is not fully adequate for a Shiba chain owing to the slow $1/r$ decay of the Shiba wavefunctions away from the impurity. This has some interesting consequences [35, 36]. Here, we restrict our discussion to the simplified model in (9.59).

The Hamiltonian (9.59)—sometimes referred to as the Kitaev chain—is just a lattice version of the spinless p-wave superconductor in (9.7). We just need to identify the Shiba-state energy E_0 as playing the role of the chemical potential, $E_0 \to -\mu$. It is not difficult to diagonalize the Kitaev chain and confirm that it has a topological state at finite Δ whenever the chemical potential is situated in the normal-state band. The phase diagram of the model is shown in Fig. 9.7. For completeness, the Kitaev chain,

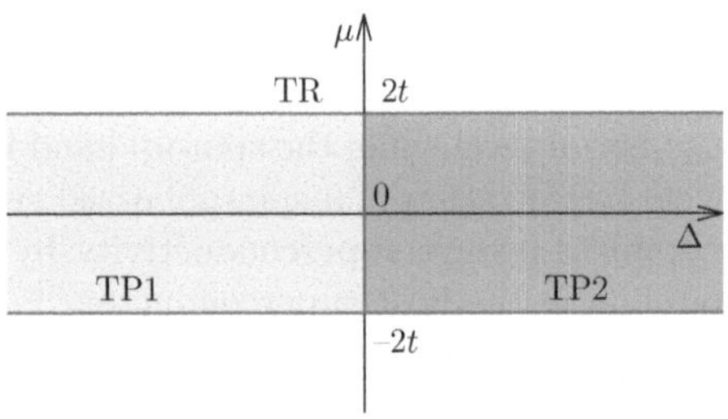

Fig. 9.7 [Colour online] Phase diagram of the Kitaev chain as function of p-wave pairing strength Δ and chemical potential μ. There is a topological superconducting phase when the chemical potential is within the band of the normal-state Hamiltonian, while the system is topologically trivial when the chemical potential is outside the band.

[14] The BdG Hamiltonian had two Shiba states per adatom, with energies symmetric about zero. The fact that there are two states is a consequence of the doubling of the number of degrees of freedom in the BdG formalism. This is why in the second-quantization representation of (9.59), each site supports only a single pair of annihilation and creation operators.

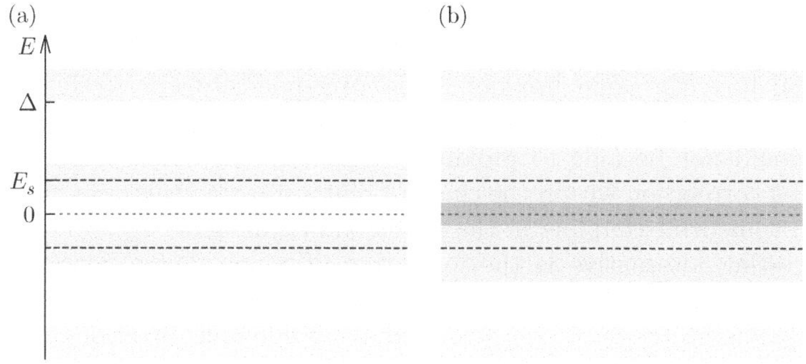

Fig. 9.8 [Colour online] Illustration of the subgap Shiba bands in the excitation spectrum of a dilute adatom chain. (a) For weak hybridization of deep Shiba states, the positive- and negative-energy Shiba bands do not cross the Fermi energy at the centre of the gap, and the system is non-topological. (b) For stronger hybridization, the Shiba bands overlap at the centre of the gap. In this case, the pairing correlations open a gap that is of p-wave nature owing to the spin polarization of the subgap states. This realizes a topological superconducting phase that hosts Majorana bound states at its ends.

including this phase diagram, is discussed in more detail in Section 9.5.3. Here, we continue with the discussion of the Shiba chain not to break the flow of the argument, drawing on basic aspects of the phase diagram of the Kitaev chain.

This provides the following somewhat simplified picture of Shiba chains; see Fig. 9.8. As the adatoms are placed closer together, the hybridization increases and with it the bandwidth of the subgap Shiba bands. Initially, the Shiba bands (including the BdG partner with an energy of opposite sign) do not cross the chemical potential at the centre of the gap. This is analogous to the Fermi energy lying outside the normal-state band for the Kitaev chain. Hence, the system is in a topologically trivial phase. Eventually, the two Shiba bands cross the centre of the gap. Now, the pairing correlations Δ within the Shiba bands will again open a gap at the Fermi energy. This is a p-wave gap, unlike the larger gap of the host superconductor! The Shiba chain is in a topological phase and hosts zero-energy Majorana bound states at its ends.

We have simply assumed that the effective Kitaev-chain Hamiltonian for the Shiba chain contains pairing terms, but have not discussed their microscopic origin. This question is closely related to the collective behaviour of the impurity spins. So far, we have just noted that the Shiba states are spin-polarized along the direction of the corresponding impurity spin. But we have ignored the question of how the impurity spins of different adatoms are oriented with respect to one another, a question that is of obvious importance for the physics of the adatom chain. Indeed, we expect that the impurity spins interact through the familiar RKKY interaction mediated by the host superconductor and may thus order magnetically. Two such orderings have been predominantly discussed, which actually involve somewhat different physics of the pairing terms [26, 37–40].

One plausible possibility is that the chain orders ferromagnetically, with all impurity spins aligning along a certain direction. In that case, the Shiba states are all spin-polarized along the same direction. This corresponds to a perfectly spin-polarized system, and consequently the spin-singlet Cooper pairs of a pure s-wave host superconductor would not be able to proximity-couple to the chain of Shiba states. To induce pairing correlation within the chain of Shiba states in this case, we need to rely on (Rashba) spin–orbit coupling in the superconducting host.

An interesting alternative is the formation of a spin helix, with adatom spins rotating along the chain. In this case, neighbouring impurity spins are not aligned and the corresponding Shiba states are polarized along different directions. A spin-singlet Cooper pair of the host superconductor can effectively tunnel into the chain as long as its spin-up and spin-down electrons enter on different sites of the chain. Thus, the effective pairing correlations that result from these processes are just of the spinless p-wave type that are included in the Kitaev chain (9.59).

In the remainder of this section, we will assume that the adatom spins order ferromagnetically, as suggested by experiment. However, it is useful to make two comments:

1. The assumption of ferromagnetic order combined with (Rashba) spin–orbit coupling in the host superconductor is less restrictive that it may appear. The reason is that this situation can be mapped on a Hamiltonian with helical spin ordering by a simple unitary transformation [41].
2. The stability of ordering in one dimension is obviously subtle. It depends on specifics of the microscopic Hamiltonian such as the presence or absence of continuous spin symmetries or the range of the substrate-induced interaction between adatom spins. Thermal fluctuations may well preclude long-range spin order. In that case, we assume that there is at least short-range order on a scale that is large compared with the length of the adatom chains.

9.5.2.2 Dense adatom chains

We now consider the situation where hopping between the Anderson impurity levels of neighbouring adatoms is sufficiently strong that the spin-down band crosses the Fermi energy. At the same time, the spin-up band is fully occupied. We further assume that the adatom chain is ferromagnetically ordered, as observed experimentally [15]. We show here that this system is prone to be in a topological superconducting phase. A more complete theoretical treatment along these lines can be found in the literature [42].

To understand the low-energy physics in this limit, it is sufficient to consider the spin-down band that crosses the Fermi level. In the vicinity of the Fermi energy, we can linearize its dispersion so that the main characteristics are its Fermi wavevector k_0 and Fermi velocity v_F. A subgap excitation in this band cannot decay into the superconducting host, owing to the superconducting gap, but some of its spectral weight will be transferred. In the absence of spin–orbit coupling, the superconducting substrate will not be able to induce a gap within this spin-polarized band, but there can be p-wave-like correlations in the presence of spin–orbit coupling. These superconducting correlations can be read off from the Green function for the adatom bands once we have accounted for the coupling to the superconductor through the appropriate self-energy. The self-energy can be computed exactly when treating the on-site Hubbard term in mean-field theory and assuming ferromagnetic order of the chain from the outset.

The BdG Hamiltonian of the adatom chain on top of a spin–orbit-coupled s-wave superconductor can be written as

$$H = \begin{pmatrix} H_s & H_T \\ H_T^T & H_d \end{pmatrix}. \tag{9.60}$$

Here, the adatoms are described as a chain of Anderson impurities, i.e. a Hamiltonian H_d with matrix elements

$$H_d^{ij} = [(\epsilon_d - \mu)\delta_{ij} - W_{ij}]\tau_z - B\sigma_z\delta_{ij} \tag{9.61}$$

in site space. We have already used a mean-field decoupling of the on-site Hubbard terms that results in the exchange field B governing the spin splitting. The principal new ingredient compared with the individual adatom is the direct hopping term

$$W_{ij} = -w(\delta_{i,j+1} + \delta_{i,j-1}) \tag{9.62}$$

between adatom orbitals, which leads to the formation of the adatom spin-up and spin-down bands with band width $2w$. As usual, τ_i and σ_i with $i = x, y, z$ are Pauli matrices in Nambu and spin space. We choose the chain to be aligned along the x direction.

The host superconductor obeys the BdG Hamiltonian

$$H_s = \left[\frac{(\mathbf{p} + k_h\sigma_x\hat{\mathbf{x}})^2}{2m} - \mu\right]\tau_z + \Delta\tau_x. \tag{9.63}$$

Here, Δ is the superconducting order parameter and k_h denotes the strength of spin–orbit coupling. We make no attempt at a microscopic description of the spin–orbit coupling appropriate for real materials, but rather retain only the relevant term needed for inducing p-wave pairing in the adatom chain. This term couples spin to the momentum along the adatom chain. The adatoms hybridize with the superconductor through H_T, which includes tunnelling amplitudes t between the adatom impurity level and the superconductor at the positions $\mathbf{R}_j = ja\hat{\mathbf{x}}$ of the impurities (a is the lattice spacing along the chain direction).

We denote the Green functions of the adatom chain before and after coupling to the superconductor by $g_d(E)$ and $G_d(E)$, respectively. These two Green functions are related by the Dyson equation

$$G_{d,ij}(E) = g_{d,ij}(E) + \sum_{mn} g_{d,im}(E)\Sigma_{d,mn}(E)G_{d,nj}(E). \tag{9.64}$$

where m, n, i, j are site indices along the chain and $\Sigma_d(E)$ is the self-energy accounting for the coupling to the superconductor,

$$\Sigma_{mn}(E) = t^2 \tau_z g_{s,mn}(E) \tau_z. \tag{9.65}$$

This self-energy describes hopping from the chain into the superconductor and back, with free propagation in the superconductor in between, as described by the Green function $g_{s,mn}(E)$.

The Green function of the superconductor is readily computed. First, we consider off-diagonal elements in site space, $m \neq n$. Then,

$$\begin{aligned} g_{s,mn}(E) &= \langle \mathbf{R}_m | (E - H_s)^{-1} | \mathbf{R}_n \rangle \\ &= \langle \mathbf{R}_m | \left\{ E - \left[\frac{(\mathbf{p} + k_h a \sigma_x \hat{\mathbf{x}})^2}{2m} - \mu \right] \tau_z - \Delta \tau_x \right\}^{-1} | \mathbf{R}_n \rangle \\ &= \langle \mathbf{R}_m | e^{-ik_h x \sigma_x} \left[E - \left(\frac{\mathbf{p}^2}{2m} - \mu \right) \tau_z - \Delta \tau_x \right]^{-1} e^{ik_h x \sigma_x} | \mathbf{R}_n \rangle \\ &= e^{-ik_h (m-n) a \sigma_x} \frac{1}{V} \sum_{\mathbf{k}} \frac{e^{i\mathbf{k}\cdot(\mathbf{R}_m - \mathbf{R}_n)}}{E - \xi_k \tau_z - \Delta \tau_x} \\ &= e^{-ik_h (m-n) a \sigma_x} [(E + \Delta \tau_x) P_0(|m-n|a) + \tau_z P_1(|m-n|a)]. \end{aligned} \tag{9.66}$$

Here, we have introduced the integrals

$$\begin{aligned} P_0(r) &= \frac{\nu_0}{2} \int d\xi_k \int_{-1}^{1} dx \, \frac{e^{ikrx}}{E^2 - \xi_k^2 - \Delta^2} \\ &= -\frac{\pi \nu_0}{\sqrt{\Delta^2 - E^2}} \frac{\sin k_F r}{k_F r} e^{-r/\xi_E}, \end{aligned} \tag{9.67}$$

$$\begin{aligned} P_1(r) &= \frac{\nu_0}{2} \int d\xi_k \int_{-1}^{1} dx \, \frac{\xi_k e^{ikrx}}{E^2 - \xi_k^2 - \Delta^2} \\ &= -\pi \nu_0 \frac{\cos k_F r}{k_F r} e^{-r/\xi_E}, \end{aligned} \tag{9.68}$$

with $\xi_E = \hbar v_F / \sqrt{\Delta^2 - E^2}$. An explicit evaluation of these integrals can be found in Appendix A of [35]. Then, we obtain

$$\begin{aligned} &g_{s,mn}(E) = \\ &\quad - \pi \nu_0 e^{-ik_h x_{mn} \sigma_x} \left(\frac{E + \Delta \tau_x}{\sqrt{\Delta^2 - E^2}} \frac{\sin k_F r_{mn}}{k_F r_{mn}} e^{-r_{mn}/\xi_E} + \tau_z \frac{\cos k_F r_{mn}}{k_F r_{mn}} e^{-r_{mn}/\xi_E} \right), \end{aligned} \tag{9.69}$$

with $x_{mn} = x_m - x_n = (m-n)a$ and $r_{mn} = |x_{mn}|$. An analogous calculation for the diagonal elements in site space, $m = n$, yields (see also (9.203))

$$g_{s,mm}(E) = -\pi\nu_0 \frac{E + \Delta\tau_x}{\sqrt{\Delta^2 - E^2}}. \tag{9.70}$$

Thus, the self-energy takes the form

$$\Sigma_{d,mn}(E) = \begin{cases} -\Gamma \dfrac{E - \Delta\tau_x}{\sqrt{\Delta^2 - E^2}}, & m = n, \\ \dfrac{-\Gamma}{k_F r_{mn}} e^{-ik_h x_{mn}\sigma_x} e^{-r_{mn}/\xi_E} \left[\dfrac{E - \Delta\tau_x}{\sqrt{\Delta^2 - E^2}} \sin k_F r_{mn} + \tau_z \cos k_F r_{mn} \right], & m \neq n. \end{cases} \tag{9.71}$$

This result for the self-energy has several important physical consequences, as we will discuss in the following.

In general, the self-energy is a 4×4 matrix in Nambu and spin space:

$$\Sigma_d(E) = \begin{pmatrix} \Sigma^{ee}_{\uparrow\uparrow} & \Sigma^{ee}_{\uparrow\downarrow} & \Sigma^{eh}_{\uparrow\downarrow} & \Sigma^{eh}_{\uparrow\uparrow} \\ \Sigma^{ee}_{\downarrow\uparrow} & \Sigma^{ee}_{\downarrow\downarrow} & \Sigma^{eh}_{\downarrow\downarrow} & \Sigma^{eh}_{\downarrow\uparrow} \\ \Sigma^{he}_{\downarrow\uparrow} & \Sigma^{he}_{\downarrow\downarrow} & \Sigma^{hh}_{\downarrow\downarrow} & \Sigma^{hh}_{\downarrow\uparrow} \\ \Sigma^{he}_{\uparrow\uparrow} & \Sigma^{he}_{\uparrow\downarrow} & \Sigma^{hh}_{\uparrow\downarrow} & \Sigma^{hh}_{\uparrow\uparrow} \end{pmatrix}, \tag{9.72}$$

where each block is still a matrix in site space. In the absence of spin–orbit coupling, the self-energy decomposes into two independent 2×2 blocks, as several matrix elements vanish:

$$\Sigma_d(E) = \begin{pmatrix} \Sigma^{ee}_{\uparrow\uparrow} & 0 & \Sigma^{eh}_{\uparrow\downarrow} & 0 \\ 0 & \Sigma^{ee}_{\downarrow\downarrow} & 0 & \Sigma^{eh}_{\downarrow\uparrow} \\ \Sigma^{he}_{\downarrow\uparrow} & 0 & \Sigma^{hh}_{\downarrow\downarrow} & 0 \\ 0 & \Sigma^{he}_{\uparrow\downarrow} & 0 & \Sigma^{hh}_{\uparrow\uparrow} \end{pmatrix}. \tag{9.73}$$

The diagonal entries in particle–hole space describe the renormalization of the quasiparticle weight and the dispersion, while the off-diagonal entries describe the proximity-induced s-wave correlations. This is considered in more detail in Appendix 9.B.

For the ferromagnetically ordered adatom chain, we assumed that only the spin-down band crosses the Fermi energy. For a description of the low-energy physics, we can therefore project the Green functions and the self-energy onto this subspace; for example,

$$\Sigma_d(E) \to \begin{pmatrix} \Sigma^{ee}_{\downarrow\downarrow} & \Sigma^{eh}_{\downarrow\downarrow} \\ \Sigma^{he}_{\downarrow\downarrow} & \Sigma^{hh}_{\downarrow\downarrow} \end{pmatrix}. \tag{9.74}$$

The diagonal entries describe the renormalization of the quasiparticle weight and the dispersion of the spin-down band, while the off-diagonal entries describe the proximity-induced pairing correlations. As these pairing correlations are induced in a spin-polarized band, they require non-zero spin–orbit coupling. They are necessarily of p-wave nature and thus odd in momentum as well as off-diagonal in site space.

In the original 4×4 scheme, the p-wave pairing terms correspond to the $\tau_x \sigma_x$ entries. Moreover, we focus on subgap energies, $E \ll \Delta$. Then, the projected self-energy takes the explicit form

$$\Sigma_{d,mn}(E) \simeq -\frac{\Gamma E}{\Delta}\delta_{mn} - i\frac{\Gamma e^{-r_{mn}/\xi_0}}{k_F r_{mn}} \sin k_h x_{mn} \sin k_F r_{mn}\, \tau_x (1 - \delta_{mn}). \tag{9.75}$$

This is already written in the 2×2 matrix notation after projection, where $\tau_x \sigma_x$ becomes τ_x. The first term renormalizes the band dispersion, while the second describes the induced p-wave pairing. Note that it is indeed odd in site space, as expected for p-wave correlations, and vanishes in the absence of spin–orbit coupling, i.e. when $k_h = 0$. Here, we will not pursue a detailed evaluation of these pairing correlations, but simply assume that they are finite:

$$\Sigma_{d,mn}(E) \simeq -\frac{\Gamma E}{\Delta}\delta_{mn} + \Delta_{mn}\tau_x. \tag{9.76}$$

The interested reader can find a more complete discussion in [42].

Then, the Dyson equation for the projected Green function in momentum space becomes

$$G_d^{-1}(k, E) \simeq E(1 + \Gamma/\Delta) - v_F(k - k_0)\tau_z + \Delta(k)\tau_x, \tag{9.77}$$

where we have explicitly linearized the dispersion of the spin-down band and $\Delta(k)$ denotes the Fourier transform of Δ_{mn} to momentum space. Besides the induced p-wave correlations, this expression includes a strong renormalization of the quasiparticle weight when $\Gamma \gg \Delta$. Indeed, Γ measures the strength of hybridization between adatoms and superconductor. In adatom experiments, this coupling is essentially determined by atomic physics and is expected to be large compared with the gap. In this situation, excitations in the adatom band will have much spectral weight in the superconductor, leading to a strong renormalization of the quasiparticle weight:

$$Z = \frac{1}{1 + \Gamma/\Delta}. \tag{9.78}$$

This renormalization affects the dispersion and the induced gap of the adatom band at subgap energies:

$$G_d(k, E) \simeq \frac{Z}{E - Zv_F(k - k_0)\tau_z + Z\Delta(k)\tau_x}. \tag{9.79}$$

We observe that the effective Fermi velocity is strongly renormalized:

$$v_F \to \tilde{v}_F = Zv_F. \tag{9.80}$$

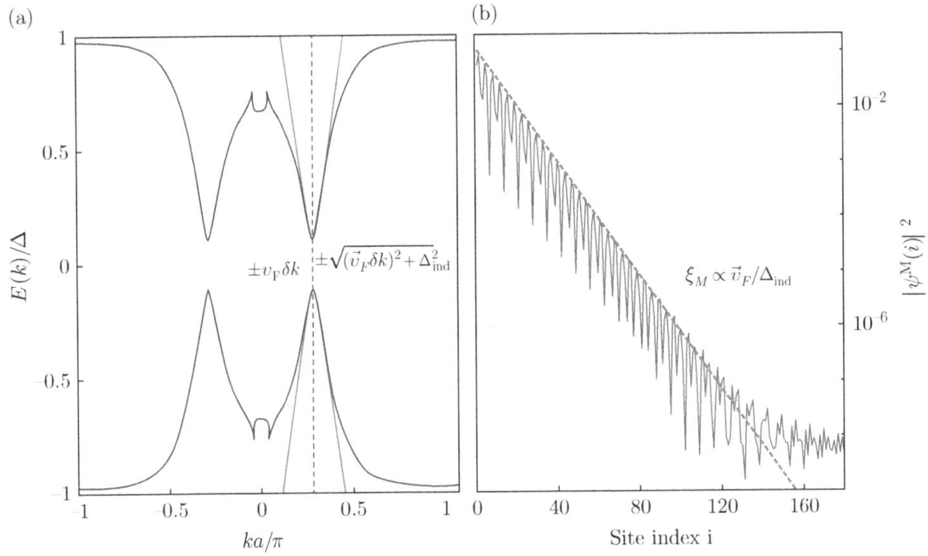

Fig. 9.9 [Colour online] (a) Excitation spectrum of a dense adatom chain coupled to a host superconductor. Only the spin-down adatom band (shown by the black dashed lines) crosses the Fermi level. The subgap dispersion accounting for the coupling to the superconductor is shown by the full curves [blue online]. The approximate theory presented in the text is shown by the dotted curves [red online] and accurately reproduces the numerically exact results near the Fermi wavevector of the adatom band. Note the strong renormalization of the Fermi velocity compared with the bare dispersion, as well as the proximity-induced p-wave gap. (b) Numerically exact results for a Majorana wavefunction, showing that it is localized on a scale that is small compared with the coherence length of the host superconductor.

Loosely speaking, this renormalization can be understood by noting that the excitations 'spend little time in the wire and propagate along the wire only during these intervals. Indeed, to a good approximation, the non-pairing contributions to the self-energy are local in site space. Similarly, the physical induced gap involves the same renormalization factor:

$$\Delta_{\rm ind} = Z|\Delta(k_0)|. \tag{9.81}$$

Notice, however, that $\Delta(k)$ is itself proportional to the large coupling Γ, making $\Delta_{\rm ind}$ independent of Γ at strong coupling.

This renormalization is confirmed [42] by more detailed calculations of the subgap spectrum for all momenta k. The results of such a calculation are shown in Fig. 9.9(a). The figure shows the original dispersion of the adatoms as black dashed lines, while the true dispersion accounting for the coupling to the superconductor is shown by the full curves [blue online]. These dispersion curves exhibit the minimal gap at the wavevector k_0 where the adatom band crosses the Fermi energy, but the slopes are dramatically reduced in accordance with the downward renormalization of the

Fermi velocity. Indeed, we can quantitatively compare these exact excitation spectra with the results of the approximate theory presented here. Equation (9.79) predicts a low-energy dispersion

$$E_k = \pm\sqrt{[Zv_F(k-k_0)]^2 + \Delta_{\text{ind}}^2}, \tag{9.82}$$

which is shown by the dotted curves [red online] in Fig. 9.9(a).

The most important consequence of this renormalization concerns the localization length of the Majorana bound states. We can extract a characteristic length from (9.82), which is given by

$$\xi_M = Z\frac{\hbar v_F}{\Delta_{\text{ind}}}. \tag{9.83}$$

This length describes the coherence length of the induced superconducting correlations. It is important to note that owing to the renormalization of the Fermi velocity, this can be significantly smaller than the coherence length of the host superconductor, $\xi = \hbar v_F/\Delta$. This is important because ξ_M also governs the localization of the Majorana end states. The Majorana states can thus be much more strongly localized than the coherence length of the host superconductor when the coupling to the superconductor is strong! This might have been observed in a recent experiment [15]. This scenario for the Majorana localization length is also confirmed by more detailed numerical calculations, as illustrated in Fig. 9.9(b).

9.5.3 Kitaev chain

9.5.3.1 Finite chain and Majorana end states

Let us first consider

$$\mathcal{H} = -t\sum_j (c_{j+1}^\dagger c_j + c_j^\dagger c_{j+1}) + \Delta\sum_j (c_{j+1}c_i + c_j^\dagger c_{j+1}^\dagger) - \mu\sum_j c_j^\dagger c_j, \tag{9.84}$$

(i.e. the Hamiltonian in (9.59) with the replacement $E_0 \to -\mu$) for a finite chain of N sites [18]. It turns out that there is a particularly simple and instructive solution of the finite chain for the special point $t = \Delta$ and $\mu = 0$. We start by writing the fermionic operator

$$c_j = \tfrac{1}{2}(\gamma_{Bj} + i\gamma_{Aj}) \tag{9.85}$$

in terms of two Majorana operators γ_{Aj} and γ_{Bj}, with $\gamma_{Aj} = \gamma_{Aj}^\dagger$ and $\gamma_{Bj} = \gamma_{Bj}^\dagger$. Using the inverse relations $\gamma_{Aj} = -i(c_j - c_j^\dagger)$ and $\gamma_{Bj} = c_j + c_j^\dagger$, as well as the usual fermionic anticommutation relations for c_j, it is easy to check that the operators γ_{Aj} and γ_{Bj} do indeed satisfy the Majorana relation (9.3). We can now express the Hamiltonian in terms of these new operators. At the special point, this yields

$$\mathcal{H} = -it\sum_{j=1}^{N-1} \gamma_{Bj}\gamma_{A,j+1}. \tag{9.86}$$

This Hamiltonian can be diagonalized by introducing $N-1$ new (conventional) fermionic operators through

$$d_j = \frac{1}{2}(\gamma_{Bj} - i\gamma_{A,j+1}) \tag{9.87}$$

for $j = 1, \ldots, N-1$. Note that these new operators combine Majorana operators that derive from neighbouring sites. If we now express the Hamiltonian in terms of these new operators, we find

$$\mathcal{H} = 2t \sum_{j=1}^{N-1} (d_j^\dagger d_j - \tfrac{1}{2}). \tag{9.88}$$

Thus, the d_j are fermionic quasiparticle (Bogoliubov) operators of the superconductor with energy t.

It is important to realize that we started with N fermionic operators c_j. In contrast, we seem to have only $N-1$ quasiparticle operators d_j. Where did we lose one of the fermionic operators? If we look back at the Hamiltonian (9.86) written in terms of the Majorana operators, we realize that two of the Majorana operators actually appear neither in the Hamiltonian nor in the quasiparticle operators d_j, namely γ_{A1} and γ_{BN}!

To understand what this means, we note that these operators commute with the Hamiltonian $\mathcal{H}$. Thus, they are eigenoperators of the Hamiltonian with *zero* energy. There is one such zero-energy Majorana excitation localized at each end of the chain. We can combine these two Majorana operators into an additional (highly non-local) conventional fermion

$$d_0 = \frac{1}{2}(\gamma_{BN} - i\gamma_{A1}). \tag{9.89}$$

This fermionic operator does not appear in the Hamiltonian and thus has zero energy. Let us assume that we find a many-body ground state $|\text{gs}\rangle$ of the chain with the additional condition $d_0|\text{gs}\rangle = 0$. Then there is necessarily a second ground state

$$d_0^\dagger|\text{gs}\rangle. \tag{9.90}$$

Indeed, since the quasiparticle excitation generated by $d_0^\dagger$ has zero energy, this state has exactly the same energy as $|\text{gs}\rangle$. Thus, we find that the ground state of a finite Kitaev chain is doubly degenerate and that this degeneracy is associated with the existence of Majorana end states at the two ends of the chain.

9.5.3.2 Bulk properties and phase diagram

To compute the bulk properties of the Kitaev chain for arbitrary parameters, we consider the Hamiltonian (9.84) for N sites with periodic boundary conditions, i.e. we identify

$$c_1 = c_{N+1}. \tag{9.91}$$

By translational invariance, this can be diagonalized by introducing a_k through

$$c_j = \frac{1}{\sqrt{N}} \sum_j e^{ikj} a_k. \tag{9.92}$$

Indeed, this yields

$$\mathcal{H} = \sum_k \xi_k a_k^\dagger a_k + \Delta \sum_k (e^{ik} a_k a_{-k} + e^{-ik} a_{-k}^\dagger a_k^\dagger), \tag{9.93}$$

where

$$\xi_k = -2t \cos k - \mu \tag{9.94}$$

is the normal-state dispersion.

To find the quasiparticle spectrum, we pass to Nambu space by introducing the two-component Nambu operator

$$\phi_k = \begin{pmatrix} a_k \\ a_{-k}^\dagger \end{pmatrix}. \tag{9.95}$$

Note that we define ϕ_k for $k > 0$ only to avoid double counting.[15] Indeed, if one remembers this condition, it is straightforward to show that the ϕ_k satisfy the usual fermionic anticommutation relations

$$\begin{aligned} \{\phi_{k\alpha}, \phi_{k'\alpha'}^\dagger\} &= \delta_{kk'}\delta_{\alpha\alpha'}, \\ \{\phi_{k\alpha}, \phi_{k'\alpha'}\} &= 0, \\ \{\phi_{k\alpha}^\dagger, \phi_{k'\alpha'}^\dagger\} &= 0. \end{aligned} \tag{9.96}$$

We can now write the Hamiltonian as

$$\mathcal{H} = \sum_{k>0} \phi_k^\dagger H_k \phi_k + \text{const}, \tag{9.97}$$

with the BdG Hamiltonian

$$H_k = \begin{pmatrix} \xi_k & 2i\Delta \sin k \\ -2i\Delta \sin k & -\xi_k \end{pmatrix}. \tag{9.98}$$

Note that the off-diagonal terms in the BdG Hamiltonian are odd functions of k, which is a direct signature of the p-wave nature of the pairing. We can now obtain the excitation spectrum

$$E_k = \pm\sqrt{\xi_k^2 + 4\Delta^2 \sin^2 k} \tag{9.99}$$

in the usual way by diagonalizing the BdG Hamiltonian.

[15] We can leave out the $k = 0$ term since there is no pairing term in this case.

The connection with the continuum model for a spinless p-wave superconductor (see Section 9.2.1) is readily made explicit. Consider a Fermi energy close to the lower band edge so that the relevant wavevectors are small. Then we can expand both the dispersion ξ_k and the pairing $\Delta \sin k$ for small k. In this limit, we simply recover both the BdG Hamiltonian (9.8) and the spectrum (9.9) for the spinless p-wave superconductor.

The excitation spectrum (9.99) of the Kitaev chain is mostly fully gapped. The normal-state dispersion ξ_k vanishes only for $k = \pm k_F$, where the Fermi wavevector k_F is determined by the condition $-2t \cos k_F = \mu$. Similarly, the pairing term is non-zero except when $k = 0$ and $k = \pm\pi$. Thus, the system becomes gapless only when the Fermi wavevector becomes equal to 0 or $\pm\pi$. This happens when the chemical potential is just at the band edges of the normal-state dispersion, i.e. when $\mu = -2t$ ($k_F = 0$) or $\mu = +2t$ ($k_F = \pm\pi$).

The lines $\mu = \pm 2t$ where the excitation spectrum becomes gapless correspond to phase boundaries between two topologically distinct phases. The corresponding phase diagram as a function of Δ and μ (both measured in units of t) is shown in Fig. 9.7. The Kitaev chain is in a topological phase when the chemical potential lies within the band of the normal-state band. Conversely, the system becomes topologically trivial when the chemical potential is outside the band. In the latter case, the system is adiabatically connected to the vacuum ($\mu \to -\infty$) or a fully occupied band ($\mu \to \infty$). The special point $t = \Delta$ and $\mu = 0$ discussed above is well within the topological phase.

9.6 Non-Abelian statistics

Perhaps the most fascinating property of Majorana zero modes is their non-Abelian quantum statistics. Non-Abelian statistics of Majorana zero modes was first discussed in two-dimensional systems, where Majoranas appear as zero-energy bound states associated with vortices in spinless $p + ip$ superconductors [36, 43–45]. At first sight, it might not be obvious whether Majoranas in one-dimensional systems as discussed here would also obey the same quantum statistics. In fact, the Majoranas in one dimension are associated with domain walls rather than vortices, while physical arguments for non-Abelian statistics in two dimensions rely heavily on the phase structure of the order parameter associated with a vortex. More generally, quantum statistics is not well defined in strictly one-dimensional systems, since exchanging particles cannot be disentangled from interactions because the particles necessarily pass one another in the exchange process (or, formally, one can pass between, for example, boson and fermion representations by means of a Jordan–Wigner transformation). The second point can be readily circumvented by considering wire networks rather than strictly one-dimensional systems [46].

9.6.1 Manipulation of Majorana bound states

A necessary prerequisite for probing non-Abelian statistics of Majorana zero modes in experiment is the ability to manipulate the Majorana zero modes. The most direct way of performing braiding operations is by explicitly moving the Majorana states in real space. Using the quantum-wire scenario as an example, let us briefly discuss how this might be achieved in practice. To move the Majorana zero mode along the wire, we need to move the domain wall with which it is associated. As we saw in Section 9.4, we can induce the domain wall by spatially varying parameters in such a way that the topological gap changes sign.

Consider first the Kitaev limit of the quantum-wire Hamiltonian (9.35), i.e. the limit of large Zeeman splitting. In this limit, the topological phase transition occurs when the chemical potential moves through the bottom of the band, and in the vicinity of the phase transition, the gap is given by the chemical potential μ. Thus, we can tune through the phase transition by changing the chemical potential, and induce a domain wall by changing the chemical potential along the wire. This can in principle be achieved by changing the local electrostatic potential through a series of gate electrodes along the wire.

Next consider the topological insulator limit of the quantum wire Hamiltonian (9.35). In Section 9.4, we considered the case of zero chemical potential, $\mu = 0$. In that case, the topological gap is given by the difference of Zeeman field and gap, $B - \Delta$. While, in principle, B and Δ can be varied along the wire, these are not conveniently controlled experimentally. It turns out that, also in this limit, we can tune through the transition by varying the chemical potential. Indeed, one can solve for the spectrum of the Hamiltonian (9.35) in this limit at finite chemical potential by squaring the Hamiltonian, along the same lines as described in Section 9.4. In this way, one finds that the gap is given by

$$B - \sqrt{\Delta^2 + \mu^2}\,. \tag{9.100}$$

This shows that again one can change the topological gap by tuning μ [46].

An alternative method relies on changing the superconducting order parameter of the proximity-coupled superconductor [47]. While it may be inconvenient to change the magnitude of Δ, the phase of the order parameter is readily manipulated. Indeed, there is a phase gradient associated with a supercurrent flowing along the superconductor. Incorporating this phase gradient into the Hamiltonian (9.35), one readily finds that the phase boundary between topological and non-topological phases depends on the phase gradient. In fact, it turns out that in quantum wires in the topological insulator limit, a finite phase gradient pushes the system towards the topological phase. In this limit, the topological gap is given by $B - \Delta$. Roughly, the effect of the supercurrent can be viewed as reducing the superconducting correlations, pushing the system towards the topological phase. As a result, one can induce Majorana-carrying domain walls in the quantum wires by having supercurrents of different strengths flowing along different segments of the wire. These domain walls can be moved by changing these supercurrents as function of time.

Based on such methods of manipulating domain walls, one can explicitly establish the non-Abelian statistics of the Majorana bound states within the quantum-wire platform [46]. Below, we will see that, in principle, Majoranas can also be effectively braided merely by varying the pairwise couplings between a number of Majoranas on a Y-junction [48]. These couplings can be manipulated by, for example, moving pairs of Majoranas closer together, increasing their spatial overlap, or changing the magnitude of the topological gap in between, affecting the localization length and hence the spatial overlap. Another interesting method relies on charging physics [49].

9.6.2 Non-Abelian Berry phase

The basis for analysing the braiding of Majorana bound states is the non-Abelian Berry phase. An essential assumption in the standard derivation of the Berry phase [50] is that the instantaneous spectrum is non-degenerate at all times. The existence of zero-energy Majorana modes implies that the ground state is degenerate and this degeneracy persists during the entire braiding process. The adiabatic evolution in the presence of degeneracies was first analysed by Wilczek and Zee [51] in a classic paper. They found that in this case, the adiabatic dynamics is not simply described by a geometric phase associated with a conventional vector potential (Berry connection) but rather by one associated with a geometric unitary transformation on the subspace of degenerate states, which can be expressed in terms of a non-Abelian vector potential or Berry connection. This is sometimes referred to as a non-Abelian Berry phase. For completeness, we briefly recapitulate the derivation of the non-Abelian Berry phase since it provides the basis for describing a simple model for the braiding of Majorana zero modes in Section 9.6.3.

Consider a Hamiltonian $H(\boldsymbol{\lambda}(t))$ that depends on time through a set of parameters $\boldsymbol{\lambda} = (\lambda_1, \lambda_2, \ldots)$, with the instantaneous spectrum

$$H(\boldsymbol{\lambda}(t))|\psi_\alpha^n(t)\rangle = E_n(t)|\psi_\alpha^n(t)\rangle. \tag{9.101}$$

This spectrum contains one or several subsets n of degenerate states. The states within each of these degenerate subspaces of dimension d_n are labelled by $\alpha = 1, \ldots, d_n$. Let us now define the adiabatic solution of the corresponding time-dependent Schrödinger equation

$$i\partial_t|\eta_\alpha^n(t)\rangle = H(\boldsymbol{\lambda}(t))|\eta_\alpha^n(t)\rangle, \tag{9.102}$$

with initial condition

$$|\eta_\alpha^n(t=0)\rangle = |\psi_\alpha^n(t=0)\rangle. \tag{9.103}$$

In the adiabatic limit, the time evolution does not take the system out of the degenerate subspace to which the initial state belongs. But, in contrast to the non-degenerate case, the time-evolved state need not remain parallel to $|\psi_\alpha^n(t)\rangle$ at later times. Instead, the time-evolved state can be a linear combination of all the states within the degenerate subspace:

$$|\eta_\alpha^n(t)\rangle = \sum_{\beta=1}^{d_n} U_{\alpha\beta}^n(t)|\psi_\beta^n(t)\rangle. \tag{9.104}$$

Note that the $U_{\alpha\beta}^n(t)$ are just prefactors in a linear combination and not Hilbert-space operators.

To deduce the $U^n_{\alpha\beta}(t)$, we insert this expansion into the time-dependent Schrödinger equation, which yields

$$i\sum_\beta [\partial_t U^n_{\alpha\beta}(t)]|\psi^n_\beta(t)\rangle + i\sum_\beta U^n_{\alpha\beta}(t)|\partial_t\psi^n_\beta(t)\rangle = E_n(t)\sum_\beta U^n_{\alpha\beta}(t)|\psi^n_\beta(t)\rangle. \quad (9.105)$$

Multiplying this equation from the left by $\langle\psi^n_\gamma(t)|$ and dropping the subspace index n for simplicity of notation, we find

$$i\partial_t U_{\alpha\gamma} + i\sum_\beta U_{\alpha\beta}\langle\psi_\gamma|\partial_t\psi_\beta\rangle = EU_{\alpha\gamma}. \quad (9.106)$$

We can now define the non-Abelian Berry connection

$$A_{\alpha\beta}(t) = i\langle\psi_\beta|\partial_t\psi_\alpha\rangle, \quad (9.107)$$

so that we obtain, in matrix notation,

$$i\partial_t U = U(E - A). \quad (9.108)$$

Note that this equation has the same structure as the Schrödinger equation of the time-evolution operator for a time-dependent Hamiltonian, except that on the right-hand side, the analogue of the Hamiltonian stands to the right of the time-evolution operator. Thus, as is familiar for the time-evolution operator, this equation can be solved formally in terms of the anti-time-ordering operator $\tilde{T}$ that orders factors according to ascending time from left to right. Thus, we find the explicit solution

$$U^n(t) = \exp\left[-i\int_0^t dt'\, E_n(t')\right]\tilde{T}\exp\left[i\int_0^t dt'\, A^n(t')\right], \quad (9.109)$$

where we have restored the subspace index n.

To bring out the geometric nature of the time evolution, we can introduce a non-Abelian vector potential that replaces time derivatives by derivatives with respect to the parameters λ_j:

$$\mathbf{A}^n_{\alpha\beta}(t) = i\langle\psi^n_\beta|\nabla_{\boldsymbol{\lambda}}\psi^n_\alpha\rangle. \quad (9.110)$$

Then, U^n can be written in terms of an anti-path-ordered integral in parameter space:

$$U^n(t) = \exp\left[-i\int_0^t dt'\, E_n(t')\right]\tilde{P}\exp\left[i\int d\boldsymbol{\lambda}\cdot\mathbf{A}^n(\boldsymbol{\lambda})\right]. \quad (9.111)$$

The path-ordered exponential of the non-Abelian vector potential generalizes the familiar Berry phase. It depends only on the path in parameter space, not on the way in which the path is being traversed, and is a purely geometric object. Now, we can also express the time-evolution operator

$$\mathcal{U}(t) \simeq \sum_n\sum_\alpha |\eta^n_\alpha(t)\rangle\langle\eta^n_\alpha(0)| = \sum_n\sum_{\alpha\beta} U^n(t)|\eta^n_\beta(t)\rangle\langle\psi^n_\alpha(0)| \quad (9.112)$$

within the adiabatic approximation.

Finally, we briefly discuss how the vector potential transforms under a change of basis of the degenerate subspace,

$$|\psi'_\alpha(t)\rangle = \sum_\beta \Omega_{\alpha\beta}(t)|\psi_\beta(t)\rangle. \tag{9.113}$$

From the definition of the vector potential, one readily finds

$$\mathbf{A} \to \mathbf{A}' = i(\nabla_\lambda \Omega)\Omega^\dagger + \Omega A \Omega^\dagger, \tag{9.114}$$

i.e. Ω transforms just like a regular non-Abelian vector potential. In some cases, this gauge freedom can be used to diagonalize the non-Abelian vector potential. In this basis, the path ordering is no longer necessary and the exponent becomes a diagonal matrix whose non-zero entries just take the form of a standard Berry phase.

9.6.3 Braiding Majorana zero modes

As illustrated in Fig. 9.10, a minimal model for non-Abelian braiding starts from a Y-junction of three one-dimensional topological superconductors, labelled wires 1, 2, and 3 [46, 48, 49]. If all three arms are in the topological phase, there are four Majorana bound states in this system. Three of these are located at the outer ends of the three wires, with Bogoliubov operators labelled γ_j for wire j, and a fourth Majorana mode γ_0 is located at the junction of the three wires. As long as the three arms have a finite

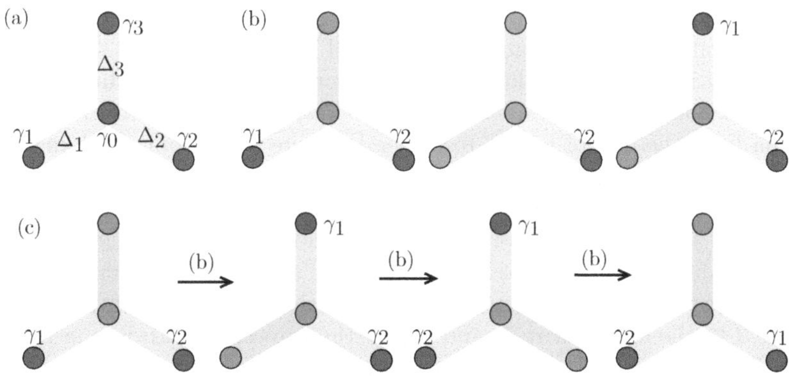

Fig. 9.10 [Colour online] (a) Y-junction with a central Majorana γ_0 and three Majoranas γ_j ($j = 1, 2, 3$) at the outer ends. The outer Majoranas are coupled to the inner Majorana with strength Δ_j. (b) Basic step of the braiding procedure, which moves a zero-energy Majorana from the end of wire 1 to the end of wire 3 by tuning the Δ_j. Blue (yellow) wires indicate zero (non-zero) couplings Δ_j. Dark red circles correspond to zero-energy Majoranas, while green circles indicate Majoranas with finite energy due to coupling. In the intermediate step, the zero-energy Majorana is delocalized over the three pink Majoranas along the yellow wires. (c) Three steps as in (b) braid the zero-energy Majoranas γ_1 and γ_2.

length, these outer Majorana bound states hybridize with the central Majorana and the system is described by the Hamiltonian

$$H = i \sum_{j=1}^{3} \Delta_j \gamma_0 \gamma_j. \tag{9.115}$$

This Hamiltonian couples the central Majorana γ_0 to a linear combination of the outer three Majoranas, $\gamma_\Sigma = (1/E)\sum_{j=1}^{3} \Delta_j \gamma_j$, with proper normalization by $E = (\Delta_1^2 + \Delta_2^2 + \Delta_3^2)^{1/2}$. Thus, the eigenenergies of H are $\pm E$. There are also two linearly independent combinations of the outer Majoranas that do not appear in the Hamiltonian and thus remain true *zero-energy* Majoranas for any (time-independent) choice of the couplings Δ_j. Owing to the presence of these zero-energy modes, the two eigenvalues of H are each doubly degenerate. These zero-energy Majoranas are particularly simple when just one of the couplings Δ_j is non-zero. In this case, the two zero-energy Majoranas are simply the Majoranas located at the ends of those wires with zero coupling.

The couplings Δ_j can be changed as a function of time. For instance, this can be achieved by varying the length of the topological section in each arm: the shorter the topological section, the stronger the overlap and hence the coupling between the outer and the central Majorana. As discussed above, this can be done, say in quantum-wire based realizations, by driving part of the arm into the non-topological phase by the application of a gate voltage or a supercurrent in the adjacent s-wave superconductor. Alternatively, we can leave the length unchanged but vary parameters (such as the Zeeman field, induced superconducting pairing correlations, or chemical potential) such that the topological gap of the arm varies. The smaller the topological gap, the larger the spatial extent and hence the overlap of the Majorana end states.

We can now imagine the following braiding procedure [46, 49]. Initially, only Δ_3 in non-zero. Then, γ_1 and γ_2 are zero-energy Majoranas. In a first step, we move a Majorana from the end of wire 1 to the end of wire 3, without involving the zero-energy Majorana γ_2. To this end, we first increase Δ_1 to a finite value. The zero-energy Majorana originally located at the end of wire 1 is now delocalized and we have a linear combination of γ_0, γ_1, and γ_3. We then localize the Majorana zero mode at the end of wire 3 by reducing Δ_3 to zero, leaving only Δ_1 non-zero. The braiding process is completed by two analogous moves: we first move the zero-energy Majorana from the end of wire 2 to the end of wire 1, and finally the zero-energy Majorana from wire 3 to wire 2. The combined effect of this procedure is to exchange the initial zero-energy Majoranas at the ends of wires 1 and 2.

We will now calculate the adiabatic evolution of the initial state under this braiding protocol. To do so, we introduce conventional fermionic operators through

$$c_1 = \frac{1}{2}(\gamma_1 - i\gamma_2), \qquad c_2 = \frac{1}{2}(\gamma_0 - i\gamma_3). \tag{9.116}$$

Using the inverse relations

$$\gamma_1 = c_1 + c_1^\dagger, \qquad \gamma_2 = i(c_1 - c_1^\dagger), \qquad \gamma_3 = i(c_2 - c_2^\dagger), \qquad \gamma_0 = c_2 + c_2^\dagger, \tag{9.117}$$

we can write H in terms of c_1 and c_2. We can now write the Hamiltonian in the basis $\{|00\rangle, |11\rangle, |10\rangle, |01\rangle\}$, where the basis states are defined as

$$|11\rangle = c_1^\dagger c_2^\dagger|00\rangle, \qquad |10\rangle = c_1^\dagger|00\rangle, \qquad |01\rangle = c_2^\dagger|00\rangle, \tag{9.118}$$

with $c_1|00\rangle = c_2|00\rangle = 0$. This yields

$$H = \begin{pmatrix} \Delta_3 & i\Delta_1 - \Delta_2 & 0 & 0 \\ -i\Delta_1 - \Delta_2 & -\Delta_3 & 0 & 0 \\ 0 & 0 & \Delta_3 & -i\Delta_1 - \Delta_2 \\ 0 & 0 & i\Delta_1 - \Delta_2 & -\Delta_3 \end{pmatrix}. \tag{9.119}$$

The block-diagonal structure is a consequence of fermion number parity conservation. In fact, it is easy to show that the Hamiltonian H commutes with the fermion number parity operator

$$P = -\gamma_0\gamma_1\gamma_2\gamma_3. \tag{9.120}$$

The top-left block $H_{\text{even}} = \Delta_3\tau_z - \Delta_1\tau_y - \Delta_2\tau_x$ corresponds to even fermion parity, while the bottom-right block $H_{\text{odd}} = \Delta_3\tau_z + \Delta_1\tau_y - \Delta_2\tau_x$ has odd fermion parity. Here we have defined Pauli matrices τ_i within the even and odd subspaces. If we also define Pauli matrices π_j in the even–odd subspace, then we can write

$$H = \Delta_3\tau_z - \Delta_1\tau_y\pi_z - \Delta_2\tau_x \tag{9.121}$$

for the overall Hamiltonian H. Writing H_{even} and H_{odd} in terms of Pauli matrices makes it obvious that these Hamiltonians take the form of a spin Hamiltonian in magnetic fields $B_{\text{even}} = (-\Delta_2, -\Delta_1, \Delta_3)$ and $B_{\text{odd}} = (-\Delta_2, \Delta_1, \Delta_3)$, respectively. The degeneracy due to the presence of the Majorana modes implies that the two subspaces have the same eigenvalues. At the same time, the spectrum for each subspace by itself is non-degenerate.

Thus, in the present basis, the non-Abelian vector potential is diagonal and, in line with (9.111), its diagonal entries can be computed just as with conventional Berry phases. For a spin in a magnetic field, we know that the Berry phase is just half the solid angle subtended by the unit vector along the magnetic-field direction during the closed trajectory, with opposite signs for the spin-up and spin-down states [50]. Thus, we can now read off the non-Abelian Berry phases for the braiding procedure described above. Let us start with the odd subspace. Then the analogue of the Zeeman field is the vector $(-\Delta_2, \Delta_1, \Delta_3)$. The Berry phase is independent of the basis in spin space, and thus we temporarily rotate the basis in $\boldsymbol{\tau}$ space by $-\frac{1}{2}\pi$ around the z axis, so that $\tau_x \to -\tau_y$, $\tau_y \to \tau_x$, and $\tau_z \to \tau_z$. In this rotated basis, the effective magnetic field becomes $\boldsymbol{\Delta} = (\Delta_1, \Delta_2, \Delta_3)$. At the beginning of the braiding process, this field points along the positive z direction. We first increase Δ_1 and subsequently reduce Δ_3 to zero. Thus, we rotate the unit vector $\boldsymbol{\Delta}$ in the xz plane to the equator. Next, we increase Δ_2 and reduce Δ_1 to zero. This rotates $\boldsymbol{\Delta}$ by $\frac{1}{2}\pi$ around the equator. Finally, we increase Δ_3 and reduce Δ_2 to zero, which rotates $\mathbf{\Delta}$ back towards the pole. In total, this procedure encloses one quarter of the upper hemisphere, i.e. a solid angle of $\frac{1}{2}\pi$, yielding a Berry phase of $\frac{1}{4}\pi$, with opposite signs for the spin-up and spin-down states.

To obtain the corresponding phase in the even subspace, we note that the effective Zeeman field in this subspace merely differs in the sign of the y component. In effect, this implies that the corresponding $\mathbf{\Delta}$ encloses the same solid angle but encircles it in the opposite direction. Hence, the Berry phases for the even and odd subspaces are equal in magnitude, but opposite in sign, and we find from (9.111) that

$$U_{12} = e^{i\frac{1}{4}\pi\tau_z\pi_z} \tag{9.122}$$

for the exchange of Majoranas 1 and 2. Here, we have dropped the dynamic phase, which is the same for all relevant states. Finally, we can re-express this in terms of the original Majorana operators using the identity $i\gamma_1\gamma_2 = \tau_z\pi_z$ which yields the basis-independent representation

$$U_{12} = e^{-\frac{1}{4}\pi\gamma_1\gamma_2} \tag{9.123}$$

of the effect of braiding Majoranas γ_1 and γ_2.

We can check explicitly that the Majorana braiding matrices (9.123) satisfy the defining relations

$$\sigma_i\sigma_j = \sigma_j\sigma_i, \qquad |i-j|\geq 2, \tag{9.124}$$

$$\sigma_i\sigma_{i+1}\sigma_i = \sigma_{i+1}\sigma_i\sigma_{i+1}, \qquad i = 1,\ldots,N-1, \tag{9.125}$$

of the braid group. Here, we imagine an N-particle system, with the particles ordered and enumerated in some arbitrary fashion as $1, 2, \ldots, N$. Then, σ_i denotes one of $N-1$ generators of the braid group, describing a counterclockwise exchange of particles i and $i+1$. Thus, we can identify σ_i with $U_{i,i+1}$. It is also not difficult to show that the Majorana braiding matrices (9.123) are indeed non-Abelian by showing that $\sigma_i\sigma_{i+1} \neq \sigma_{i+1}\sigma_i$.

9.7 Experimental signatures

There have been many proposals for detecting Majorana bound states experimentally. In this section, we briefly introduce signatures that have been used in experiments.

9.7.1 Conductance signatures

9.7.1.1 Normal-metal lead

A simple and direct method of detecting bound states in superconductors relies on measurements of the tunnelling conductance. The differential conductance is non-zero whenever a state in the sample is energetically aligned with the Fermi level in the normal-metal lead. Tuning the bias voltage between lead and sample effectively shifts the Fermi level in the lead and enables measurements as a function of energy. While tunnelling into a superconductor is typically suppressed at subgap energies, subgap bound states appear as sharp resonances in the differential conductance.

To understand the nature of transport through such bound states, consider an isolated subgap state that gives rise to a sharp singularity in the density of states. A single electron may tunnel from the lead and occupy the quasiparticle bound state. In the absence of coupling to other degrees of freedom, the quasiparticle cannot relax into the superconductor, thus blocking single-particle transport. In contrast, current can flow by Andreev reflection, when an electron entering from the lead is reflected as a hole, creating a Cooper pair in the sample.

Now, let us consider a proximity-coupled wire in a topological phase, terminated at one end by a tunnel barrier and connected to a normal-state lead. We calculate the Andreev current from the normal-state lead to the proximity-providing superconductor by scattering theory. The amplitude for an electron in the lead to tunnel through the tunnel barrier and Andreev-reflect from the superconductor as a hole and for the hole to tunnel back into the lead is $t_h r_{he} t_e$, where r_{he} (and r_{eh}) are amplitudes for Andreev reflections and t_e (t_h) is the barrier transmission amplitude for electrons (holes). Note that we leave the energy dependence of these amplitudes implicit. In addition, the total current also comprises processes in which the Andreev reflection is followed by a reflection at the barrier and further Andreev reflections. The total amplitude for Andreev reflection is the sum of all of these processes (see Fig. 9.11(a)):

$$A_{he} = t_h[1 + r_{he}r_e r_{eh}r_h + (r_{he}r_e r_{eh}r_h)^2 + \ldots]r_{he}t_e = \frac{t_h r_{he} t_e}{1 - r_{he}r_e r_{eh}r_h}. \tag{9.126}$$

To obtain the tunnelling current, we multiply the Andreev reflection probability $|A_{he}|^2$ by the Fermi distribution of incoming electrons and outgoing holes, $n_F(\omega - eV)[1 - n_F(\omega + eV)]$, and integrate over all energies. Note that the electron and hole reservoirs in the lead are shifted relative to the Fermi energy of the sample by $\mp eV$. We further add the contribution due to Andreev reflections of incoming holes and divide by two to prevent double counting. This yields the Andreev current

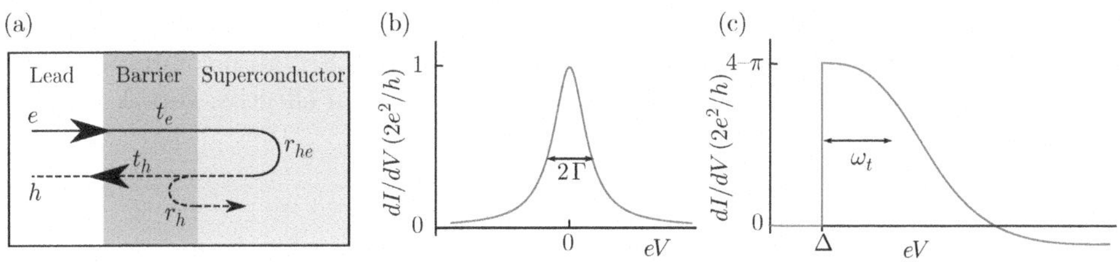

Fig. 9.11 (a) The total amplitude for Andreev reflection is the sum of the amplitudes for a number of processes, as described in the text. (b) Differential conductance as a function of bias voltage for a normal-metal lead. (c) Differential conductance as a function of bias voltage for a superconducting lead.

$$I = \frac{1}{2} 2e \int \frac{d\omega}{2\pi\hbar} |A_{he}|^2 [n_F(\omega - eV) - n_F(\omega + eV)] \tag{9.127}$$

for a spinless superconductor, where the charge $2e$ accounts for the fact that a Cooper pair is transmitted during each Andreev reflection.

At subgap energies, the transmission through the superconductor vanishes and the reflection matrix

$$r = \begin{pmatrix} r_{ee} & r_{eh} \\ r_{he} & r_{hh} \end{pmatrix} \tag{9.128}$$

must be unitary. Particle–hole symmetry relates the matrix elements through

$$\tau_x r(-E) \tau_x = r^*(E). \tag{9.129}$$

Specifically, we find $r_{ee} = r_{hh}^*$ and $r_{eh} = r_{he}^*$ at the Fermi energy. This implies that $\det r(E = 0)$ is real. When combined with unitarity, this demands that the determinant of the reflection matrix take on only two possible values:

$$\det r = \pm 1. \tag{9.130}$$

This corresponds to the following two cases, making $\det r$ a topological index: (i) reflection from the trivial phase with perfect normal reflection $|r_{ee}| = 1$ and zero Andreev reflection $r_{eh} = 0$, corresponding to $\det r = 1$, and (ii) reflection from the topological phase with perfect Andreev reflection $|r_{eh}| = 1$ and zero normal reflection $r_{ee} = 0$. Note that it is impossible to tune smoothly between the two cases.

Before returning to the conductance signatures, we briefly note that this result allows for an alternative derivation of the existence of Majorana bound states. In the topological phase, we find $r_{he} r_{eh} = 1$ at the Fermi level. Consider a topological superconductor terminated by a short normal section and a hard wall. An electron at the Fermi energy impinging on the superconductor is Andreev reflected as a hole, the hole undergoes normal reflection at the hard wall (with phase π) and Andreev reflection from the superconductor, and finally the electron is normally reflected from the hard wall, closing the trajectory. At the Fermi energy, the reflection phases add to a multiple of 2π, implying the formation of a bound state by Bohr–Sommerfeld quantization. This zero-energy bound state is just the Majorana.

To obtain A_{he} also at non-zero energies, we first recall the Andreev reflection amplitudes $r_{he} = r_{eh} = \exp[-i \arccos(\omega/\Delta)]$ for an s-wave superconductor with real order parameter $\Delta > 0$. In a p-wave superconductor (see (9.8)), an incoming electron and outgoing hole (both with momentum p_F) experience an effective gap $\Delta = \Delta' p_F$ and thus

$$r_{he} = \exp[-i \arccos(\omega/\Delta)], \tag{9.131}$$

as for the s-wave superconductor. The gap has the opposite sign for the reverse process, since both the incoming hole and the reflected electron have momentum $-p_F$. The Andreev reflection amplitude is thus

$$r_{eh} = \exp[-i \arccos(\omega/\Delta) + i\pi] = \exp[i \arccos(-\omega/\Delta)] \tag{9.132}$$

In the vicinity of the Fermi level, we can expand

$$r_{he}r_{eh} \simeq 1 + 2i\omega/\Delta, \tag{9.133}$$

and at weak tunnelling through the barrier, we can approximate $r_{e/h} \simeq 1 - \frac{1}{2}t_{e/h}^2$ (assuming real $r_{e/h}$ and $t_{e/h}$). Using these approximations, we arrive at the Breit–Wigner form

$$|A_{he}|^2 = \frac{t_h^2 t_e^2}{4\omega^2/\Delta^2 + \frac{1}{4}(t_e^2 + t_h^2)^2} \tag{9.134}$$

for the Andreev reflection amplitude, and the Andreev current becomes

$$I = e \int \frac{d\omega}{2\pi\hbar} \frac{\Gamma_e \Gamma_h}{\omega^2 + \frac{1}{4}(\Gamma_e + \Gamma_h)^2} [n_F(\omega - eV) - n_F(\omega + eV)], \tag{9.135}$$

where we have introduced the electron and hole tunnelling rates $\Gamma_{e/h} = \frac{1}{2}\Delta t_{e/h}^2$ through the barrier. These can be evaluated at the Fermi level, where they are equal by particle–hole symmetry: $\Gamma_e = \Gamma_h = \Gamma$. The resonance of the integrand (9.135) at energy $\omega = 0$ reflects the Majorana bound state at the junction.

Using (9.135) to compute the differential conductance, we find

$$\frac{dI}{dV} = \frac{2e^2}{h} \frac{\Gamma^2}{eV^2 + \Gamma^2}. \tag{9.136}$$

The differential conductance is a Lorentzian as a function of bias voltage, with quantized height $2e^2/h$ and peak width determined by the tunnelling rate through the barrier; see Fig. 9.11(b) [53, 54]. This quantized zero-bias conductance peak can serve as a robust fingerprint of an isolated Majorana bound state. For other subgap states such as regular Andreev bound states, the conductance is not restricted to quantized values. Moreover, such resonances will typically shift in energy as functions of gate voltage or magnetic field.

Nevertheless, it remains a challenging task to resolve this quantized conductance peak in experiment. First, temperature broadening of the distribution function in the normal lead limits the energy resolution. Once temperature exceeds the intrinsic broadening Γ, the width of the conductance peak is determined by temperature T, and the conductance peak is correspondingly reduced by a factor of order Γ/T. The situation may be particularly unfavourable in multichannel wires, where the coupling of the topological channel to the lead is typically very weak [52].

Second, the zero-bias Majorana peak is also broadened by inelastic quasiparticle transitions in the superconductor. At finite temperatures, there will be inelastic (e.g., phonon-assisted) transitions of quasiparticles between the zero-energy state and other subgap states or the quasiparticle continuum. These quasiparticle-poisoning processes reduce the lifetime of the zero-energy excitation and can be accounted for phenomenologically in (9.136) by including an additional rate Γ_{qp} in the broadening:

$$\frac{dI}{dV} = \frac{2e^2}{h} \frac{\Gamma^2}{eV^2 + (\Gamma + \Gamma_{qp})^2}. \tag{9.137}$$

Thus, quasiparticle poisoning also destroys the conductance quantization at zero bias. In addition, such relaxation processes of quasiparticles allow for an additional single-particle current that is non-quantized and adds to the Andreev current [55, 56].

9.7.1.2 Superconducting lead

An alternative experiment tunnels into the Majorana bound state from a (non-topological) superconducting tip. One advantage of this setup is that the gap exponentially suppresses finite-temperature broadening. It is important to understand that for a superconducting electrode, Majorana bound states are no longer signalled by zero-bias peaks. The threshold for electron tunnelling corresponds to the Majorana bound state overlapping with the BCS singularity in the density of states of the electrode. Thus, the Majorana bound state is signalled by differential conductance peaks at bias voltages $eV = \pm\Delta$, where Δ denotes the gap of the electrode.

Heuristically, we can derive the tunnelling current from (9.135) by noting that the tunnelling rates Γ_e and Γ_h are proportional to the density of states in the lead electrode. In a superconductor, we thus expect $\Gamma_{e/h} = \Gamma\rho(\omega \mp eV)$, with the dimensionless BCS density of states $\rho(\omega) = \theta(|\omega| - \Delta)|\omega|/\sqrt{\omega^2 - \Delta^2}$ normalized to the normal-state density of states. It can indeed be shown that this is the result of a more formal calculation [57].

Up to exponentially small corrections in $\Delta/T \gg 1$ and for $eV \simeq \Delta$, we can set $n_F(\omega - eV) - n_F(\omega + eV) \simeq 1$ (reflecting the above-mentioned insensitivity to temperature). Then, the current becomes

$$I = e\int_{-(eV-\Delta)}^{eV-\Delta} \frac{d\omega}{2\pi\hbar} \frac{\Gamma^2\rho(\omega - eV)\rho(\omega + eV)}{\omega^2 + \frac{1}{4}\Gamma^2[\rho(\omega - eV) + \rho(\omega + eV)]^2}. \tag{9.138}$$

In this low-temperature limit, the current vanishes for $eV < \Delta$. We measure voltage from the threshold, $\eta = eV - \Delta$, so that for $\eta \simeq 0$, the bound state is energetically aligned with the BCS singularities of the lead. For $|\omega| < \eta$, we can approximate

$$\rho(\omega \pm eV) \simeq \sqrt{\Delta/2(\eta \pm \omega)} \gg 1. \tag{9.139}$$

This yields a current

$$I = \frac{4e}{h}\int_{-\eta}^{\eta} \frac{d\omega}{\sqrt{\eta^2 - \omega^2}} \frac{\omega_t^3}{\omega^2 + \omega_t^3\left(\frac{1}{\sqrt{\eta - \omega}} + \frac{1}{\sqrt{\eta + \omega}}\right)^2}, \tag{9.140}$$

where we have introduced the effective tunnel coupling $\omega_t = \frac{1}{2}(\Delta\Gamma^2)^{1/3}$. We finally arrive at

$$I = \frac{4e}{h}\eta\int_{-1}^{1} \frac{dx}{\sqrt{1 - x^2}} \frac{1}{x^2\left(\frac{\eta}{\omega_t}\right)^3 + \left(\frac{1}{\sqrt{1 - x}} + \frac{1}{\sqrt{1 + x}}\right)^2} \tag{9.141}$$

for $\eta > 0$, with the limiting cases

$$\frac{dI}{dV} = \begin{cases} 0, & eV - \Delta < 0, \\ \frac{2e^2}{h}(4 - \pi), & eV - \Delta = 0, \\ -\frac{2e^2}{h}\frac{\omega_t^3}{(eV - \Delta)^3} \times \text{const}, & eV - \Delta \gg \omega_t. \end{cases} \tag{9.142}$$

The conductance is shown in Fig. 9.11(c). At $eV = \Delta$, the conductance jumps from zero to the maximum value $(4 - \pi)(2e^2/h)$ and then decreases on the scale ω_t, eventually developing a shallow negative differential conductance dip.

This result has several remarkable implications [57]:

1. The peak conductance is universal, independent of tunnelling strength, Majorana wavefunction, and the sign of the voltage. This parallels the conductance quantization for a normal-metal tip. For both normal-state and superconducting electrodes, conventional subgap states exhibit non-universal behaviour, so they can in principle be distinguished from Majorana resonances.
2. The peak width depends on a lower power of the tunnelling strength for a superconducting electrode, $\sim \Gamma^{2/3}$, than for a normal-state electrode, $\sim \Gamma$. This weak dependence on the junction transmission allows one to distinguish bound-state resonances from competing multiple Andreev peaks, whose width scales as Γ^2.
3. Thermal broadening is practically irrelevant for a superconducting lead at $T \ll \Delta$.
4. For a superconducting lead, the peak conductance is less vulnerable to quasiparticle poisoning. At the threshold $eV = \Delta$, the tunnelling rates $\Gamma_{e/h}$ diverge owing to the BCS singularity, making additional broadening due to inelastic transitions ineffective.

This leads to a striking Majorana signature when a superconducting STM tip is used to map out the conductance in the vicinity of the bound state. For a Majorana state, the threshold conductance is independent of the location of the tip, and the Majorana appears as a plateau of height $(4 - \pi)(2e^2/h)$. The extension of the plateau is limited only by experimental resolution.

9.7.2 4π-periodic Josephson effect

So far, we have considered junctions of a topological superconductor with a normal metal or a conventional superconductor. Junctions of two topological superconductors, harbouring two Majorana bound states γ_L and γ_R, provide additional signatures. The coupling across the junction fuses the two Majoranas into a conventional fermion with non-zero energy. Nevertheless, the junction retains important signatures of the topological phase in the two superconductors.

This is rooted in the bound-state spectrum of the junction, which can be obtained from the tunnelling Hamiltonian

$$H_T = tc_L^\dagger c_R + t^* c_R^\dagger c_L, \tag{9.143}$$

which couples the topological superconductors. Here, t is the tunnelling matrix element and the operator $c_{L/R}$ annihilates an electron at the junction in the left/right superconductor. When the two banks of the junction have the same superconducting phase, the projection of the tunnelling Hamiltonian onto the low-energy Majorana excitations reads

$$H_T = (t + t^*)u_L u_R P, \tag{9.144}$$

where $P = i\gamma_L\gamma_R$ is the parity operator of the fermion formed from the two Majoranas. Here, we have used the fact that the electron operators project as $c_L \simeq u_L\gamma_L$ and $c_R \simeq iu_R\gamma_R$, where $u_{L/R}$ are real Majorana wavefunctions in the left and right banks. Indeed, these expressions are consistent with the results for the Kitaev chain in Section 9.5.3 for $t = \Delta$ and $\mu = 0$, where the low-energy projections of the two end fermions are $c_1 \simeq i\gamma_{A1}$ and $c_N \simeq \gamma_{BN}$.

In an appropriate gauge, a phase difference φ across the junction can be incorporated entirely into the tunnelling amplitude, $t = t_0 e^{i\varphi/2}$, with t_0 real. Thus, we find the phase-dependent subgap spectrum

$$E = \pm 2\cos\left(\frac{1}{2}\varphi\right) t_0 u_L u_R, \tag{9.145}$$

where the sign corresponds to the parity eigenvalue ± 1. This is illustrated in Fig. 9.12. Remarkably, E is 4π-periodic for fixed parity, and tuning the phase by 2π changes the energy of the system! Of course, the entire spectrum, including both fermion parity sectors, is 2π-periodic, as required by gauge invariance.

For fixed fermion parity, this result predicts a 4π-periodic Josephson current, quite unlike the 2π-periodic Josephson current of conventional Josephson junctions. This remarkable consequence of Majorana physics follows when it is recalled that the Josephson current can be obtained from the subgap spectrum through

$$I = 2e\frac{dE}{d\varphi}, \tag{9.146}$$

as can be readily established from the tunnelling Hamiltonian (9.143) with $t = t_0 e^{i\varphi/2}$. For fixed fermion parity, only one of the subgap states contributes and the Josephson current has period 4π. We also observe that the Josephson current has the same magnitude, but opposite signs, for the two fermion parities.

An important point of this argument is that there is a degeneracy of the two states at $\varphi = \pi$. The Josephson behaviour would revert to the conventional 2π periodicity,

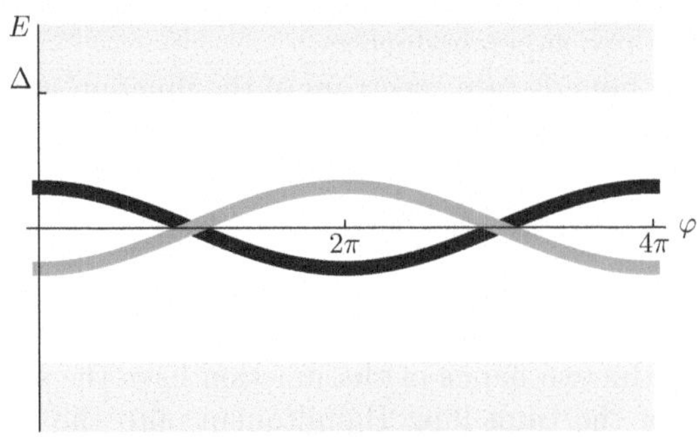

Fig. 9.12 Schematic spectrum of a Josephson junction harbouring two Majorana bound states. The shaded areas represent the quasiparticle continuum above the gap.

should this crossing turn into an anticrossing. However, this cannot happen, since the crossing is protected by fermion parity! The crossing of the subgap states at $\varphi = \pi$ implies a ground-state degeneracy and can be viewed as a quantum phase transition, at which the fermion parity of the many-body ground state changes.

It is also interesting to connect these considerations with our discussion of symmetry classes in Section 9.2.1. For $\varphi = \pi$, the pairing gaps have opposite signs in the two banks. As discussed in Section 9.2.1, each bank is described by a Hamiltonian in class BDI, albeit with opposite topological indices ± 1. Hence, the topological index jumps by 2 across the junction, which necessitates the presence of two Majorana bound states, in agreement with our findings. The protection of these two Majoranas relies on chiral symmetry, which is broken by the complex order parameter away from $\varphi = \pi$, and hence the energy levels split.

Experimental observation of the 4π-periodic Josephson effect requires that fermion parity be preserved. If the phase difference is varied too slowly, parity may change by quasiparticle poisoning, masking the 4π periodicity. One way of varying the phase difference swiftly is via the ac Josephson effect in the presence of a finite bias voltage across the junction. The 4π periodicity generates an ac current at half the usual Josephson frequency, a clear signature of topological superconductivity. However, one needs to keep in mind that the time-dependent phase difference may induce diabatic transitions between the low-energy bound states and the quasiparticle continuum above the gap. Such transitions are most likely in the vicinity of the phase difference where the bound-state energy becomes maximal, and cause switching between the fermion parities. This also masks the fractional Josephson frequency. Even in the presence of these transitions, however, a signature remains present in the finite-frequency current noise, which has a peak at half the Josephson frequency [58]. This effect is particularly prominent at low bias voltages, when transitions occur only after many cycles.

An alternative route, which requires only static measurements, is based on Shapiro steps. In conventional junctions, the combination of a dc voltage $V_{\rm dc}$ and an ac voltage $V_{\rm ac} \sin \omega t$ generates a Josephson current

$$I = I_J \sin\left(\varphi + 2eV_{\rm dc}t - \frac{eV_{\rm ac}}{\omega} \cos \omega t\right). \tag{9.147}$$

Expanding this expression in Bessel functions, one can show that the current exhibits steps as a function of bias voltage. These steps originate from resonances between the ac voltage and the phase winding due to the dc voltage, which occur when $2eV_{\rm dc} = n\omega$, with n an integer. Clearly, this condition is modified when the current–phase relation is 4π-periodic and the steps occur instead at $eV_{\rm dc} = n\omega$. Thus, Majoranas only contribute to every second Shapiro step, predicting a prominent even–odd asymmetry of the Shapiro steps as a strong signatures of Majorana states.

9.8 Conclusions

In these lecture notes, we have provided an introduction to the physics of one-dimensional topological superconductivity and Majorana bound states. This field is currently attracting significant theoretical and experimental attention, fuelled by the prospect of not only establishing the existence of these exotic quasiparticles, but also observing a new type of quantum statistics. Condensed matter has already enriched physics through the (Abelian) anyonic statistics of the quasiparticles in the fractional quantum Hall effect. Observing non-Abelian statistics would take this to yet another level. It is quite remarkable that such fundamentally new physics is lurking in material systems as mundane as hybrids of semiconductors and superconductors, with the relevant phases accessible to a standard mean-field analysis. This contrasts sharply with the Abelian anyons that occur in the strongly correlated fractional quantum Hall states.

Beyond the non-Abelian statistics, the field has been energized by its potential for topological quantum information processing. One can envision the remarkable properties of Majorana bound states being exploited to store and process quantum information in an intrinsically fault-tolerant manner. However, it turns out that it is impossible to construct a universal topological quantum computer based on braiding Majorana bound states. Two possible workarounds are being discussed in the literature. The less ambitious but perhaps more realistic approach is to complement the topologically protected braiding operations by additional gate operations that are unprotected. In addition to the topologically protected gate operations based on Majorana braiding, it would suffice to include two unprotected operations, namely the one-qubit operation $\exp(\frac{1}{8}\pi\gamma_0\gamma_1)$ and the two-qubit operation $\exp(i\frac{1}{4}\pi\gamma_0\gamma_1\gamma_2\gamma_3)$ [59]. Even when such non-topological gate operations are included, it is still possible to gain significantly from the topological protection of information storage and the partial protection of information processing. The more ambitious programme tries to find platforms that realize yet more exotic quasiparticles such as Fibonacci anyons with a richer braid group and the capacity to realize a universal topological quantum computer.

There are many aspects of Majorana physics that have not been discussed in these lecture notes, such as effects of disorder and interactions, alternative experimental platforms, and numerous proposals for experimental Majorana signatures. Most importantly, we have not discussed the existing experiments in any detail. However, we hope that these notes have provided sufficient detail for readers to develop their own informed opinion on these and forthcoming experiments.

9.A Pairing Hamiltonians: BdG and second quantization

Up to a constant, a general second-quantized pairing Hamiltonian $\mathcal{H}$ can be brought into BdG form by doubling the degrees of freedom:

$$\mathcal{H} = \frac{1}{2}\int d^d x \, \Psi^\dagger(x) H \Psi(x) + \text{const}, \tag{9.148}$$

where H is the first-quantized BdG Hamiltonian. Throughout these lecture notes, we choose the Nambu spinor ordered as

$$\Psi(x) = (\psi_\uparrow(x),\ \psi_\downarrow(x),\ \psi_\downarrow^\dagger(x),\ -\psi_\uparrow^\dagger(x))^T. \tag{9.149}$$

As an example, we consider the second-quantized Hamiltonian for the topological insulator edge:

$$\mathcal{H} = \int dx \,\{- iv_F[\psi_\uparrow^\dagger(x)\partial_x\psi_\downarrow(x) + \psi_\downarrow^\dagger(x)\partial_x\psi_\uparrow(x)] \\ - B[\psi_\uparrow^\dagger(x)\psi_\uparrow(x) - \psi_\downarrow^\dagger(x)\psi_\downarrow(x)] + \Delta[\psi_\uparrow^\dagger(x)\psi_\downarrow^\dagger(x) + \text{h.c.}]\}. \tag{9.150}$$

Introducing the Nambu spinor and using the anticommutation relations of the electronic operators, we can bring this Hamiltonian into the BdG form (9.148) with the BdG Hamiltonian H given by (9.24).

With the definition (9.149) of the spinor, time reversal is effected by

$$T = i\sigma_y K, \tag{9.151}$$

where K denotes complex conjugation, and charge conjugation by

$$C = -i\tau_y. \tag{9.152}$$

Owing to the doubling of the degrees of freedom, the BdG Hamiltonian acquires the constraint

$$CT\Psi = \Psi, \tag{9.153}$$

known as particle–hole symmetry. For the BdG Hamiltonian, particle–hole symmetry implies

$$\{H, CT\} = 0, \qquad CT\Psi = \Psi. \tag{9.154}$$

We denote the eigenfunctions of H as

$$\Phi_n(x) = (u_{\uparrow,n}(x),\ u_{\downarrow,n}(x),\ \ v_{\uparrow,n}(x),\ v_{\downarrow,n}(x))^T, \tag{9.155}$$

with eigenvalues E_n, satisfying

$$H\Phi_n(x) = E_n\Phi_n(x). \tag{9.156}$$

This equation is known as the BdG equation. Particle–hole symmetry implies

$$HCT\Phi_n = -CTH\Phi_n = -CTE_n\Phi_n = -E_nCT\Phi_n, \tag{9.157}$$

i.e. for every BdG eigenspinor Φ_n with energy E_n, there is an eigenspinor $\Phi_{-n} = CT\Phi_n$ with energy $-E_n$. The eigenspinors are orthonormal:

$$\int dx\, \Phi_n^\dagger(x)\Phi_m(x) = \delta_{nm}, \tag{9.158}$$

where δ_{nm} is the Kronecker symbol when $\Phi_n(x)$ and $\Phi_m(x)$ are normalizable and a Dirac δ-function when they are scattering states. In addition, the completeness can be written as

$$\sum_n \Phi_n(x)\Phi_n^\dagger(y) = \delta(x-y). \tag{9.159}$$

Using the BdG eigenspinors, the second-quantized Hamiltonian (9.148) can be written as (up to a constant)

$$\begin{aligned}\mathcal{H} &= \frac{1}{2}\int dx\, \Psi^\dagger(x) H \Psi(x) \\ &= \int dx \int dy\, \Psi^\dagger(y) H \delta(x-y)\Psi(x) \\ &= \sum_n \int dx \int dy\, \Psi^\dagger(y) H \Phi_n(y)\Phi_n^\dagger(x)\Psi(x) \\ &= \sum_n E_n \gamma_n^\dagger \gamma_n,\end{aligned} \tag{9.160}$$

where the γ_n are the Bogoliubov quasiparticle operators. They can be expressed in terms of the original electron operators as

$$\gamma_n = \int dx\, \Phi_n^\dagger(x)\Psi(x), \tag{9.161}$$

$$\gamma_n^\dagger = \int dx\, \Psi^\dagger(x)\Phi_n(x). \tag{9.162}$$

Using (9.154), (9.157), and the unitarity CT, we have

$$\begin{aligned}\gamma_{-n}^\dagger &= \int dx\, \Psi^\dagger(x)\Phi_{-n}(x) \\ &= \int dx\, [CT\Psi(x)]^\dagger [CT\Phi_n(x)] \\ &= \int dx\, \Phi_n^\dagger(x)\Psi(x) \\ &= \gamma_n.\end{aligned} \tag{9.163}$$

One can readily check that the Bogoliubov operators satisfy the fermionic anticommutation relations.

We can also write the electronic operators in terms of Bogoliubov operators using (9.159):

$$\begin{aligned}\Psi(x) &= \sum_n \Phi_n(x)\gamma_n \\ &= \sum_{n>0}[\Phi_n(x)\gamma_n + \Phi_{-n}(x)\gamma_{-n}] \\ &= \sum_{n>0}[\Phi_n(x)\gamma_n + CT\Phi_n(x)\gamma_n^\dagger]. \end{aligned} \tag{9.164}$$

These equations have to be complemented by the Majorana mode when there is an isolated zero-energy eigenspinor with associated Bogoliubov operator $\gamma_0 = \gamma_0^\dagger$; see Section 9.3.2.

9.B Proximity-induced pairing

All realizations of topological superconducting phases discussed in these lecture notes are based on proximity-induced superconductivity. However, we have never explicitly discussed the s-wave superconductor that induces the superconducting correlations in the one-dimensional system. Instead, we have directly included a pairing term in the BdG Hamiltonian of the one-dimensional system. In this appendix, we want to briefly discuss the proximity effect more explicitly for a one-dimensional wire proximity-coupled to a BCS superconductor, not accounting for Zeeman fields or spin–orbit coupling.

The s-wave superconductor is described by the pairing Hamiltonian

$$\mathcal{H}_s = \frac{1}{2}\int d^3r\, \psi^\dagger(\mathbf{r})H_s\psi(\mathbf{r}), \tag{9.165}$$

$$H_s = \xi_{\mathbf{p}}\tau_z + \Delta\tau_x, \tag{9.166}$$

$$\xi_{\mathbf{p}} = \frac{\mathbf{p}^2}{2m} - \mu, \tag{9.167}$$

written in terms of the Nambu spinor

$$\psi(\mathbf{r}) = (\psi_\uparrow(\mathbf{r}),\ \psi_\downarrow(\mathbf{r}),\ \psi_\downarrow^\dagger(\mathbf{r}),\ -\psi_\uparrow^\dagger(\mathbf{r}))^T. \tag{9.168}$$

Here, Δ is the s-wave gap of the proximity-providing superconductor and the τ_i denote Pauli matrices in Nambu space. The one-dimensional wire can be modelled by the BdG Hamiltonian

$$\mathcal{H}_d = \frac{1}{2}\int dx\, d^\dagger(x)H_d d(x), \tag{9.169}$$

where

$$d(x) = (d_\uparrow(x),\ d_\downarrow(x),\ d_\downarrow^\dagger(x),\ -d_\uparrow^\dagger(x))^T. \tag{9.170}$$

At low energies in the vicinity of the Fermi energy, we can linearize the dispersion of the wire so that its BdG Hamiltonian becomes

$$H_d = v_F p_x \tau_z, \tag{9.171}$$

with the momentum p_x measured from the respective Fermi point. The hybridization between the adatom chain and the superconductor is modelled by

$$\mathcal{H}_T = -\frac{t}{2}\int d^3r \left[\psi^\dagger(\mathbf{r})\tau_z d(x) + d^\dagger(x)\tau_z\psi(\mathbf{r})\right]\delta(y)\delta(z), \tag{9.172}$$

which describes local tunnelling between wire and superconductor.

In order to describe the effect of the superconductor on the wire, we consider the Green function of the electrons in the wire and account for the coupling to the superconductor through a self-energy. Since we are dealing with a quadratic problem, this self-energy can be computed exactly:

$$\Sigma(x_1 - x_2, E) = t^2\tau_z g_s(\mathbf{x}_1 - \mathbf{x}_2, E)\tau_z, \qquad \mathbf{x}_i = (x_i, 0, 0), \tag{9.173}$$

where $g_s(\mathbf{r})$ is the real-space Green function of the uncoupled BCS superconductor. We have also used the translational invariance of the adatom chain along the x direction so that the self-energy depends only on the distance between the two positions. The Green function of the superconductor can now be evaluated as

$$\begin{aligned} g_s(\mathbf{x}_1 - \mathbf{x}_2, E) &= \int \frac{d^3p}{(2\pi)^3}\frac{e^{i(x_1-x_2)p_x}}{E - \Delta\tau_x - \xi_p\tau_z} \\ &\simeq \nu_0^{2D}\int \frac{dp_x}{2\pi}\, e^{i(x_1-x_2)p_x}\int d\xi\,\frac{E + \Delta\tau_x}{E^2 - \Delta^2 - \xi^2} \\ &= -\pi\nu_0^{2D}\frac{E + \Delta\tau_x}{\sqrt{\Delta^2 - E^2}}\,\delta(x_1 - x_2). \end{aligned} \tag{9.174}$$

Here, ν_0^{2D} is a two-dimensional density of states at the Fermi level. Note that to a good approximation, the self-energy is strictly local. Thus, it becomes independent of momentum when Fourier transforming to momentum space along the x direction:

$$\Sigma(k, E) = -\Gamma_{2D}\frac{E - \Delta\tau_x}{\sqrt{\Delta^2 - E^2}}, \qquad \Gamma_{2D} = \pi\nu_0^{2D}t^2 \tag{9.175}$$

We thus find the dressed Green function

$$G_s(k, E) = \left[g_s^{-1}(k, E) - \Sigma(k, E)\right]^{-1} \tag{9.176}$$

describing the propagation of electrons in the quantum wire.

To understand this self-energy better, it is instructive to consider various limits. For $E \gg \Delta$, i.e. for energies far above the gap, we find the retarded self-energy

$$\Sigma^R(k, E) \simeq -i\Gamma_{2D}. \tag{9.177}$$

The self-energy is purely imaginary and describes the fact that high-energy excitations in the wire can decay into the superconductor with rate $2\Gamma_{2D}$. Indeed, $2\Gamma_{2D} = 2\pi\nu_0^{2D}t^2$ just coincides with a simple golden-rule result for this process at fixed k, since the density of states of the superconductor is unaffected by pairing at high energies.

At subgap energies, $E \ll \Delta$, the self-energy is purely real, reflecting the fact that the superconductor is gapped and excitations in the wire can only virtually enter the superconductor, but not decay into it. We can now expand the self-energy for small E:

$$\Sigma(k, E) \simeq -\frac{\Gamma_{2D}}{\Delta}E - \Gamma_{2D}\tau_x. \tag{9.178}$$

At first sight, the induced gap is given by Γ_{2D}. However, this cannot be the case at strong hybridization $\Gamma_{2D} \gg \Delta$. In fact, in this limit, the term linear in E becomes important and induces a significant renormalization of the quasiparticle weight:

$$\begin{aligned} G_s(k, E) &= [E(1 + \Gamma_{2D}/\Delta) - v_F k\tau_z - \Gamma_{2D}\tau_x]^{-1} \\ &= \frac{Z}{E - Zv_F k\tau_z + Z\Gamma_{2D}\tau_x}, \end{aligned} \tag{9.179}$$

where

$$Z = \frac{1}{1 + \Gamma_{2D}/\Delta}. \tag{9.180}$$

This renormalization of the quasiparticle weight reflects the fact that even at subgap energies, excitations of the quantum wire have appreciable spectral weight in the superconductor. We can now identify the induced gap

$$\Delta_{\text{ind}} = Z\Gamma_{2D} \simeq \begin{cases} \Gamma_{2D}, & \Gamma_{2D} \ll \Delta, \\ \Delta, & \Gamma_{2D} \gg \Delta. \end{cases} \tag{9.181}$$

We observe that as a result of the renormalization of the quasiparticle weight, the induced gap indeed saturates at the host gap, as expected.

However, this renormalization also has other important consequences. The dispersion is obtained from the poles of the Green function:

$$\det G_s^{-1}(k, E) = 0 \tag{9.182}$$

for each k. This yields

$$E(k) = \pm\sqrt{(Zv_F k)^2 + (Z\Gamma_{2D})^2}\,. \tag{9.183}$$

Thus, at strong hybridization, $\Gamma_{2D} \gg \Delta$, there is also a significant renormalization of the Fermi velocity v_F:

$$v_F \to \tilde{v}_F = v_F \frac{\Delta}{\Gamma_{2D}}. \tag{9.184}$$

This implies that also the effective coherence length of the proximity-induced superconducting correlations in the wire can be quite different from the coherence length of the superconductor. If we assume that both wire and superconductor have bare Fermi velocities of the same order (as is presumably the case in the adatom scenario), the correlation length of the proximity-induced superconductivity is

$$\xi = \frac{\hbar \tilde{v}_F}{\Delta_{ind}} = \frac{\hbar v_F}{\Gamma_{2D}}. \tag{9.185}$$

At strong hybridization, this is much smaller than the coherence length of the bulk superconductor, $\hbar v_F/\Delta$.

9.C Shiba states

In this appendix, we outline the derivation of the Shiba states for a single magnetic impurity. We first consider the approach in which the adatom is described as a classical magnetic moment. Subsequently, we describe the adatom as a spin-$\frac{1}{2}$ Anderson impurity, treating the on-site interaction within a mean-field approximation.

9.C.1 Adatom as a classical magnetic impurity

Our starting point is the Hamiltonian (9.48). When choosing the impurity spin $\mathbf{S}$ to point along the z direction, this 4×4 Hamiltonian separates into independent 2×2 blocks $\mathcal{H}_\pm$ for spin-up (+) and spin-down (−) electrons:

$$\mathcal{H}_\pm = \xi_\mathbf{p}\tau_z + [V\tau_z \mp JS]\delta(\mathbf{r}) + \Delta\tau_x. \tag{9.186}$$

To solve for the bound-state spectrum, we isolate the impurity terms on the right-hand side,

$$(E - \xi_\mathbf{p}\tau_z - \Delta\tau_x)\psi(\mathbf{r}) = (V\tau_z \mp JS)\delta(\mathbf{r})\psi(\mathbf{0}), \tag{9.187}$$

and pass to the momentum representation, $\psi(\mathbf{r}) = \int [d\mathbf{p}/(2\pi)^3]\, \psi_\mathbf{p}$. This yields

$$(E - \xi_\mathbf{p}\tau_z - \Delta\tau_x)\psi_\mathbf{p} = (V\tau_z \mp JS)\psi(\mathbf{0}), \tag{9.188}$$

and hence

$$\psi_\mathbf{p} = \frac{1}{E - \xi_\mathbf{p}\tau_z - \Delta\tau_x}(V\tau_z \mp JS)\psi(\mathbf{0}). \tag{9.189}$$

We can now obtain an equation for the spinor $\psi(\mathbf{0})$ evaluated at the position of the impurity only:

$$\psi(\mathbf{0}) = \int \frac{d\mathbf{p}}{(2\pi)^3} \frac{E + \xi_\mathbf{p}\tau_z + \Delta\tau_x}{E^2 - \xi_\mathbf{p}^2 - \Delta^2}(V\tau_z \mp JS)\psi(\mathbf{0}). \tag{9.190}$$

For subgap energies $E < \Delta$, the integral can be readily performed. This yields

$$\left[\mathbf{1} + \frac{E + \Delta\tau_x}{\sqrt{\Delta^2 - E^2}}(\beta\tau_z \mp \alpha)\right]\psi(\mathbf{0}) = 0. \tag{9.191}$$

Here, we have introduced the dimensionless measures $\alpha = \pi\nu_0 JS$ and $\beta = \pi\nu_0 V$ of the exchange coupling and the potential scattering, respectively. ν_0 denotes the normal-phase density of states.

Setting the determinant of the prefactor of $\psi(\mathbf{0})$ in (9.191) equal to zero, we find that $\mathcal{H}_\pm$ has a subgap solution with energies

$$E = \pm\Delta\frac{1 - \alpha^2 + \beta^2}{\sqrt{(1 - \alpha^2 + \beta^2)^2 + 4\alpha^2}}. \tag{9.192}$$

The positive (negative) sign corresponds to the spin-up (spin-down) sector. The energies of the two Shiba states cross at zero when $\alpha^2 = 1 + \beta^2$. For stronger exchange coupling, the ground state changes from even to odd electron number. Notice also that $E_0 \to \Delta$ for vanishing exchange coupling, $\alpha \to 0$, in accordance with Anderson's theorem.

Inserting these eigenenergies into (9.191), we can obtain the corresponding eigenspinors,

$$\begin{aligned}\psi(\mathbf{0}) &= \begin{pmatrix} u(\mathbf{0}) \\ v(\mathbf{0}) \end{pmatrix} \\ &= C\begin{pmatrix} \sqrt{1 + (\alpha \pm \beta)^2} \\ \pm\sqrt{1 + (\alpha \mp \beta)^2} \end{pmatrix},\end{aligned} \tag{9.193}$$

where C is a normalization constant. The normalization constant can be found from the condition

$$\int \frac{d\mathbf{p}}{(2\pi)^3}\left(|u_\mathbf{p}|^2 + |v_\mathbf{p}|^2\right) = 1, \tag{9.194}$$

where

$$\psi_\mathbf{p} = \begin{pmatrix} u_\mathbf{p} \\ v_\mathbf{p} \end{pmatrix} \tag{9.195}$$

is given in (9.189). A somewhat lengthy, but elementary, evaluation of this condition yields

$$\psi_\mathbf{p} = \frac{\sqrt{2\pi\alpha\nu_0\Delta}}{\{[1 + (\alpha - \beta)^2][1 + (\alpha + \beta)^2]\}^{3/4}}\begin{pmatrix} \sqrt{1 + (\alpha \pm \beta)^2} \\ \pm\sqrt{1 + (\alpha \mp \beta)^2} \end{pmatrix}. \tag{9.196}$$

A couple of interesting comments can be made here:

1. Even when $E = 0$, i.e. for $\alpha^2 - \beta^2 = 1$, the electron and hole wavefunctions of the Shiba state are in general different from one another. This is quite distinct from the case of Majorana zero-energy states, for which electron and hole wavefunctions are necessarily complex conjugates of one another.
2. Electron and hole wavefunctions do become equal up to a sign in the absence of potential scattering.

Finally, we can also give the 4-spinor results for the Shiba state wavefunctions at the position of the impurity:

$$\psi_+(\mathbf{0}) = \frac{\sqrt{2\pi\alpha\nu_0\Delta}}{\{[1+(\alpha-\beta)^2][1+(\alpha+\beta)^2]\}^{3/4}} \begin{pmatrix} \sqrt{1+(\alpha+\beta)^2} \\ 0 \\ \sqrt{1+(\alpha-\beta)^2} \\ 0 \end{pmatrix}, \tag{9.197}$$

$$\psi_-(\mathbf{0}) = \frac{\sqrt{2\pi\alpha\nu_0\Delta}}{\{[1+(\alpha-\beta)^2][1+(\alpha+\beta)^2]\}^{3/4}} \begin{pmatrix} 0 \\ \sqrt{1+(\alpha-\beta)^2} \\ 0 \\ -\sqrt{1+(\alpha+\beta)^2} \end{pmatrix}. \tag{9.198}$$

9.C.2 Adatom as a spin-$\frac{1}{2}$ Anderson impurity

Following the main text, we focus on the case where the impurity level is fully spin-polarized along the z axis, so $\langle n \rangle = 1$ and $\langle m \rangle = 1$. Using the Nambu spinor notation

$$d = (d_\uparrow,\ d_\downarrow,\ d_\downarrow^\dagger,\ -d_\uparrow^\dagger)^T, \tag{9.199}$$

we can write down the BdG Hamiltonian for the impurity level as

$$\mathcal{H}_d = \frac{1}{2} d^\dagger H_d d, \tag{9.200}$$

$$H_d = \left(\epsilon_d - \mu + \frac{1}{2}U\right)\tau_z + \frac{1}{2}U\sigma_z. \tag{9.201}$$

The adatom induces localized subgap states once the the hybridization with the superconductor is included. The spectrum of subgap states can be found from the poles of the local Green function $G(E)$ of the superconductor at the impurity position (chosen at the origin). Owing to the local nature of the tunnelling, the latter obeys a purely multiplicative Dyson equation

$$G(E) = g(E) + g(E)\Sigma(E)G(E), \tag{9.202}$$

where $g(E)$ is the local Green function of the homogeneous superconductor in the absence of coupling to the adatom:

$$g(E) = \int \frac{d\mathbf{p}}{(2\pi)^3} \frac{E + \xi_\mathbf{p}\tau_z + \Delta\tau_x}{E^2 - \xi_p^2 - \Delta^2} = -\pi\nu_0 \frac{E + \Delta\tau_x}{\sqrt{\Delta^2 - E^2}}. \tag{9.203}$$

The effect of the adatom is included through the self-energy

$$\Sigma(E) = t^2(E - H_d)^{-1}. \tag{9.204}$$

The poles of $G(E)$ and hence the subgap spectrum can be found from the condition

$$\det G^{-1}(E) = \det\left[g^{-1}(E) - \Sigma(E)\right] = 0. \tag{9.205}$$

This separates into separate equations for the two spin components:

$$\det\left\{\frac{E - \Delta\tau_x}{\sqrt{\Delta^2 - \omega^2}} - \frac{\Gamma}{\left[\left(\omega \pm \frac{1}{2}U\right) - \left(\epsilon_d - \mu + \frac{1}{2}U\right)\tau_z\right]}\right\} = 0, \tag{9.206}$$

with $\Gamma = \pi\nu_0 t^2$. In the limit $U, \Gamma \gg \Delta$, one finds subgap states with energies

$$E = \pm\Delta\frac{\Gamma^2 + \left(\epsilon_d - \mu + \frac{1}{2}U\right)^2 - \left(\frac{1}{2}U\right)^2}{\sqrt{[\Gamma^2 + \left(\epsilon_d - \mu + \frac{1}{2}U\right)^2 - \left(\frac{1}{2}U\right)^2]^2 + \Gamma^2U^2}}. \tag{9.207}$$

When written in term of $E_{d\uparrow} = \epsilon_d - \mu$ and $E_{d\downarrow} = \epsilon_d - \mu + U$, this yields (9.56).

Acknowledgements

We would like to acknowledge our collaborators on this subject, from whom we learned much of what we know. These are Jason Alicea, Erez Berg, Arne Brataas, Piet Brouwer, Matthias Duckheim, Matthew Fisher, Katharina Franke, Leonid Glazman, Arbel Haim, Bert Halperin, Benjamin Heinrich, Liang Jiang, Torsten Karzig, Graham Kells, Yuval Oreg, David Pekker, Armin Rahmani, Gil Refael, Alessandro Romito, Michael Ruby, Ady Stern, and Yuval Vinkler. We would also like to acknowledge financial support through the Helmholtz Virtual Institute 'New States of Matter and Their Excitations', as well as the DFG Priority Programme 'Topological Insulators'.

References

[1] J. Alicea, *Rep. Prog. Phys.* **75**, 076501 (2012).
[2] C. W. J. Beenakker, *Annu. Rev. Condens. Matter Phys.* **4**, 113 (2013).
[3] S. Das Sarma, M. Freedman, and C. Nayak, *NPJ Quantum Inf.* **1**, 15001 (2015).
[4] C. W. J. Beenakker, *Rev. Mod. Phys.* **87**, 1037 (2015).
[5] E. Majorana, *Nuovo Cim.* **5**, 171 (1937).
[6] F. Wilczek, *Nat. Phys.* **5**, 614 (2009).
[7] A. Kitaev, *Ann. Phys. (NY)* **303**, 2 (2003).
[8] C. Nayak, S. H. Simon, A. Stern, M. Freedman, and S. Das Sarma, *Rev. Mod. Phys.* **80**, 1083 (2008).
[9] V. Mourik, K. Zuo, S. M. Frolov, S. R. Plissard, E. P. A. M. Bakkers, and L. P. Kouwenhoven, *Science* **336**, 1003 (2012).

[10] A. Das, Y. Ronen, Y. Most, Y. Oreg, M. Heiblum, and H. Shtrikman, *Nat. Phys.* **8**, 887 (2012).
[11] H. O. H. Churchill, V. Fatemi, K. Grove-Rasmussen, M. T. Deng, P. Caroff, H. Q. Xu, and C. M. Marcus, *Phys. Rev. B* **87**, 241401(R) (2013).
[12] M. T. Deng, C. L. Yu, G. Y. Huang, M. Larsson, P. Caroff, and H. Q. Xu, *Nano Lett.* **12**, 6414 (2012).
[13] L. P. Rokhinson, X. Liu, and J. K. Furdyna, *Nat. Phys.* **8**, 795 (2012).
[14] A. D. K. Finck, D. J. Van Harlingen, P. K. Mohseni, K. Jung, and X. Li, *Phys. Rev. Lett.* **110**, 126406 (2013).
[15] S. Nadj-Perge, I. K. Drozdov, J. Li, H. Chen, S. Jeon, J. Seo, A. H. MacDonald, B. A. Bernevig, and A. Yazdani, *Science* **346**, 602 (2014).
[16] W. Chang, S. M. Albrecht, T. S. Jespersen, F. Kuemmeth, P. Krogstrup, J. Nygard, and C. M. Marcus, *Nat. Nanotechnol.* **10**, 232 (2015).
[17] M. Ruby, F. Pientka, Y. Peng, F. von Oppen, B. W. Heinrich, and K. J. Franke, *Phys. Rev. Lett.* **115**, 197204 (2015).
[18] A. Y. Kitaev, *Phys. Usp.* **44**, 131 (2001).
[19] R. Jackiw and C. Rebbi, *Phys. Rev. D* **13**, 3398 (1976).
[20] S. Ryu, A. P. Schnyder, A. Furusaki, and A. W. W. Ludwig, *New J. Phys.* **12**, 065010 (2010).
[21] L. Fu and C. L. Kane, *Phys. Rev. Lett.* **100**, 096407 (2008).
[22] L. Fu and C. L. Kane, *Phys. Rev. B* **79**, 161408R (2009).
[23] L. Jiang, D. Pekker, J. Alicea, G. Refael, Y. Oreg, A. Brataas, and F. von Oppen, *Phys. Rev. B* **87**, 075438 (2013).
[24] R. M. Lutchyn, J. D. Sau, and S. Das Sarma, *Phys. Rev. Lett.* **105**, 077001 (2010).
[25] Y. Oreg, G. Refael, and F. von Oppen, *Phys. Rev. Lett.* **105**, 177002 (2010).
[26] S. Nadj-Perge, I. K. Drozdov, B. A. Bernevig, and A. Yazdani, *Phys. Rev. B* **88**, 020407(R) (2013).
[27] L. Yu, *Acta Phys. Sin.* **21**, 75 (1965).
[28] H. Shiba, *Prog. Theor. Phys.* **40**, 435 (1968).
[29] A. I. Rusinov, *Zh. Eksp. Teor. Fiz. Pisma* **9**, 146 (1968) [*JETP Lett.* **9**, 85 (1969)].
[30] A.ZV. Balatsky, I. Vekhter, and J.-X. Zhu, *Rev. Mod. Phys.* **78**, 373 (2006).
[31] A. Yazdani, B. A. Jones, C. P. Lutz, M. F. Crommie, and D. M. Eigler, *Science* **275**, 1767 (1997).
[32] A. Yazdani, C. M. Howald, C. P. Lutz, A. Kapitulnik, and D. M. Eigler, *Phys. Rev. Lett.* **83**, 176 (1999).
[33] H. Shiba, *Prog. Theor. Phys.* **50**, 50 (1973).
[34] P. W. Anderson, *Phys. Rev.* **124**, 41 (1961).
[35] F. Pientka, L. Glazman, and F. von Oppen, *Phys. Rev. B* **88**, 155420 (2013).
[36] F. Pientka, L. I. Glazman, and F. von Oppen, *Phys. Rev. B* **89**, 180505(R) (2014).
[37] J. Klinovaja, P. Stano, A. Yazdani, and D. Loss, *Phys. Rev. Lett.* **111**, 186805 (2013).
[38] B. Braunecker and P. Simon, *Phys. Rev. Lett.* **111**, 147202 (2013).
[39] M. M. Vazifeh and M. Franz, *Phys. Rev. Lett.* **111**, 206802 (2013).
[40] J. Li, H. Chen, I. K. Drozdov, A. Yazdani, B. A. Bernevig, and A. H. MacDonald, *Phys. Rev. B* **90**, 235433 (2014).

[41] B. Braunecker, G. I. Japaridze, J. Klinovaja, and D. Loss, *Phys. Rev. B* **82**, 045127 (2010).
[42] Y. Peng, F. Pientka, L. I. Glazman, and F. von Oppen, *Phys. Rev. Lett.* **114**, 106801 (2015).
[43] N. Read and D. Green, *Phys. Rev. B* **61**, 10267 (2000).
[44] D. A. Ivanov, *Phys. Rev. Lett.* **86**, 268 (2001).
[45] A. Stern, F. von Oppen, and E. Mariani, *Phys. Rev. B* **70**, 205338 (2004).
[46] J. Alicea, Y. Oreg, G. Refael, F. von Oppen, and M. P. A. Fisher, *Nat. Phys.* **7**, 412 (2011).
[47] A. Romito, J. Alicea, G. Refael, and F. von Oppen, *Phys. Rev. B* **85**, 020502(R) (2012).
[48] J. D. Sau, D. J. Clarke, and S. Tewari, *Phys. Rev. B* **84**, 094505 (2011).
[49] B. van Heck, A. R. Akhmerov, F. Hassler, M. Burrello, and C. W. J. Beenakker, *New J. Phys.* **14**, 035019 (2012).
[50] M. V. Berry, *Proc. R. Soc. Lond. Ser. A* **392**, 45 (1984).
[51] F. Wilczek and A. Zee, *Phys. Rev. Lett.* **52**, 2111 (1984).
[52] F. Pientka, G. Kells, A. Romito, P. W. Brouwer, and F. von Oppen, *Phys. Rev. Lett.* **109**, 227006 (2012).
[53] K. T. Law, P. A. Lee, T. K. Ng, *Phys. Rev. Lett.* **103**, 237001 (2009).
[54] K. Flensberg, *Phys. Rev. B* **82**, 180516 (2010).
[55] I. Martin and D. Mozyrsky, *Phys. Rev. B* **90**, 100508 (2014).
[56] M. Ruby, F. Pientka, Y. Peng, F. von Oppen, B. W. Heinrich, and K. J. Franke, *Phys. Rev. Lett.* **115**, 087001 (2015).
[57] Y. Peng, F. Pientka, Y. Vinkler-Aviv, L. I. Glazman, and F. von Oppen, *Phys. Rev. Lett.* **115**, 266804 (2015).
[58] D. Badiane, M. Houzet, and J. S. Meyer, *Phys. Rev. Lett.* **107**, 177002 (2011).
[59] S. B. Bravyi and A. Yu. Kitaev, *Ann. Phys. (NY)* **298**, 210 (2002).

10
Transport of Dirac surface states

D. CARPENTIER

Laboratoire de Physique
Ecole Normale Supérieure de Lyon
Lyon, France

Chapter Contents

10.1 Introduction

10.1.1 Purpose of the lectures

The occurrence of robust states at their surfaces is the most salient feature of three-dimensional topological insulators [29, 47]. Indeed, it is their existence in angle-resolved photoemission spectroscopy (ARPES) experiments that is used as a signature of the topological property of the bulk bands. These surface states turn out to be described by a relativistic two-dimensional Dirac equation at low energy. In these lectures, we focus on the transport properties of these Dirac surface states. While transport may not be the ideal probe of the existence of Dirac-like electronic excitations, it remains a tool of choice in condensed matter. In the following, we survey some of the transport properties of Dirac excitations and the techniques appropriate to their study. For the sake of pedagogy, we will focus on simplest transport properties, neglecting in particular transport in hybrid structures with superconductors, which would deserve a lecture course of their own.

Naturally, there is a strong overlap between the study of transport properties of graphene and surface states of topological insulators. Indeed, the low-energy electronic excitations of graphene are also described as two-dimensional Dirac particles. The discovery of graphene has led to a large amount of work on the associated transport properties: there are already a number of textbooks and extensive reviews on the subject, including [10, 19, 20, 24, 35] and [14, 65] on related issues. In the context of topological insulator surface states, the review [12] focuses on the quantum coherent transport properties. In these lectures, we start with a survey of classical transport properties of Dirac fermions at high carrier concentration and the inherent anisotropic scattering using the Boltzmann equation. The minimum conductivity of evanescent Dirac states in a short junction is described within the Landauer formalism. Then the quantum coherent regime is approached within the context of diagrammatic perturbation theory and the Kubo formula, in the spirit of [4]. This technique also allows us to recover the results on classical transport at zero and high chemical potentials.

10.1.2 Dirac surface states of topological insulators

10.1.2.1 Generic Hamiltonian

We will neglect the consequence of the possible presence of conducting bulk states in the insulating gap. Although these states are present in contemporary experiments, we prefer to focus on the simpler ideal situation for pedagogical reasons. A more realistic treatment should include the coupling between the surface and these bulk states. While the surface of topological insulators is generically characterized by an odd number of Dirac species [29, 47], we consider in the following the simplest situation described by a single Dirac cone. For energies lying in the bulk gap, electronic states are described as eigenstates of the low-energy Bloch Hamiltonian

$$H(\vec{k}) = \hbar v_F \vec{\sigma} \cdot \vec{k}, \tag{10.1}$$

where $\vec{k}$ is the two-dimensional momentum along the surface and $\vec{\sigma}$ represents two Pauli matrices for an effective spin $\frac{1}{2}$. The momentum–spin locking for eigenstates of the Hamiltonian (10.1) is reminiscent of the bulk spin–orbit coupling at the origin of the bulk band inversion. The Hamiltonian (10.1) is invariant under a so-called symplectic time-reversal symmetry T that satisfies $T^2 = -\mathbf{I}$ [21, 66, 71]:

$$TH(\vec{k})T^{-1} = H(-\vec{k}), \qquad \text{with} \quad T = i\sigma_y \mathcal{C}, \tag{10.2}$$

where $\mathcal{C}$ denotes the complex conjugation operator acting on the right. We choose to write the eigenstates of (10.1) as

$$|u(\vec{k} = ke^{i\theta})\rangle = \frac{1}{\sqrt{2}}\begin{pmatrix} 1 \\ \pm e^{i\theta} \end{pmatrix}, \qquad \text{with} \quad \epsilon(\vec{k}) = \pm v_F \hbar k. \tag{10.3}$$

Note that the Hamiltonian (10.1) is also relevant to discussion of transport at the surfaces of weak topological insulators or crystalline topological insulators characterized by an even number of Dirac cones but with a symplectic time-reversal symmetry [48], quantum wells close to a topological transition (in which case a small mass term $m\sigma_z$ should be added [60]), and other realizations (see [63] for a recent discussion of Dirac matter).

10.1.2.2 Hexagonal warping

For energies far from the Dirac point, the linearized Hamiltonian (10.1) has to be complemented by higher-order terms, leading to warping of the Fermi surface. In the case of the surface states of Bi_2Te_3, this warping corresponds to hexagonal deformation of the Fermi surface, and is due to an additional term in the Hamiltonian for surface states:

$$H_w = \frac{1}{2}\lambda\sigma_z(k_+^3 + k_-^3), \tag{10.4}$$

with $k_\pm = k_x \pm ik_y$. The resulting hexagonal symmetry of the Fermi surface originates from the combination of a trigonal discrete C_3 lattice symmetry with time-reversal symmetry [28, 39]. The corresponding dispersion relation $\epsilon^2(\vec{k} = ke^{i\theta}) = \hbar^2 v_F^2 k^2 + \lambda^2 k^6 \cos^2 3\theta$ leads to the snowflake shape of constant-energy surfaces [3]. Defining $\epsilon_F = \hbar v_F k_F$ and $k = k_F \tilde{k}(\theta)$, the shape of the Fermi surface at energy ϵ_F is conveniently parametrized by the dimensionless parameter $b = \lambda E_F^2/(2\hbar^3 v_F^3)$ as

$$1 = \tilde{k}^2(\theta) + 4b^2\tilde{k}^6(\theta)\cos^2 3\theta. \tag{10.5}$$

While this parameter takes reasonable small values $0.04 < b < 0.09$ for energies $0.05\,\text{eV} < \epsilon_F < 0.15\,\text{eV}$ in Bi_2Se_3, it ranges from $b = 0.13$ for $\epsilon_F = 0.13\,\text{eV}$ to $b = 0.66$ for $\epsilon_F = 0.295\,\text{eV}$ in Bi_2Te_3 and leads to sizeable consequences on transport at high chemical potential [3].

10.1.2.3 Disorder

A description of transport amounts to a consideration of the scattering of electronic excitations, in particular on impurities. In the following, we adopt a statistical description of these impurities: we describe them by a continuous field corresponding to an additional term $V(\hat{r})\mathbf{1}$ in the Hamiltonian. This field is random, and its realizations are chosen according to a characteristic distribution $P[V]$. For simplicity, we adopt the simplest convention, corresponding to a Gaussian distribution, with vanishing average $\langle V(\hat{r})\rangle_V = 0$ and variance

$$\langle V(\hat{r})V(\hat{r}')\rangle_V = \gamma_{\mathrm{V}}(\hat{r} - \hat{r}'), \tag{10.6}$$

where $\langle \ldots \rangle_V$ corresponds to an average over disorder configurations and the correlation $\gamma_{\mathrm{V}}(\hat{r})$ is exponentially decaying over a short distance ξ. We will often approximate it by a δ function in the continuum limit.

This Gaussian-distributed potential can be recovered as the continuum limit of the Edwards model of localized impurities [4]. Indexing independent impurities by j, the corresponding random potential is written as

$$\hat{V}(\hat{r}) = \sum_j \hat{v}(\vec{r} - \vec{R}_j), \tag{10.7}$$

where $\hat{v}(\vec{r}) = v(\vec{r})\mathbf{I}$ couples only to the density of Dirac fermions. The averaged matrix elements of this potential between Dirac eigenstates are

$$\langle |\langle \vec{k}|V|\vec{k}'\rangle|^2\rangle_V = n_i|v(\vec{k},\vec{k}')|^2\, |\langle u(\vec{k}')|u(\vec{k})\rangle|^2 \equiv \gamma_{\mathrm{V}}(\vec{k},\vec{k}')|\langle u(\vec{k}')|u(\vec{k})\rangle|^2, \tag{10.8}$$

where n_i is the impurity concentration. In the limit $n_i \to \infty, v(\vec{k},\vec{k}') \to 0$, while, keeping $\gamma_{\mathrm{V}}(\vec{k},\vec{k}') = n_i|v(\vec{k},\vec{k}')|^2$ constant, we recover a Gaussian continuous random field. A more realistic treatment of the disorder encountered at the surface of topological insulators, along the lines of [38], goes beyond the scope of these lectures.

10.1.3 Graphene

10.1.3.1 Low-energy Bloch Hamiltonian

Graphene consists of a hexagonal lattice of carbon atoms whose electronic properties can be described by considering a single p_z atomic orbital per lattice site. The electronic Bloch wavefunctions are naturally decomposed on the two sublattices of the hexagonal lattice according to

$$\psi(\vec{k},\vec{x}) = e^{i\vec{k}\cdot\vec{x}}[u_A(\vec{k},\vec{x}) + u_B(\vec{k},\vec{x})]. \tag{10.9}$$

The corresponding Bloch Hamiltonian acting on the functions $u_{A/B}(\vec{k},\vec{x})$ is written as

$$H(\vec{k}) = \begin{pmatrix} g(\vec{k}) & f(\vec{k}) \\ \overline{f}(\vec{k}) & g(\vec{k}) \end{pmatrix}. \tag{10.10}$$

At low energy, only nearest-neighbour hopping integrals can be kept, imposing vanishing amplitudes diagonal in the sublattice, $g(\vec{k}) = 0$, and hopping between different sublattices $f(\vec{k})$, which vanishes at the two Dirac points $\vec{K}$ and $\vec{K}' = -\vec{K}$. The existence of two cones, associated with states of opposites chiralities at a given energy, is a consequence of the Nielsen–Ninomiya theorem [44], which states the impossibility of realizing a lattice model with realistic couplings but a net chirality among its excitations [33]. Hence for small energies, eigenstates are labelled by a quasimomentum close to either $\vec{K}$ or $\vec{K}'$: it is thus convenient to introduce a 'valley index' and write an effective Hamiltonian in this extended basis. By using $f(\pm\vec{K} + \vec{q}) = \hbar v_F(\pm q_x - iq_y)$, we write the corresponding Hamiltonian as

$$H(\vec{q}) = \begin{pmatrix} H(\vec{K} + \vec{q}) & 0 \\ 0 & H(\vec{K}' + \vec{q}) \end{pmatrix} = \begin{pmatrix} \hbar v_F \vec{\sigma} \cdot \vec{q} & 0 \\ 0 & \hbar v_F \vec{\sigma} \cdot \vec{q} \end{pmatrix} \tag{10.11}$$

where $H(\vec{q})$ acts on vectors of states $(|u_{\vec{K},A}(\vec{k})\rangle, |u_{\vec{K},B}(\vec{k})\rangle, -|u_{\vec{K}',B}(\vec{k})\rangle, |u_{\vec{K}',A}(\vec{k})\rangle)$ with the definition $|u_{A/B}(\pm\vec{K} + \vec{q})\rangle = |u_{\vec{K}/\vec{K}',A/B}(\vec{q})\rangle$.

10.1.3.2 Time-reversal symmetry

The Hamiltonian (10.10) describes spinless fermions on a hexagonal lattice: the spectrum for the electrons is spin-degenerate and is described neglecting the spin degree of freedom. Hence, this Hamiltonian is invariant under time-reversal symmetry for spinless electrons: if $\psi(\vec{k}, \vec{x})$ is an eigenstate of energy $\epsilon_{\vec{k}}$, then $\overline{\psi}(\vec{k}, \vec{x})$ is also an eigenstate of the same energy, where we use the notation $\overline{\psi}(\vec{k}, \vec{x}) = \mathcal{C}\psi(\vec{k}, \vec{x})$ for the complex conjugate of $\psi(\vec{k}, \vec{x})$. This symmetry manifests itself as $\overline{H}(-\vec{k}) = H(\vec{k})$ on the Bloch Hamiltonian (10.10). Expressed in the valley/sublattice Hilbert space, it is written as

$$TH(\vec{q})T^{-1} = H(-\vec{q}), \qquad T = (i\tau_y \otimes i\sigma_y)\, \mathcal{C}. \tag{10.12}$$

This antiunitary time-reversal operator satisfies $T^2 = \mathbf{I}$, as expected for spinless particles. Owing to the emergence of the pseudospin $\frac{1}{2}$ in sublattice space, the low-energy Hamiltonian $H(\vec{q})$ possesses a second time-reversal symmetry acting in each valley on spin-$\frac{1}{2}$ fermions [14]:

$$\tilde{T}H(\vec{q})\tilde{T}^{-1} = H(-\vec{q}), \qquad \tilde{T} = (\mathbf{I} \otimes i\sigma_y)\, \mathcal{C}, \tag{10.13}$$

which is a symplectic symmetry: $\tilde{T}^2 = -\mathbf{I}$. Two time-reversal symmetries, which are defined as antiunitary operators commuting with the Hamiltonian, necessarily differ by a unitary operator that commutes with the Hamiltonian: a standard symmetry [50]. Here, this symmetry emerges in the low-energy regime and consists of the exchange of valleys (without reversal of momenta $\vec{q}$): $U = i\tau_y \otimes \mathbf{I}$. The presence of these two time-reversal symmetries, an orthogonal and a symplectic one, leads to a possible crossover between universality classes of phase-coherent weak localization physics: this crossover is controlled by the correlation of the disorder, and more precisely whether the U symmetry is statistically preserved, i.e. whether disorder correlation is diagonal in the valley index [59] (see also [5, 42] for more realistic and complex descriptions at low energy). In the present lectures, we focus on transport properties of a single Dirac cone corresponding to the situation where the total Hamiltonian including disorder is valley-diagonal, i.e. is invariant under the symmetry U.

10.1.4 Overview of transport properties

The typical behaviour of the electrical conductivity of Dirac fermions is represented in Fig. 10.1. A remarkable feature is the existence of a non-vanishing conductivity at the Dirac point even in the absence of disorder [25, 26, 34, 46, 54, 62, 69]. We expect the transport in this limit to be unconventional and quantum in nature as the Fermi wavelength becomes increasingly large close to the Dirac point. Indeed, this conductivity at the Dirac point was shown to correspond to a 'pseudo-diffusive' regime, with the statistics of the transmission coefficients being characteristic of diffusive transport in conventional metals [62]. We will discuss this minimal conductivity in the clean limit in a tunnel barrier geometry where it is related to the transport through evanescent Dirac states [62], and its possible relation with the so-called Zitterbewegung of Dirac fermions [34]. When disorder is increased, both the density of states at the Dirac point and the associated conductivity increase. This increase can be described using a self-consistent Born approximation [54], or alternatively a self-consistent Boltzmann approach that can be extended to the the regime at high Fermi energies [2]. In this latter approach, the density of states is renormalized by the fluctuations of disorder or chemical potential, which become dominant at very low Fermi energy $\tilde{\epsilon}_F = [\langle(\epsilon + V)^2\rangle_V]^{1/2}$. The quantum regime of weak disorder is difficult to describe accurately within this Boltzmann approach [1]. The behaviour at stronger

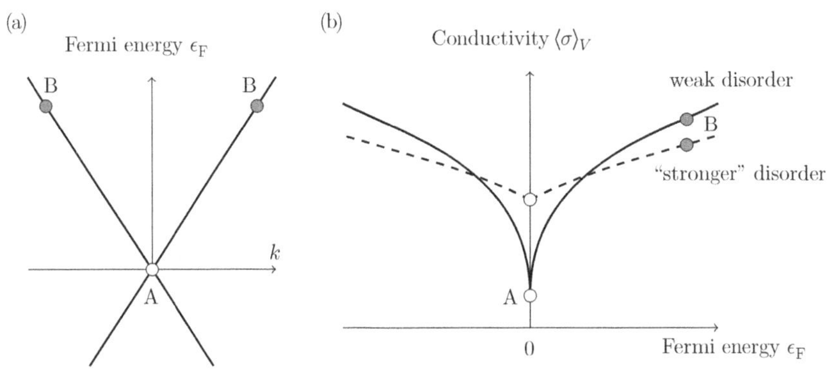

Fig. 10.1 Schematic behaviour of the conductivity as a function of the Fermi energy (b) for a Dirac dispersion relation (a). A characteristic feature is the existence of a finite conductivity at the Dirac point (point A), which increases with the disorder amplitude. At higher energies, a more common diffusive metallic behaviour is recovered (point B), with intrinsically anisotropic scattering properties.

disorder in this quantum low-energy regime is a manifestation of the absence of Anderson localization for a model of a single Dirac cone of fermions [13, 45, 51]. These Dirac fermions are the signature of the bulk topological property of valence bands: they cannot be gapped out, in particular by disorder (provided the bulk gap does not close) [53]. This property was later used to obtain a classification of topological phases, identifying those that allowed surface states robust towards Anderson localization [50].

At higher Fermi energies (point B in Fig. 10.1), we recover a standard situation where the Fermi wavelength is much smaller than characteristic lengths for transport, including the mean free path: a semiclassical approach via the Boltzmann equation is possible. As we will see, in this regime, the manifestation of the Dirac nature of the electrons lies in the anisotropy of scattering, even in the presence of 'isotropic impurities'. Naturally, this property requires the use of a transport time, different from the elastic scattering time, to define the diffusion constant. For small samples in which transport can remain phase-coherent over sizeable distances, quantum corrections to this diffusive transport have to be taken into account. The standard description of this quantum regime was extended to the case of Dirac diffusion in the context of graphene [5, 6, 36, 42, 43]. In this context, the result depends on the type of disorder and its symmetry with respect to valley indices: we obtain a standard weak-localization physics (orthogonal class), or weak antilocalization (symplectic class). The situation of Dirac surface states of topological insulators is simpler since no crossover is allowed without magnetic disorder. We will describe this regime, following the diagrammatic approach in the spirit of [4]. We will not discuss the Altshuler–Aronov effect. The interested reader can turn to [52] for an alternative and interesting description of the semiclassical regime for Dirac fermions as a propagation along classical trajectories.

10.2 Minimal conductivity close to the Dirac point

10.2.1 Zitterbewegung

The transport in the limit $\epsilon_F \to 0$ has been related to the peculiar nature of Dirac fermions [34, 62]. In particular, the occurrence of a finite conductivity in the clean limit was discussed in relation with Zitterbewegung, i.e. an intrinsic agitation of Dirac fermions [34]. Indeed, the current operator $\vec{j} = ev_F\vec{\sigma}$ associated with the Hamiltonian (10.1) does not commute with it. Hence, $\langle\vec{j}\rangle$ is not a constant of motion for eigenstates of the Hamiltonian, which signals the existence of a 'trembling motion', or Zitterbewegung, around the centre of motion [64]. This Zitterbewegung was claimed to play the role of an intrinsic disorder manifesting itself in a finite conductivity at $\epsilon_F = 0$ [34, 35]. Of particular interest is the geometry of a large 'tunnel' junction of Dirac material at the Dirac point. Let us express the current operator in the eigenstates basis (10.3) of the Hamiltonian (10.1) with $\vec{k} = ke^{i\theta}$:

$$j_x = ev_F \begin{pmatrix} \cos\theta & i\sin\theta\, e^{-i\theta} \\ -i\sin\theta\, e^{i\theta} & -\cos\theta \end{pmatrix}, \qquad j_y = ev_F \begin{pmatrix} \sin\theta & -i\cos\theta\, e^{-i\theta} \\ i\cos\theta\, e^{i\theta} & -\sin\theta \end{pmatrix}. \tag{10.14}$$

In this basis, the non-commutativity of $\vec{j}$ with the Hamiltonian originates from the off-diagonal terms describing transitions between the $\pm v_F\hbar k$ eigenstates. It is natural to expect this Zitterbewegung to manifest itself strongly close to the Dirac point and in the presence of broadening of the eigenstates originating from either disorder or confinement. This is indeed what occurs.

10.2.2 Clean large tunnel junction

Following [62], we consider a wide barrier at the surface of a topological insulator, such that the length of the barrier L is much smaller than its circumference W around the sample, as shown in Fig. 10.2. The confinement of the tunnel junction is described by the potential $eV(x)\mathbf{1}$ added to the Hamiltonian (10.1), with $V(x) = 0$ exactly at the Dirac point for $0 < x < L_x$ and $V(x) = V_\infty$ outside the barrier ($x < 0$ and $x > L$). The periodic boundary condition in the y direction around the sample implies the quantization of momentum along y: $k_n = n2\pi/W$, with $-\frac{1}{2}W \leq n \leq \frac{1}{2}W$. The conductance of the junction can then be deduced from the Landauer formula [32]

$$G = \frac{W}{L}\sigma = \frac{e^2}{h}\sum_n T_n, \tag{10.15}$$

where the T_n denote the transmission coefficients of the current carried by the different modes of the junction. At $\epsilon = 0$, only evanescent states carry current in the junction. They are described by the eigenfunctions

$$\psi(x,y) = \frac{1}{\sqrt{2}}\begin{pmatrix} ae^{k_n x} \\ be^{-k_n x}\end{pmatrix} e^{ik_n y}. \tag{10.16}$$

We recover here a crucial property of the Dirac Hamiltonian (10.1): at each boundary $x = 0$ or $x = L$, these evanescent states are entirely polarized in either the $\uparrow$ or $\downarrow$ state

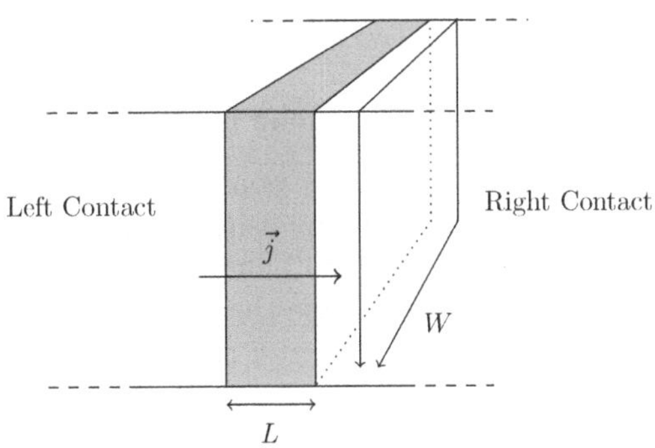

Fig. 10.2 Schematic representation of a tunnel junction at the surface of a topological insulator. In the ideal situation, for a chemical potential inside the bulk gap, only the Dirac surface states transport the current between the contacts.

(corresponding to localization in a single sublattice in the case of graphene). This is a consequence of the chiral symmetry of the Dirac Hamiltonian [30]: the operator $C = \sigma_z$ anticommutes with the Hamiltonian (10.1). Hence, C relates eigenstates at $+\epsilon(\vec{k}\,)$ to eigenstates at $-\epsilon(\vec{k}\,)$. However, at $\epsilon(\vec{k}\,) = 0$, this chiral symmetry implies that all eigenstates of the Hamiltonian are also eigenstates of C. The conductance at $\epsilon = 0$ directly probes transport properties of these chirality eigenstates, although in their evanescent form.

We can now solve the standard diffusion problem through the potential well and find the transmission coefficients

$$\begin{aligned} T_n(\vec{k} = ke^{i\theta}) &= \frac{\cos^2\theta}{\cosh^2 k_n L - \sin^2\theta} \\ &\simeq \frac{1}{\cosh^2 k_n L} \qquad \text{for large } V_\infty, \text{ i.e. } k_x \gg k_n. \end{aligned} \tag{10.17}$$

In the limit of a wide and narrow junction, $W \gg L$, the ensemble of transmission coefficients T_n samples accurately the underlying distribution function $\rho(T)$ and we find a dimensionless conductance

$$g = \frac{G}{e^2/h} = \sum_{n=-\infty}^{+\infty} T_n \simeq \frac{W}{2\pi L}\int_{-\pi L}^{+\pi L} \frac{1}{\cosh x}\,dx = \frac{W}{\pi L}. \tag{10.18}$$

This result corresponds to a minimal conductivity $\sigma_{\min} = e^2/\pi h$. Quite remarkably, the transmission coefficients are distributed according to the law [62]

$$\rho(T) = \frac{g}{2T\sqrt{1-T}}, \tag{10.19}$$

characteristic of the conventional orthogonal diffusive metallic regime. Accordingly, the tunnel transport though a wide Dirac junction has been denoted a pseudodiffusive regime. The occurrence of the diffusive distribution function of transmissions explains the identification between this tunnel conductivity and the diffusive conductivity of a long wide conductor in the presence of weak disorder presented in Section 10.2.3.

10.2.3 Minimal conductivity from linear response theory

The above minimum conductance at the Dirac point in the clean limit can be recovered in the case of a large sample of size $L = W$ by using the Kubo formula. We follow the approach of [49] (see also the earlier work [40]) and consider the Kubo formula for the conductivity calculated within linear response theory:

$$\begin{aligned} &\sigma_{ij}(\omega, \beta, \tau) = \\ &\quad \frac{\hbar}{4\pi L^2}\int d\epsilon\, \frac{f_\beta(\epsilon+\hbar\omega) - f_\beta(\epsilon)}{\hbar\omega}\, \mathrm{Tr}[\mathrm{Im}\,\hat{\mathcal{G}}^A(\epsilon,\tau)\,\hat{\jmath}_i\,\mathrm{Im}\,\hat{\mathcal{G}}^A(\epsilon+\hbar\omega,\tau)\,\hat{\jmath}_j], \end{aligned} \tag{10.20}$$

where $f_\beta(\epsilon)$ is the Fermi–Dirac distribution function, the current density operator reads $\vec{\jmath} = ev_F\vec{\sigma}$, and the trace runs over the quantum numbers (spin and momentum) of electronic states. In this expression, $\mathrm{Im}\,\hat{\mathcal{G}}^A(\epsilon,\tau) = \hat{\mathcal{G}}^A(\epsilon,\tau) - \hat{\mathcal{G}}^R(\epsilon,\tau)$, where $\hat{\mathcal{G}}^{R,A}$ correspond to the retarded and advanced Green's functions for the Hamiltonian (10.1), with or without disorder potential V:

$$\hat{\mathcal{G}}^{R/A}(\vec{k},\epsilon,\tau_\phi) = \left[\left(\epsilon \pm i\frac{\hbar}{2\tau_\phi}\right)\mathbf{I} - H(\vec{k}) - V\right]^{-1}. \tag{10.21}$$

Here τ stands either for an elastic mean free time τ_e in the disorder case or for a phenomenological phase coherence time $\tau_\phi(T)$ for the Bloch states accounting for the inelastic interactions of the electron with the phonons, other electrons, or magnetic Kondo impurities [32]. In the presence of disorder, we approximate the average of the conductivity over disorder $\langle\sigma\rangle_V$ by replacing the Green's functions in (10.20) by their average over disorder (see Section 10.3.2 for a discussion of this point and refinements). This simply amounts to replacing the phase coherence time τ_ϕ by the shorter elastic mean free time τ_e (see (10.48) and (10.50)). The (averaged) Green's functions for the Dirac fermions are written as

$$\mathcal{G}^{R/A}(\vec{k},\epsilon,\tau) = \frac{(\epsilon \pm i\hbar/2\tau)\mathbf{I} + \hbar v_F \vec{k}\cdot\vec{\sigma}}{(\epsilon \pm i\hbar/2\tau)^2 - \epsilon^2(\vec{k})}. \tag{10.22}$$

The trace over spin space in the expression (10.20) can now be performed. The remaining order of limits is crucial: we keep τ finite and using

$$\lim_{\beta\to\infty}\lim_{\omega\to 0} \frac{f_\beta(\epsilon+\hbar\omega) - f_\beta(\epsilon)}{\hbar\omega} = -\delta(\epsilon), \tag{10.23}$$

we obtain the minimal conductivity (using $\eta^{-1} = 2\tau$):

$$\lim_{\beta\to\infty}\lim_{\omega\to 0}\sigma_{ij}(\omega,\beta,\tau) = \frac{2}{\hbar}\left(\frac{ev_F}{2\pi}\right)^2 \int_0^\infty dx\, \frac{\eta^2}{(\eta^2 + v_F^2 x)^2} = \frac{e^2}{\pi h}, \tag{10.24}$$

which is precisely the result found for the wide and narrow junction using the Landauer formula. Note that a finite η or τ was crucial in deriving this result: its presence is related in the clean case either to a dephasing time in the large sample geometry or to a lifetime in the sample owing to the presence of the absorbing boundaries for a narrow strip considered in Section 10.2.2. The independence of the result (10.24) on τ and thus on a weak disorder breaks down as the disorder strength is increased, as shown in numerical studies [1], and in agreement with contribution of the quantum correction (weak antilocalization) described in Section 10.4.

10.3 Classical conductivity at high Fermi energy

At higher Fermi energies, represented schematically as the region of point B in Fig. 10.1, we recover a conventional situation of a metal with a Fermi wavelength $2\pi/k_F$ much smaller than lengthscales characteristics of transport (the mean free path l_e or the size of the sample L). This regime is conveniently described using a semiclassical description. First, we will identify the signature of the Dirac nature of excitations within the Boltzmann equation approach, before resorting to the Kubo approach previously introduced.

10.3.1 Boltzmann equation

Classical phase space is spanned by variables $\vec{r}_c, \vec{p}_c$: a statistical description of a ensemble of particles amounts to defining a density of states $f(\vec{r}_c, \vec{p}_c, t)$ at time t. The Boltzmann equation states that the evolution of this density of states is the sum of three terms:

$$\frac{\partial f}{\partial t} = -\frac{d\vec{r}_c}{dt} \cdot \vec{\nabla}_{\vec{r}_c} f - \frac{d\vec{p}_c}{dt} \cdot \vec{\nabla}_{\vec{p}_c} f + I[f], \tag{10.25}$$

where $I[f]$ is a collision integral defined below in (10.34), which describes the evolution of the density f due to scattering. To proceed in this semiclassical description of electrons in crystals, we need equations of motions for $d\vec{r}_c/dt$ and $d\vec{p}_c/dt$. It has recently been understood that these equations depend not only on the band structures $\epsilon(\vec{k})$ but also on geometrical properties of the field of eigenvectors associated with these bands [67]. Although these geometrical characteristics do not enter the simplest transport properties addressed in these lectures, it is interesting to introduce them for extensions to, for example, magnetotransport. Let us sketch briefly the derivation of these semiclassical equations of motion [41, 67].

10.3.1.1 Semiclassical equations of motion

We want to describe the time evolution of a semiclassical wavepacket restricted to a single band indexed by n (or more generally a subset of bands). This amounts to considering a wavepacket

$$|\psi^{(n)}_{\vec{r}_c,\vec{k}_c}\rangle = \int \frac{d^2\vec{k}}{(2\pi)^2} \chi(\vec{k} - \vec{k}_c) e^{-i[\vec{k}+(e/\hbar c)\vec{A}(\vec{r}_c)]\cdot\vec{r}_c} |\psi(\vec{k}, n)\rangle, \tag{10.26}$$

where $\langle \vec{r}\,|\psi(\vec{k}, n)\rangle = e^{i\vec{k}\cdot\vec{r}} \langle \vec{r}\,|u(\vec{k}, n)\rangle$ are eigenstates associated with the band n, and the vector potential $\vec{A}$ originates from the possible presence of a magnetic field. Imposing the localization of the wavepacket around $\vec{r}_c$, i.e.

$$\left\langle \psi^{(n)}_{\vec{r}_c,\vec{k}_c} \middle| \hat{r} \middle| \psi^{(n)}_{\vec{r}_c,\vec{k}_c} \right\rangle = \vec{r}_c, \tag{10.27}$$

imposes that the phase of $\chi(\vec{k} - \vec{k}_c)$ is related to the Berry connection in band n:

$$\chi(\vec{k} - \vec{k}_c) = |\chi(\vec{k} - \vec{k}_c)| e^{i(\vec{k}-\vec{k}_c)\cdot\vec{\mathcal{A}}^{(n)}(\vec{k}_c)}. \tag{10.28}$$

In this expression, $\vec{\mathcal{A}}^{(n)}(\vec{k}_c)$ is a connection defined not on the field of electronic states $|\psi(\vec{k}, n)\rangle$, but rather on the states $|u(\vec{k})\rangle = e^{-i\vec{k}\cdot\hat{r}} |\psi(\vec{k}\,)\rangle$ invariant under translations on the lattice. These states are eigenstates of the $\vec{k}$-dependent Bloch Hamiltonian $H(\vec{k}\,) = e^{i\vec{k}\cdot\hat{r}} H e^{-i\vec{k}\cdot\hat{r}}$. Following Berry [17], we can define a connection $\vec{\mathcal{A}}^{(n)}(\vec{k}_c)$ associated with the parallel transport within the space of eigenvectors $|u(\vec{k}, n)\rangle = e^{-i\vec{k}\cdot\hat{r}} |\psi(\vec{k}, n)\rangle$. It is this connection that naturally occurs in the expression (10.28):

it should not be confused with other 'projected' connections that can be defined in terms of Bloch eigenstates [16, 27].

Following Sundaram and Niu [58], we write down a classical Lagrangian

$$\mathcal{L} = \left\langle \psi^{(n)}_{\vec{r}_c,\vec{k}_c} \middle| i\hbar\partial_t \middle| \psi^{(n)}_{\vec{r}_c,\vec{k}_c} \right\rangle - \left\langle \psi^{(n)}_{\vec{r}_c,\vec{k}_c} \middle| H \middle| \psi^{(n)}_{\vec{r}_c,\vec{k}_c} \right\rangle, \tag{10.29}$$

with

$$\left\langle \psi^{(n)}_{\vec{r}_c,\vec{k}_c} \middle| i\hbar\partial_t \middle| \psi^{(n)}_{\vec{r}_c,\vec{k}_c} \right\rangle = \frac{e}{c}\vec{r}_c \cdot \frac{d\vec{A}(\vec{r}_c)}{dt} + \hbar\vec{k}_c \cdot \frac{d\vec{r}_c}{dt} + \hbar\frac{d\vec{k}_c}{dt} \cdot \vec{\mathcal{A}}^{(n)}(\vec{k}_c), \tag{10.30}$$

$$\left\langle \psi^{(n)}_{\vec{r}_c,\vec{k}_c} \middle| H \middle| \psi^{(n)}_{\vec{r}_c,\vec{k}_c} \right\rangle = \epsilon(\vec{k}_c) - \vec{B}\cdot\vec{m}(\vec{k}_c) - eV(\vec{r}_c), \tag{10.31}$$

where $\vec{m}(\vec{k}_c)$ is an orbital magnetic moment [67], which is neglected below. The Lagrange equations on $\mathcal{L}$ provide the required classical equations of motion:

$$\hbar\frac{d\vec{k}_c}{dt} = -e\vec{E} - \frac{e}{c}\frac{d\vec{r}_c}{dt} \times \vec{B}(\vec{r}_c), \qquad \text{with} \quad \vec{B} = \vec{\nabla}_{\vec{r}} \times \vec{A}(\vec{r}), \tag{10.32}$$

$$\frac{d\vec{r}_c}{dt} = \frac{1}{\hbar}\vec{\nabla}_{\vec{k}}\epsilon(\vec{k}_c) - \frac{d\vec{k}_c}{dt} \times \mathcal{F}(\vec{k}_c), \qquad \text{with} \quad \vec{\mathcal{F}}(\vec{k}_c) = \vec{\nabla}_{\vec{k}} \times \vec{\mathcal{A}}(\vec{k}_c). \tag{10.33}$$

Note that in the presence of time-reversal symmetry, the Berry curvature $\mathcal{F}(\vec{k}_c)$ vanishes and we recover the standard classical equation of motion in a crystal. Focusing on situations in the absence of a magnetic field, we will thus forget these Berry terms in the following.

10.3.1.2 Linear homogeneous response

We focus on the charge response of Dirac surface states to a homogeneous field: this amounts to considering the homogeneous solutions $f(\vec{p}_c = \hbar\vec{k}_c, t)$ of the equation (10.25). This simplification does not hold when considering, for example, the thermoelectric current with a spatially varying temperature. In the homogeneous case, the collision integral occurring in (10.25) is simply defined as

$$\begin{aligned} I[f] &= \int \frac{d^2\vec{k}'}{(2\pi)^2} \{f(\vec{k}')[1-f(\vec{k})] - f(\vec{k})[1-f(\vec{k}')]\}\mathcal{M}(\vec{k},\vec{k}') \\ &= \int \frac{d^2\vec{k}'}{(2\pi)^2} [f(\vec{k}') - f(\vec{k})]\mathcal{M}(\vec{k},\vec{k}'), \end{aligned} \tag{10.34}$$

where $\mathcal{M}(\vec{k},\vec{k}')$ is a transition amplitude specified below in (10.41). By definition, the equilibrium distribution, which is stationary without any external perturbing field, satisfies $I[f_{\text{eq}}] = 0$. Within linear response theory, we expand the stationary homogeneous distribution to first order in the perturbing field $\vec{E}$ around the equilibrium distribution:

$$f(\vec{k}) = f_{\text{eq}}(\vec{k}) + f^{(1)}(\vec{k}), \tag{10.35}$$

where $f_{\text{eq}}(\vec{k}) = n_F(\epsilon(\vec{k}) - \epsilon_F)$, with n_F and ϵ_F being respectively the Fermi–Dirac distribution function and the Fermi energy. Here and in the following, we use the simpler notation $\vec{k}$ for the momentum parametrizing the semiclassical wavepacket. The Boltzmann equation then simplifies to $-e\vec{E} \cdot \vec{\nabla}_{\vec{k}} f = I[f]$. The linearity in f of (10.34) allows us to write, to lowest order in $\vec{E}$,

$$-e\vec{E} \cdot \vec{\nabla}_{\vec{k}} f_{\text{eq}} = I[f^{(1)}]. \tag{10.36}$$

10.3.1.3 Transport-time approximation

The standard transport-time ansatz for a solution of the Boltzmann equation (10.25) amounts to replacing the collision integral (10.34) by $I[f] = \tau_{\text{tr}}^{-1} f$: if the external field $\vec{E}$ driving the system out of equilibrium is turned off, then τ_{tr} describes the characteristic time of relaxation of the distribution f towards equilibrium. We will discuss the validity of this ansatz below. Introducing the group velocity $\vec{v}(\vec{k}) = \vec{\nabla}_{\vec{k}} \epsilon(\vec{k})$, we can rewrite (10.36) using the transport-time ansatz as

$$f^{(1)}(\vec{k}) = -e\vec{E} \cdot [\vec{v}(\vec{k})\tau_{\text{tr}}]\partial_\epsilon n_F(\epsilon(\vec{k})) \simeq e\vec{E} \cdot \vec{\Lambda}_{\text{tr}}(\vec{k})\delta(\epsilon(\vec{k}) - \epsilon_F), \tag{10.37}$$

where we have introduced the vector transport length $\vec{\Lambda}_{\text{tr}}$ [55–57]. Equation (10.37) expresses that the transport-time ansatz accounts for the application of an electric field $\vec{E}$ by a translation of the Fermi surface according to

$$\begin{aligned} f(\vec{k}) &= f_{\text{eq}}(\vec{k}) - \frac{e\tau_{\text{tr}}}{\hbar}\vec{E} \cdot \vec{\nabla}_{\vec{k}} f_{\text{eq}} \\ &\simeq f_{\text{eq}}\left(\vec{k} - \frac{e\tau_{\text{tr}}}{\hbar}\vec{E}\right) \\ &= n_F(\epsilon(\vec{k}) + e\vec{\Lambda}_{\text{tr}}(\vec{k}) \cdot \vec{E}). \end{aligned} \tag{10.38}$$

For an isotropic Fermi surface, it is natural to expect the response to a homogeneous electric field $\vec{E}$ to be independent of the direction of application: a single transport time is necessary to describe this response. However, for an anisotropic Fermi surface with several symmetry axis, we expect different transport times or transport vectors $\vec{\Lambda}(\vec{k}, \vec{E})$ to be necessary to describe the response to different orientations of the applied electric field with respect to the Fermi surface. This is the case for the hexagonally warped Fermi surface occurring *e.g.* in Bi_2Te_3 and introduced in eq.(10.5), as described in [3]. Note that this anisotropy of the Fermi surface leading to the existence of different transport vectors should not be confused with the anisotropy of scattering by disorder which manifests itself as a discrepancy between transport and elastic mean free time. In the case of Dirac surface states, the scattering is anisotropic, but the Fermi surface remains isotropic when warping is neglected.

10.3.1.4 Conductivity

The current density can be deduced from (10.38) by using $\vec{j} = e\int d^2\vec{k}\,[f(\vec{k}) - f_{\text{eq}}(\vec{k})]\vec{v}(\vec{k})$. The conductivity tensor $\sigma_{\alpha\beta}$ defined by $j_\alpha = \sigma_{\alpha\beta}E_\beta$ satisfies the Einstein relation $\sigma_{\alpha\beta} = e^2\rho(\epsilon_F)D_\alpha\delta_{\alpha\beta}$. To express the diffusion coefficients D_α, we introduce the coordinate $k_\parallel$ along constant-energy contours and $\rho(\epsilon, k_\parallel)$ the corresponding density of states satisfying $d^2\vec{k}/(2\pi)^2 = \rho(\epsilon, k_\parallel)\, d\epsilon\, dk_\parallel$. We obtain

$$D_\alpha = \frac{1}{\rho(\epsilon_F)} \oint dk_\parallel\, \rho(\epsilon_F, k_\parallel) v_\alpha(k_\parallel)\Lambda_\alpha(k_\parallel). \tag{10.39}$$

In the case of an isotropic two-dimensional Fermi surface, we recover the usual form $D_x = D_y = D = \tau_{\text{tr}} v_F^2/2$, corresponding to

$$\sigma_{xx}(\epsilon) = e^2\rho(\epsilon_F)\frac{\tau_{\text{tr}} v_F^2}{2} = \frac{e^2}{h}\frac{\epsilon_F}{2}\frac{\tau_{\text{tr}}}{\hbar}, \tag{10.40}$$

with $\rho(\epsilon_F) = \epsilon_F/(2\pi\hbar^2 v_F^2)$ for Dirac fermions as opposed to $\rho(\epsilon) = m/(2\pi\hbar^2)$ and $\sigma = (e^2/h)(v_F^2\tau m/\hbar)$ for parabolic bands.

10.3.1.5 Transport time

The conductivity depends on the phenomenological transport time in (10.40), which we will now express in terms of the amplitude of the scattering potential. Using the Born approximation, the transition amplitude of scattering is expressed in terms of the matrix elements of the disorder potential introduced in (10.8):

$$\mathcal{M}(\vec{k}, \vec{k}') = \frac{2\pi}{\hbar}\langle|\langle\vec{k}|V|\vec{k}'\rangle|^2\rangle_V \delta(\epsilon(\vec{k}) - \epsilon(\vec{k}')). \tag{10.41}$$

The corresponding collision integral (10.34) satisfies the required condition $I[f_{\text{eq}}] = 0$ for any equilibrium distribution parametrized by the energy $f_{\text{eq}}(\epsilon(\vec{k}))$. In (10.8), we identify a contribution specific to the Dirac fermion originating from the last term, which expresses a strong backscattering reduction: scattering is much less efficient for Dirac fermions than for non-relativistic electrons. This property is a signature of time-reversal symmetry for a single Dirac cone: the states $|\psi(\vec{k})\rangle$ and $|\psi(-\vec{k})\rangle$ carry a spin $\frac{1}{2}$ and constitute a Kramers pair: they are thus orthogonal. This property implies that scattering for Dirac fermions is intrinsically anisotropic: we thus have to resort to the standard description of transport properties in the presence of anisotropic scattering. Let us plug back the expression (10.41) in the Boltzmann equation (10.36) with the transport-time ansatz:

$$\begin{aligned} &- e\vec{E}\cdot\vec{v}(\vec{k})\delta(\epsilon(\vec{k}) - \epsilon_F) = I[f^{(1)}] = \frac{1}{\tau_{\text{tr}}} f^{(1)}(\vec{k}) \\ &\Longrightarrow \quad \frac{\hbar}{\tau_{\text{tr}}} = 2\pi \int d\theta'\, \rho(\epsilon_F, \theta')[1 - \hat{v}(\theta)\cdot\hat{v}(\theta')]\cos^2\left(\frac{\theta - \theta'}{2}\right)\gamma_{\text{V}}(\epsilon_F, \theta, \theta'). \end{aligned} \tag{10.42}$$

For Dirac fermions, $\hat{v}(\theta)\cdot\hat{v}(\theta') = \cos(\theta - \theta')$: we obtain for an isotropic Dirac Fermi sea $\rho(\epsilon_F, \theta) = \rho(\epsilon_F)/2\pi$ and a transport time independent of the incident direction:

$$\frac{\hbar}{\tau_{\text{tr}}} = \rho(\epsilon_F)\int d\theta'\, \frac{1 - \cos^2\theta'}{2}\gamma_{\text{V}}(\epsilon_F, \theta'). \tag{10.43}$$

The disorder amplitude γ_V was defined in (10.6) and(10.8). This expression for the transport time has to be contrasted with the definition of the elastic scattering time that appears in, for example, the Dingle factor for Shubnikov–de Haas oscillations:

$$\frac{\hbar}{\tau_e} = \rho(\epsilon_F) \int d\theta' \, \gamma_{\mathrm{V}}(\epsilon_F, \theta'). \tag{10.44}$$

The discrepancy between the transport and elastic scattering times is a consequence of the anisotropic nature of scattering, which originates in the nature of the Dirac fermions in the present case. For an isotropic disorder for which $\gamma_{\mathrm{V}}(\epsilon_F, \theta) = \gamma_{\mathrm{V}}(\epsilon_F)$ independent of θ, we recover the standard result $\tau_{\mathrm{tr}} = 2\tau_e$: it takes twice as long for Dirac fermions to diffuse isotropically than for conventional electrons.

Note that the expression (10.43) also implies that

$$\sigma = \frac{e^2 v_F^2}{2} \rho(\epsilon_F) \tau_{\mathrm{tr}}, \qquad \text{with} \quad \rho(\epsilon_F)\tau_{\mathrm{tr}} = \frac{2\hbar}{\pi \gamma_{\mathrm{V}}(\epsilon_F)}. \tag{10.45}$$

As a consequence of this result, for Dirac fermions in the classical regime, the energy dependence of the Boltzmann conductivity originates only from the disorder correlations. Corrections to this behaviour can be attributed to a renormalization of the density, requiring a self-consistent treatment beyond the Born approximation [38], or to quantum corrections as described below.

10.3.2 Linear response approach

We aim at recovering the previous classical conductivity for the Dirac fermions within a linear response approach that will allow us later to incorporate quantum corrections. We start from the Kubo formula, introduced in Section 10.2.3 when studying the minimal conductivity at the Dirac point.

10.3.2.1 Kubo formula

We consider the longitudinal conductivity $\sigma = \sigma_{xx}$ of a sample of typical size L. This conductivity is calculated within linear response theory from the Kubo formula introduced in (10.20) in terms of the Green's function defined in (10.21). Focusing on the zero-temperature and $\omega = 0$ longitudinal conductivity, we can consider the expression

$$\sigma = \frac{\hbar}{2\pi L^2} \operatorname{Re} \operatorname{Tr}[\hat{j}_x \hat{\mathcal{G}}^R(\epsilon_F) \hat{j}_x \hat{\mathcal{G}}^A(\epsilon_F)], \tag{10.46}$$

where the trace runs over the quantum numbers (spin and momentum) of electronic states and we have neglected contributions proportional to $\mathcal{G}^R\mathcal{G}^R$ or to $\mathcal{G}^A\mathcal{G}^A$, which are systematically of lower order than the terms we have kept in the following perturbative expansion in $1/k_F l_e$ [4]. This conductivity depends on disorder through the Green's functions (10.21). We focus on the diffusive regime, corresponding to the semiclassical regime where λ_F is small compared with the mean scattering length l_e. The natural small parameter is $1/(k_F l_e)$. In this regime, we do not expect the conductivity to depend on the exact configuration of disorder, but only on its strength.

Such a quantity is called a self-averaging observable: its typical value is identified with its average over disorder configurations, which is easier to calculate.

Working perturbatively in the disorder allows us to expand the retarded and advanced Green's functions as follows:

$$\hat{\mathcal{G}}^{R/A} = [(\hat{\mathcal{G}}_0^{R/A})^{-1} - V]^{-1} = \hat{\mathcal{G}}_0^{R/A} \sum_{n=0}^{\infty} (V \hat{\mathcal{G}}_0^{R/A})^n, \qquad (10.47)$$

where the Green's functions for the pure Hamiltonian are defined in (10.22) with $\tau = \tau_\phi$ and no disorder potential. Averaging any combination of these Green's functions over a Gaussian distribution for V amounts to pairing all occurrences of the disorder potential V. When this task is performed on the conductivity (10.46), two different pairings appear: the first consists in pairing potentials V within the expansions of $\hat{\mathcal{G}}^R$ and $\hat{\mathcal{G}}^A$ independently of each other, while the second pairing is of occurrences of V in the expansion of $\hat{\mathcal{G}}^R$ with occurrences in the expansion of $\hat{\mathcal{G}}^A$. The former amounts to replacing $\hat{\mathcal{G}}^R$ and $\hat{\mathcal{G}}^A$ by their averages over disorder, while the latter corresponds to the cooperon and diffuson contributions discussed below.

10.3.2.2 Averaged Green's function and self-energy

Averaging of the expansion (10.47) over disorder can be accounted for by introducing a self-energy Σ defined as

$$\langle \hat{\mathcal{G}}^R \rangle_V^{-1} = \hat{\mathcal{G}}_0^{-1} - \Sigma, \qquad (10.48)$$

where, to lowest order in γ_V,

$$\Sigma(\epsilon) = \int \frac{d\vec{k}'}{(2\pi)^2} \langle V(\vec{k}')V(-\vec{k}')\rangle_V \mathcal{G}_0^R(\vec{k} - \vec{k}', \epsilon) = \gamma_V \int \frac{d^2\vec{k}}{(2\pi)^2} \mathcal{G}_0^R(\vec{k}, \epsilon). \qquad (10.49)$$

The real part of this self-energy is incorporated in a redefinition of the arbitrary origin of energies, while its imaginary part defines the elastic scattering time

$$-\mathrm{Im}(\Sigma) = \frac{\hbar}{2\tau_e} = \pi \gamma_V \rho(\epsilon_F). \qquad (10.50)$$

Hence, the averaging procedure of the Green's function amounts to replace the dephasing rate of the bare Green's function by $\tau_\phi^{-1} \to \tau_\phi^{-1} + \tau_e^{-1}$. In practice, τ_ϕ^{-1} is often negligible compared with τ_e^{-1} in this Mathiessen rule, and averaged Green's functions are simply given by (10.22) with $\tau = \tau_e$. We can now use these expressions in the average of the Kubo expression (10.46) by approximating $\langle \mathcal{G}^R \mathcal{G}^A \rangle_V$ by the product of averages $\langle \mathcal{G}^R \rangle_V \langle \mathcal{G}^A \rangle_V$. Performing the remaining trace, we recover an Einstein formula for the conductivity,

$$\langle \sigma \rangle_0 = \frac{v_F^2 \tau_e}{2} = D_0 e^2 \rho(E_F), \qquad (10.51)$$

but with a diffusion coefficient D_0 that is half the correct Boltzmann expression (10.40). This discrepancy is a consequence of the inherent anisotropy of scattering for Dirac fermions, which manifests itself in the difference between the transport and elastic scattering times. In the present perturbative expansion, it occurs as the contribution of an additional class of diagrams.

10.3.2.3 The dominant contributions: cooperon and diffuson

Standard diagrammatic theory of the diffusive regime amounts to summing an infinite set of dominant diagrams perturbative in disorder strength [37]. These contributions are conveniently represented by the diffusive propagation of pseudoparticles, the so-called diffusons and cooperons (see [4] for a recent pedagogical presentation). We can resort to a simple physical argument to gain an intuitive understanding of the origin of these contributions. This is most conveniently done by considering the probability to transfer an electron across the sample from position $\vec{r}_1$ to $\vec{r}_2$, which is closely related to the conductivity. In a semiclassical description, this probability is related to the amplitude $\mathcal{A}_{\vec{r}_1\vec{r}_2}$ of diffusion from $\vec{r}_1$ to $\vec{r}_2$, which itself can be summed *à la Feynman* over contributions labelled by classical diffusive paths:

$$\begin{aligned} P(\vec{r}_1,\vec{r}_2) &= |\mathcal{A}_{\vec{r}_1\vec{r}_2}|^2 \\ &= \left|\sum_{\mathcal{C}:\vec{r}_1\to\vec{r}_2} \mathcal{A}_{\mathcal{C}}\right|^2 \\ &= \sum_{\mathcal{C},\mathcal{C}'} \mathcal{A}_{\mathcal{C}}\mathcal{A}^*_{\mathcal{C}'}, \end{aligned} \tag{10.52}$$

where $\mathcal{C}$ and $\mathcal{C}'$ are two diffusive paths (or scattering sequences for discrete impurities), from $\vec{r}_1$ to $\vec{r}_2$. $\mathcal{A}_{\mathcal{C}}$ and $\mathcal{A}_{\mathcal{C}'}$ represent the corresponding diffusion amplitudes along these paths. Note that we can view $\mathcal{A}^*_{\mathcal{C}'}$ as the amplitude of diffusion for a hole in the Fermi sea.

Electrons are described at the point $\vec{r}_1$ by a Bloch state, and when evolving along a given path $\mathcal{C}$, their phase is incremented by $k_F\mathcal{L}(\mathcal{C})$, where $\mathcal{L}(\mathcal{C})$ is the length of $\mathcal{C}$ (we neglect any geometrical Berry contribution in this argument). In a good metal, which is the situation considered in this section, the Fermi wavelength $2\pi/k_F$ is typically of the order of the crystal lattice constant. Hence, this phase $k_F\mathcal{L}(\mathcal{C})$ varies by an amount of the order of 2π as soon as the path is modified over a (few) lattice spacing(s). Hence, for two different paths $\mathcal{C} \neq \mathcal{C}'$, the relative phase $\mathcal{L}(\mathcal{C}) - \mathcal{L}(\mathcal{C}')$ appearing in (10.52) will vary by 2π over neighbouring paths for which the amplitude $|\mathcal{A}_{\mathcal{C}}\mathcal{A}^*_{\mathcal{C}'}|$ can be assumed constant. Hence, this term will vanish upon the summation over the paths $\mathcal{C}$ and $\mathcal{C}'$ (corresponding to a disorder average). The only terms surviving this summation are those for which $\mathcal{L}(\mathcal{C}) = \mathcal{L}(\mathcal{C}')$. This property is naturally associated with pairs of identical paths $\mathcal{C} = \mathcal{C}'$, which can be viewed as the propagation of an electron and a hole correlated by the disorder: this statistical coupling of paths by disorder is conveniently viewed as the propagation of a pseudoparticle called a *diffuson*; see Fig. 10.3(a). A second solution exists for a path $\mathcal{C}$ containing a loop; see Fig. 10.3(b). In this case, the path $\mathcal{C}'$ is identified with $\mathcal{C}$ except around the loop, along which the direction of propagation is reversed. $\mathcal{C}'$ has approximately the same length as $\mathcal{C}$. The corresponding contribution can be viewed as the diffusion of a diffuson up to and from the loop, and the counterpropagation of a particle and a hole around the loop. This counterpropagation can also be interpreted as the correlated propagation of two particles along the same loop and called a *cooperon* by analogy with Cooper pairs in

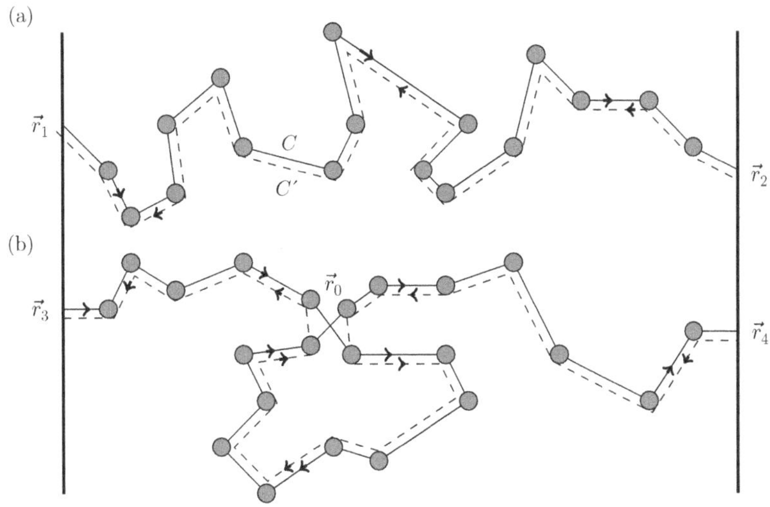

Fig. 10.3 (a) Contribution corresponding to an electron and hole moving along the same path $\mathcal{C} = \mathcal{C}'$, described as the propagation of a diffuson. (b) Contribution corresponding to an electron-like and a hole-like excitation moving in opposite directions along a loop of the path. Along this loop, this corresponds to the propagation of two particles in the same direction, accounted for by the diffusive propagation of a cooperon. Note that in this case, quantum crossing between paths occurs at the point $\vec{r}_0$.

superconductors. The existence of the cooperon is closely related to the time-reversal symmetry of transport, which identifies the amplitude $\mathcal{A}_{\mathcal{C}'}$ with $\mathcal{A}_{\mathcal{C}}$: it will disappear upon the application of a small magnetic field. Note that in Fig. 10.3(b), a crossing of paths appears when the reconnection of a diffuson with a cooperon is drawn. The number of such crossings will turn out to be the correct parameter for the perturbative theory.

10.3.2.4 Classical or quantum?

In the presence of a dynamical environment accounting for the inelastic interaction of the propagating electrons with other electrons, phonons, etc., we realize that the contribution corresponding to the diffuson is not affected: both the electron and the hole contributions encounter the same environment during their evolution, and are not dephased with respect to each other. Their possible interference contributions are not affected by this fluctuating environment: this is the signature of a classical contribution. On the other hand, the cooperon contribution corresponds to an electron and a hole propagating in opposite directions along the loop: they encounter different dynamical environments during their propagation and are dephased with respect to each other during this evolution. This is the manifestation of a quantum contribution: we expect the cooperon contribution to correspond to loops of length smaller than the typical dephasing length L_ϕ (with $L_\phi^2 \simeq D\tau_\phi$). When we study the conductivity

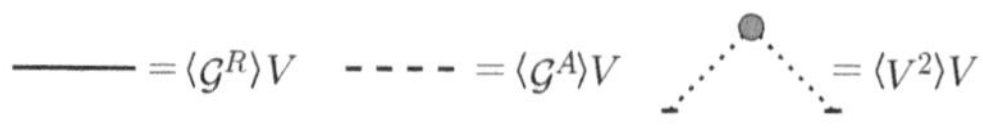

Fig. 10.4 Conventions for the diagrammatic perturbative theory.

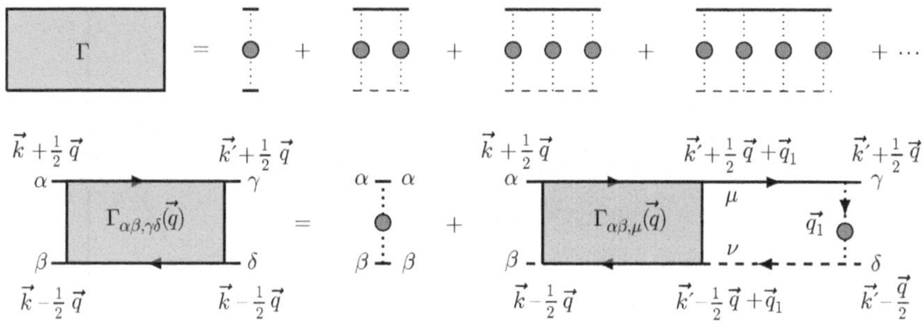

Fig. 10.5 Diagrammatic representation of the recursive calculation satisfied by the diffuson structure factor.

fluctuations, we will encounter different cooperons and diffusons, which correspond to the propagations correlated by the disorders of a particle and a hole evolving in different thermal environments: in this situation, both cooperon and diffuson are affected by a dephasing due to their environment, and correspond to quantum corrections to the conductivity fluctuations.

10.3.2.5 Diffuson contribution to the conductivity

Let us focus for now on the classical contribution to the conductivity. For simplicity, in the following we will consider a Dirac cone without warping (see [3] for a non-perturbative treatment of warping using the diagrammatic formalism). According to the previous discussion, the correction to the expression (10.51) of the conductivity comes from contributions of the diffuson (Fig. 10.3(a)). The corresponding term requires a summation over a geometric series of diagrams of the same perturbative order [4, 37]. It is best represented diagrammatically: we will use the convention of Fig. 10.4 to represent averaged Green's functions and disorder correlations (second cumulants). The real-space picture of the diffuson contribution can be represented as a contribution to the conductivity: it corresponds to the insertion in the trace occurring in the Kubo formula (10.46) of a sequence (diffusion path) of retarded and advanced Green's functions representing the evolution of a particle and hole, correlated by the disorder. It turns out that sequences (or paths) of all lengths contribute to the final result: the summation over all sequences amounts to considering a 'diffuson structure factor' Γ obtained from the algebraic sum of terms shown in Fig. 10.5.

The recursive nature of this algebraic sum represented in Fig. 10.5 can be expressed by the relation

$$\Gamma_{\alpha\beta,\gamma\delta}(\vec{q}) = \gamma_{\mathrm{V}} \mathbf{I}_{\alpha\gamma} \otimes \mathbf{I}_{\beta\delta} + \gamma_{\mathrm{V}} \Gamma_{\alpha\beta,\mu\nu}(\vec{q})\ \Pi_{\mu\nu,\gamma\delta}(\vec{q}), \tag{10.53}$$

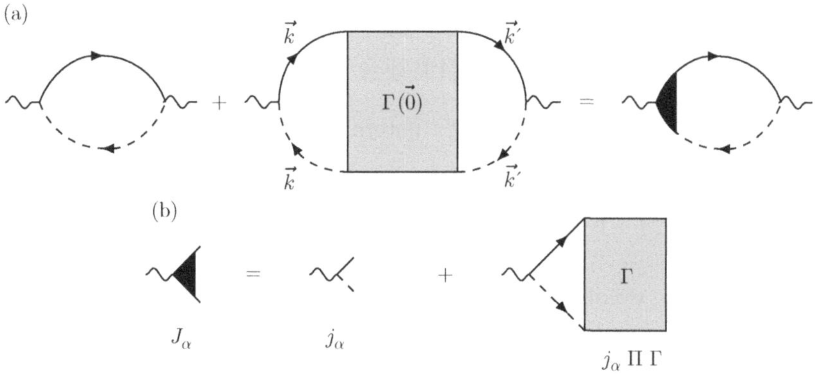

Fig. 10.6 Diagrammatic representation of (a) the bare and diffuson contributions to the averaged conductivity and (b) the renormalization of the vertex current operator accounting for the contribution of the diffuson.

where we have explicitly written the dependence on spin indices and Π is the quantum diffusion probability [4]:

$$\Pi_{\mu\nu,\gamma\delta}(\vec{q}) = \frac{1}{L^2}\sum_{\vec{q}_1}\langle \mathcal{G}^R_{\mu\gamma}(\vec{q}_1+\vec{q})\rangle_V \langle \mathcal{G}^A_{\nu\delta}(\vec{q}_1)\rangle_V. \tag{10.54}$$

By using the expression (10.22) for the Green's functions, we can perform the integral over momentum in the diffusive limit and obtain

$$\Pi(\vec{q}) = \frac{1}{2\gamma_V}[(1-2w_e^2)\mathbf{I}\otimes\mathbf{I} + \tfrac{1}{2}(1-w_e^2)\vec{\sigma}\otimes\vec{\sigma} \\ - iw_e(\hat{q}\cdot\vec{\sigma}\otimes\mathbf{I} + \mathbf{I}\otimes\hat{q}\cdot\vec{\sigma}) - w_e^2\hat{q}\cdot\vec{\sigma}\otimes\hat{q}\cdot\vec{\sigma}] + \mathcal{O}(w_e^2), \tag{10.55}$$

with $w_e = \tau_e v_f q/2 \ll 1$ in the diffusive limit and where we have used the notation $\vec{\sigma} = (\sigma_x, \sigma_y)$ and $\hat{q} = \vec{q}/|q|$. Note that in the present case, the 'complexity' of the structure factor Γ, i.e. its spin content, originates from the free Green's functions of the Dirac particles embedded in the probability Π and not the symmetry of the disorder correlations γ_V as is standard for quadratic bands [4].

The sum of the bare and diffuson contributions to the conductivity represented in Fig. 10.6 is expressed as

$$\sigma = \frac{\hbar}{2\pi L^2}\,\mathrm{Re}\,\mathrm{Tr}[j_x\Pi j_x] + \frac{\hbar}{2\pi L^2}\,\mathrm{Re}\,\mathrm{Tr}[j_x\Pi\Gamma\Pi j_x], \tag{10.56}$$

where we have used the following condensed notation for the spin contractions: $\mathrm{Tr}[j_x\Pi j_x] = j_{\beta\alpha}\Pi_{\alpha\beta,\gamma\delta}j_{\gamma\delta}$. These two contributions can be recast as a modification or renormalization of the vertex current operator using (10.53), written in condensed form as $\gamma_V^{-1}\Gamma = \mathbf{I} + \Gamma\Pi$:

$$\sigma = \frac{\hbar}{2\pi L^2}\gamma_V^{-1}\,\mathrm{Re}\,\mathrm{Tr}[j_x\Pi\Gamma j_x] = \frac{\hbar}{2\pi L^2}\,\mathrm{Re}\,\mathrm{Tr}[j_x\Pi J_x], \tag{10.57}$$

with the 'renormalized' vertex current operator represented in Fig. 10.6 and defined by

$$J_x = j_x + \Gamma\Pi j_x = \gamma_V^{-1}\Gamma j_x. \tag{10.58}$$

Note that only the limit $\vec{q} \to \vec{0}$ of $\Pi(\vec{q})$ enters this expression, which from (10.55) reduces to

$$\Pi(\vec{q} = \vec{0}\,) = \frac{1}{2\gamma_V}(\mathbf{I} \otimes \mathbf{I} + \tfrac{1}{2}\sigma_x \otimes \sigma_x + \tfrac{1}{2}\sigma_y \otimes \sigma_y). \tag{10.59}$$

From this expression, we obtain the renormalized vertex

$$\begin{aligned} J_\alpha &= (ev_F)\gamma_V^{-1}\Gamma(\vec{q} = \vec{0}\,)\sigma_\alpha \\ &= (ev_F)[\mathbf{I} \otimes \mathbf{I} - \gamma_V\Pi(\vec{q} = \vec{0}\,)]^{-1}\sigma_\alpha \\ &= (ev_F)2(\mathbf{I} \otimes \mathbf{I} - \tfrac{1}{2}\sigma_x \otimes \sigma_x - \tfrac{1}{2}\sigma_y \otimes \sigma_y)^{-1}\sigma_\alpha \\ &= 2(ev_F)\sigma_\alpha = 2j_\alpha \qquad \text{for } \alpha = x, y. \end{aligned} \tag{10.60}$$

The final contraction can be done without further algebra: we obtain twice the result of (10.51). This result correspond to a renormalization by 2 of the current operator: $J_\alpha = 2j_\alpha$. Hence we recover the Boltzmann result of (10.45): the anisotropic scattering inherent to the Dirac nature of the particles leads to a doubling of the transport time with respect to the elastic scattering time.

Note that the notation of the first line of (10.60) is misleading and should be read as $J_\alpha = (ev_F)\gamma_V^{-1}\lim_{q\to 0}\Gamma(\vec{q})\sigma_\alpha$. Indeed, from the discussion around Fig. 10.3, we expect $\Gamma(\vec{q})$ to possess long-wavelength diffusive modes, corresponding to eigenenergies $\simeq 1/(Dq^2)$ of $\Gamma(\vec{q})$. These diffusive modes encode the quantum corrections to transport. As a consequence, in the limit $\vec{q} \to \vec{0}$, the operator $\mathbf{I} - \gamma_V\Pi(\vec{q})$ is no longer invertible and $\Gamma(\vec{q})$ becomes ill defined. However, we can explicitly check that the contraction $\Gamma(\vec{q}) \cdot \sigma_\alpha$ remains well defined in this limit, which justifies a posteriori the above notation. Indeed, the renormalization of the elastic scattering time into a transport time occurs on a short distance scale, and is expected to be independent from the long-distance physics of the diffusive quantum modes. This is demonstrated by the above property: the vertex renormalization of (10.60) does not depend on the vanishing modes of $\mathbf{I} - \gamma_V\Pi(\vec{q})$, but only on its (non-universal) massive modes.

10.4 Quantum transport of Dirac fermions

The last three decades have seen the exploration of electronic transport in conductors below the micrometre scale [9, 18]. Owing to the interaction with its environment, the phase ϕ of an electron is randomly incremented during its evolution: this phase evolution on the unit circle is characterized by the rate of increase of the fluctuations $\overline{(\delta\phi)^2}$. As the electron evolves in real space, its phase ϕ spreads over the unit circle. Beyond a characteristic length L_ϕ, the variance of the phase is of order $\overline{(\delta\phi)^2} \simeq (2\pi)^2$:

the statistical uncertainty on the electron phase due to the coupling with the environment forbids any measurable interference effect. By lowering the temperature (density of phonons), we increase the corresponding phase coherence length $L_\phi(T)$ for the electrons (at lower temperatures, other mechanisms such as electron–electron interactions and the Kondo effects on magnetic impurities take over). At temperatures $T \simeq 100\,\text{mK}$, the typical order of magnitude of $L_\phi(T)$ is a few micrometres. The study of such small conductors at low temperature has led to the appearance of a new domain of research: mesoscopic quantum physics [9, 18]. Many features of transport of such mesoscopic conductors are remarkable: the quantum corrections to the conductance of a mesoscopic conductor depend on the precise locations of impurities in a given sample: different mesoscopic samples prepared according to exactly the same protocol (or successive annealing of a given sample allowing for disorder reorganizations) display different values of conductance. In this regime, the phase-coherent conductance is said to be a non-self-averaging observable: it fluctuates from sample to sample for sizes $L \leq L_\phi(T)$. In the limit $L \gg L_\phi(T)$, the conductor can be viewed as an incoherent collection of pieces of size $L_\phi(T)$ and relative fluctuations are statistically reduced: we recover the previous classical regime. The description of the conductance of a phase-coherent conductor requires the use of a distribution function, which for weak disorder is a Gaussian characterized by two cumulants. Moreover, the conductance fluctuates as a function of a weak transverse magnetic flux threading the sample. These fluctuations should not be confused with noise: they are perfectly reproducible for a given sample and do not fluctuate in time like typical $1/f$ noise. Indeed, the whole magnetoconductance trace is modified as the sample is annealed: each curve appears as a unique signature of the location of the impurities in the sample. It is a real fingerprint of the configuration of disorder.

The nature of this magnetoconductance and its quantum origin can be understood as follows. As hinted at in Section 10.3.2, along a given diffusive path, the phase of an electronic state is incremented by $\delta\phi_{\mathcal{L}} = k\mathcal{L}$, where $\mathcal{L}$ is the length of the path. For electrons at the Fermi level, $k \simeq k_F$, and this phase $\delta\phi_{\mathcal{L}} \simeq 2\pi\mathcal{L}/\lambda_F$ is extremely sensitive on the lengthscale $\mathcal{L}$, which is typically much larger than the Fermi wavelength λ_F. Between different samples the positions of these impurities are different, and all the lengths $\mathcal{L}$ are modified by at least λ_F, and correspondingly the phases $\delta\phi_{\mathcal{L}}$ are redistributed randomly. The conductivity being a *non-self-averaging* quantity, its value thus differs from sample to sample. A different procedure allows redistribution of these phases along the diffusive path, namely the application of a transverse magnetic field. The presence of such a field can be accounted for by an extra dephasing $e\int_{\mathcal{L}} \vec{A} \cdot d\vec{l}$ along each path $\mathcal{L}$, where $\vec{A}$ is the vector potential. The shapes of these paths $\mathcal{L}$ and thus the associated magnetic phases are random: similarly to a change in impurity positions, the magnetic field redistributes the phases associated with each path in a sample and accordingly changes the value of the conductivity. Whenever a new quantum of magnetic flux is added though the sample, the typical phase shift between two paths crossing the sample is of order 2π, and we obtain a different value of the conductivity. Moreover this function $G(B)$, called a magnetoconductance trace, provides an invaluable access to the statistics of conductance in the quantum regime. Since both the magnetic field and the change in disorder amount to redistribution of the phases in a random manner, we expect both perturbations to lead to the same statistics of the conductance. This is the so-called ergodic hypothesis, which turns out to be quantitatively valid for the first two moments of the conductivity distribution [61].

Let us now consider the coherent regime of transport, relevant for transport on timescales shorter than the dephasing time $\tau_\phi(T)$, defined from the imaginary part $\hbar/(2\tau_\phi)$ of the self-energy for electrons; see (10.21). In this regime, and for weak enough disorder, which is the case experimentally, the conductivity is Gaussian-distributed, and is fully characterized by its first two cumulants. The first cumulant $\langle\delta\sigma\rangle_V$ describes the so-called weak (anti-)localization correction to the averaged conductivity, while the second cumulant $\langle(\delta\sigma)^2\rangle_V$ is associated with the universal fluctuations of the conductivity from sample to sample or as a function of the magnetic field. We have already guessed in the discussion in Section 10.3 that the origin of these quantum corrections to the conductivity lies in the existence of long-wavelength statistical correlations conveniently viewed as propagating diffuson and cooperon modes. In the following, we will briefly review the description of the corresponding diagrammatic contributions.

10.4.1 Quantum correction to the conductivity: weak antilocalization

The two contributions represented in Fig. 10.3 to the average conductivity correspond to diagrams similar to that of Fig.10.6 with either a diffuson or a cooperon structure factor. We have already seen in Section 10.3 that the diffusive mode of the diffuson does not contribute to the average conductivity: only short-distance contributions renormalize the vertex operator. Hence, we shall focus only on the contribution of a cooperon structure factor represented in Fig. 10.7, where $\tilde{\Gamma}$ is a structure factor, analogous to that in Fig.10.5, and accounting for an infinite series of maximally crossed diagrams:

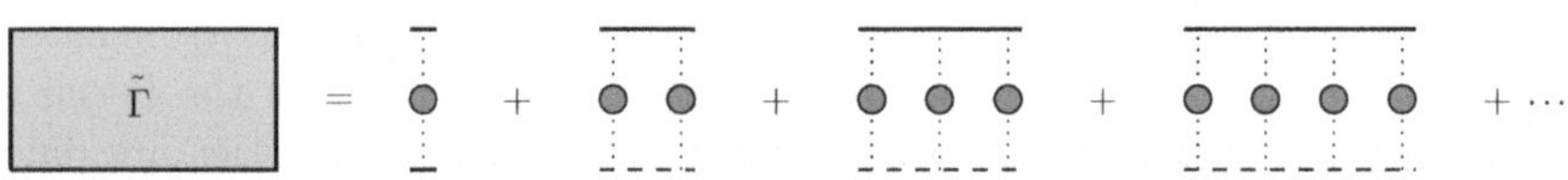

This series of terms can be recast as a geometric series by time reversal of the advanced branch: similarly to the diffuson, the cooperon structure factor satisfies a recursive equation:

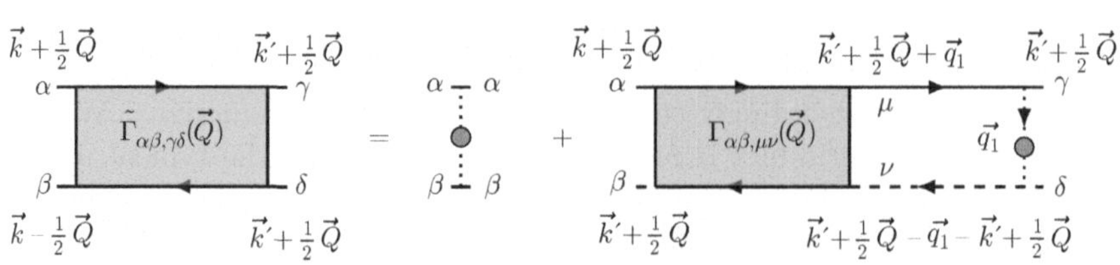

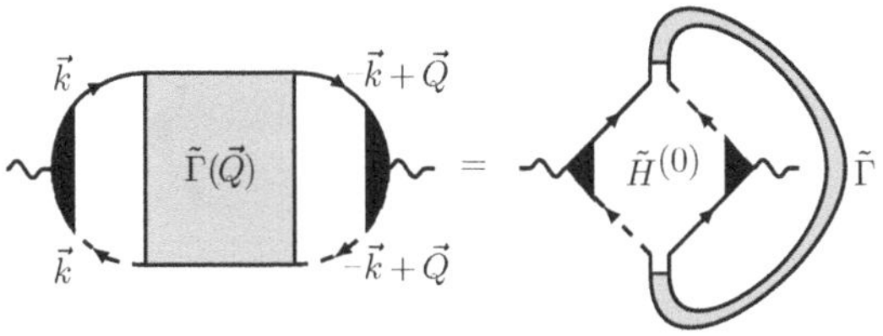

Fig. 10.7 Diagrammatic representation of the first quantum correction to the averaged conductivity (left) and equivalent representation in terms of a contraction between a Hikami box and a cooperon structure factor (right).

or, equivalently,

$$\tilde{\Gamma}_{\alpha\beta,\gamma\delta}(\vec{Q}) = \gamma_V \mathbf{I}_{\alpha\gamma} \otimes \mathbf{I}_{\delta\beta} + \gamma_V \tilde{\Gamma}_{\alpha\beta,\mu\nu}(\vec{Q})\tilde{\Pi}_{\mu\nu,\gamma\delta}(\vec{Q}), \tag{10.61}$$

with

$$\tilde{\Pi}_{\mu\nu,\gamma\delta}(\vec{Q}) = \frac{1}{V}\sum_{\vec{q}_1} \langle \mathcal{G}^R_{\mu\gamma}(\vec{q}_1)\rangle_V \langle \mathcal{G}^A_{\nu\delta}(\vec{Q}-\vec{q}_1)\rangle_V . \tag{10.62}$$

The value of $\tilde{\Pi}$ can be deduced by time reversal of the advanced branch of the expression (10.55) for $\Pi(\vec{q})$. From inversion of (10.61) we obtain the long-wavelength behaviour of the cooperon structure factor $\tilde{\Gamma}(\vec{Q})$. Only one eigenmode is diffusive, with an eigenvalue $1/(DQ^2)$ for $Q \to 0$, corresponding to a singlet mode. Hence, in the diffusive limit, the cooperon structure factor reduces to a projector onto this singlet mode:

$$\begin{aligned}\tilde{\Gamma}(\vec{Q}) &= \frac{\gamma_V}{\tau_e}\frac{1}{DQ^2}\frac{1}{4}(\mathbf{I}\otimes\mathbf{I} - \sigma_x\otimes\sigma_x - \sigma_y\otimes\sigma_y - \sigma_z\otimes\sigma_z)\\ &= \frac{\gamma_V}{\tau_e}\frac{1}{DQ^2}|S\rangle\langle S|,\end{aligned} \tag{10.63}$$

where $D = v_F^2\tau_e$ is the diffusion constant and $|S\rangle$ is a singlet state as defined in Section 10.4.3.

The weak antilocalization correction represented on the left side of Fig. 10.7 is conveniently viewed as a contraction of a cooperon structure factor and a Hikami box, as shown on the right side of Fig. 10.7. The corresponding contribution is

$$\langle \delta\sigma_0 \rangle = \frac{\hbar}{2\pi L^2}\,\mathrm{Tr}[\mathcal{G}^A(\vec{k})J_x\mathcal{G}^R(\vec{k})\tilde{\Gamma}(\vec{Q})\mathcal{G}^R(\vec{Q}-\vec{k})J_x\mathcal{G}^A(\vec{Q}-\vec{k})]. \tag{10.64}$$

In this expression, the $\vec{Q}$ integral (occurring in the trace over quantum numbers) is dominated by the small $\vec{Q}$ contribution originating from the diffusive pole of the cooperon. This justifies a posteriori the projection on the single diffusive mode in (10.63). Focusing on the most dominant part of this expression, we can set $Q \to 0$ except in the cooperon propagator, the Green's functions being regular in $\vec{k}$: this amounts to setting $\vec{k} = \vec{0}$ in the expression of the Hikami box.

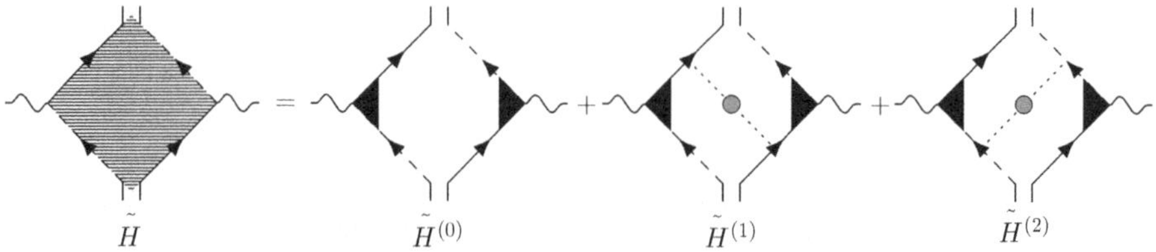

Fig. 10.8 First renormalized Hikami box as the sum of three contributions.

Let us now turn to the expression for this Hikami box: it turns out that besides the contribution depicted in Fig. 10.7, two other terms of the same order have to be included. This is a standard mechanism when scattering is anisotropic [4] and follows from the nature of the Dirac Green's functions. The three contributions consist in considering a renormalized Hikami box as the sum of three terms represented in Fig. 10.8. The integrals corresponding to the three terms are performed according to

$$\begin{aligned}\tilde{H}^{(0)}_{xx} &= 4\int_{\vec{k}} [\mathcal{G}^A(\vec{k}\,)\sigma_x\mathcal{G}^R(\vec{k}\,)] \otimes [\mathcal{G}^R(-\vec{k}\,)\sigma_x\mathcal{G}^A(-\vec{k}\,)] \\ &= \rho(\epsilon_F)\left(\frac{2\tau_e}{\hbar}\right)^3 \frac{\pi}{16}(-4\mathbf{I}\otimes\mathbf{I} + 3\sigma_x\otimes\sigma_x + \sigma_y\otimes\sigma_y), \end{aligned} \tag{10.65}$$

$$\begin{aligned}\tilde{H}^{(1)}_{xx} &= 4\gamma_V \int_{\vec{k}}\int_{\vec{q}_1} [\mathcal{G}^A(\vec{k}\,)\sigma_x\mathcal{G}^R(\vec{k}\,)\mathcal{G}^R(-\vec{q}_1)] \otimes [\mathcal{G}^R(-\vec{k}\,)\mathcal{G}^R(\vec{q}_1)\sigma_x\mathcal{G}^A(\vec{q}_1)] \\ &= \frac{\pi}{16}\rho(\epsilon_F)\left(\frac{2\tau_e}{\hbar}\right)^3 (\mathbf{I}\otimes\mathbf{I} - \sigma^x\otimes\sigma^x), \end{aligned} \tag{10.66}$$

$$\begin{aligned}\tilde{H}^{(2)}_{xx} &= 4\gamma_V \int_{\vec{k}}\int_{\vec{q}_1} [\mathcal{G}^A(-\vec{q}_1)\mathcal{G}^A(\vec{k}\,)\sigma_x\mathcal{G}^R(\vec{k}\,)] \otimes [\mathcal{G}^R(\vec{q}_1)\sigma_x\mathcal{G}^A(\vec{q}_1)\mathcal{G}^A(-\vec{k}\,)] \\ &= \tilde{H}^{(1)}_{xx} \end{aligned} \tag{10.67}$$

Summing these three contributions, we obtain the renormalized Hikami box:

$$\begin{aligned}\tilde{H}_{xx} &= \tilde{H}^{(0)}_{xx} + \tilde{H}^{(1)}_{xx} + \tilde{H}^{(2)}_{xx} \\ &= \rho(\epsilon_F)\left(\frac{2\tau_e}{\hbar}\right)^3 \frac{\pi}{16}(-2\mathbf{I}\otimes\mathbf{I} + \sigma^x\otimes\sigma^x + \sigma^y\otimes\sigma^y). \end{aligned} \tag{10.68}$$

The resulting weak antilocalization correction is obtained via the final contraction with a cooperon structure factor (10.63) as shown in Fig.10.7 (with $\tilde{H}^{(0)}$ replaced by $\tilde{H}$), and leads to the result

$$\langle\delta\sigma\rangle_V = \frac{e^2}{\pi\hbar}\frac{1}{L^2}\sum_{\mathbf{Q}}\frac{1}{Q^2} = \frac{e^2D}{\pi\hbar}\int_{\tau_{\rm tr}}^{\tau_\phi}\frac{dt}{4\pi Dt} \tag{10.69}$$

$$= \frac{e^2D}{\pi\hbar}\int_{\tau_{\rm tr}}^{\tau_\phi}P(0,t)\,dt \tag{10.70}$$

$$\simeq \frac{e^2}{\pi h}\ln\frac{L_\phi}{l_e}. \tag{10.71}$$

The form of this correction appears reminiscent of its physical origin: it is due to the cooperon interference at a point before and after a diffusive travel around a loop. Thus, this correction is proportional to the probability $P(0,t)$ that this cooperon diffuses back to its origin on loops smaller than L_ϕ,

10.4.2 Universal conductance fluctuations

We now consider the second cumulant $\langle(\delta\sigma)^2\rangle_V$ of the distribution function of conductivity, with $\delta\sigma = \sigma - \langle\sigma\rangle_V$. From the relation (10.40), $\sigma = e^2\rho(\epsilon_F)D$, we expect the fluctuations of conductance $\langle(\delta\sigma)^2\rangle_V$ to originate either from fluctuations of the diffusion coefficient $e^2\rho(\epsilon_F)\langle\delta D\rangle_V$ or fluctuations of the density of states $e^2D\langle\delta\rho(\epsilon_F)\rangle_V$. These two physical sources of fluctuations correspond to two different types of diagrams (Figs. 10.9 and 10.10). Their identification proceeds along the same lines as for the average conductivity: similarly to a Lego game, we need to assemble the elementary blocks we have identified: the cooperon and diffuson structure factors and the Hikami boxes. The diffuson structure factor is deduced from that of the cooperon (10.63) by time-reversal symmetry of the advanced branch: it reduces again to a projector on a single state, denoted the diffuson singlet as defined in Section 10.4.3:

$$\begin{aligned}\Gamma(\vec{q}=\vec{0}) &= \frac{\gamma_{\rm V}}{\tau_e}\frac{1}{Dq^2}\frac{1}{4}(\mathbf{I}\otimes\mathbf{I}+\sigma_x\otimes\sigma_x-\sigma_y\otimes\sigma_y+\sigma_z\otimes\sigma_z)\\ &= \frac{\gamma_{\rm V}}{\tau_e}\frac{1}{Dq^2}\,|\tilde{S}\rangle\langle\tilde{S}|.\end{aligned} \tag{10.72}$$

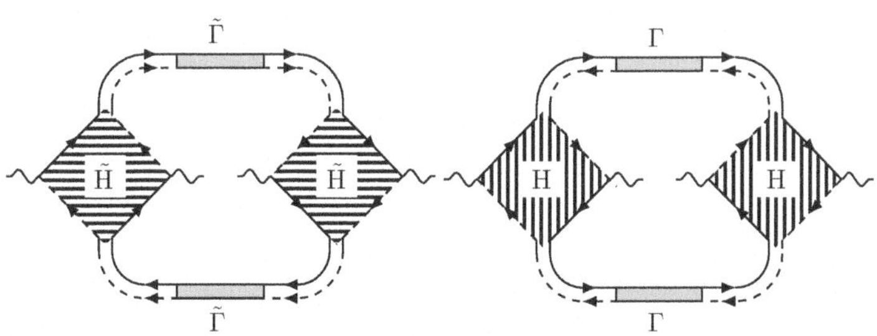

Fig. 10.9 Diagrams describing the contributions to the conductivity fluctuations accounting for the fluctuations of the diffusion coefficient.

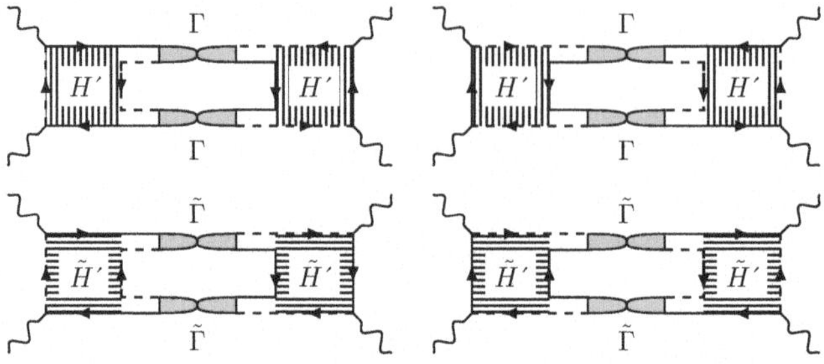

Fig. 10.10 Diagrams describing the contributions to the conductivity fluctuations accounting for the fluctuations of the density of state.

Following a similar approach to that adopted in the case of the cooperon, we identify a second renormalized Hikami box for diagrams involving diffuson structure factors in Fig. 10.9:

$$H = \rho(E_F)\left(\frac{2\tau_e}{\hbar}\right)^3 \frac{\pi}{16}(2\mathbf{I}\otimes\mathbf{I} + \sigma^x\otimes\sigma^x + \sigma^y\otimes\sigma^y). \tag{10.73}$$

The resulting contractions of the diagrams of Fig. 10.9 between two diffuson or cooperon structure factors lead to the result

$$\Delta\sigma_1^2 = 8\left(\frac{e^2}{h}\right)^2 \sum_{\vec{q}} \frac{1}{(L^2q^2)^2}. \tag{10.74}$$

This contribution can be interpreted as describing the fluctuations of the diffusion coefficient D [4].

A second contribution to the conductance fluctuations originates from diagrams with a different topology, represented in Fig. 10.10. They describe the fluctuations of the density of states $\rho(\epsilon_F)$ [4]. Their determination requires two additional Hikami boxes:

$$H' = \rho(E_F)\left(\frac{2\tau_e}{\hbar}\right)^3 \frac{\pi}{16}(\mathbf{I}\otimes\mathbf{I} + \sigma^x\otimes\sigma^x), \tag{10.75}$$

$$\tilde{H}' = \rho(E_F)\left(\frac{2\tau_e}{\hbar}\right)^3 \frac{\pi}{16}(\mathbf{I}\otimes\mathbf{I} - \sigma^x\otimes\sigma^x). \tag{10.76}$$

The final result after contraction in spin space of these diagrams is

$$\Delta\sigma_2^2 = 4\left(\frac{e^2}{h}\right)^2 \sum_{\vec{q}} \frac{1}{(L^2q^2)^2}. \tag{10.77}$$

Summing the two contributions (10.74) and (10.77), we finally get the amplitude of the universal conductance fluctuations (UCF):

$$\begin{aligned}\langle(\delta\sigma)^2\rangle_V &= 12\left(\frac{e^2}{h}\right)^2\sum_{\vec{q}}\frac{1}{(L^2q^2)^2}\\ &= \frac{12}{\pi^4}\left(\frac{e^2}{h}\right)^2\sum_{n_x\neq 0,n_y}\frac{1}{(n_x^2+n_y^2)^2}.\end{aligned} \tag{10.78}$$

The results (10.69) and (10.78) together define the quantum corrections to the diffusive transport of Dirac fermions in $d = 2$. They correspond exactly to known results for the symplectic class in $d = 2$, and the previous calculations appear as a tedious way to recover these results for the specific case of Dirac fermions. In the next subsection, we will discuss how this is indeed the case by introducing the notion of universality class for quantum corrections to diffusive transport.

10.4.3 Notion of universality class

10.4.3.1 Universality class and number of diffusive modes

In deriving diagrammatically the perturbation theory of weak localization, we have identified the building blocks as diffusive modes (either cooperon or diffuson) and the Hikami boxes that reconnect these propagating modes to the current vertices. Moreover, we have found that the quantum corrections to the two first cumulants of the conductivity probability distribution function depend on the number of such Goldstone modes. This unusual universality strongly points towards an effective field theory approach underlying the above perturbation theory. This is indeed the case, and the nonlinear sigma model developed to analyse the Anderson localization transitions turns out to be a very elegant framework not only to understand the universal results in the perturbative regime and classify all the possible universality classes [70], but also to derive in a systematic manner the higher cumulants of the conductivity probability distribution function [8]. In particular, the number of diffusive modes responsible for the quantum corrections to the conductance appears as the 'dimension' of the target space of this effective field theory. This classification of symmetry classes of quantum transport has been used to analyse the occurrence of topological order in a gapped phase from the stability of their surface states with respect to disorder [50]. In this case, this stability manifests itself as a topological term allowed in the field theory action, which forbids Anderson localization. Such topological terms are irrelevant within the regime perturbative in disorder on which we focus here, and will not be discussed further.

A discussion of the field theory approach to the weak localization of electrons goes beyond the scope of the present lectures. A thorough discussion of the construction of the generating functional for the conductance moments can be found in [8] (see also [22]), while a recent pedagogical introduction can be found in the textbook [7]. The general idea of this approach, inspired by the diagrammatic perturbative expansion

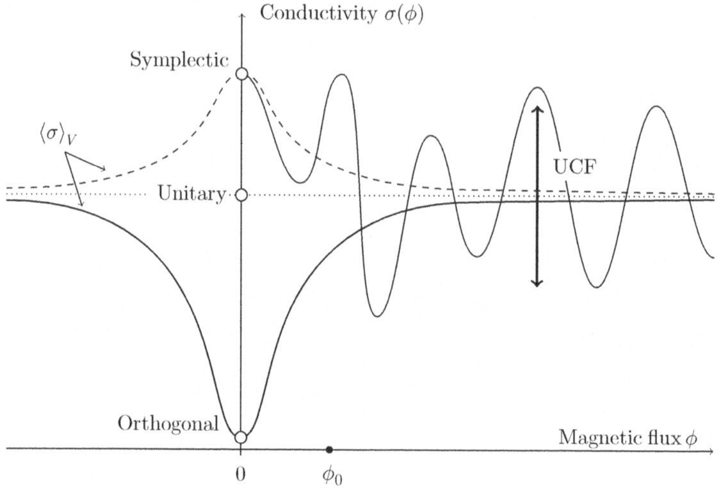

Fig. 10.11 Schematic behaviour of the quantum contribution to the averaged conductivity, observed in samples of size large compared with the phase-coherent lengthscale $L_\phi(T)$, as a function of a weak magnetic field for different symmetry classes. (i) In the orthogonal class, corresponding for example to parabolic bands with charged impurities, a weak localization behaviour is present, which vanishes when a magnetic field is applied, leading to the characteristic behaviour represented by the thick full curve. (ii) In the symplectic class, corresponding for example to Dirac bands with charged impurities, the quantum correction at $B = 0$ corresponds to a weak antilocalization, leading to the behaviour represented by the dashed curve. (iii) When time-reversal symmetry is broken (unitary class), no dependence on a weak magnetic field is observed (dotted line). For short samples of size comparable to $L_\phi(T)$, fluctuations of the conductivity occur as a functions of the magnetic field, characteristic of the unitary symmetry class.

presented in Section 10.3.2, amounts to considering cumulants of pairs of Green's functions $\mathcal{G}^R\mathcal{G}^A$ occurring in the Kubo formula (10.46). When doing so, the pairings between fermionic fields corresponding to both the cooperon and the diffuson modes are treated on an equal footing. This amounts, after standard field theory techniques, to considering an action for the field

$$Q = \begin{pmatrix} d_{\uparrow\uparrow} & d_{\uparrow\downarrow} & -c_{\uparrow\downarrow} & c_{\uparrow\uparrow} \\ d_{\downarrow\uparrow} & d_{\downarrow\downarrow} & -c_{\downarrow\downarrow} & c_{\downarrow\uparrow} \\ c^*_{\downarrow\uparrow} & c^*_{\downarrow\downarrow} & d^*_{\downarrow\downarrow} & -d^*_{\downarrow\uparrow} \\ -c^*_{\uparrow\uparrow} & -c^*_{\uparrow\downarrow} & -d^*_{\uparrow\downarrow} & d^*_{\uparrow\uparrow} \end{pmatrix}, \tag{10.79}$$

where c and d correspond to the modes in the cooperon and diffuson pairing channels (we consider spin-$\frac{1}{2}$ particles with no additional quantum number, as opposed to, for example, graphene). In a typical Landau approach, the dominant terms of this action in the long-wavelength limit can be determined from symmetry constraints: the Goldstone modes of the resulting nonlinear sigma model, corresponding to the diffusive cooperon and diffuson modes, are thus entirely determined by the dimension of space and the statistical symmetry of the disordered model. We summarize the results of this approach below.

The spin structure of the Dyson equation (10.53) reflects the construction of the diffuson as the tensor product of two spins $\frac{1}{2}$ associated with the retarded and advanced Green's functions. It is thus naturally diagonalized by using the basis of singlet and triplet states for a spin $\frac{1}{2}$ and a time-reversed spin (with $T|\uparrow\rangle = |\downarrow\rangle$ and $T|\downarrow\rangle = -|\uparrow\rangle$):

$$\begin{gathered} |S\rangle = \tfrac{1}{\sqrt{2}}(|\uparrow\uparrow\rangle + |\downarrow\downarrow\rangle), \quad |T_1\rangle = \tfrac{1}{\sqrt{2}}(|\uparrow\uparrow\rangle - |\downarrow\downarrow\rangle), \\ |T_2\rangle = |\uparrow\downarrow\rangle, \quad |T_3\rangle = |\downarrow\uparrow\rangle. \end{gathered} \tag{10.80}$$

Similarly, the cooperon's structure factor is diagonalized in the basis of two spins $\frac{1}{2}$:

$$\begin{gathered} |\tilde{S}\rangle = \tfrac{1}{\sqrt{2}}(|\uparrow\downarrow\rangle - |\downarrow\uparrow\rangle), \quad |\tilde{T}_1\rangle = \tfrac{1}{\sqrt{2}}(|\uparrow\downarrow\rangle + |\downarrow\uparrow\rangle), \\ |\tilde{T}_2\rangle = |\uparrow\uparrow\rangle, \quad |\tilde{T}_3\rangle = |\downarrow\downarrow\rangle. \end{gathered} \tag{10.81}$$

In these bases, (10.53) and the equivalent equation for the cooperon are diagonalized for each state: the various diffuson and cooperon modes propagate either diffusively or not. We can write formally

$$\Gamma_{S/T}(\vec{q}) = \frac{\gamma_V}{\tau_e} \frac{1}{Dq^2 + \eta_\phi + \eta^{D,S/T}}, \tag{10.82}$$

$$\tilde{\Gamma}_{S/T}(\vec{Q}) = \frac{\gamma_V}{\tau_e} \frac{1}{DQ^2 + \eta_\phi + \eta^{C,S/T}}, \tag{10.83}$$

with $\eta_\phi = \hbar/\tau_\phi$. The various dephasing rates account for the possible decay over short lengthscales of the respective mode: $\eta^{C/D,S/T} = 0$ for the diffusive mode, while it is finite for modes contributing only on short lengthscales (crossover between different universality classes can be described along these lines [23]).

The general expression for the quantum correction to the averaged conductivity is the following sum of contributions:

$$\langle \delta\sigma \rangle_V = -\frac{e^2 D}{\pi\hbar} \left(-\frac{1}{4}\frac{1}{L^d} \sum_{\mathbf{Q}} \frac{1}{DQ^2 + \eta_S^{(C)} + \eta_\phi} + \frac{1}{4} \sum_\alpha \frac{1}{L^d} \sum_{\mathbf{Q}} \frac{1}{DQ^2 + \eta_{T_\alpha}^{(C)} + \eta_\phi} \right), \tag{10.84}$$

where d is the dimensionality of diffusion. In this expression, the cooperon singlet mode occurs as a negative correction, while all triplet modes contribute a positive correction: this quantum correction thus depends solely on the number of modes of the cooperon structure factor. The different symmetry classes, initially identified through random matrix considerations (see [15] for a general review), correspond to different numbers of cooperon modes:

- In situations with spin-rotation symmetry, corresponding to a spinless Hamiltonian with a time-reversal symmetry satisfying $T^2 = \mathbf{I}$, all cooperon modes are present and contribute to the quantum correction, which is negative. This corresponds to the weak-localization situation.
- When spin–momentum locking occurs, due either to the pure Hamiltonian (Dirac case) or to the disorder type (spin–orbit disorder), all triplet modes are affected and cannot diffuse on long distances. Only the singlet modes contribute to (10.84) and we find a weak antilocalization. This corresponds to a situation where time-reversal symmetry satisfies $T^2 = -\mathbf{I}$. Note that the $d = 2$ situation of random spin–orbit is special since disorder affects only the z components of spins: one triplet and one singlet mode remain unaffected and we obtain a "pseudo-unitary' class [31].
- Finally, when spin symmetry is broken, either by magnetic impurities or by a magnetic field, no cooperon modes diffuse and we obtain a vanishing quantum correction.

These three cases are summarized in Table 10.1.

Similarly, the amplitude of universal conductance fluctuations can be written as the sum of the contributions from the different diffusive modes:

$$\begin{aligned}\langle(\delta\sigma)^2\rangle_V = F(\eta_m^{D,S} + \eta_\phi) + \sum_{\alpha=1,2,3} F(\eta_m^{D,T_\alpha} + \eta_\phi) \\ + F(\eta_m^{C,S} + \eta_\phi) + \sum_{\alpha=1,2,3} F(\eta_m^{C,T_\alpha} + \eta_\phi),\end{aligned} \tag{10.85}$$

where

$$F(\eta) = 6\sum_{\vec{q}} \frac{1}{[(Lq)^2 + \eta]^2}. \tag{10.86}$$

Hence the conductance fluctuations depend linearly on the number of diffusive modes. This is summarized in Table 10.2.

Table 10.1 Summary of the symmetry classes for the different types of disorder for quadratic and Dirac Hamiltonians.

	SCALAR DISORDER	RANDOM SPIN–ORBIT	MAGNETIC DISORDER
Quadratic dispersion	Orthogonal	Symplectic ($d = 3$)	Unitary
		Pseudo-unitary ($d = 2$)	
Dirac dispersion	Symplectic	Symplectic	Unitary

Table 10.2 Summary of the numbers of singlet (S) and triplet (T) modes for the cooperon and diffuson structure factors. The weak-localization and universal conductance fluctuations (UCF) contributions are represented as respective integer factors depending solely on the numbers of diffusive modes. The corresponding proportionality factors are defined in (10.84) and (10.85)

SYMMETRY CLASS	TIME-REVERSAL SYMMETRY	DIFFUSON MODES	COOPERON MODES	WEAK LOCALIZATION	UCF
Orthogonal	$T^2 = \mathbf{I}$	1 S + 3 T	1 S + 3 T	−2	8
Symplectic	$T^2 = -\mathbf{I}$	1 S	1 S	+1	2
Unitary	0	1 S	0	0	1

10.4.4 Effect of a magnetic field

10.4.4.1 Transverse magnetic field

In the presence of a magnetic field, the probability of return to the origin during time t of a diffusive walk is modified to [4]:

$$Z_c(t, B) = \frac{\phi/\phi_0}{\sinh(4\pi BDt/\phi_0)} \tag{10.87}$$

where $\phi = BL^2$ and the argument of the sinh function is the dimensionless magnetic flux through the region $4\pi l_t^2$ typically explored by the diffusive path during time t: $l_t^2 = Dt$. The corresponding quantum contribution to the conductivity can be written as

$$\begin{aligned}\langle\delta\sigma(B)\rangle_V &= (\#\ \mathrm{C,S}\ - \#\ \mathrm{C,T})\,\frac{e^2 D}{\pi\hbar}\int_{\tau_{\mathrm{tr}}}^{\tau_\phi} Z_c(t,B)\\ &= (\#\ \mathrm{C,S}\ - \#\ \mathrm{C,T})\left[\Psi\left(\frac{1}{2}+\frac{\hbar}{4eDB\tau_{\mathrm{tr}}}\right)-\Psi\left(\frac{1}{2}+\frac{\hbar}{4eDB\tau_\phi}\right)\right],\end{aligned} \tag{10.88}$$

where $\Psi(x)$ is the digamma function. This formula is commonly used as a fit to extract the phase-coherent time τ_ϕ from experimental transport measurements. Note that in this case, the expression (10.88) describes a crossover from the orthogonal or symplectic class at $B = 0$ to the unitary class at larger magnetic field, as represented in Fig. 10.11.

10.4.4.2 Aharonov–Bohm-like oscillations

Finally, we consider a cylinder made out of a topological insulating material, with a metallic Dirac metal at its surface (see Fig. 10.12). The cylinder is assumed to be long compared with the dephasing lengthscale $L_\phi(T)$ so that conductance fluctuations are negligible (they are statistically reduced by the incoherent combination of contributions of domains of size $L_\phi(T)$). This conductivity is thus well described by its

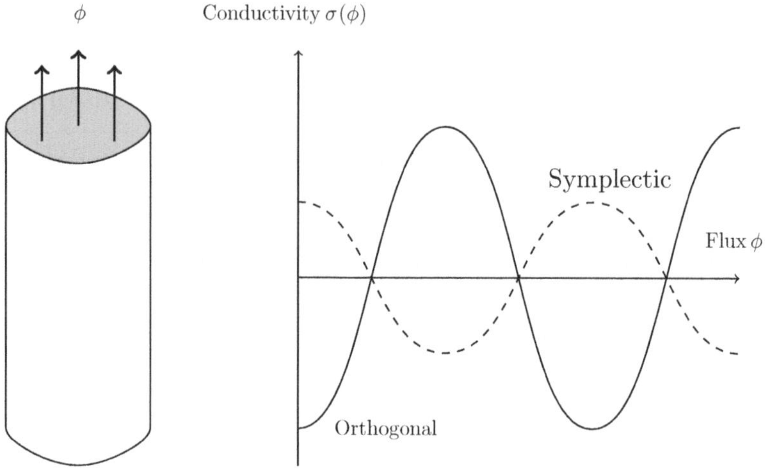

Fig. 10.12 Schematic representation of the Aharonov–Bohm oscillations of a metal at the surface of a cylinder. In the case of Dirac particles, we expect behaviour predicted by the symplectic symmetry class.

average $\langle\sigma\rangle_V$, which consist of both the classical contribution and the quantum correction. This quantum correction is due to the contribution of cooperon diffusive modes. When a magnetic flux is threaded through the section of the cylinder, this cooperon, which carries a charge, $2e$ acquires an Aharonov–Bohm phase, leading to oscillations of the quantum contribution to the conductivity with a period $\tilde{\phi}_0 = h/2e$. As depicted in Fig. 10.12, the phase of these oscillations is fixed by the sign of the quantum correction at $\phi = 0$, and thus by the symmetry class. In the case of Dirac fermions, these oscillations are reminiscent of the π phase acquired owing to momentum–spin locking by particles when winding around the cylinder in the ballistic limit. However, the absolute phase of these ballistic oscillations is very sensitive to energy, which renders it hard to measure experimentally. See [11, 68] for discussions of this effect in the context of the surface states of topological insulators and [12] for a detailed discussion.

Acknowledgements

I warmly thank P. Adroguer, J. Cayssol, A. Fedorenko, G. Montambaux, and E. Orignac, with whom I learned most of what is included in these lectures. I am particularly indebted to P. Adroguer for a recent stimulating discussion on the pseudo-unitary class in two dimensions and E. Orignac for proofreading the manuscript of these lecture notes.

References

[1] Adam, S., Brouwer, P. W., and Das Sarma, S. (2009). Crossover from quantum to Boltzmann transport in graphene. *Phys. Rev. B*, **79**, 201404(R).

[2] Adam, S., Hwang, E. H., Galitski, V. M., and Das Sarma, S. (2007). A self-consistent theory for graphene transport. *Proc. Natl Acad. Sci. USA*, **104**, 18392.

[3] Adroguer, P., Carpentier, D., Cayssol, J., and Orignac, E. (2012). Diffusion at the surface of topological insulators. *New J. Phys.*, **14**, 103027.

[4] Akkermans, E. and Montambaux, G. (2007). *Mesoscopic Physics of Electrons and Photons.* Cambridge University Press, Cambridge.

[5] Aleiner, I. L. and Efetov, K. B. (2006). Effect of disorder on transport in graphene. *Phys. Rev. Lett.*, **97**, 236801.

[6] Altland, A. (2006). Low-energy theory of disordered graphene. *Phys. Rev. Lett.*, **97**, 236802.

[7] Altland, A. and Simons, B. (2010). *Condensed Matter Field Theory.* Cambridge University Press, Cambridge.

[8] Altshuler, B.L., Kravtsov, V.E., and Lerner, I.V. (1991). Distribution of mesoscopic fluctuations and relaxation properties in disordered conductors. In *Mesoscopic Phenomena in Solids* (ed. B. L. Altshuler, P. A. Lee, and R. A. Webb), p. 449. North-Holland, Amsterdam.

[9] Altshuler, B. L., Lee, P. A., and Webb, R. A. (eds.) (1991). *Mesoscopic Phenomena in Solids.* North-Holland, Amsterdam.

[10] Ando, T. (2008). Physics of graphene. *Prog. Theor. Phys. Suppl.*, **176**, 203.

[11] Bardarson, J. H., Brouwer, P.W., and Moore, J. E. (2010). Aharonov–Bohm oscillations in disordered topological insulator nanowires. *Phys. Rev. Lett*, **105**, 156803.

[12] Bardarson, J. H. and Moore, J.E. (2013). Quantum interference and Aharonov–Bohm oscillations in topological insulators. *Rep. Prog. Phys.*, **76**, 056501.

[13] Bardarson, J. H., Tworzydlo, J., Brouwer, P. W., and Beenakker, C. W. J. (2007). One-parameter scaling at the Dirac point in graphene. *Phys. Rev. Lett.*, **99**, 106801.

[14] Beenakker, C.W.J. (2008). Colloquium: Andreev reflection and Klein tunneling in graphene. *Rev. Mod. Phys.*, **80**, 1337.

[15] Beenakker, C. W. J. (1997). Random-matrix theory of quantum transport. *Rev. Mod. Phys.*, **69**, 731.

[16] Bena, C. and Montambaux, G. (2009). Remarks on the tight-binding model of graphene. *New J. Phys.*, **11**, 095003.

[17] Berry, M. V. (1984). Quantal phase factors accompanying adiabatic changes. *Proc. R. Soc. Lond., Ser. A*, **392**, 45.

[18] Bouchiat, H., Guéron, S., Montambaux, G., and Dalibard, J. (eds.) (2004). *Nanophysics: Coherence and Transport. Les Houches Summer School, Session LXXXI.* Elsevier, Amsterdam.

[19] Castro Neto, A. H., Guinea, F., Peres, N. M. R., Novoselov, K. S., and Geim, A. K. (2009). The electronic properties of graphene. *Rev. Mod. Phys.*, **81**, 109.

[20] Das Sarma, S., Adam, S., Hwang, E. H., and Rossi, E. (2011). Electronic transport in two-dimensional graphene. *Rev. Mod. Phys.*, **83**, 407.

[21] Dyson, F. J. (1962). The threefold way. Algebraic structure of symmetry groups and ensembles in quantum mechanics. *J. Math. Phys.*, **3**, 1199.

[22] Efetov, K. B., Larkin, A. I., and Khmelnitskii, D. E. (1980). Interaction of diffusion modes in the theory of localization. *Sov. Phys. JETP*, **52**, 568.

[23] Fedorenko, A. A. and Carpentier, D. (2009). Magnetic dephasing in a mesoscopic spin glass. *Europhys. Lett.*, **88**, 57009.

[24] Foa Torres, L. E. F., Roche, S., and Charlier, J.-C. (2014). *Introduction to Graphene-Based Nanomaterials.* Cambridge University Press, Cambridge.

[25] Fradkin, E. (1986). Critical behaviour of disordered degenerate semiconductors. I. Models, symmetries, and formalism. *Phys. Rev. B*, **33**, 3257.

[26] Fradkin, E. (1986). Critical behaviour of disordered degenerate semiconductors. II. Spectrum and transport properties in mean-field theory. *Phys. Rev. B*, **33**, 3263.

[27] Fruchart, M., Carpentier, D., and Gawedzki, K. (2014). Parallel transport and band theory in crystals. *Europhys. Lett.*, **106**, 60002.

[28] Fu, L. (2009). Hexagonal warping effects in the surface states of the topological insulator Bi_2Te_3. *Phys. Rev. Lett.*, **103**, 266801.

[29] Hasan, M. Z. and Kane, C. L. (2010). Colloquium: Topological insulators. *Rev. Mod. Phys.*, **82**, 3045.

[30] Hatsugai, Y., Morimoto, T., Kawarabayashi, T., Hamamoto, Y., and Aoki, H. (2013). Chiral symmetry and its manifestation in optical responses in graphene: interaction and multilayers. *New J. Phys.*, **15**, 035023.

[31] Hikami, S., Larkin, A., and Nagaoka, Y. (1980). Spin–orbit interaction and magnetoresistance in the two dimensional random system. *Prog. Theor. Phys.*, **63**, 707.

[32] Imry, Y. (2002). *Introduction to Mesoscopic Physics.* Oxford University Press, Oxford.

[33] Itzykson, C. and Drouffe, J.-M. (1991). *Statistical Field Theory*, Volume 1. Cambridge University Press, Cambridge.

[34] Katsnelson, M. I. (2006). Zitterbewegung, chirality, and minimal conductivity in graphene. *Eur. Phys. J. B*, **51**, 157.

[35] Katsnelson, M. I. (2012). *Graphene: Carbon in Two Dimensions.* Cambridge University Press, Cambridge.

[36] Khveshchenko, D. V. (2006). Electron localization properties in graphene. *Phys. Rev. Lett.*, **97**, 036802.

[37] Langer, J. L. and Neal, T. (1966). Breakdown of the concentration expansion for the impurity resistivity of metals. *Phys. Rev. Lett.*, **16**, 984.

[38] Li, Q., Rossi, E., and Das Sarma, S. (2012). Two-dimensional electronic transport on the surface of three-dimensional topological insulators. *Phys. Rev. B*, **86**, 235443.

[39] Liu, C.-X., Qi, X.-L., Zhang, H., Dai, X., Fang, Z., and Zhang, S.-C. (2010). Model Hamiltonian for topological insulators. *Phys. Rev. B*, **82**, 045122.

[40] Ludwig, A. W. W., Fisher, M. P. A., Shankar, R., and Grinstein, G. (1994). Integer quantum Hall transition: an alternative approach and exact results. *Phys. Rev. B*, **50**, 7526.

[41] Marder, M. P. (2010). *Condensed Matter Physics*, 2nd edn. Wiley, Hoboken, NJ.

[42] McCann, E., Kechedzhi, K., Fal'ko, V.I., Suzuura, H., Ando, T., and Altshuler, B.L. (2006). Weak-localization magnetoresistance and valley symmetry in graphene. *Phys. Rev. Lett.*, **97**, 146805.

[43] Morpurgo, A. F. and Guinea, F. (2006). Intervalley scattering, long-range disorder, and effective time-reversal symmetry breaking in graphene. *Phys. Rev. Lett.*, **97**, 196804.

[44] Nielsen, H. B. and Ninomiya, M. (1981). A no-go theorem for regularizing chiral fermions. *Phys. Lett. B*, **105**, 219.

[45] Nomura, K., Koshino, M., and Ryu, S. (2007). Topological delocalization of two-dimensional massless Dirac fermions. *Phys. Rev. Lett.*, **99**, 146806.

[46] Peres, N. M. R., Guinea, F., and Castro Neto, A. H. (2006). Electronic properties of disordered two-dimensional carbon. *Phys. Rev. Lett.*, **73**, 125411.

[47] Qi, X.-L. and Zhang, S.-C. (2011). Topological insulators and superconductors. *Rev. Mod. Phys.*, **83**, 1057.

[48] Ringel, Z., Kraus, Y., and Stern, A. (2012). Strong side of weak topological insulators. *Phys. Rev. B*, **86**, 045102.

[49] Ryu, S., Mudry, C., Furusaki, A., and Ludwig, A. W. W. (2007). Landauer conductance and twisted boundary conditions for Dirac fermions in two space dimensions. *Phys. Rev. B*, **75**, 205344.

[50] Ryu, S., Schnyder, A. P., Furusaki, A., and Ludwig, A. W. W. (2010). Topological insulators and superconductors: tenfold way and dimensional hierarchy. *New J. Phys.*, **12**, 065010.

[51] San-Jose, P., Prada, E., and Golubev, D. S. (2007). Universal scaling of current fluctuations in disordered graphene. *Phys. Rev. B*, **76**, 195445.

[52] Schneider, M. and Brouwer, P. W. (2014). Quantum corrections to transport in graphene: a trajectory-based semiclassical analysis. *New J. Phys.*, **16**, 073015.

[53] Schuessler, A., Ostrovsky, P. M., Gornyi, I. V., and Mirlin, A. D. (2009). Analytic theory of ballistic transport in disordered graphene. *Phys. Rev. B*, **79**, 075405.

[54] Shon, N. H. and Ando, T. (1998). Quantum transport in two-dimensional graphite system. *J. Phys. Soc. Jpn*, **67**, 2421.

[55] Sondheimer, E. H. (1962). The Boltzmann equation for anisotropic metals. *Proc. R. Soc. Lond., Ser. A*, **268**, 100.

[56] Sorbello, R. S. (1974). On the anisotropic relaxation time. *J. Phys. F*, **4**, 503.

[57] Sorbello, R.S. (1975). Effects of anisotropic scattering on electronic transport properties. *Phys. Condens. Matter*, **19**, 303.

[58] Sundaram, G. and Niu, Q. (1999). Wave-packet dynamics in slowly perturbed crystals: gradient corrections and Berry-phase effects. *Phys. Rev. B*, **59**, 14915.

[59] Suzuura, H. and Ando, T. (2002). Crossover from symplectic to orthogonal class in a two-dimensional honeycomb lattice. *Phys. Rev. Lett.*, **89**, 2666603.

[60] Tkachov, G. and Hankiewicz, E. M. (2011). Weak antilocalization in HgTe quantum wells and topological surface states: massive versus massless Dirac fermions. *Phys. Rev. B*, **84**, 035444.

[61] Tsyplyatyev, O., Aleiner, I. L., Fal'ko, V. I., and Lerner, I. V. (2009). Applicability of the ergodic hypothesis to mesoscopic fluctuations. *Phys. Rev. B*, **68**, 121301.

[62] Tworzydlo, J., Trauzettel, B., Titov, M., Rycerz, A., and Beenakker, C. W. J. (2006). Sub-Poissonian shot noise in graphene. *Phys. Rev. Lett.*, **96**, 246802.

[63] Vafek, O. and Vishwanath, A. (2014). Dirac fermions in solids—from high T_c cuprates and graphene to topological insulators and Weyl semimetals. *Annu. Rev. Condens. Matter Phys.*, **5**, 83.

[64] Zawadzki, W. and Rusin, T. M. (2011). Zitterbewegung (trembling motion) of electrons in semiconductors: a review. *J. Phys. Condens. Matter*, **23**, 143201.

[65] Wakabayashi, K., Takane, Y., Yamamoto, M., and Sigrist, M. (2009). Electronic transport properties of graphene nanoribbons. *New J. Phys.*, **11**, 095016.

[66] Wigner, E. P. (1959). *Group Theory and its Application to the Quantum Mechanics of Atomic Spectra.* Academic Press, New York.

[67] Xiao, D., Chang, M.-C., and Niu, Q. (2010). Berry phase effects on electronic properties. *Rev. Mod. Phys.*, **82**, 1959.

[68] Zhang, Y. and Vishwanath, A. (2010). Anomalous Aharonov–Bohm conductance oscillations from topological insulator surface states. *Phys. Rev. Lett*, **105**, 206601.

[69] Ziegler, K. (1998). Delocalization of 2D Dirac fermions: the role of a broken supersymmetry. *Phys. Rev. Lett.*, **80**, 3113.

[70] Zirnbauer, M. R. (1996). Riemannian symmetric superspaces and their origin in random-matrix theory. *J. Math. Phys.*, **37**, 4986.

[71] Zirnbauer, M. (2011). Symmetry classes. In *The Oxford Handbook of Random Matrix Theory* (ed. G. Akemann, J. Baik, and P. Di Francesco), Chap. 3. Oxford University Press, Oxford.

11
Spin textures in quantum Hall systems

Benoît Douçot

Sorbonne Universités, Université Pierre and Marie Curie Paris 6, and CNRS
Laboratoire de Physique Théorique et Hautes Énergies
4 place Jussieu
75252 Paris Cedex 05, France

Chapter Contents

Colour figures. For those figures in this chapter that use colour, please see the version of these lecture notes at http://topo-houches.pks.mpg.de. These figures are indicated by '[Colour online]' at the start of the caption.

11.1 Introduction

The subject of these lectures combines several different manifestations of topology in a condensed matter system. The most classical one is through the notion of texture. By this, we mean any non-singular and topologically non-trivial spatial configuration of some relevant order parameter. Textures are therefore qualitatively different from defects, for which the order-parameter field exhibits a point-like singularity in two-dimensional (2D) space, or more generally a codimension-2 surface of singularity in D-dimensional space. In this latter situation, the set of points where the order-parameter field is smooth exhibits a non-trivial topology, equivalent to that of a circle S^1. Denoting by M the order-parameter manifold, defects are naturally classified by the group $\pi_1(M)$ of homotopy classes of smooth maps from S^1 to M [1–3]. By contrast, textures with a finite energy correspond to configurations in which the order parameter is uniform at infinity, which allows us to compactify physical space into a D-dimensional sphere S^D. Textures are then classified according to the higher homotopy group $\pi_D(M)$. In most systems, textures appear as finite-energy excitations above an ordered ground state. A remarkable aspect of quantum Hall ferromagnets is that non-trivial textures have been predicted to form, if the electronic g factor is not too large, as soon as electrons are added to or removed from a filled Landau level [4]. Spin textures on a 2D system are classified by $\pi_2(S^2) = \mathbb{Z}$, so they carry an integer topological charge N_{top}. A striking prediction of Sondhi et al. [4] is that N_{top} is identical to the electric charge: it is equal to $+1$ for a hole (skyrmion) and to -1 for an electron (antiskyrmion). This picture has been confirmed experimentally, in particular thanks to NMR measurements of the electronic spin susceptibility [5] and nuclear spin relaxation [6]. Experimentally, it is easier to control the skyrmion density $1 - \nu$ than their total number. Here ν denotes, as usual, the filling factor of the lowest Landau level. For a small but finite skyrmion density, it has been predicted that the long-range Coulomb interaction between the charges bound to skyrmions will favour their ordering into a 2D periodic lattice [7, 8]. Several experiments have provided substantial evidence for the existence of skyrmion lattices in 2D electron gases close to $\nu = 1$. Let us mention for example specific-heat measurements [9, 10], NMR relaxation [11], Raman spectroscopy [12], and microwave pinning-mode resonances [13].

More recently, the physics of quantum Hall ferromagnets has been stimulated by the discovery of new systems that can provide access to more than two internal states for each electron. The first of these is the quantum Hall bilayer [14], in which, besides the physical electronic spin, the additional bilayer degree of freedom can be viewed as a kind of isospin. Skyrmions in these systems have been studied in great detail [15]. Unfortunately, bilayers are far from the maximal $SU(4)$ symmetry that one may expect in a system with four possible internal states. The discovery of graphene opened a very promising way to achieve such a large symmetry. In graphene, the isospin degree of freedom is implemented thanks to the existence of two inequivalent Dirac points. It has been shown that in the presence of an external magnetic field B, the couplings that break $SU(4)$ symmetry are smaller than the symmetry-preserving ones by a factor a/l, where a is the lattice spacing and $l = \sqrt{\hbar/eB}$ is the magnetic length [16, 17]. Other possible examples of systems with more than two internal states are semiconductors with valley degeneracy [18–20] and cold atoms [21, 22]. Theoretical work has been dedicated to the elucidation of phase diagrams for skyrmionic matter in the presence of various physically relevant interactions and anisotropies [21–23] and to the computation of the associated collective mode spectrum [24]. These latter calculations have been partly motivated by NMR relaxation-rate measurements on bilayer systems [25, 26]. Recently, we have revisited these questions for fermions with d internal states and for $SU(d)$-symmetric effective Hamiltonians [27]. This high symmetry allowed us to set up an accurate variational calculation for the optimal wavefunction describing a periodic lattice of skyrmions, for which a simple analytic expression has been obtained. Because these periodic states fully break the underlying $SU(d)$ symmetry, we expect a collective mode spectrum composed of $d^2 - 1$ Goldstone branches and one magnetophonon branch. These expectations have been confirmed by explicit calculations based on a time-dependent Hartree–Fock treatment of our $SU(d)$-symmetric effective Hamiltonian.

The goal of these lectures is to provide a theory-oriented introduction to the physics of textures in quantum Hall ferromagnets, so they do not attempt to review this already rich subject, and many important aspects will not be mentioned. To give an idea, the APS website records 600 citations for the paper by Sondhi et al. [4]. Our recent approach on periodic textures will be presented in Section 11.3, and the associated collective modes will be the subject of Section 11.4. But before discussing our contributions, I try to show in some detail how to derive the effective models that we use from microscopic models of interacting fermions in the lowest Landau level. Establishing this connection is the goal of Section 11.2. Most of the results there are already quite old and are due to many researchers [4, 15, 28–32]. I have tried to give a unified presentation of these seminal works using the framework of coherent-state quantization [33, 34]. This formalism appears at two stages, with different manifestations and purposes. The first is to associate a Slater determinant $|\mathcal{S}_\psi\rangle$ composed of single-electron orbitals in the lowest Landau level with a prescribed texture, described in terms of a smooth d-component spinor field $\psi_a(r)$ $(1 \leq a \leq d)$. Coherent-state quantization is used to construct precisely $|\mathcal{S}_\psi\rangle$ and to compute the expectation values of some physical observables such as the particle density and the interaction energy. The key remark here is that projection onto the lowest Landau level turns the physical plane into a two-dimensional phase space, in which each single-particle quantum state occupies an area equal to $2\pi l^2$. In the strong-field limit, this area goes to zero as $1/B$, so we have a kind of classical limit, in which we can neglect the non-commutation between the two guiding-centre coordinates $\hat{R}_x$ and $\hat{R}_y$. Going away from this limit yields naturally a gradient expansion in which the small parameter is nl^2, where n is the average topological charge density. The second use of coherent-state quantization is at the many-particle level. We can indeed view the Slater determinants $|\mathcal{S}_\psi\rangle$ as coherent states for the many-fermion problem, spanning a low-energy subspace within the fermionic Fock space. Although the mathematical aspects here are not as clear to us as for the single-particle level, this viewpoint can still provide a useful intuition for the many-body problem [35, 36], and will motivate our subsequent treatment of collective dynamics in Section 11.4.

11.2 Physical properties of spin textures

11.2.1 Intuitive picture

Before beginning our discussion of textures in quantum Hall systems, it is useful to recall a few aspects of the physics of a single charge e particle moving on a 2D plane in a strong uniform magnetic field. In this limit, the particle undergoes a fast cyclotron motion with characteristic frequency $\omega_c = eB/m$. In the absence of disorder, the corresponding energy spectrum is quantized according to $E_n = \hbar\omega_c(n + \frac{1}{2})$, with n a non-negative integer. Each of these Landau levels is infinitely degenerate in the thermodynamic limit. The physical origin of this degeneracy can be understood in classical terms: in the absence of disorder, the centre of cyclotron orbits can be located anywhere on the plane, and all these locations give the same energy. The coordinates R_x and R_y of this so-called guiding centre are

$$R_x = \frac{x}{2} + \frac{p_y}{eB}, \qquad R_y = \frac{y}{2} - \frac{p_x}{eB}. \tag{11.1}$$

Quantum mechanically, they become operators, $\hat{R}_x$ and $\hat{R}_y$, whose commutator is

$$[\hat{R}_r, \hat{R}_y] = -il^2. \tag{11.2}$$

In the limit of a very strong magnetic field, the cyclotron gap $\hbar\omega_c$ is larger than other energy scales, in particular the scale e^2/l associated with Coulomb interactions, so we may project all the single-particle states onto the highest occupied Landau level. To simplify the discussion, we shall assume that the electronic filling factor ν is less than d, so this Landau level is the lowest one, corresponding to $n = 0$ and often denoted by the acronym LLL. Physically, the only remaining degrees of freedom for a given electron are, besides its d internal levels, its guiding-centre coordinates $\hat{R}_x$ and $\hat{R}_y$. Equation (11.2) shows that they behave as a canonically conjugate pair of observables. This turns the physical plane into a 2D phase-space, with an effective Planck's constant $\hbar_{\text{eff}} = l^2$. Heisenberg's uncertainty principle suggests then that each quantum state in the LLL occupies an area $2\pi l^2$, so the degeneracy of the LLL should be equal to the system area divided by $2\pi l^2$, which is equal to the total magnetic flux through the system divided by the flux quantum $\Phi_0 = h/e$: we may also say that each single-particle state in the LLL occupies the area corresponding to one flux quantum. This intuitive estimate turns out to give the exact degeneracy for each Landau level.

On physical grounds, the fact that a single charge added to or removed from a fully polarized state at $\nu = 1$ binds a spin texture is relatively easy to understand. To be specific, let us remove one electron. The remaining $N - 1$ electrons would lower their electrostatic energy if they could benefit from the created hole by moving slightly away from each other. However, as we have just discussed, the physical plane becomes like a phase space, with each single-particle state occupying an area threaded by one flux quantum. The natural way to create a smooth distribution of the remaining $N - 1$ electrons would be to remove one flux quantum through the whole plane. In the absence of a Zeeman spin anisotropy, this can be done with a small energy cost by twisting the spins slowly. The Berry phase associated with the spin texture creates an artificial effective magnetic field, which adds to the physical external field.

To formulate the previous remarks in more precise language, we describe the spin texture by a smooth d-component spinor field $\psi_a(r)$ $(1 \leq a \leq d)$, which is expected to have only small variations on the scale of the magnetic length. In this presentation, I will often denote this spinor field by $|\psi(r)\rangle$. This notation is suggestive and convenient, although it has a potential risk of confusion, even at the single-particle level, because of the presence of the r variable inside the ket. I trust the reader not to be misled by this choice of notation. We suppose that the quantum state of the $(N-1)$-electron system can be described by a Slater determinant obtained from a collection of single-particle orbitals $|\Phi_\alpha\rangle$ $(1 \leq \alpha \leq N-1)$ that have the form

$$\Phi_{\alpha,a}(r) = \chi_\alpha(r)\psi_a(r), \tag{11.3}$$

where $\chi_\alpha(r)$ describes the remaining orbital degree of freedom, once the spins are constrained to follow the prescribed texture $\psi_a(r)$. We wish first to minimize the kinetic energy of this state. A good guide is to minimize separately the kinetic energy of the individual orbitals $|\Phi_\alpha\rangle$. A simple calculation shows that

$$\langle\Phi_\alpha|(P-eA)^2|\Phi_\alpha\rangle = \langle\chi_\alpha|(P-eA_{\text{eff}})^2 + V_{\text{eff}}|\chi_\alpha\rangle, \tag{11.4}$$

with

$$A_{\text{eff}} = A - \frac{\Phi_0}{2\pi}\mathcal{A}, \qquad \mathcal{A} = \frac{1}{i}\langle\psi|\nabla|\psi\rangle, \tag{11.5}$$

and

$$V_{\text{eff}} = \langle\nabla\psi|\nabla\psi\rangle - \langle\nabla\psi|\psi\rangle\langle\psi|\nabla\psi\rangle. \tag{11.6}$$

$\mathcal{A}$ is often referred to as the *Berry connection* associated with the spin texture $|\psi(r)\rangle$. The above expressions (11.5) and (11.6) are valid provided the spinor field $|\psi(r)\rangle$ is everywhere normalized to unity, i.e. we impose $\langle\psi(r)|\psi(r)\rangle = 1$ for all r. The total kinetic energy is minimized by putting all the effective orbitals $\chi_\alpha(r)$ in the lowest Landau level of the effective magnetic field corresponding to A_{eff}, in the presence of the potential V_{eff}. Because V_{eff} is small for a slowly varying texture, we expect that its effect is subleading compared with A_{eff}. This level contains N_{eff} states, where N_{eff} is simply equal to the number of flux quanta of the effective magnetic field through the system. Now, it is an important fact that the total flux associated with the Berry connection is 2π times an integer N_{top} called the *total topological charge.* To be more precise, this sharp 'quantization' holds when the spin texture goes to a constant value $|\psi(\infty)\rangle$ far from the origin in all directions. This allows us to compactify the plane, which is topologically trivial (it can be shrunk smoothly into a point), into a two-dimensional sphere S^2 , which has a non-trivial topology.

At this point, it is important to note that there is clearly a gauge freedom in the definition of the spinor field $|\psi(r)\rangle$. As shown by (11.3), a change of $|\psi(r)\rangle$ into $e^{i\theta(r)}|\psi(r)\rangle$, where the phase $\theta(r)$ is an arbitrary function, can be compensated by the related change of $\chi_\alpha(r)$ into $e^{-i\theta(r)}\chi_\alpha(r)$, so that the physical orbitals $|\Phi_\alpha\rangle$ remain unchanged. It is then more appropriate to view $|\psi(r)\rangle$ as a representative of the complex line it generates in the complex space $\mathbb{C}^d$. In mathematical terms, the spinor field $|\psi(r)\rangle$ should be viewed as a map from S^2 to the complex projective space $\mathbb{C}P^{d-1}$. This map can be used to define a line bundle over S^2, whose Chern number is the topological charge $N_{\rm top}$. Coming back to the main discussion, we therefore have the very important relation

$$N_{\rm eff} = N - N_{\rm top}. \tag{11.7}$$

If we wish to describe a system with one hole, we need to get $N_{\rm eff} = N - 1$, so that $N_{\rm top} = 1$. Likewise $N_{\rm top} = -1$ for an added electron. This shows that topologically non-trivial textures (called skyrmions) are bound to extra charges, the driving force being the Coulomb interaction.

Before going further, let us give an illustration of a single-skyrmion texture in the $d = 2$ case. Then, we can associate with each spinor (ψ_1, ψ_2) the expectation value $\mathbf{n}(r) = \langle\psi(r)|\boldsymbol{\sigma}|\psi(r)\rangle$, where $\boldsymbol{\sigma} = (\sigma_x, -\sigma_y, \sigma_z)$ denotes the usual Pauli spin matrices. We can use this map from $\mathbb{C}P^1$ to S^2 to construct a spin texture if we choose a map from the physical plane to $\mathbb{C}P^1$. For $r = (x, y)$ in the plane, let us pick $\psi_1 = z$ and $\psi_2 = 1$, where $z = x + iy$. At the origin $z = 0$, $\mathbf{n}(0) = -\mathbf{e}_z$, and, far from it, $\mathbf{n}(\infty) \to \mathbf{e}_z$. Explicitly, we get

$$\mathbf{n}(x,y) = \left(\frac{2x}{|r|^2+1}, \frac{2y}{|r|^2+1}, \frac{|r|^2-1}{|r|^2+1}\right), \qquad |r|^2 = x^2 + y^2. \tag{11.8}$$

This map from the plane to S^2 is nothing but the inverse of the stereographic projection of S^2 to the $z = 0$ plane from the north pole. To find $\mathbf{n}(x, y)$, we simply draw the line joining the point $(x, y, 0)$ to the north pole $(0, 0, 1)$. This line intersects the unit sphere $x^2 + y^2 + z^2 = 1$ at a second point, besides the north pole, which is simply $\mathbf{n}(x, y)$. This construction is depicted on Fig. 11.1. It shows clearly that each spin value

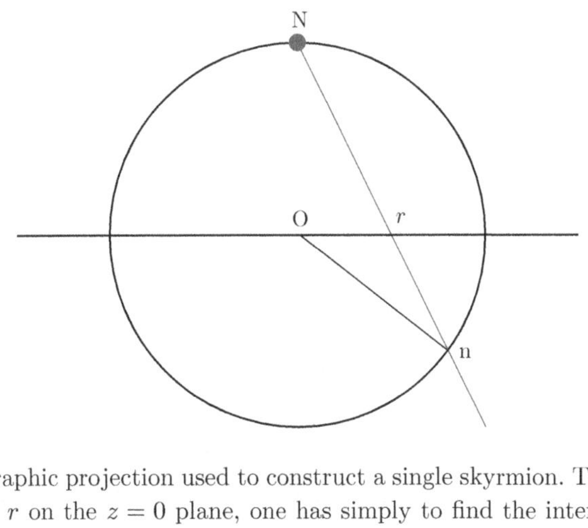

Fig. 11.1 Stereographic projection used to construct a single skyrmion. To find the spin orientation $\mathbf{n}$ at a point r on the $z = 0$ plane, one has simply to find the intersection between the unit sphere and the line joining r to the north pole N.

on S^2 (with the exception of the north pole) is reached exactly once under the inverse stereographic projection. We also note that when $|r| = 1$, $\mathbf{n}(x, y) = (x, y, 0)$, so the spin configuration, restricted to the unit circle, coincides with a 2π vortex. However, unlike the vortex, which has a singular core and no well-defined limit as $|r| \to \infty$, the texture is everywhere smooth and reaches the north pole if $|r| \to \infty$ along all possible directions. This qualitative difference with a 2π vortex is obtained by allowing spins to move away from the $z = 0$ plane.

Although physically appealing, the previous discussion is not totally satisfactory. Its main problem is that minimizing the expectation value $\langle \chi_\alpha | (P - eA_{\text{eff}})^2 + V_{\text{eff}} | \chi_\alpha \rangle$ doesn't imply that the single-particle states $|\Phi_\alpha\rangle$ belong to the lowest Landau level (corresponding to the *physical* magnetic field). This problem disappears if one considers, like Pasquier [29, 30], the case of a positive topological charge and a subset of textures in which the components $\psi_a(r)$ of $|\psi(r)\rangle$ are *analytic* functions of $z = x + iy$. Because the lowest Landau level corresponds (in the circular gauge) to wavefunctions of the form $\psi(r) = f(z) \exp(-|z|^2/4l^2)$ with $f(z)$ analytic, we see that if the orbital parts $\chi_\alpha(r)$ are of this form, then multiplying them by an analytic spinor $\psi_a(r)$ will produce single-particle states $\Phi_{\alpha,a}(r)$ that are also in the lowest Landau level. As shown by MacDonald, Fertig, and Brey [28], Slater determinants associated with such textures are exact ground states for a model with $N_{\text{el}} < N$ electrons and a point-like interaction. However, it is not always sufficient to restrict ourselves to analytic textures. The first reason is that in the presence of spin anisotropies, or even isotropic but long-range interactions, the optimal textures are no longer analytic. Another reason is that we wish to keep the possibility of studying excited states, that live outside the analytic subspace. In spite of these restrictions, these works by Pasquier have brought important ideas, in particular the realization that we can use tools from geometric quantization to analyse physical properties of quantum Hall textures. In these

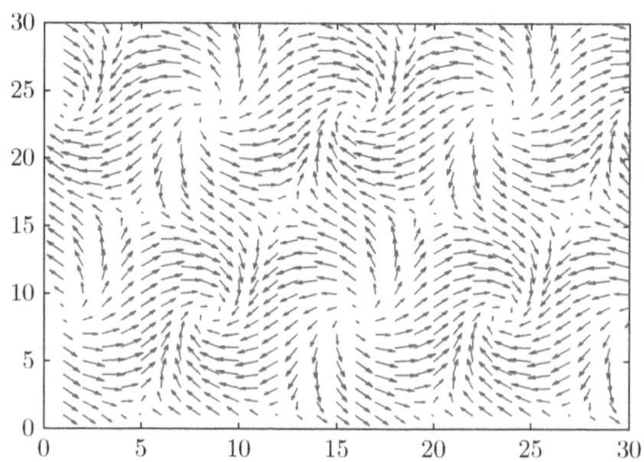

Fig. 11.2 A periodic spin texture for the $SU(2)$ case ($d = 2$). The projection of the spins on the horizontal plane is shown. This triangular pattern corresponds to the optimal variational state described in Section 11.3.4.

lectures, I will show that geometric quantization is also a very useful tool to construct a large class of textures, not limited to analytic functions.

11.2.2 Construction of spin textures

Guided by the previous discussion, we would like to find a way to associate a Slater determinant $|\mathcal{S}_\psi\rangle$ for N_e electrons in the lowest Landau level with a smooth spinor field $\psi_a(r)$ $(1 \leq a \leq d)$. On physical grounds, this could be achieved by switching on the following auxiliary Zeeman-like Hamiltonian acting on a single-particle wavefunction $\phi_a(r)$:

$$(\hat{H}_{\psi,\mathrm{cl}}\phi)_a(r) = -\psi_a(r)\sum_{b=1}^{d}\psi_b^*(r)\phi_b(r). \tag{11.9}$$

The ground state of $\hat{H}_{\psi,\mathrm{cl}}$ is infinitely degenerate, being composed of all the single-particle states of the form $\phi_a(r) = f(r)\psi_a(r)$, where $f(r)$ is an arbitrary function with complex values. Excited eigenstates of $\hat{H}_{\psi,\mathrm{cl}}$ are also infinitely degenerate, since they correspond to wavefunctions $\phi_a(r)$ that are orthogonal at every point r to the local spinor $\psi_a(r)$; that is, $\sum_{b=1}^{d}\psi_b^*(r)\phi_b(r) = 0$ everywhere. In the following discussion, it will be convenient to view single-particle Hamiltonians as $d \times d$ matrices whose entries are operators acting on the orbital part of the wavefunction. With this convention, we may write $\hat{H}_{\psi,\mathrm{cl}}$ as

$$(\hat{H}_{\psi,\mathrm{cl}})_{ab} = -\psi_a(r)\psi_b^*(r), \qquad 1 \leq a, b \leq d. \tag{11.10}$$

It turns out that this is not quite the construction that we need, because it doesn't take into account the projection on the lowest Landau level. Denoting by $\mathcal{P}_{\mathrm{LLL}}$ the self-adjoint projector onto this level, a natural candidate to create a texture would come from looking at the ground state of $\hat{H}_\psi = \mathcal{P}_{\mathrm{LLL}}\hat{H}_{\psi,\mathrm{cl}}\mathcal{P}_{\mathrm{LLL}}$, or, with the previous notation,

$$(\hat{H}_\psi)_{ab} = -\mathcal{P}_{\mathrm{LLL}}\psi_a(r)\psi_b^*(r)\mathcal{P}_{\mathrm{LLL}}, \qquad 1 \leq a, b \leq d. \tag{11.11}$$

As we explained in Section 11.2.1, the projection on the lowest Landau level turns the physical plane into a two-dimensional phase space, in which coordinates R_x and R_y become canonically conjugated. In this quantization process, the role of Planck's constant $\hbar$ is played by the square of the magnetic length l, in tune with the general picture that each quantum state occupies an area $2\pi\hbar$ in phase space. The limit of large magnetic fields, where l is much smaller than the characteristic length associated with the spatial variations of the texture field $\psi_a(r)$, can therefore be viewed as a *semiclassical limit*. In this limit, we expect the following qualitative properties for the spectrum of $\hat{H}_\psi$ [37]: the two degenerate levels with eigenvalues -1 and 0 are replaced by two bands whose widths are at most proportional to l^2, containing $N - N_{\mathrm{top}}$ and $(d-1)N + N_{\mathrm{top}}$ states, respectively, and separated by a well-defined gap. It is then natural to define the quantum state associated with the smooth texture $\psi_a(r)$ as the Slater determinant $|\mathcal{S}_\psi\rangle$ composed of the $N - N_{\mathrm{top}}$ single-particle states lying in the lowest band of $\hat{H}_\psi$. The goal of this subsection is to investigate the physical properties of such states. With this purpose in mind, all the relevant information is encoded in the projector $\hat{P}_\psi$ on this lowest band. Our first task is then to write explicit expressions for $\hat{P}_\psi$ in terms of the texture $\psi_a(r)$.

The two equations that we wish to solve are

$$[\hat{H}_\psi, \hat{P}_\psi] = 0, \tag{11.12}$$

$$\hat{P}_\psi \hat{P}_\psi = \hat{P}_\psi. \tag{11.13}$$

The main difficulty here is that we wish to diagonalize a $d \times d$ matrix $(\hat{H}_\psi)_{ab}$ whose elements are themselves operators, rather than numbers. A very important remark is that these operators do commute in the classical limit $l^2 \to 0$, so the usual diagonalization methods (designed to work with commuting numbers) can be applied there. One can hope then that the non-commuting nature of the elements $(\hat{H}_\psi)_{ab}$ can be dealt with in a semiclassical expansion. That this is indeed possible is well known in the mathematical literature, where it has been shown that the projector $\hat{P}_\psi$ can be constructed as a formal power series in l^2 [38, 39].

In doing these calculations, it will be useful to make use of the correspondence principle between classical and quantum mechanics, to represent operators such as $(\hat{H}_\psi)_{ab}$ and $(\hat{P}_\psi)_{ab}$ in terms of functions of the underlying phase-space coordinates x and y. It is a general fact that there are many ways to represent operators by functions over classical phase space in such a way that standard classical mechanics emerges as a limit of this quantization process when $\hbar \equiv l^2 \to 0$. Rather than being a problem, this ambiguity has positive aspects, because it allows us to choose the precise correspondence which simplifies our calculations. In the present problem, we shall use the so-called covariant symbol in Berezin's terminology [34], also called the Husimi distribution in the quantum optics and quantum chaos communities.

For a given operator $\hat{f}$ acting in the lowest Landau level, the associated covariant symbol $f(z, \bar{z})$ is simply the expectation value of $\hat{f}$ taken on the normalized coherent state $|\Phi_{\bar{z}}\rangle$ centred at (x, y), with $z = x + iy$. To bring more fluidity to this presentation, the precise definition and elementary properties of these coherent states are relegated to Appendix 11.A. There, we also show that a given operator is uniquely determined by its covariant symbol, from which it can be explicitly constructed through the normal-ordering procedure.

The non-commuting algebra of quantum operators can be represented as a deformation of the commuting algebra of functions over phase space. To see this, we need to know the covariant symbol of the product of two operators. It is common to write $\hat{f}\hat{g} = \widehat{f \star g}$, where the explicit formula for the star product is

$$f \star g = f \exp\left[-i\frac{l^2}{2}\left(\overleftarrow{\partial_x}\overrightarrow{\partial_y} - \overleftarrow{\partial_y}\overrightarrow{\partial_x}\right) + \frac{l^2}{2}\left(\overleftarrow{\partial_x}\overrightarrow{\partial_x} + \overleftarrow{\partial_y}\overrightarrow{\partial_y}\right)\right] g. \tag{11.14}$$

This may be represented as a power series in l^2:

$$f \star g = \sum_{n=0}^{\infty} l^{2n} f \star_n g \tag{11.15}$$

with

$$f \star_0 g = fg, \tag{11.16}$$

$$f \star_1 g = -\frac{1}{2}i\{f,g\} + \frac{1}{2}\nabla f \cdot \nabla g. \tag{11.17}$$

The Poisson bracket is defined as usual by

$$\{f,g\} = \frac{\partial f}{\partial x}\frac{\partial g}{\partial y} - \frac{\partial f}{\partial y}\frac{\partial g}{\partial x}. \tag{11.18}$$

Equations (11.16) and (11.17) imply that

$$[\hat{f},\hat{g}] = -il^2\widehat{\{f,g\}} + \mathcal{O}(l^4), \tag{11.19}$$

which is an expression of the correspondence principle between classical and quantum mechanics.

With this choice of correspondence between functions and operators, it is natural to modify accordingly the definition of the single-particle Hamiltonian $\hat{H}_\psi$ and to replace (11.11) by

$$(\hat{H}_\psi)_{ab} = -\widehat{\psi_a(r)\psi_b^*(r)}, \qquad 1 \le a,b \le d. \tag{11.20}$$

This new definition does not have any effect on the physics, because it simply modifies the way we parametrize Slater determinants $|\mathcal{S}_\psi\rangle$ in terms of classical spinor fields $\psi_a(r)$.

Let us write then $\hat{P}_\psi = \hat{P}_0 + l^2\hat{P}_1 + \mathcal{O}(l^4)$, and compute the first two terms $\hat{P}_0$ and $\hat{P}_1$ in the semiclassical expansion of $\hat{P}_\psi$. We have dropped the ψ subscript on $\hat{P}_0$ and $\hat{P}_1$ to lighten the notation. From now on, we shall replace operators like $\hat{H}_\psi$ and $\hat{P}_\psi$ by their $d \times d$ matrix symbols H_ψ and P_ψ. For two such matrix symbols $A \equiv A_{ij}(r)$ and $B \equiv B_{ij}(r)$, the matrix star product $A \star B$ is defined by the usual matrix multiplication rule, but with the ordinary product replaced by the star product, i.e.

$$(A \star B)_{ik} = \sum_j A_{ij} \star B_{jk} \tag{11.21}$$

Similarly, we define

$$[A,B]_\star = A \star B - B \star A. \tag{11.22}$$

To zeroth order in l^2, (11.12) and (11.13) give

$$[H_\psi, P_0] = 0, \tag{11.23}$$

$$P_0 P_0 = P_0. \tag{11.24}$$

This is the standard diagonalization problem for a Hermitian matrix. Because we are interested in the eigenvalue of $\hat{H}_\psi$ that goes to -1 in the classical limit, we choose

$$(P_0)_{ij} = \psi_i \psi_j^*. \tag{11.25}$$

Here, we have assumed that the local spinor field is everywhere normalized, i.e. $\sum_j |\psi_j(r)|^2 = 1$ for any r.

The first-order terms in (11.12) and (11.13) read

$$[H_\psi, P_1] + [H_\psi, P_0]_1 = 0, \tag{11.26}$$

$$P_0 P_1 + P_1 P_0 + P_0 \star_1 P_0 = P_1. \tag{11.27}$$

Here, we have used the notation

$$[A, B]_\star = \sum_{n=0}^{\infty} l^{2n} [A, B]_n. \tag{11.28}$$

These equations can be seen as ordinary matrix equations for P_1:

$$[H_\psi, P_1] = -[H_\psi, P_0]_1, \tag{11.29}$$

$$P_1 - P_0 P_1 - P_1 P_0 = P_0 \star_1 P_0. \tag{11.30}$$

Because $P_0 = -H_\psi$, $[H_\psi, P_0]_\star = 0$, which implies $[H_\psi, P_0]_1 = 0$. Therefore, P_1 has to commute with P_0. Multiplying (11.30) by P_0 on both sides and subtracting the two results gives a necessary condition for the existence of P_1:

$$P_0(P_0 \star_1 P_0) = (P_0 \star_1 P_0) P_0. \tag{11.31}$$

This condition is always satisfied, because the star product is associative: starting from $P_0 \star (P_0 \star P_0) = (P_0 \star P_0) \star P_0$ and keeping the first-order term in l^2 gives exactly (11.31). Using this property, it is easy to find that P_1 is given by

$$P_1 = (1 - 2P_0)(P_0 \star_1 P_0). \tag{11.32}$$

This is the most important result of this section. We emphasize that it holds for an arbitrary smooth spinor field $\psi_a(r)$. The small parameter in the expansion of P_ψ is just the ratio between the typical length over which $\psi_a(r)$ varies and the magnetic length l. Let us check that this result allows us to compute the local charge density bound to the texture. From (11.32) and (11.17), using the classical expression (11.25) for P_0 gives

$$\begin{aligned}(P_1)_{jk} = &-\frac{1}{2} i(\{\psi_j, \psi_k^*\} + \{\psi_j, \psi_l\}\psi_l^* \psi_k^* + \psi_j \psi_l \{\psi_l^*, \psi_k^*\} + \psi_j \{\psi_l, \psi_l^*\} \psi_k^*) \\ &+ \frac{1}{2} \nabla \psi_j \cdot \nabla \psi_k^* + \frac{1}{2} (\nabla \psi_j \cdot \nabla \psi_l) \psi_l^* \psi_k^* + \frac{1}{2} \psi_j \psi_l (\nabla \psi_l^* \cdot \nabla \psi_k^*) \\ &- \frac{1}{2} \psi_j [\nabla \psi_l . \nabla \psi_l^* - 2\psi_l (\nabla \psi_m \cdot \nabla \psi_l^*) \psi_m^*] \psi_k^*.\end{aligned} \tag{11.33}$$

It can be checked that this expression is invariant under local gauge transformations. A direct consequence is

$$\mathrm{Tr}[P_1(r)] = -\mathcal{B}(r) = -2\pi Q(r). \tag{11.34}$$

Here, $\mathcal{B} = \partial_x \mathcal{A}_y - \partial_y \mathcal{A}_x$ is the curvature of the Berry connection, and $Q(r)$ is the topological charge density associated with the ψ texture. The local particle density

$$\rho(r) = \sum_a \langle \psi_a^+(r)\psi_a(r)\rangle \tag{11.35}$$

is equal to $(1/2\pi l^2)\mathrm{Tr}(P_\psi(r))$, as can be seen from (11.136) in Appendix 11.C. Therefore,

$$\rho(r) = \frac{1}{2\pi l^2} - Q(r) + \mathcal{O}(l^2). \tag{11.36}$$

This is a local form of (11.7), because $1/2\pi l^2 = B/\Phi_0$ is the particle density in a filled Landau level. Because the integrated result, (11.7), is a relation between integers, we expect that the spatial integral of all higher-order corrections in l^2 to $\rho(r)$ are equal to zero.

11.2.3 Energetics of spin textures

We consider the usual $SU(d)$-symmetric Hamiltonian with two-body potential $V(r-r')$:

$$H = \frac{1}{2}\sum_{ab}\int d^2r \int d^2r'\, V(r-r')\Psi_a^+(r)\Psi_b^+(r')\Psi_b(r')\Psi_a(r), \tag{11.37}$$

where the single-particle creation and annihilation operators are projected onto the lowest Landau level, as detailed in Appendix 11.C. In the Slater determinant associated with the classical texture $\psi_a(r)$, we can use Wick's theorem to evaluate the expectation value of H, and we get $\langle H\rangle_\psi = \langle H\rangle_{H,\psi} + \langle H\rangle_{F,\psi}$, with

$$\langle H\rangle_{H,\psi} = \frac{1}{2}\sum_{ab}\int d^2r \int d^2r'\, V(r-r')\langle\Psi_a^+(r)\Psi_a(r)\rangle_\psi\langle\Psi_b^+(r')\Psi_b(r')\rangle_\psi, \tag{11.38}$$

$$\langle H\rangle_{F,\psi} = -\frac{1}{2}\sum_{ab}\int d^2r \int d^2r'\, V(r-r')\langle\Psi_a^+(r)\Psi_b(r')\rangle_\psi\langle\Psi_b^+(r')\Psi_a(r)\rangle_\psi. \tag{11.39}$$

Using (11.138) from Appendix 11.C, we can express these energies in terms of the matrix symbol $P_\psi(r)$ as

$$\langle H\rangle_{H,\psi} = \frac{1}{2(2\pi l^2)^2}\int d^2r \int d^2r'\, V(r-r')\,\mathrm{Tr}[P_\psi(r)]\,\mathrm{Tr}[P_\psi(r')], \tag{11.40}$$

$$\langle H\rangle_{F,\psi} = -\frac{1}{2(2\pi l^2)^2}\int d^2r \int d^2r'\, V(r-r')\,\mathrm{Tr}[P_\psi(s(r,r'))P_\psi(s(r',r))] \times \exp\left[-\frac{(r-r')^2}{2l^2}\right]. \tag{11.41}$$

Here, we have set

$$s(r, r') = \frac{r + r'}{2} + \frac{i}{2}\hat{z} \times (r' - r). \tag{11.42}$$

Let us first consider the case of a point-like interaction $V(r - r') = W\delta(r - r')$. We get

$$\langle H \rangle_\psi = \frac{W}{2(2\pi l^2)^2} \int d^2r \, (\{\mathrm{Tr}[P_\psi(r)]\}^2 - \mathrm{Tr}[P_\psi(r)^2]). \tag{11.43}$$

For a fully polarized system, we can choose a basis in internal space such that $(P_\psi)_{ab}(r) = \delta_{a1}\delta_{b1}f(r)$, and $\langle H \rangle_\psi = 0$ as expected, because the orbital wavefunction is completely antisymmetric, so it is impossible for two particles to be at the same point. Expanding the integrand in (11.43), we get

$$\{\mathrm{Tr}[P_\psi(r)]\}^2 - \mathrm{Tr}[P_\psi(r)^2] = 2l^2[\mathrm{Tr}\, P_1 - \mathrm{Tr}(P_0 P_1)] + \mathcal{O}(l^4). \tag{11.44}$$

From (11.33), we find

$$\mathrm{Tr}(P_0 P_1) = -\frac{1}{2}\mathcal{B} - \frac{1}{2}(\langle \nabla\psi|\nabla\psi\rangle - \langle \nabla\psi|\psi\rangle\langle\psi|\nabla\psi\rangle). \tag{11.45}$$

Putting everything together gives

$$\langle H \rangle_\psi = -\frac{W}{4\pi l^2} N_{\mathrm{top}} + \frac{W}{8\pi^2 l^2} \int d^2r \, (\langle \nabla\psi|\nabla\psi\rangle - \langle \nabla\psi|\psi\rangle\langle\psi|\nabla\psi\rangle). \tag{11.46}$$

We now turn to the physically important case of Coulomb interaction, $V(r) = e^2/4\pi\epsilon r$. In fact, the Hartree contribution is easily written for an arbitrary interaction. Equation (11.38) becomes, taking (11.36) into account,

$$\langle H \rangle_{H,\psi} = \frac{1}{2}\int d^2r \int d^2r' \, V(r - r') \left[\frac{1}{2\pi l^2} - Q(r) + \mathcal{O}(l^2)\right] \left[\frac{1}{2\pi l^2} - Q(r') + \mathcal{O}(l^2)\right]. \tag{11.47}$$

Therefore,

$$\langle H \rangle_{H,\psi} = \frac{N - 2N_{\mathrm{top}}}{4\pi l^2} \tilde{V}(k = 0) + \frac{1}{2}\int d^2r \int d^2r' \, V(r - r') Q(r) Q(r'). \tag{11.48}$$

We may object that the last term is not the only $\mathcal{O}(l^0)$ contribution to $\langle H \rangle_{H,\psi}$. We should in principle include $\mathcal{O}(l^2)$ terms in the local particle density. However, because the density in a filled Landau level is spatially uniform, such terms will appear in $\langle H \rangle_{H,\psi}$ only through their integral over the whole plane. As we have discussed at the end of Section 11.2.2, these integrals are expected to vanish because the total number of electrons, $N_e = N - N_{\mathrm{top}}$, contains only the first $\mathcal{O}(l^0)$ correction due to the topological charge of the texture.

Let us now consider the Fock term. Because the Gaussian kernel in (11.41) is sharply peaked around the origin, with a characteristic length equal to l, it is natural to expand the trace $\mathrm{Tr}[P_\psi(s(r,r'))P_\psi(s(r',r))]$ around $r = r'$. The following integrals are useful:

$$\int dx \int dy \frac{1}{\sqrt{x^2+y^2}} \exp\left(-\frac{x^2+y^2}{2l^2}\right) = \pi\sqrt{2\pi}\, l, \tag{11.49}$$

$$\int dx \int dy \frac{x^2}{\sqrt{x^2+y^2}} \exp\left(-\frac{x^2+y^2}{2l^2}\right) = \pi\sqrt{\frac{\pi}{2}}\, l^3. \tag{11.50}$$

Keeping the first two terms in the semiclassical expansion of $\langle H\rangle_{F,\psi}$ gives

$$\langle H\rangle_{F,\psi} = -\frac{e^2}{16\pi\sqrt{2\pi}\,\epsilon l^3}\int d^2r \left[1 + 2l^2\,\mathrm{Tr}(P_0P_1) + \frac{1}{4}l^2\,\mathrm{Tr}(\nabla P_0\cdot\nabla P_0)\right]. \tag{11.51}$$

Finally, using (11.45), the Fock contribution for the Coulomb interaction reads

$$\langle H\rangle_{F,\psi} = -\frac{e^2}{8\sqrt{2\pi}\,\epsilon l}(N - N_{\mathrm{top}}) + \frac{e^2}{32\pi\sqrt{2\pi}\,\epsilon l}\int d^2r\,(\langle\nabla\psi|\nabla\psi\rangle - \langle\nabla\psi|\psi\rangle\langle\psi|\nabla\psi\rangle). \tag{11.52}$$

Note that the first term has the expected form for the Fock contribution in a fully polarized system, which is equivalent to a system of spinless fermions. It is negative and proportional to the particle number, in agreement with the physical interpretation that it removes the Coulomb self-interaction of all particles present in the system. The value of the stiffness in the second term is consistent with the value given in the literature [4, 15], in which the Coulomb interaction potential is often written as $V(r) = e^2/\epsilon r$ (Gaussian units).

11.2.4 Choice of an effective model

So far, we have computed properties of a single Slater determinant $|\mathcal{S}_\psi\rangle$ associated with a smooth spinor ψ. Here, we would like to construct an effective model by considering the family $\mathcal{M}$ of such Slater determinants as an (overcomplete) basis of low-energy states. An essential ingredient is the overlap $\langle\mathcal{S}_\psi|\mathcal{S}_{\psi'}\rangle$ between two states in $\mathcal{M}$. The reader is invited to check that the modulus of this overlap can be simply expressed in terms of the corresponding projectors $\hat{P}_\psi$ and $\hat{P}_{\psi'}$:

$$|\langle\mathcal{S}_\psi|\mathcal{S}_{\psi'}\rangle|^2 = \mathrm{Det}(I - \hat{P}_\psi\hat{P}_{\psi'}\hat{P}_\psi). \tag{11.53}$$

This implies that

$$\log|\langle\mathcal{S}_\psi|\mathcal{S}_{\psi'}\rangle|^2 = \int \frac{d^2r}{2\pi l^2}\,\mathrm{Tr}_{\mathbb{C}^d}[\text{Cov. Symb. }\log(I - \hat{P}_\psi\hat{P}_{\psi'}\hat{P}_\psi)]. \tag{11.54}$$

The leading order in l^2 is easily extracted by taking the principal symbols, that is, by making the approximation $P_\psi \simeq |\psi(r)\rangle\langle\psi(r)|$. At this leading order, the star product becomes the ordinary product, and we have

$$\log|\langle\mathcal{S}_\psi|\mathcal{S}_{\psi'}\rangle|^2 = \int \frac{d^2r}{2\pi l^2}\,\mathrm{Tr}_{\mathbb{C}^d}\log\left\{I - [1 - |\langle\psi(r)|\psi'(r)\rangle|^2]|\psi(r)\rangle\langle\psi(r)|\right\} + \mathcal{O}(l^2). \tag{11.55}$$

Finally,

$$\log |\langle \mathcal{S}_\psi | \mathcal{S}_{\psi'} \rangle|^2 = \int \frac{d^2r}{2\pi l^2} \log |\langle \psi(r) | \psi'(r) \rangle|^2 + \mathcal{O}(l^2). \tag{11.56}$$

This form of the overlap is compatible with a lattice model where each of its N sites hosts a quantum degree of freedom with d independent internal states. Furthermore, Slater determinants associated with textures can be seen as coherent states. To substantiate this claim, let us consider the Hilbert space that is a tensor product of N copies of $\mathbb{C}^d$. Coherent states can be defined as factorizable states of the form $|\mathcal{C}_\psi\rangle = |\psi(1)\rangle \otimes |\psi(2)\rangle \otimes \cdots \otimes |\psi(N)\rangle$, where $|\psi(i)\rangle \in \mathbb{C}^d$ for all i. Note that we have here a redundancy because changing $|\psi(j)\rangle$ into $e^{i\theta(j)}|\psi(j)\rangle$ modifies only the global phase of $|\mathcal{C}_\psi\rangle$. There are least two ways to deal with this. The first is to lift this ambiguity by fixing a gauge. For example, one may impose that $\psi_1(j)$ be real and positive. The problem with this prescription is that it is useless when $\psi_1(j)$ is equal to zero. To cover the whole projective space $\mathbb{C}P^{d-1}$, we have to consider for each site j at least d open subsets characterized by $\psi_i(j) \neq 0$ for $1 \leq i \leq d$ and patch them together. This procedure is mathematically clean, although it may not be the most convenient for practical calculations. The second way, which we will follow, is to work with unconstrained spinors, but keep in mind that all physical properties are invariant under local gauge transformations.

The overlap between two such states is given by

$$\langle \mathcal{C}_\psi | \mathcal{C}_{\psi'} \rangle = \prod_{i=1}^{N} \langle \psi(i) | \psi'(i) \rangle. \tag{11.57}$$

Taking the squared modulus and transforming slightly, we get

$$\log |\langle \mathcal{C}_\psi | \mathcal{C}_{\psi'} \rangle|^2 = \sum_{i=1}^{N} \log |\langle \psi(i) | \psi'(i) \rangle|^2. \tag{11.58}$$

In the semiclassical limit of slowly varying textures, the discrete texture $|\psi(i)\rangle$ becomes a smooth one $|\psi(r)\rangle$ and we can approximate the sum by an integral, which gives exactly the same leading term as in (11.56) previously derived.

This discussion motivates the following semiclassical model to describe a system containing $N_{\text{eff}} = N - N_{\text{top}}$ electrons in the lowest Landau level. The semiclassical limit is reached when the typical distance between nearby skyrmions is large compared with the magnetic length, that is, when $N_{\text{top}} << N$. The expectation value of the Hamiltonian on Slater determinants detailed in Section 11.2.3 suggests the following form for the energy in this effective model:

$$\langle H \rangle_\psi \equiv \langle \mathcal{C}_\psi | H | \mathcal{C}_\psi \rangle = \langle H \rangle_{\text{ex},\psi} + \langle H \rangle_{\text{el},\psi}, \tag{11.59}$$

where

$$\langle H \rangle_{\text{ex},\psi} = E_{\text{ex}} \int d^2r \left(\frac{\langle \nabla\psi | \nabla\psi \rangle}{\langle \psi | \psi \rangle} - \frac{\langle \nabla\psi | \psi \rangle \langle \psi | \nabla\psi \rangle}{\langle \psi | \psi \rangle^2} \right) \tag{11.60}$$

expresses the short-range part of the Coulomb interactions (exchange energy) and

$$\langle H\rangle_{\mathrm{el},\psi} = \frac{1}{2}\sum_{ab}\int d^2r \int d^2r'\, V(r-r')Q(r)Q(r') \tag{11.61}$$

is the residual long-range part due to the spatial variations of the topological charge density $Q(r)$. Equation (11.60) has been written for the most general (not necessarily normalized) spinor field $|\psi(r)\rangle$. The presence of the second term in the integrand ensures the expected local gauge invariance of $\langle H\rangle_{\mathrm{ex},\psi}$.

How should we use $\langle H\rangle_\psi$? The first viewpoint is that the family $\mathcal{M}$ of Slater determinants $|\mathcal{S}_\psi\rangle$ associated with smooth spinor fields provides a good starting point for a variational calculation of the ground state for a system of $N - N_{\mathrm{top}}$ electrons (with $N_{\mathrm{top}} << N$). This is a very useful approach indeed, which is strongly supported by the fact that it gives the exact ground states for a model with point-like repulsive interactions when $0 \leq N_{\mathrm{top}} << N$. We shall dedicate a substantial part of these lectures to such studies, with emphasis on spatially periodic textures (Sections 11.3.3 and 11.3.4).

But the use of $\langle H\rangle_\psi$ extends beyond the variational determination of ground states and their properties. Viewing $\mathcal{M}$ as a family of coherent states embedded in the electronic Fock space, $\langle H\rangle_\psi$ can be regarded as the covariant symbol of an effective Hamiltonian operator from which the low-energy dynamics can, in principle, be reconstructed. This second viewpoint is in tune with a long tradition in many-body physics. The general idea is that most variational approximations involve a particular continuous family of trial states (such as Slater determinants for Hartree–Fock or general BCS states for Hartree–Fock–Bogoliubov approximations), which can often be regarded as a classical phase space. The underlying quantum dynamics in electronic Fock space can then be conveniently analysed from the perspective of coherent-state quantization. For several illustrations of this viewpoint, the reader is invited to consult [35, 36, 40]. We shall refer only briefly to coherent-state quantization of spin textures, in a discussion of the quantum zero-point correction to the variational energy $\langle H\rangle_\psi$ (Section 11.3.2).

A third way to use $\langle H\rangle_\psi$ is somewhat intermediate between the two previous ones. In the spirit of the time-dependent Hartree–Fock approximation, it regards $\langle H\rangle_\psi$ as a classical Hamiltonian. The associated dynamics on $\mathcal{M}$ is expected to approximate rather well the full quantum dynamics. The situation is reminiscent of quantum antiferromagnets. There, strictly speaking, the semiclassical limit is reached when the spin S is large. But it is well known that, at least for magnetically ordered systems, observations on real systems with a small value of S and calculations on a semiclassical expansion in $1/S$ can show surprisingly good agreement. Most likely, classicality is an emerging property in such systems. Even starting from 'extremely quantum' spins $\frac{1}{2}$ at the microscopic level, coarse graining leads to effective spins that tend to behave more and more classically as the spatial scale grows. This analogy with quantum magnets is especially natural for the periodic textures considered in Sections 11.3.3 and 11.3.4. Their spontaneously broken $SU(d)$ symmetry makes them similar to non-collinear long-range ordered antiferromagnets.

Let us now derive the classical equations of motion generated by $\langle H\rangle_\psi$. The first important consideration here is that the time-dependent Schrödinger equation $i(\partial/\partial t)|\Psi\rangle = H|\Psi\rangle$ can be derived from the following variational principle:

$$\delta \int_{t_i}^{t_f} \left(i\langle \Psi | \tfrac{\partial \Psi}{\partial t} \rangle - \langle \Psi | H | \Psi \rangle \right) dt = 0. \tag{11.62}$$

In other words, Schrödinger's equation can be regarded as a classical Hamilton equation in Hilbert space! The corresponding Hamiltonian function takes the value $\langle \Psi | H | \Psi \rangle$ on the state $|\Psi\rangle$. To specify the dynamics, we also need to know the underlying symplectic structure (or Poisson brackets). If M is the dimension of the Hilbert space, the Schrödinger equation reads

$$\frac{\partial \Psi_a}{\partial t} = -i \frac{\partial \langle \Psi | H | \Psi \rangle}{\partial \bar{\Psi}_a}, \qquad \frac{\partial \bar{\Psi}_a}{\partial t} = i \frac{\partial \langle \Psi | H | \Psi \rangle}{\partial \Psi_a}, \qquad 1 \leq a \leq M. \tag{11.63}$$

Writing $\Psi_a = (q_a + ip_a)/\sqrt{2}$, with q_a and p_a real variables, we get exactly the usual form of Hamilton's equations:

$$\frac{\partial q_a}{\partial t} = \frac{\partial \langle \Psi | H | \Psi \rangle}{\partial p_a}, \qquad \frac{\partial p_a}{\partial t} = -\frac{\partial \langle \Psi | H | \Psi \rangle}{\partial q_a}. \tag{11.64}$$

A great advantage of variational principles is their ability to deal with constraints. The previous discussion suggests that a reasonable approximation would be to constrain the time evolution to take place within the manifold $\mathcal{M}$ of coherent states $|\mathcal{S}_\psi\rangle$. The variational principle then becomes

$$\delta \int_{t_i}^{t_f} \left[i \sum_{j=1}^{N} \langle \psi(j) | \tfrac{\partial \psi(j)}{\partial t} \rangle - \langle H \rangle_\psi \right] dt = 0. \tag{11.65}$$

To simplify the writing, we have assumed here that the local spinors are normalized, that is $\langle \psi(j) | \psi(j) \rangle = 1$. Taking the continuum limit $N \to \infty$, this becomes

$$\delta \int_{t_i}^{t_f} \left[i \int \frac{d^2 r}{2\pi l^2} \langle \psi(r) | \tfrac{\partial \psi(r)}{\partial t} \rangle - \langle H \rangle_\psi \right] dt = 0. \tag{11.66}$$

We shall consider this Hamiltonian dynamics on $\mathcal{M}$ further in Sections 11.3.1 and 11.4.1. A slight modification of it will be useful, in order to deal with the local gauge symmetry sending $|\psi(r)\rangle$ into $f(r,t)|\psi(r)\rangle$, with $f(r,t)$ an arbitrary function. Allowing functions such that $|f|^2 \neq 1$ can be important if we deal with spinors that are not normalized by construction (such as spinors with holomorphic components). Again, this symmetry expresses the invariance of physical quantities with respect to changes in the global factor in front of the wavefunction. The original variational principle (11.62) can be modified slightly to make this generalized gauge symmetry manifest, and this gives

$$\delta \int_{t_i}^{t_f} \left(i \frac{\langle \Psi | \frac{\partial \Psi}{\partial t} \rangle}{\langle \Psi | \Psi \rangle} - \frac{\langle \Psi | H | \Psi \rangle}{\langle \Psi | \Psi \rangle} \right) dt = 0. \tag{11.67}$$

The corresponding evolution equation is

$$i\left(\left|\frac{\partial\Psi}{\partial t}\right\rangle - \frac{\langle\Psi|\frac{\partial\Psi}{\partial t}\rangle|\Psi\rangle}{\langle\Psi|\Psi\rangle}\right) = H|\Psi\rangle - \frac{\langle\Psi|H|\Psi\rangle}{\langle\Psi|\Psi\rangle}|\Psi\rangle. \tag{11.68}$$

This is equivalent to $i\left|\frac{\partial\Psi}{\partial t}\right\rangle = H|\Psi\rangle + f(t)|\Psi\rangle$, where $f(t)$ is an arbitrary function. This modified equation has the same physical content as the usual one. To see this, let us diagonalize H in an orthonormal basis $|\alpha\rangle$, so $H|\alpha\rangle = \omega_\alpha|\alpha\rangle$, $1 \leq \alpha \leq M$. If $|\Psi(t)\rangle = \sum_\alpha^M \Psi_\alpha(t)|\alpha\rangle$, then we have

$$\Psi_\alpha(t) = g(t)e^{-i\omega_\alpha t}\Psi_\alpha(0), \qquad g(t) = \exp\left[-i\int_0^t f(t')dt'\right]. \tag{11.69}$$

Adapting this to the case of the constrained dynamics on the coherent-state manifold $\mathcal{M}$ gives

$$\delta\int_{t_i}^{t_f}\left[i\int\frac{d^2r}{2\pi l^2}\frac{\left\langle\psi(r)\left|\frac{\partial\psi(r)}{\partial t}\right.\right\rangle}{\langle\psi(r)|\psi(r)\rangle} - \langle H\rangle_\psi\right]dt = 0. \tag{11.70}$$

We shall use this variational principle in Section 11.4.1. It is very close in spirit to the time-dependent Hartree–Fock approximation, with the difference that the wavefunction of the system is not allowed to be the most general Slater determinant, but one of the form $|\mathcal{S}_\psi\rangle$.

11.2.5 Classical ground states of the $\mathbb{C}P^{d-1}$ model

The previous discussion shows that it is very useful to look for the spinor fields that minimize the variational energy $\langle H\rangle_\psi$. Since the energy scale E_{ex} is of order $e^2/\epsilon l$, we see that $\langle H\rangle_{\text{el},\psi}/\langle H\rangle_{\text{ex},\psi}$ is proportional to $ln^{1/2}$, where n is the average topological charge density. In our semiclassical limit, $ln^{1/2} << 1$ and the term $\langle H\rangle_{\text{el},\psi}$ can be treated as a perturbation with respect to the leading term $\langle H\rangle_{\text{ex},\psi}$. It is therefore natural to concentrate on the $\mathbb{C}P^{d-1}$ model defined by the energy functional $\langle H\rangle_{\text{ex},\psi}$. This model has been studied in great detail by field theorists, starting in the 1970s. A very pedagogical presentation of many of its properties can be found in Rajaraman's book [41]. Our first concern is to find the minima of $\langle H\rangle_{\text{ex},\psi}$ with the constraint of a fixed topological charge. To achieve this, we use the Bogomol'nyi–Prasad–Sommerfield (BPS) inequality, which states that

$$\langle H\rangle_{\text{ex},\psi}/E_{\text{ex}} \geq 2\pi|N_{\text{top}}|. \tag{11.71}$$

To prove this bound in the present case, we start from the following expressions:

$$\frac{\langle H\rangle_{\text{ex},\psi}}{2E_{\text{ex}}} = I + J, \tag{11.72}$$

$$\pi N_{\text{top}} = I - J, \tag{11.73}$$

with

$$I = \int d^2r \left(\frac{\langle \partial_{\bar{z}}\psi | \partial_z \psi \rangle}{\langle \psi | \psi \rangle} - \frac{\langle \partial_{\bar{z}}\psi | \psi \rangle \langle \psi | \partial_z \psi \rangle}{\langle \psi | \psi \rangle^2} \right), \tag{11.74}$$

$$J = \int d^2r \left(\frac{\langle \partial_z\psi | \partial_{\bar{z}} \psi \rangle}{\langle \psi | \psi \rangle} - \frac{\langle \partial_z\psi | \psi \rangle \langle \psi | \partial_{\bar{z}} \psi \rangle}{\langle \psi | \psi \rangle^2} \right). \tag{11.75}$$

Since both I and J are positive, we have $I + J \geq |I - J|$, thereby proving the inequality (11.71). This proof also shows that this lower bound is reached if and only if $J = 0$ for $N_{\text{top}} \geq 0$ or $I = 0$ for $N_{\text{top}} \leq 0$. A sufficient condition for this to hold is $|\partial_{\bar{z}}\psi\rangle = 0$ for $N_{\text{top}} \geq 0$ or $|\partial_z\psi\rangle = 0$ for $N_{\text{top}} \leq 0$. So the variational exchange energy is minimal for analytic textures when holes ares added ($N_{\text{top}} \geq 0$) or for anti-analytic ones when particles are added ($N_{\text{top}} \leq 0$). This is striking, because these spaces of analytic (or anti-analytic) textures are quite large. If we fold the plane onto a finite torus, using periodic boundary conditions, they form a dN_{top}-dimensional complex vector space.

It is interesting to apply the inequality (11.71) to the case of point-like interactions. Keeping the first two terms in the l^2 expansion of $\langle H \rangle_\psi$ given in (11.46), it gives

$$\langle H \rangle_\psi \geq \frac{W}{4\pi l^2}(|N_{\text{top}}| - N_{\text{top}}). \tag{11.76}$$

This becomes an equality for analytic spinors when $N_{\text{top}} \geq 0$, so that $\langle H \rangle_\psi = 0$. This is consistent with the form of the Slater determinant $|\mathcal{S}_\psi\rangle$ in this case. Its wavefunction, in first quantization, reads

$$\Psi(r_1 a_1, \ldots, r_{N_e} a_{N_e}) = \prod_{i<j}(z_i - z_j) \prod_{i=1}^{N_e} \psi_{a_i}(z_i) e^{-|z_i|^2/4l^2}. \tag{11.77}$$

The first factor prevents two particles from occupying the same position, so this wavefunction is an eigenstate of the point-like interaction Hamiltonian with eigenvalue zero [28–30].

11.3 Periodic textures

11.3.1 Perturbation theory for degenerate Hamiltonians

We have just seen that in the manifold $\mathcal{M}$, the ground states of $\langle H \rangle_{\text{ex}}$ form a rather large submanifold $\mathcal{D}$ composed of analytic textures for $N_{\text{top}} \geq 0$ (and anti-analytic ones when $N_{\text{top}} \leq 0$). What is the effect of the residual interaction $\langle H \rangle_{\text{el}}$ on such a system? We know that some care has to be taken while perturbing degenerate systems. The first task is to recast the unperturbed $\langle H \rangle_{\text{ex}} \equiv H_0$ in a form that is convenient for a perturbative analysis. This raises immediate difficulties for degenerate systems, because we have to understand how the ground-state manifold $\mathcal{D}$ behaves with respect to the symplectic structure of $\mathcal{M}$.

Let us pick a point on $\mathcal{D}$ and perform a linear analysis of the equations of motion around it. If H_0 is positive and vanishes on $\mathcal{D}$, then the Williamson theorem [42] states that we can find canonical coordinates in a neighbourhood of this point (taken as the origin) such that the second-order Taylor expansion of H_0 at the origin reads

$$H_0 = \tfrac{1}{2} \sum_{j=N_0+1}^{N_0+N_d} p_j^2 + \tfrac{1}{2} \sum_{j=N_0+N_d+1}^{N} \omega_j (p_j^2 + q_j^2), \tag{11.78}$$

with $\omega_j > 0$. Here N_0, N_d, and N_m are non-negative integers such that $N = N_0 + N_d + N_m$ is the total number of degrees of freedom, and hence the dimension of $\mathcal{M}$ is $2N$. In our system of textures, the number of degrees of freedom is dN, where N denotes the number of flux quanta through the system. The conflict of notation should not be a problem, because the discussion in this subsection is mostly conceptual.

Near the origin, $\mathcal{D}$ is defined by the equations

$$p_{N_0+1} = \ldots = p_N = 0 = q_{N_0+N_d+1} = \ldots = q_N \tag{11.79}$$

This leaves $q_1, \ldots, q_{N_0+N_d}, p_1, \ldots, p_{N_0}$ as independent coordinates on $\mathcal{D}$ near the origin, so the dimension of $\mathcal{D}$ is $2N_0 + N_d$. N_0, N_d, and N_m will be referred to as the numbers of zero modes, drift motions, and massive modes, respectively. The important remark here is that the dimension of $\mathcal{D}$ doesn't fix separately the values of N_0 and N_d. These values can be extracted by a different procedure [43], which we outline here.

Phase-space geometry is not Euclidean (because there is no distance invariant under canonical transformations) but symplectic. Its basic object is not a metric, but a rank-two antisymmetric form $\omega = \sum_{j=1}^{N} dp_j \wedge dq_j$. This form assigns to two infinitesimal vectors $(\delta p_i, \delta q_i)$ and $(\delta' p_i, \delta' q_i)$ the number $\omega(\delta, \delta') = \sum_{j=1}^{N} (\delta p_j \, \delta' q_j - \delta q_j \, \delta' p_j)$. This form is invariant under canonical transformations and in particular, under Hamiltonian evolutions. If $N = 1$, this conservation law is just Liouville's theorem on the conservation of phase-space volume. The invariance of ω shows that the notion of orthogonality of two infinitesimal vectors (in the sense of ω) has an intrinsic meaning. A simple inspection shows that the tangent vectors at the origin that are orthogonal to all tangent vectors along $\mathcal{D}$ correspond to $\delta p_1 = \ldots = \delta p_{N_0+N_d} = 0 = \delta q_1 = \ldots = \delta q_{N_0}$. This implies that the restriction of ω to the tangent space of $\mathcal{D}$ at the origin contains a d-dimensional subspace of vectors that are orthogonal to all tangent vectors along $\mathcal{D}$. This subspace (the kernel of the restriction of ω) is spanned by the drift motions, for which the only non-zero components of the velocity are $\dot{q}_{N_0+1}, \ldots, \dot{q}_{N_0+N_d}$. Such motions are generated by Hamiltonians of the form $\sum_{j=N_0+1}^{N_0+N_d} a_j p_j$. Their physical importance is that, unlike small oscillations associated with massive modes, they are in principle unbounded, which makes them qualitatively different from both zero modes and massive modes.

This local analysis is interesting, but it raises immediately the question whether it can be extended to a larger region (i.e. an open subset in $\mathcal{M}$) intersecting $\mathcal{D}$. The answer is positive, with the assumption that N_d should be constant along the intersection of $\mathcal{D}$ with this region. If this condition is satisfied, we can find canonical coordinates such that a normal form similar to (11.78) holds, with two differences. First, the massive-mode frequencies ω_j may vary along $\mathcal{D}$ so that ω_j becomes a function of $q_1, \dots, q_{N_0+N_d}, p_1, \dots, p_{N_0}$. Second, the quadratic kinetic term associated with drift motion is not necessarily diagonal, and is replaced by $\frac{1}{2}\sum_{i=N_0+1}^{N_0+N_d}\sum_{j=N_0+1}^{N_0+N_d}(g^{-1})_{ij}p_ip_j$, where $g_{ij}(q_1, \dots, q_{N_0+N_d}, p_1, \dots, p_{N_0})$ is a metric tensor on $\mathcal{D}$.

I am not going to discuss the proof of this result here, because I fear that it may be of limited interest to most readers. We expect that this should be a direct consequence of the relative Darboux theorem for a submanifold $\mathcal{D}$ of a symplectic manifold $\mathcal{M}$ [44]. [1] One version of this theorem states that if coordinates (p'_i, q'_i) exist such that $\mathcal{D}$ is defined by the equations $p'_{N_0+1} = \dots = p'_N = 0 = q'_{N_0+N_d+1} = \dots = q'_N$, and that ω takes its canonical form on $\mathcal{D}$, then there exists a smooth one-to-one transformation from (p'_i, q'_i) to *canonical coordinates* (p_i, q_i) on $\mathcal{M}$ that acts like the identity on $\mathcal{D}$. Given this relative Darboux theorem, the main task seems to be to prove the existence of the coordinates (p'_i, q'_i) with the desired properties. This is easy in two particular cases. When $N_d = 0$, the restriction of ω to $\mathcal{D}$ is non-degenerate, so the existence of these coordinates is provided by the usual Darboux theorem applied to the submanifold $\mathcal{D}$. When $N_0 = 0$, all pairs of tangent vectors along $\mathcal{D}$ at an arbitrary point of $\mathcal{D}$ are orthogonal, so the restriction of ω to $\mathcal{D}$ is zero. The existence of the required coordinates follows because we impose only the value of ω on $\mathcal{D}$. For intermediate values of N_d (between 1 and $\dim \mathcal{D} - 1$), we have to prove that the distribution of N_d-dimensional subspaces on $\mathcal{D}$ obtained by taking the kernel of the restriction of ω is integrable, i.e. that this kernel coincides with the tangent space to the submanifolds defined by fixing the values of all coordinates, except $q'_{N_0+1}, \dots, q'_{N_0+N_d}$. I believe that this is true,[2] and probably well known to mathematicians, although I haven't been able to locate a proof in the literature.

Coming back to our problem, we have rather good numerical evidence (but no conceptual proof, unfortunately) that in the case where $\mathcal{D}$ is the submanifold of analytic textures, we have $N_d = 0$. In this situation, we may therefore write

$$H_0 = \frac{1}{2}\sum_{j=N_0+1}^{N}\omega_j(p_j^2 + q_j^2) \equiv \sum_{j=1}^{N_m}\omega_j(p_s, q_s)J_j. \tag{11.80}$$

Here, we have introduced the slow variables $(p_s, q_s) \equiv (p_1, \dots, p_{N_0}, q_1, \dots, q_{N_0})$, which are good canonical coordinates on $\mathcal{D}$, and action variables J_j for massive modes.

For this Hamiltonian, the J_j's associated with massive modes are integrals of motion. When they are non-zero, they can induce a motion along $\mathcal{D}$, because Hamilton's equations read as follows for $1 \le j \le N_0$:

$$\dot{q}_j = \sum_{k=1}^{N_m}\frac{\partial\omega_k}{\partial p_j}(p_s, q_s)J_j, \tag{11.81}$$

$$\dot{p}_j = -\sum_{k=1}^{N_m}\frac{\partial\omega_k}{\partial q_j}(p_s, q_s)J_j. \tag{11.82}$$

[1] I am grateful to San Vu Ngoc for a discussion on this point and for directing my attention to the relative Darboux theorem.

[2] B. Douçot, Handwritten notes in French, available upon request.

At the classical level, we may take $J_k = 0$ for all k, and no motion along $\mathcal{D}$ is generated. Quantum-mechanically, this is no longer true, and we can no longer ignore the quantum zero-point-energy correction H_{qzpc} coming from the massive modes. Its effect is to lift the degeneracy between the classical states lying on $\mathcal{D}$, and to favour the minima of H_{qzpc}. A more detailed discussion of this quantum correction is given in Appendix 11.D.

Let us now switch on a small perturbation H_1, which it is natural to expand in powers of the massive-mode action-angle coordinates $p_{N_0+j} = \sqrt{2J_j}\cos\theta_j$, $q_{N_0+j} = \sqrt{2J_j}\sin\theta_j$:

$$H_1 = H_1^{(0)} + \sum_{j=1}^{N_m} J_j^{1/2}\left[a_j(p_s,q_s)e^{i\theta_j} + \bar{a}_j(p_s,q_s)e^{-i\theta_j}\right] + \mathcal{O}(J_j), \tag{11.83}$$

where $H_1^{(0)}$ and a_j are of order ϵ.

At first order in ϵ, we may keep only the term $H_1^{(0)}$ in H_1. Indeed, the term proportional to $J_j^{1/2}$ corresponds to a constant driving force that pulls the massive coordinates away from their unperturbed equilibrium value at $J_j = 0$. This shift induces a change in the energy of the massive mode j of order $|a_j|^2/\omega_j$ that is proportional to ϵ^2/ω_j and therefore of higher order in ϵ than the $H_1^{(0)}$ term. We see also that this induced ϵ^2 contribution to the effective Hamiltonian is smaller when the characteristic frequencies associated with massive modes are large. To summarize, the first corrections due to the H_1 perturbation are captured by the effective Hamiltonian

$$H_{\text{eff}} = H_{\text{qzpc}}(p_s,q_s) + H_1^{(0)}(p_s,q_s). \tag{11.84}$$

We emphasize once again that it acts on the degenerate manifold $\mathcal{D}$, which is symplectic because $N_d = 0$, so it constitutes a good classical phase space. On this manifold, H_{eff} is the sum of the quantum zero-point correction due to the massive modes of H_0 and the perturbation H_1, restricted to $\mathcal{D}$.

These considerations motivate us to discuss briefly the spectrum of massive oscillators in the vicinity of a periodic analytic texture, which therefore belongs to $\mathcal{D}$, and was chosen to minimize the perturbation $H_1^{(0)}(p_s,q_s) \equiv \langle H_{\text{el}}\rangle_\psi$. The construction of this optimal texture will be presented in Section 11.3.3.

11.3.2 Remarks on the Hessian of the exchange energy

To be specific, we assume a positive topological charge. The discussion of Section 11.2.5 shows that the variational energy can be conveniently be written as

$$\langle H\rangle_{\text{ex},\psi}/E_{\text{ex}} = 2\pi N_{\text{top}} + 4J, \tag{11.85}$$

where J is defined by (11.75) Consider now small deviations $|\psi\rangle \to |\psi\rangle + \sqrt{\langle\psi|\psi\rangle}\,|\phi\rangle$ away from the analytic spinor $|\psi\rangle$. Inserting this ansatz into (11.85) gives directly the second derivative (Hessian operator) of $\langle H\rangle_{\text{ex},\psi}$:

$$\langle H\rangle_{\text{ex},\psi}/E_{\text{ex}} = 2\pi|N_{\text{top}}| + 4\langle\phi|M^{+}PM|\phi\rangle + \ldots \tag{11.86}$$

Here, we have introduced two operators M and P acting on the spinor fields ϕ describing small deviations. Explicitly,

$$M|\phi\rangle = |\partial_{\bar{z}}\phi\rangle + \frac{1}{2}\frac{\langle\partial_{\bar{z}}\psi|\psi\rangle}{\langle\psi|\psi\rangle}|\phi\rangle, \tag{11.87}$$

$$P|\phi\rangle = |\phi\rangle - \frac{|\psi(z)\rangle\langle\psi(z)|}{\langle\psi(z)|\psi(z)\rangle}|\phi\rangle. \tag{11.88}$$

A very important property of the operator M is that

$$[M, M^{+}] = \tfrac{1}{2}\mathcal{B}(r) = \pi Q(r). \tag{11.89}$$

Here, $\mathcal{B} = \partial_x\mathcal{A}_y - \partial_y\mathcal{A}_x$ is the gauge-invariant flux density associated with the Berry connection. If $\mathcal{B}(r)$ is constant, then the spectrum of $M^{+}M$ is $\{\frac{1}{2}\mathcal{B}n, n = 0, 1, 2, \dots\}$. As we are going to show in the next subsection, the spatial variations of $\mathcal{B}(r)$ are quite small for the optimal periodic texture. The residual inhomogeneities are not expected to close the gaps in this Landau-level-like spectrum. What is more difficult to analyse is the effect of the projector P. It imposes the local variation $|\phi(r)\rangle$ to be orthogonal to the reference spinor $|\psi(r)\rangle$. This is certainly a serious perturbation for small values of d. At large d, the probability for two randomly chosen spinors to be orthogonal becomes large, so we may expect that the effect of P is small in the large-d limit. Most likely, the Hessian of the $\mathbb{C}P^{d-1}$ model is gapped, with an energy gap of order $(e^2/4\pi\epsilon l)nl^2$, where $\overline{Q(r)} = n$. Good numerical evidence that this is indeed true has been obtained recently by Dima Kovrizhin, using the dynamics given by (11.70). His results for $d = 3$, with the reference spinor described as in Sections 11.3.3 and 11.3.4, are shown in Fig. 11.3.

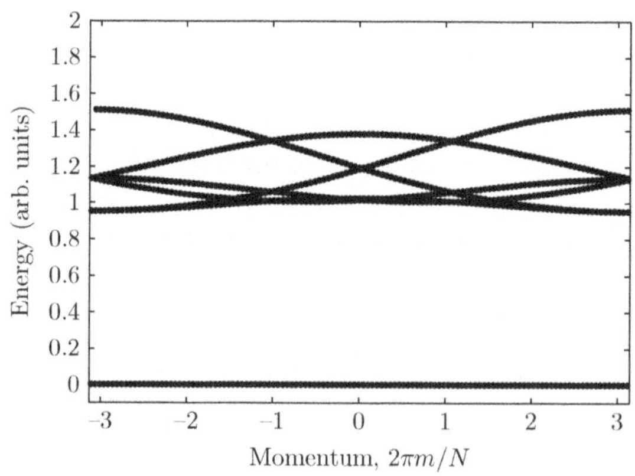

Fig. 11.3 The spectrum of the Hessian in the case $d = 3$. The flat branch at zero frequency corresponds to variations within the analytic subspace. The first Landau-like level is broadened by the combined effect of the periodic modulation of the Berry flux $\mathcal{B}(r)$ and of the projector P. Nevertheless, there is a clear gap between the degenerate zero-energy level and this broadened first Landau level.

In view of (11.84), we need to estimate the quantum zero-point-energy correction due to the massive modes of $\langle H \rangle_{\text{ex}}$. The form of the Hessian (11.86) shows that it is bilinear in $\phi_a(r)$ and $\bar{\phi}_{a'}(r')$. The discussion in Appendix 11.D suggests then that H_{qzpc} should vanish, because it is only sensitive to squeezing operators, which would be detected by the presence of quadratic terms of the form $\phi_a(r)\phi_{a'}(r')$ or $\bar{\phi}_a(r)\bar{\phi}_{a'}(r')$ in the Taylor expansion of the covariant symbol $\langle H \rangle_{\text{ex}}$. But such terms are clearly absent from (11.86). At this stage, we therefore conjecture that $H_{\text{qzpc}} = 0$. A sound mathematical theory of the family $|\mathcal{S}_\psi\rangle$, viewed as coherent states in the fermionic Fock space, doesn't seem to be available, in part because of the infinite dimensionality of the space of possible smooth textures. Physically, we expect the presence of a spatial cutoff below the magnetic length, so we might be able to get back to the more familiar situation of a finite-dimensional family. Besides the previous formal argument, our conjecture seems to be supported by two different observations. It has indeed been found by a combination of numerical and analytical studies that quantum corrections to the effective energy functional introduced in Section 11.2.4 are small [45]. Second, as observed long ago, analytic textures are exact zero-energy eigenstates for a model with point-like repulsive interaction [28–30], and therefore their degeneracy is preserved to all orders in quantum fluctuations. Of course, it is not clear whether this conclusion, valid for the model with point-like repulsion, can be transferred to the quantum $\mathbb{C}P^{d-1}$ model. In Section 11.2.3, we have computed to $\mathcal{O}(l^2)$ the correction to the expectation value of the interaction energy, and this led to the $\mathbb{C}P^{d-1}$ energy functional (11.46). I don't know if the next corrections arising from both Hartree and Fock terms will cancel. Unfortunately, the computation of the Fock term requires knowledge of P_2 in the series expansion of P_ψ, whose expression is a priori quite complicated. So the precise relationship between the quantum $\mathbb{C}P^{d-1}$ model and the model with point-like repulsion remains an interesting open question.

11.3.3 Variational procedure for energy minimization

Let us consider the case $0 < N_{\text{top}} << N$, so that $\langle H \rangle_{\text{ex},\psi}$ is minimized for analytic textures. Neglecting the quantum zero-point correction coming from the finite-frequency modes of the Hessian of $\langle H \rangle_{\text{ex}}$, we have then to minimize $\langle H \rangle_{\text{el},\psi}$ with the constraint of a fixed topological charge. Intuitively, the Coulomb interaction being repulsive, one would like to make the topological charge density as uniform as possible. This motivates a variational search within the class of periodic textures. Let us then pick two independent vectors γ_1 and γ_2 on the plane. Our goal is to construct holomorphic spinor fields $\psi_a(z)$ such that all physical properties of the corresponding texture are periodic under translations by γ_1 and γ_2. This seems at first impossible, because the only holomorphic functions that have such double periodicity are constants. However, we should view $|\psi(z)\rangle$ as a representative of the complex line it generates, so the appropriate notion of periodicity is that for any lattice vector $\gamma = n_1\gamma_1 + n_2\gamma_2$ (with n_1 and n_2 integers), there should be a holomorphic function $f_\gamma(z)$ such that

$$|\psi(z + \gamma)\rangle = f_\gamma(z)|\psi(z)\rangle. \tag{11.90}$$

A particularly important class of functions satisfying this is the family of θ functions, which satisfy $f_\gamma(z) = \exp(a_\gamma z + b_\gamma)$, a_γ and b_γ being complex numbers, which are functions of the lattice vector γ. These two functions are called the type of the θ function. A remarkable mathematical result states that any periodic holomorphic map from the complex plane to $\mathbb{C}P^{d-1}$ is obtained, up to a gauge transformation, from a holomorphic spinor field whose d components are θ functions of the same type [46]. An important example of θ functions is

$$\theta_p(z) = \sum_n \exp\left\{i\left[\pi\tau d\left(n-\frac{p}{d}\right)\left(n-1-\frac{p}{d}\right)+2\sqrt{d}\left(n-\frac{p}{d}\right)z\right]\right\}. \tag{11.91}$$

Here, we have chosen $\gamma_1 = \pi\sqrt{d}$ and $\gamma_2 = \pi\sqrt{d}\tau$, with $\operatorname{Im}\tau > 0$, and p is an integer. These functions satisfy

$$\theta_p(z+\gamma_1) = \theta_p(z), \tag{11.92}$$

$$\theta_p(z+\gamma_2) = e^{-i2\sqrt{d}z}\theta_p(z). \tag{11.93}$$

In particular, we have $a_{\gamma_1} = 0$ and $a_{\gamma_2} = -i2\sqrt{d}$. Let us compute the topological charge of such a texture enclosed in the parallelogram $\mathcal{P}(\gamma_1,\gamma_2)$ spanned by the two basis vectors. This is done easily because

$$\left(\frac{\langle\psi|\partial_z\psi\rangle}{\langle\psi|\psi\rangle}\right)(z+\gamma) = a_\gamma + \left(\frac{\langle\psi|\partial_z\psi\rangle}{\langle\psi|\psi\rangle}\right)(z). \tag{11.94}$$

From this, we deduce that

$$\frac{1}{2\pi}\oint_{\mathcal{P}(\gamma_1,\gamma_2)} \mathcal{A}\cdot dl = \frac{1}{2\pi i}(a_{\gamma_1}\gamma_2 - a_{\gamma_2}\gamma_1). \tag{11.95}$$

The topological charge enclosed in $\mathcal{P}(\gamma_1,\gamma_2)$ is then equal to d. A similar calculation shows that the phase of these θ functions winds by $2\pi d$ when one goes counterclockwise around $\mathcal{P}(\gamma_1,\gamma_2)$. This domain therefore contains d zeros for each θ function of the above type. Another remarkable result is the special case of the Riemann–Roch theorem for a complex torus. It states that the θ functions of a given type form a complex vector space, whose dimension is finite and is equal to the topological charge within $\mathcal{P}(\gamma_1,\gamma_2)$. For the type given by (11.92) and (11.93), a possible basis is the set of θ_p's for $0 \leq p \leq d-1$. The positions of zeros in $\mathcal{P}(\gamma_1,\gamma_2)$ are illustrated on Fig. 11.4.

These properties are very reminiscent of the problem of a quantum particle on a torus in the presence of a uniform magnetic field [47]. This is not a coincidence, because quantum wavefunctions in the lowest Landau level are analytic functions multiplied by $\exp(-|z|^2/4l^2)$. To fold the plane onto a torus, periodic boundary conditions are imposed by fixing the eigenvalues of the magnetic translation operators along γ_1 and γ_2, which commute only if $\mathcal{P}(\gamma_1,\gamma_2)$ encloses a finite number of flux quanta Φ/Φ_0. These boundary conditions impose that the analytic factors in the admissible wavefunctions are θ functions of a prescribed type, with the corresponding topological charge equal to Φ/Φ_0.

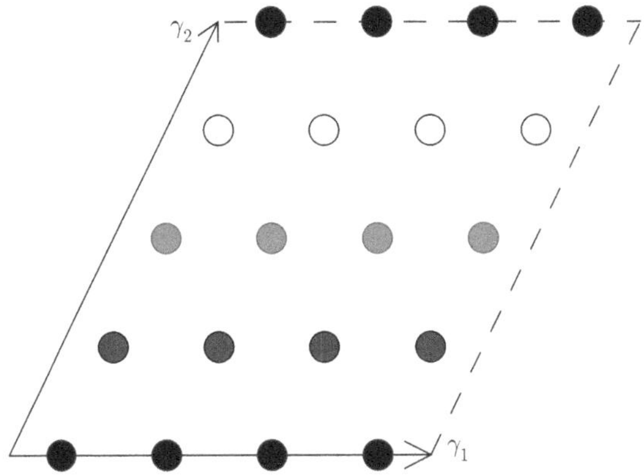

Fig. 11.4 [Colour online] Location of zeros of the basis θ_p functions for $d = 4$: black [blue] circles, $p = 0$; dark grey [green], $p = 1$, light grey [red], $p = 2$; white [yellow], $p = 3$.

It is also useful for our purposes to understand the action of translations on θ functions. We have

$$\theta(z - w + \gamma) = \exp[a_\gamma z + (b_\gamma - a_\gamma w)]\, \theta(z - w). \tag{11.96}$$

This shows that the translated function $z \to \theta(z - w)$ is again a θ function. While a_γ is preserved in this operation, b_γ is in general modified. To preserve b_γ as well, we have to multiply the transformed function $\theta(z - w)$ by an exponential factor. We make then the following definition:

$$\mathcal{T}_w \theta(z) = e^{\mu(w) z} \theta(z - w). \tag{11.97}$$

This new transformation preserves b_γ if and only if

$$\mu(w)\gamma - a_\gamma w \in 2\pi\mathbb{Z}. \tag{11.98}$$

Because this holds in particular for both γ_1 and γ_2, the possible values of w are discrete, and they correspond to the points of a lattice generated by γ_1/d and γ_2/d:

$$w(n_1, n_2) = \frac{n_1}{d}\gamma_1 + \frac{n_2}{d}\gamma_2, \tag{11.99}$$

$$\mu(w) = \frac{n_1}{d} a_{\gamma_1} + \frac{n_2}{d} a_{\gamma_2}. \tag{11.100}$$

As in the case of a particle in the lowest Landau level, this is a projective representation of translations, and, more precisely,

$$\mathcal{T}_w \mathcal{T}_{w'} = e^{i(2\pi/d)(m_1 m_2' - m_2 m_1')} \mathcal{T}_{w'} \mathcal{T}_w. \tag{11.101}$$

Note that the phase factor $(m_1m'_2 - m_2m'_1)/d$ has a simple meaning: it is the topological charge inside the parallelogram delimited by w and w'. These translations have a simple action on the basis function θ_p:

$$\mathcal{T}_{\frac{\gamma_1}{d}}\theta_p = e^{i(2\pi p/d)}\theta_p, \tag{11.102}$$

$$\mathcal{T}_{\frac{\gamma_2}{d}}\theta_p = \lambda\theta_{p+1}, \tag{11.103}$$

where $\lambda = e^{-i\pi\tau(d+1/d)}$.

Given all these preparations, we see that the general periodic texture with periods γ_1 and γ_2 takes the form

$$\psi_a(r) = \sum_{b=0}^{d-1} M_{ab}\theta_b, \tag{11.104}$$

where M_{ab} are the complex entries of a $d \times d$ matrix M. The global $SU(d)$ symmetry manifests itself through the invariance of $\langle H\rangle_{\mathrm{el}}$ under the transformation $M \to UM$, where U is an arbitrary unitary matrix.

11.3.4 Properties of periodic textures

After a substantial amount of numerical work, due to Dima Kovrizhin, it appears that the variational energy $\langle H\rangle_{\mathrm{el}}$ is minimal for $\tau = e^{i\pi/3}$, and $M = I$ (the identity matrix).

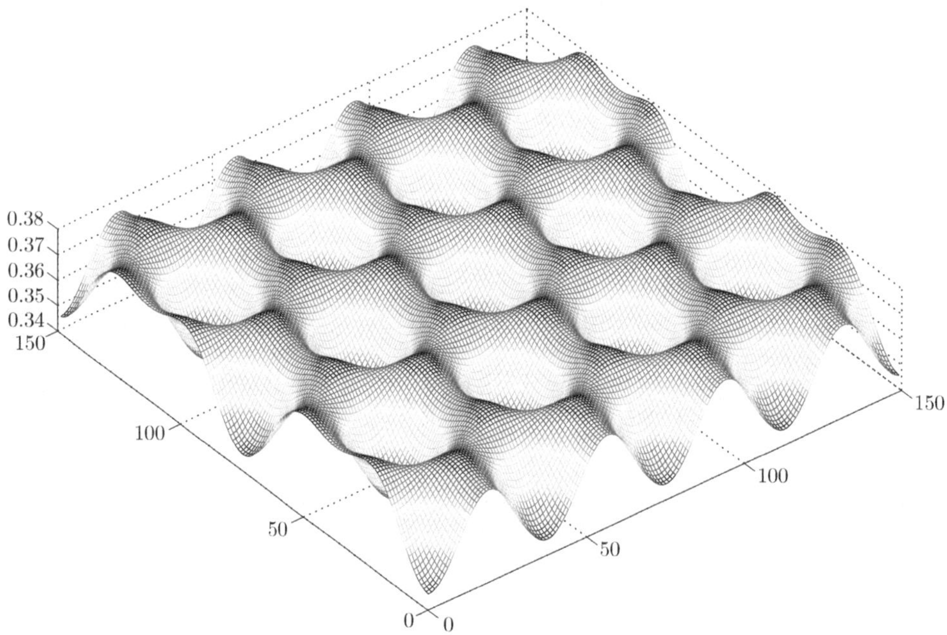

Fig. 11.5 [Colour online] Spatial variations of the topological charge density for the optimal periodic crystal with $d = 4$.

This corresponds to a spontaneous symmetry breaking of the global $SU(d)$ symmetry. In fact, all unitary matrices M are ground states of $\langle H\rangle_{\rm el}$. The optimal value of τ corresponds to a triangular lattice of skyrmions. Such a state has been represented for $d=2$ on Fig. 11.2.

An interesting property of these periodic textures is that the spatial modulation of the topological charge density is more periodic than the ansatz $M=I$ suggests at first glance, in the sense that its elementary periods are γ_1/d and γ_2/d. At large d, the modulation contains mostly the lowest harmonic, and its amplitude decays exponentially with d. The large-d behaviour can be computed explicitly for a square lattice:

$$Q(x,y)\simeq\frac{2}{\pi}-4de^{-\pi d/2}[\cos(2\sqrt{d}\,x)-2e^{-\pi d/2}\cos^2(4\sqrt{d}\,x)+(x\leftrightarrow y)]+\dots \tag{11.105}$$

A picture of these spatial modulations is shown in Fig. 11.5.

Only the triangular lattice seems to yield a true local energy minimum. This is most directly seen by computing eigenfrequencies of small-deformation modes, using the method to be described below.

11.4 Collective excitations around periodic textures

11.4.1 Time-dependent Hartree–Fock equations

Let us now turn to collective excitations around such textures. What we are going to do is very reminiscent of the traditional linear spin-wave theory in quantum magnetic systems. We shall use the variational formulation of the quantum dynamics given by (11.70), where the variation of $|\psi(r,t)\rangle$ has to be taken within the subspace of analytic spinors. To achieve this, it is convenient to work on a system with finite volume, to keep the phase-space dimension finite. One way to do this is to introduce a large supercell $\mathcal{P}(N_1\gamma_1,N_2\gamma_2)$, containing dN_1N_2 topological charges. We fix the type for the θ functions across this supercell. The corresponding allowed translations take the form

$$w(m_1,m_2)=\frac{m_1}{dN_2}\gamma_1+\frac{m_2}{dN_1}\gamma_2, \tag{11.106}$$

$$\mu(m_1,m_2)=\frac{m_1}{dN_2}a_{\gamma_1}+\frac{m_2}{dN_1}a_{\gamma_2}. \tag{11.107}$$

These can be used to construct a basis of dN_1N_2 θ functions:

$$\chi_{p,m_1,m_2}=\mathcal{T}_{w(m_1,m_2)}\theta_p,\qquad 1\leq m_1\leq N_2-1,\qquad 1\leq m_2\leq N_1-1. \tag{11.108}$$

Now, it is important to note that the type of these θ functions across the elementary cell $\mathcal{P}(\gamma_1,\gamma_2)$ depends on (m_1,m_2) since

$$\frac{\chi_{p,m_1,m_2}(z+\gamma)}{\chi_{p,m_1,m_2}(z)}=e^{a_\gamma z+b_\gamma+c_\gamma} \tag{11.109}$$

and, for example,

$$c_{\gamma_1} = -i2\pi \frac{m_2}{N_1}, \qquad c_{\gamma_2} = i2\pi \frac{m_1}{N_2}. \tag{11.110}$$

We see that the type associated with a $\mathcal{P}(\gamma_1, \gamma_2)$ cell is N_2-periodic in m_1 and N_1-periodic in m_2: somehow, we recover a notion of *Brillouin zone*, although we do not have periodicity in the usual sense (as for plane waves or Bloch functions), because we are dealing with analytic functions.

We are now looking for small deviations away from the optimal texture, assuming that

$$\psi_a(z,t) = \theta_a(z,t) + \sum_{p,m_1,m_2} M_{a,p,m_1,m_2}(t)\chi_{p,m_1,m_2}(z). \tag{11.111}$$

In the spirit of spin-wave theory, we assume here that the amplitudes $M_{a,p,m_1,m_2}(t)$ are infinitesimals of order one. Using the fact that the topological charge density in the reference texture is γ_1- and γ_2-periodic, we deduce that the linearized equations of motion couple (m_1, m_2) only to itself and to $(-m_1, -m_2)$. This is very similar to what we would get in a superfluid or in a quantum antiferromagnet if we could identify (m_1, m_2) with the momentum of the excitation. The same structure has been obtained for the Bogoliubov theory of collective modes in superfluids in the presence of a vortex lattice [48]. The new feature here is the presence of internal degrees of freedom, leading to matrix eigenvalue equations of size $2d^2 \times 2d^2$, but because of the high symmetry of the $Q(r)$ profile, and in particular its γ_1/d- and γ_2/d-periodicity, this large matrix structure breaks into small blocks of size 2 by 2, which makes the computation of the collective mode spectrum much easier!

11.4.2 Collective-mode spectrum

To analyse the corresponding collective-mode spectrum, let us first concentrate on the zero-momentum sector $(m_1, m_2) = (0, 0)$, which yields a Hamiltonian system with $N = d^2$ degrees of freedom. As we have seen, this sector exhibits a d^2-dimensional ground-state manifold obtained by letting the unitary group act on the reference texture, given by $M = I$. To proceed further, it is useful to return to the general analysis of Section 11.3.1 concerning the classical dynamics in the vicinity of a degenerate ground-state manifold $\mathcal{D}$. There, we introduced a normal form characterized by three integers N_0, N_d, and N_m. Recall that the dimension of $\mathcal{D}$ is $2N_0 + N_d$. What is the effect of changing the value of N_d while keeping the dimensions of $\mathcal{M}$ and $\mathcal{D}$ fixed? Let us show, by a few examples that it affects qualitatively the structure of the linearized equations of motion.

The simplest case is $\dim \mathcal{D} = 1$. This is realized for $(N_0, N_d, N_m) = (0, 1, 0)$. An example of this is provided by a free particle in one dimension with Hamiltonian $H = \frac{1}{2}P^2$. Here $\mathcal{D}$ is the X axis, and the equations of motion read

$$\begin{pmatrix} \dot{X} \\ \dot{P} \end{pmatrix} = \begin{pmatrix} 0 & 1 \\ 0 & 0 \end{pmatrix} \begin{pmatrix} X \\ P \end{pmatrix}. \tag{11.112}$$

An important feature is that this matrix is not diagonalizable—it forms a Jordan block with eigenvalue zero. In physical terms, this means that moving away by ϵ along the P axis generates drift motion parallel to $\mathcal{D}$ with velocity ϵ.

Next, let us consider $\dim \mathcal{D} = 2$. The first possibility is to have $(N_0, N_d, N_m) = (0, 2, 0)$. Let us assume that $\mathcal{D}$ is the (X_1, X_2) plane. A simple choice for H is $H = \frac{1}{2}P_1^2 + \frac{1}{2}P_2^2$, with the corresponding equations of motion

$$\begin{pmatrix} \dot{X}_1 \\ \dot{P}_1 \\ \dot{X}_1 \\ \dot{P}_1 \end{pmatrix} = \begin{pmatrix} 0 & 1 & 0 & 0 \\ 0 & 0 & 0 & 0 \\ 0 & 0 & 0 & 1 \\ 0 & 0 & 0 & 0 \end{pmatrix} \begin{pmatrix} X_1 \\ P_1 \\ X_2 \\ P_2 \end{pmatrix}. \tag{11.113}$$

There are now two Jordan blocks, one for each flat direction along $\mathcal{D}$. This situation is characterized by the fact that generating functions of drift motions, P_1 and P_2, commute everywhere, and in particular on the ground-state subspace.

A second possibility with two degrees of freedom is $(N_0, N_d, N_m) = (1, 0, 1)$. As an example, we may take $\mathcal{D}$ to be the (X_1, P_1) plane and the Hamiltonian as $H = \frac{1}{2}\omega\left(X_2^2 + P_2^2\right)$, with the equations of motion

$$\begin{pmatrix} \dot{X}_1 \\ \dot{P}_1 \\ \dot{X}_1 \\ \dot{P}_1 \end{pmatrix} = \begin{pmatrix} 0 & 0 & 0 & 0 \\ 0 & 0 & 0 & 0 \\ 0 & 0 & 0 & \omega \\ 0 & 0 & -\omega & 0 \end{pmatrix} \begin{pmatrix} X_1 \\ P_1 \\ X_2 \\ P_2 \end{pmatrix}. \tag{11.114}$$

Here, we have only one zero eigenvector for each flat direction and there is no Jordan block. Besides, there is a finite-frequency oscillator. The qualitative difference with the previous case is that the generating functions of drift motions, X_1 and P_1, do not commute on $\mathcal{D}$.

It is instructive to see how these two possibilities can be realized in a simple system with two classical spins. We choose $H = \vec{S}_1 \cdot \vec{S}_2$, with the constraints $||\vec{S}_1||^2 = s_1$, and $||\vec{S}_2||^2 = s_2$. The ground-state manifold $\mathcal{D}$ is easily obtained. It is parametrized by a unit vector $\vec{n}$ such that $\vec{S}_1 = s_1\vec{n}$ and $\vec{S}_2 = -s_2\vec{n}$. So $\mathcal{D}$ is a two-dimensional sphere. Because of the global spin-rotation symmetry, we can always induce motions along $\mathcal{D}$ by taking the components of the total angular momentum operator as generators. The classical equations of motion,

$$\frac{d\vec{S}_i}{dt} = (\vec{S}_1 + \vec{S}_2) \times \vec{S}_i, \tag{11.115}$$

exhibit the eigenfrequencies $\{0, 0, s_1 - s_2, s_2 - s_1\}$. We see that something special happens when the two spins have the same length. In the general case, $s_1 \neq s_2$, so that $\vec{S}_1 + \vec{S}_2 \neq 0$ on $\mathcal{D}$. Because of this, the generators of global rotations do not commute on $\mathcal{D}$, which shows that we are in the second case, according to the previous discussion. This is consistent with the fact that we have a massive mode here.

The case $s_1 = s_2$ is special in that the generators of global rotations commute on $\mathcal{D}$, so we are in the first case with two Jordan blocks. Physically, these drift motions correspond to a precession of both spins along $\vec{S}_1 + \vec{S}_2$, which is small but non-zero for small deviations away from $\mathcal{D}$ in the system phase space. In mathematical terms, in the vicinity of $\mathcal{D}$, the symplectic structure on $\mathcal{M}$ is similar to the familiar one on the cotangent bundle over $\mathcal{D}$. In this situation, physicists usually prefer the Lagrangian formulation, which here takes the form

$$L \propto (\partial_t \vec{n})^2. \tag{11.116}$$

This case is very close to what happens in a Néel antiferromagnet, and is illustrated in Fig. 11.6

Let us now come back to the collective dynamics around the optimal texture. We can show that we are dealing with a situation very much like the latter case, with rotations replaced by $SU(d)$ transformations. We consider the infinitesimal one corresponding to the anti-Hermitian matrix ξ. It is easy to show that the generator of this transformation is the functional

$$\Phi_\xi = i \int d^2 r \sum_{a,b} \frac{\bar{\psi}_a(r)\xi_{ab}\psi_b(r)}{\langle \psi(r)|\psi(r)\rangle}. \tag{11.117}$$

Using the symmetries between the basis θ functions $\theta_p(r)$, such as (11.102) and (11.103), it is easy to show that on the optimal texture for which $\psi_p(r) = \theta_{p-1}(r)$, the integral gives a contribution proportional to $\sum_{a,b} \delta_{ab}\xi_{ab} = \mathrm{Tr}\,\xi = 0$ for ξ in the Lie algebra of $SU(d)$. So we have $N_d = d^2 - 1$, and therefore we expect exactly $d^2 - 1$ Jordan blocks for the linearized equations of motion in the zero-momentum sector. This has been confirmed by detailed numerical studies for $d \in \{2, 3, 4\}$. The missing degree of freedom corresponds to diagonal generators, which make the difference between the Lie algebras of $U(d)$ and $SU(d)$. But the equations of motion are ill defined in this two-dimensional block, in agreement with the fact that it corresponds to uniform gauge transformations.

Let us now consider a small but finite momentum (in the sense of (m_1, m_2)), which may be treated as a small perturbation. When it acts on any of the $d^2 - 1$ Jordan blocks obtained for zero momentum, we observe that this block disappears, and

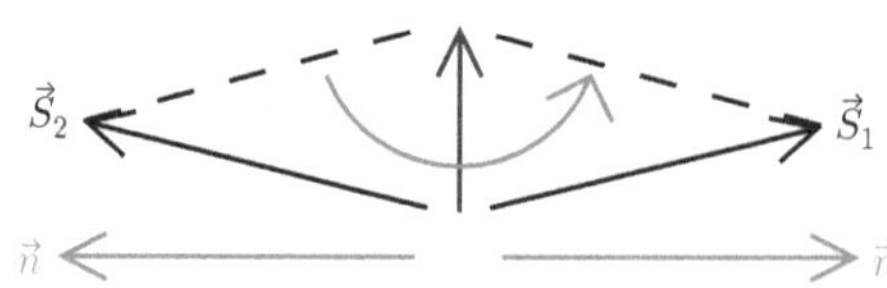

Fig. 11.6 [Colour online] Illustration of drift motion for a spin configuration (thick black arrows) close to the antiferromagnetic ground state. Spins precess around the total magnetization (vertical white [blue] arrow) at an angular velocity that is proportional to the deviation from collinearity.

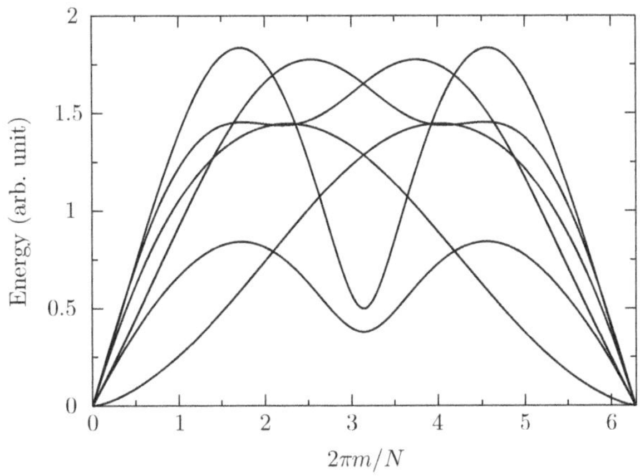

Fig. 11.7 Collective mode spectrum of an $SU(3)$ skyrmion lattice with Coulomb interactions. The spectrum is shown along the diagonal cut through the Brillouin zone, $k_1 = k_2$. The lowest-lying mode at small wavevectors is the magnetophonon. All remaining modes are linear gapless excitations. Some of theses branches appear degenerate, because of the particular symmetry of the chosen wavevectors. The total number of Goldstone branches is $3^2 - 1 = 8$.

gives rise to a pair of opposite eigenfrequencies, which grow linearly with momentum. Putting all the perturbed blocks together, we have $d^2 - 1$ Goldstone branches that disperse linearly. There remains another mode, which originates from the diagonal generators at zero momentum, whose dispersion is not always linear in momentum, and depends on the shape of the long-range interaction. For a repulsive potential $V(r) \propto r^{-\alpha}$, the corresponding dispersion relation is $\omega \propto k^{1+\alpha/2}$. This is reminiscent of the dispersion of magnetophonons in a 2D Wigner crystal [49]. All these features can be clearly identified on Fig. 11.7.

11.4.3 Towards an effective sigma model description

Can we go beyond this linearized dynamics? The analogy with antiferromagnets suggests that it should be possible. It is well known that in these systems, the key to this upgrade beyond linear spin-wave theory is to replace an expansion in *small deviations* away from the reference ordered state by a *gradient expansion*, which is, in a nonlinear setting, reminiscent of the perturbation theory of Jordan blocks by a small momentum performed in the linear case. In the case of Néel order, we get the $O(3)$ nonlinear sigma model, whose Lagrangian density reads

$$L \propto (\partial_t \vec{n})^2 - (\partial_x \vec{n})^2 - (\partial_y \vec{n})^2. \tag{11.118}$$

Clearly, this generalizes (11.116) to slowly varying configurations. In the $SU(d)$ case, we should expect an expression of the form

$$L \propto \mathrm{Tr}\big(\partial_t g\, A\, \partial_t g^+ - \partial_x g\, B_x\, \partial_x g^+ - \partial_y g\, B_y\, \partial_y g^+\big)\,, \tag{11.119}$$

where A, B_x, and B_y are fixed $d \times d$ matrices and $g(r,t) \in SU(d)$ is slowly varying in space and time, in perfect analogy with non-collinear antiferromagnets [50].

The connection between this slowly varying matrix field $g(r,t)$ and the actual time-dependent spinor $\psi_a(r,t)$ is expected to be something like

$$\psi_a(r,t) = \sum_{b=1}^{d} g_{ab}(r,t)\theta_{b-1}(r). \tag{11.120}$$

The problem that we face here is that this effective model is supposed to capture the low-energy dynamics in the classical manifold of *analytic* textures. The previous relation would force $g(r,t)$ to be holomorphic, that is, $\partial_{\bar{z}} g = 0$, but then it would run into conflict with the unitarity of $g(r,t)$. We may speculate that a similar effective theory to that in (11.119) could continue to hold but that (11.120) would have to be replaced by something compatible with the analyticity constraint. A natural idea would be to write

$$\psi_a(r,t) = \mathcal{P}_{\text{hol}}\left(\sum_{b=1}^{d} g_{ab}(r,t)\theta_{b-1}(r)\right), \tag{11.121}$$

where $\mathcal{P}_{\text{hol}}$ is a projector (to be precisely defined) on the space of holomorphic functions. Note that we may have a similar deformation of the algebra of functions over the plane as that considered in Section 11.2.2, given the strong resemblance between the lowest Landau level and the subspace of holomorphic functions. In particular, a notion of star product should emerge, through the relation

$$\mathcal{P}_{\text{hol}}(f\mathcal{P}_{\text{hol}}(g\theta)) = \mathcal{P}_{\text{hol}}((f \star g)\theta). \tag{11.122}$$

These remain open questions at present. Even if this effective description of the residual interaction between the $d^2 - 1$ Goldstone branches in terms of a nonlinear sigma model on a 'non-commutative plane' is validated by future studies, another framework is likely to be needed to account for the interactions between the magnetophonon and these Goldstone modes. So the subject is far from being closed!

11.A Coherent states in the lowest Landau level

Let us choose the circular gauge. It is well known that the lowest Landau level corresponds in this gauge to wavefunctions of the form $\psi(r) = f(z)\exp(-|z|^2/4l^2)$, where $z = x + iy$ if $r = (x,y)$. An orthonormal basis on the infinite plane is given by

$$\langle r|n\rangle = \frac{(z/l)^n}{\sqrt{2\pi 2^n n!}\, l} e^{-|z|^2/4l^2}. \tag{11.123}$$

The projector on the lowest Landau level is then constructed as

$$\mathcal{P}_{\text{LLL}} = \sum_{n=0}^{\infty} |n\rangle\langle n| \tag{11.124}$$

Explicitly,

$$\langle r|\mathcal{P}_{\rm LLL}|r'\rangle = \frac{1}{2\pi l^2}\exp\left(\frac{z\bar{z}'}{2l^2} - \frac{|z|^2+|z'|^2}{4l^2}\right). \tag{11.125}$$

The state $\mathcal{P}_{\rm LLL}|r'\rangle$ belongs to the lowest Landau level. It has the distinctive property that it is orthogonal to the codimension-1 subspace of all the wavefunctions ψ that vanish at r'. Therefore, we expect it to be strongly localized around r', which is confirmed by a direct inspection of the behaviour of $|\langle r|\mathcal{P}_{\rm LLL}|r'\rangle|$ when r gets close to r'. This is then a natural candidate to define the coherent state centred at r'. Let us normalize this state. We have

$$\langle r'|\mathcal{P}^2_{\rm LLL}|r'\rangle = \langle r'|\mathcal{P}_{\rm LLL}|r'\rangle = \frac{1}{2\pi l^2}. \tag{11.126}$$

The normalized coherent state centred at r' is thus $|\Phi_{\bar{z}'}\rangle = \sqrt{2\pi}\, l\mathcal{P}_{\rm LLL}|r'\rangle$, and we have

$$\langle r|\Phi_{\bar{z}'}\rangle = \Phi_{\bar{z}'}(r) = \frac{1}{\sqrt{2\pi}\, l}\exp\left(\frac{z\bar{z}'}{2l^2} - \frac{|z|^2+|z'|^2}{4l^2}\right). \tag{11.127}$$

This expression motivates the notation because r' is involved only through $\bar{z}'$ in the analytical part (i.e. function of z) of the wavefunction $\Phi_{\bar{z}'}(r)$. The overlap between two coherent states is given by

$$\langle \Phi_{\bar{z}}|\Phi_{\bar{z}'}\rangle = \exp\left(\frac{z\bar{z}'}{2l^2} - \frac{|z|^2+|z'|^2}{4l^2}\right). \tag{11.128}$$

In particular, since $|\langle \Phi_{\bar{z}}|\Phi_{\bar{z}'}\rangle| = \exp(-|z-z'|^2/4l^2)$, this overlap decreases very quickly beyond a spatial scale given by the magnetic length. We also see that when this length goes to zero, coherent states become orthogonal, which fits nicely with our intuition of the classical limit.

11.B From covariant symbols on a two-dimensional plane to operators

It is convenient to introduce a raising operator b^+ and a lowering operator b, both acting in the lowest Landau level according to

$$b^+|n\rangle = \sqrt{n+1}\,|n+1\rangle, \qquad b|n\rangle = \sqrt{n}\,|n-1\rangle. \tag{11.129}$$

They satisfy the usual commutation relation $[b, b^+] = 1$. From this definition and the expression (11.127) for coherent states, we see that

$$|\Phi_{\bar{z}}\rangle = e^{-|z|^2/4l^2}\sum_{n=0}^{\infty}\frac{1}{\sqrt{n!}}\left(\frac{\bar{z}}{\sqrt{2}\, l}\right)^n |n\rangle = e^{-|z|^2/4l^2}\exp\left(\frac{\bar{z}}{\sqrt{2}\, l}b^+\right)|0\rangle. \tag{11.130}$$

From this, we check the important relations

$$b|\Phi_{\bar{z}}\rangle = \frac{\bar{z}}{\sqrt{2}\,l}\,|\Phi_{\bar{z}}\rangle, \qquad \langle\Phi_{\bar{z}}|b^+ = \frac{z}{\sqrt{2}\,l}\,\langle\Phi_{\bar{z}}|. \tag{11.131}$$

This shows that the covariant symbol is very easy to extract for *normal-ordered* operators (or *Wick-ordered* operators in Berezin's terminology), i.e. linear combinations of monomials of the form $(b^+)^m b^n$. The previous relations show indeed that

$$\widehat{z^m \bar{z}^n} = (\sqrt{2}\,l)^{m+n}(b^+)^m b^n. \tag{11.132}$$

11.C Single-particle density matrix in a texture Slater determinant

In a large magnetic field, and for a filling factor less than d, a very good approximation in which to study the low-energy physics is to restrict the single-particle states to the lowest Landau level. It is then convenient to introduce projected creation operators $\Psi_a^+(r)$ and annihilation operators $\Psi_a(r)$ for a single electron in internal state a at position r. Using the orthogonal basis $|n\rangle$ for $n = 0, 1, \ldots$, these operators are given by

$$\Psi_a(r) = \sum_{n=0}^{\infty} \langle r|n\rangle c_{a,n}, \qquad \Psi_a^+(r) = \sum_{n=0}^{\infty} \langle n|r\rangle c_{a,n}^+. \tag{11.133}$$

Here, $c_{a,n}$ destroys an electron in the internal state a and the basis orbital state $|n\rangle$, and $c_{a,n}^+$ is the corresponding creation operator. Assuming the canonical anticommutation rules $\{c_{a,n}, c_{a',n'}^+\} = \delta_{aa'}\delta_{nn'}$, we have the less conventional

$$\{\Psi_a(r), \Psi_{a'}^+(r')\} = \delta_{aa'}\langle r|\mathcal{P}_{\mathrm{LLL}}|r'\rangle. \tag{11.134}$$

The effect of the projection is to smear the expected delta function $\delta(r - r')$ on the right-hand side and replace it by the matrix elements of the projector $\mathcal{P}_{\mathrm{LLL}}$. As in the main text, let us denote by $\hat{P}_\psi$ the projector on the occupied subspace in the Slater determinant associated with the classical texture $\psi_a(r)$, and let $P_\psi(z, \bar{z}) \equiv P_\psi(r)$ be the corresponding covariant matrix symbol. A very useful fact is that we can express the single-particle density matrix in this Slater determinant, in terms of the covariant symbol $P_\psi(r)$. Indeed,

$$\langle\Psi_b^+(r')\Psi_a(r)\rangle = \langle r|\mathcal{P}_{\mathrm{LLL}}(\hat{P}_\psi)_{ab}\mathcal{P}_{\mathrm{LLL}}|r'\rangle. \tag{11.135}$$

So,

$$\langle\Psi_b^+(r')\Psi_a(r)\rangle = \frac{1}{2\pi l^2}\,\langle\Phi_{\bar{z}}|(\hat{P}_\psi)_{ab}|\Phi_{\bar{z}'}\rangle = \frac{1}{2\pi l^2}(P_\psi)_{ab}(r, r'). \tag{11.136}$$

The last term has to be understood as the result of a continuation process from the diagonal part $(P_\psi)_{ab}(r)$ to the non-diagonal one $(P_\psi)_{ab}(r, r')$. For any operator $\hat{A}$

acting on the lowest Landau level, with covariant symbol $A(z, \bar{z}) = A(r)$, we have the useful relation

$$A(r, r') = A\left(\frac{r + r'}{2} + \frac{i}{2}\hat{z} \times (r' - r)\right) \exp\left[-i\frac{r \times r'}{2l^2} - \frac{(r - r')^2}{4l^2}\right]. \qquad (11.137)$$

Finally, our main result is

$$\langle \Psi_b^+(r')\Psi_a(r)\rangle = \frac{1}{2\pi l^2}(P_\psi)_{ab}\left(\frac{r + r'}{2} + \frac{i}{2}\hat{z} \times (r' - r)\right) \exp\left[-i\frac{r \times r'}{2l^2} - \frac{(r - r')^2}{4l^2}\right]. \qquad (11.138)$$

To prove (11.137), let us consider the operator $\hat{f}_{\lambda,\mu}$, whose covariant symbol is $f_{\lambda,\mu}(r) \equiv f_{\lambda,\mu}(z, \bar{z}) = \exp(\lambda z + \mu \bar{z})$. By taking derivatives with respect to the external variables λ and μ, we can reach any monomial in z and $\bar{z}$. From (11.132), we see that

$$\hat{f}_{\lambda,\mu} = \exp(\sqrt{2}\, l\lambda b^\dagger) \exp(\sqrt{2}\, l\mu b). \qquad (11.139)$$

The extended symbol is then easily computed:

$$f_{\lambda,\mu}(r, r') \equiv \langle \Phi_{\bar{z}}|\hat{f}_{\lambda,\mu}|\Phi_{\bar{z}'}\rangle = \exp(\lambda z + \mu \bar{z}')\, \langle \Phi_{\bar{z}}|\Phi_{\bar{z}'}\rangle. \qquad (11.140)$$

Now, if an analytic function of x and y, $g(x, y) = g(z, \bar{z})$, is continued to form the function $g(z, \bar{z}')$, the variable $x = \frac{1}{2}(z + \bar{z})$ has to be changed into $x_{\text{new}} = \frac{1}{2}(z + \bar{z}')$. Likewise, $y = \frac{1}{2}i(\bar{z} - z)$ has to be changed into $y_{\text{new}} = \frac{1}{2}i(\bar{z}' - z)$. Explicitly, we get

$$x_{\text{new}} = \frac{1}{2}(x + x') - \frac{1}{2}i(y' - y), \qquad (11.141)$$

$$y_{\text{new}} = \frac{1}{2}(y + y') + \frac{1}{2}i(x' - x). \qquad (11.142)$$

This is conveniently expressed as

$$r_{\text{new}} = \frac{r + r'}{2} + \frac{i}{2}\hat{z} \times (r' - r). \qquad (11.143)$$

Using the expression (11.128) for the overlap between coherent states, we see that

$$f_{\lambda,\mu}(r, r') = f_{\lambda,\mu}\left(\frac{r + r'}{2} + \frac{i}{2}\hat{z} \times (r' - r)\right) \exp\left[-i\frac{r \times r'}{2l^2} - \frac{(r - r')^2}{4l^2}\right], \qquad (11.144)$$

which establishes (11.137).

11.D Hamiltonians with quadratic covariant symbol

To keep the discussion simple, let us consider a single degree of freedom, whose associated phase space is the plane. It is convenient to use the notations of Sections 11.A and 11.B, with the replacement $l^2 = \hbar$. Suppose that for some Hamiltonian $\hat{H}$, the covariant symbol $H(z,\bar{z}) \equiv \langle \Phi_{\bar{z}} | \hat{H} | \Phi_{\bar{z}} \rangle$ is minimal at some point z_0. By a translation, it is possible to shift the origin in the plane, so we may assume that $z_0 = 0$. The Taylor expansion of $H(z,\bar{z})$ around the origin contains no linear term, so it reads

$$H(z,\bar{z}) = E_0 + \frac{\omega_0}{2}\bar{z}z + \frac{\Delta}{4}z^2 + \frac{\bar{\Delta}}{4}\bar{z}^2 + \dots \qquad (11.145)$$

Here ω_0 is real and positive, and Δ is complex. The corresponding quantum Hamiltonian $\hat{H}$ can be written in terms of b and b^+ operators using the normal-ordering prescription, as shown in Appendix 11.B:

$$\hat{H} = E_0 + \hbar\omega_0 b^+ b + \frac{\hbar\Delta}{2}(b^+)^2 + \frac{\hbar\bar{\Delta}}{2}b^2 + \dots \qquad (11.146)$$

After a $U(1)$ rotation $b \to be^{i\alpha}$, $b^+ \to b^+ e^{-i\alpha}$, it is possible to eliminate the phase of Δ, which will then be assumed to be real. This quadratic Hamiltonian is diagonalized by a Bogoliubov transformation: $b = (\cosh\theta)\beta - (\sinh\theta)\beta^+$. Terms of the form β^2 or $(\beta^+)^2$ disappear if we choose $\tanh 2\theta = \Delta/\omega_0$. We then get

$$\hat{H} = E_0 + \frac{1}{2}\hbar(\Omega - \omega_0) + \hbar\Omega b^+ b. \qquad (11.147)$$

Here $\Omega = \sqrt{\omega_0^2 - \Delta^2}$ is the oscillation frequency of the harmonic mode. The quantum zero-point correction to the variational estimate E_0 of the ground-state energy is

$$H_{\mathrm{qzpc}} = \frac{1}{2}\hbar(\Omega - \omega_0). \qquad (11.148)$$

This correction is always negative and it vanishes only when $\Delta = 0$. This is in agreement with the fact that the normal-ordered term $\hbar\omega_0 b^+ b$ is diagonal in the standard oscillator basis and gives zero when applied to the coherent state $|\Phi_0\rangle = |0\rangle$. Only the $(b^+)^2$ term and its Hermitian conjugate can change the ground-state energy, but it is purely off-diagonal, so the corresponding energy shift has to be negative.

This discussion may seem surprising to experienced readers who expect $\frac{1}{2}\hbar\Omega$ to be the quantum zero-point energy of a harmonic oscillator. In fact, there is no contradiction between the two viewpoints, which differ only in a different choice of the reference energy. If we start from the traditional harmonic-oscillator Hamiltonian $\hat{H} = \frac{1}{2}\omega_0(\hat{p}^2 + \hat{q}^2)$, we get $\hat{H} = \hbar\omega_0 b^+ b + \frac{1}{2}\hbar\omega_0$. The corresponding covariant symbol is $H(z,\bar{z}) = E_0 + \frac{1}{2}\omega_0|z|^2$, so the traditional zero-point energy is incorporated in the constant $E_0 = \frac{1}{2}\hbar\omega_0$. In physical terms, H_{qzpc} measures the additional energy shift due to squeezing phenomena induced by non-diagonal operators such as $(b^+)^2$ and b^2. If the coherent states initially chosen in our variational procedure are the correct ones, there are no such squeezing terms, and the traditional zero-point energy is already correctly taken into account by computing the expectation value of the quantum-mechanical operator $\hat{H}$ in the chosen coherent-state family. In our approach, a non-zero H_{qzpc} signals that residual quantum fluctuations will dress the initial optimal coherent state and introduce a finite amount of squeezing. Most of the above remarks can be generalized to quadratic Hamiltonians with an arbitrary finite number of oscillators, but to keep the discussion simple, we won't elaborate further on this issue.

Acknowledgements

I would like to thank my friends and collaborators who introduced me to this beautiful field and shared many insights with me: Pascal Lederer, Mark Görbig, Roderich Moessner, and Dima Kovrizhin. I am also grateful to Frédéric Faure, Laurent Charles, and San Nu Ngoc for several discussions on the theme of semiclassical analysis and geometric quantization, which helped me to understand this subject a lot better. Finally, I wish to thank the organizers of this wonderful Les Houches summer school and the students for their very stimulating questions and comments.

References

[1] G. Toulouse and M. Kléman, *J. Phys. (Paris) Lett.* **37**, L149 (1976).
[2] G. E. Volovik and V. P. Mineev, *Sov. Phys. JETP* **45**, 1186 (1977).
[3] N. D. Mermin, *Rev. Mod. Phys.* **51**, 591 (1979).
[4] S. L. Sondhi, A. Karlhede, S. A. Kivelson, and E. H. Rezayi, *Phys. Rev. B* **47**, 16419 (1993).
[5] S. E. Barrett, G. Dabbagh, L. N. Pfeiffer, K. W. West, and R. Tycko, *Phys. Rev. Lett.* **74**, 5112 (1995).
[6] R. Tycko, S. E. Barrett, G. Dabbagh, L. N. Pfeiffer, and K. W. West, *Science* **268**, 1460 (1995).
[7] L. Brey, H. A. Fertig, R. Côté, and A. H. MacDonald, *Phys. Rev. Lett.* **75**, 2562 (1995).
[8] R. Côté, A. H. MacDonald, L. Brey, H. A. Fertig, S. M. Girvin, and H. T. C. Stoof, *Phys. Rev. Lett.* **78**, 4825 (1997).
[9] V. Bayot, E. Grivei, S. Melinte, M. B. Santos, and M. Shayegan, *Phys. Rev. Lett.* **76**, 4584 (1996).
[10] V. Bayot, E. Grivei, J.-M. Beuken, S. Melinte, and M. Shayegan, *Phys. Rev. Lett.* **79**, 1718 (1997).
[11] G. Gervais, H. L. Stormer, D. C. Tsui, P. L. Kuhns, W. G. Moulton, A. P. Reyes, L. N. Pfeiffer, K. W. Baldwin, and K. W. West, *Phys. Rev. Lett.* **94**, 196803 (2005).
[12] Y. Gallais, J. Yan, A. Pinczuk, L. N. Pfeiffer, and K. W. West, *Phys. Rev. Lett.* **100**, 086806 (2008).
[13] H. Zhu, G. Sambandamurthy, Y. P. Chen, P. Jiang, L. W. Engel, D. C. Tsui, L. N. Pfeiffer, and K. W. West, *Phys. Rev. Lett.* **104**, 226801 (2010).

[14] S. Q. Murphy, J. P. Eisenstein, G. S. Boebinger, L. N. Pfeiffer, and K. W. West, *Phys. Rev. Lett.* **72**, 728 (1994).
[15] K. Moon, H. Mori, K. Yang, S. M. Girvin, A. H. MacDonald, L. Zheng, D. Yoshioka, and S.-C. Zhang, *Phys. Rev. B* **51**, 5138 (1995).
[16] J. Alicea and M. P. A. Fisher, *Phys. Rev. B* 74, 075422 (2006).
[17] M. O. Goerbig, R. Moessner, and B. Douçot, *Phys. Rev. B* **74**, 161407 (2006).
[18] S. Prabhu-Gaunkar, S. Birner, S. Dasgupta, C. Knaak, and M. Grayson, *Phys. Rev. B* **84**, 125319 (2011).
[19] K. Eng, R. N. McFarland, and B. E. Kane, *Phys. Rev. Lett.* **99**, 016801 (2007).
[20] F. Herzog, M. Bichler, G. Koblmüller, S. Prabhu-Gaunkar, W. Zhou, and M. Grayson, *Appl. Phys. Lett.* **100**, 192106 (2012).
[21] R. W. Cherng and E. Demler, *Phys. Rev. A* **83**, 053613 (2011).
[22] R. W. Cherng and E. Demler, *Phys. Rev. A* **83**, 053614 (2011).
[23] J. Bourassa, B. Roostaei, R. Côté, H. A. Fertig, and K. Mullen, *Phys. Rev. B* **74**, 195320 (2006).
[24] R. Côté, D. B. Boisvert, J. Bourassa, M. Boissonneault, and H. A. Fertig, *Phys. Rev. B* **76**, 125320 (2007).
[25] I. B. Spielman, L. A. Tracy, J. P. Eisenstein, L. N. Pfeiffer, and K. W. West, *Phys. Rev. Lett.* **94**, 076803 (2005).
[26] N. Kumada, K. Muraki, K. Hashimoto, and Y. Hirayama, *Phys. Rev. Lett.* **94**, 096802 (2005).
[27] D. L. Kovrizhin, B. Douçot, and R. Moessner, *Phys. Rev. Lett.* **110**, 186802 (2013).
[28] A. H. MacDonald, H. A. Fertig, and L. Brey, *Phys. Rev. Lett.* **76**, 2153 (1996).
[29] V. Pasquier, *Phys. Lett. B* **490**, 258 (2000).
[30] V. Pasquier, *Phys. Lett. B* **513**, 241 (2001).
[31] Z. F. Ezawa, G. Tsitsishvili, and K. Hasebe, *Phys. Rev. B* **67**, 125314 (2003).
[32] Z. F. Ezawa, and G. Tsitsishvili, *Phys. Rev. D* **72**, 085002 (2005).
[33] R. J. Glauber, *Phys. Rev.* **131**, 2766 (1963).
[34] F. A. Berezin, *Math. USSR Izv.*, **8**, 1109 (1974).
[35] D. J. Rowe and A. G. Ryman, *Phys. Rev. Lett.* **45**, 406 (1980).
[36] D. J. Rowe, A. G. Ryman, and G. Rosensteel, *Phys. Rev. A* **22**, 2362 (1980).
[37] F. Faure, *Lett. Math. Phys.* **55**, 219 (2001).
[38] J. Sjöstrand, *C. R. Acad. Sci. Paris, Ser. I* **317**, 217 (1993).
[39] C. Emmrich and A. Weinstein, *Commun. Math. Phys.* **176**, 710 (1996).
[40] F. A. Berezin, *Commun. Math. Phys.* **63**, 131 (1978).
[41] R. Rajaraman, *Solitons and Instantons*, North-Holland, Amsterdam (1982).
[42] V. I. Arnold, *Mathematical Methods of Classical Mechanics*, 2nd edn, Springer-Verlag, Berlin (1989), see Appendix 6.
[43] B. Douçot and P. Simon, *J. Phys. A* **31**, 5855 (1998).
[44] V. I. Arnold and A. B. Givental, Symplectic geometry, in *Dynamical Systems IV: Symplectic Geometry and its Applications* (ed. V. I. Arnold and S. P. Novikov), pp. 1–136, Encyclopaedia of Mathematical Sciences, Volume 4, Springer-Verlag, Berlin (2001), see p. 24.

[45] M. Abolfath, J. J. Palacios, H. A. Fertig, S. M. Girvin, and A. H. MacDonald, *Phys. Rev. B* **56**, 6795 (1997).
[46] O. Debarre, *Complex Tori and Abelian Varieties*, AMS/SMF, Providence, RI/Paris (2005), see Chapter IV.
[47] F. D. M. Haldane, and E. H. Rezayi, *Phys. Rev. B* **31**, 2529, (1985).
[48] S. I. Matveenko and G. V. Shlyapnikov, *Phys. Rev. A* **83**, 033604 (2011).
[49] R. Côté and A. H. MacDonald, *Phys. Rev. Lett.* **65**, 2662 (1990).
[50] T. Dombre and N. Read, *Phys. Rev. B* **39**, 6797 (1989).

12
Out-of-equilibrium behaviour in topologically ordered systems on a lattice: fractionalized excitations and kinematic constraints

Claudio Castelnovo

TCM Group
Cavendish Laboratory, University of Cambridge
Cambridge CB3 0HE, UK

Topological Aspects of Condensed Matter Physics. First Edition. C. Chamon et al.

Chapter Contents

Colour figures. For those figures in this chapter that use colour, please see the version of these lecture notes at http://topo-houches.pks.mpg.de. These figures are indicated by '[Colour online]' at the start of the caption.

Preface

These lecture notes touch upon aspects of out-of-equilibrium behaviour in topologically ordered systems, broadly interpreted. It should be noted that the selection of topics reflects a personal choice and it is not intended as a systematic review. Hopefully, these notes will stimulate the appetite of the interested reader to pursue further study in this area of research.

12.1 Topological order, broadly interpreted

First, it should be pointed out that these lectures do not aim to introduce or adhere to a specific and accurate definition of *topological order*. Other lectures at this school may serve that purpose. Here, 'topologically ordered systems' refers very loosely to systems that do not develop a local order parameter at low temperature (e.g. via the spontaneous breaking of a symmetry) and yet exhibit non-trivial global properties.

Within this broad definition, which encompasses classical statistical- mechanical systems as well as quantum-mechanical systems, we shall take topologically ordered systems to be characterized by the following properties:

1. There is a lack of a local order parameter characterizing the low-temperature phase; rather, these systems remain in a disordered 'liquid' state with non-trivial non-local correlations.
2. The collective excitations of the low-temperature phase take the form of point-like quasiparticles that carry a fraction of the microscopic degrees of freedom in the system.

It is often the case that the low-temperature phase can be effectively interpreted as a special *vacuum*, capable of hosting the emergent collective excitations as its *elementary particles*.

The properties of the emergent excitations and those of the vacuum are closely related. From a dynamical perspective, the vacuum determines both at the local and at the global (topological) levels the rules of motion of the excitations. Vice versa, as the quasiparticles move across the system, they change it. The excitations act indeed as dynamical facilitators, since it is through their motion that the system can respond to external perturbations and/or relax to equilibrium.

The close interplay between excitations and their vacuum is often responsible for non-trivial and interesting dynamical properties, in particular when the system is driven out of equilibrium. This is a rich and interesting regime, controlled by the interplay of many (often independently tunable) factors, such as the interactions between the emergent quasiparticles, the local and global kinematic constraints imposed by the vacuum, and, in quantum-mechanical systems, the mutual statistics of the excitations.

For the reader who may be familiar with these models, examples include lattice dimer models, vertex models (e.g. the six- and eight-vertex models in two dimensions (2D) and spin ices in 3D), and the toric code model and Kitaev's model.

In these lectures, we will discuss specifically the case of classical spin ice (Section 12.2) and the quantum toric code (Section 12.3). The combination of strongly correlated physics, topological order, and far-from-equilibrium behaviour makes the study of these systems a tall order. For classical systems, we will see that one can make substantial progress in understanding the dynamics, in particular following thermal and field quenches, thanks to an effective modelling of the vacuum and its emergent excitations. At the quantum-mechanical level, a similar modelling is not readily available and the depth of our present understanding is limited to the study of leading energy barriers and asymptotic behaviour. We close with the discussion of an intriguing parallel that can be drawn between the toric code Hamiltonian and a class of lattice systems known as kinetically constrained models, which were designed to achieve trivially disordered low-temperature phases with emergent *long relaxation timescales* (Section 12.3.4).

12.2 Example 1: (classical) spin ice

As shown in Chapter 3 of this volume, the behaviour of spin-ice models and materials at low temperature can be understood as a spin-liquid 'vacuum' with an emergent gauge symmetry (inherited from the 2in–2out local constraint that minimizes the energy). This vacuum hosts classical fractionalized excitations that take the form of free magnetic charges in 3D, or emergent magnetic monopoles. (For a review, see e.g. [1, 2].)

As a first approximation, the collective behaviour of these systems at low temperature (Fig. 12.1) resembles that of a Coulomb liquid or pair plasma (i.e. a gas of positive and negative Coulomb-interacting point charges, which is overall neutral) [3].

Such an effective description goes a long way to capture the low-temperature behaviour of spin ice, with far less effort than would otherwise be required by conventional theoretical approaches for strongly correlated magnetic systems on a 3D lattice.

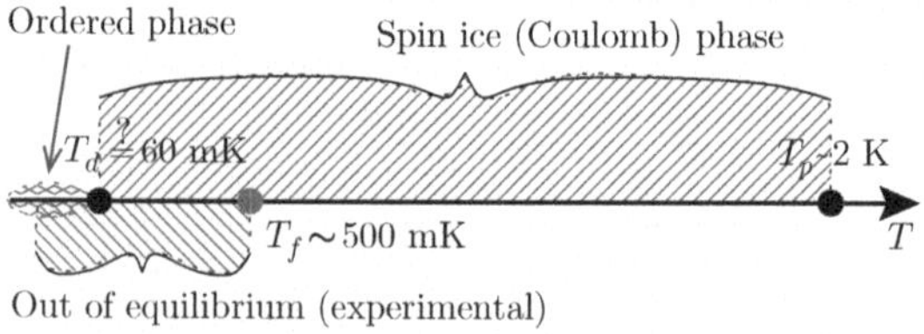

Fig. 12.1 Schematic illustration of the different temperature regimes in spin ice. The theoretically predicted ordering transition at T_d appears to be prevented in experiments by freezing of the magnetic degrees of freedom below a threshold temperature T_f, as evidenced for example by a discrepancy between field-cooled and zero-field-cooled magnetization. The 2 in–2out spin-ice regime undergoes a continuous crossover to trivial paramagnetic behaviour around T_p.

An example can be found in the use of Debye–Hückel theory to obtain the low-temperature heat capacity of spin ice. This is discussed in detail in [3] and we only report here a brief outline of the approach for illustrative purposes. In order to compute the heat capacity, one often looks for ways to approximate the free energy of the system. With strongly correlated localized spins, it is customary for instance to use appropriate truncated expansions. Rather than working with the spins directly, however, in spin ice one can choose to work with the effective description in terms of a gas of Coulomb-interacting charges, focusing on the nature of the elementary excitations and neglecting, to a first approximation, the 2in–2out spin background. One can then assemble the free energy of the system in this new language:

$$F = F_{\text{chem pot}} + F_{\text{charge entropy}} + F_{\text{el}}, \tag{12.1}$$

where $F_{\text{chem pot}}$ is the contribution due to the fact that emergent excitations cost energy (chemical potential); $F_{\text{charge entropy}}$ is the entropic contribution of distributing point charges on a lattice; and F_{el} is the electrostatic (or, better, magnetostatic) contribution.

The first two terms are straightforward. $F_{\text{chem pot}} \propto \rho\Delta$, where ρ is the monopole density and Δ is their bare energy cost (i.e. their cost in a generic 2in–2out spin-ice configuration infinitely far from any other monopoles); and $F_{\text{charge entropy}} \propto -TS_{\text{mixing}}$, where the mixing entropy takes the usual form $S_{\text{mixing}} \propto -\rho \ln \rho - (1-\rho)\ln(1-\rho)$ (for a more detailed expression accounting for positive and negative charges separately, see [3]).

The third term is a tall order and an exact expression is not known. However, several analytical approximations are readily available in the literature on Coulomb liquids and charged plasmas. One of the simplest approximations goes under the name of Debye–Hückel theory (see e.g. [4]). It provides an analytical expression for F_{el} in terms of the ratio between the Coulomb interaction strength at nearest-neighbour distance, E_{nn}, and the temperature T as

$$F_{\text{el}} \propto -T\left[\tfrac{1}{2}x^2 - x + \ln(1+x)\right], \qquad x \propto \sqrt{\frac{E_{\text{nn}}}{T}\rho}\,. \tag{12.2}$$

We now have an expression for the full free energy F as a function of the monopole density ρ. All the relevant parameters (Δ, E_{nn}) and the proportionality constants that have been omitted above can in fact be obtained independently from microscopic details about spin ice. Therefore, by minimization with respect to ρ, one can solve for the thermodynamic equilibrium value of the monopole density as a function of temperature. From the latter, one then obtains the free energy of the system and, using known thermodynamic relations, the heat capacity. In the non-interacting limit $F_{\text{el}} = 0$, analytical expressions can be obtained, whereas in the Debye–Hückel case above, one has to resort to a recursive set of equations that can be solved numerically to the desired accuracy.

The outcome is in excellent agreement with numerical simulations as well as with the experimentally measured behaviour of the heat capacity of the system at low temperature [5]—far better than one can achieve with conventional approaches for strongly interacting localized spin systems. The Debye–Hückel approach also allows one to obtain further insight into the system, for instance the behaviour of the density of monopoles and their screening length.

Exercise 12.1 Using the parameters in [3], compute $F(\rho)$ in the Debye–Hückel approximation and obtain the recursive equations for the equilibrium monopole density.

The benefit of a Coulomb-liquid description is not limited to thermodynamic properties. It is also key to understanding response and equilibration in these systems. A generic spin in a 2in–2out background can only flip if a thermal fluctuation allows it to overcome the energy barrier to create a monopole–antimonopole pair, $\Delta_s = 2\Delta - E_{\rm nn}$. This is unlikely to happen when the temperature is appreciably smaller than the barrier, since the spin-flip process takes a correspondingly long time $\sim e^{\Delta_s/T}$. On the contrary, three of the four spins next to an isolated monopole can flip without incurring such large energy barrier. Their reversal results in the monopole hopping from one tetrahedron to the next (see Fig. 12.2), whereby the number of monopoles remains unchanged (we disregard here the weaker energetic contribution due to long-range interactions with other monopoles).

The first process can take place at a density $\sim (1-\rho)$ of sites on the lattice. Therefore, the associated timescale is $\tau \sim e^{\Delta_s/T}/(1-\rho) \sim e^{\Delta_s/T}$, at the regime of interest of sufficiently low temperatures where $\rho \sim e^{-\Delta/T} \ll 1$. The second process does not

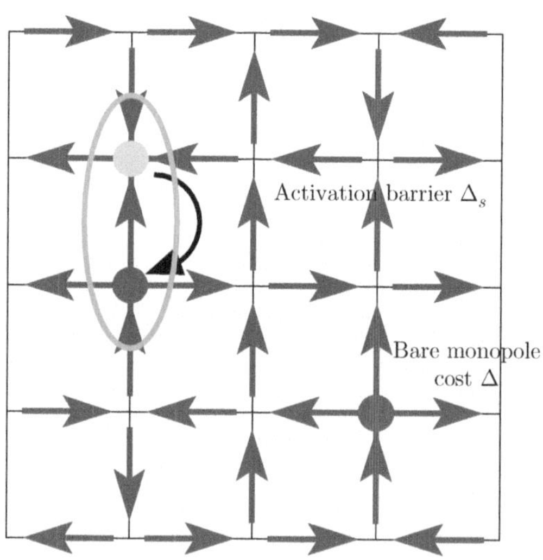

Fig. 12.2 On the left, a generic spin reversal in 2in–2out spin ice incurs a large energy barrier due to the creation of a monopole–antimonopole pair. On the right, a monopole acts as a spin facilitator, in that it allows three of the four neighbouring spins to flip without such barrier. Flipping one of those spins results in the monopole hopping to a neighbouring tetrahedron and no energy change in the system (up to long-range Coulombic interactions).

incur an energy barrier, but it can only take place next to an existing monopole and therefore $\tau \sim 1/\rho \sim e^{\Delta/T}$. Which of the two processes dominates is determined by the smallest of the two energy scales, Δ and $\Delta_s = 2\Delta - E_{\rm nn}$. In known spin-ice materials, $\Delta > E_{\rm nn}$ and spin flip via monopole hopping is exponentially preferred at low temperatures.

From these observations, we conclude that magnetic monopoles act as facilitators for the spin-flip dynamics in the system. Therefore, they play a key role in the way the system responds to external perturbations and equilibrates. Typical timescales for macroscopic response are proportional to the inverse density of monopoles, $\tau \sim \tau_0/\rho$, where τ_0 is some characteristic single-spin-flip timescale. A hydrodynamic theory that demonstrates this relationship for non-interacting monopoles is presented in [6]. This result captures well the leading order divergence of magnetic relaxation timescales observed in AC susceptibility measurements [7, 8]. Corrections due to the Coulomb interactions between monopoles were investigated numerically in [8]. (See also [9–15] for other factors playing a role in the dynamical slowing down at low temperatures, and potentially interesting open issues.)

Suggested reading See [6] for a derivation of the linear response dynamics in the non-interacting monopole approximation using hydrodynamics of irreversible processes.

In general, a system where point-like excitations freely moving in three dimensions are responsible for bulk magnetic response is bound to exhibit an interesting separation of timescales. On the one hand, monopoles are only created and annihilated in pairs. Therefore, monopole density relaxation processes involve monopole motion over distances of the order of the average monopole–monopole separation $\xi \sim \rho^{-1/3}$. In a ballistic regime where positive monopoles are driven towards negative monopoles, the corresponding timescale is of the order of ξ monopole hops. At sufficiently high temperatures and/or beyond the screening length, the monopole motion is diffusive and the timescale varies as $\xi^2 \sim \rho^{-2/3}$. Finally, any changes in the bulk magnetization and other observables that depend on the local spin orientations require the monopoles to visit a finite fraction of the spins in the systems. Therefore, on average, they have to move across a finite fraction of ξ^3 spins per monopole, corresponding to a timescale of the order of $\xi^3 \sim \rho^{-1}$.

The Coulomb-liquid nature of low-temperature excitations leads to other unusual features in the dynamical behaviour of driven spin-ice systems close to equilibrium. For example, response properties typical of electrolytes have been argued to apply to spin ices in the form of an increase in the monopole density upon switching on an external magnetic field, in an analogue of the second *Wien effect* [16–18] in electrolytes. Experimental work testing this hypothesis has been presented [17], but no consensus has yet been reached on the observation of a magnetic Wien effect.

Suggested reading It is insightful to revisit Onsager's theory for electrolytes (see e.g. [18]) and translate it into the magnetic language appropriate for spin ices [17].

These examples illustrate the close interplay between the nature of the fractionalized excitations in spin ice and its dynamical properties. Such interplay is bound to be reflected, if not enhanced, when the system is brought strongly out of equilibrium. In the following, we shall discuss a couple of examples in some detail. Specifically, we shall consider sudden quenches from a high- to a low-monopole-density state, triggered by tuning either the temperature or an applied magnetic field.

We mention in passing that the phase diagram of spin ice includes a critical end point in presence of a magnetic field. 'Slow quenches' (i.e. continuous variations of the parameters as functions of time) to or across the critical point should therefore give rise to out-of-equilibrium scaling behaviour à la Kibble–Zurek, in the novel context of a system with emergent gauge symmetry and emergent Coulomb-interacting quasi-particles. Theoretical and experimental work investigating this possibility is currently underway.

12.2.1 Thermal quenches

One way to cause the system to evolve from a state with high monopole density to one with low monopole density is by lowering its temperature. Here we consider for simplicity the case where the system is initially at infinite temperature (Ising paramagnet) and it is suddenly quenched to a target (low) temperature [19]. In effective Coulomb-liquid terms, this is equivalent to quenching a plasma where positive and negative charges can be created (and annihilated) only in pairs, and each charge costs some finite amount of energy Δ.

Immediately after the quench, the system is strongly out of equilibrium (e.g. the monopole density is much larger than its thermodynamic value at the target temperature). When coupled to a thermal bath, it will relax to equilibrium via the available dynamical processes, namely monopole motion and monopole–antimonopole creation/annihilation.

Note the stark contrast with conventional magnets, where thermal quenches are usually described in terms of domain nucleation, growth, and coarsening [20]. In spin ice, it is clear that this language is unlikely the right one to understand the evolution of the system. Rather, we see that the language of reaction–diffusion processes is more befitting.

Whereas monopole–antimonopole annihilation events lower the energy of the system, pair-creation events face a finite energy cost Δ_s. Detailed balance (i.e. the requirement that the dynamical processes are compatible with thermodynamic equilibrium) imposes that creation is statistically suppressed with respect to annihilation by a Boltzmann factor $e^{-\Delta_s/T}$.

For the sake of the discussion below, we limit ourselves to the case where the target temperature is much smaller than the pair creation energy cost and $e^{-\Delta_s/T} \sim 0$. We can thus neglect creation processes altogether. Within this assumption, the equilibrium density of monopoles at the target temperature is also vanishingly small and we shall set it to zero (recall that $\rho \sim e^{-\Delta/T}$ and $\Delta < \Delta_s < 2\Delta$; therefore, $\Delta_s \gg T$ implies $\Delta \gg T$ and $\rho \sim 0$). The equations of motion for the monopole density can be generically written as [21]

$$\frac{\partial \rho_\pm(r,t)}{\partial t} + \nabla \cdot \mathbf{J}_\pm = -\kappa\, \rho_+(r,t)\rho_-(r,t), \tag{12.3}$$

$$\mathbf{J}_\pm = -D\nabla \rho_\pm(r,t) - \mu q_\pm \rho_\pm(r,t)\nabla V(r,t), \tag{12.4}$$

where $\rho_\pm$ and $\mathbf{J}_\pm$ are the densities and currents of positive and negative monopoles, respectively, κ is the annihilation reaction constant, and the two current terms are due to diffusion (constant D) and deterministic drift caused by (long-range) interactions (mobility μ, interaction potential $V(r)$). In spin ice systems, it is often the case that the relevant constants can be estimated analytically or obtained from independent comparisons with simulations or experiments, remarkably leaving few to no fitting parameters in the equations!

12.2.1.1 Nearest-neighbour spin ice

Let us focus first on the case where spin–spin interactions are truncated at nearest-neighbour distance and correspondingly the charges in the Coulomb liquid language are non-interacting ($V(r) = 0$). Incidentally, in this case, $\Delta_s = 2\Delta$.

Within this approximation, the dynamical processes in the system are limited to diffusion of non-interacting charges and monopole–antimonopole annihilation events. At mean-field level (uniform system, no spatial dependence), the diffusive part is irrelevant and we are left with a straightforward reaction equation:

$$\frac{d\rho}{dt} = -\kappa \rho^2(t). \tag{12.5}$$

The right-hand side is determined by the rate of monopole annihilation events, which is proportional to the probability of finding a monopole–antimonopole pair in the system ($\sim \rho^2$) divided by the characteristic timescale for a single annihilation event to take place (namely, the characteristic spin-flip timescale τ_0). The constant $\kappa \propto 1/\tau_0$ depends on details of the underlying microscopic lattice through a combinatorial factor accounting for the ways to arrange two monopoles next to one another across a bond of the lattice.

Exercise 12.2 Provide an estimate of κ for spin ice on the pyrochlore lattice and compare your answer with the two values provided in [19] and reported in the caption of Fig. 12.3.

The mean field equation (12.5), complemented by the initial condition $\rho(t=0) = \rho_0$, can be solved straightforwardly to find

$$\rho(t) = \frac{\rho_0}{1 + \kappa \rho_0 t}, \tag{12.6}$$

and the long-time decay in the monopole density goes as t^{-1}. The accuracy of the mean-field solution in describing the behaviour of a nearest-neighbour spin ice depends crucially on how uniform the initial charge distribution is, to ensure that diffusion timescales are indeed irrelevant.

Spatial variations in the initial distribution of monopoles and antimonopoles can, however, alter the behaviour significantly [22]. For instance, if the charges are distributed entirely at random with density ρ_0, then the net charge fluctuations in a volume of linear size ℓ scale as $\sqrt{\rho_0 \ell^3}$. Given that annihilation processes conserve the local net charge (they always remove one positive and one negative monopole at the same time), they cannot remove these fluctuations. After a time t sufficient for monopoles to diffuse over the length ℓ (i.e. $\ell = \sqrt{Dt}$), all possible annihilation events within the volume of size ℓ will have taken place, leaving behind a number $\sim \sqrt{\rho_0 \ell^3}$ of monopoles of the same charge owing to the statistical net charge fluctuations. The density of leftover monopoles scales as $\sqrt{\rho_0 \ell^3}/\ell^3 = \rho_0^{1/2} (Dt)^{-3/4}$; it decays with time more slowly than the mean-field behaviour ($\sim t^{-1}$) and therefore dominates at long times. (We refer the reader to [22] for a more detailed derivation and discussion of this result.)

However, none of this in fact applies to spin ice. As a monopole travels along a given path across the system, it modifies the underlying spin-ice vacuum by polarizing the spins along the path. Another monopole of the same charge *cannot* follow the same path in the same direction. Equivalently, we can drive at most ℓ^2 monopoles of equal charge across a system of linear size ℓ before the system becomes fully polarized and no more monopoles of the same charge can travel in that same direction. This means that the most net charge that can accumulate in a volume ℓ^3 of a spin-ice system is of the order of ℓ^2. The density of leftover monopoles in spin ice therefore scales as $\sqrt{\ell^2}/\ell^3 = (Dt)^{-1}$ rather than $(Dt)^{-3/4}$, which is no longer asymptotically slower than mean-field behaviour.

These observations are confirmed by the excellent agreement between Monte Carlo simulations of thermal quenches in nearest-neighbour spin ice and the solution of the mean-field equation, (12.6), illustrated in Fig. 12.3. Notice that the agreement is achieved without any fitting parameters!

12.2.1.2 Dipolar spin ice

Let us now consider the case of dipolar spin ice, where monopole excitations are coupled by long-range Coulomb interactions. In general, they introduce an additional energetic term that has a smoothing effect on spatial fluctuations of the net charge. Therefore, the naive expectation from this coarse-grained picture is that the monopole density decay following a thermal quench in dipolar spin ice is at least as fast as in the nearest-neighbour case. (We refer the reader to [21] for a discussion of annihilation–diffusion reaction processes with long-range interactions.)

Monte Carlo simulations confirm this expectation at short times, as illustrated in Fig. 12.4(c). However, for sufficiently low target temperatures, a long-lived metastable plateau develops in the time evolution of the monopole density. This new and unexpected feature is due to a curious interplay between long-range emergent physics

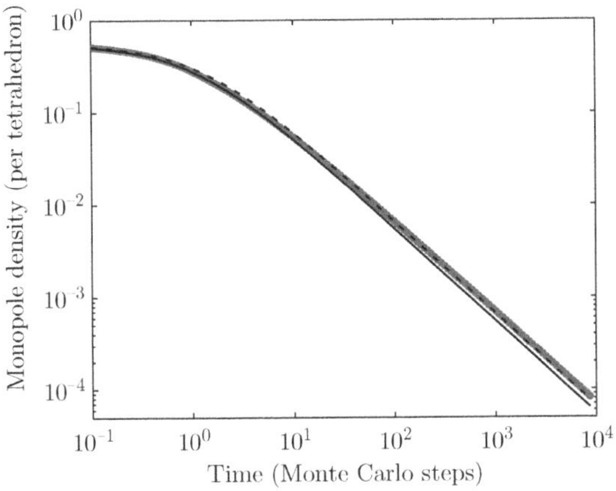

Fig. 12.3 Monopole density evolution in nearest-neighbour spin ice, after a thermal quench from $T = 10\,\mathrm{K}$ to $T = 0\,\mathrm{K}$, for system sizes $L = 32$, 64, and 128. The analytical mean-field result (12.6) is shown for $\kappa = 3/2\tau_0$ (dashed black line) and $\kappa = 9/5\tau_0$ (solid black line); see [19]. (Reprinted with permission from [19]. Copyright 2010 by the American Physical Society.)

(the Coulomb-liquid description) and lattice-scale physics (related to how monopole motion changes the underlying spin-ice vacuum).

When a positive and a negative monopole meet in spin ice, the spin between them can sometimes be the minority spin rather than one of the three majority ones, as illustrated in Fig. 12.4(a). In this case, flipping the spin does not annihilate the two monopoles but rather creates an even more energetically costly defect: a 4in and a 4out pair of tetrahedra. At low temperatures, the likelihood of such process is so low that it is effectively forbidden. We shall dub such monopole–antimonopole pairs *non-contractible*. Once they meet at the 'wrong' spin (i.e. the minority spin), the two monopoles of a non-contractible pair are bound together, held by their reciprocal Coulomb attraction. This is a direct consequence of the long-range nature of the dipole–dipole interaction.

The monopoles forming non-contractible pairs do not necessarily have to separate in order to be able to annihilate. It can also happen that another (free) monopole collides with the pair, whereby it can annihilate one of the monopoles in the pair (that with opposite charge to the free monopole) and free up the other one. Pictorially, one can think of this as *radioactive decay*, triggered by the absorption of a monopole, in contrast to *spontaneous decay* of the pair, where the monopole and antimonopole separate and annihilate elsewhere on the lattice. The radioactive process straightforwardly reduces the energy of the system, whereas the spontaneous process incurs a finite energy barrier.

Which of the two processes controls the long-time decay of the monopole density depends on the relative population of free monopoles and non-contractible pairs. If free monopoles are abundant, then one can expect that nearly all non-contractible pairs will decay radioactively (vanishing energy barrier, fast relaxation channel). If, instead, most monopoles in the system form non-contractible pairs, then their

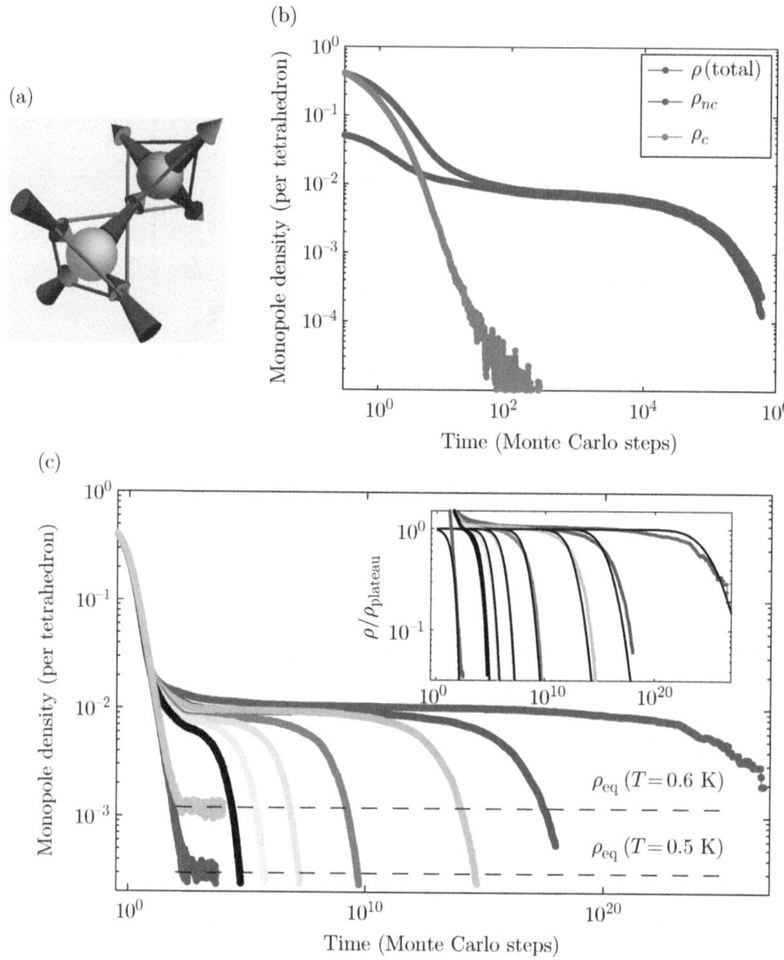

Fig. 12.4 [Colour online] (a) Spin configuration of two adjacent tetrahedra hosting a non-contractible monopole–antimonopole pair. (b) Monte Carlo simulation of dipolar spin ice, showing the total monopole density [red], non-contractible pair density [blue], and free-monopole density [magenta], following a thermal quench from $T = \infty$ to $T = 0.125\,\text{K}$, with system size $L = 8$ and $Dy_2Ti_2O_7$ parameters. (c) Decay of the total monopole density in Monte Carlo simulations of dipolar spin ice, following thermal quenches from infinite temperature to different finite target temperatures (see the caption to Fig.3 in [19] for details). Inset: comparison of the long-time tail of the monopole density with the Poissonian modelling of the spontaneous decay of non-contractible pairs discussed in the text. (Reprinted with permission from [19]. Copyright 2010 by the American Physical Society.)

annihilation must occur via spontaneous decay (slow relaxation channel, due to the finite activation energy barrier).

At high temperature, when the system is nearly paramagnetic and the defects are dense, one can readily verify that the density of free monopoles is statistically larger (by about one order of magnitude) than the density of non-contractible pairs—as reflected in the initial conditions that can be inferred from Fig. 12.4(c). Therefore, we see that a *population inversion* is required to cause the system to relax via the slow channel and to develop a long-lived metastable plateau at low temperature.

Exercise 12.3 Obtain an estimate of the density of non-contractible pairs in the paramagnetic limit (i.e. randomly oriented Ising spins), and compare it with the total density of monopoles in the same state.

Once again, the long-range Coulomb interaction plays a crucial role in determining how the free versus non-contractible monopoles evolve with time. Indeed, free monopoles and antimonopoles are drawn together by Coulomb forces that are stronger than the attraction between free monopoles and non-contractible pairs (charge–dipole interaction). Naively, one would thus expect that the long-range interactions favour direct annihilation of free monopoles over the radioactive decay of non-contractible pairs. If the bias is sufficiently pronounced, it can eventually cause the density of free monopoles to become vanishingly small, whereas the density of non-contractible pairs remains finite, leading to the above-mentioned population inversion.

Numerical Monte Carlo simulations of dipolar spin ice suggest that this understanding of the behaviour of the system in terms of a Coulomb-liquid picture of monopoles and non-contractible pairs is in fact correct. In particular, the population inversion does take place and the density of non-contractible pairs is solely responsible for the long-lived metastable plateau (see Fig. 12.4(b)).

As in the case of a nearest-neighbour spin ice, one can use differential equations for reaction–diffusion processes to model the evolution of the monopole density following a quench and to confirm the qualitative understanding presented above. The processes that ought to be included are

(1) monopole–antimonopole annihilation;
(2) non-contractible pair formation;
(3) radioactive and spontaneous decay of non-contractible pairs.

They are qualitatively illustrated (with the exception of the spontaneous decay) in Fig. 12.5(a). In contrast to the nearest-neighbour case, one has to introduce an additional density variable to represent non-contractible pairs (a new 'species' of particles whose evolution is directly related to that of the free monopoles).

Exercise 12.4 Write the differential equations corresponding to the diagrams in Fig. 12.5(a) in the mean-field limit. Consider explicitly the limit where the temperature is much smaller than the energy barrier for spontaneous decay and non-contractible pairs can only decay radioactively. Solve the equations either analytically or numerically and compare the free versus non-contractible monopole densities (you should use initial conditions similar to those in Fig. 12.4(b)). Try to identify qualitatively the range of parameters in the differential equations for which a population inversion takes place and comment whether spin ice is likely to fall within this range or not. (The differential equations coefficients for spin ice can be estimated from microscopic probabilistic arguments akin to the derivation of κ in the nearest-neighbour case.)

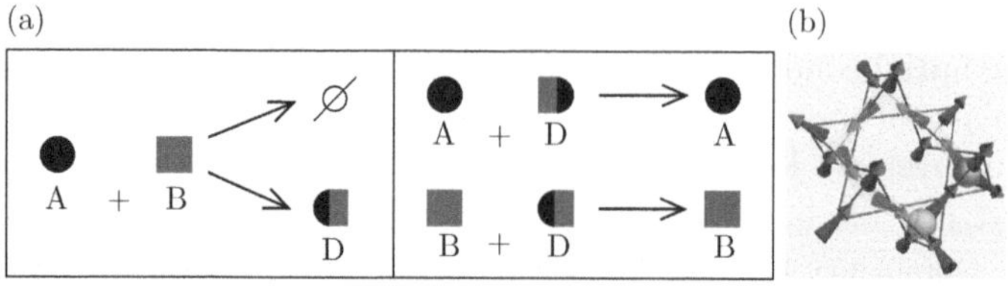

Fig. 12.5 [Colour online] (a) Qualitative illustration of the dynamical processes involved in the monopole density evolution following a thermal quench in dipolar spin ice. (A = positive and B = negative monopole; D = non-contractible pair). (b) Example of a hexagonal path for the spontaneous decay of a non-contractible pair. (Reprinted with permission from [19]. Copyright 2010 by the American Physical Society.)

Here we limit ourselves to modelling in some detail the long time tail of the monopole density decay. As discussed above, it is evident from Fig. 12.4(b) that the non-contractible pairs are largely responsible for this tail. Moreover, in this regime, we expect that spontaneous decay of non-contractible pairs is the leading dynamical process in the system.

First, we ought to estimate the typical energy barrier $\Delta E_{\rm nc}$ of a spontaneous decay process. This is determined by the largest distance that a monopole and an antimonopole in a non-contractible pair need to be separated by before they are able to annihilate elsewhere in the lattice. The shortest possible path is illustrated in Fig. 12.5(b). It requires separating the two monopoles up to third-neighbour distance, before they are brought together again to annihilate. From the value of the magnetic charge of a spin-ice monopole, using the known lattice spacing and the formula for the magnetic Coulomb interaction, one can readily obtain

$$\Delta E_{\rm nc} = -\frac{\mu_0}{4\pi} Q_{\rm m}^2 \left(\frac{1}{d_{3\rm n}} - \frac{1}{d_{\rm nn}} \right). \tag{12.7}$$

Exercise 12.5 Using spin-ice parameters from the literature, compute the value of $\Delta E_{\rm nc}$ for $Dy_2Ti_2O_7$ and $Ho_2Ti_2O_7$.

Now that we have an estimate of the energy barrier, we can proceed with modelling the spontaneous decay of non-contractible pairs. (Notice that the existence of a hexagonal decay path for each non-contractible pair is far from obvious and ought to be regarded as a working assumption here; it will be confirmed a posteriori by comparison with simulations.) We shall assume that the spontaneous decay events are uncorrelated and they obey a Poissonian distribution, with decay probability per unit time $\mathcal{P}(t) = e^{-t/\tau_{\rm nc}}/\tau_{\rm nc}$. The timescale for the activated process is $\tau_{\rm nc} = \tau_0 e^{\Delta E_{\rm nc}/T}$, where τ_0 is the characteristic microscopic spin-flip timescale ($\tau_0 = 1$ in Monte Carlo simulations). Finally, the non-contractible pair density at time t is determined by the number of pairs that have not annihilated via spontaneous decay at any $t' \leq t$, i.e.

$$\rho(t) \propto 1 - \int_0^t \mathcal{P}(t')\, dt' \propto e^{-t/\tau_{\rm nc}}. \tag{12.8}$$

In the Coulomb-liquid description, the value of $\Delta E_{\rm nc}$ is well defined. However, one should recall that it is in fact the result of a resummation of the dipolar interactions between spins that neglects quadrupolar corrections [23]. Therefore, it is subject to (small) statistical fluctuations in Monte Carlo simulations of dipolar spin ice, which are reasonably fitted by a Gaussian distribution. In the case of $\mathrm{Dy_2Ti_2O_7}$ for example, the peak of the distribution occurs at $\Delta E_{\rm nc} \simeq 1.47\,\mathrm{K}$ and the variance is $0.01\,\mathrm{K}^2$ [19]. The value of $\rho(t)$ in (12.8) ought to be averaged over such Gaussian distribution before comparing with simulations:

$$\mathcal{G}(\Delta E) \propto \exp\left[-\frac{(\Delta E - \Delta E_{\rm nc})^2}{2\sigma^2}\right], \tag{12.9}$$

$$\begin{aligned} \langle \rho(t) \rangle_{\rm dis} &\propto \int \mathcal{G}(\Delta E) \rho(t; \Delta E)\, d\Delta E \\ &\propto \int \exp\left[-\frac{(\Delta E - \Delta E_{\rm nc})^2}{2\sigma^2}\right] \exp\left(-\frac{t}{\tau_0 e^{\Delta E/T}}\right) d\Delta E. \end{aligned} \tag{12.10}$$

Notice that (12.8) has only one fitting parameter left: the proportionality constant, i.e. the height of the metastable plateau induced by the long-lived non-contractible pairs. The comparison between theory and simulations is illustrated in the inset in Fig. 12.4(c). We note the good agreement over *more than 20 orders of magnitude*(!), demonstrating that the qualitative understanding in terms of a Coulomb liquid and non-contractible pairs is indeed correct, and that the choice of single-hexagon paths for the spontaneous decay is justified.

We close by stressing the role played by the long-range Coulomb interactions between the monopoles in determining the strikingly different behaviour in dipolar versus nearest-neighbour spin ice. On the one hand, they are responsible (at short range) for the existence of metastable non-contractible pairs. On the other hand, their long-range nature contributes to the population inversion that is key to the long-time plateau in the monopole density at low temperatures.

12.2.2 Field quenches

An alternative protocol to drive a spin-ice system from high to low monopole density involves the use of an applied magnetic field pointing in one of the global [111] crystallographic directions. For intermediate and large field strength, the field maps in the Coulomb liquid language onto a (staggered) chemical potential for the monopoles [23]. The resulting phase diagram is that typical of a liquid–gas system, with a first-order transition line ending at a critical endpoint (see Fig. 12.6).

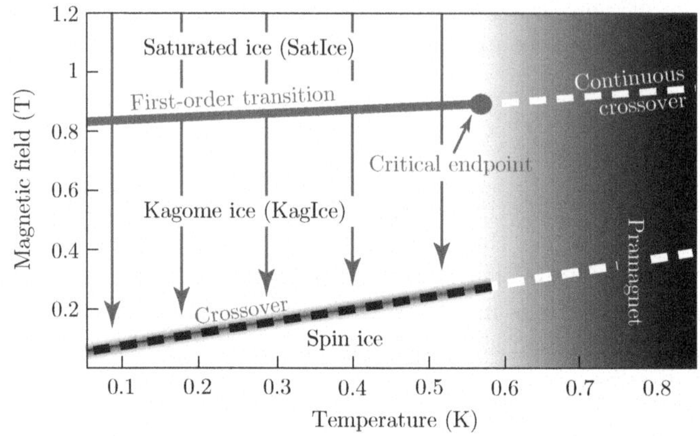

Fig. 12.6 Phase diagram of spin ice in the presence of a [111] field. The vertical arrows represent field quenches from saturated ice (high monopole density) to kagome ice (low monopole density), as discussed in the text. (Reprinted with permission from [24].)

To understand this phase diagram, it is convenient to divide the spin lattice (pyrochlore, or corner-sharing tetrahedral lattice) into alternating kagome and triangular layers perpendicular to the field direction, as illustrated in Fig. 12.7(a). In the limit of strong fields (the saturated ice regime), all of the spins point along the field direction while respecting the local easy axes (Fig. 12.7(b)). The ice rules are violated everywhere and each tetrahedron hosts a monopole; the monopoles form an 'ionic crystal' of alternating positive and negative charges. As the field strength is reduced,

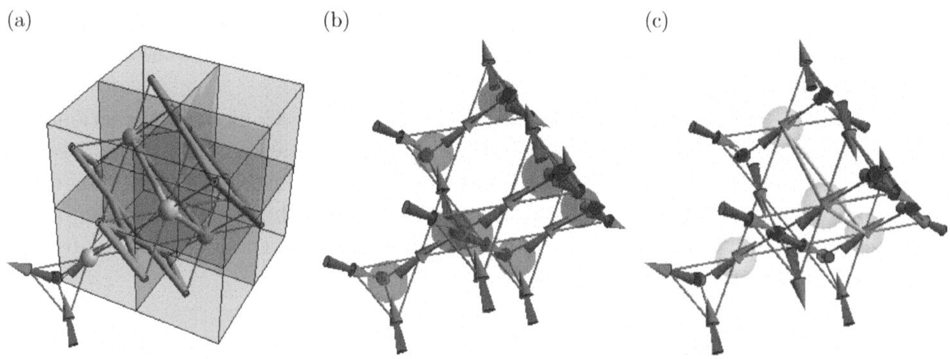

Fig. 12.7 [Colour online] (a) With respect to the global [111] direction identified by the field, the pyrochlore lattice can be seen as a stack of triangular [yellow] and kagome [green] layers perpendicular to the field direction. The easy axis of the triangular spins is parallel to the field, whereas the kagome easy axes are canted, all with the same projection factor $\frac{1}{3}$ onto the field direction. (b) Saturated spin-ice state. (c) Example of a kagome-ice spin configuration. (Reprinted with permission from [24].)

violations of the ice rules are no longer sufficiently offset by a gain in Zeeman energy, and a regime where most tetrahedra obey the ice rule is recovered (at low temperature). This necessarily requires some of the spins to point against the applied field. At intermediate field strengths, these are mostly spins in the kagome planes, because their Zeeman energy is smaller by a factor of three compared with the spins in the triangular planes. This leads to an extensively degenerate regime known as kagome ice, illustrated in Fig. 12.7(c). At low field strengths, the kagome-ice regime becomes entropically unstable to the conventional spin-ice regime, i.e. the ensemble of all configurations satisfying the ice rules irrespective of the polarization of the triangular spins. All of these regimes cross over at sufficiently large temperatures into a conventional paramagnetic regime.

The range of behaviours that can be investigated in quenches involving an applied field is far richer than in thermal quenches [24]. For instance, the presence of phase transitions can lead to qualitatively different responses, including the possibility of critical slowing down and universal scaling à la Kibble–Zurek [42]. The fact that triangular and kagome spins couple differently to the applied field can be used to tune the dimensionality of the system (2D $\leftrightarrow$ 3D). Moreover, the ability to tune both temperature and Zeeman energy against the long-range Coulomb interaction allows us to control the dynamical processes at play and even to alter the characteristic monopole hopping timescales.

Here we focus for simplicity on field quenches across the first-order transition, while the temperature is held constant. Our initial condition is the large-field (saturated-ice) state, where each spin has positive projection in the direction of the field (Fig. 12.7(b)). Every 'upward-pointing' tetrahedron is occupied by a positive monopole and every 'downward-pointing' tetrahedron by a negative monopole. Further, we only consider temperatures and target field values whereby the thermal equilibrium state after the quench is that of kagome ice. Here, the Zeeman energy of the triangular spins is sufficiently larger than the temperature that they remain effectively fully polarized in the field direction. On the other hand, the Zeeman energy of the kagome spins is comparable to the temperature, and they are therefore disordered (in so far as the ice rules due to exchange and dipolar interactions allow). We note that this choice of temperature and field after the quench typically corresponds to a negligibly small equilibrium monopole density—hence the quenches can be regarded once again to be from high to zero monopole density, although the starting configuration is very different from the initial paramagnetic state used in thermal quenches. For a more detailed discussion of [111] field quenches in spin ice, we refer the reader to [24].

12.2.2.1 Initial decay

Immediately following a field quench from saturated ice at low temperature, the monopole density is far greater than its thermodynamic equilibrium value. Therefore, dynamical spin-flip processes that lead to monopole–antimonopole annihilation become favoured.

Notice that the triangular spins do not participate in the initial decay of the monopole density. Not only are they pinned by a larger Zeeman energy than the kagome spins, but also—and more importantly—they are akin to the intervening spin in a

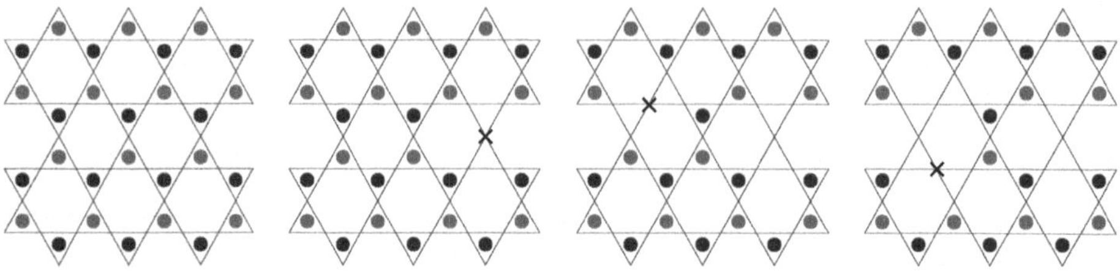

Fig. 12.8 [Colour online] Pictorial representation of the initial monopole annihilation processes within a kagome plane, from left to right. Positive and negative monopoles are represented by grey [red] and black [blue] dots; the spins are not shown for simplicity. The [green] crosses indicate the spins that have flipped in going from one configuration to the next (left to right panels).

non-contractible pair. Flipping a triangular spin in saturated ice leads to the creation of a 4in and a 4out defect rather than to the annihilation of two monopoles.

The initial dynamics of a field quench is thus confined to the 2D kagome planes. Here, flipping a spin between two monopoles leads to their straightforward annihilation, which lowers the energy of the system. The process continues so long as there are kagome spins available between neighbouring monopoles, akin to a random dimer deposition process on the bonds of the dual honeycomb lattice (see Fig. 12.8).

Notice that dimers can sometimes 'desorb' during the initial decay when thermal fluctuations lead to a second reversal of the same spin, thus creating anew the two monopoles that had previously annihilated. The desorption rate can be controlled by tuning the value of the target field as well as the temperature. Here we focus for simplicity on the regime where the desorption rate is negligible.

Ignoring the long-range Coulomb interaction between the monopoles, one should expect to be able to model the initial decay process with reasonable accuracy at the mean field level, given the uniformity of the charge distribution in the initial (saturated) state. The equation of motion is thus the same as for nearest-neighbour thermal quenches, (12.5). The agreement with Monte Carlo simulations of field quenches in dipolar spin ice is excellent without fitting parameters (Fig. 12.9), suggesting that the Coulomb interactions do not have a measurable effect on the reaction process.

The solution of the mean-field equations is temperature-independent. As time passes, we see from Fig. 12.9 that the results of the simulations eventually depart from the mean-field behaviour and become strongly temperature-dependent. This signals the end of the initial (dimer-deposition-like) regime. Randomly selected neighbouring monopoles have straightforwardly annihilated until only isolated ones are left behind, and they need to diffuse across the system before their density can decay further.

Exercise 12.6 Implement a numerical algorithm (e.g. Monte Carlo) to estimate the density of leftover monomers following a dimer deposition process on the honeycomb lattice. Compare it with the density of monopoles in dipolar spin-ice simulations at the end of the initial decay (from Fig. 12.9). Comment on the comparison in light of the fact that spin-ice simulations include a small but non-zero desorption probability and long-range Coulomb interactions.

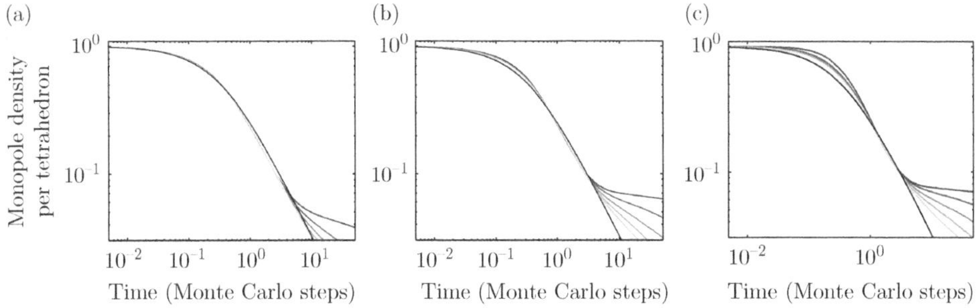

Fig. 12.9 [Colour online] Monte Carlo simulations of field quenches in dipolar spin ice for different values of the target field H: (a) 0.2 T; (b) 0.3 T; (c) 0.35 T. Only the initial (short-time) decay of the monopole density is shown. The differently shaded [coloured] curves correspond to different values of the temperature, and the superposed black line is the solution of the mean-field equation (12.5), without any fitting parameters. (Parts (a) and (b) reprinted with permission from [24].)

When the value of the target field becomes sufficiently large, it is no longer possible to disregard desorption events. This is the likely cause of the departure from mean-field behaviour at short times, which begins to appear in Fig. 12.9(c).

12.2.2.2 Intermediate regime

The initial decay ends when there are no more monopoles and antimonopoles next to one another that can be annihilated by flipping the intervening spin. Monopoles are now required to travel across the lattice before their density can be further reduced.

Figure 12.10 illustrates the behaviour over a large time window, for different fields and temperatures. In general, we observe that the relaxation timescales in the system become substantially longer after the initial decay discussed in Section 12.2.2.1. The new timescales show a clear temperature dependence (the lower the temperature, the slower the decay), as one would expect in the presence of activation-energy barriers obstructing the relaxation. This scenario is similar to that observed in thermal quenches in dipolar spin ice (Fig. 12.4(c)). However, we see that the behaviour in field quenches is far richer, with intermediate-time regimes that appear to be distinct from both the initial and the asymptotically long-time decay.

These intermediate regimes are controlled by finite-size, finite-time processes and are rather challenging to model analytically. The comparison between analytics and numerics is not as straightforward when we do not have access to some asymptotic limit (e.g. short or long times). This interesting and unique regime of an emergent reaction–diffusion process in presence of long-range Coulomb interactions and kinematic constraints, which can in principle be accessed experimentally in spin-ice materials, lacks a proper understanding to date.

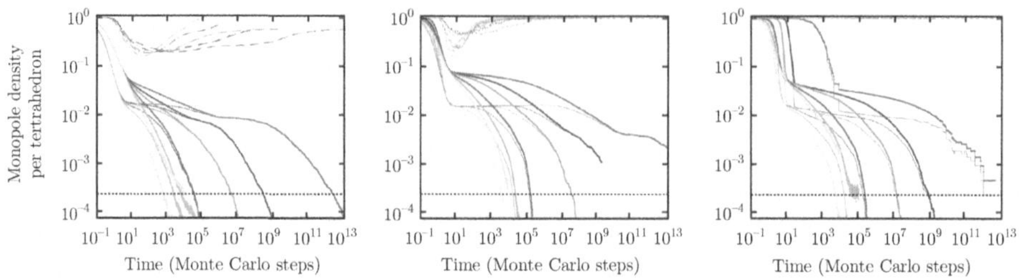

Fig. 12.10 [Colour online] Monopole density (thick lines), density of triangular spins in the direction of the initial magnetization (thin dotted-dashed lines), and density of non-contractible pairs (thin solid lines) from Monte Carlo simulations for a system of size $L = 8$ (8192 spins) at temperatures $T = 0.1$, 0.15, 0.2, 0.3, 0.4, and 0.5 K [red, blue, green, magenta, cyan, and yellow curves, respectively] for different fields H: (a) 0.2 T; (b) 0.4 T; (c) 0.6 T. At intermediate times, some of the triangular spins reverse, as shown by the dip in their density; the latter has been magnified by factors of 100 and 1000 in (a) and (b), respectively, to aid visualization. In (c), the density of triangular spins in the direction of the applied field remains very nearly 1 throughout the simulations; the triangular spins remain polarized throughout the quench and the monopole motion is effectively 2D. (The black dotted horizontal line in each plot indicates the density threshold of one monopole in the entire Monte Carlo system.) (Reprinted with permission from [24].)

12.2.2.3 Long-time behaviour

At long times, the monopole density decay becomes increasingly dominated by the longest relaxation timescale in the system. We should therefore be able to capture the physics of this regime by modelling its asymptotic behaviour analytically.

Whereas at small and intermediate target field values (Fig. 12.10(a) and (b)), most of the monopoles at long times form non-contractible pairs, this is clearly not the case at larger fields (Fig. 12.10(c)). Hereinafter, for simplicity, we shall focus only on the latter case.

For large field values, the long relaxation times cannot be ascribed to long-lived non-contractible pairs. Rather, it must be that an energy barrier impedes the diffusion and annihilation of free monopoles. The origin of this barrier can be understood if we recall that monopole diffusion at large fields and low temperatures takes place nearly exclusively within each kagome plane, whilst the triangular spins remain fully polarized (Fig. 12.10(c)).

Under these conditions, a positive monopole in a kagome plane has a lower Zeeman energy when it sits in an upward-pointing tetrahedron than in a downward-pointing tetrahedron (vice versa for a negative monopole, as illustrated in Fig. 12.11). If we were to make a monopole hop across the kagome lattice, then at every other step it would have to overcome a Zeeman energy barrier $dE \simeq 4.48H$ K (where H is the value of the target field measured in tesla).

Alternatively, the system can create a monopole–antimonopole pair next to the existing monopole and then annihilate the existing monopole with the oppositely

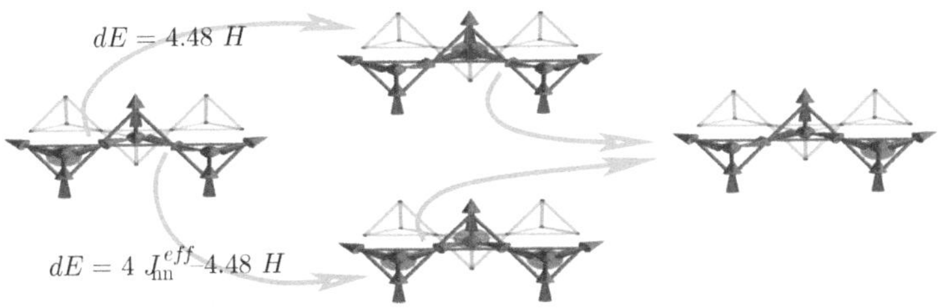

Fig. 12.11 [Colour online] Schematic representation of monopole motion in a kagome plane, via ordinary hopping (upper intermediate diagram) and via pair-assisted hopping (lower intermediate diagram). Both processes result in a negative monopole being transferred from a downward-pointing tetrahedron (left diagram) to one of the four nearest downward-pointing tetrahedra (right diagram). Each process encompasses two spin flips but, according to the order in which they are executed, the two processes face different energy barriers dE with opposite field dependence. The figure shows the value of the barriers for nearest-neighbour spin ice. In the text, we discuss how they are modified in the presence of long-range dipolar interactions. The field dependence, however, remains unchanged. The tails of the curved [green] arrows originate from the spin being flipped in going from one diagram to the next. Only the spins in the front three tetrahedra are drawn for convenience. The triangular spins remain polarized throughout. (Reprinted with permission from [24].)

charged member of the pair. The outcome is equivalent to moving a monopole from one Zeeman-favoured tetrahedron to another Zeeman-favoured tetrahedron two lattice spacings away from the first. This process costs interaction energy (monopole pair creation + Coulomb interactions), but it can be done while gaining Zeeman energy. The corresponding barrier, using the Coulomb-liquid description, can be estimated as $dE \simeq 2\Delta - 2E_{\text{nn}} + E_{2\text{n}} - 4.48H$ K, where Δ is the bare monopole cost and E_{nn} ($E_{2\text{n}}$) is the strength of the Coulomb interaction between two nearest-neighbour (next-nearest-neighbour) monopoles.

Exercise 12.7 Using spin-ice parameters for $Dy_2Ti_2O_7$ and $Ho_2Ti_2O_7$ from the literature, estimate the corresponding values of the pair-assisted hopping barrier.

Notice that the two dynamical processes have opposite dependence on the applied field strength. Using spin-ice parameters appropriate for $Dy_2Ti_2O_7$, the second process (i.e. pair-assisted hopping) becomes energetically favoured with respect to the first for $H \gtrsim 0.5$ T. When $H = 0.6$ T (Fig. 12.10(c)), the barrier to pair-assisted hopping is of the order of 2 K, whereas the barrier to ordinary hopping is approximately 3 K.

In order to confirm our understanding of the slowing down of the monopole hopping, we attempt to collapse the long-time tails of the Monte Carlo simulations of dipolar spin ice by rescaling time using the characteristic activated timescale $e^{dE/T}$. For the target field $H = 0.6$ T, we find a good collapse when we choose $dE = 2.4$ K, in reasonable agreement with the estimated value for pair-assisted hopping (Fig. 12.12).

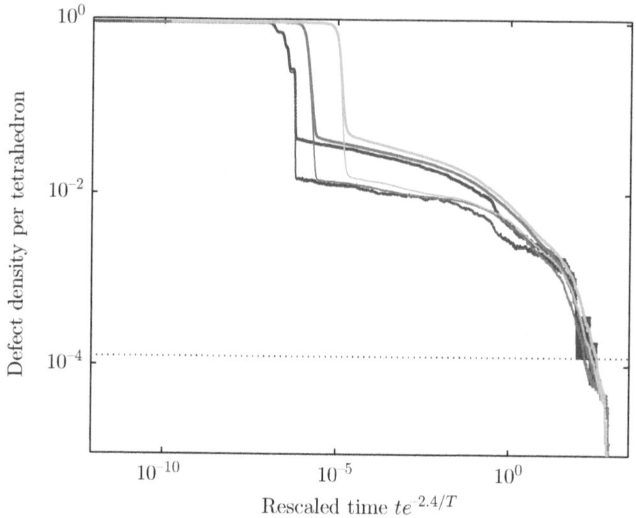

Fig. 12.12 [Colour online] Collapse of the long-time decay of the monopole density (thick lines) and of the non-contractible monopole density (thin lines) after rescaling the time axis by a factor $e^{-2.4/T}$. The Monte Carlo simulations are for $L = 10$ with $H = 0.6\,\mathrm{T}$ and $T = 0.13$, 0.15, and $0.18\,\mathrm{K}$ [red, blue, and green curves, respectively]. The good quality of the collapse indicates that the simulated systems are large enough for the energy scale of 2.4 K not to exhibit any appreciable dependence on system size. (Reprinted with permission from [24].)

However, larger system sizes and longer simulation times are required for a more discerning and conclusive comparison [24].

In summary, field quenches in spin ices offer a realization of several paradigmatic concepts in non-equilibrium dynamics: dimer adsorption, Coulombic reaction–diffusion physics, and kinetically constrained slow dynamics. There is an unusually high degree of tunability, since one is able to control, say, the timescale of the elementary dynamical move through a Zeeman energy barrier, the dimensionality of the final state ($d = 2$ for kagome versus $d = 3$ for spin ice), and the relative importance of dimer desorption compared with Coulomb interactions between the monomers.

Given the availability of a range of experimental probes for magnetic systems and the ability to apply time-dependent fields of the strength required for spin ice materials, we can expect that it will be possible to study some of these out-of-equilibrium phenomena experimentally in the near future.

Further reading on recent experimental work in this direction—including a curious interplay between magnetic and thermal degrees of freedom leading to magnetic deflagration effects [25]—can be found in [11, 26, 27] and references therein.

12.3 Example 2: Kitaev's toric code

So far, we have considered out-of-equilibrium phenomena in classical topologically ordered systems. The observed interplay between the non-local nature of the phase, the emergent excitations, and local kinematic constraints is likely to give rise to interesting properties and phenomena also in related quantum-mechanical systems. However, the study of strongly correlated quantum systems in two or higher dimensions far from equilibrium is in general a tall order. For this reason, we shall focus here on one of the simplest examples of quantum systems that exhibit topological order in 2D: Kitaev's toric code [28].

12.3.1 The model

The toric code is a system of spin-$\frac{1}{2}$ degrees of freedom living on the bonds of a square lattice, subject to the Hamiltonian

$$H = -\lambda_A \sum_s A_s - \lambda_B \sum_p B_p, \tag{12.11}$$

where λ_A, $\lambda_B > 0$ are two coupling constants, and the star and plaquette operators

$$A_s = \prod_{j \in \text{star}(s)} \sigma_j^x, \qquad B_p = \prod_{j \in \partial p} \sigma_j^z \tag{12.12}$$

are defined as in Fig. 12.13. (The experienced reader will recognize this as a gauge-fixed lattice gauge theory.) The beauty of the model lies in its simplicity. Every term in the Hamiltonian commutes with every other, that is, $[A_s, A_{s'}] = 0$, $[B_p, B_{p'}] = 0$, and $[A_s, B_p] = 0$, for all s, s', p, p'. One can therefore diagonalize simultaneously all these operators. Every eigenstate of H is also an eigenstate of each A_s and B_p (which have eigenvalue ± 1). With the choice of λ_A and λ_B both positive, the ground state $|\psi_0\rangle$ satisfies the equations

$$A_s|\psi_0\rangle = |\psi_0\rangle, \qquad B_p|\psi_0\rangle = |\psi_0\rangle, \qquad \forall s, p.$$

Despite its simplicity, the model and its ground state are far from trivial. Let us assume periodic boundary conditions for the system (i.e. it is defined on a torus—hence the first part of its name). The number of star (plaquette) operators equals the

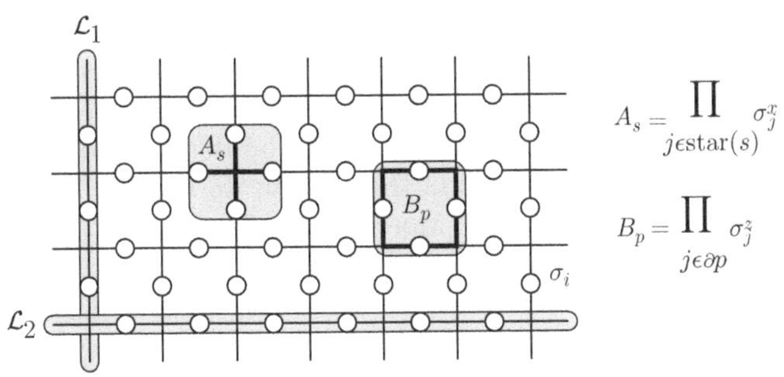

Fig. 12.13 Illustration of the star A_s and plaquette B_p operators in the toric-code Hamiltonian. The figure also shows the support of two winding loop operators on the direct lattice, $\mathcal{L}_1$ and $\mathcal{L}_2$ (periodic boundary conditions are assumed).

number of lattice sites N. However, not all of them are independent. The product of all star (plaquette) operators is always 1 because it is the trivial product of squares of spin-$\frac{1}{2}$ (Pauli matrix) operators. (Notice, for instance, that every time we take the product of two neighbouring star operators, the σ_i^x operator shared between them is squared, and similarly for plaquette operators.) Therefore, the number of independent star and plaquette operators is $2N - 2$, whereas the number of degrees of freedom in the system is $2N$. Specifying the values of all A_s and B_p determines uniquely the energy of the system (H) but it *does not* identify a unique state. Rather, it identifies a fourfold-degenerate manifold of states.

In order to resolve the ground-state degeneracy, we need to find two additional spin-$\frac{1}{2}$-like operators that commute with all A_s and B_p and yet are not directly dependent on them. We can verify that there are no such operators with local support, and we need to consider instead products of ensembles of spin operators that span the entire width of the system. For instance, we can use the two loops $\mathcal{L}_1$ and $\mathcal{L}_2$ illustrated in Fig. 12.13 and take the products

$$\Gamma_1 = \prod_{i \in \mathcal{L}_1} \sigma_i^z, \qquad \Gamma_2 = \prod_{i \in \mathcal{L}_2} \sigma_i^z. \tag{12.13}$$

It is straightforward to see that the new operators Γ_1 and Γ_2 commute with one another and with all B_p (trivially) and all A_s operators (the latter is a consequence of the fact that $\Gamma_{1,2}$ share either two or no spins with any of the star operators). Their eigenvalues ± 1 completely resolve the degeneracy.

The choice of paths $\mathcal{L}_1$ and $\mathcal{L}_2$ is immaterial so long as their respective winding numbers are preserved ($\mathcal{L}_1$ winds around the torus in one direction once; $\mathcal{L}_2$ winds once along the other direction). Given two different choices for $\mathcal{L}_1$, the product of the two corresponding Γ_1 operators is equivalent to the product of all the plaquette operators in between them. In the ground state, the latter are all equal to 1 and so is their product; hence, the two Γ_1 operators must have the same eigenvalue. Similarly for $\mathcal{L}_2$. (These additional operators are equivalent to winding Wilson loops that the reader may be familiar with from lattice gauge theory.)

Notice that the nature of the degeneracy is not related to the breaking of a symmetry. Indeed, one can show that all local operators have trivial (namely, zero-range) correlations. The degeneracy depends on the topology of the system: it is fourfold on a torus (genus $g = 1$) and it is in general 2^{2g}-fold on a surface of genus g. The information that distinguishes one ground state from another is contained in the eigenvalues of the operators Γ_1 and Γ_2. These values cannot be determined from knowledge of the state of any finite subset of spins in the system; we need to know their state for a subset that spans the entire system. Moreover, we have seen that the eigenvalues of Γ_1 and Γ_2 are independent of the microscopic choice of paths $\mathcal{L}_1$ and $\mathcal{L}_2$; they depend only on their global properties, namely how they wind around the torus. As such, we say that the system is topologically ordered, and the different degenerate states are dubbed topological sectors.

The ability to store quantum information non-locally as a superposition of ground states of this system, inherently protected from local perturbations, is responsible for the great interest that Kitaev's toric code has received in recent years from the quantum information and quantum computing communities (hence the second part of its name).

Exercise 12.8 Check that the choice of σ^z operators in (12.13) is arbitrary and one could equally use σ^x operators upon replacing the paths $\mathcal{L}_1$ and $\mathcal{L}_2$ on the direct lattice with equivalent paths on the dual lattice. Discuss the action of the new winding operators with respect to the old ones (equivalently, their commutation relations).

Although the toric code is indeed very different from spin ice, an interesting parallel can be drawn between the two systems. Let us consider the eigenvalues ± 1 of the σ^x operators and let us represent them as arrows pointing from one sublattice to the other $(+)$ and vice versa $(-)$. The star operators in the Hamiltonian favour a ground state where the product of the σ^x operators around each site of the lattice is $+1$. In the language of the arrows, this corresponds to enforcing an even number of arrows pointing into (equivalently, out of) each site. This is the same as the 2in–2out ice rules in spin ice, with the addition of 4in and 4out states. The plaquette operators in the Hamiltonian are kinetic terms with respect to the arrow representation, introducing quantum dynamics into an otherwise classical vertex model. In summary, the toric code 'looks like' a quantum spin-ice model in 2D with the addition of low-energy 4in and 4out vertices. This addition is, however, responsible for a major difference in their properties, whereby one system is in a Coulomb phase with an emergent gauge symmetry and the other is in a $\mathbb{Z}_2$ topologically ordered state.

12.3.2 Elementary excitations

In order to understand the nature of the elementary excitations over the ground state of the toric code, let us consider the action of a σ_i^x operator applied to a given spin i. While it trivially commutes with the star operators, the value of the σ^z component of the spin is changed and therefore the two plaquette operators that share this spin acquire a negative eigenvalue (Fig. 12.14(a)). The energy of the system is correspondingly raised by $4\lambda_B$.

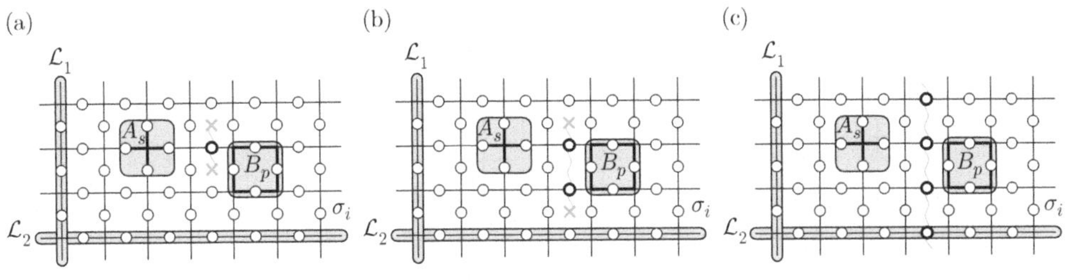

Fig. 12.14 [Colour online] Qualitative illustration of the creation and separation of plaquette-type defects in the toric code. A single spin flip creates two negative plaquettes (a), which can then separate via the action of other spin-flip operations without incurring further energy barriers (b). If the two defective plaquettes wind around the entire system before annihilating, they change the topological sector of the state (c).

In a conventional spin-$\frac{1}{2}$ ground state, a single spin reversal is typically the lowest-energy excitation. Acting with further flipping operators costs increasingly more energy. However, much like spin ice, this is not the case for the toric code. Consider the action of another σ_j^x at a site j that belongs to one of the two negative plaquettes created by σ_i^x. Having now two spins flipped, the eigenvalue of that plaquette reverts to its lowest-energy (positive) state. On the other hand, there is a new plaquette that shares spin j but not spin i, and its eigenvalue becomes negative. In a nutshell, the action of σ_j^x is to separate the negative plaquettes without introducing any further energetic defects (akin to how appropriate spin flips separate monopoles in nearest-neighbour spin ice without any energy cost). This is illustrated in Fig. 12.14(b). Therefore, the elementary excitations in the toric code are deconfined plaquettes (equivalently, stars) with negative eigenvalue. Each defect costs an energy $2\lambda_B$ (equivalently, $2\lambda_A$). They can only be created or annihilated in pairs.

In contrast to classical spin ice, we have here two types of excitations: defective stars and defective plaquettes. Although they do not interact, they have non-trivial reciprocal statistics. Indeed, let us consider two negative plaquettes p and p' on the lattice (they can only be created in pairs and therefore it is not useful to consider only one of them in isolation). In order to create these two excitations from the ground state of the system, one has to choose a path from p to p' on the dual lattice and act with the product of all σ_i^x operators along the path (see Fig. 12.14). One can check that the choice of path is immaterial, since any two different paths differ from one another by products of star operators (assuming that there are no star defects in between them). Similar considerations apply to negative star operators at s and s', with respect to paths on the direct lattice from s to s' and products of σ_i^z operators. We can now imagine having two plaquette and two star defects in the system; we keep three of them fixed and we drag, say, one of the plaquettes around one of the star defects (but *not* around the other) and back to its initial position (Fig. 12.15). The initial and final states are the same in terms of the positions of the defects. However, the braiding operation of moving one plaquette around a star necessarily changes the parity of the number of times that the dual path p–p' intersects the direct path s–s'. This results in the state of the system acquiring an overall phase factor $e^{i\pi} = -1$. Plaquette and star defects have relative *semionic* statistics!

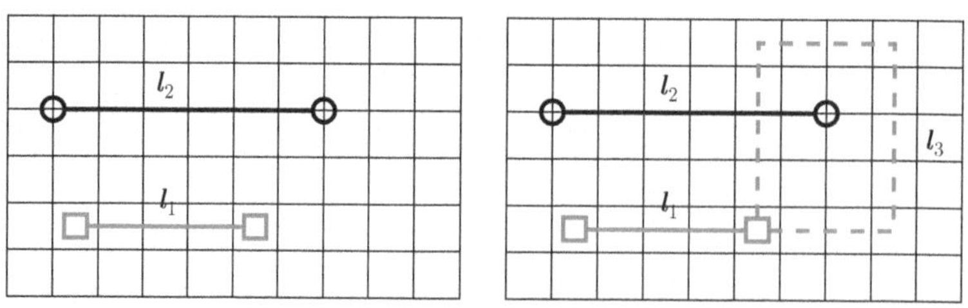

Fig. 12.15 [Colour online] Qualitative illustration of the braiding of a plaquette defect around a star defect.

Exercise 12.9 Construct the wavefunctions of the two states represented in Fig. 12.15, namely

$$|\psi_{\text{left}}\rangle = \prod_{i\in l_1} \sigma_i^x \prod_{j\in l_2} \sigma_j^z |\psi_0\rangle,$$

$$|\psi_{\text{right}}\rangle = \prod_{i\in(l_1\cup l_3)} \sigma_i^x \prod_{j\in l_2} \sigma_j^z |\psi_0\rangle.$$

Using the well-known (anti-)commutation relations between Pauli matrices, show that the two states are indeed identical up to an overall minus sign.

The emergence of quasiparticles with fractional statistics with respect to the microscopic degrees of freedom in the system is another instance of fractionalization in topologically ordered systems.

12.3.3 Dynamics

Once defective stars and plaquettes are created in the system, they are static in so far as the action of the Hamiltonian is concerned. None of the operators in (12.11) can alter their positions or values. Defects acquire dynamics only if we assume that either thermal or quantum fluctuations are present, which generally couple to σ^x and σ^z operators (as well as σ^y, but we shall not discuss that case in these notes). In the presence of such fluctuations, the defects are able to move freely across the system. Once again, similarly to spin ice, defects act as dynamical facilitators for the system's response and relaxation. Spin flips that result in the hopping of a defect do not incur an energy barrier; whereas generic spin flips away from existing defects must overcome the energy barrier to create two new excitations.

In contrast to spin ice, where the role of the defects as dynamical facilitators is readily reflected in the magnetic response of the system (e.g. its susceptibility), the case of the toric code is more subtle owing to the lack of any local correlations. Here, we discuss how the dynamics of the excitations relates to the topological properties of the system.

Let us prepare the system in a given topological sector. The creation, say, of a pair of defective plaquettes only disrupts the spins along the path that was chosen to generate them. Away from this path, the eigenvalues of the winding loop operators in (12.13) remain unaltered. Once fluctuations allow the defective plaquettes to move around, the eigenvalues of the winding loop operators are statistically well defined only if the two plaquettes remain close to one another. Once they separate and wander across the system, the information about the initial topological sector is lost. Indeed, if we create a pair of negative plaquettes, wind them around the system, and then annihilate them, the outcome is that all winding loop operators crossing the winding path of the plaquettes change sign and the topological sector of the system changes (Fig. 12.14(c)).

In order to understand how a defect-driven change in a topological sector takes place dynamically, let us take a look at the shape of the relevant energy barrier. Starting from the ground state, the system faces an energy increase for the creation of

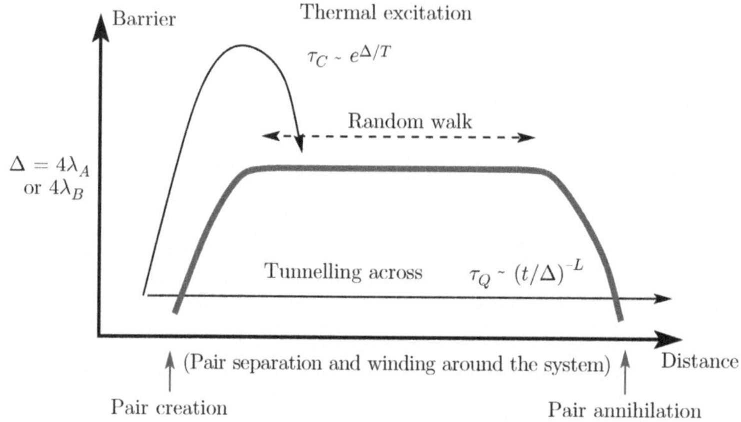

Fig. 12.16 Qualitative illustration of the shape of the energy barrier to changing topological sector in the toric code via defect creation, diffusion, and annihilation. The arrows represent thermal and quantum processes, with their relevant timescales.

two defects ($\Delta = 4\lambda_A$ or $\Delta = 4\lambda_B$, depending on the type of defect). The energy then remains constant as the defects move about. In order to change topological sector, one of the defects has to separate from the other and wind around the system before they annihilate. As a result, the width of the barrier is at least of the order of the system size L, after which the energy decreases again to the ground-state value upon annihilating the two defects. The shape of such barrier is depicted in Fig. 12.16.

Local quantum fluctuations can induce a change in topological sector by exciting a virtual pair of defects and making them hop (while the system is in a virtual excited state) across the entire lattice before they annihilate and the energy is finally lowered. If the strength of the quantum fluctuations that couple to the σ_i^x operators is t, then the height Δ and width $\sim L$ of the barrier imply that the quantum tunnelling under the barrier is a perturbative process of order L in t/Δ. The tunnelling rate $\sim (t/\Delta)^L$ is therefore exponentially suppressed in the size of the system. Correspondingly, the relaxation timescale from one topological sector to another due to local quantum fluctuations grows exponentially with system size, $\tau_Q \sim \exp[L \ln(\Delta/t)]$.

This is to be contrasted with the analogous process in the presence of thermal fluctuations. Once a pair of (real) defects has been thermally excited against the energy cost Δ, they are free to diffuse across the system. The probability that they wind around the system and then annihilate is related to the first-passage probability of a random walk to come back to the origin after winding around the torus an odd number of times, which is polynomial in L. The overall probability of the process is therefore given by the product of the activation probability $\exp(-\Delta/T)$ times a factor that does not depend on Δ or T and that scales polynomially in system size, $\tau_C \sim e^{\Delta/t}\,\mathrm{Poly}(L)$.

Whether thermal or quantum processes are the dominant contribution to the relaxation of the system to equilibrium is thus a matter of order of limits. In a system of finite size, there is a temperature below which the exponential slowing down of thermal processes due to the activation barrier exceeds the exponential suppression in system size of quantum tunnelling, and the latter becomes the faster process ($\tau_Q \ll \tau_C$). On the other hand, if the system becomes larger and larger at fixed temperature, then the protection from quantum fluctuations is bound to become far greater than the protection from thermal fluctuations, and the latter become the faster relaxation channel ($\tau_Q \gg \tau_C$). A more general discussion of quantum versus thermal relaxation processes and their relation to (topological) quantum glassiness can be found in [29].

It is worth commenting that relaxation times that scale polynomially in system size are unusual in ordered phases and signal a remarkable weakness. Whereas quantum topological order in the toric code is highly robust to quantum perturbations, it is immediately lost (in the thermodynamic limit) when the system is coupled to a thermal bath. This issue is discussed in detail in [30–33]. Freeman et al. [34] use numerical simulations to investigate the relaxation dynamics of the toric code coupled to a thermal bath and discuss its connection to thermal fragility.

12.3.4 Intriguing comparison: kinetically constrained models

As illustrated in the examples above, the appearance of topologically ordered phases (in lattice models) is closely related to the presence of dominant energy terms that enforce local constraints (cf. dimer/vertex/plaquette constraints). Although insufficient to drive the system into a conventionally ordered phase, these terms are directly responsible for the non-trivial global properties of the system.

It is interesting to draw a parallel between the role of local constraints in topologically ordered systems and another area of research, namely that of *kinetically constrained models* [35], where local constraints are used instead to induce non-trivial dynamical properties (i.e. unusually slow response and equilibration while the thermodynamic properties remain altogether trivial).

Kinetically constrained models have received much attention in the literature as an attempt to understand the emergence of long relaxation timescales and glassiness in systems without disorder.

Here, we briefly review two examples and comment on their analogies and differences with respect to the topologically ordered systems considered earlier. We limit our discussion to classical 2D systems, although higher-dimensional [36] as well as quantum [29, 37] examples are also available.

12.3.4.1 Square-lattice plaquette model

The first model we consider is an Ising model on a square lattice (with spins living on the sites, not the bonds) and Hamiltonian [38]

$$H = -J \sum_p \prod_{i \in p} S_i \qquad (J > 0), \tag{12.14}$$

where p labels the plaquettes on the lattice and $\prod_{i \in p} S_i$ is the product of the four spins at the corners of plaquette p. It belongs to a broader class of models known as *gonihedric* models and is also directly mappable onto Baxter's eight-vertex model (notice the direct correspondence with the toric code).

The system does not exhibit any phase transitions as a function of temperature, and the high-temperature paramagnetic phase is continuously connected to the low-temperature phase where all plaquettes have the same sign ($\prod_{i \in p} S_i = +1$ for $J > 0$). This is most straightforwardly seen in the language of the dual variables $\tau_p \equiv \prod_{i \in p} S_i$, defined on the centres of the plaquettes of the original lattice, where the Hamiltonian reduces to that of a trivial paramagnet,

$$H = -J \sum_p \tau_p. \tag{12.15}$$

Notice that the Hamiltonian in (12.14) is invariant under transformations that flip straight lines of spins on the direct lattice, spanning the entire system (notice the analogies and differences with the winding loops introduced in the discussion of the toric code). This invariance has two important consequences. First, the zero-temperature limit when all plaquettes are polarized is sub-extensively degenerate (namely, the number of degenerate configurations scales with the exponential of the linear size L of the system rather than the exponential of the volume L^2). Second, all two-spin correlators

$$\langle S_i S_j \rangle = \sum_{\{S_k\}} S_i S_j \frac{e^{-\beta H}}{Z}, \qquad Z = \sum_{\{S_k\}} e^{-\beta H}, \tag{12.16}$$

vanish identically at all temperatures. This is because there is always at least one straight line (horizontal or vertical or both) that goes through spin i but not spin j. Therefore, the correlators vanish by symmetry (so long as the system remains ergodic).

If the duality transformation trivializes the thermodynamics of the system, the dynamical processes then become non-trivial. At low temperature, in order to transition from one lowest-energy configuration to another, the system must overcome an energy barrier that is similar to that encountered in the toric code. First, a thermally excited spin creates four defective plaquettes ($\Delta = 8J$). Then, neighbouring spins can flip to annihilate two defective plaquettes and create two new ones, thus effectively separating the four defective plaquettes in pairs without changing the energy of the system (see Fig. 12.17). In contrast to the toric code, however, the motion must follow a straight line. If the pairs wind around the system before they annihilate, the system ends in a new lowest-energy configuration. Once again, we expect relaxation timescales that are exponential in the height of the barrier over the temperature, $e^{\Delta/T}$, times a temperature-independent factor that scales polynomially with the system size.

Although we have considered the timescale for the system to relax from one lowest-energy state to another, similar arguments apply to the relaxation timescales in the system as the temperature is progressively reduced (see e.g. [38]).

12.3.4.2 Triangular-lattice plaquette model

The second model that we consider is similar to the former, but defined on the triangular lattice (again with Ising spins living on the sites). The Hamiltonian of the system can be written as [39]

$$H = J \sum_{\nabla} \prod_{i \in \nabla} S_i \qquad (J > 0), \tag{12.17}$$

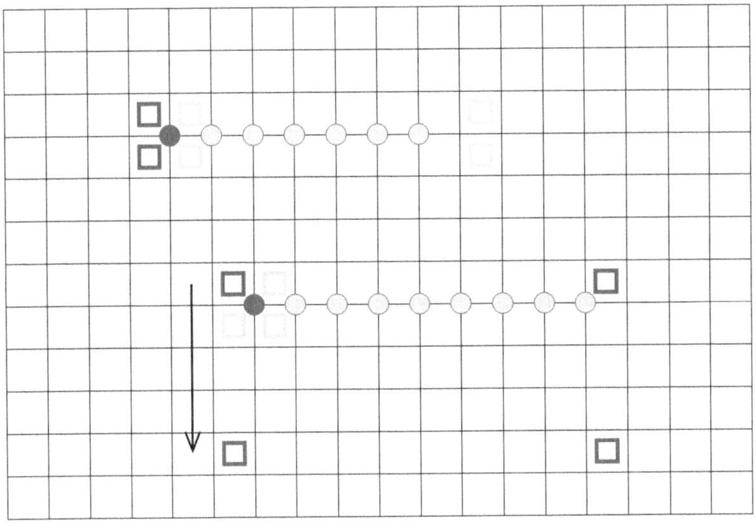

Fig. 12.17 [Colour online] Illustration of defects in the square-plaquette model. A single spin flip changes the sign of the four plaquettes to which it belongs, which are then pairwise free to move along straight lines across the system. The bottom portion of the figure illustrates how isolated defects are brought together to eventually annihilate, as the system attempts to reach one of its defect-less ground states via the allowed (constrained) defect dynamics (see [38] for details).

where $\bigtriangledown$ labels the downward-pointing triangular plaquettes and $\prod_{i\in\bigtriangledown} S_i$ is the product of the three Ising spins at their three vertices. (This is similar to—but not to be confused with—the Baxter–Wu model, which includes upward- as well as downward-pointing triangles.)

The thermodynamic properties of the system are best understood in terms of dual variables $\tau_{\bigtriangledown} \equiv \prod_{i\in\bigtriangledown} S_i$, which live on the triangular lattice formed by the centres of the downward-pointing triangles in the original lattice (Fig. 12.18(a)). If the linear dimension of the system is a power of 2 (i.e. $L = 2^n$, $\exists\, n \in \mathbb{N}$) and periodic boundary conditions are assumed, one can show that there is a one-to-one correspondence between the two representations of the system [39]. In the new language, the Hamiltonian becomes

$$H = J \sum_{\bigtriangledown} \tau_{\bigtriangledown}, \tag{12.18}$$

i.e. that of an ensemble of non-interacting spins in an applied magnetic field. In the dual language, it is straightforward to write the partition function of the system and use the mapping to obtain correlation functions of the original degrees of freedom [40]. The system does not undergo a phase transition as a function of temperature and the lowest-energy configuration (where all $\tau_{\bigtriangledown} = -1$) is continuously connected to the trivial paramagnetic phase.

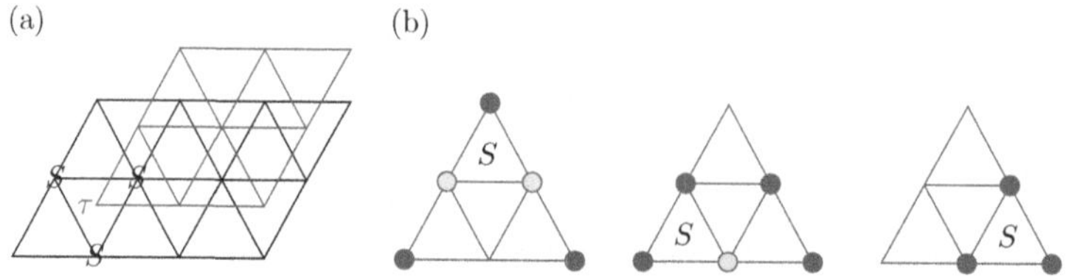

Fig. 12.18 [Colour online] (a) Illustration of the original triangular lattice of the S spins (light grey lattice) and the dual triangular lattice of the τ spins (dark grey [blue] lattice). (b) The steps (from left to right) to annihilate three defective plaquettes at the corners of a dual triangle of side 2, by flipping three S spins in sequence. Notice that in the process we cannot avoid creating one additional defect.

Whereas the duality transformation allows one to demonstrate straightforwardly the trivial thermodynamics of the system, the dynamical processes become non-trivial. Flipping an individual spin of the original system (S) now leads to changing the sign (i.e. flipping) the three plaquette variables (τ) that share the spin (Fig. 12.18). This should be contrasted with the corresponding defect dynamics in the toric code (where, for example, flipping a bond spin changes the sign of the two adjacent star or plaquette operators) and in the square-lattice plaquette model (where flipping a site spin changes the sign of four plaquettes). Notice that the S spins live only in the *upward-pointing* triangular plaquettes of the τ spin lattice.

The presence of such dynamical constraints plays a crucial role in the response and equilibration properties of the system, which become drastically different from those expected for a trivial paramagnet in an applied field. Similarly to the square-plaquette model, Monte Carlo simulations show the emergence of unusually long relaxation timescales and glassiness at low temperatures [39]. However, the behaviour in this case is remarkably different from the activated behaviour encountered in the toric code and in the square-plaquette model, since the characteristic timescale grows exponentially with the *square of the inverse temperature* [39–41].

In order to understand this behaviour, let us consider how the system approaches the lowest-energy state as the temperature is lowered. For this purpose, it is sufficient to consider the lowest-energy excitations above the ground state where all $\tau_{\bigtriangledown} = -1$. It is possible to show that these excitations take the form of equilateral triangles of linear size $\ell = 2^k$, with k integer, that have single isolated defects $\tau_{\bigtriangledown} = +1$ at each of their three vertices, as illustrated in Fig. 12.18. (The proof is given in detail in [39] and will not be reported here.)

These defect structures are metastable in that they cannot be removed (or moved) without incurring an energy cost. The steps towards the annihilation of a structure with $k = 1$ are shown explicitly in Fig. 12.18. They require flipping three original (S) spins, which in turn flip three plaquette (τ) spins each. In the process, we generate one more defect than the three we started with, and the overall energy barrier is therefore $2J$.

The same process can be iterated for larger defect structures: to annihilate a structure of linear size 2^k, one has to annihilate the three structures of linear size 2^{k-1}

within it, which requires overcoming the barrier to create one extra defect, $2J$. The situation is similar for each of the structures of linear size 2^{k-1}, etc., until we arrive at $k = 1$, where the process above applies. The overall barrier is thus $\Delta = 2Jk = 2J \log_2 \ell$, where ℓ is the initial separation between defects.

Exercise 12.10 Follow the discussion in the text to prove that the smallest number of additional intermediate defects that one ought to create in order to annihilate three defective plaquettes at the corners of an equilateral triangle of side 2^k is k.

In thermodynamic equilibrium, the average separation between defects scales as the inverse square root of their density, namely $\ell \sim e^{J/T}$, since from (12.18) we see that the energy cost of a defect is $2J$, whence their density is $\sim e^{-2J/T}$. Therefore, $\Delta = 2J \log_2 e^{J/T}$ and the corresponding relaxation timescale is $\tau \sim e^{\Delta/T} \sim \exp[2J^2/(T^2 \ln 2)]$.

Even though it is still the case that relaxation timescales diverge only in the limit of zero temperature, the plaquette energy terms on the triangular lattice exhibit a qualitatively different behaviour from those on the square lattice. They give rise to an unusually strong slowing down that is exponential in the square of the inverse temperature. This is a substantial improvement in robustness to thermal fluctuations. One might thus wonder whether new lattice models can be designed where an appropriate combination of plaquette energy terms manages to achieve, say, topological order as in the toric code and exponential inverse temperature squared protection from thermal fluctuations, as in the triangular-plaquette model. In this case, the enhanced protection would not be thermodynamic (in the sense of topological order surviving up to a finite-temperature phase transition) but rather dynamical, slowing down the destabilizing thermal defects into a nearly glassy state.

12.3.4.3 Quantum kinetically constrained models

Quantum versions of kinetically constrained models also exist, although their discussion is beyond the scope of these short lectures. They are in general less studied and less well understood than their classical counterparts. Some examples are discussed in [29], covering both 2D and 3D cases that exhibit energy barriers and quantum relaxation rates akin to those illustrated in Fig. 12.16.

Similarly to the case of the toric code model, one finds that large but low-energy barriers are effective at slowing down quantum tunnelling processes (exponentially suppressed in the width of the barrier), yet they are rather ineffective with respect to thermal fluctuations (exponentially suppressed in the height of the barrier but only polynomially suppressed in the width). This leads to an interesting parallel between classical and quantum glassiness, and the fact that (topological) quantum glassiness can be a behaviour inherent to zero temperature, which disappears immediately at any finite temperature [29].

As suggested in these notes, it is often found that a rich dynamical phenomenology in quantum kinetically constrained models is accompanied by the emergence of quantum topological order. In their dual description, quantum kinetically constrained models can be seen once again as models of point-like particles that move on a lattice according to allowed and disallowed processes. Other models that typically exhibit topological properties are those where particle hopping processes are accompanied by non-trivial phase factors (see e.g. Haldane's model, fractional Chern insulators, and the recent artificial gauge fields in ultracold atomic systems). It will be interesting to investigate how quantum kinetically constrained models behave when similar phase factors are present in the allowed dynamical processes.

12.4 Conclusions

In summary, we have discussed a few examples of how systems with topological properties behave out of equilibrium. The topological nature of the low-energy state in these systems is closely related to the fractionalized character of its elementary excitations. In turn, we have seen that these excitations are directly responsible for the response and equilibration behaviour. This intriguing interplay gives rise to a rich variety of exciting phenomena that we are just beginning to understand and classify.

In the context of statistical-mechanical models such as classical spin ice, we have shown how the nature of the low-temperature phase and its excitations is reflected in reaction–diffusion relaxation processes with local and global kinematic constraints as well as emergent long-range Coulomb interactions. Spin ice thus offers a realization of several paradigmatic concepts in non-equilibrium dynamics, with an unusually high degree of tunability.

We have also discussed how a similar interplay between the topological ground state and its fractionalized excitations leads to interesting equilibration properties in quantum-mechanical systems, in the presence of both quantum and thermal perturbations. However, the additional complexity of out-of-equilibrium quantum mechanics in strongly interacting systems limits the discussion at present to somewhat simple examples (e.g. the toric code) and achieves a far less detailed understanding than its classical counterpart (e.g. spin ice and kinetically constrained models). Notwithstanding this, one encounters interesting scenarios demonstrating the interplay of topological order and glassiness, which raise intriguing questions. Quantum topological order appears to be more susceptible to thermal fluctuations than conventional (local) types of order, at least in 2D and 3D. Could topological protection be improved by slowing down (thermal) defects, i.e. freezing them into a glassy state? Can this be achieved without disorder, using fractal structures as for example in the kinetically constrained triangular-plaquette model?

Overall, this is an exciting and timely research direction, also thanks to recent material and technological developments that are producing an increasing number of experimental results on systems with topological properties out of equilibrium.

Acknowledgements

This work was supported in part by EPSRC Grant EP/K028960/1.

References

[1] S. T. Bramwell and M. J. P. Gingras, *Science* **294**, 1495 (2001).

[2] C. Castelnovo, R. Moessner, and S. L. Sondhi, *Annu. Rev. Condens. Matter Phys.* **3**, 35 (2012).

[3] C. Castelnovo, R. Moessner, and S. L. Sondhi, *Phys. Rev. B* **84**, 144435 (2011).

[4] Y. Levin, *Rep. Prog. Phys.* **65**, 1577 (2002).

[5] D. J. P. Morris, D. A. Tennant, S. A. Grigera, B. Klemke, C. Castelnovo, R. Moessner, C. Czternasty, M. Meissner, K. C. Rule, J.-U. Hoffmann, K. Kiefer, S. Gerischer, D. Slobinsky, and R. S. Perry, *Science* **326**, 411 (2009).

[6] I. A. Ryzhkin, *JETP* **101**, 481 (2005).

[7] J. Snyder, B. G. Ueland, J. S. Slusky, H. Karunadasa, R. J. Cava, and P. Schiffer, *Phys. Rev. B* **69**, 064414 (2004).

[8] L. D. C. Jaubert and P. C. W. Holdsworth, *Nat. Phys.* **5**, 258 (2009).

[9] J. A. Quilliam, I. R. Yaraskavitch, H. A. Dabkowska, B. D. Gaulin, and J. B. Kycia, *Phys. Rev. B* **83**, 094424 (2011).

[10] K. Matsuhira, C. Paulsen, E. Lhotel, C. Sekine, Z. Hiroi, and S. Takagi, *J. Phys. Soc. Jpn* **80**, 123711 (2011).

[11] S. R. Giblin, S. T. Bramwell, P. Holdsworth, D. Prabhakaran, and I. Terry, *Nat. Phys.* **7**, 252 (2011).

[12] L. R. Yaraskavitch, H. M. Revell, S. Meng, K. A. Ross, H. M. L. Noad, H. A. Dabkowska, B. D. Gaulin, and J. B. Kycia, *Phys. Rev. B* **85**, 020410 (2012).

[13] H. M. Revell, L. R. Yaraskavitch, J. D. Mason, K. A. Ross, H. M. L. Noad, H. A. Dabkowska, B. D. Gaulin, P. Henelius, and J. B. Kycia, *Nat. Phys.* **9**, 34 (2013).

[14] H. Takatsu, K. Goto, H. Otsuka, R. Higashinaka,K. Matsubayashi, Y. Uwatoko, and H. Kadowaki, *J. Phys. Soc. Jpn* **82**, 104710 (2013).

[15] G. Sala, M. J. Gutmann, D. Prabhakaran, D. Pomaranski, C. Mitchelitis, J. B. Kycia, D. G. Porter, C. Castelnovo, and J. P. Goff, *Nat. Mater.* **13**, 488 (2014).

[16] L. Onsager, *J. Chem. Phys.* **2**, 599 (1934).

[17] S. T. Bramwell, S. R. Giblin, S. Calder, R. Aldus, D. Prabhakaran, and T. Fennell, Nature **461**, 956 (2009).

[18] V. Kaiser, S. T. Bramwell, P. C. W. Holdsworth, and R. Moessner, *Nat. Mater.* **12**, 1033 (2013).

[19] C. Castelnovo, R. Moessner, and S. L. Sondhi, *Phys. Rev. Lett.* **104**, 107201 (2010).

[20] A. J. Bray, *Adv. Phys.* **43**, 357 (1994).

[21] V. V. Ginzburg, L. Radzihovsky, and N. A. Clark, *Phys. Rev. E* **55**, 395 (1997).

[22] D. Toussaint and F. Wilczek, *J. Chem. Phys.* **78**, 2642 (1983)

[23] C. Castelnovo, R. Moessner, and S. L. Sondhi, *Nature* **451**, 42 (2008).

[24] S. Mostame, C. Castelnovo, R. Moessner, and S. L. Sondhi, *Proc. Natl Acad. Sci. USA* **111**, 640 (2014).

[25] D. Slobinsky, C. Castelnovo, R. A. Borzi, A. S. Gibbs, A. P. Mackenzie, R. Moessner, and S. A. Grigera, *Phys. Rev. Lett.* **105**, 267205 (2010).

[26] C. Paulsen, M. J. Jackson, E. Lhotel, B. Canals, D. Prabhakaran, K. Matsuhira, S. R. Giblin, and S. T. Bramwell, *Nat. Phys.* **10**, 135 (2014).

[27] M. J. Jackson, E. Lhotel, S. R. Giblin, S. T. Bramwell, D. Prabhakaran, K. Matsuhira, Z. Hiroi, Q. Yu, and C. Paulsen, *Phys. Rev. B* **90**, 064427 (2014).
[28] A. Y. Kitaev, *Ann. Phys. (NY)* **303**, 2 (2003).
[29] C. Castelnovo and C. Chamon, *Phil. Mag.* **92**, 304 (2011).
[30] C. Castelnovo and C. Chamon, *Phys. Rev. B* **76**, 184442 (2007).
[31] C. Castelnovo and C. Chamon, *Phys. Rev. B* **78**, 155120 (2008).
[32] Z. Nussinov and G. Ortiz, *Proc. Natl Acad. Sci. USA* **106**, 16944 (2009).
[33] Z. Nussinov and G. Ortiz, *Ann. Phys. (NY)* **324**, 977 (2009).
[34] C. D. Freeman, C. M. Herdman, D. J. Gorman, and K. B. Whaley, *Phys. Rev. B* **90**, 134302 (2014).
[35] F. Ritort and P. Sollich, *Adv. Phys.* **52**, 219 (2003).
[36] D. A. Johnston, A. Lipowski, and R. P. K. C. Malmini, in *Rugged Free-Energy Landscapes: Common Computational Approaches to Spin Glasses, Structural Glasses and Biological Macromolecules* (ed. W. Janke), pp. 173–199. Lecture Notes in Physics, Vol. 736, Springer-Verlag, Berlin (2008).
[37] C. Chamon, *Phys. Rev. Lett.* **94**, 040402 (2005).
[38] A. Lipowski, *J. Phys. A: Math. Gen.* **30**, 7365 (1997).
[39] M. E. J. Newman and C. Moore, *Phys. Rev. E* **60**, 5068 (1999).
[40] J. P. Garrahan and M. E. J. Newman, *Phys. Rev. E* **62**, 7670 (2000).
[41] J. P. Garrahan, *J. Phys. Condens. Matter* **14**, 1571 (2002).
[42] J. Hamp, A. Chandran, R. Moessner, and C. Castelnovo *Phys. Rev. B* **92**, 075142 (2015). DOI: 10.1103/PhysRevB.92.075142

13
What is life?—70 years after Schrödinger

Antti J. Niemi

Department of Physics and Astronomy, Uppsala University
PO Box 803, S-75108, Uppsala, Sweden
Laboratoire de Mathematiques et Physique Théorique
CNRS UMR 6083, Fédération Denis Poisson, Université de Tours
Parc de Grandmont, F37200, Tours, France
Department of Physics, Beijing Institute of Technology
Haidian District, Beijing 100081, China

Topological Aspects of Condensed Matter Physics. First Edition. C. Chamon et al.

Chapter Contents

13 What is life?—70 years after Schrödinger 567
Antti J. NIEMI

Colour figures. For those figures in this chapter that use colour, please see the version of these lecture notes at http://topo-houches.pks.mpg.de and arXiv.1412.8321 [cond-mat.soft]. These figures are indicated by '[Colour online]' at the start of the caption.

Preface

This chapter is dedicated to the 70th anniversary of Schrödinger's 1944 Lectures in Dublin.

There are at present fundamental problems in theoretical physics awaiting solution, e.g. the relativistic formulation of quantum mechanics and the nature of atomic nuclei (to be followed by more difficult ones such as the problem of life). (P. A. M. Dirac, 1931)

In February 1943, Erwin Schrödinger presented a series of lectures at the Dublin Institute for Advanced Studies with the title *What is Life? The Physical Aspect of the Living Cell.* The lectures were subsequently published in the form of a small book [1]. According to Google Scholar, this book has been cited even more frequently than Schrödinger's articles on quantum mechanics. In particular, it was credited by Crick and Watson as the source of inspiration that led them to reveal the double-helix structure of DNA.

My lectures at Les Houches were a celebration of the anniversary of Schrödinger's lectures, and for that reason I decided to share a title.

Besides Crick and Watson, Schrödinger's book has been, and continues to be, a source of inspiration to generations of physicists. But it seems to me that its full dimensionality might not yet have been fully comprehended, or appreciated. In particular, the book has many parallels to Philip Anderson's highly inspiring 1972 article *More is different. Broken symmetry and the nature of the hierarchical structure of science* [2]. Schrödinger clearly realized that *Bol'she* (*больше*) makes a difference when he wrote that

...living matter, while not eluding the 'laws of physics' as established up to date, is likely to involve 'other laws of physics' hitherto unknown, which, however, once they have been revealed, will form just as integral a part of this science as the former.

The time should be ripe to accept a universal formal definition of life in terms of proteins and their dynamics: proteins are the workhorses of all living organisms, they are true nanomachines that participate in all the metabolic activities that constitute life as we know it. From this perspective, life is indeed something that can be modelled and understood using both known and still to be revealed laws that govern the subcellular physics of living matter. For a physicist like me, this is an exciting way to try and answer Schrödinger's question.

The underlying theme in my lectures is to view proteins as an exciting example of a physical system where much of Anderson's *Bol'she* can be found. Indeed, proteins seem to bring together *most* of the contemporary lines of research in modern theoretical physics: geometry of string-like structures, topological solitons, spin chains, integrable models, equilibrium and non-equilibrium statistical physics, quest for entropy, emergent phenomena, ...and much, much more. Moreover, the tools that are needed to fully understand proteins range from highly formal to extensively numerical, and for a theorist there are almost endless opportunities to address questions with direct and important experimental relevance: physical, chemical, biological, medical, ... The amount of data is *overwhelming* and it is readily accessible. Experiments can be done and directly compared with theoretical calculations and numerical computations; the subject continues to grow at a highly exponential rate.

These lectures were prepared for students in condensed matter physics, both theoretical and experimental. I assumed the students did not really have any prior knowledge of proteins, that they did not even know what a protein looks like at the atomic level. Thus, I begin with a *protein minimum*. It explains the basic facts that I think one needs to know to get started in research on the physics of proteins. The rest of the lectures address proteins from the point of view of a physicist, from a perspective that I hope appeals to the way a physicist thinks. As such, the presentation could be somewhat intimidating to chemists and biologists, who might find the concepts and techniques that I introduce as foreign, something they have not seen and are not accustomed to, in the context of a problem that is traditionally viewed as theirs. However, I assure you that everything I describe is very simple. A good command of basic algebra is all that it really takes to follow these lectures. Indeed, proteins brings physics together in a unique fashion with biology, chemistry, applied mathematics, even medical research, Theory and expriments? with the goal of understanding matter that is alive. I hope you 'catch the bug' too!

13.1 A protein minimum

13.1.1 Why proteins?

Proteins are nanoscale machines that control and operate all metabolic processes in all living organisms. They often have to function with extreme precision: like most machines, those made of proteins need to have their parts and pieces in the right place, in a good shape, and finely tuned. How else could these self-producing nanomachines work in such great harmony, cooperate over an enormous range of scales, and uphold something as complex as life? Indeed, it is widely understood that the biological function of a protein depends critically on its three-dimensional geometry. From this perspective, the so-called *protein folding problem*, which aims to explain and derive the shape of a biologically active protein using laws of physics, addresses the origin of life itself [3, 4].

Furthermore, a wrong fold is a common cause for a protein to lose its function. A wrongly folded protein can be dangerous, even fatal, to a biological organism. It is now widely understood that diverse neurodegenerative diseases, including various forms of dementia such as Alzheimer's disease and Parkinson's disease, type 2 diabetes, and about half of all cancers are caused by wrong folds in certain proteins [5]. At the same time, bacteria are on the rampage and emergent resistance through evolutionary processes is rendering existing antibiotics ineffective at a rapid pace [6–8]. No effective methods and treatments have been found to prevent or cure viral maladies like HIV, Ebola, or respiratory syndromes such as SARS and MERS. Our future protection against these and various other harmful and deadly pathogens depends on our skills and knowledge to develop conceptually new, protein-level approaches to fight and eliminate our enemies. Research on proteins is really about 'Saving the Planet' as much as in any video game or movie ever made. But it is for real: by doing research on proteins you have a change to become a real-life 'Gordon Freeman'.

For all these and many other reasons, the ability to accurately describe the physics of proteins, their structure and dynamics, would have an enormous impact on biology, pharmacy, and health sciences. It would provide huge benefits to society by paving ways to prevent and cure many tormenting diseases. In particular, it would provide us an answer to *What is life?* along the lines foreseen by Schrödinger.

In the following, I will give a short introduction to proteins, what you need to get your research started as a physicist. For those who are really seriously interested in the biological aspects, I recommend the textbook *Molecular Biology of the Cell* [9].

13.1.2 Protein chemistry and the genetic code

Proteins are one-dimensional linear polymers. They are composed of 20 different amino acids that share a number of structural properties: there are the *backbone* atoms that are common to all amino acids and there are the *residues* or *side chains* that are different for each of the 20 amino acids.

In Fig. 13.1, we show the chemical composition of a generic amino acid. When two amino acids meet, a chemical process can take place that joins them together into a dipeptide plus water, as shown in the figure. When this process repeats itself, we eventually arrive at a long polypeptide chain , a.k.a. a protein, as shown in Fig. 13.2. Once the protein attains the correct shape, it becomes ready for biological action.

Note the carbon atoms that are denoted C_α in Figs. 13.1 and 13.2. These are called the α-carbons, and they have a central role in protein structure. As shown in the figures, the α-carbons connect the residues to the backbone; the C_α forms the

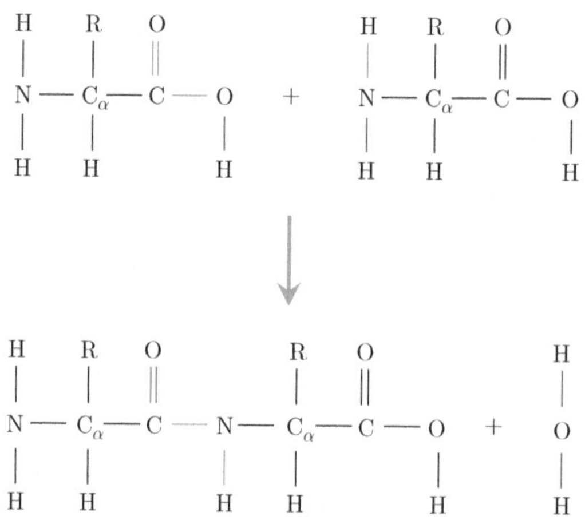

Fig. 13.1 [Colour online] Amino acids have a common backbone with heavy-atom pattern –N–C–C–O–, but there are also 20 different residues (side chains), which we denote here by R. When two amino acids combine together we obtain a dipeptide, in addition to a water molecule.

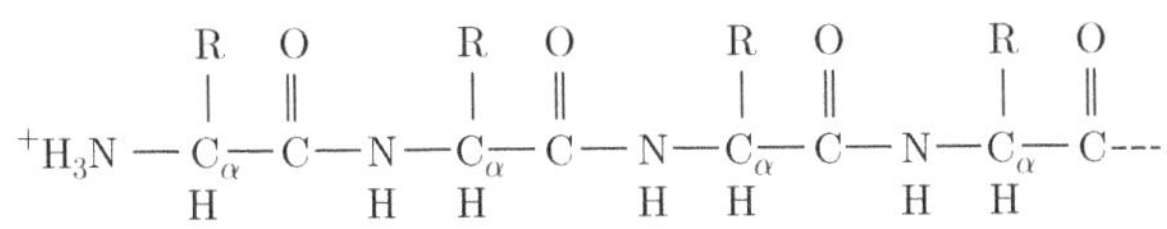

Fig. 13.2 [Colour online] Proteins are long linear chains of amino acids, each with a similar backbone structure but 20 different residue structures (R).

centre of an sp^3-hybridized tetrahedron that subjects it to strong steric constraints and holds it rigidly in place relative to the other atoms. As we shall find, the α-carbons largely determine the shape of the protein. Thus, much of our subsequent analysis of protein structure and dynamics is based on the central role of the C_α, for reasons that become increasingly apparent as we proceed.

In a living organism like you and me, the instructions for making proteins are stored in our *genome*. At the level of DNA, the *genetic code* consists of a sequence of nucleobases that connect the two strands of DNA. A group of three nucleobases corresponds to a single amino acid; there is a segment of DNA for each protein. The genetic code is copied from DNA to RNA in a process called *transcription*. This process, like all other processes in our bodies, is driven by various proteins. Particular proteins called enzymes act as catalysts to help and control complex biological reactions.

Our DNA consists of four different nucleobases. Hence, there are $4 \times 4 \times 4 = 64$ different combinations. But two of them are instructions to *stop* the process of transcription. Thus, we have a total of 62 combinations of nucleobases that encode the 20 amino acids—the genetic code is degenerate.

Research project 13.1 From the point of view of physics, we have an appetizing similarity between the genetic code, where a group of three nucleobases corresponds to an amino acid, and the Standard Model of particle physics, where baryons are made of three quarks. Can you find a symmetry principle akin to the Eightfold Way that relates the 62 codons to the 20 amino acids? *Hint:* A good way to start trying to do this is to follow [10].

Once formed, the RNA has the mission of carrying the genetic code to a ribosome. A ribosome is essentially a nanoscale three-dimensional printer. It is made of proteins, and it has the duty to produce new proteins according to the instructions given to it by RNA. The process where a ribosome combines amino acids into a protein chain is called *translation*.

13.1.3 Data banks and experiments

The amount of data and information available on the Internet is enormous, both on the genetic code and on proteins. There are various open-access libraries both on the sequences and on the structures of proteins; the amount of data is already more than any single person can possibly ever analyse, and it continues to increase at an exponential rate.

For those who are mainly interested in biology and related bioinformatics, an excellent resource on protein sequences and their biological function is UniProt:

http://www.uniprot.org/

This data bank contains presently almost 90 million different protein sequences. As shown in Fig. 13.3(a) the number of known sequences grows at a very high, exponential, rate.

For those who are mainly interested in physics of proteins, the Protein Data Bank (PDB) is an excellent resource:

http://www.pdb.org/

As shown in Fig. 13.3(b), the number of known protein structures in the PDB is around 100 000 and growing—but not at all as fast, since only about 0.1% of known protein sequences have a known structure.

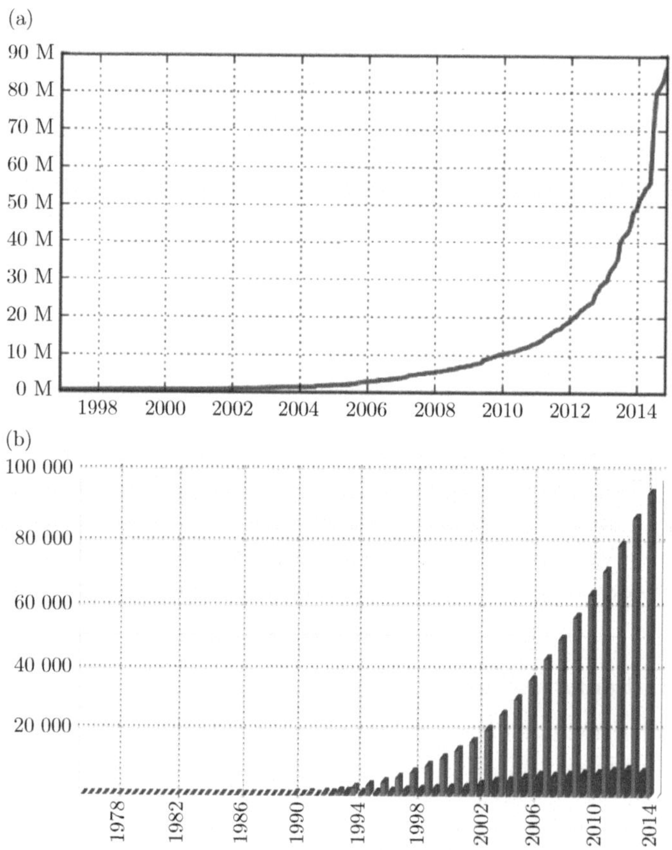

Fig. 13.3 (a) The increase in the number of sequences in UniProt, as a function of year. (Taken from http://www.uniprot.org/.) (b) The increase in the number of structures in the PDB as a function of year. Both annual increase and accumulated total are shown. (Taken from http://www.pdb.org/.)

Numerous other good sources of information exist and can be found on the Internet. For example, the PSI Structural Biology Knowledgebase is a comprehensive database for various structural aspects of proteins. It can be found at

http://sbkb.org/

Most of the 100 000 structures in the PDB have been resolved using X-ray crystallography. But other techniques are also being used. In particular, the number of NMR structures is increasing. Until now, it has been very difficult to resolve long protein sequences using NMR techniques. Most NMR structures are quite short, and those with more than 100 amino acids are rare. The advantage of NMR over X-ray crystallography is that, because no crystallization is needed, NMR can more easily provide dynamical information. It is possible to follow proteins in motion using NMR, while crystallized structures have problems moving. However, X-ray techniques such as small-angle X-ray scattering (SAXS) and wide-angle X-ray scattering (WAXS) are now being developed that can observe proteins in motion. In the near future, further techniques, such as those using free-electron lasers, will be able to provide detailed structural and dynamical information at very short timescales. Various other methods are also in use and under development. The experimental study of protein structure and dynamics is still very much in its infancy. This makes the study of proteins an exciting field to enter for those who are experimentally minded, as well as for theoreticians.

The protein crystals in the PDB are *ordered* and they are commonly presumed to display a crystallized conformation that is close to the biologically active one. But *most* proteins are apparently intrinsically unstructured. Such proteins cannot be crystallized into any kind of biologically unique conformation. When these proteins are biologically active, they do not have any single conformation. Instead, they change their form, perpetually. Most proteins in our bodies are like this, intricate nanomachines that are in constant activity. Very little—in fact next-to-nothing—is known about the structural aspects of intrinsically unstructured proteins. In these lectures, we shall look at examples of both ordered and intrinsically disordered proteins.

Our experimental considerations will mainly make use of a subset of crystallographic PDB structures that have been measured with ultra-high precision, with a resolution better than 1.0 Å. The reason why we prefer to use these ultra-high-resolution structures is because of a process called *refinement* that commonly takes place during experimental data validation and model building [11]. During refinement and validation, one iteratively improves the parameters of an approximate trial structure from experimental observations, until some kind of a best fit between the trial structure and the observed diffraction pattern is obtained. As an ansatz, and as reference, the process utilizes known experimental crystallographic structures. Widely used experimentally determined, and highly accurate, template libraries of small molecules include that of Engh and Huber [12]. Thus, the process of refinement *might* introduce a bias towards structures that are already known. In particular, it is not clear to what extent the structure of a small molecule persists in the complex, highly interactive environment of a large protein.

Indeed, it is important to recognize and keep in mind that the PDB data files are prone to all kinds of errors [11, 13–15]. The data should be used with care. MolProbity is an example of a web server that can be used to analyse the quality of an experimental protein structure. It can be found at

http://molprobity.biochem.duke.edu/

One might presume that in the case of ultra-high-resolution structures—those that have been measured with better than 1.0 Å resolution—the quality of observed diffraction patterns should be very good. These structures should have much less need for refinement during model building. Thus, they should be much less biased towards known structures. The number of misplaced atoms should be relatively low.

In order to minimize radiation damage, crystallographic structures are often measured at temperatures near that of liquid nitrogen i.e. around 80–90 K. Thus, the thermal fluctuations in the atomic coordinates should be small. In the PDB data, the experimental uncertainty in the atomic coordinates is estimated by the (temperature) B-factors. Besides the thermal fluctuations, these B-factors also summarize all the other uncertainties that the experimentalist thinks affects the precision. In Fig. 13.4, we show the distribution of the Debye–Waller fluctuation distance in our subset of ultra-high-precision structures for the C_α atom coordinates. The fluctuation distance can be estimated using the Debye–Waller relation

$$\sqrt{\langle \mathbf{x}^2 \rangle} \approx \sqrt{\frac{B}{8\pi^2}}. \tag{13.1}$$

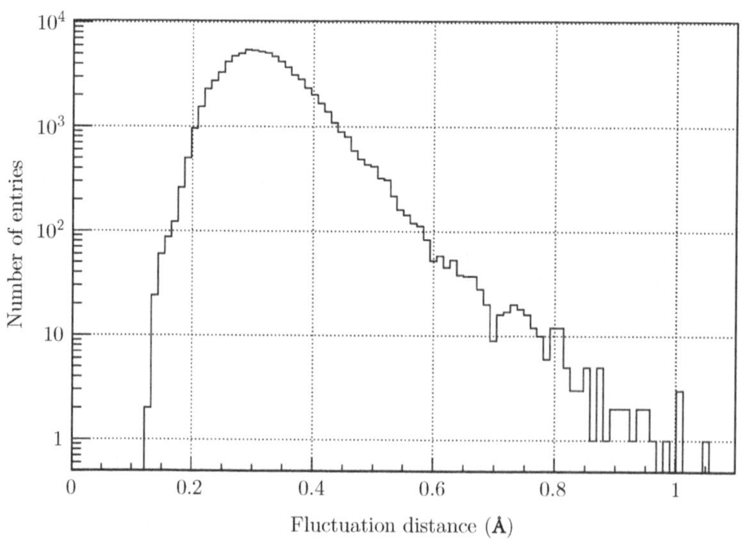

Fig. 13.4 The distribution of the Debye–Waller fluctuation distance for the C_α atoms among those crystallographic PDB structures that have been measured with better than 1.0 Å resolution.

ATOM	98	N	HIS	A	12	37.554	22.049	−3.039	1.00	7.00	N
ATOM	99	CA	HIS	A	12	38.052	23.320	−2.567	1.00	7.99	C
ATOM	100	C	HIS	A	12	37.862	23.459	−1.070	1.00	6.88	C
ATOM	101	O	HIS	A	12	38.654	24.106	−0.390	1.00	7.05	O
ATOM	102	CB	HIS	A	12	37.387	24.460	−3.314	1.00	11.49	C
ATOM	103	CG	HIS	A	12	38.195	24.796	−4.556	1.00	15.84	C
ATOM	104	ND1	HIS	A	12	39.254	25.596	−4.535	1.00	18.10	N
ATOM	105	CD2	HIS	A	12	38.101	24.186	−5.778	1.00	17.75	C
ATOM	106	CE1	HIS	A	12	39.815	25.479	−5.722	1.00	18.99	C
ATOM	107	NE2	HIS	A	12	39.122	24.635	−6.465	1.00	19.38	N

Fig. 13.5 An example of a PDB file, in this case the amino acid histidine (HIS). The second column lists the atom number (98–107) along the backbone. The third column lists the type of atom; CA stands for C_α and entries 98–101 are the backbone N-C_α-C-O atoms of HIS. Entries 102–107 are side chain atoms. In this case, HIS is the 12th amino acid along the backbone. The (x, y, z) coordinates are listed in the following three columns. The B-factors are listed in the last column, before a list of the chemical symbols. Note that in this list, the hydrogen atoms are absent (hydrogens can be difficult to observe).

This corresponds roughly to the one-standard-deviation uncertainty in the experimentally measured coordinate values. Figure 13.5 is an example of a generic PDB entry that shows how the B-factors are listed, together with the atomic coordinates.

According to Fig. 13.4, among our ultra-high resolution PDB structures the one-standard-deviation error distance in the C_α atomic positions peaks in the range of 0.3–0.5 Å. The lower bound is around 0.15–0.2 Å, and in these lectures we shall adopt ~0.15 Å, i.e. around 20% of the radius of the carbon atom, as the lower-bound estimate for the size of quantum-mechanical zero-point fluctuations in the C_α positions. Note that historically ~0.2 Å has been considered as the boundary between X-rays and γ-rays.

13.1.4 Phases of proteins

Like most linear polymers, proteins have a highly complex phase structure that can depend on a multitude of factors, including the chemical structure of a polymer and its solvent, temperature, pressure, changes in the solvent's acidity, and many other environmental factors [16, 17]. In a good solvent environment, the interactions between a polymer segment and the solvent molecules usually cause the polymer to expand, and the polymer behaves like a self-avoiding random walk. In a poor solvent environment, such as the water that surrounds proteins in our cells, the polymer–polymer self-interactions dominate and the polymer tends to collapse into a space-filling conformation. These two phases are separated by a θ-regime, where the repulsive and attractive interactions cancel each other and the polymer has the geometric character of a random walk (Brownian motion).

In the limit where the number N of monomers is very large, many aspects of the phase structure become *universal* [18–20]. An example of a universal quantity in the case of a linear polymer such as a protein is the compactness index ν. It is defined in terms of the radius of gyration R_g [16, 17, 21–23]:

$$R_g^2 = \frac{1}{2N^2} \sum_{i,j} (\mathbf{r}_i - \mathbf{r}_j)^2. \tag{13.2}$$

Here, $\mathbf{r}_i$ are the coordinates of all the atoms in the polymer. In the case of a protein, with no loss of generality, we may restrict $\mathbf{r}_i$ to the coordinates of the backbone C_α atoms only. The compactness index ν governs the large-N asymptotic form of (13.2). When the number N of monomers becomes very large, we have [23]

$$R_g^2 \xrightarrow{N \text{ large}} R_0^2 N^{2\nu}(1 + R_1 N^{-\delta_1} + \ldots). \tag{13.3}$$

It should be obvious that ν coincides with the inverse Hausdorff dimension of the structure. Besides the compactness index ν, the critical exponents δ_1 etc. are also universal quantities. But the form factor R_0 that characterizes the effective distance between the monomers in the large-N limit, and the subsequent amplitudes R_1 etc. that parametrize the finite-size corrections, are not universal [23]. These parameters can in principle be computed from the chemical structure of the polymer and solvent, in terms of environmental factors such as temperature and pressure.

As a universal quantity, ν is independent of the detailed atomic structure. Different values of ν correspond to different phases of polymer. The four commonly accepted mean-field values of ν are

$$\nu = \begin{cases} \frac{1}{3}, \\ \frac{1}{2}, \\ \frac{3}{5}, \\ 1. \end{cases} \tag{13.4}$$

Under poor solvent conditions such as in the case of the proteins in our cells, a linear single-chain polymer collapses into the space-filling conformation and we have the mean-field exponent $\nu = \frac{1}{3}$. For a fully flexible ideal chain, the mean-field value is $\nu = \frac{1}{2}$. This phase takes place in the θ-regime that separates the collapsed phase from the high-temperature self-avoiding random-walk phase, for which we have the mean-field Flory value $\nu = \frac{3}{5}$. Finally, when $\nu = 1$, the polymer is like a rigid stick. A number of proteins are like this, some collagens for example.

Research project 13.2 Three-dimensional dynamical systems such as the Lorenz equation provide numerous examples of space curves with attractors that have all kind of Hausdorff dimensions. Can you find physical examples of polymers (proteins) where ν takes values that correspond to phases that are different from the four listed in (13.4)?

The mean-field values of the critical exponents ν, δ_1, etc. in (13.3) may be corrected by fluctuations. In particular, in the universality class of the self-avoiding random walk, the improved values are [24, 25]

$$\begin{aligned} \nu &= 0.5880 \pm 0.0015, \\ \delta_1 &= 0.47 \pm 0.03. \end{aligned} \tag{13.5}$$

The computation of (13.5) in [24, 25] utilizes the concept of universality to argue that the three-dimensional self-avoiding random walk is in the same universality class as the $O(n)$-symmetric scalar field theory with a quartic self-interaction, in the limit where the number of components $n \to 0$. A subsequent numerical Monte Carlo evaluation of the critical exponents (13.5), computed using a self-avoiding random-walk model on a square lattice [23], gave very similar values:

$$\begin{aligned} \nu &= 0.5877 \pm 0.0006, \\ \delta_1 &= 0.56 \pm 0.03. \end{aligned} \tag{13.6}$$

In the case of crystallographic PDB protein structures, we may evaluate the dependence of the radius of gyration on the number of residues using the coordinates of the C_α atoms. The result is shown in Fig. 13.6.

A least-squares fit to the data gives

$$R_g \approx R_0 N^\nu \approx 2.280 N^{0.375} \text{ Å}. \tag{13.7}$$

Note that proteins are not homopolymers. But when N increases, the detailed amino acid structure of a protein should become increasingly irrelevant in determining the

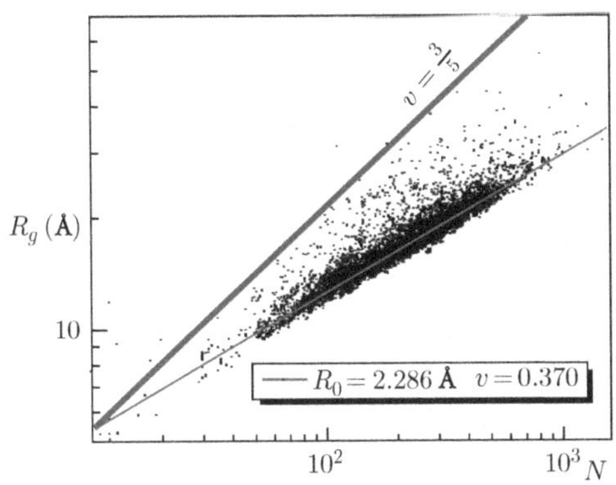

Fig. 13.6 [Colour online] The C_α radius of gyration as a function of residues, in the case of those monomeric crystallographic PDB proteins that have been measured with better than 2.0 Å resolution. The lower [red] line is the least-squares linear fit, and the upper [blue] line is for the Flory value $\nu = \frac{3}{5}$.

relation between the radius of gyration and the number of residues. For long protein chains, the inhomogeneity due to amino acids should be treated as a finite-size correction in (13.3), when N becomes very large. Indeed, in the limit of a very large number of residues, a generic protein is like a chain along which the 20 residues have been quite randomly distributed. It should be like a spin chain embedded in $\mathbb{R}^3$ where each residue is a spin variable, with 20 different (random) values. Thus, when the ratio $20/N$ becomes very small, the effect of an individual residue becomes small in an average, statistical sense. The protein approaches a homopolymer that is equipped with an 'averaged' residue.

Research project 13.3 Develop a theory of spin chains embedded in $\mathbb{R}^3$.

13.1.5 Backbone geometry

According to Fig. 13.6, those proteins that can be crystallized are in the collapsed $\nu \approx \frac{1}{3}$ phase. To describe the properties of their thermodynamical phase state, we need to identify a proper set of *order parameters* in the sense of Landau, Ginzburg, and Wilson; the concept of order parameter is described in numerous textbooks.[1] A local order parameter is a systematically constructed effective dynamical variable that describes collectively a set of elemental degrees of freedom such as atoms and molecules in a system that is subject to the laws of statistical physics. Examples of order parameters include the magnetization in the case of a ferromagnet, the director in a nematic crystal, the condensate wavefunction in superfluid helium-4, and Cooper pair in a BCS superconductor. The concept of an order parameter is often intimately related to the concept of symmetry breaking and emergent phenomena. For example, in each of the cases that we mentioned, we have a symmetry that becomes broken, and this symmetry breaking gives rise to emergent structures. In particular, the breaking of the symmetry is described in terms of the properties of the pertinent order parameter in each case.

In the case of a protein, we have already concluded that the phase structure relates to aspects of the protein geometry. The different phases of a protein are characterized by different Hausdorff dimensions (13.4). Moreover, proteins have an apparent symmetry that has become broken.

Amino acids are chiral molecules. An amino acid can be either left-handed (`L`) or right-handed (`D`). The only exception is glycine, which has no chirality. For two amino acids to form a dipeptide as shown in Fig. 13.1, they must have the same chirality; you can't easily shake someone's left hand with your own right. For some reason, the symmetry between `L` and D is broken in Nature—practically all amino acids that appear in proteins of living organisms from prokaryotes such as bacteria to eukaryotes like us, are left-handed chiral. This symmetry breaking apparently reflects itself in the higher-level geometric structures of proteins: as a polymer chain, the proteins that are found in living organisms are more often twisted in a *right-handed* manner than

[1] In the context of protein research, order parameters are sometimes called *reaction coordinates*.

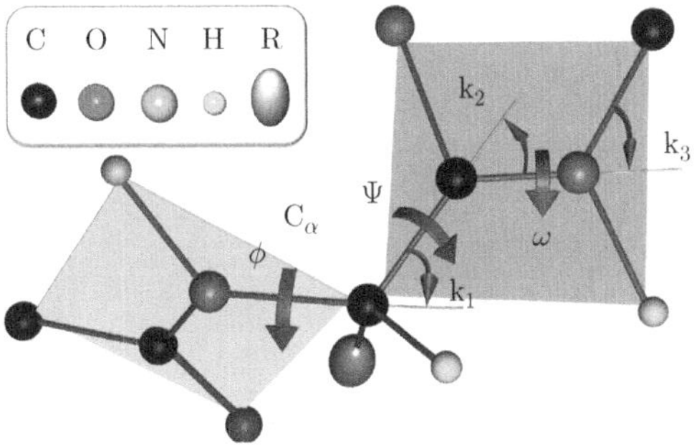

Fig. 13.7 [Colour online] The atoms along the protein backbone form peptide planes; the two covalent bonds C=O and N–H are antiparallel. The definitions of the various bond and torsion angles between the covalent bonds are also shown. The two torsion angles (ϕ, ψ) are the Ramachandran angles.

the opposite. Thus, any local order parameter that describes the phase properties of proteins in a living organism should somehow capture the helical aspects of protein geometry.

Figure 13.7 details the local geometry of a protein. In this figure, we identify a C_α atom together with its covalently bonded N, C, and H atoms, and the residue R, which starts with a covalently bonded carbon atom called C_β. The covalent bonds between these five atoms form an sp^3-hybridized tetrahedron, with C_α at the centre. Take the C atom to be the top of the tetrahedron, and N, H, and R as the three bottom vertices. Consider the axis of the tetrahedron that runs along the covalent bond from the C to C_α and look down this axis from C towards C_α. If the H atom is in the clockwise direction from the residue R, then the amino acid is left-handed; this is the case in Fig. 13.7. Otherwise, the amino acid is right-handed.

In Fig. 13.7, we also have two *peptide planes*, one prior to the C_α atom and the other after C_α. The C_α atom is located at the joint vertex of the two adjacent peptide planes. We proceed to analyse in detail the properties of the peptide plane geometry.

It turns out that the geometry of the peptide planes is indeed *very* rigidly planar. The planarity is measured by the angle ω shown in Fig. 13.7; this is the angle between the C=O covalent bond and the N–C_α covalent bond (or N–H covalent bond) The values of ω are found to fluctuate *very* little around $\omega = \pi$, which corresponds to the *trans* conformation of the backbone and is shown in the figure; there are a few entries, mainly involving the amino acid proline, where the backbone is in the *cis* conformation where ω vanishes. The *cis* conformation is equally planar, and the fluctuations around $\omega = 0$ are minimal. Figure 13.8 shows the distribution of the ω angles in our ultra-high-resolution subset of crystallographic protein structures—those that have been measured with better than 1.0 Å resolution.

The values of the three covalent bond angles $(\kappa_1, \kappa_2, \kappa_3)$ defined in Fig. 13.7 are also shown in Fig. 13.8. Their values are likewise subject to relatively small variations.

The various covalent bond lengths along the protein backbone also have values that fluctuate very little around their average values. Figure 13.9 shows the various

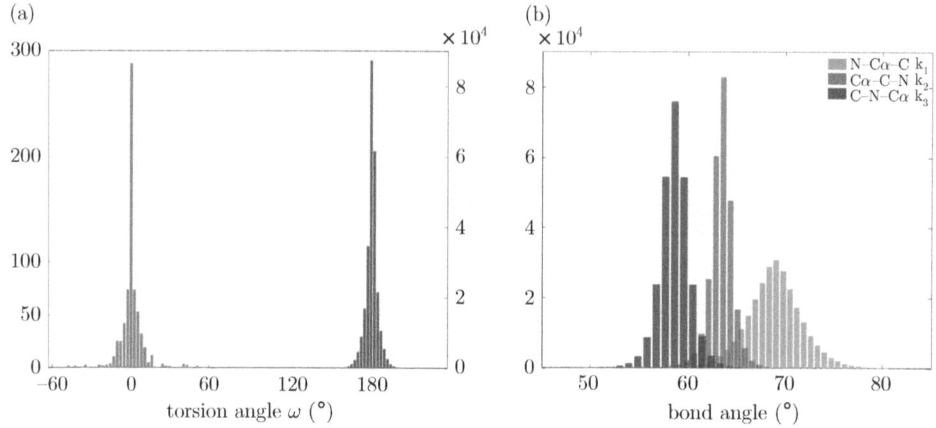

Fig. 13.8 [Colour online] (a) Distribution of the torsion angle ω between the covalent bonds C=O and N–H from Fig. 13.6 in our ultra-high-resolution (1.0 Å) resolution PDB subset. The result shows that the geometry of the peptide planes is indeed planar, with very high precision. For *trans*, we have $\omega \approx \pi$ and for *cis*, we have $\omega \approx 0$. (b) Distribution of the three bond angles $(\kappa_1, \kappa_2, \kappa_3)$ defined in Fig. 13.7 in our data set. The variation around the average values is relatively small.

bond lengths between the heavy atoms along the backbone. We also show the distance between two consecutive C_α atoms, which is also subject to very small fluctuations.

13.1.6 Ramachandran angles

Figures 13.8 and 13.9 show that the three covalent bond angles $(\kappa_1, \kappa_2, \kappa_3)$, the torsion angle ω, and the various covalent bond lengths reveal no dependence on local geometry. Each of these variables has a fairly uniform distribution, which is apparently quite insensitive to variations in local backbone geometry. Thus, we are left with only the two Ramachandran angles (ϕ, ψ) in Fig. 13.7 as the potential local order parameters to characterize local geometry along the backbone. Indeed, it turns out that the variations in their values are *not* small, and in particular they appear to depend on backbone geometry. This is shown by the Ramachandran map in Fig. 13.10, which displays the (ϕ, ψ) distribution in PDB structures that have been measured with better than 2.0 Å resolution.

Note the asymmetry, both in ϕ and in ψ; this asymmetry translates into helicity of the protein backbone. Regular right-handed helical structures (right-handed α-helices) are quite common, while left-handed helical structures are very rare.

Since the phase structure of a protein relates to its geometry, we may expect that the set of the two Ramachandran angles could be utilized as the local order parameters to describe the phase structure of proteins. However, it turns out that this is not the case [26]: the Ramachandran angles form an incomplete set of local order parameters. To show this, we consider all the PDB structures in our data set of ultra-high-resolution structures, those that have been measured with better than 1.0 Å resolution. We perform the following *reconstruction*. From the PDB coordinates

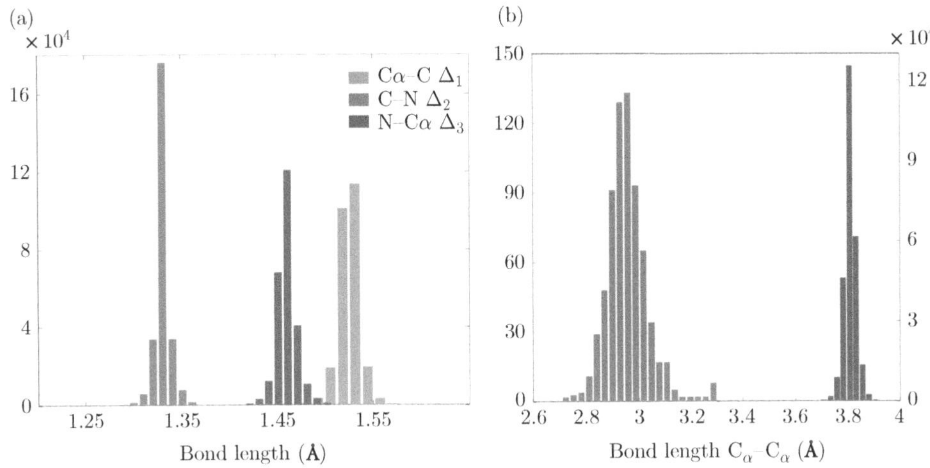

Fig. 13.9 [Colour online] (a) Distribution of the three covalent bond lengths C_α–C, C–N, and N–C_α shown in Fig. 13.7 along the protein backbone. (b) Distribution of the length between neighbouring C_α atoms along the protein backbone. The smaller value ~3.0 Å corresponds to *cis* and the larger value ~3.8 Å is for *trans*.

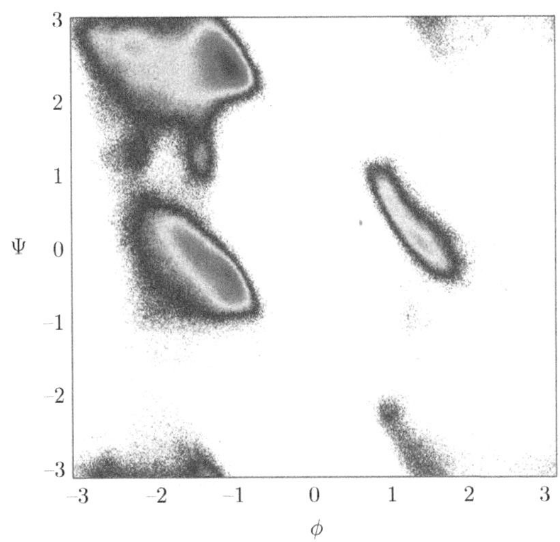

Fig. 13.10 [Colour online] Distribution of Ramachandran angles (ϕ, ψ) defined in Fig. 13.7 in radians.

of the atoms, we first compute the numerical values of all the Ramachandran angles, for each and every peptide plane. Then we continue and do the inverse: we start from the Ramachandran angles that we have computed, and we assume that all the other angles and bond lengths have their average values. We then try to reconstruct the original protein structure, by computing the C_α coordinates.

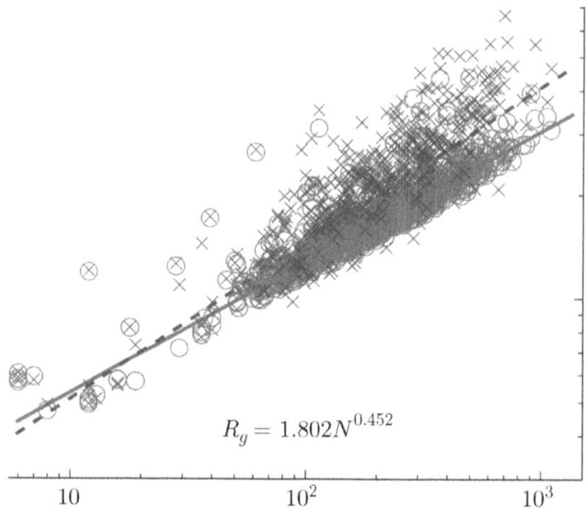

Fig. 13.11 [Colour online] Comparison of R_g between the PDB structures, and the structures obtained by a reconstruction which is based on Ramachandran angles, with all other backbone angles and bond lengths set to their ideal values. Red circles are the original PDB structures, blue crosses are the reconstructed structures.

It turns out that this reconstruction of the coordinates, going from C_α to Ramachandran and then back, usually fails. It is not possible to perform such a reconstruction in the case of a generic protein.

In Fig. 13.11, we show how the reconstructed proteins fail to reproduce even the correct fractal geometry of folded proteins [26]. Instead of (13.7), we obtain the asymptotic relation

$$R_g \approx 1.8N^{0.45} \text{ Å} \tag{13.8}$$

for the reconstructed protein structures: the inverse Hausdorff dimension $\nu = 0.45$ is incorrect. In order to reconstruct the correct fractal geometry, it turns out that we need to include *all* the angular variables in Fig. 13.7 as variables. Only for the bond lengths can the average PDB values be used.

Eventually, later in these lectures, we shall construct a *different* set of local order parameters and show their completeness. However, we first address the modelling of proteins in terms of their (primary) full atom-level description.

13.1.7 Homology modelling

As shown in Fig. 13.3(a), the number of sequences in UniProt grows at a rate that is much higher than that of structures in the PDB, shown in Fig. 13.3(b). The gap is enormous and keeps on growing: sequencing is now fast, cheap and routine, while crystallographic protein structure determination is difficult, time-consuming, and often very expensive; apparently the average cost of a PDB structure is around US$100 000.

Thus, it is impossible to close the gap between sequences and structures by purely experimental methods. To close this gap, we need to develop efficient and accurate computational methods.

Homology[2] modelling [27–29], together with other comparative modelling techniques, is presently the most reliable and effective approach to generate a three-dimensional model of a protein's structure from knowledge of its amino acid sequence. These methods are consistently the top performers in the biannual *Critical Assessment of Protein Structure Prediction* (CASP) tests; see http://predictioncenter.org/.

Homology modelling techniques aim to construct the atomic level structure of a given protein (target), by comparing its amino acid sequence with various libraries of known, homologically related protein structures (templates). Apparently, the reason why this kind of method works is as follows. It seems to be the case that the number of possible protein folds in nature is limited, and that the three-dimensional protein structures seem to be better conserved than the amino acid sequences [30]. However, the quality of a model obtained for the target is largely dictated by the evolutionary distance between the target and the available template structures. If no closely homologous template can be found, these approaches typically fail.

From the point of view of physics, any template-based approach has the disadvantage that there is no energy function. Thus, no energetic analyses of structure and dynamics can be performed. For these, other techniques need to be introduced: to understand the physical properties of a protein, we need to know the energy function.

Research project 13.4 Try to develop a structure prediction approach that uses the best of both worlds—one that finds an initial ansatz structure using templates, and then develops it using techniques of physics. For hints, continue reading ...

13.1.8 All-atom models

... if we were organisms so sensitive that a single atom, or even a few atoms, could make a perceptible impression on our senses—Heavens, what would life be like!

(E. Schrödinger)

We present a short overview of all-atom molecular dynamics, where impressive progress is being made. The aim of an all-atom approach is to model the way a protein folds, by following the time evolution of each and every atom involved, including those of the surrounding solvent (water). But despite impressive progress, the subject remains very much under development and provides enormous challenges to those brave enough to face them: both conceptual and technical problems remain to be resolved, involving issues relevant to physics, chemistry, and computer science, as well as problems for those interested in optimization and efficient algorithm development.

[2] Homology between two proteins is commonly measured on the basis of amino acid sequence similarity. High sequence similarity is a sign of shared ancestry, but for short proteins it can also be due to chance.

There are many examples of all-atom energy functions, called force fields in this context. The most widely used are CHARMM (http://www.charmm.org/) and Amber (http://ambermd.org/); we also mention Gromacs (http://www.gromacs.org/), which is a popular platform for performing molecular dynamics (MD) simulations using different force fields.

A typical all-atom force field that describes a protein chain with N atoms has the following generic form:

$$E(r^N) = \sum_{\text{bonds}} \frac{1}{2} k_b (l - l_0)^2 + \sum_{\text{angles}} \frac{1}{2} k_a (\kappa - \kappa_0)^2 + \sum_{\text{torsions}} \frac{1}{2} V_n [1 + \cos(n\omega - \gamma)] \tag{13.9}$$

$$+ \sum_{j=1}^{N-1} \sum_{i=j+1}^{N} \left\{ \epsilon_{i,j} \left[\left(\frac{r_{0ij}}{r_{ij}} \right)^{12} - 2 \left(\frac{r_{0ij}}{r_{ij}} \right)^{6} \right] + \frac{q_i q_j}{4\pi\epsilon_0 r_{ij}} \right\}. \tag{13.10}$$

The first two contributions in (13.9) describe harmonic oscillations of the bond lengths and bond angles around certain *ideal values* (l_0, κ_0). The third contribution involves torsion angles and evokes a mathematical pendulum, similarly with *ideal-value* ground state(s) given by γ. Torsion angles ω are often found to be much more flexible than bond angles κ, and thus the numerical values of V_n are commonly orders of magnitude smaller than those of k_a. Moreover, a mathematical pendulum that is used for the torsion angles in lieu of a harmonic oscillator allows for larger-amplitude motions and multiple equilibrium states, which is consistent with their more flexible character.

The second contribution (13.10) involves the 6–12 Lennard-Jones potential that approximates the interaction between a pair of neutral atoms; the form is chosen for computational efficiency. Finally, we have the Coulomb interaction. In practice, long-range interactions (13.10) are cut off beyond a distance around 10 Å, again for computational efficiency. The 'ideal' values of the various parameters are usually determined by a process of optimization, using comparative simulations. For parameter fine tuning, one may use very short peptides that have accurately known experimental structures. Such structures can be found for example in the Engh–Huber library [12].

In a full all-atom MD simulation, one solves Newton's equation of motion with (13.9) and (13.10) in an environment of water, which in a full all-atom approach is also treated explicitly. We note that, for example, the r^{-12} term in (13.10) models short-range Pauli repulsion due to overlapping electron orbitals, and the r^{-6} term emerges from long-range van der Waals interactions. Thus, these terms have a quantum-mechanical origin, and the proper interpretation of the ensuing Newton's equation is in terms not of a classical equation but rather of a semiclassical one.

An MD simulation solves Newton's equation iteratively, with a time step $\Delta\tau$. This time step should be short in comparison with the shortest timescale $\Delta t_{\min}$ that characterizes the fastest atomic-level oscillations. The ratio of the two defines a dimensionless expansion parameter. For good convergence of the iterative, discretized Newton's equation, we should have

$$\frac{\Delta\tau}{\Delta t_{\min}} \ll 1. \tag{13.11}$$

This implies that $\Delta\mathcal{T}$ should be no more than a few femtoseconds: the modelling of protein folding in an all-atom MD is conceptually a weak-coupling expansion in terms of the dimensionless ratio (13.11).

Research project 13.5 Newton's equation for the mathematical pendulum,

$$\ddot{\omega} = V \sin\omega,$$

is integrable. Its naive discretization

$$\ddot{\omega} \rightarrow \frac{1}{\epsilon^2}(\omega_{n+1} - 2\omega_n + \omega_{n-1})$$

yields the so-called standard map [31]

$$\omega_{n+1} - 2\omega_n + \omega_{n-1} = \epsilon^2 V \sin\omega,$$

which is not integrable. However, an integrable discretization of the mathematical pendulum is known. See for example Chapter VIII of [32]. Find which one describes protein folding more accurately.

13.1.9 All-atom simulations

Enormous computational resources have been developed and dedicated to solve the *protein folding problem* [3, 4]. From the point of view of molecular dynamics this amounts to a numerical simulation of the all-atom time evolution in a protein, from a random chain initial condition to the natively folded conformation using a force field such as (13.9), (13.10). For example, the Blue Gene family of supercomputers was originally designed by IBM to address the problem of protein folding and gene development, which explains the name. Subsequently, special purpose computers have been constructed to address the folding problem, at all-atom level of scrutiny. Thus far, the most powerful is the *Anton* supercomputer, built by D. E. Shaw Research [33–35]. In the case of relatively short proteins, *Anton* can produce a few microseconds of *in vitro* folding trajectory per day *in silico* [34]; this is about three to four orders of magnitude more than a Blue Gene can achieve. In a series of MD simulations of 12 fast-folding proteins, from chignon with 10 residues to genetically modified λ-repressor with 80 residues, and with each protein capable of folding within a number of microseconds *in vitro*, *Anton* was able to produce dynamical trajectories that reproduced the experimentally observed folded structures, in some examples with an impressive precision [35]. At the moment, these results set the benchmark for all-atom protein folding simulations. But despite the encouraging results obtained by *Anton*, several issues remain to be overcome before proteins can be routinely folded at an all-atom level, starting from a knowledge of the amino acid sequence only. We name a few, as a challenge for future research:

- For the majority of proteins, it takes much longer than a millisecond or so to fold. For example myoglobin, which is probably the most widely studied protein and one that we shall fold in the sequel, needs about 2.5 s *in vitro* to reach its native state when it starts from a random chain initial condition. Thus, it would take at least 1000 years to simulate the folding of myoglobin at the all-atoms level using presently available computers.

- In an all-atom MD simulation with explicit water, an increase in the number of water molecules quickly exhausts the capacity of presently available computers. For example, in the case of the 80-residue λ-repressor mutant simulation using *Anton*, only around 11 000 explicit water molecules could be included. This should be contrasted with real physiological conditions: the normal pH of blood plasma is around 7.4. Since pH is defined as the $\log_{10}$ of the reciprocal of the hydrogen ion activity in a solution, this translates to an average of one proton per $10^{7.4}$ water molecules. Protein folding is strongly affected by pH; proteins have different natively folded states at different pH values. Thus, it remains a formidable task to describe physiologically relevant pH environments in a truly all-atom manner.
- All-atom force fields utilize a quadratic, harmonic-oscillator approximation around the ideal values of the bond lengths and angles (13.9), and a mathematical pendulum in the case of torsional angles. As long as the atomic fluctuations around the ideal values remain *very* small, this is a decent approximation. But whenever the atoms deviate from their ideal positions more than 'just a little', higher order nonlinear corrections are inevitably present and cannot be ignored. The existing all-atom force fields are not designed to account for this. The force fields are not built to describe protein conformations in a realistic manner, whenever the lengths and angles are not *very* close to their ideal values.

13.1.10 Thermostats

This section describes a technical aspect that is not needed in the rest of the lectures. We include it here, however, since we feel it addresses a highly important yet still open theoretical issue that needs to be addressed by any all-atom modelling approach.

Finally, we have the theoretically highly interesting problem of *thermostatting*. An all-atom simulation solves Newton's equation. As a consequence, energy is conserved and we have a microcanonical ensemble. But in a living cell the energy is not conserved; instead, temperature is fixed. Accordingly, proteins in living organisms should be described in terms of a canonical ensemble. One needs to convert the microcanonical ensemble that is described by the all-atom Newton's equation into a canonical one. To achieve this conversion, one adds *thermostats* to the system. A thermostat models an environment that maintains its own temperature constant while interacting with the system of interest: we refer to [36] for a detailed description of thermostats.

The Langevin equation is an example of a thermostat that is well grounded in physical principles. In the case of uniform damping, we write it as follows:

$$\ddot{\mathbf{x}}_i = -\nabla_i E(\mathbf{x}) - \lambda \dot{\mathbf{x}}_i + \mathbf{F}_i(t), \tag{13.12}$$

$$\langle \mathbf{F}_i(t) \rangle = 0, \qquad \langle \mathbf{F}_i(t) \cdot \mathbf{F}_j(s) \rangle = \lambda k_B T \delta_{ij}(t-s). \tag{13.13}$$

The derivation of the Langevin equation assumes two sets of variables, a.k.a. particles: there are light, small particles that are fast, and there are heavy particles that are slow. The Langevin equation describes the dynamics of the latter, in the limit where the ensuing particles are much slower and heavier than the small and fast ones. The derivation is based on standard arguments of statistical physics. Thus, the Langevin equation is a priori a well-grounded and rigorous method to introduce temperature into a Newtonian system.

In the case of all-atom MD, the Langevin equation cannot be used. There are no small and fast variables around. The oscillating atoms are themselves the small and fast variables. Moreover, from a conceptual point of view, the presence of a white noise (13.13) de facto converts the ensuing numerical algorithm into a Monte Carlo process, albeit an elaborate one.

Many alternatives to Langevin thermostat have been introduced. An example is the Gaussian thermostat [36]. Unlike the Langevin one, it is *deterministic*. Instead of small and fast background variables, one couples the variables $\mathbf{x}_i$ of interest to explicit thermostat variables $\mathbf{X}_k$, with equations of motion

$$m_i \ddot{\mathbf{x}}_i = -\nabla_i E(\mathbf{x}) - \nabla_i \mathcal{E}(\mathbf{x}, \mathbf{X}), \tag{13.14}$$

$$M_k \ddot{\mathbf{X}}_k = -\nabla_k \mathcal{U}(\mathbf{X}) - \nabla_k \mathcal{E}(\mathbf{x}, \mathbf{X}) - \alpha_k \dot{\mathbf{X}}_k. \tag{13.15}$$

The thermostatting effect is modelled by the last term in (13.15); the α_k is determined by subjecting the auxiliary variables to the non-holonomic constraint

$$\frac{1}{2} M_k \dot{\mathbf{X}}_k^2 = \frac{3}{2} k_B T, \tag{13.16}$$

$$\Longrightarrow \quad \alpha_k = \left(\frac{3}{2} k_B T \right)^{-1} \left\{ \frac{1}{2} M_k \dot{\mathbf{X}}_k^2 + \dot{\mathbf{X}}_k [\nabla_k \mathcal{U}(\mathbf{X}) + \nabla_k \mathcal{E}(\mathbf{x}, \mathbf{X})] \right\}, \tag{13.17}$$

for each of the thermostat variables. The disadvantage of a Gaussian thermostat is in the lack of a Hamiltonian character in the equations of motion (13.14) and (13.15).

A Hamiltonian, albeit singular, thermostatting has been proposed by Nosé and Hoover [37–40]. Their thermostat is probably the most widely used in the context of all-atom protein folding simulations. In the simplest variant, the all-atom phase space is extended by a single *ghost* particle with a singular, logarithmically divergent, potential that provides the temperature for all the rest.[3] The singular character of the potential introduces an inexhaustible heat reservoir, in essence.

Following [41], we consider the toy-model case of a single canonical degree of freedom (p, q) in the presence of a single Nosé–Hoover thermostat degree of freedom (P, Q). The classical action is [37–40]

$$S = \int_{-\mathrm{T}}^{\mathrm{T}} dt \left\{ p\dot{q} + P\dot{Q} - \frac{p^2}{2mQ^2} - V[q] - \frac{P^2}{2M} - \frac{1}{\beta_0} \ln Q \right\}. \tag{13.18}$$

[3] We remark that this ghost particle is a little like a Higgs particle that gives the mass for other particles in subatomic physics, except that instead of mass, it gives temperature and it cannot be observed.

We may assume that $V[q]$ has the double-well profile

$$V[q] = \lambda(q^2 - m^2)^2. \tag{13.19}$$

We are interested in the tunnelling amplitude between the two minimum-energy configurations $q = \pm m$:

$$\langle +m, \mathrm{T} \mid -m, -\mathrm{T}\rangle = \int\int_{q(-\mathrm{T})=-m}^{q(+\mathrm{T})=+m} [dp]\,[dq]\, e^{iS} \tag{13.20}$$

The integration over p is Gaussian but yields a Jacobian factor that depends on Q. This Jacobian has the same functional form as the last term in (13.18), and thus it can be absorbed by a redefinition (renormalization) of $\beta_0 \to \beta$. As usual, we evaluate the transition amplitude using the Euclidean (imaginary-time) formalism, obtained by sending $t \to it$. The Euclidean action is

$$S = \int_{-\mathrm{T}}^{\mathrm{T}} dt \left\{ \frac{1}{2} Q^2 q_t^2 + V[q] + \frac{1}{2} Q_t^2 + \frac{1}{\beta} \ln Q \right\}, \tag{13.21}$$

and we search for a finite-action instanton trajectory that connects the two states $q = \pm m$; note that the continuation to imaginary time 'inverts' the potential term, as shown in Fig. 13.12.

The instanton is a solution to the equations of motion

$$Q^2 q_{tt} = V_q - 2QQ_t q_t \simeq V_q - \gamma q_t, \tag{13.22}$$

$$QQ_{tt} = Q^2 q_t^2 + \frac{1}{\beta}. \tag{13.23}$$

Note how the coupling between q and the thermostat variable Q gives rise to an effective friction-like coefficient $\gamma(t)$.

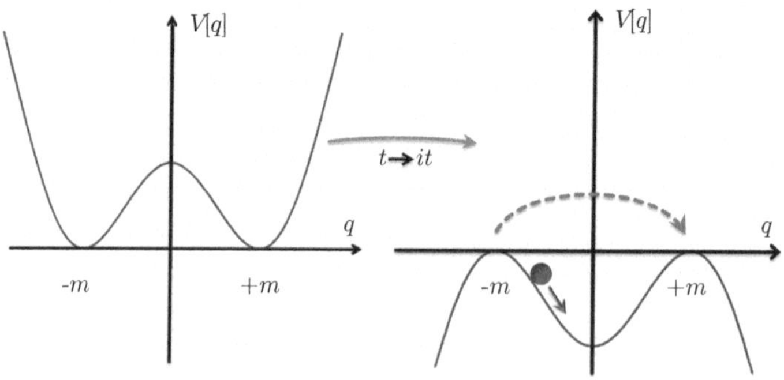

Fig. 13.12 Analytic continuation to Euclidean time has the effect of inverting the potential $V[q]$. The instanton is a trajectory that starts at (Euclidean) time $-\mathrm{T}$ from the local maximum at $q = -m$ and reaches the local maximum at $q = -m$ at time $+\mathrm{T}$, as shown on the right.

We first consider the case where the thermostat field Q is absent. This amounts to setting $Q \equiv 1$ in (13.21) and (13.22) and removing (13.23), leaving us with the equation

$$q_{tt} = V_q = 4\lambda q(q^2 - m^2). \tag{13.24}$$

By adjusting the initial velocity, we conclude that a solution exist that starts from $q = -m$ at time $-\mathrm{T}$, and ends at $q = +m$ at time $+\mathrm{T}$. This solution is the instanton that gives rise to a finite tunnelling amplitude between $\pm m$; in the limit $\mathrm{T} \to \infty$, the instanton has the familiar double-well topological soliton profile

$$q(t) = m \tanh[\sqrt{2\lambda}\, m(t - a)]. \tag{13.25}$$

Now suppose that the thermostat field is present. We first consider a scenario where at $\pm\mathrm{T}$ the system reaches a stationary state where $q = \pm q_0 \neq \pm m$. Since the action (13.21) should remain finite as $\mathrm{T} \to \infty$, we arrive at the Gibbsian relation

$$Q(\mathrm{T}) \xrightarrow{\mathrm{T}\to\pm\infty} Q_\pm = e^{-\beta V[q_\pm]}. \tag{13.26}$$

This proposes that β is indeed the Boltzmannian temperature factor, when positive.

Next, we integrate (13.23) and then take the limit $\mathrm{T} \to \infty$; since the Euclidean action should be finite, we may assume that Q_t vanishes as $\mathrm{T} \to \infty$, which removes the surface term. We find

$$\int_{-\infty}^{\infty} dt \, (Q_t^2 + Q^2 q_t^2) = - \int_{-\infty}^{\infty} dt \, \frac{1}{\beta}. \tag{13.27}$$

For a non-trivial tunnelling configuration and with a finite Euclidean action (13.21), the integral on the left-hand-side of (13.27) should be finite and non-vanishing. But since the left-hand side is manifestly non-negative, the Boltzmann temperature factor β cannot be positive as it should. Thus, we conclude that the presence of the thermostat field suppresses tunnelling. We note that a suppression of the tunnelling amplitude by a Nosé–Hoover thermostat in the case of double-well potential has been observed in numerical simulations [40].

Proteins regularly need to tunnel over various different potential barriers in their presumably highly rugged energy landscape, when proceeding from a random initial configuration to the natively folded state. Thus, we suspect that simulations using Nosé–Hoover thermostats can cause complications whenever we have a protein for which we can expect that the folding pathway goes through various intermediates and molten globules.

Research project 13.6 Investigate how the suppression of tunnelling amplitudes in the case of a properly modified Nosé–Hoover thermostat could be avoided.

Other kind of thermostats have also been introduced; in particular, we mention the Berendsen thermostat [42]. These thermostats, which are often convenient in numerical simulations, are designed to approach canonical ensembles in a limit. But they commonly lack a Hamiltonian interpretation, i.e. they are non-physical, and thus a physical interpretation of the results is absent. While this might not be an issue when the goal is *simply* to find a local energy minimum of an all-atom force field such as (13.9), (13.10), a non-physical thermostat cannot be used to model the dynamics of proteins.

13.1.11 Other physics-based approaches

The all-atom approaches where the discretized Newton's equation is solved iteratively are conceptually weak-coupling expansions in the dimensionless parameter (13.11). To describe the folding of most proteins, one needs to be able to extend this expansion and ensure its convergence over some 15 orders of magnitude or even more. From the perspective of subatomic physics, this is like extending perturbative Standard Model calculations all the way to the Planck scale.

Several approaches have been developed with the goal of introducing an expansion parameter that corresponds to a timescale clearly larger than the femtosecond scale. Such *coarse-grained* force fields average over the very short-timescale atomic fluctuations. If the denominator in (13.11) can be increased, so can the numerator, and it becomes possible to develop expansions that reach longer *in vivo* timescales with no increase in the *in silico* time. In practice, a carefully crafted coarse-grained force field can cover up to around three orders of magnitude longer folding trajectories than all-atom approaches, while still maintaining good overall quality. Here, we mention in particular UNRES as an example of such a carefully crafted coarse grained force field [43–45]. See the homepage http://www.unres.pl/.

Finally, we comment on the various versions of the Gõ model [46] and the closely related elastic (Gaussian) network models [47]. These approaches were historically important in obtaining insight into protein folding at a time when computer power was insufficient for any kind of serious all-atom folding simulation. In these models, the folded configuration is presumed to be known; the individual atomic coordinates of the folded protein chain appear as an input. A simple energy function is then introduced, tailored to ensure that the known folded configuration is a minimum-energy ground state. In the Gõ model, the energy could be as simple as a square-well potential centred at the native conformation. In elastic network models, the atoms are connected by harmonic oscillators, with energy minima that correspond to the natively folded state. Since the positions of all the relevant atoms appear as parameters in these models, they contain more parameters than unknowns and so no predictions can be made: from the point of view of a system of equations, these models are overdetermined. In any *predictive* energy function, the number of adjustable parameters must remain *smaller* than the number of independent atomic coordinates. Otherwise, no predictions can be made, and no physical principles can be tested.

13.2 Bol'she

In 1972, Anderson wrote an article [2] entitled *More is different* that has been inspirational to many. In particular, he argued for the importance of emergent phenomena. But the call for *Bol'she* is already present in Schrödinger's 1944 book:

... living matter, while not eluding the 'laws of physics' as established up to date, is likely to involve 'other laws of physics' hitherto unknown, which, however, once they have been revealed, will form just as integral a part of this science as the former.

However, Anderson makes a *crucially* important point that cannot be found in Schrödinger's book. Anderson's article has this point even as a subtitle: *Broken symmetry and the nature of the hierarchical structure of science.* Anderson realized that in order for emergent phenomena to give rise to structural self-organization, one needs a symmetry that becomes broken.

We have already pointed out that a broken symmetry is present in the case of proteins. Individual amino acids can be either left-handed chiral or right-handed chiral, equally. But, for some reason, living matter is built almost exclusively from amino acids that are left-handed chiral. We note that, apparently as a consequence, protein chains are predominantly chiral with right-handed helicity.

13.2.1 The importance of symmetry breaking

Consider a fluid-dynamical system such as water, the atmosphere, or any other scenario that can be described by the Navier–Stokes or Euler equation or their many descendants. These are fundamentally atomic systems, but with an enormous number of constituents. Their macroscopic properties are governed, often with very high precision, by the properties of a local order parameter that computes the fluid velocity. Structures such as vortices and tornadoes, and solitonic waves like the one that emerges from the Korteweg–de Vries equation, are all described by a solution that breaks an underlying symmetry.

A fluid-dynamical vortex line is a familiar example of a highly regular collective state of individual atoms, with a topological character. It is an example of an emergent structure. At the atomic level of scrutiny, the individual atoms and molecules that constitute the vortex are subject to random, Brownian thermal motion. By following a single individual water molecule, you cannot really conclude whether a vortex is present. The vortex materializes only at the macroscopic level, when the individual haphazard atomic motions become collectively self-organized into a regular pattern.

It would be inconceivable to construct a macroscopic vortex line such as a tornado in the atmosphere from purely atomic-level considerations, even in principle. A vortex is an example of a soliton, and solitons cannot be constructed simply by adiabatically building up individual atomic-level interactions as small perturbations around a ground state consisting of free individual atoms. A (topological) soliton emerges when nonlinear interactions combine elementary constituents into a localized collective excitation that is stable against small perturbations and cannot decay, unwrap, or disentangle.

13.2.2 An optical illusion

We start to describe the importance of symmetry breaking at an intuitive level. We consider a simple optical illusion, *not* a physical example. But it nevertheless demonstrates how *Bol'she* gives rise to a symmetry that becomes broken, and the illusion of breaking symmetry leads to the formation of structure, in our eyes.

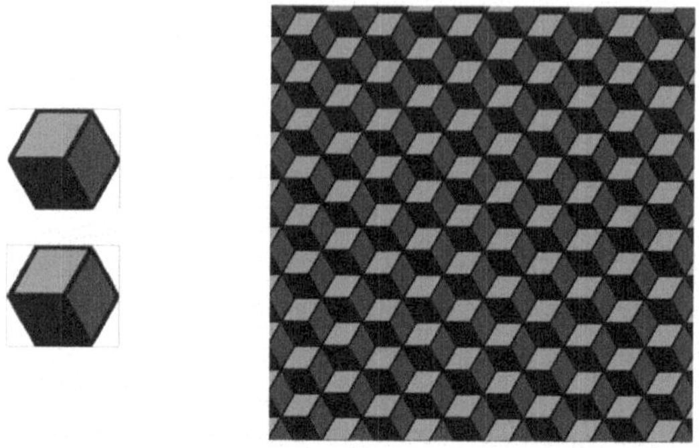

Fig. 13.13 Rotate the figure, slowly, by 180°. When you focus your eyes on the two individual cubes on the left, nothing much happens. However, if you focus on the set of cubes on the right, you see an abrupt effect like a phase transition between two different but identical ground states; we have a $\mathbb{Z}_2$ symmetry that has become broken by visual perception.

We then proceed to consider two physically relevant examples, where a very similar broken symmetry is present, but now in a physically relevant fashion. Most importantly, each of the two examples we present describes a simple physical scenario that shares many features with proteins.

In Fig. 13.13, we have two sets of cubes. On the left, we have two individual cubes; on the right, there are more cubes. If we look at the cubes on the right, we should be able to visualize some order—for example, light grey steps that come down from the left. Now we rotate the figure, slowly. When we keep our eyes focused on the two individual cubes on the left, nothing really happens besides an overall rotation of the two cubes. But if we instead focus on the set of cubes on the right, we should observe a rapid transition: there is a point at which the direction of the steps changes, abruptly, so that, after a rotation by 180°, we still have the same light grey steps, still coming down from the left.

The system on the right is *Bol'she*, with a discrete $\mathbb{Z}_2$ symmetry under a rotation by 180°. There are two ground states, and our mind chooses one, thus breaking the symmetry. In fact, the scenario is very much like that in Fig. 13.12, with two identical ground states. In this sense, Anderson's *Bol'she* is present in Fig. 13.13. No similar optical illusory effect is observed when the two individual cubes on the left are rotated.

We now proceed to describe two physical examples where a similar kind of $\mathbb{Z}_2$ symmetry becomes broken, with equally dramatic *physical*—not illusory—consequences.

13.2.3 Fractional charge

Polyacetylene in *trans* conformation is like a simplified protein. Figure 13.14(a) shows the structure. There is a 'backbone' consisting entirely of carbon atoms, and at each carbon atom there is a 'side chain' with a single hydrogen—much like in a protein, but simpler. In Fig. 13.14(b), we depict the (*trans*) polyacetylene chain by combining each (CH) unit into a single vertex. A double line describes the σ-bond and a single

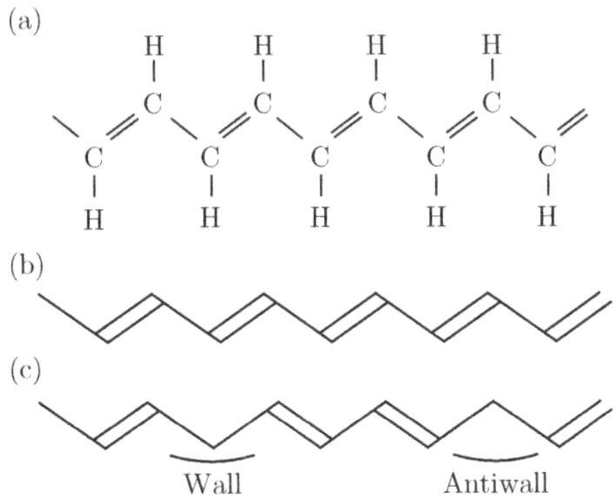

Fig. 13.14 (a) The *trans* conformation of polyacetylene. (b) One of the two degenerate ground states in a *trans*-polyacetylene chain. The other ground state is obtained by a reflection that interchanges the double bonds and the single bonds. (c) A state of a *trans*-polyacetylene chain with one of the double bonds converted into a single bond and then transported to form two domain walls carrying fractional electric charges.

line the π-bond between two consecutive C atoms. Owing to a Peierls instability, the asymmetry of the chain causes the ground state to be doubly degenerate. The free energy acquires the double-well profile that we have depicted in Fig. 13.12. The two ground states are related to each other by a $\mathbb{Z}_2$ reflection (parity) symmetry of the polyacetylene chain about a (CH) site, which exchanges the σ- and π-bonds.

When we choose one of the ground states, we break the symmetry. But we may introduce domain walls that interpolate between two different ground states. In Fig. 13.14(c), we show an example. Here, we have two domain walls, each of which interpolates between two different ground states; between the two domain walls, we have a region where the the σ- and π-bonds have become interchanged. Quantitatively, in terms of the double-well potential shown in Fig. 13.12, each of the two domain walls corresponds to a topological soliton (instanton) profile (13.25).

We now argue that we have *Bol'she*, which makes things different: The chain in Fig. 13.14(c) is obtained from the chain in Fig. 13.14(b) by removal of a single electron. There are 15 bonds in (b) and 14 in (c). The removal of a single electron converts a double bond into a single bond, and we have simply moved the bonds around to make the two identical domain walls. Since the structures in Fig. 13.14(b) and those in Fig. 13.14(c) differ from each other only by the removal of a single bond and since the two domain walls are identical, related to each other by parity, the two domain walls must share equally the quantum numbers of the missing bond: the two domain walls each have electric charge $\frac{1}{2}$.

This phenomenon of fermion number (charge) fractionalization demonstrates how Anderson's *Bol'she* really makes a difference: such exotic quantum number assignments could never be obtained simply by linearly superposing an integer number of initially non-interacting electrons and holes adiabatically, in a continuous manner, into a weakly interacting system: *a charge-$\frac{1}{2}$ state simply cannot be made by combining together any finite number of particles with an integer charge*—unless something *Bol'she* is involved.

Fine Points A bond line in polyacetylene corresponds to two electrons, one with spin up and the other with spin down. Thus, an isolated domain wall must have a net electric charge that is equal in magnitude but opposite in sign to that of a single electron. But since the spins of the electrons that have been removed are paired, the domain wall carries no spin. This unusual spin–charge assignment has been observed experimentally and it constitutes the essence of the fermion number fractionalization [48–50] that gives rise to electrical conductivity along polyacetylene. In the absence of the spin doubling, we would observe a domain wall that carries half the electric charge of one electron. Note that if we add a single electron to a domain wall, we obtain a state that is charge-neutral but carries the spin of the electron. Alternatively, if we remove a single electron from a domain wall, we obtain a state with spin $\frac{1}{2}$ and a charge that equals (minus) twice that of the electron. Neither of these states is possible without *Bol'she*.

13.2.4 Spin–charge separation

We shall eventually argue that proteins are *very* much like one-dimensional spin chains; the side chains are akin to spin variables along the backbone. Thus, our second example is a one-dimensional spin chain. For conceptual clarity, we may assume that the spin variables are single electrons (or maybe protons like H^+). We assume that the background lattice prefers a ground state that is an antiferromagnetic Néel state where all the spins point into an opposite direction from their nearest neighbours. Furthermore, we assume that in the ground state, all the sites have single occupancy and that there is a very strong repulsive force between the electrons that prevents a double occupancy. Such a ground state is a Mott insulator, and we have depicted the structure in Fig. 13.15.

As in the case of polyacetylene, the ground state is doubly degenerate: the $\mathbb{Z}_2$ symmetry transformation operates by reversing the direction of the spin at every single lattice site. By choosing one of the two ground states, we break the $\mathbb{Z}_2$ symmetry.

If we reverse the direction of a single spin along the chain, we form a localized configuration with three parallel neighbouring spins; see Fig. 13.16(a).

By successively reversing the direction of neighbouring spins but without changing the total spin, we can decompose this configuration into two separate domain walls, each consisting of two parallel spins. These domain walls interpolate between the two ground states of the spin chain, as shown in Fig. 13.16(b). Since the lattices in

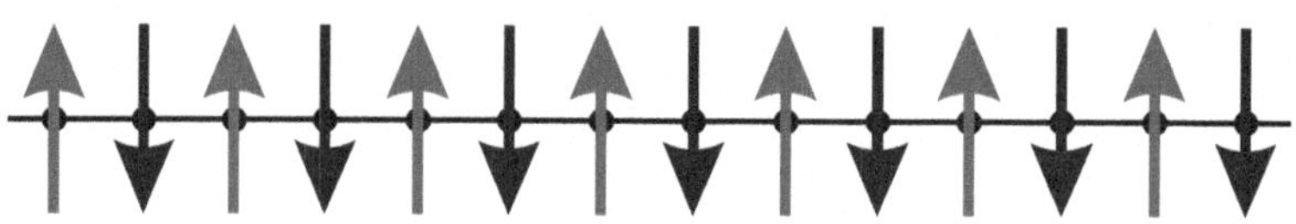

Fig. 13.15 [Colour online] One the two degenerate ground states in a Néel antiferromagnet, with alternating spin directions along a one-dimensional lattice of electrons. The $\mathbb{Z}_2$-symmetric ground state is obtained by reversing the direction of every spin.

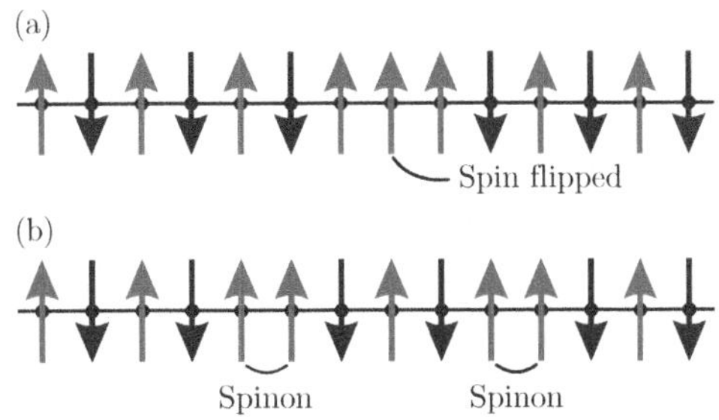

Fig. 13.16 [Colour online] (a) The same as in Fig. 13.15, but with the spin of only one electron reversed. (b) The state in (a) is decomposed into two domain walls representing the spinons.

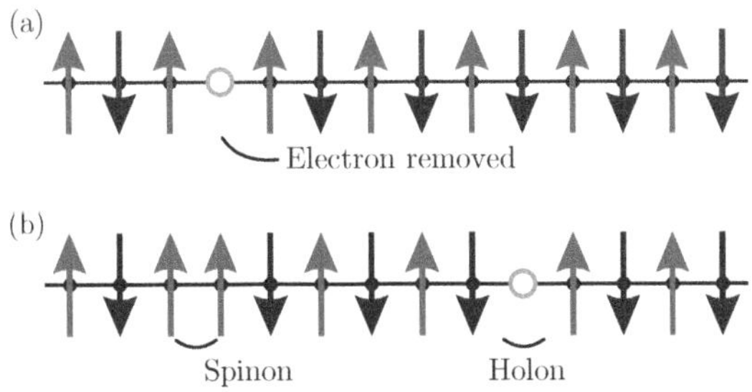

Fig. 13.17 [Colour online] (a) Spinon (left) and chargon (right) states in the underdoped state. (b) The state in (a) is decomposed into two domain walls representing the spinons.

Fig. 13.16(a) and (b) differ from each other only by the flip of a single spin, the total change in the spin between the two lattices is 1. The two domain walls in Fig. 13.16(b) are also identical. Thus each must have a spin equal to $\frac{1}{2}$.

Since we have made the two domain walls without adding or removing any electrons, each of them must be charge-neutral. We conclude that the domain walls are *spinons*—they describe the same spin degree of freedom as a single electron but with no charge [51]. The two domain walls interpolate between the two different ground states of the antiferromagnetic chain, very much like the domain walls in the case of polyacetylene.

Now we consider the same antiferromagnetic lattice but with one of the electrons removed as shown in Fig. 13.17.

This corresponds to an underdoped case: we have a hole in the spin chain. When we move the hole to the right, we arrive at the situation depicted in Fig. 13.17(b). In addition to the hole, we have here another domain wall that is similar to the two domain walls that we have in Fig. 13.16(b). This domain wall is a spinon, it has spin $\frac{1}{2}$ but carries no charge. Since we have removed one electron and since the spinor does not carry charge, the hole must carry a charge that is opposite to that of one electron. But no spin is available for this hole—it describes a spinless *holon*.

The ARPES experiment at Lawrence Berkeley Laboratory has confirmed that such spinons and holons do exist: they have been observed in a one-dimensional copper oxide ($SrCuO_2$) wire [52].

Note that in Fig. 13.17(a), the two spins on the left and on the right of the hole are parallel to each other, but in Fig. 13.17(b), the two spins are opposite to each other. This confirms that, like the spinon, the spinless holon is a domain wall that interpolates between the two different ground states of the chain.

Finally, in Fig. 13.18(a), we have added a single electron to the lattice. This example is of particular conceptual interest since it allows us to directly address what happens to a (pointlike) electron when it enters the antiferromagnetic environment: the presence of the electron introduces a single site with a double bond ($\uparrow\downarrow$). The chain is now overdoped. As before, we transport the doubly filled state, for example to the right, so that we arrive at the situation depicted in Fig. 13.18(b). Note that owing to Pauli exclusion, the transport occurs so that we move alternatively either a spin-up or a spin-down state one step to the right. The final configuration shown in Fig. 13.18(b) describes two separate domain walls that both interpolate between the two distinct ground states of the spin chain. One of these domain walls is again a spinon. The other one is a *doublon*. Since one electron has been added and since the spinon has spin but no charge, we conclude that the doublon does not have any spin but it carries the entire charge of one electron. The charge of the doublon is opposite to that of the holon.

The two examples we have discussed—fermion number fractionalization and spin–charge separation—make it plain and clear how much difference *Bol'she* can make: it would be impossible to form states with the spin of an electron but no charge, or states with the charge of an electron but no spin, simply by superposing an integer number of non-interacting electrons and then adiabatically switching on their mutual interactions. For states with such exotic quantum numbers, we need to have an environment with a symmetry that has become broken.

13.2.5 All-atom is Landau liquid

The Landau (Fermi) liquid is a paradigm on which much of our understanding of many-body systems like metals is based. This paradigm states that in a physical system with

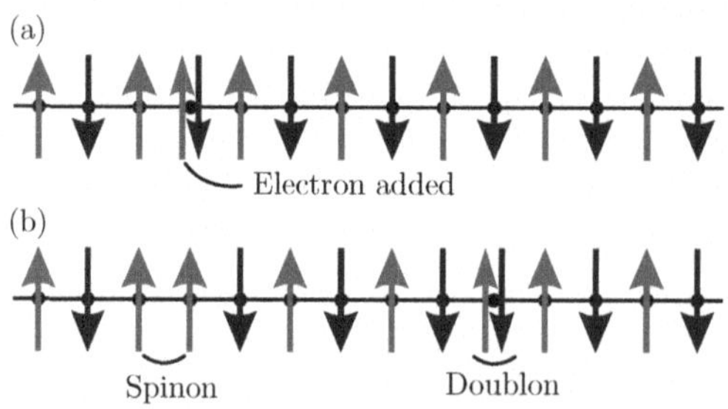

Fig. 13.18 [Colour online] (a) The Néel state with one electron added (overdoped case). (b) Spinon (left) and doublon (right) states in the overdoped state.

a large number of atoms, each atom retains its individual integrity. The properties of a material system that is described by a Landau liquid can be understood by superposing its individual constituents in a weak-coupling expansion around a given ground state. In particular, a Landau liquid that is made of electrons and protons can only have spin and charge assignments that are obtained by superposing the individual spins and charges. This is the case when the properties of the system can be understood using the notion of adiabaticity: we imagine that we start from an initial condition where the elemental constituents have no mutual interactions. The interactions are then turned on, adiabatically, in a continuous manner. Accordingly, the ground state of the original non-interacting system becomes continuously deformed into the ground state of the interacting system.

The two examples that we have described—polyacetylene and an antiferromagnetic spin chain—show that the Landau liquid paradigm is not a universal one. In a Landau-liquid system, it would be impossible to have states with exotic quantum numbers such as an electric charge that is half of that of a single electron. The Landau-liquid paradigm can fail whenever we have emergent structures that display symmetries that become broken. In such scenarios, there are often collective excitations such as topological solitons that cannot be built simply by adding together small adiabatic perturbations around a ground state of non-interacting elemental constituents.

The all-atom description (13.9) and (13.10) of a protein force field implicitly assumes the Landau-liquid paradigm. According to (13.9) and (13.10), the individual atoms oscillate around their ideal values under the influence of a potential that is either a harmonic oscillator or a mathematical pendulum. The Lennard-Jones and Coulomb potentials introduce continuously evolving deformations around the ideal atomic positions, in a manner that can be modelled by a weak-coupling expansion of the iterative Newton's equation in powers of (13.11); in practical simulations, these long-range interactions are turned off beyond a distance of around 10 Å.

It remains to be seen whether an all-atom Landau-liquid description of proteins breaks down. But the basal ingredient, that of a broken $\mathbb{Z}_2$ symmetry, which also appears in our examples of polyacetylene and an antiferromagnetic spin chain, is certainly present: the amino acids are left-handed chiral, and as a consequence proteins that constitute living matter prefer right-handed helicity along their backbone.

13.3 Strings in three space dimensions

Thus we have come to the conclusion that an organism and all the biologically relevant processes that it experiences must have an extremely ''many-atomic' structure and must be safeguarded against haphazard, 'single-atomic' events attaining too great importance.

(E. Schrödinger)

We start our search for broken symmetry and the ensuing *Bol'she* that makes us alive by considering differentiable (class $\mathcal{C}^3$) strings in $\mathbb{R}^3$.

13.3.1 Abelian Higgs model and the limit of slow spatial variations

The Abelian Higgs model (AHM) is the paradigmatic framework to describe vortices as solitons. Solitonic vortices are important to many physical phenomena, from cosmic strings in the Early Universe to type II superconductors. In particular, the Weinberg–Salam model of electroweak interactions with its Higgs boson is a non-Abelian extension of the AHM.

The AHM involves a single complex scalar (Higgs) field ϕ and a vector field A_i. These fields are subject to the $U(1)$ gauge transformation

$$\begin{aligned} \phi &\rightarrow e^{ie\vartheta}\phi, \\ A_i &\rightarrow A_i + \partial_i \vartheta, \end{aligned} \tag{13.28}$$

where ϑ is a function and e is a parameter. The standard AHM Hamiltonian is

$$\mathcal{H} = \frac{1}{4}G_{ij}^2 + |(\partial_i - ieA_i)\phi|^2 + \lambda\left(|\phi|^2 - v^2\right)^2 \tag{13.29}$$

where

$$G_{ij} = \partial_i A_j - \partial_j A_i \tag{13.30}$$

When the space dimension D is odd, a Chern–Simons term (ChS) can be added to (13.29). Explicitly,

$$\begin{aligned} D = 1: &\qquad ChS \sim A, \\ D = 3: &\qquad ChS \sim AdA, \\ D = 5: &\qquad ChS \sim AdAdA, \\ &\qquad \vdots \end{aligned} \tag{13.31}$$

The Chern–Simons term is the paradigmatic way to break parity.

In a material system, (13.29) with (13.31) is the Kadanoff–Wilson energy in the limit where the fields have slow spatial variations [18, 19]. To describe this limit, we start from the full free energy of a material system that is based on the AHM field multiplet; we denote it by

$$\mathcal{F}(\phi, A_i) \tag{13.32}$$

This free energy is in general a non-local functional of the field variables. But it must be gauge-invariant. Thus, it can only depend on manifestly gauge-invariant combinations of the fields such as

$$|\phi|^2, \quad |(\partial_i - ieA_i)\phi|^2, \quad \ldots \tag{13.33}$$

Consider the limit where the lengthscale that is associated with spatial variations of the field variables becomes very large in comparison with other characteristic lengthscales of the system. In this limit, we can expand the free energy in powers of the gauge-covariant derivatives of the fields. The expansion looks like this [53]:

$$\mathcal{F}(\phi, A_i) = V(|\phi|) + Z(|\phi|)|(\partial_i - ieA_i)\phi|^2 + ChS(A) + W(|\phi|)G_{ij}^2 + \ldots \tag{13.34}$$

The leading term is called the effective potential. The higher-derivative terms are multiplied by functions $Z(|\phi|)$, $W(|\phi|)$, etc.

The AHM energy (13.29) with (13.31) constitutes the leading-order non-trivial contribution to (13.34), in powers of fields and their covariant derivatives.

We introduce a set of new variables (J_i, ρ, θ), obtained from (A_i, ϕ) by the following change of variables:

$$\begin{aligned}\phi &= \rho e^{i\theta} \\ A_i \to J_i &= \frac{i}{2e|\phi|^2}\left[\phi^*(\partial_i - ieA_i)\phi - \text{c.c.}\right].\end{aligned} \tag{13.35}$$

We can introduce these new variables whenever $\rho \neq 0$. Note that both ρ and J_i are gauge-invariant under the gauge transformation (13.28). But

$$\theta \to \theta + \vartheta. \tag{13.36}$$

When we write (13.29) and (13.31) in terms of these new variables (13.35), we have

$$\mathcal{H} = \frac{1}{4}\left(J_{ij} + \frac{2\pi}{e}\sigma_{ij}\right)^2 + (\partial_i\rho)^2 + \rho^2 J_i^2 + \lambda\left(\rho^2 - \eta^2\right)^2 + ChS, \tag{13.37}$$

where

$$J_{ij} = \partial_i J_j - \partial_j J_i, \tag{13.38}$$

$$\sigma_{ij} = \frac{1}{2\pi}[\partial_i, \partial_j]\theta. \tag{13.39}$$

We observe that (13.37) involves only variables that are manifestly $U(1)$ gauge-invariant. In particular, unlike in the case of (13.29), in (13.37), the local gauge invariance is entirely removed by a change of variables [54].

The term σ_{ij} is a string current. It has a Dirac δ-like support that coincides with the worldsheet of the cores of vortices. When (13.37) describes a finite-energy vortex, (13.39) subtracts a singular string contribution that appears in J_{ij}. Since J_i is singular in the presence of a vortex line, it makes a divergent contribution to the third term on the right-hand side of (13.37). But the divergence is removed, provided the density ρ vanishes on the worldsheet of the vortex core. Thus, the vanishing of ρ along a string-like line in space is a *necessary* condition for the presence of finite-energy vortex lines.

13.3.2 The Frenet equation

Proteins are string-like objects. Thus, to understand proteins, we need to develop the formalism of strings in $\mathbb{R}^3$, at least to some extent.

The geometry of a class $\mathcal{C}^3$ differentiable string $\mathbf{x}(z)$ in $\mathbb{R}^3$ is governed by the Frenet equation, described widely in elementary courses of differential geometry. The parameter $z \in [0, L]$ where L is the length of the string in $\mathbb{R}^3$. We can compute the length from

$$L = \int_0^L dz\, ||\mathbf{x}_z|| = \int_0^L dz\, \sqrt{\mathbf{x}_z \cdot \mathbf{x}_z} \equiv \int_0^L dz\, \sqrt{g}\,. \tag{13.40}$$

Here, we recognize the static version of the standard Nambu–Goto action, with generic parameter $z \in [0, L]$. We reparametrize the string to express it in terms of the arclength $s \in [0, L]$ in the ambient $\mathbb{R}^3$, by the change of variables

$$s(z) = \int_0^z ||\mathbf{x}_z(z')||\, dz'. \tag{13.41}$$

In the following, we use the arclength parametrization exclusively. We consider a single open string that does not self-cross. We introduce the unit-length tangent vector

$$\mathbf{t} = \frac{d\mathbf{x}(s)}{ds} \equiv \mathbf{x}_s, \tag{13.42}$$

the unit-length binormal vector

$$\mathbf{b} = \frac{\mathbf{x}_s \times \mathbf{x}_{ss}}{||\mathbf{x}_s \times \mathbf{x}_{ss}||}, \tag{13.43}$$

and the unit-length normal vector

$$\mathbf{n} = \mathbf{b} \times \mathbf{t}. \tag{13.44}$$

The three vectors $(\mathbf{n}, \mathbf{b}, \mathbf{t})$ define the right-handed orthonormal Frenet frame. We may introduce this framing at every point along the string, whenever

$$\mathbf{x}_s \times \mathbf{x}_{ss} \neq 0. \tag{13.45}$$

We proceed, for the moment, by assuming this to be the case. The Frenet equation then computes the frames along the string as follows:

$$\frac{d}{ds}\begin{pmatrix} \mathbf{n} \\ \mathbf{b} \\ \mathbf{t} \end{pmatrix} = \begin{pmatrix} 0 & \tau & -\kappa \\ -\tau & 0 & 0 \\ \kappa & 0 & 0 \end{pmatrix}\begin{pmatrix} \mathbf{n} \\ \mathbf{b} \\ \mathbf{t} \end{pmatrix}. \tag{13.46}$$

Here

$$\kappa(s) = \frac{||\mathbf{x}_s \times \mathbf{x}_{ss}||}{||\mathbf{x}_s||^3} \tag{13.47}$$

is the curvature of the string on the osculating plane that is spanned by $\mathbf{t}$ and $\mathbf{n}$, and

$$\tau(s) = \frac{(\mathbf{x}_s \times \mathbf{x}_{ss}) \cdot \mathbf{x}_{sss}}{||\mathbf{x}_s \times \mathbf{x}_{ss}||^2} \tag{13.48}$$

is the torsion that measures how the string deviates from its osculating plane. Both $\kappa(s)$ and $\tau(s)$ are extrinsic geometric quantities, i.e. they depend only on the shape of the string in the ambient space $\mathbb{R}^3$. Conversely, if we know the curvature and torsion, we can construct the string. For this, we first solve for $\mathbf{t}(s)$ from the Frenet equation. We then solve for the string by integration of (13.42). The solution is unique, modulo a global translation and rotation of the string.

Finally, we note that both the curvature (13.47) and the torsion (13.48) transform as scalars, under reparametrizations of the string. For this, we introduce an infinitesimal local diffeomorphism along the string, by deforming s as follows:

$$s \to s + \epsilon(s). \tag{13.49}$$

Here, $\epsilon(s)$ is an arbitrary infinitesimally small function such that

$$\epsilon(0) = \epsilon(L) = 0 = \epsilon_s(0) = \epsilon_s(L). \tag{13.50}$$

Under this reparametrization of the string, the curvature and torsion transform as follows:

$$\begin{aligned} \delta\kappa(s) &= -\epsilon(s)\kappa_s, \\ \delta\tau(s) &= -\epsilon(s)\tau_s, \end{aligned} \tag{13.51}$$

which is how scalars transform. The Lie algebra of diffeomorphisms (13.49) is the classical Virasoro (Witt) algebra.

13.3.3 Frame rotation and Abelian Higgs multiplet

In order to relate the Abelian Higgs multiplet with extrinsic string geometry, we observe that the normal and binormal vectors do not appear in (13.42). As a consequence an $SO(2)$ rotation around $\mathbf{t}(s)$ (Fig. 13.19),

$$\begin{pmatrix} \mathbf{n} \\ \mathbf{b} \end{pmatrix} \to \begin{pmatrix} \mathbf{e}_1 \\ \mathbf{e}_2 \end{pmatrix} = \begin{pmatrix} \cos\eta(s) & \sin\eta(s) \\ -\sin\eta(s) & \cos\eta(s) \end{pmatrix} \begin{pmatrix} \mathbf{n} \\ \mathbf{b} \end{pmatrix}. \tag{13.52}$$

has no effect on the string. For the Frenet equation, this rotation gives

$$\frac{d}{ds}\begin{pmatrix} \mathbf{e}_1 \\ \mathbf{e}_2 \\ \mathbf{t} \end{pmatrix} = \begin{pmatrix} 0 & \tau + \eta_s & -\kappa\cos\eta \\ -(\tau + \eta_s) & 0 & \kappa\sin\eta \\ \kappa\cos\eta & -\kappa\sin\eta & 0 \end{pmatrix} \begin{pmatrix} \mathbf{e}_1 \\ \mathbf{e}_2 \\ \mathbf{t} \end{pmatrix}. \tag{13.53}$$

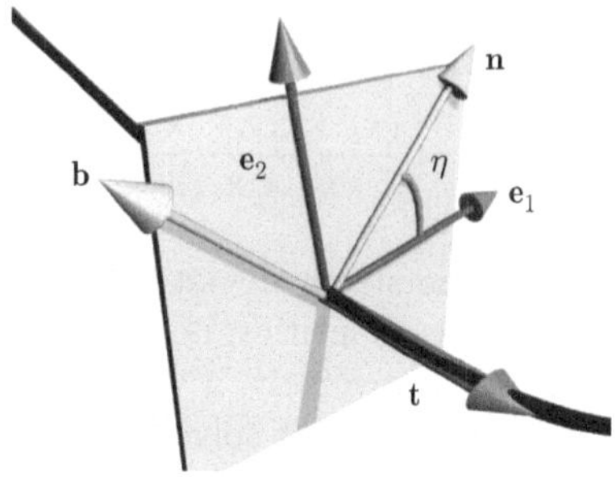

Fig. 13.19 [Colour online] Rotation between the Frenet frame and a generic frame, on the normal plane of the string.

We may utilize the κ-dependent terms in (13.53) to promote κ into a complex quantity, with a modulus that coincides with the manifestly frame-independent geometric curvature (13.47):

$$\kappa \xrightarrow{\eta} \kappa(\cos\eta + i\sin\eta) \equiv \kappa e^{i\eta}. \tag{13.54}$$

This enables us to interpret the transformation of (κ, τ) in (13.53) in terms of a one-dimensional version of the $U(1)$ gauge transformation (13.28). We identify the curvature as the Higgs field and the torsion as the $U(1)$ gauge field [55]:

$$\begin{aligned} \kappa &\to \kappa e^{-i\eta} \equiv \phi, \\ \tau &\to \tau + \eta_s \equiv A_i. \end{aligned} \tag{13.55}$$

Note that when we choose

$$\eta(s) \to \eta_B(s) = -\int_0^s \tau(\tilde{s})\, d\tilde{s}, \tag{13.56}$$

we arrive at the *unitary* gauge in terms of the abelian Higgs multiplet. This defines Bishop's parallel transport framing [56]. The Bishop-frame vectors $\mathbf{e}_{1,2}^B$ do not rotate around the tangent vector:

$$\frac{d}{ds}\left(\mathbf{e}_1^B + i\mathbf{e}_2^B\right) = -\kappa e^{-i\eta_B}\mathbf{t}. \tag{13.57}$$

Thus, unlike the Frenet framing, which that cannot be introduced when the curvature $\kappa(s)$ vanishes, the Bishop framing can be introduced and defined in an unambiguous and continuous manner even in that case. However, it turns out that in the case of proteins, which is the subject we are interested in, the Bishop frames do not work very well [57].

13.3.4 The unique string Hamiltonian

The curvature and torsion are the *only* available geometric quantities for constructing energy functions of strings, while (13.29) with (13.31) is the *unique* energy of the Abelian Higgs multiplet in the Kadanoff–Wilson sense of universality.

Consider a string in the limit where the curvature and torsion are slowly varying functions along it. The shape of the string cannot depend on the framing, and thus its energy can only involve combinations of the curvature and torsion in a manifestly frame-independent fashion.

On the other hand, (13.29) with (13.31) is the unique $SO(2) \sim U(1)$-invariant energy function that involves a complex Higgs field and a gauge field. It emerges from general arguments and symmetry principles alone, in the limit where the lengthscale that is associated with spatial variations of the field variables becomes very large in comparison with other characteristic lengthscales of the system.

Thus, the *only* Hamiltonian that can describe a string and its dynamics in the infrared limit is [55]

$$H = \int_0^L ds \left[|(\partial_s + ie\tau)\kappa|^2 + \lambda(|\kappa|^2 - m^2)^2 \right] + a \int_0^L ds\, \tau. \tag{13.58}$$

We have included here the one-dimensional version of the Chern–Simons term (13.31). It introduces net helicity along the string, breaking the $\mathbb{Z}_2$ symmetry between strings that are twisted in the right-handed and left-handed sense.

In (13.58), both κ and τ are expressed in a generic, arbitrary framing of the string. The corresponding gauge-invariant variables (13.35) are the curvature (13.47) and torsion (13.48) that characterize the extrinsic string geometry. In terms of these gauge-invariant variables, which from now on we denote by (κ, τ) the Hamiltonian (13.58) is

$$H = \int_0^L ds \left[(\partial_s \kappa)^2 + e^2\kappa^2\tau^2 + \lambda\,(\kappa^2 - m^2)^2 \right] + a \int_0^L ds\, \tau, \tag{13.59}$$

where we have simply followed the steps that gave us (13.37). Thus, (13.59) is the unique energy of a string, in terms of geometrically defined curvature and torsion, and in the limit where the spatial variations of curvature and torsion along the string are small.

13.3.5 Integrable hierarchy

A relation exist between (13.59), the integrable hierarchy of the nonlinear Schrödinger (NLS) equation, and the Heisenberg spin chain of ferromagnetism. For this, we introduce the Hasimoto variable

$$\psi(s) = \kappa(s) e^{ie \int_0^s ds'\, \tau(s')} \tag{13.60}$$

that combines the curvature and torsion into a single gauge-invariant complex variable. In terms of (13.60), we obtain the Hamiltonian of the integrable NLS equation [58, 59, 62] as follows:

$$\kappa_s^2 + e^2\kappa^2\tau^2 + \lambda\kappa^4 = \bar{\psi}_s\psi_s + \lambda(\bar{\psi}\psi)^2 = \mathcal{H}_3. \tag{13.61}$$

The lower-level conserved densities in the integrable NLS hierarchy are the helicity $\mathcal{H}_{-2}$, length (i.e. Nambu–Goto) $\mathcal{H}_{-1}$, number operator $\mathcal{H}_1$, and momentum $\mathcal{H}_2$:

$$\begin{aligned} \mathcal{H}_{-2} &= \tau, \\ \mathcal{H}_{-1} &= L, \\ \mathcal{H}_1 &= \kappa^2, \\ \mathcal{H}_2 &= \kappa^2\tau. \end{aligned} \tag{13.62}$$

The energy (13.59) is a combination of $\mathcal{H}_{-2}$, $\mathcal{H}_1$, and $\mathcal{H}_3$. From the perspective of the NLS hierarchy, the momentum $\mathcal{H}_2$ should also be included for completeness. If higher-order corrections are desired, the natural candidate is the modified Korteweg–de Vries (mKdV) density

$$\mathcal{H}_4 = \kappa\kappa_{ss}\tau + 4\kappa^2\tau^3 - 4e^2\kappa_s^2\tau + 3\lambda, \kappa^4\tau \tag{13.63}$$

which appears as the next-level conserved density in the NLS hierarchy.

We note that the Heisenberg spin chain is obtained from $\mathcal{H}_1$ using the Frenet equation:

$$\int_0^L ds\, \mathcal{H}_1 = \int_0^L ds\, \kappa^2 = \int_0^L ds\, |\mathbf{t}_s|^2. \tag{13.64}$$

The combination of $\mathcal{H}_{-1}$ and $\mathcal{H}_1$ leads to Polyakov's rigid-string action [60]. This combination also coincides with the Kratky–Porod model of polymers [61].

In [60], the concept of perturbative-level Wilsonian universality is employed to argue that no additional terms besides $\mathcal{H}_{-1}$ and $\mathcal{H}_1$ should appear in the infrared limit. But, in the presence of non-perturbative structures, any perturbative argument becomes incomplete: The NLS Hamiltonian (13.61) supports solitons that do not coexist with perturbative arguments.

13.3.6 Strings from solitons

Solitons are the paradigm structural self-organizers in Nature. They materialize in diverse scenarios [58, 59, 62, 63]; we have already seen that solitons conduct electricity in organic polymers. But solitons can also transmit data in transoceanic cables, and they can transport chemical energy along proteins. Solitons explain the Meissner effect in superconductivity, and they model dislocations in liquid crystals. Solitons are used to describe hadronic particles, cosmic strings, and magnetic monopoles in high-energy physics.

We argue that solitons also describe life. We argue that each of us has some 10^{20} solitons in our body. These solitons are the building blocks of folded proteins, they are the essential ingredients in all the metabolic processes that make us alive.

The NLS equation that we obtain from (13.61) is the paradigm equation that supports solitons [58, 59, 62, 63]. Depending on the sign of λ, the soliton is either dark ($\lambda > 0$) or bright ($\lambda < 0$). The torsion-independent contribution to (13.59),

$$H = \int_{-\infty}^{\infty} ds \, \left[\kappa_s^2 + \lambda\,(\kappa^2 - m^2)^2\right], \tag{13.65}$$

reproduces our previous instanton equation (13.22) with $Q = \sqrt{2}$. The Hamiltonian (13.65) supports the double-well soliton, a.k.a. the paradigmatic *topological* soliton: when m^2 is positive and when κ can take both positive and negative values, the equation of motion

$$\kappa_{ss} = 2\lambda\kappa(\kappa^2 - m^2) \tag{13.66}$$

is solved by (see (13.25))

$$\kappa(s) = m \tanh[m\sqrt{\lambda}(s - s_0)]. \tag{13.67}$$

We have concluded that the energy function

$$\mathcal{E} = \int ds \left[\kappa_s^2 + \lambda(\kappa^2 - m^2)^2 + \frac{1}{2}d\kappa^2\tau^2 - b\kappa^2\tau - a\tau + \frac{1}{2}c\tau^2\right] \tag{13.68}$$

is the most general one, a linear combination of *all* the densities (13.61) and (13.62). In (13.68), we have also included the Proca mass; this is the last term. Even though the Proca mass does not appear in the integrable NLS hierarchy, it does have a claim to be gauge-invariant [64, 65]. Eventually, we shall present a topological argument why the Proca mass should be included.

The energy (13.68) is quadratic in the torsion. Thus, we can eliminate τ using its equation of motion

$$\frac{\delta\mathcal{E}}{\delta\tau} = d\kappa^2\tau - b\kappa^2 - a + c\tau = 0. \tag{13.69}$$

This gives

$$\tau[\kappa] = \frac{a + b\kappa^2}{c + d\kappa^2} \equiv \frac{a}{c}\,\frac{1 + (b/a)\kappa^2}{1 + (d/c),\kappa^2} \tag{13.70}$$

and we obtain the following *effective* energy for the curvature:

$$\mathcal{E}_\kappa = \int ds \left(\frac{1}{2}\kappa_s^2 + V[\kappa]\right), \tag{13.71}$$

with the equation of motion

$$\frac{\delta\mathcal{E}_\kappa}{\delta\kappa} = -\kappa_{ss} + V_\kappa = 0, \tag{13.72}$$

where

$$V[\kappa] = -\left(\frac{bc - ad}{d}\right)\frac{1}{c + d\kappa^2} - \left(\frac{b^2 + 8\lambda m^2}{2b}\right)\kappa^2 + \lambda\kappa^4. \tag{13.73}$$

This is a deformation of the potential in (13.65); the two share the same large-κ asymptotics. When we select the parameters properly, we can expect that (13.69), (13.72), and (13.73) continue to support topological solitons. But we do not know their explicit profile, in terms of elementary functions. In the sequel, we shall construct these solitons numerically.

Once we have constructed the soliton of (13.72), we evaluate $\tau(s)$ from (13.70). We substitute these profiles into the Frenet equation (13.46) and solve for $\mathbf{t}(s)$. We then integrate (13.42) to obtain the string $\mathbf{x}(s)$ that corresponds to the soliton.

13.3.7 Anomaly in the Frenet frames

When the curvature of a string vanishes, we have an anomaly in the Frenet framing. It turns out that the origin of the anomaly is the *raison d'être* for a topological soliton to reside on a string.

Up to now, we have assumed (13.45) so that the curvature (13.47) does not vanish. But we have also observed that in the case of the Abelian Higgs model (13.37), it is natural for the density ρ to vanish on the worldsheet of a vortex core. Thus, in the context of the AHM, the vanishing of ρ relates to important, physically significant effects.

Furthermore, the explicit soliton profile (13.67) displays both positive and negative values, and in particular (13.67) vanishes when $s = s_0$. Consequently, we should consider the possibility that κ may vanish, even become negative, along a string.

We start by extending the curvature (13.47) so that it has both positive and negative values. According to (13.54), the negative values of κ are related to the positive ones by a $\eta = \pm\pi$ frame rotation,

$$\kappa \xrightarrow{\eta = \pm\pi} e^{\pm i\pi}\kappa = -\kappa. \tag{13.74}$$

Hence, we compensate for an extension of (13.47) to negative values by introducing the discrete $\mathbb{Z}_2$ symmetry [57]

$$\left.\begin{aligned} \kappa &\leftrightarrow -\kappa, \\ \eta &\leftrightarrow \eta \pm \pi \end{aligned}\right\} \quad \Longleftrightarrow \quad \kappa e^{i\eta} \leftrightarrow \kappa e^{i\eta}. \tag{13.75}$$

An (isolated) point where $\kappa(s)$ vanishes is called an inflection point. Figure 13.20 shows an example of an inflection point. As shown in this figure, in the limit of a plane curve, we obtain a discontinuity in the Frenet frames, when the string goes through the inflection point: the zweibein $(\mathbf{n}, \mathbf{b})$ becomes reflected according to

$$\mathbf{n} + i\mathbf{b} \to -(\mathbf{n} + i\mathbf{b}) = e^{\pm i\pi}(\mathbf{n} + i\mathbf{b}) \tag{13.76}$$

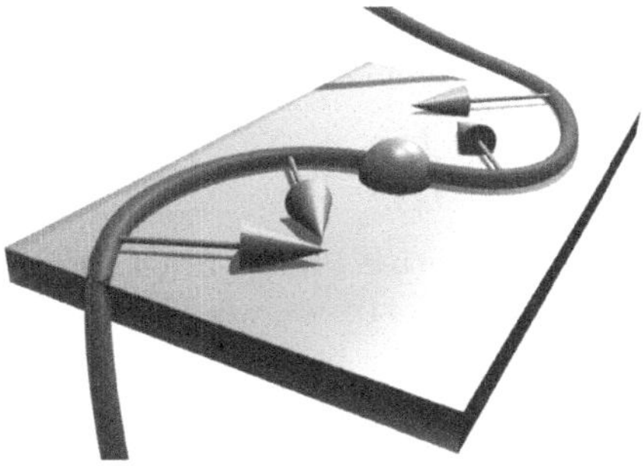

Fig. 13.20 [Colour online] A string with inflection point (ball). At each point, the Frenet-frame normal vector points towards the centre of the osculating circle. At the inflection point, we have a discontinuity in the direction of the normal vectors: the radius of the osculating circle diverges, and the normal vectors are reflected in the osculating plane, from one side to the other side of the string.

when we have a simple inflection point along the string. At the inflection point itself, the Frenet frames cannot defined. Thus, we cannot deduce whether we have a jump by $\eta = +\pi$ or by $\eta = -\pi$ at the inflection point. We cannot conclude whether the Frenet-frame vectors $(\mathbf{n}, \mathbf{b})$ become rotated clockwise or counterclockwise by an angle π along the tangent vector. There is a $\mathbb{Z}_2$ anomaly in the definition of Frenet framing, due to inflection points.

To analyse the anomaly, consider a string $\mathbf{x}(s)$ that has a simple inflection point when $s = s_0$. Thus, $\kappa(s_0) = 0$ but $\kappa_s(s_0) \neq 0$, as shown in Fig. 13.20. To simplify the notation, we redefine the parameter s so that the inflection point occurs at $s_0 = 0$.

We can always remove the inflection point by a generic deformation of the string: a deformation that is restricted to the plane as in Fig. 13.20 only slides the inflection point along the string without removing it. In order to remove the inflection point, we need to deform the string away from its instantaneous tangent plane: the codimension of the inflection point in $\mathbb{R}^3$ is 2, and the inflection point is not generic along a string.

Consider two different generic deformations,

$$\mathbf{x}(s) \to \mathbf{x}(s) + \delta\mathbf{x}_{1,2}(s) = \mathbf{x}_{1,2}(s). \tag{13.77}$$

In the case shown in Fig. 13.20, these two deformations amount to moving the string either slightly up from the plane or slightly down from the plane, around the inflection point. We assume that the deformations are very small and compactly supported, so that

$$\delta\mathbf{x}_{1,2}(\pm\varepsilon_\pm) = 0. \tag{13.78}$$

Here $\varepsilon_\pm > 0$ are small and determine the parameter values where the deformations $\mathbf{x}_{1,2}(s)$ bifurcate.

Imagine now a closed string, denoted by γ, that starts from $\mathbf{x}(-\varepsilon_-)$, follows along $\mathbf{x}_1$ to $\mathbf{x}(+\varepsilon_+)$, and then returns along $\mathbf{x}_2$ back to $\mathbf{x}(-\varepsilon_-)$. Introducing the Frenet-frame normal vectors of γ and shifting γ slightly into the direction of these normals, we obtain a second closed string, which we call $\tilde{\gamma}$. Let $\mathbf{t}$ and $\tilde{\mathbf{t}}$ be the corresponding tangent vectors. The Gauss linking number of γ and $\tilde{\gamma}$ is

$$\text{Lk} = \frac{1}{4\pi} \oint_{\gamma} \oint_{\tilde{\gamma}} ds\, d\tilde{s}\, \frac{\mathbf{x} - \tilde{\mathbf{x}}}{|\mathbf{x} - \tilde{\mathbf{x}}|^3} \cdot (\mathbf{t} \times \tilde{\mathbf{t}}). \tag{13.79}$$

Proceeding along $\mathbf{x}_{1,2}(s)$, the respective Frenet frames are continuously rotated by $\eta_{1,2} \approx \pm\pi$; in the limit where $\delta\mathbf{x}_{1,2} \to 0$, we would obtain the discontinuous jump (13.76). By continuity of Frenet framing in the complement of inflection points, the linking number has values $\text{Lk} = \pm 1$ when the $\eta_{1,2}$ change in the same direction; we recall that γ proceeds 'backwards' along $\mathbf{x}_2$. But if the framings along $\mathbf{x}_1(s)$ and $\mathbf{x}_2(s)$ rotate in the opposite directions, we have $\text{Lk} = 0$.

Accordingly, the relative sign of $\eta_{1,2}$ depends on the way in which the inflection point is circumvented: we have a *frame anomaly* in the Frenet framing as $\delta\mathbf{x}_{1,2} \to 0$, and the value of Lk depends on how we define $\delta\mathbf{x}_{1,2}(s)$.

Exercise 13.1 Analyse the framing of a string in the presence of an inflection point, using the Bishop frames (13.56).

13.3.8 Perestroika

An inflection point together with the corresponding Frenet-frame anomaly can be given an interpretation in terms of a string specific bifurcation, which is called *inflection-point perestroika* [66–70]. It explains why a *uniquely* defined Frenet framing across the inflection point, or any other framing that rotates around the tangent vector, is not possible:

Consider a segment of a string, along which the torsion $\tau(s)$ vanishes. Accordingly the string segment is constrained on a plane, as in Fig. 13.20. When a string is constrained on a plane, a simple isolated inflection point is generic. This follows since for a string on plane the inflection point has codimension 1. Moreover, in the case of a string on a plane, a single simple inflection point is a topological invariant. It can only be moved around the plane, but not made to disappear unless it escapes the plane, which we now assume not to be the case. If we have two simple inflection points along a string on plane, we can bring them together to deform the string so that no inflection point remains. Thus, the inflection point is a mod 2 topological invariant of a string on a plane.

Consider now a generic string in $\mathbb{R}^3$; a generic string is not constrained on a plane. The codimension of a single simple inflection point is 2, and thus a generic string does not have any inflection points. But along a string that moves freely in $\mathbb{R}^3$, an isolated simple inflection point appears generically, at some point, at some moment, during the motion. When this infection-point perestroika takes place along the moving string, it leaves a trail behind: the inflection-point perestroika changes the number of *flattening points*, which are points along the string where the torsion $\tau(s)$ vanishes [69, 70].

At a simple flattening point, the curvature $\kappa(s)$ is generically non-vanishing, but the torsion $\tau(s)$ changes its sign. Accordingly the inflection-point perestroika can only change the number of simple flattening points by 2. Apparently, it always does [69, 70].

Unlike the inflection point, a flattening point where $\tau(s) = 0$ is generic along a static space string. Furthermore, unlike a simple inflection point, a single simple flattening point that occurs in a one-parameter family of strings in $\mathbb{R}^3$ is a topological invariant. It cannot disappear on its own, under local deformations that leave the ends of the string intact. A pair of flattening points can be combined together, into a single bi-flattening point, which can then dissolve. When this happens, a second string-specific bifurcation that is called a *bi-flattening perestroika* takes place.

Apparently, inflection-point and bi-flattening perestroikas are the only two bifurcations where the number of flattening points can change [70]. The number of simple flattening points is a *local* invariant of the string. Besides the flattening number, and the self-linking number in the case of a framed string, a generic smooth string does not possess any other essential local invariants [69]. The two are also mutually independent, even though they often appear together.

For example, one can deduce that the self-linking number of a string increases by 1 if the torsion is positive when the string approaches its simple inflection point, and if two simple flattening points disappear after the passage of the inflection point. Moreover, if the torsion is negative, the self-linking number decreases by 1 when two flattening points disappear after the passage [69]. But when two simple flattening points dissolve in a bi-flattening perestroika, the self-linking number in general does not change.

A bifurcation is the paradigmatic cause of structural transitions, including phase transitions, in any dynamical system. Inflection-point and bi-flattening perestroikas are the only bifurcations that are string-specific. Accordingly, these two perestroikas must have a profound influence on determining the physical behaviour and phase structure of string-like configurations. In particular, they must be responsible for any string-specific structural reorganization that can take place when the value of the compactness index ν (13.4) changes. Since perestroikas are related to the creation and disappearance of topological solitons such as (13.67) along a string, it is clear that perestroikas and topological solitons, with the ensuing physical effects, are commonplace whenever we have a string with an energy function of the form (13.68).

Example A good example of the interplay between inflection points and flattening points is given by (13.67) or, more generally, by a soliton solution of (13.72) and (13.70). For a regular string, the denominator of (13.70) should not vanish. Thus, in the case of an inflection point, the ratio d/c should be positive. When b/a is negative, we have a symmetric pair of inflection points around the inflection point. Thus, starting from a one-parameter family of strings $\kappa(s, v)$ with v the parameter, if initially $\kappa(s, v)$ is sufficiently large and, for example, positive, and we are not close to an inflection point, then there are no flattening points either. When v is varied so that the inflection point is approached, a pair of flattening points emerges and remains whenever the curvature has the profile (13.67). In particular, we conclude that it is important to retain the Proca mass term, even a very small one, as a regulator.

13.4 Discrete Frenet frames

Proteins are not like continuous differentiable strings. Rather, they are like piecewise linear polygonal strings. Thus, to understand the physical properties of proteins, we need to develop the formalism of discrete strings. Accordingly, we proceed to generalize the Frenet-frame formalism to the case of polygonal strings that are piecewise-linear [57].

Let $\mathbf{r}_i$ with $i = 1, \ldots, N$ be the vertices of a piecewise-linear discrete string. At each vertex we introduce the unit tangent vector

$$\mathbf{t}_i = \frac{\mathbf{r}_{i+1} - \mathbf{r}_i}{|\mathbf{r}_{i+1} - \mathbf{r}_i|}, \tag{13.80}$$

the unit binormal vector

$$\mathbf{b}_i = \frac{\mathbf{t}_{i-1} - \mathbf{t}_i}{|\mathbf{t}_{i-1} - \mathbf{t}_i|}, \tag{13.81}$$

and the unit normal vector

$$\mathbf{n}_i = \mathbf{b}_i \times \mathbf{t}_i. \tag{13.82}$$

The orthonormal triplet $(\mathbf{n}_i, \mathbf{b}_i, \mathbf{t}_i)$ defines a discrete version of the Frenet frames (13.42)–(13.44) at each position $\mathbf{r}_i$ along the string, as shown in Fig. 13.21.

In lieu of the curvature and torsion, we have the bond angles and torsion angles, defined as in Fig. 13.22.

When we know the Frenet frames at each vertex, we can compute the values of these angles: the bond angles are

$$\kappa_i \equiv \kappa_{i+1,i} = \arccos(\mathbf{t}_{i+1} \cdot \mathbf{t}_i) \tag{13.83}$$

and the torsion angles are

$$\tau_i \equiv \tau_{i+1,i} = \text{sign}\{\mathbf{b}_{i-1} \times \mathbf{b}_i \cdot \mathbf{t}_i\} \arccos(\mathbf{b}_{i+1} \cdot \mathbf{b}_i) \tag{13.84}$$

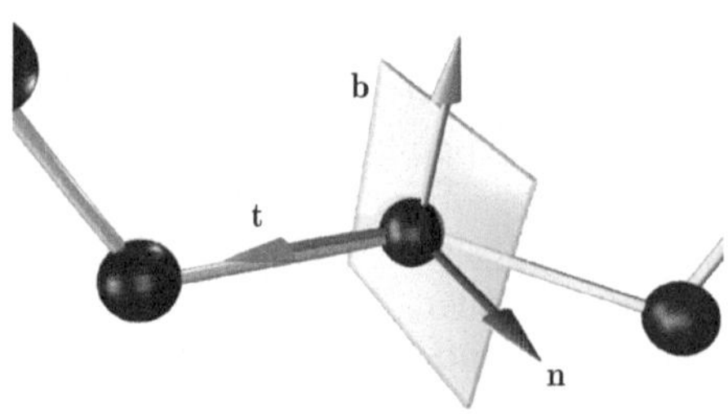

Fig. 13.21 [Colour online] Discrete Frenet frames along a piecewise-linear discrete string.

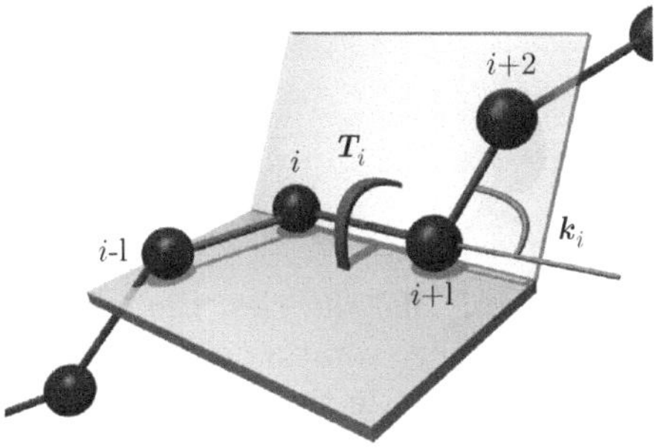

Fig. 13.22 [Colour online] Definition of bond (κ_i) and torsion (τ_i) angles, along the piecewise linear discrete string.

Conversely, when the values of the bond and torsion angles are all known, we can use the discrete Frenet equation

$$\begin{pmatrix} \mathbf{n}_{i+1} \\ \mathbf{b}_{i+1} \\ \mathbf{t}_{i+1} \end{pmatrix} = \begin{pmatrix} \cos\kappa\cos\tau & \cos\kappa\sin\tau & -\sin\kappa \\ -\sin\tau & \cos\tau & 0 \\ \sin\kappa\cos\tau & \sin\kappa\sin\tau & \cos\kappa \end{pmatrix}_{i+1,i} \begin{pmatrix} \mathbf{n}_i \\ \mathbf{b}_i \\ \mathbf{t}_i \end{pmatrix} \tag{13.85}$$

to compute the frame at position $i + i$ from the frame at position i. Once all the frames have been constructed, the entire string is given by

$$\mathbf{r}_k = \sum_{i=0}^{k-1} |\mathbf{r}_{i+1} - \mathbf{r}_i| \cdot \mathbf{t}_i. \tag{13.86}$$

Without any loss of generality, we may choose $\mathbf{r}_0 = 0$, make $\mathbf{t}_0$ point into the direction of the positive z axis, and let $\mathbf{t}_1$ lie on the y–z plane.

The vectors $\mathbf{n}_i$ and $\mathbf{b}_i$ do not appear in (13.86). Thus, as in the case of continuum strings, a discrete string remains intact under frame rotations of the $(\mathbf{n}_i, \mathbf{b}_i)$ zweibein around $\mathbf{t}_i$. This local $SO(2)$ rotation acts on the frames as follows:

$$\begin{pmatrix} \mathbf{n} \\ \mathbf{b} \\ \mathbf{t} \end{pmatrix}_i \to e^{\Delta_i T^3} \begin{pmatrix} \mathbf{n} \\ \mathbf{b} \\ \mathbf{t} \end{pmatrix}_i = \begin{pmatrix} \cos\Delta_i & \sin\Delta_i & 0 \\ -\sin\Delta_i & \cos\Delta_i & 0 \\ 0 & 0 & 1 \end{pmatrix} \begin{pmatrix} \mathbf{n} \\ \mathbf{b} \\ \mathbf{t} \end{pmatrix}_i . \tag{13.87}$$

Here Δ_i is the rotation angle at vertex i and T^3 is one of the SO(3) generators

$$T^1 = \begin{pmatrix} 0 & 0 & 0 \\ 0 & 0 & -1 \\ 0 & 1 & 0 \end{pmatrix}, \quad T^2 = \begin{pmatrix} 0 & 0 & 1 \\ 0 & 0 & 0 \\ -1 & 0 & 0 \end{pmatrix}, \quad T^3 = \begin{pmatrix} 0 & -1 & 0 \\ 1 & 0 & 0 \\ 0 & 0 & 0 \end{pmatrix} \tag{13.88}$$

that satisfy the Lie algebra

$$[T^a, T^b] = \epsilon^{abc} T^c. \tag{13.89}$$

Using these matrices, we can write the effect of frame rotation on the bond and torsion angles as follows:

$$\kappa_i T^2 \to e^{\Delta_i T^3}(\kappa_i T^2)e^{-\Delta_i T^3}, \tag{13.90}$$

$$\tau_i \to \tau_i + \Delta_{i-1} - \Delta_i. \tag{13.91}$$

From the point of view of lattice gauge theories, the transformation of bond angles is like an adjoint $SO(2) \in SO(3)$ gauge rotation of a Higgs triplet around the Cartan generator T^3, when the Higgs triplet is in the direction of T^2. The transformation of torsion angle coincides with that of the $SO(2)$ lattice gauge field. Since the $\mathbf{t}_i$ remain intact under (13.87), the gauge transformation of (κ_i, τ_i) has no effect on the geometry of the discrete string.

A priori, the fundamental range of the bond angle is $\kappa_i \in [0, \pi]$, while for the torsion angle the range is $\tau_i \in [-\pi, \pi)$. Thus, we identify (κ_i, τ_i) as the canonical latitude and longitude angles of a two-sphere $\mathbb{S}^2$. In parallel with the continuum case, we find it useful to extend the range of κ_i into negative values $\kappa_i \in [-\pi, \pi] \bmod 2\pi$. As in (13.74), we compensate for this twofold covering of $\mathbb{S}^2$ by a $\mathbb{Z}_2$ symmetry, which now takes the form

$$\begin{aligned} \kappa_k &\to -\kappa_k \qquad \text{for all} \quad k \geq i, \\ \tau_i &\to \tau_i - \pi. \end{aligned} \tag{13.92}$$

This is a special case of (13.90) and (13.91), with

$$\begin{aligned} \Delta_k &= \pi \quad \text{for} \quad k \geq i+1, \\ \Delta_k &= 0 \quad \text{for} \quad k < i+1. \end{aligned} \tag{13.93}$$

13.4.1 The C_α trace reconstruction

We have already concluded that the Ramachandran angles are not sufficient for reconstructing the protein backbones: As shown in Fig. (13.11) ,the reconstructed backbones are not in the same universality class as folded proteins. The value of the compactness index ν is different. For a correct reconstruction, we need to utilize all the bond and torsion angles that we have defined in Fig. 13.7. Only for the bond lengths can the average values be used.

We now consider the protein backbone reconstruction in terms of the virtual C_α backbone. We identify the vertices in Fig. 13.22 with the C_α atoms, so that (κ_i, τ_i) are the virtual C_α backbone bond and torsion angles. For the virtual C_α–C_α bond lengths, we use the average PDB value

$$|\mathbf{r}_{i+1} - \mathbf{r}_i| \sim 3.8\,\text{Å}. \tag{13.94}$$

It turns out that, *unlike* in the case of the Ramachandran angles, the ensuing reconstructed C_α backbones reproduce the original crystallographic structures, with a very high precision; the difference is mostly within the range of experimental errors, as measured by the B-factor (13.1).

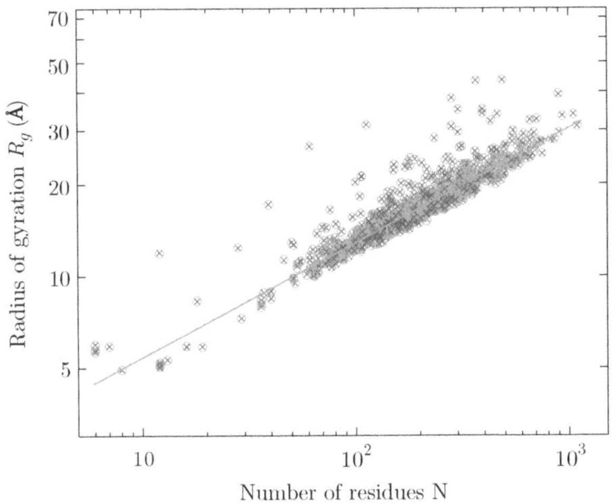

Fig. 13.23 [Colour online] Comparison of the C_α–C_α radius-of-gyration data between the original PDB structures and those reconstructed in terms of variable virtual bond and torsion angles in combination with optimal C_α–C_α virtual bond lengths. The [blue] crosses are the original PDB structures, and the [red] circles are the reconstructed ones. The line shows the fits of the radius of gyration. We see no visual difference between the two cases.

In Fig. 13.23, we compare the radius-of-gyration values in our ultra-high-resolution protein structures for the original PDB structures and those that have been reconstructed using the virtual C_α backbone bond and torsion angles when we use the constant virtual bond length value (13.94). Unlike in Fig. (13.11), we now observe no visual difference. For the reconstructed data, we obtain the relation [26]

$$R_g \approx 2.281 N^{0.375} \text{ Å}. \tag{13.95}$$

This is remarkably close to (13.7)—the difference is immaterial. Thus, we conclude that in the case of crystallographic protein structures, the virtual C_α trace bond and torsion angles (κ_i, τ_i) form a complete set of geometrical local order parameters.

13.4.2 Universal discretized energy

The goal is to describe the structure and dynamics of proteins *beyond* the limitations of an expansion in a small coupling like (13.11). For this, we propose to start with an energy function where the virtual C_α backbone bond and torsion angles appear as local order parameters; these variables form a complete set of local order parameters for reconstruction.

Let F be the thermodynamic Helmholtz free energy of a protein. Its minimum-energy configuration describes a folded protein, under thermodynamic equilibrium conditions. The free energy is the sum of the internal energy U and the entropy S at temperature T:

$$F = U - TS. \tag{13.96}$$

It is a function of all the interatomic distances:

$$F = F(r_{\alpha\beta}), \qquad r_{\alpha\beta} = |\mathbf{r}_\alpha - \mathbf{r}_\beta|, \tag{13.97}$$

where the indices $\alpha, \beta, \ldots$ extend over all the atoms in the protein system, including those of the solvent environment.

We assume that the characteristic lengthscales of spatial deformations along the protein backbone around its thermal equilibrium configuration are large in comparison with the covalent bond lengths; there are no abrupt wrenches and buckles, only gradual bends and twists. We also assume that the C_α virtual bond length oscillations have a characteristic timescale that is short in comparison with the timescale we consider; we adopt (13.94) as a time-averaged value for all the virtual bond lengths. The completeness of the C_α bond and torsion angles then suggests that in the vicinity of the free-energy minimum we should utilize these angles as the local order parameters. Accordingly, we consider the response of the interatomic distances to variations in these angles:

$$r_{\alpha\beta} = r_{\alpha\beta}(\kappa_i, \tau_i) \tag{13.98}$$

Suppose that at a local extremum of the free energy, the C_α bond and torsion angles have the values

$$(\kappa_i, \tau_i) = (\kappa_{i0}, \tau_{i0}). \tag{13.99}$$

Consider a conformation where the (κ_i, τ_i) deviate from these extremum values. The deviations are

$$\begin{aligned} \Delta\kappa_i &= \kappa_i - \kappa_{i0}, \\ \Delta\tau_i &= \tau_i - \tau_{i0}. \end{aligned} \tag{13.100}$$

Taylor-expand the infrared-limit Helmholtz free energy (13.96) around the extremum:

$$\begin{aligned} F[r_{\alpha\beta} &= r_{\alpha\beta}(\kappa_i, \tau_i)] \equiv F(\kappa_i, \tau_i) \\ &= F(\kappa_{i0}, \tau_{i0}) + \sum_k \left(\frac{\partial F}{\partial \kappa_k}\bigg|_0 \Delta\kappa_k + \frac{\partial F}{\partial \tau_k}\bigg|_0 \Delta\tau_k \right) \\ &+ \sum_{k,l} \left(\frac{1}{2} \frac{\partial^2 F}{\partial \kappa_k \partial \kappa_l}\bigg|_0 \Delta\kappa_k \, \Delta\kappa_l + \frac{\partial^2 F}{\partial \kappa_k \tau_l}\bigg|_0 \Delta\kappa_k \, \Delta\tau_l \right. \\ &\left. + \frac{1}{2} \frac{\partial^2 F}{\partial \tau_k \partial \tau_l}\bigg|_0 \Delta\tau_k \, \Delta\tau_l \right) + \mathcal{O}(\Delta^3). \end{aligned} \tag{13.101}$$

The first term in the expansion evaluates the free energy at the extremum. Since (κ_{i0}, τ_{i0}) correspond to the extremum, the second term vanishes and we are left with the following expansion of the averaged free energy:

$$F(\kappa_i, \tau_i) = F(\kappa_{i0}, \tau_{i0}) + \sum_{k,l} \left(\frac{1}{2} \frac{\partial^2 F}{\partial \kappa_k \partial \kappa_l} \bigg|_0 \Delta\kappa_k \, \Delta\kappa_l + \frac{\partial^2 F}{\partial \kappa_k \tau_l} \bigg|_0 \Delta\kappa_k \, \Delta\tau_l + \frac{1}{2} \frac{\partial^2 F}{\partial \tau_k \partial \tau_l} \bigg|_0 \Delta\tau_k \, \Delta\tau_l \right) + \ldots \tag{13.102}$$

In the limit where the characteristic scale of the extent of spatial deformations around a minimum-energy configuration is large in comparison with a covalent bond length, and the amplitude of these deformations remains small, we may *rearrange* the expansion (13.102) in terms of of the differences in the angles as follows. First come local terms. Then come terms that connect the nearest neighbours. Then come terms that connect the next-to-nearest neighbours. And so forth ... This reordering of the expansion ensures that we recover the derivative expansion (13.34) at leading order when we take the continuum limit where the virtual bond length vanishes. Moreover, since the free energy *must* remain invariant under the local frame rotations (13.90) and (13.91), we conclude [55, 71–78] that to leading order the expansion of the free energy *must* coincide with a discretization of the AHM energy (13.29) with (13.31):

$$F = -\sum_{i=1}^{N-1} 2\kappa_{i+1}\kappa_i + \sum_{i=1}^{N} \left[2\kappa_i^2 + \lambda(\kappa_i^2 - m^2)^2 + \frac{1}{2} q \kappa_i^2 \tau_i^2 - p\tau_i + \frac{1}{2} r \tau_i^2 \right] + \ldots \tag{13.103}$$

The corrections include next-to-nearest-neighbour couplings and so forth, which are higher-order terms from the point of view of our systematic expansion. The approximation (13.103) is valid, as long as there are no abrupt wrenches and buckles but only gradual bends and twists along the backbone. In particular, long-range interactions are accounted for as long as they don't introduce any localized buckling of the backbone.

In (13.103), λ, q, p, r, and m depend on the atomic-level physical properties and the chemical microstructure of the protein and its environment. In principle, these parameters can be computed from this knowledge.

We note the following. The free energy (13.103) is a deformation of the standard energy function of the discrete nonlinear Schrödinger (DNLS) equation [58, 59]. The first sum together with the first three terms in the second sum is the energy of the standard DNLS equation, in terms of the discretized Hasimoto variable (13.60). The fourth (p) term is the conserved helicity, which breaks the $\mathbb{Z}_2$ parity symmetry and is responsible for the right-handed helicity of the C_α backbone. The last (r) term is the Proca mass that we again add as a 'regulator'. Observe in particular the explicit presence of the nonlinear, quartic contribution to the (virtual) bond angle energy. This is the familiar double-well potential, shown in Fig. 13.12. Its $\mathbb{Z}_2$ symmetry eventually becomes broken. The breaking of this symmetry is essential for the emergence of structure in the case of proteins. It is the source of *Bol'she* that makes us alive. We note that this kind of explicit nonlinearity is absent in (13.9).

We summarize. The expression (13.103) for the free energy describes the small-amplitude fluctuations around the local extremum (κ_{i0}, τ_{i0}) in the space of bond and torsion angles. It can be identified as the long-wavelength (infrared) limit of the full free energy, in the sense of Kadanoff and Wilson. To the present order of the expansion in powers of (κ_i, τ_i), the functional form (13.103) is the *most general* long-wavelength free energy that is compatible with the principle of gauge invariance. This *fundamental* symmetry principle ensures that no physical quantity depends on our choice of coordinates (framing) along the backbone.

Research project 13.7 Develop a method to compute the parameters in (13.103) from an all-atom energy function.

13.4.3 Discretized solitons

The energy (13.103) is a deformation of the integrable energy of the DNLS equation [58, 59, 62]: The first term together with the λ- and d-dependent terms constitute the (naively) discretized Hamiltonian of the NLS model in the Hasimoto variable. The conventional DNLS equation is known to support solitons. Thus, we can try and find soliton solutions of (13.103).

As in (13.70), we first eliminate the torsion angle:

$$\tau_i[\kappa] = \frac{a + b\kappa_i^2}{c + d\kappa_i^2} = a\frac{1 + b\kappa_i^2}{c + d\kappa_i^2} \equiv a\hat{\tau}_i[\kappa]. \tag{13.104}$$

For bond angles, we then have

$$\kappa_{i+1} = 2\kappa_i - \kappa_{i-1} + \frac{dV[\kappa]}{d\kappa_i^2}\kappa_i \qquad (i = 1, \ldots, N), \tag{13.105}$$

where we set $\kappa_0 = \kappa_{N+1} = 0$, and $V[\kappa]$ is given by (13.73). This equation is a deformation of the conventional DNLS equation, and it is not integrable, a priori. For a numerical solution, we extend (13.105) to the following iterative equation [73]:

$$\kappa_i^{(n+1)} = \kappa_i^{(n)} - \epsilon\{\kappa_i^{(n)}V'[\kappa_i^{(n)}] - (\kappa_{i+1}^{(n)} - 2\kappa_i^{(n)} + \kappa_{i-1}^{(n)})\}. \tag{13.106}$$

Here $\{\kappa_i^{(n)}\}_{i\in N}$ denotes the nth iteration of an initial configuration $\{\kappa_i^{(0)}\}_{i\in N}$, and ϵ is some sufficiently small but otherwise arbitrary numerical constant; we often choose $\epsilon = 0.01$ in practical computations. The fixed point of (13.106) is clearly independent of the value chosen. But the radius and rate of numerical convergence in a simulation towards the fixed point depend on the value of ϵ: The fixed point is clearly a solution of (13.105).

Once the numerically constructed fixed point is available, we calculate the corresponding torsion angles from (13.104). Then, we obtain the frames from (13.85) and can proceed to the construction of the discrete string, using (13.86).

At the moment, we do not know of an analytical expression for the soliton solution of (13.105). But we have found [72, 74, 76] that an *excellent* approximate solution can be obtained by discretizing the topological soliton (13.67):

$$\kappa_i \approx \frac{m_1 e^{c_1(i-s)} - m_2 e^{-c_2(i-s)}}{e^{c_1(i-s)} + e^{-c_2(i-s)}}. \tag{13.107}$$

Here, $(c_1, c_2, , m_1, m_2, s)$ are parameters. m_1 and m_2 specify the asymptotic κ_i-values of the soliton. Thus, these parameters are entirely determined by the character of the regular, constant bond and torsion angle structures that are adjacent to the soliton. In particular, these parameters are not specific to the soliton per se, but to the adjoining regular structures. The parameter s defines the location of the soliton along the string. This leaves us with only two loop-specific parameters, c_1 and c_2. These parameters quantify the length of the bond angle profile that describes the soliton.

For the torsion angle, (13.104) involves one parameter (a) that we have factored out as the overall relative scale between the bond angle and torsion angle contributions to the energy; this parameter determines the relative flexibility of the torsion angles with respect to the bond angles. Then, there are three additional parameters $(b/a, c/a, d/a)$ in the remainder $\hat{\tau}[\kappa]$. Two of these are again determined by the character of the regular structures that are adjacent to the soliton. As such, these parameters are not specific to the soliton. The remaining single parameter specifies the size of the regime where the torsion angle fluctuates.

On the regions adjacent to a soliton, we have constant values of (κ_i, τ_i). In the case of a protein, these are the regions that correspond to the standard regular secondary structures: in a rough sense, proteins are made of right-handed α-helices, β-strands, and loops. For example, the standard right-handed α-helix is

$$\alpha\text{-helix:} \qquad \begin{cases} \kappa \approx \frac{1}{2}\pi, \\ \tau \approx 1, \end{cases} \tag{13.108}$$

and the standard β-strand is

$$\beta\text{-strand:} \qquad \begin{cases} \kappa \approx 1, \\ \tau \approx \pi. \end{cases} \tag{13.109}$$

All the other standard regular secondary structures such as 3/10 helices and left-handed helices are similarly described by definite constant values of κ_i and τ_i.

Protein loops correspond to regions where the values of (κ_i, τ_i) are variable, while protein loops are the soliton proper: a soliton is a configuration that interpolates between two regular structures, with constant values of (κ_i, τ_i).

13.4.4 Proteins out of thermal equilibrium

When a protein folds towards its native state, it is out of thermal equilibrium. Several studies propose that in the case of a small protein, the folding takes place in a manner that is consistent with Arrhenius' law; we recall that this law states that the reaction rate depends exponentially on the ratio of activation energy E_A and temperature:

$$r \propto \exp\left(-\frac{E_A}{k_B T}\right). \tag{13.110}$$

On the other hand, in the case of simple spin chains, it has been found that the *Glauber dynamics* [79, 80] describes the approach to thermal equilibrium in a manner that follows Arrhenius's law. Since we have argued that proteins can be viewed as spin chains, with residues corresponding to the spin variables, it is natural to try and model the way in which a protein approaches thermal equilibrium by using Glauber dynamics.

Glauber proposed to model non-equilibrium dynamics in terms of a Markovian Monte Carlo (MC) time evolution, defined by the heat bath probability distribution [79, 80]

$$\mathcal{P} = \frac{x}{1+x}, \qquad \text{with} \quad x = \exp\left(-\frac{\Delta E}{kT}\right). \tag{13.111}$$

Here, ΔE is the energy difference between consecutive MC time steps ($\sim$activation energy). We compute it from (13.103) in the case of a protein. In addition, in the case of a protein, we need to account for steric constraints: analysis of PDB structures reveals that the distance between two C_α atoms that are *not* nearest neighbours along the backbone, is always larger than (13.94):

$$|\mathbf{r}_i - \mathbf{r}_k| > 3.8\,\text{Å} \quad \text{for} \quad |i-k| \geq 2. \tag{13.112}$$

Note the apparent similarity between Arrhenius' law and Glauber's algorithm. We also note that the scale of units of kT, which appears in (13.111) as a temperature factor, should not be directly identified with the Boltzmannian temperature factor $k_B T$. The scale of units depends on the overall scale of the energy function (13.103), and in particular by our choice of the normalization factor in the first, nearest-neighbour interaction term. To determine the unit, we need a renormalization condition. For this, we need to perform a proper experimental measurement(s), and compare the predictions of our model with those of the protein that it describes, at that temperature. One suitable renormalization point could be to try and identify the experimentally measured θ-transition temperature by comparison with the properties of our model.

13.4.5 Temperature renormalization

This subsection is somewhat technical. The details are not needed in the rest of the lectures. We include it because we feel that good understanding of temperature renormalization of parameters is relevant to the physics of proteins [78]. For example, thus far this has not been really addressed in any other approach we are aware of. You might find the subject described here to be an inspiration for your future research.

In the probability distribution (13.111) the nearest-neighbour coupling contribution in (13.103) becomes normalized as follows:

$$-\frac{2}{kT}\sum_{i=1}^{N}\kappa_{i+1}\kappa_i. \tag{13.113}$$

Thus, the temperature factor kT depends on the physical temperature factor k_BT in a non-trivial fashion. That is, we should really write

$$\frac{2}{kT} \to \frac{J(T)}{k_BT} \tag{13.114}$$

where $J(T)$ is the strength of the nearest-neighbour coupling at Boltzmannian k_BT. Its numerical value depends on the temperature in a manner that is governed by the standard renormalization group equation

$$T\frac{dJ}{dT} = \beta_J(J;\lambda,m,q,p,r) \sim \beta_J(J) + \ldots \tag{13.115}$$

For simplicity, we may assume that to leading order the dependence of β_J on the other couplings can be ignored. Note that the parameters, and thus their β-functions, depend on the properties of the environmental factors: the properties and pH of the solvent, the pressure, etc.

In the low-temperature limit, we can expand the nearest-neighbour coupling as follows:

$$J(T) \approx J_0 - J_1T^{\alpha} + \ldots \qquad \text{as } T \to 0. \tag{13.116}$$

Here, J_0 is non-vanishing, and the critical exponent α controls the low-temperature behaviour of $J(T)$. The asymptotic expansion (13.116) corresponds to a β-function (13.115) that in the $T \to 0$ limit approaches

$$\beta_J(J) = \alpha(J - J_0) + \ldots \tag{13.117}$$

Consequently, at low temperatures, we have

$$kT \approx \frac{2k_B}{J_0}T. \tag{13.118}$$

In terms of the temperature factor, (13.115) translates into

$$T\frac{d}{dT}\left(\frac{1}{kT}\right) = -\frac{1}{kT} + \frac{1}{2k_BT}\beta_J\left(\frac{2k_BT}{kT}\right). \tag{13.119}$$

We try to find an approximate solution in the collapsed phase, when the temperature T is below the critical θ-point temperature T_θ at which the transition between the collapsed phase and the random-walk phase takes place This coincides with the physical temperature value that corresponds to the unfolding transition temperature factor value kT_Θ. Let

$$\beta_J\left(\frac{2k_BT}{kT}\right) = \frac{2k_BT}{kT} + F\left(\frac{2k_BT}{kT}\right) \tag{13.120}$$

and define

$$y = \frac{1}{kT}, \qquad x = \frac{1}{2k_BT}. \tag{13.121}$$

Equation (13.119) then translates into

$$\frac{dy}{dx} = -F\left(\frac{y}{x}\right), \tag{13.122}$$

with the solution

$$\ln(cx) = -\int^{y/x} \frac{du}{F(u)+u}, \tag{13.123}$$

where c is an integration constant. We shall assume that the leading nonlinear corrections are logarithmic, which is often the case. To leading order, we then have

$$F(u) = (\eta - 1)u + \alpha u \ln u + \ldots \tag{13.124}$$

Note that in general there are higher-order corrections. When we reintroduce the original variables and set

$$\eta = -\alpha \ln J_0, \tag{13.125}$$

we get for the temperature factor

$$kT \approx \frac{2}{J_0} k_BT \exp\left(\frac{J_1}{J_0} T^\alpha\right) \tag{13.126}$$

$$\approx \frac{2}{J_0} k_BtT + \frac{2J_1}{J_0^2} k_BT^{\alpha+1} + \ldots \qquad \text{as} \quad T \to 0 \qquad (\alpha > 0), \tag{13.127}$$

where we have chosen the integration constant so that in the low-temperature limit we obtain (13.118).

For the nearest-neighbour coupling, (13.126) yields

$$J(T) \approx J_0 \exp\left(-\frac{J_1}{J_0} T^\alpha\right). \tag{13.128}$$

Thus, the coupling between bond angles becomes weak, at an exponential rate, when the temperature approaches the transition temperature T_θ between the collapsed phase and the random-walk phase.

Similarly, all the other couplings that are present in (13.103) are temperature-dependent. Each has its own renormalization group equation. For example, the quartic κ_i self-coupling λ in (13.103) flows according to a renormalization group equation of the form

$$T\frac{d\lambda}{dT} = \beta_\lambda(\lambda). \tag{13.129}$$

For simplicity, we again assume that to leading order β_λ depends only on λ.

It is natural to interpret λ as a measure of the strength of hydrogen bonds: structures such as α-helices and β-strands become stable owing to the presence of hydrogen bonds. At the same time, the value of λ controls the affinity of κ_i towards the ground-state value of the quartic potential in (13.103). The hydrogen bonds are presumed to become vanishingly weak when the protein unfolds. This can take place when the protein approaches the transition temperature T_θ that separates the collapsed phase from the random-walk phase. This suggest that, asymptotically,

$$\lambda(T) \to \lambda_\theta|T - T_\theta|^{\gamma_\lambda} \qquad \text{as } T \to T_\theta \text{ from below.} \tag{13.130}$$

Here, γ_λ is a critical exponent that controls the way in which the strengths of (effective) hydrogen bonds vanish. More generally, we may send $T_\theta \to T_H$, which is the temperature at which hydrogen bonds disappear even between the solvent molecules; in the case of water under atmospheric conditions, $T_H \approx 100°\text{C}$. In general, we expect the value of T_H to be higher than that of T_θ.

Above $T > T_\theta$, when the hydrogen bonds become vanishingly weak, we expect that effectively $\lambda \approx 0$ (or $m \approx 0$) in (13.103). On the other hand, we expect that as the temperature decreases, the value of $\lambda(T)$ increases, so that in the low-temperature limit we have

$$\lambda(T) \to \lambda_0 - \lambda_1 T^{\gamma_0} + \dots \qquad \text{as } T \to 0. \tag{13.131}$$

Thus,

$$\beta_\lambda(\lambda) \approx \gamma_0(\lambda - \lambda_0) + \mathcal{O}((\lambda - \lambda_0)^2). \tag{13.132}$$

Here, λ_0 should be close to the value we obtain from PDB when we compute the parameters in (13.103) from the crystallographic low-temperature structure.

Research project 13.8 Try to evaluate the β-function (13.115) numerically in the case of a simple protein, such as villin headpiece (PDB code 1YRF), using results from detailed experimental measurements.

Research project 13.9 Determine how the parameters in (13.9) and (13.10) depend on temperature.

13.5 Solitons and ordered proteins

Various taxonomy schemes such as CATH (http://www.cathdb.info/) and SCOP (http://scop.mrc-lmb.cam.ac.uk/scop/) have revealed that folded proteins are build in a modular fashion, from a relatively small number of building blocks. Despite an essentially exponentially increasing number of new crystallographic protein structures being discovered, novel fold topologies are now rarely found and some authors think that most modular building blocks of proteins are already known [30, 81]. This convergence in protein architecture demonstrates that protein folding should be a process that is driven by some kind of universal structural self-organization principle.

We know that solitons are the paradigmatic structural self-organizers. Thus, we argue that the soliton solution of the DNLS equation (13.105) with (13.104) must be the universal modular building block from which folded proteins can be constructed. Indeed, we know that the energy function (13.103) is unique, in the limit where it becomes applicable. Moreover, it has already been shown that over 92% of all C_α-traces of PDB proteins can be described in terms of no more than 200 different parametrizations of the DNLS soliton (13.107), with a root-mean-square distance (RMSD) precision that is better than 0.5 Å [75].

Accordingly, we set out to describe the modular building blocks of proteins in terms of various parametrizations of the DNLS soliton profile, which is described by (13.106), (13.104), (13.85), and (13.86).

13.5.1 λ-repressor as a multisoliton

In order to identify the soliton structure of a given protein, we start by computing the C_α virtual backbone bond and torsion angles from the PDB data. We *initially* fix the $\mathbb{Z}_2$ gauge in (13.92) so that all the bond angles take positive values $\kappa_i \in [0, \pi]$. A generic protein profile consists of a set of κ_i with values that are typically between $\kappa_i \approx 1$ and $\kappa_i \approx \pi/2$; the upper bound can be estimated using steric constraints. The torsion angle values τ_i are commonly much more unsettled, and their values extend more widely over the entire range $\tau \in [-\pi, +\pi]$.

As an example we consider the λ-repressor, which is a protein that controls the lysogenic-to-lytic transition in bacteriophage λ-infected *E. coli* cells. The transition between the lysogenic and lytic phases in bacteriophage λ-infected *E. coli* is the paradigmatic example of a genetic switch mechanism, which has been described in numerous molecular biology textbooks and review articles (see e.g. [82, 83]). The interplay between the lysogeny-maintaining λ-repressor protein and the regulator protein that controls the transition to the lytic state is a simple model for more complex regulatory networks, including those that can lead to cancer in humans.

The λ-repressor structure that we consider has PDB code 1LMB. It is a homodimer with 92 residues in each of the two monomers. It maintains the lysogenic state by binding to DNA with a helix–turn–helix motif that is located between residue sites 33 and 51. The λ-repressor is a fast-folding protein. In [35], an 80-residue-long mutant of the λ-repressor was studied in an all-atom simulation.

In Fig. 13.24 (left column) we show the (κ_i, τ_i) spectrum of 1LMB, with the convention (i.e. $\mathbb{Z}_2$ gauge fixing) that κ_i is positive. We display the segments between

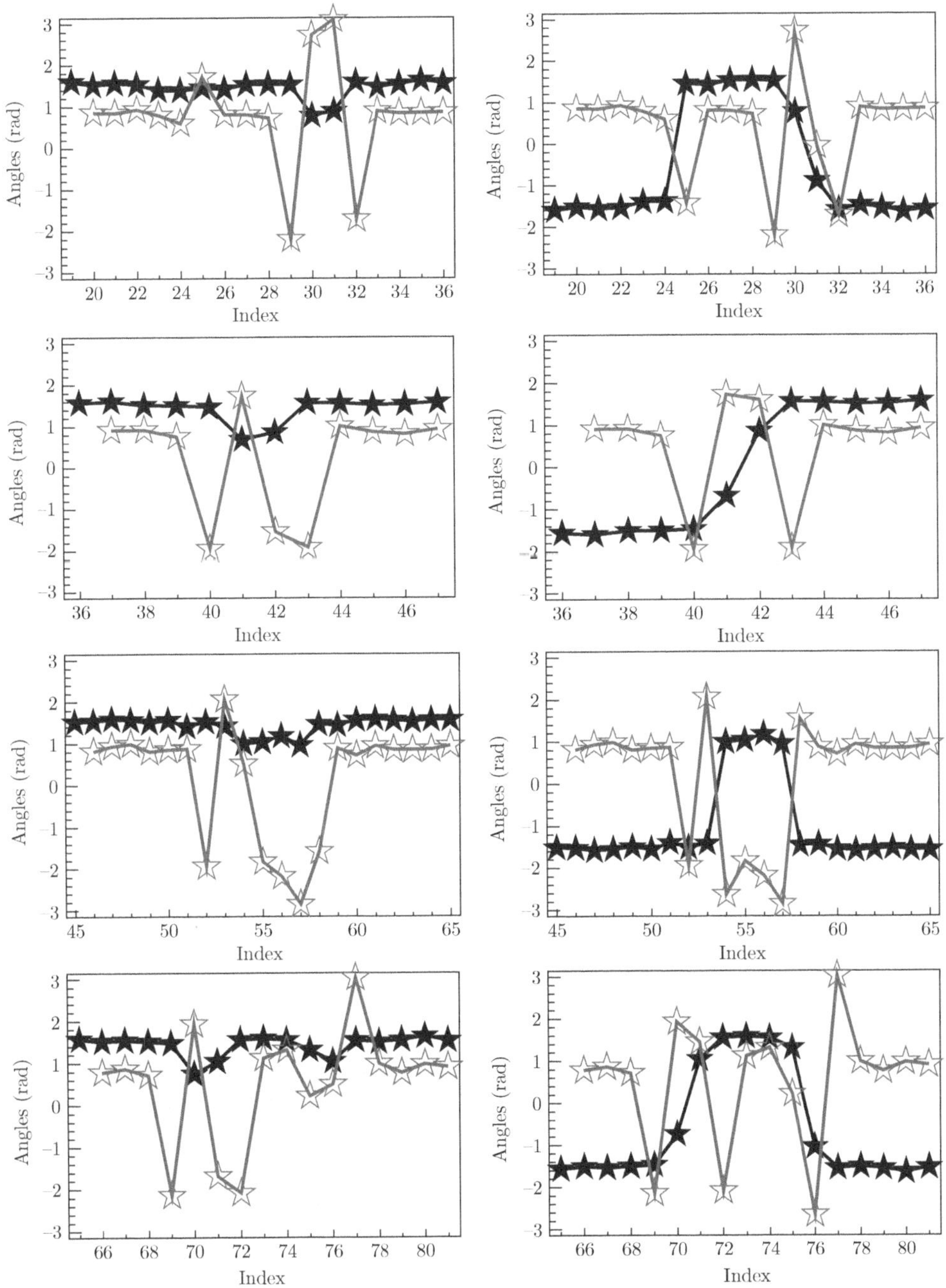

Fig. 13.24 Bond angle κ and torsion angle τ spectrum of the λ-repressor 1LMB, with indexing that follows PDB. Left column: spectrum in the $\mathbb{Z}_2$ gauge where all $\kappa_i > 0$. Right column: spectrum after implementation of the Z_2 gauge transformations that identify the solitons.

residues 19 and 82. This spectrum is fairly typical of the PDB structures that we have analysed.

The results in Section 13.3.8 suggest that in analysing the PDB data, one should first pay attention to flattening points, i.e. points where τ_i changes its sign. The flattening points should be located near a putative inflection point where a soliton is located and perestroika takes place. Accordingly, we perform in the spectrum of Fig. 13.24 (left column) the $\mathbb{Z}_2$ gauge transformations (13.92) in the vicinity of the apparent flattening points, to identify the putative multisoliton profile of κ_i. For example, in the case of 1LMB, we observe four regions with an irregular τ_i profile as shown in Fig. 13.24. By a judicious choice of $\mathbb{Z}_2$ gauge transformations, we identify seven different solitons in κ_i. The profiles are shown in the right column of Fig. 13.24. Each of the soliton profiles is clearly accompanied by putative flattening points. Note that τ_i is multivalued, mod 2π. Thus, the large fluctuations in the values of τ_i are deceptive. Once we account for the multivaluedness, we find that τ_i is actually quite regular. This is in full accordance with the observed, much higher, flexibility of the torsion angles in relation to the bond angles that is known to occur in proteins.

On the basis of the general considerations in Section 13.3.8, we argue that protein folding from a regular unfolded configuration with no solitons to the biologically active natively folded configuration with its solitons is a process that is driven by inflection and flattening-point perestroikas. The initial configuration with no solitons can be chosen to coincide with the minimum of the second sum in (13.103). It could also be, for example, a uniform right-handed α-helix (13.108), or β-strand (13.109), or polyproline II-type conformation. When the protein folds, it proceeds from this initial configuration towards the final configuration, through successive perestroikas, i.e. bifurcations. These perestroikas deform the C_α backbone, creating DNLS-like solitons along it, thus causing the backbone to enter the space-filling $\nu \sim \frac{1}{3}$ collapsed phase.

In the case of 1LMB, we identify seven soliton profiles as shown in Fig. 13.24 (right column). We proceed to determine the parameters in the energy function (13.103). For this, we train the energy function so that it describes the seven individual solitons in terms of a solution of (13.105) with (13.104) for each of them individually. The training is performed by demanding that the fixed point of the iterative equation (13.106) models the ensuing C_α backbone structure as a soliton solution, with a prescribed precision.

We have developed a program *GaugeIT* that implements the $\mathbb{Z}_2$ gauge transformations to identify the background, and a program *PropoUI* that trains the energy so that it has an extremum that models the background in terms of solitons. These programs are described at

http://www.folding-protein.org

In the case of a protein for which the PDB structure is determined with ultra-high resolution, typically below 1.0 Å, *PropoUI* routinely constructs a soliton solution that describes the C_α backbone with a precision comparable to the accuracy of the experimentally measured crystallographic structure; recall that the accuracy of the experimental PDB structure is estimated by the B-factors using the Debye–Waller relation (13.1).

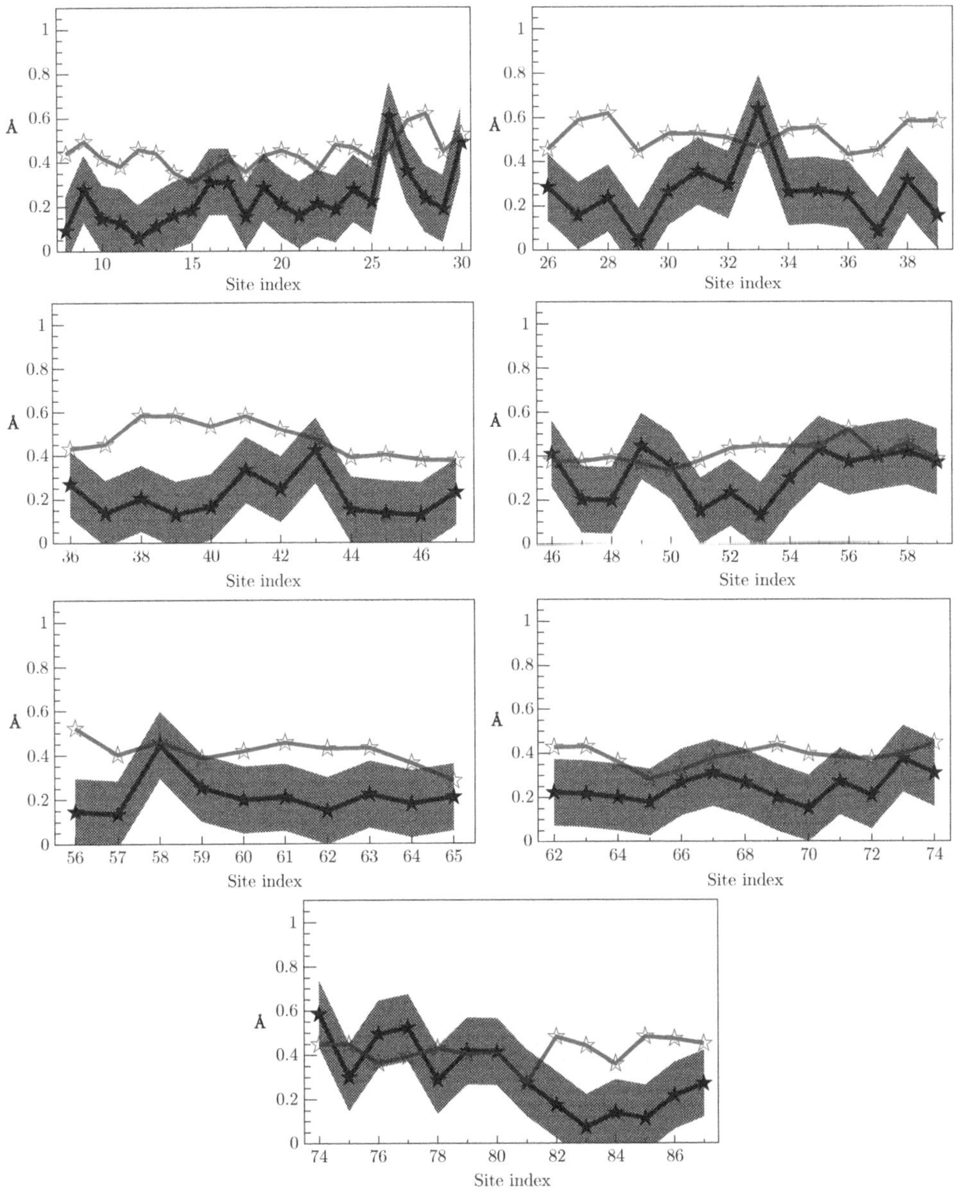

Fig. 13.25 [Colour online] Distance between the PDB backbone of the first 1LMB chain and its seven solitons. The black line denotes the distance between the PDB structure and the corresponding soliton. The grey area around the black line describes the lower-bound estimate of 15 pm (quantum-mechanical) zero-point fluctuation distance around each soliton, obtained from Fig. (13.4). The grey [red] line denotes the Debye–Waller fluctuation distance (13.1).

In Fig. 13.25, we compare the distance between the C_α backbone and the seven individual soliton solutions for 1LMB. The B-factor fluctuation distance in the experimental structure 1LMB, evaluated from the Debye–Waller relation, is also displayed. As shown in the figure, the solitons describe the loops with a precision that is fully comparable to the experimental uncertainties. The grey zone around the soliton profile denotes our best estimate for the extent of quantum-mechanical zero-point fluctuations; according to Fig. 13.4, there are practically no Debye–Waller values less than 15 pm, which we have chosen here as the zero-point fluctuation distance, in the figure.

13.5.2 Structure of myoglobin

Myoglobin [9] is the primary oxygen-carrying protein in the muscle cells of mammals. It is closely related to haemoglobin, which is the oxygen-binding protein in blood. Myoglobin gives meat its red colour; the more red, the more myoglobin. It also allows organisms to hold their breath for a period of time: diving mammals such as whales and seals have a high myoglobin concentration in their muscles.

Myoglobin was the first protein to have its three-dimensional structure determined by X-ray crystallography. Subsequently, it has remained among the most actively studied proteins. But, theoretically, the investigation of myoglobin, for example in an all-atom MD simulation, remains a formidable task: the experimentally measured folding time from a random chain to the natively folded state is around 2.5 s [84]. At the same time, the fastest special purpose MD supercomputer ever built, *Anton*, can produce at most a few microseconds of *in vitro* folding trajectory per day *in silico*—in the case of proteins that are much shorter and simpler than myoglobin. Accordingly, it would take at least a million or so days to reproduce a *single* myoglobin folding trajectory *in silico*, at an all-atom level, even with *Anton*. A good convergence of Newton's iteration with the energy function (13.9), (13.10) over such a long timescale would be truly amazing.

We have already noted that all-atom MD simulation is conceptually a weak-coupling expansion, appropriate for describing phenomena over very short time periods only: the time ratio (13.11) is the small, iterative expansion parameter. In the case of long-time trajectories, such a weak-coupling expansion cannot be expected to be very effective, not even convergent. Alternative approximate methods need to be introduced to model myoglobin and the large majority of proteins that fold much more slowly than the microsecond to millisecond scale.

From the perspective of quantum field theory, this means that we need a non-perturbative approach. For example, in quantum chromodynamics (QCD), we do not expect standard perturbation theory to be capable of describing hadrons. On the other hand, lattice QCD is designed for modelling hadrons. But it can hardly describe the scattering of quarks and gluons.

Indeed, we have argued that in the case of proteins, the energy function (13.103) is an example of such a non-perturbative approach. It avoids altogether the need to introduce a time step ratio such as (13.11) as a weak-coupling expansion parameter. Instead, the C_α geometry is modelled in terms of small variations in the angular variables around their equilibrium conformations. Since the approximation does not involve time directly, it becomes in principle possible to describe folding phenomena that can be very difficult, even impossible, to model in terms of conventional and presently available all-atom MD.

We proceed to apply the energy function (13.103) to investigate detailed properties of myoglobin. Here, we analyse the static structure, and in Section 13.5.3, we consider out-of-equilibrium dynamics.

We first construct the multisoliton solution that models the myoglobin backbone. We use the crystallographic PDB structure 1ABS as a decoy, to train the energy function. This structure has been measured at very low liquid helium temperatures $\sim$20 K. Thus, the thermal B-factor fluctuations (13.1) are small.

At this point, we suggest that the reader should download the PDB structure 1ABS, to make it easier to follow details of our analysis. We propose to use the Java interface provided at the PDB site (http://www.pdb.org/) for visualization of the backbone and side-chain atoms. We also recommend the analysis tools available on the Molprobity website (http://molprobity.biochem.duke.edu/), which we shall refer to in the following. We shall provide the values for all the parameters in the energy function (13.103), which the reader can use as input in the programs *Propro* and *GaugeIT*, which are described at our website (http://www.folding-protein.org). This enables a detailed analysis of the multisoliton that describes 1ABS, and should help the reader to start independent research.

We proceed to the construction of the multisoliton: 1ABS has 154 amino acids, and the PDB index runs over $i = 0, \ldots, 153$. Conventionally, one identifies the structure as a bundle of eight α-helices (A,B,C, ...,H), which are separated by seven loops as shown in Fig. 13.26.

Here, we shall limit the construction of the multisoliton to the sites with PDB index N between 8 and 149. That is, we include all the named helices but not include the flexible tails at the ends of the backbone. These tails could be included, but without much additional insight into the issues that we address.

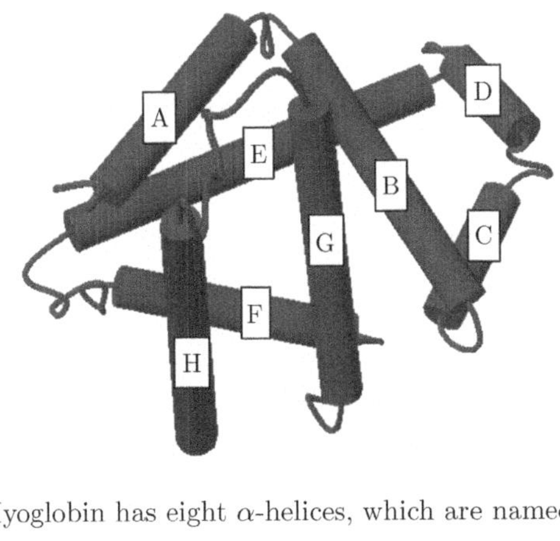

Fig. 13.26 Myoglobin has eight α-helices, which are named A,B,C, ..., H.

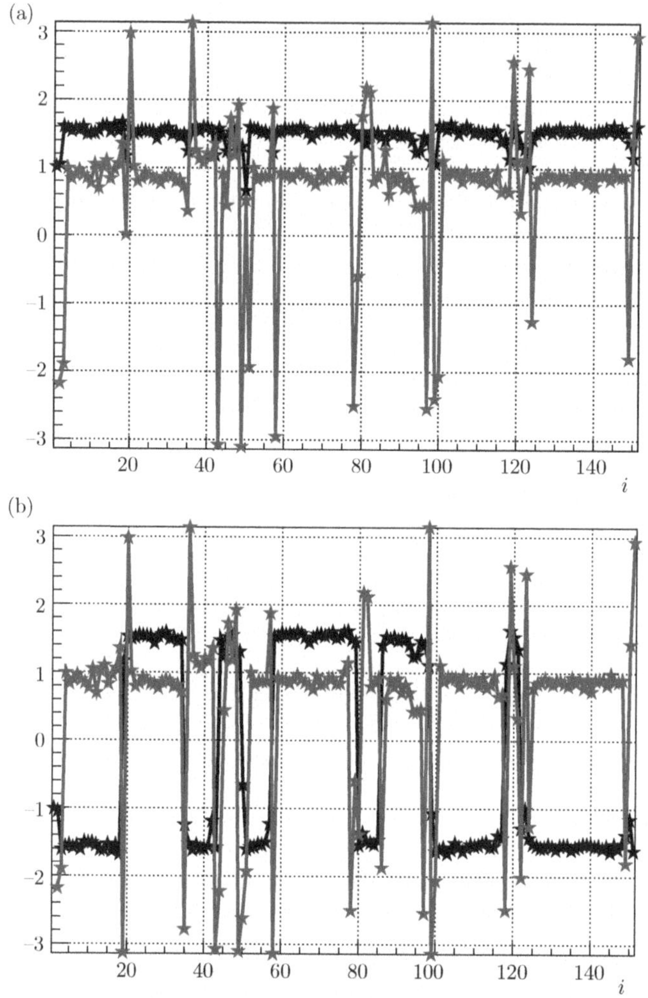

Fig. 13.27 [Colour online] (a) The κ_i (black) and τ_i (grey [red]) profiles of 1ABS using the standard differential geometric convention that bond angles are positive. (b) The soliton structure becomes visible in the κ_i profile once we implement the transformations (13.92).

In Fig. 13.27(a), we show the backbone bond and torsion angle spectra, with the convention that all κ_i are positive. In Fig. 13.27(b), we show the spectra after we have implemented the $\mathbb{Z}_2$ transformations (13.92) to putatively identify the multisoliton profile; in our analysis, we use the program packages *Propro* and *GaugeIT* described at our website (http://www.folding-protein.org).

We recall that both Fig. 13.27(a) and (b) correspond to the same intrinsic backbone geometry. The $\mathbb{Z}_2$ transformation is a symmetry of the discrete string that is obtained by solving the discrete Frenet equation.

Table 13.1 The solitons along the 1ABS C_α-backbone, with indexing starting from the N-terminus. We have left out the end sites that correspond to monotonic helices, and the N- and C-terminal segments. The type identifies whether the soliton corresponds to a loop that connects α-helices and (or) 3/10-helices.

SOLITON	1	2	3	4	5
Sites	15–27	30–41	39–49	47–57	54–66
Type	α–α	α–3/10	3/10–3/10	3/10–3/10	3/10–α
Soliton	6	7	8	9	10
Sites	72–87	83–92	94–106	110–123	121–135
Type	α–α	α–α	α–α	α–α	α–α

We conclude from the κ_i profile Fig. 13.27(b) that the myoglobin backbone has 11 helices that are separated by 10 single soliton loops. The numbers of loops and helices are more or less unambiguously determined by the number of inflection points that we identify visually in Fig. 13.27, in the manner explained in Section 13.3.8. The PDB sites of the 10 individual soliton profiles that we use for our construction are identified in Table 13.1. We emphasize that our geometry-based identification of the loops and helices along the 1ABS backbone does not necessarily coincide with the conventional one used, for example, in crystallography, which is based on inspection of hydrogen bonds. In particular, according to the conventional classification, the soliton pair 3 and 4, the soliton pair 6 and 7, and the soliton pair 9 and 10, are all interpreted as a single loop.

From our geometric point of view, the PDB data reveal that in 1ABS, we have four different types of solitons: those that connect two α-helices, those that connect an α-helix with a 3/10-helix or vice versa, and those that connect two 3/10-helices; see Table 13.1.

In Table 13.2, we give our parameter values for the multisoliton solution. This describes the 1ABS backbone with 0.78 Å RMSD accuracy.

Note that in those terms in (13.103) that involve the torsion angles, the numerical parameter values are consistently much smaller than in terms that contain only the bond angles. This is in line with the known fact that in proteins the torsion angles, i.e. dihedrals, are usually quite flexible while the bond angles are relatively stiff.

Note also that our energy function has 80 parameters, while there are 153 amino acids in the entire myoglobin backbone. Thus, the energy function (13.103) is *highly predictive*: the number of free parameters is even *less than the number of amino acids*. This shows that myoglobin displays structural redundancy in its amino acids.

The predictive power of (13.103) can alternatively be characterized as follows: When we assume that all the bond lengths have the constant value (13.94), we are left with 282 C_α angular coordinate values in our truncated backbone. These we need to

Table 13.2 Parameter values in energy (13.103) for the multisoliton solution that describes 1ABS.

SOLITON	λ_1	λ_2	m_1	m_2
1	9.923	2.232	1.54097	1.54548
2	6.48516	0.9955	1.58013	1.54058
3	2.05153	0.657	1.66032	1.60224
4	0.89676	6.74235	1.3563	1.5232
5	9.26118	0.83376	1.55206	1.5386
6	0.98018	2.1337	1.45791	1.54653
7	1.37667	3.16891	1.47151	1.04128
8	10.3168	4.2801	1.18192	1.61334
9	0.80042	1.28973	1.5154	1.60278
10	3.15255	0.91475	1.55827	1.55151
SOLITON	a	b	c	d
1	−5.62412 e-08	−4.13459 e-07	1.81044 e-08	4.273 e-09
2	−6.25287 e-11	−1.68598 e-05	1.47093 e-07	2.82807 e-07
3	−9.05135 e-08	1.20232 e-06	5.10166 e-11	5.75389 e-09
4	−2.33413 e-07	−3.3991 e-07	2.36516 e-08	7.98841 e-09
5	−9.73035 e-08	4.78674 e-07	1.03189 e-10	4.88194 e-09
6	−7.25906 e-09	3.76092 e-09	6.82624 e-10	1.87212 e-14
7	−1.39052 e-13	5.97719 e-13	3.77897 e-14	5.81911 e-14
8	−1.27193 e-07	1.41736 e-06	1.07182 e-10	1.26295 e-08
9	−2.03487 e-07	1.13574 e-06	1.46007 e-11	7.82707 e-08
10	−1.07811 e-07	1.02768 e-06	7.49571 e-11	7.73639 e-09

determine from the properties of the energy function (13.103), in order to construct the backbone from (13.85) and (13.86). We have a total of 80 parameters in Table 13.2, and thus a total of 202 coordinates remain to be determined by the multisoliton solution that minimizes the energy function. Therefore, we have 202 unknowns that are to be *predicted* by the model. These predictions then directly probe the physical principles on which (13.103) has been built.

We recall that approaches such as the Gō model and its variants and various elastic network models lack this kind of predictive power. In those models, the positions of *all* the atoms are assumed to be known a priori. In addition, one needs a description how the atoms interact. Thus, there are always *more parameters* than degrees of freedom, and no predictions can be made.

In Fig. 13.28, we interlace the 1ABS backbone with the multisoliton; the difference is very small.

In Fig. 13.29, we analyse site-wise the precision of the multisoliton configuration with the PDB structure 1ABS. The 15 pm grey-shaded region around the multisoliton profile again corresponds to the regime of zero-point fluctuations that we have

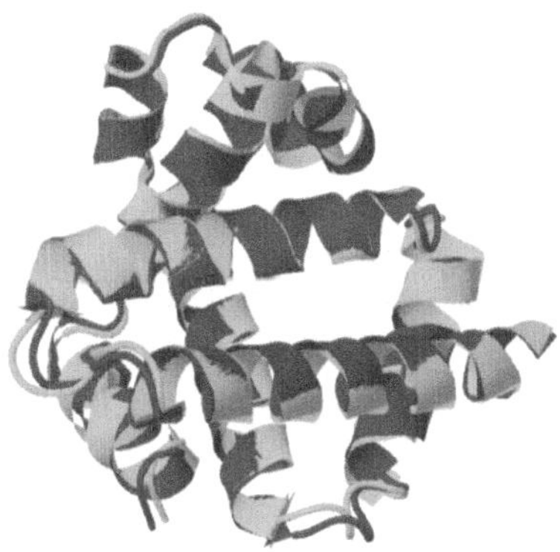

Fig. 13.28 [Colour online] Comparison between the PDB structure 1ABS (dark [purple]) and the corresponding multisoliton solution (light [blue]).

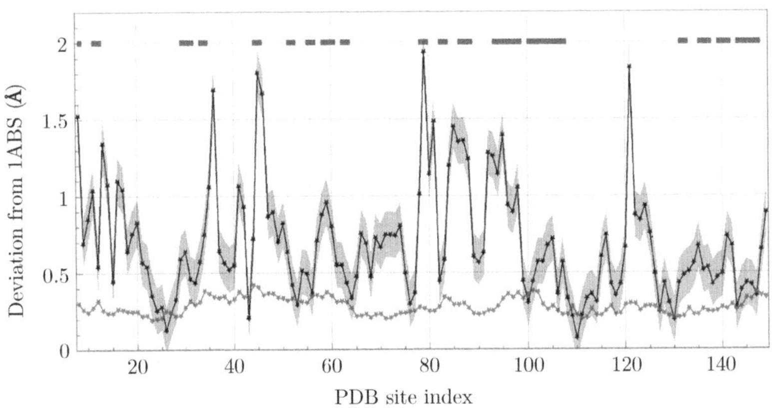

Fig. 13.29 [Colour online] Comparison of the RMSD between the 1ABS configuration and the multisoliton solution, with the Debye–Waller B-factor fluctuation distance around the 1ABS backbone. The [blue] markings at the top, along the 2.0 Å line, denote sites where *Molprobity* (http://molprobity.biochem.duke.edu/) detects imperfections.

deduced from Fig. 13.4. Conceptually, the multisoliton describes a *single* myoglobin structure in the limit of vanishing temperature. In particular, as such, the multisoliton does not account for any kind of conformational fluctuations that are due to thermal effects, lattice imperfections, or any other kind of conformational substate effects; we model thermal effects using the Glauber heath bath (13.111). On the other hand, the experimentally measured 1ABS crystal structure should not be interpreted as a single static low-temperature structure. Instead, it is an average over a large number of closely packed crystallographic structures. A comparison between Figs. 13.25 and 13.29 shows that in the latter, the distance between the PDB backbone and the multisoliton profile is larger than that between the PDB backbone and the individual solitons in Fig. 13.25. Figure 13.29 describes the *single* multisoliton solution to the equations (13.105), and (13.104), while in Fig. 13.25 we have constructed the individual solitons by solving (13.105) and (13.104) independently for each loop. A similar individual soliton construction in the case of 1ABS gives profiles that are comparable, even slightly more precise, than those in Fig. 13.25. But for energetic studies we need the full multisoliton with its energy function, and we need the local energy minimum of (13.103) for the *entire* backbone.

We presume that a multisoliton solution could be constructed with a precision even better than 0.78 Å in C_α RMSD. But the convergence of (13.106) becomes slow when we use a laptop like a MacBook Air. Thus, we have simply terminated the process when we reach the value 0.78 Å,which is in any case much better than that obtained in any other approach, using any other computer, to our knowledge.

In Fig. 13.29, we observe that the distance between the multisoliton solution and the C_α backbone of 1ABS has its largest values mainly in two regimes. These are located *roughly* between sites 35 and 45 and between sites 79 and 98; we propose that the reader inspects the structure of 1ABS using the visualization interface of the PDB. The first regime corresponds to the single soliton that models the loop between helices B and C in Fig. 13.26. The second regime corresponds to the location of helix F. This helix is part of the 'V'-shaped pocket of helices E and F, where the haem group is located. In particular, the reader can see that helix F includes the proximal histidine at site 93, which is bonded to the iron ion of the haem. Note that, in addition, our analysis detects an anomaly at about site 121.

In order to understand the origin of the observed deviations from an ideal multisoliton crystal, we check for the presence of potential structural disorders in 1ABS using *Molprobity*; we recommend that the reader perform this analysis on the *Molprobity* website (http://molprobity.biochem.duke.edu/). In Fig. 13.28, along the top, at the level of the 2.0 Å line, we have marked [blue online] those regions where, according to *Molprobity*, we have potential clashes. The *Molprobity* clash score of 1ABS is 20.32, which puts it in the 10th percentile among structures with comparable resolution (100 being the best score). The regions of potential structural clashes correlate with those regions where our multisoliton profile has the largest deviations from the 1ABS backbone (except for the vicinity of site 121, which is unproblematic according to *Molprobity*).

We first consider the difference between the multisoliton and the 1ABS backbone around sites 79–98, which was also identified by the ansatz as a potentially troublesome one. The difference appears to be largely due to a deformation of helix F. It could be caused by a bond between the proximal histidine at site 93 and the oxygen-binding haem iron. This might introduce a strain that modifies the backbone. The effect of the haem is not accounted for by our energy function in its present form. Consequently, we propose the histidine–haem interaction as the likely explanation for the relatively large deviation between our multisoliton profile and the 1ABS backbone at this position.

We proceed to consider the difference between the multisoliton and the 1ABS backbone around sites 35–45. These sites are also located very close to the haem. For example, the distance between the C_α at site 45 (arginine) and the haem oxygen 154 is 4.84 Å, and the C_α of phenylalanine at site 43 is even closer to the haem. This proximity between the backbone and the haem is reflected in the *Molprobity* clash at site 45 (between $C\delta$ and 154 HEM). We conclude that there could be strain in the backbone structure that is due to the haem, and this could explain the difference between the 1ABS backbone and the multisoliton configuration in this regime.

Finally, we note that in Fig. 13.29, we also have our previously observed anomaly at site 121 (glycine). At this point, we have no explanation for the anomaly except that glycine is flexible and that this region is on the exterior of the protein. This leaves the hydrophobic phenylalanine at nearby site 123 exposed to the solvent. Consequently, relatively strong fluctuations between several locally conformationally different but energetically degenerated substates are possible.

13.5.3 Dynamical myoglobin

Myoglobin stores O_2 by binding it to the iron atom, which is inside the myoglobin. The oxidization causes a conversion from ferrous ion (Fe^{2+}) to ferric ion (Fe^{3+}); the oxidized molecule is called oxymyoglobin. When the oxygen is absent, the molecule is called deoxymyoglobin. We propose that the reader finds examples of each from the PDB and inspects the structures using the three-dimensional Java interface.

Numerous detailed experimental investigations have been made of both oxymyoglobin and deoxymyoglobin. But to our knowledge, the understanding of the oxygen transport mechanism in myoglobin remains incomplete: We do not yet know exactly how small non-polar ligands such as O_2, CO, and NO move between the external solvent and the iron-containing haem group, which is located inside the myoglobin. From the available static crystallographic PDB structures, one cannot identify any obvious ligand pathway. It has been proposed that the process involves thermally driven large-scale conformational motions. Collective thermal fluctuations could open and close gates through which the ligands migrate. Such gates are not necessarily visible in the crystallized low-temperature structures. Computational investigations that model the dynamics of myoglobin might provide a clue as to how these gates operate.

13.5.3.1 Glauber dynamics and phase structure

We start our investigation into how ligand gates might open and close by performing heating and cooling simulations of myoglobin with the energy function (13.103) in combination with Glauber dynamics (13.111) and (13.112) [77]. We use the 1ABS multisoliton that we have constructed. The structure is a carbon-monoxy-myoglobin

(MbCO), but with the covalent bond between the CO and iron broken by photodissociation. In any case, its dynamical properties should serve as a good first-approximation model of how myoglobin behaves more generally.

We start our simulations at a low-temperature value, from the multisoliton configuration that we have constructed: A classical soliton solution is commonly interpreted as a structure that describes the limit of vanishing temperature where thermal fluctuations become very small.

We take

$$k\mathcal{T}_L = 10^{-16} \tag{13.133}$$

for the numerical value of the low-temperature factor, in terms of the dimensionless unit that is determined by our choice of the overall energy scale in (13.103). At this value of the temperature factor, thermal fluctuations are absent in our multisoliton. For the numerical high-temperature value, we choose

$$k\mathcal{T}_H = 10^{-13}. \tag{13.134}$$

By applying the renormalization group arguments in Section 13.4.5, we have related the two temperature factors $k\mathcal{T}$ and $k_B T$. Our conversion relation, which we shall justify in the sequel, is

$$k\mathcal{T} = 1.6181 \times 10^{-9} k_B T e^{0.05506T}. \tag{13.135}$$

We use CGS units, so $k_B = 1.381 \times 10^{-16}$ erg/K.

Under *in vivo* conditions, myoglobin always interacts with water. This interaction is essential for maintaining the collapsed phase. In our approach, we account for the solvent (water) implicitly, in terms of the parameter values in (13.103). In particular, as such, our model can only *effectively* take into account the highly complex phase properties of water at sub-freezing temperatures. We do not even try to address the obvious complications that appear when the temperature rises above the boiling point of water.

We start the simulation at $k\mathcal{T}_L$. The heating takes place with an exponential rate of increase during 5 million Monte Carlo steps. We model the non-equilibrium response using the Glauber protocol (13.111). According to (13.135), in terms of the physical temperature factor $k_B T$, the heating process should correspond to an adiabatically slow nearly linear rate of increase.

When we arrive at the high-temperature $k\mathcal{T}_H = 10^{-13}$, we fully thermalize the system by keeping it at this temperature value during 5 million steps. We then proceed to cool it back down to $k\mathcal{T}_L$. We use the same rate of cooling as we use for heating, i.e. we cool exponentially in $k\mathcal{T}$ during 5 million steps. Each complete heating–cooling cycle takes about 3 minutes of wall-clock time when we use a single processor in a standard laptop (MacBook Air). Consequently, time is no constraint for us and we can collect very, very good statistics. In particular, we have checked that our results, and the conclusions are quite insensitive to the rate of heating and cooling.

For statistical purposes, we have performed 100 repeated heating and cooling cycles that we have then analysed in detail; an increase in the number of cycles does not change our conclusions. Note that in an all-atom approach, a comparable simulation would take over 100 000 years even with *Anton* [33], while for us a few minutes is enough.

During the simulations, we follow the evolution of both the radius of gyration R_g and the RMSD R_{RMSD} between the simulated configuration and the folded 1ABS structure. In Fig. 13.30(a), we show the evolution of the radius of gyration and in Fig. 13.30(b) the evolution of the RMSD distance to 1ABS, as functions of steps during our 100 repeated heating and cooling cycles; Note that in these plots we have converted the temperature into the kelvin scale, using (13.135).

We make the following observations: At low temperatures, with temperature factors

$$kT < 10^{-15}, \tag{13.136}$$

the radius of gyration is essentially constant, $R_g \approx 14.6$, and only subject to very small thermal fluctuations. In the range

$$10^{-15} < kT < 10^{-14}, \tag{13.137}$$

we have a regime where the radius of gyration increases at an accelerating rate in the number of steps. The increase in R_g continues until we reach a temperature near kT_H. But for temperatures where the temperature factor is in the range

$$10^{-14} < kT < kT_H = 10^{-13}, \tag{13.138}$$

the rate of increase decelerates so that when we reach the temperature kT_H, we observe no increase in R_g. This suggests that we have reached the random-walk θ-regime. Furthermore, the radius of gyration value

$$R_g \approx 22\,\text{Å} \tag{13.139}$$

is *extremely* close to the experimentally measured value ~23.6 Å for the molten globule state of myoglobin. The difference can be entirely attributed to the 12 residues that we have excluded (7 from the N-terminus and 5 from the C-terminus) when constructing the multisoliton.

We have confirmed that the transition near kT_H is indeed a θ-transition between the collapsed phase and the random-walk phase, by heating the configuration to substantially higher temperature factor values. We have found that above this putative θ-transition, the radius of gyration remains essentially intact under temperature variations all the way to

$$kT = 10^{-8}. \tag{13.140}$$

Around this temperature value another transition takes place, presumably to the $\nu \sim \frac{3}{5}$ self-avoiding random-walk phase. In Fig. 13.30(c), we show how the RMSD between the heated configuration and 1ABS changes as a function of temperature during heating and cooling cycles between $kT_L = 10^{-16}$ and $kT = 10^{-8}$. We observe two clear transitions that are consistent with the transitions between collapsed and random-walk phases, and between random-walk and self-avoiding random-walk phases according to (13.4).

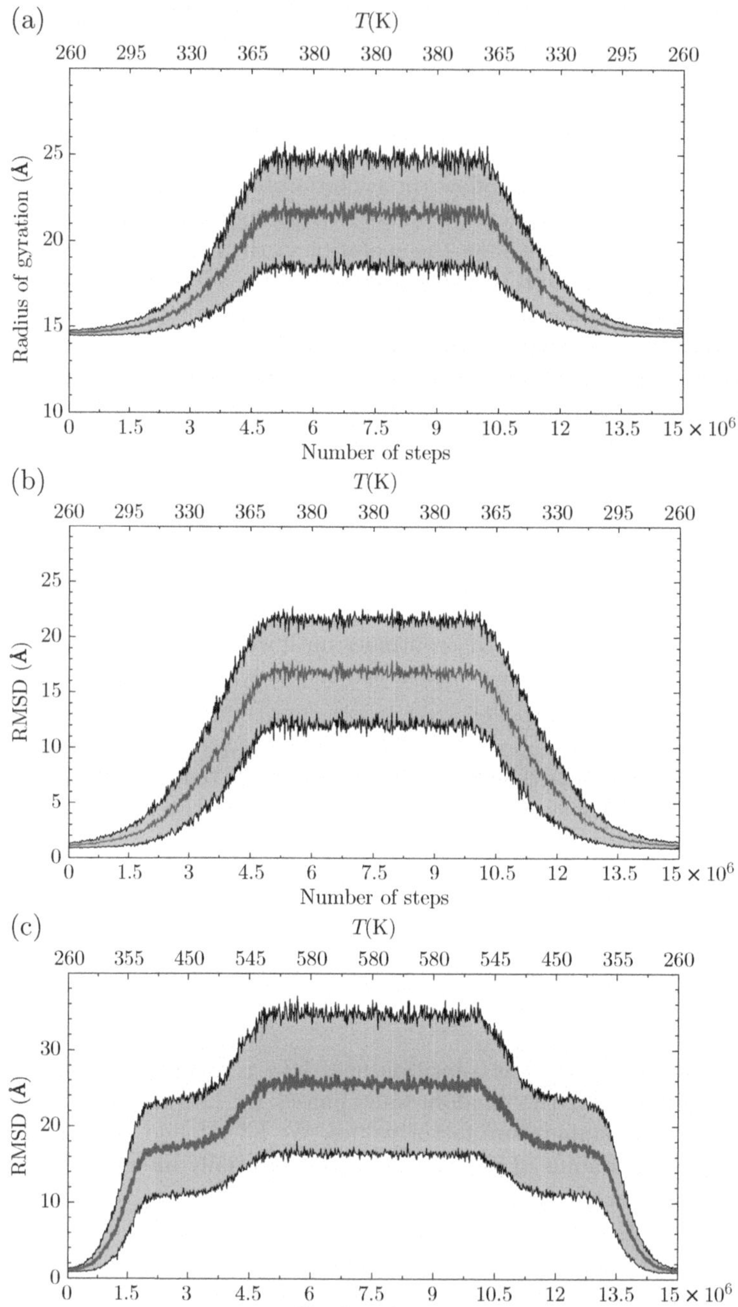

Fig. 13.30 [Colour online] (a) Evolution of the radius of gyration during 100 repeated heating and cooling cycles. (b) Evolution of the RMSD between the 1ABS backbone and the simulated configuration during 100 repeated heating and cooling cycles. (c) Evolution of the RMSD between the 1ABS backbone and the simulated configuration during 100 repeated heating and cooling cycles, to very high temperature values. In each plot, the central [blue] line denotes the average, and the shaded area around it is the extent of one standard deviation fluctuations. Along the top axis, we show the temperature on the kelvin scale, using the conversion relation (13.135).

When we decrease the temperature, the evolution of R_g becomes inverted. At the end of the cycle, when the temperature reaches $k\mathcal{T}_L$, the configuration returns back to a very close proximity of the original folded state; see Fig. 13.30(b). This demonstrates the stability of the multisoliton solution that describes the natively folded myoglobin as a local minimum of the energy (13.103).

Figure 13.31(a) shows the *average* values of R_g. These averages are evaluated at several different temperature values, over 100 runs, both during the heating period when $0 < x < 7.5$ and during the cooling period when $7.5 < x < 15$, where x is the number of MC steps in millions. The data can be approximated by a fitting function of the form

$$R_g(x) \approx a \tanh[b(x-c)] + d. \tag{13.141}$$

The parameter values are listed in Table 13.3.

In Fig. 13.31(a), we display the derivative of (13.141). We can try and use the maximum of the derivative to identify the θ-transition temperature in our model. For this, we assume that the experimentally measured [85, 86] transition temperature at $T_c \approx 348\,\text{K}$ is the one that corresponds to the θ-transition. We identify it with the maximum of the derivative of R_g, to conclude that during the heating cycle the θ-transition temperature relates to our dimensionless temperature values as follows:

$$\mathcal{T}_g^h \approx 1.63 \times 10^{-14} \approx 348\,\text{K}. \tag{13.142}$$

We use this value to determine one of the two parameters in (13.135).

During the cooling cycle, we find the slightly different

$$\mathcal{T}_g^c \approx 1.71 \times 10^{-14} \approx 349\,\text{K}. \tag{13.143}$$

We note that an asymmetry between heating and cooling has been observed experimentally [85].

The RMSD between the simulated configuration and the 1ABS backbone depends on temperature in a very similar manner. In Fig. 13.31(b), we show the comparison between simulation and the corresponding approximation (13.141),

$$R_{\text{RMSD}} \approx a \tanh[b(x-c)] + d. \tag{13.144}$$

Table 13.3 Parameter values in the fits (13.141), (13.144) for the two ranges 0–7.5 and 7.5–15 (in millions) of iteration steps

	R_g				R_{RMSD}			
RANGE	a	b	c	d	a	b	c	d
0–7.5	3.519	0.9047	3.6855	18.29	7.9	0.8318	3.5715	9.291
7.5–15	−3.486	0.9193	11.2965	18.28	−7.872	0.8327	11.4255	9.298

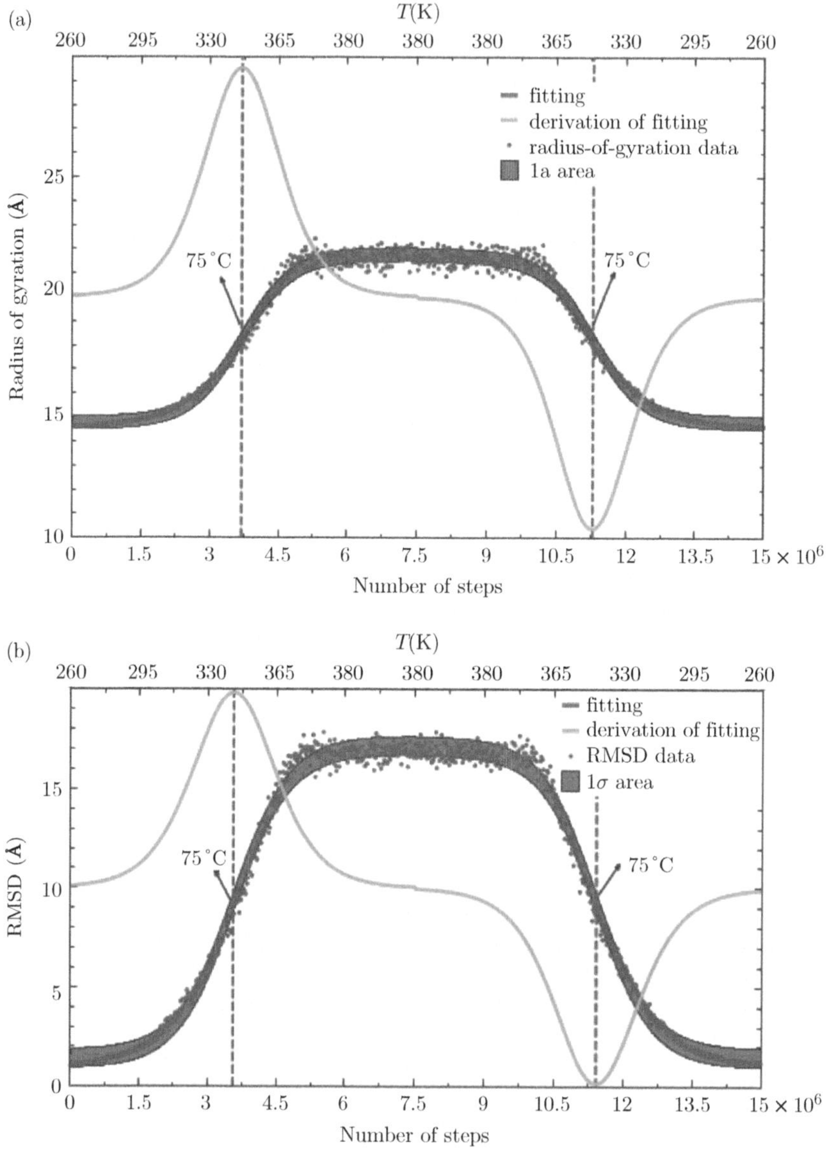

Fig. 13.31 [Colour online] The [red] line is the fitting of (13.141) to the average values of the [blue] dots, which are the numerically computed values of R_g, over the heating and cooling periods. The shaded area around the [red] fitting line is the one-standard-deviation estimate. Note that the difference between these three is so small that it is barely observable on the figure. Also shown is the derivative of (13.141) (light [blue] line). Along the top axis, we have converted the temperature into the kelvin scale.

These parameter values are also listed in Table 13.3 separately for the heating period $0 < x < 7.5$ and for the cooling period $7.5 < x < 15$. Figure 13.31(b) also shows the derivative of $R_{\mathrm{RMSD}}(x)$. As in the case of the radius of gyration, we use the maximum of the derivative to estimate the peak rate of change of the transition temperature. During the heating period, the increase in R_{RMSD} peaks at

$$\mathcal{T}^{h}_{\mathrm{RMSD}} \approx 1.35 \times 10^{-14} \approx 344\,\mathrm{K}. \tag{13.145}$$

During the cooling period, we find that the peak corresponds to a slightly higher temperature value,

$$\mathcal{T}^{h}_{\mathrm{RMSD}} \approx 1.45 \times 10^{-14} \approx 346\,\mathrm{K}. \tag{13.146}$$

These values are very close to those we observe in the case of R_g.

13.5.3.2 Backbone ligand gates

We proceed to try and identify potential thermally driven backbone ligand gates. We are interested in studying how the gates open and close as the myoglobin is heated and cooled. Moreover, thus far, we have fixed only one of the two parameters in (13.142) using the θ-transition temperature. We shall fix the second parameter in the sequel, by considering the dynamics of the backbone ligand gates.

We investigate the shape of the backbone visually, during the heating and cooling. We find that qualitatively the thermal fluctuations follow a very similar pattern. The backbone becomes unfolded in more or less the same manner, again and again, as the temperature increases. The inverse pattern is observed during cooling.

During heating, we observe a clear onset of the unfolding transition, which we characterize in terms of backbone ligand gates. We have identified three major gates, which we call Gates 1, 2, and 3 and define as follows.

Gate 1 is defined as the area between the segment that starts at PDB site 37 (Pro) and ends at 44 (Asp) and the segment that starts at 96 (Lys) and ends at 103 (Tyr). The opening of this gate takes place as the distance between the two segments increases. The open gate exposes the haem to the solvent. Figure 13.32(a) shows the location of this gate along the 1ABS backbone.

Gate 2 is located between the helical structures E and F, as shown in Fig. 13.32(b). This gate extends over the entire length of both helices E and F. Thus, in order to compare it with Gate 1, which is composed of segments with only eight residues, we select two opposing segments along helices E and F, each with eight amino acids. The first segment, located in the helical structure E, starts with site 61 (Leu) and ends with site 68 (Val). The second segment, located in the helical structure F opposite to the first segment, starts with site 89 (Leu) and ends with site 96 (Lys). We have intentionally selected these two segments to be far from the loop that connects helices E and F. This is because in our simulations, we have observed that the amplitudes of the thermal fluctuations in the segment distances tend to increase the further away the segment is located from the connecting loop: the opening and closing of the gate resemble the opening and closing of scissors, with blades formed by helices E and F

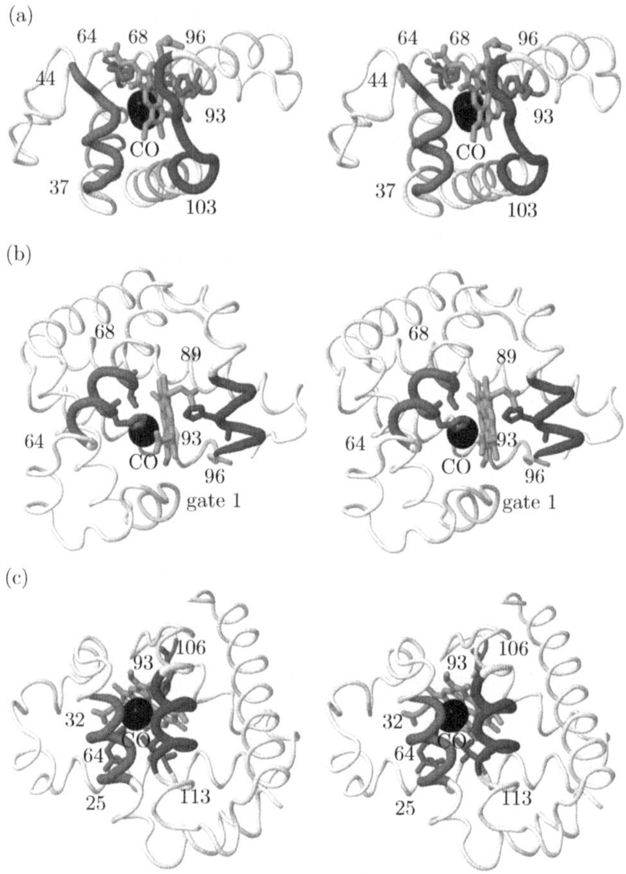

Fig. 13.32 [Colour online] Stereographic cross-eyed view of the ligand Gates 1, 2, and 3 as defined in the text. (a) Gate 1, between the segment starting at PDB site 37 (Pro) and ending at 44 (Asp) and the segment starting at 96 (Lys) and ending at 103 (Tyr). (b) Gate 2, between the segment starting at 61 (Leu) and ending at 68 (Val) and the segment starting at 89 (Leu) and ending at 96 (Lys). (c) Gate 3, between the segment starting at 25 (Gly) and ending at 32 (Leu) and the segment starting at 106 (Phe) and ending at 113 (His). We also show the location of the haem [orange], the proximal histidine (93), the valine (68), the distal histidine (64) [all green] and the CO (black ellipsoid) of 1ABS.

that are connected by the loop between these two helices. Note that the first segment along helix E includes both the distal histidine at site 64 and the valine at the end site 68. This valine is also inside the haem pocket, and it is presumed to have an important role in discrimination between CO and O_2. Similarly, the opposite segment in the helical structure F includes the proximal histidine at site 93.

Finally, Gate 3, which is shown in 13.32(c), is located between the helical structures B and G. Again, in order to compare this relatively long gate with Gate 1, we select two opposing segments, each with eight amino acids. The segment in the helical structure B starts at site 25 (Gly) and ends at site 32 (Leu). The segment in helix G starts at site 106 (Phe) and ends at site 113 (His).

During the heating and cooling cycle of the myoglobin, we follow the size of the three gates. We do this by computing the distance d_i $(i = 1, 2, 3)$ between the respective segments as a functions of temperature. We define the distance d_i between the two segments for each of the three gates as follows:

$$d_i = \sqrt{\sum_{n=1}^{8} (\mathbf{x}_n - \mathbf{y}_n)^2} \, . \tag{13.147}$$

Here $\mathbf{x}_n$ are the eight coordinates in the first segment, and $\mathbf{y}_n$ are the corresponding coordinates in the second segment, along Gate $i = 1, 2, 3$. Note that the two segments in each of the three gates are spatially oriented in an antiparallel manner with respect to PDB indexing. Consequently, in computing (13.147), we invert the indexing in one of the two segments with respect to the PDB indexing.

We start by investigating the temperature dependence of the three gates, using the experimental data available from the PDB. For this, we compute the following three gate ratios from PDB data:

$$\frac{\text{Gate 1}}{\text{Gate 2}} = \frac{d_1}{d_2}, \qquad \frac{\text{Gate 3}}{\text{Gate 2}} = \frac{d_3}{d_2}, \qquad \frac{\text{Gate 3}}{\text{Gate 1}} = \frac{d_3}{d_1}. \tag{13.148}$$

We use all the presently available myoglobin structures in PDB that have been measured with resolution 2.0 Å or better. The results are shown in Fig. 13.33. In each plot, we observe substantial fluctuations in the gate ratios in the case of PDB data that have been taken at around 100 K. But this reflects only the fact that the majority of PDB data have been collected at this temperature value. Overall, we conclude that the gate ratios show no temperature dependence for $T < 300\,\text{K}$.

We proceed to the computation of the temperature dependence of the gate ratios using our 1ABS multisoliton with the energy function (13.103). The results are shown in Fig. 13.34. We have found that Gate 3 is the first to open as the temperature increases, and the last to close as the temperature decreases. Gate 2 is the last to open , and the first to close. In the low-temperature limit, Gate 3 is about half the size of Gate 2. But its size exceeds that of Gate 2 in terms of the segment separation distance (13.147) at a temperature

$$kT_{23}^c \approx 10^{-14} \sim 340\,\text{K}. \tag{13.149}$$

The transition is very rapid, in line with the general results of [87]: when the temperature reaches the θ-transition value $\sim$348 K, Gate 3 is about twice as large as Gate 2.

Gate 1 also opens much faster than Gate 2, but more slowly than Gate 3. It also closes more slowly than Gate 2, but faster than Gate 3. In the low-temperature limit,

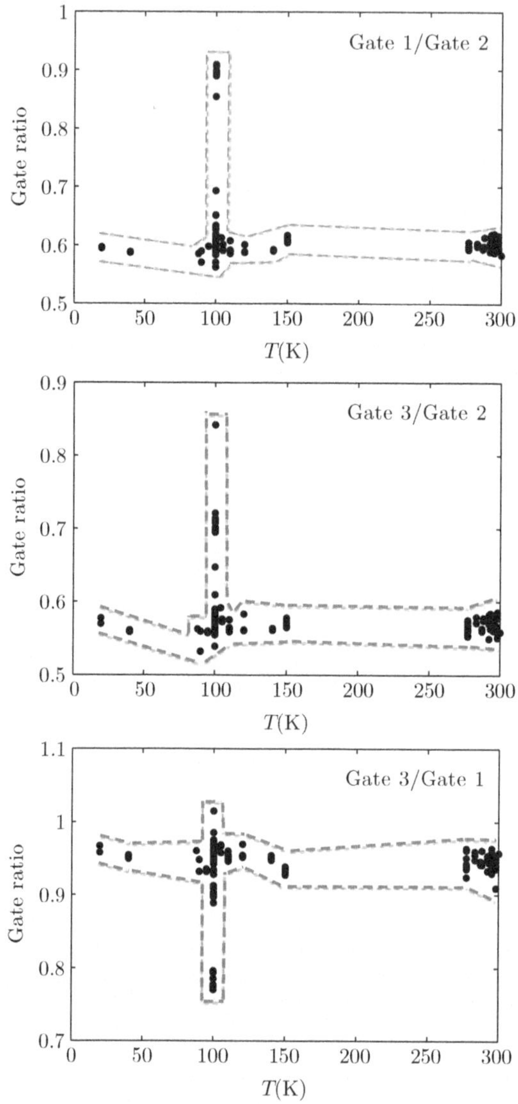

Fig. 13.33 The three gate ratios (13.148). The dashed lines are simply guides to the eye.

Gate 1 is about half as wide as Gate 2. But it becomes wider than Gate 2 when the temperature reaches a value

$$kT_{12}^{c} \approx 10^{-14}. \tag{13.150}$$

However, Gate 1 does not become quite as wide as Gate 3. This is shown in Fig. 13.34(a).

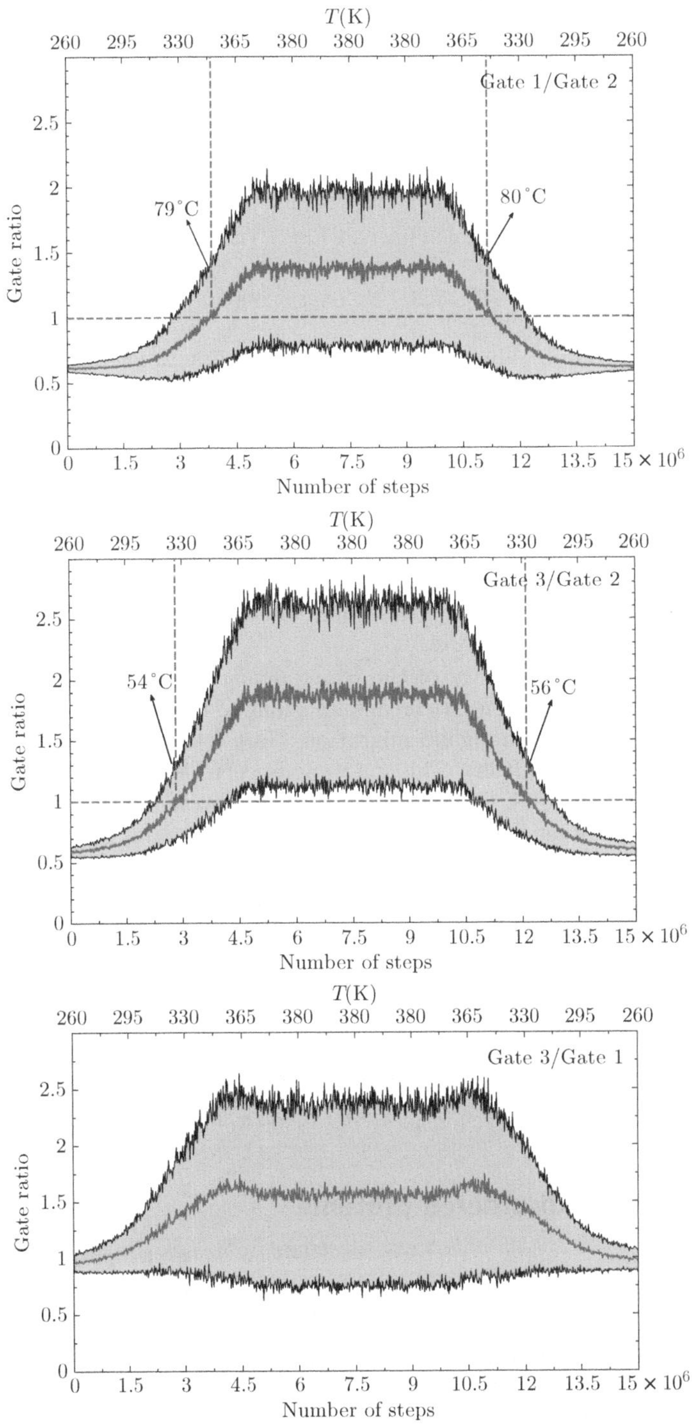

Fig. 13.34 [Colour online] Temperature dependence of the three gate ratios (13.148) during the heating and cooling cycles.

We are now in a position to determine the second parameter in (13.126) to arrive at (13.135); we recall that one of the two parameters is already determined, in (13.142). For this, we proceed as follows. When we compare the plots in Fig. 13.33, we conclude that, experimentally, the gate ratios do not display any observable temperature dependence when $T < 300\,\mathrm{K}$. Consequently, the lowest possible value of the temperature factor kT at which Fig. 13.34 can display any change in the gate ratios should correspond to a temperature above 300 K. When we read off the lowest possible kT value where we have an observable effect in Fig. 13.34, we conclude that, necessarily,

$$kT \approx 10^{-15} > k_B \times 300\,\mathrm{K}. \tag{13.151}$$

This gives a lower bound. When we adopt this lower-bound value as our estimate for the gate opening temperature, we obtain the second parameter value in (13.135).

In reality, the actual gate opening temperature can be higher, but at the moment experimental basis for choosing a higher value is lacking: the single presently existing NMR data on myoglobin in the PDB is 1MYF. It has been measured at the slightly higher temperature of 308 K. But the quality of data does not enable us to improve our estimate.

We note that a higher gate opening temperature has no qualitative effect on our conclusions, and quantitatively the differences are minor. The only effect would be a sharpening of the θ-transition onset.

We conclude with a summary of the consequences that the results in Fig. 13.34 might have for ligand migration. We have found that to the extent that backbone thermal fluctuations play a role in ligand migration, Gate 3 between the helical structures B and G can be very important. This gate opens very much like a baseball glove as the temperature is increased. Gate 1 might also play a role, but probably a lesser one than Gate 3. On the other hand, the V-shaped Gate 2 between helices E and F seems to be quite sturdy,; it does not seem to open as much as the other two gates. The presence of the distal and proximal histidines in Gate 2 and their attractive interactions with haem might have an additional stabilizing effect that is not accounted for by our model. Consequently, we do not see how the thermal backbone fluctuations that take place in Gate 2, could play a major role in ligand migration—at least to the extent that backbone fluctuations are relevant.

A recent terahertz-timescale spectroscopy experiment has detected collective thermal fluctuations in the protein, that might be related to our theoretical proposals for ligand gate dynamics [88]. However, more detailed experiments need to be performed.

13.6 Intrinsically disordered proteins

The crystallographic protein structures in PDB are *ordered* proteins. An ordered protein has an essentially unique native fold that can be determined by X-ray crystallography. But most proteins are not ordered; most do not seem to have an essentially unique native fold. Instead, the low-energy landscape of most proteins seems to comprise several states that are energetically degenerate but conformationally disparate, with local energy minima that are separated from each other by very low free-energy barriers. We call them *intrinsically disordered* proteins. Normally, these proteins cannot be crystallized, and structural data are in short supply. Our aim is to describe the properties of such proteins by extending the methods that we have developed. For this, we start by describing some formalism.

But please keep in mind that there is a grey zone between ordered and intrinsically disordered proteins: a skillful crystallographer might be able to produce a crystal out of a protein that others consider hopeless.

13.6.1 Order versus disorder

When an ordered protein is cooled to low temperatures, it should assume an essentially unique native fold that corresponds to a minimum of the low-temperature thermodynamic (Helmholtz) free energy. More specifically, in the case of a protein with an ordered native fold, cooling should produce a highly localized statistical distribution of structurally closely related conformational substates. When taken together, this ensemble constitutes the folded native state at low temperatures. But if a protein is intrinsically disordered, then instead we expect the low-temperature limit to produce a *scattered* statistical distribution of structurally disparate but energetically comparable ensembles of conformational substates. Moreover, these different substates should be separated from each other only by relatively low energy barriers. The unstructured, disordered character of the protein is a consequence of a motion around this scattered landscape: the protein swings and sways back and forth, quite freely, over the low energy barriers that separate the various energetically degenerate but structurally disparate conformations.

We may think that the state space of a ordered protein with an essentially unique native fold, consists of a set of *snapshot* states $|s\rangle$ that form a *tightly* localized and essentially Gaussian distribution around an average state $|s\rangle_{\rm ave}$. When taken together, the set of these snapshots determine the low-temperature folded native state as an ensemble of conformational substates. In fact, we expect that for an ordered protein, the extend of conformational variations around the average state $|s\rangle_{\rm ave}$ can be estimated from the crystallographic Debye–Waller B-factor.

However, in the case of an intrinsically disordered protein the low-temperature set of snapshot states $|s\rangle$ exhibits a *disperse* statistical distribution. We have no single tightly localized peak, with a clearly identifiable average value, in the statistical distribution of snapshots. Instead, the statistical distribution of the snapshot states is *scattered*. The snapshots become apportioned between several structurally disperse but energetically degenerate conformational substates.

Of particular interest are those energetically degenerate substates that are separated from each other by relatively low energy barriers. The unstructured and unsettled character of an intrinsically disordered protein is a consequence of thermally induced fluctuations that move the configuration around the structurally diverse and energetically degenerate landscape: The protein swings and sways back and forth between the disparate snapshot states $|s\rangle$. Because of the very low energy barriers, this dynamics persists even at very low temperatures, where quantum-mechanical tunnelling transitions eventually take over the thermally induced ones. Thus, the unstructured character can persists even at *very* low temperatures.

We interpret the ensemble of snapshot states in terms of a Hartree state $|\Phi\rangle$, which is a linear combination of the form

$$|\Phi\rangle \simeq \sum_i p_i \, |\mathrm{s}_i\rangle \, . \tag{13.152}$$

Here, the index set i can have both discrete and continuum portions, including various small- and large-amplitude collective coordinates. We envision that the state space spanned by $|\mathrm{s}_i\rangle$ can be endowed with a norm (metric) that enables us to select and orthonormalize the set of *eigenconformations* (snapshot structures) $|\mathrm{s}_i\rangle$. The detailed construction of a norm in the space of string-like structures will not be addressed here.

When the $|\mathrm{s}_i\rangle$ are normalized, the coefficient p_i determines the probability weight for the ensuing eigenconformation $|\mathrm{s}_i\rangle$ to contribute in the Hartree state. In particular, when it describes an intrinsically disordered protein, the Hartree state (13.152) is a *mixed* state and not a pure state; the conformational entropy is non-vanishing:

$$S = -\sum_i p_i \log p_i > 0. \tag{13.153}$$

Note that if we have a single value $p_k \approx 1$, and the other p_i with $i \neq k$ are vanishingly small, $p_i \approx 0$, then the Hartree state $|\Phi\rangle$ reduces to a pure state and describes an ordered native fold.

The eigenconformations $|\mathrm{s}_i\rangle$ are time-independent, akin to states in the Heisenberg picture of quantum mechanics. But the p_i can be time-dependent quantities; they then describe the time evolution of $|\Phi\rangle$. For sufficiently long timescales, longer than the characteristic thermal tunnelling time between different eigenconformations $|\mathrm{s}_i\rangle$, the time dependence of the p_i is governed by a Liouville equation

$$\frac{d\hat{\rho}}{dt} = \frac{\partial \hat{\rho}}{\partial t} + \{\hat{\rho}, H\} \equiv \mathcal{L}\hat{\rho}, \tag{13.154}$$

where

$$\hat{\rho} = \sum_i p_i \, |\mathrm{s}_i\rangle \, \langle \mathrm{s}_i| \tag{13.155}$$

is the density matrix. The second term in (13.154) is the Poisson bracket with the (total) semiclassical Hamiltonian H; in a quantum-mechanical version, the density matrix and the Hamiltonian are replaced by the corresponding Heisenberg operators. We note that for a non-equilibrium system, the total operator $\mathcal{L}$ becomes the (semiclassical) Lindblad superoperator.

For a system at or very near thermodynamic equilibrium, the $|\mathrm{s}_i\rangle$ concur with the extrema configurations of the low-temperature limit of the Helmholtz free energy. The probabilities p_i are evaluated from the corresponding values of the free energy, and (13.152) takes the form

$$|\Phi\rangle \simeq \frac{1}{Z} \sum_{\{\mathrm{s}\}} e^{-\beta E(\mathrm{s})} \, |\mathrm{s}\rangle \, , \tag{13.156}$$

where

$$Z = \sum_{\{s\}} e^{-\beta E(s)} \tag{13.157}$$

and $\beta = 1/kT$ is the inverse temperature. For completeness, we note that for an extremum conformation $|s_i\rangle$ that is not a local minimum of the free energy, a Maslov-index contribution needs to be included.

13.6.2 hIAPP and type 2 diabetes

We now proceed to consider an example of an intrinsically disordered protein with very extensive and important biological, medical, and pharmaceutical ramifications.

The human islet amyloid polypeptide (hIAPP), also known as amylin, is a widely studied 37-amino-acid polypeptide hormone; for extensive reviews, we refer to [89, 90]. It is processed in pancreatic β-cells from an 89-residue precursor protein, by a protease cleavage in combination of post-translational modifications. hIAPP is secreted in response to meals, and the peptide cooperates with insulin to regulate blood glucose levels. But hIAPP can also form pancreatic amyloid deposits, and their formation and buildup correlates strongly with the depletion of islet β-cells. This hIAPP amyloidosis is present in over 90% of patients with type 2 diabetes, and the deposits are considered as the hallmark of the disease in progression.

It still remains to be fully clarified whether the hIAPP amyloid aggregation is the direct cause of apoptosis in the islet β-cells. Instead, the amyloid fibres might only be a consequence of the disease, which is ultimately caused by some other yet-to-be-identified agent. Among the arguments that support the existence of a first-hand causal relationship between hIAPP fibrillation and the onset of type 2 diabetes is the observation that wild-type mice do not develop the disease, while transgenic mice that express hIAPP can fall ill with the disease. It has also been observed that direct contact between the hIAPP amyloid fibrils and the surfaces of pancreatic islet β-cells has a toxic effect on the latter.

There is a real possibility that understanding the structural landscape of hIAPP, in particular how the amyloidosis transition takes place, could be a major step towards the identification of therapeutic targets and the development of strategies to combat a potentially deadly disease that presently plagues around 5% of the world's adult population. Indeed, type 2 diabetes is arguably among the most devastating diseases to curse mankind. Its annual economic cost has been globally estimated to be in excess of 425 billion Euros, and the number of sufferers is estimated to almost double during the next 20 years.

The structure of aggregated hIAPP fibrils has been studied very extensively [89, 90]. The fibrils consist of an ordered parallel arrangement of monomers, with a zipper-like packing. Apparently, the fibril formation proceeds by nucleation, with one monomeric hIAPP molecule first assuming a hairpin-like structure. This is followed by a piling-up of several monomers, which eventually leads to the buildup of amyloid fibrils as the hallmark of the disease in progression. But the structure of full-length monomeric hIAPP, its potentially disease-causing dynamical conformational state in pancreatic cells, remains unknown, as does the reason for the occasional formation of the disease-causing hairpin-like structure. Moreover, despite the highly ordered nature of amyloid fibril aggregates, only very recently have experimental advances made it possible to obtain high-resolution models. However, we still largely lack the detailed atomic-level knowledge needed for drug development.

The monomeric form of hIAPP is presumed to be an example of an intrinsically disordered protein. When biologically active and healthy, it is in an unsettled and highly dynamic state. As such, it lacks an ordered three-dimensional folded state that could be studied by conventional X-ray crystallographic approaches. Several experimental methods based, for example, on solution and solid-state NMR and other techniques are currently under development to try and characterize the conformation of monomeric hIAPP. But the existing techniques do not yet permit a direct examination of the atomic level structure. Detergents such as sodium dodecyl sulfate (SDS) micelles are commonly introduced as stabilizing agents.

The detailed atomic-level information could in principle be extracted by theoretical means using MD simulations. However, with explicit water, presently available computer power can at best cover a dynamical *in vitro/in vivo* trajectory up to around a microsecond per a day *in silico*. But amyloid aggregation takes hours, even days. Thus, the present all-atom computational investigations are largely dependent on our ability to determine an initial conformation for the simulations. Otherwise, one might end up simulating only the initial condition.

Because of its intrinsically disordered character, the structural data of hIAPP in isolation remain sparse in the PDB. The only presently available PDB data on hIAPP consist of two NMR structures and one crystallographic structure. Both NMR structures describe hIAPP in a complex with SDS micelles; the PDB access codes are 2L86 and 2KB8. The sole available crystallographic structure describes hIAPP that has been fused with a maltose-binding protein; the PDB access code is 3G7V. We note that these three structures are all very different from each other.

The NMR structure 2L86 has been measured at pH 7.4, i.e. around the pH value of the extracellular domain where the hIAPP amyloid deposits appear. On the other hand, the NMR structure 2KB8 has been measured at pH 4.6, which is closer to the pH of around 5.5 inside the β-cell granules of the pancreas. We should point out that even though hIAPP amyloidosis is apparently an extracellular process, some evidence suggests that the aggregation might have an intracellular origin. Thus, a thorough investigation of the role of hIAPP in the onset of type 2 diabetes, should account for *both* the extracellular and the intracellular structural properties of the peptide. In addition, a detailed analysis how hIAPP interacts with cell membranes is needed; we note that, to some extent, micelles might mimic membrane effects.

We conclude by pointing out that hIAPP also affects several other organs besides the pancreas. For example, it is known to have binding sites in the brain, where it apparently has a regulatory effect on gastric emptying. Delayed gastric emptying is commonly diagnosed in patients with diabetes. But gastroparesis is also a component in a number of other disorders. Certainly, the ability of hIAPP to cross the blood–brain barrier and affect the central nervous system is related to its structure. Amyloid fibres can hardly cross the barrier. Thus, besides apparently contributing directly to type 2 diabetes, aggregation should also have a wider influence on the regulatory activity of hIAPP.

13.6.3 hIAPP as a three-soliton

We shall investigate in detail the physical properties of a 28-segment monomer of hIAPP; we propose that the reader gets access to the entry 2L86 in PDB. The segment we are interested in consists of residues 9–36 where several studies have either observed or predicted that the amyloid fibril formation starts [89, 90]. The physical properties of the short N-terminal segment that comprises residues 1–8 are not addressed here. The structure of this segment is more involved, owing to the disulfide bond that connects the cysteines located at residues 2 and 7. Moreover, it remains to be understood what is the role, if any, of residues 1–8 in hIAPP aggregation. These residues appear to have a tendency towards forming long and stable non-β-sheet fibres in solution, under the same conditions in which hIAPP aggregates into amyloid fibres.

We use the NMR structure 2L86 from the PDB as a decoy to train the energy function. We construct a multisoliton configuration as an extremum of the energy function (13.103) that accurately describes 2L86. Since 2L86 is a composite of hIAPP with SDS micelles, we propose the following biological setup. We consider the structural evolution of an isolated hIAPP in the extracellular domain, in a scenario where the polypeptide is initially in a direct but residual interaction with the cell membrane. The effect of an initial cell membrane interaction is modelled by the effect of SDS micelles in 2L86. Following the construction of the multisoliton configuration, we study the presumed disordered structural landscape of an isolated hIAPP; we try and model hIAPP as it enters the extracellular domain. For this, we subject the multisoliton to a series of heating and cooling simulations as in the case of myoglobin, using Glauber dynamics. During heating, we increase the temperature until we detect a structural change in the multisoliton, so that the configuration behaves like a random walker. We fully thermalize the configuration at the random-walk phase. We then reduce the ambient temperature, to cool the configuration to very low temperature values until it freezes into a conformation where no thermal motion prevails. Since an isolated hIAPP is intrinsically disordered, instead of a single native fold as in the case of myoglobin, we expect the low-temperature limit to produce a scattered statistical distribution of structurally disparate but energetically comparable ensembles of conformational substates (13.152). Moreover, the individual different substates should be separated from each other by relatively low energy barriers. The unstructured, disordered character of hIAPP is then a consequence of a motion around this landscape: it swings and sways back and forth, quite freely, over the low energy barriers that separate the various energetically degenerate but structurally disparate conformations. We shall find that in the case of hIAPP, the heating and cooling procedure, which in the case of myoglobin yields a single low-energy state, now produces exactly this kind of structurally scattered ensemble of conformations.

Table 13.4 shows the parameter values that we find by training the energy function (13.103) to describe 2L86. These parameters values are taken from [91].

Table 13.4 Parameter values for the three-soliton configuration that describes 2L86; soliton 1 covers the PDB segment THR 9–ASN 21, soliton 2 covers the segment ASN 22–ALA 25, and soliton 3 covers the segment ILE 26–THR 36. The value of a is fixed at $a = -10^{-7}$.

SOLITON	q_1	q_2	m_1	m_2	d/a	c/a	b/a
1	9.454	4.453	1.521	1.606	-8.164×10^{-2}	-1.402×10^{-3}	-2.568
2	2.927	2.441	1.667	1.534	-4.894×10^{-1}	-1.067×10^{-3}	-19.48
3	1.119	8.086	1.522	1.514	-3.578×10^{-2}	-5.907×10^{-3}	-1.908

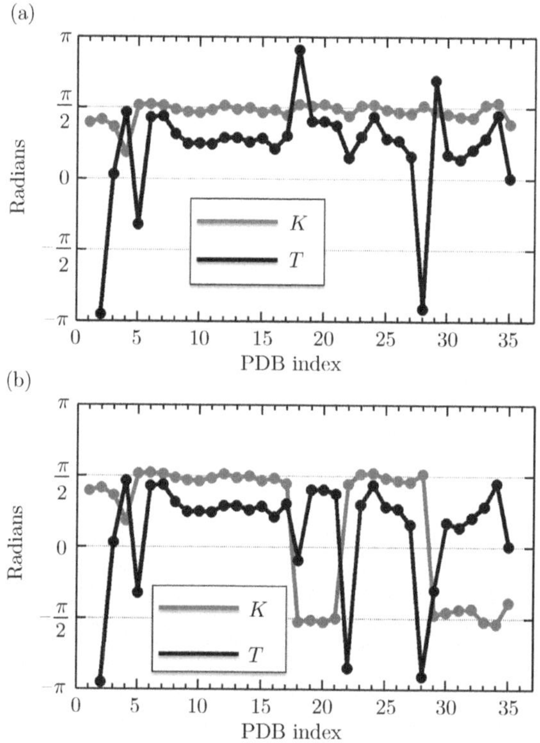

Fig. 13.35 [Colour online] (a) Spectrum of the bond and torsion angles of 2L86 (first entry) with the convention that the bond angle takes values $\kappa \in [0, \pi)$. (b) Spectrum of the bond and torsion angles that identify the soliton structures.

In Fig. 13.35(a), we show the spectrum of bond and torsion angles for the first NMR structure of 2L86, with the convention that the bond angle takes values of $\kappa \in [0, \pi)$. In Fig. 13.35(b), we have introduced the $\mathbb{Z}_2$ symmetry (13.92) to disclose three individual solitons along the backbone. The first soliton from the N-terminus is centred at site 17. The third soliton is centred at site 27. Both of these solitons correspond to clearly visible loops in the three-dimensional structure in PDB entry 2L86. The second soliton, centred at site 23, is much less obvious in the three-dimensional NMR structure. This soliton appears more like a bend in an α-helical structure, extending from the first soliton to the third. The $\mathbb{Z}_2$-transformed (κ, τ) profile shown in Fig. 13.35(b) is the background that we have used in training the energy function (13.103).

In Fig. 13.36, we compare the bond and torsion angle spectrum of our three-soliton solution with the first NMR structure of 2L86; the solution is obtained using the program *ProPro* from http://www.folding-protein.org.

The quality of our three-soliton solution is clearly very good, at the level of the bond and torsion angles.

Figure 13.37 shows the three-soliton solution, interlaced with the first NMR structure of 2L86.

The RMSD between the experimentally determined structure and the three-soliton configuration is 1.17 Å. This is somewhat large when compared with the multisoliton structures that we have found previously. But the resolution of the present

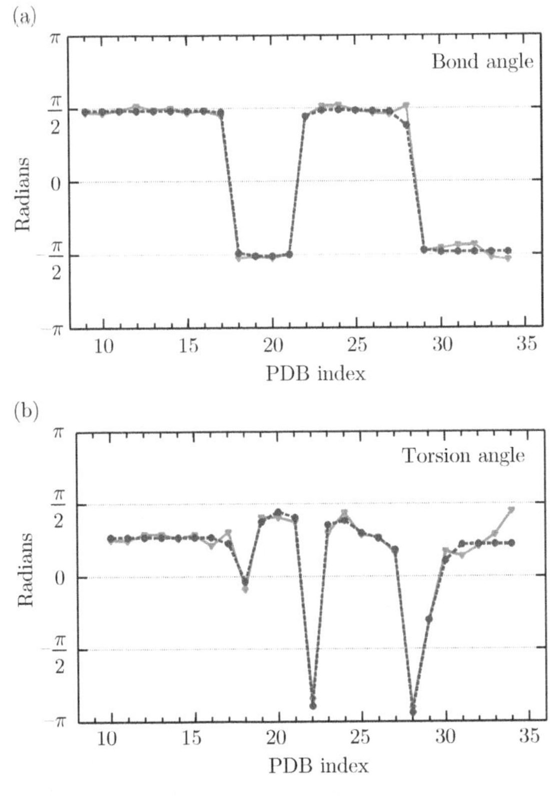

Fig. 13.36 [Colour online] Comparison of the three-soliton bond angle (circles [blue]) with the experimental 2L86 bond angle spectrum (triangles [red]). (b) Comparison of the three-soliton torsion angle (circles [blue]) with the experimental 2L86 torsion angle spectrum (triangles [red]).

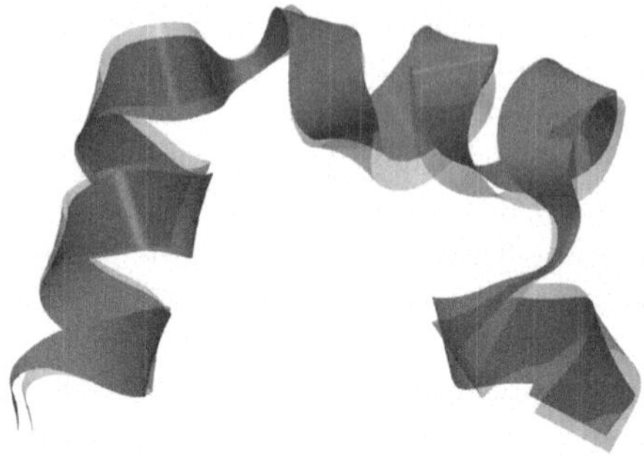

Fig. 13.37 [Cour online] The three-soliton solution (light [blue]) interlaced with the 2L86 experimental structure (dark [red]).

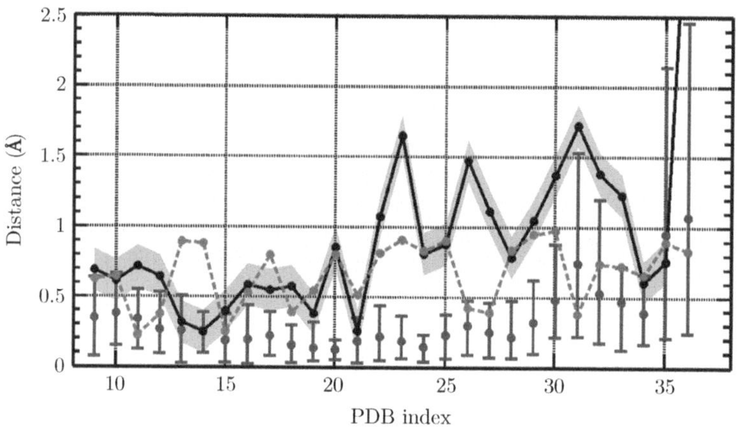

Fig. 13.38 [Colour online] The full black line denotes the C_α atom distance between the three-soliton configuration and the model 1 NMR configuration 2L86; the grey region is an estimated 0.15 Å zero-point fluctuation distance from the three-soliton configuration. The dashed [red] line denotes the B-factor Debye–Waller fluctuation distance from model 1 of 2L86. The [blue] points with error bars denote the average C_α distance between the model 1 NMR structure and the average of the remaining 19 models on 2L86; the error-bars indicate the maximum and minimum C_α distances.

experimental NMR structure is not that good, and this is reflected by the somewhat lower quality of the three-soliton solution in comparison with the case of high-resolution crystallographic structures.

Figure 13.38 compares the residue-wise C_α distances between the 20 different NMR structures in PDB entry 2L86 and our three-soliton solution. For those residues that precede the bend-like second soliton centred at site 23, the distance between the experimental structures and the numerically constructed solution is relatively small. We observe a quantitative change in the precision of the three-soliton solution, which takes place after site 23. The distance between the experimental structures and the three-soliton solution clearly increases after this residue. It could be that this change is due to the SDS micelles used in the experimental setup to stabilise hIAPP/2L86: SDS is widely used as a detergent to enable NMR structure determination in the case of proteins with high hydrophobicity. The mechanism of SDS–protein interaction is apparently not yet fully understood. But it is known that the hydrophobic tails of SDS molecules interact in particular with the hydrophobic core of a protein. These interactions are known to disrupt the native structure with the effect that the protein displays an increase in its α-helical posture; these additional α-helical structures tend to be surrounded by SDS micelles.

The residue at site 23 of hIAPP is the highly hydrophobic phenylalanine. It is followed by the very flexible glycine at site 24. Thus, the apparently abnormal bend located at site 23 and affecting the quality of our three-soliton configuration could be due to an interaction between the phenylalanine and the surrounding SDS micelles. A high sensitivity of the hIAPP conformation to the phenylalanine at site 23 is well documented.

An analysis of 2L86 structure using *Molprobity* (http://molprobity.biochem.duke.edu/) suggests a propensity towards poor rotamers between sites 23 and 36, i.e. the region where the quality of our three-soliton solution decreases.

A comparison with the statistically determined radius-of-gyration relation (13.7) reveals that for 2L86 the value of $R_g \approx 9.2$ (over residues $N = 9, \ldots, 36$). This is somewhat high. According to (13.7), we expect a value close to $R_g \approx 7.9$ when we set $N = 28$. The structure of 2L86 should be more compact.

We conclude that, most likely, the SDS–hIAPP interaction has deformed a loop that, in the absence of micelles, would be located in the vicinity of residue 23. Probably, interaction with the micelles has converted this loop into a structure resembling a bend in an α-helix. This interaction between hIAPP and SDS interferes with our construction of the three-soliton configuration, adversely affecting its precision.

13.6.4 Heating and cooling hIAPP

Following our myoglobin analysis, we proceed to investigate the properties of the three-soliton model of 2L86 under repeated heating and cooling, using the Glauber algorithm.

Figure 13.39 illustrates the evolution of the three-soliton configuration during repeated heating and cooling: (a) shows the evolution of the radius of gyration, and (b) the evolution of the RMSD from the PDB structure 2L86. Both the average value and the standard deviation from the average are shown. During the cooling period, we observe only one transition, in both the radius of gyration and the RMSD. Thus, based on our previous experience, we are confident that at high temperatures we are in the random-walk regime. The profile of each curve in Fig. 13.39 also shows that the structures are fully thermalized, in both in the high- and low-temperature regimes.

We observe that the average final value of the radius of gyration $R_g \approx 7.8$ is an excellent match with the prediction obtained from (13.7). In particular, the final configurations are quite different from the initial one: The RMSD between the initial configuration and the average final configuration is around 4.8 Å.

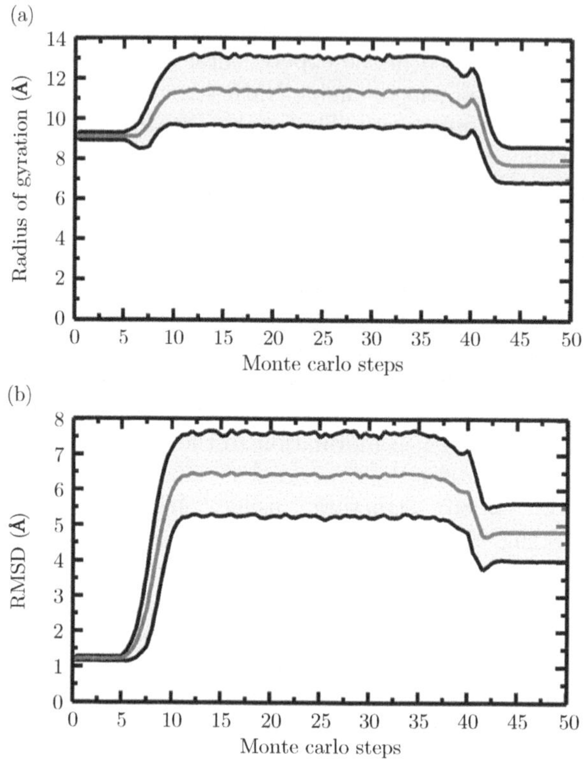

Fig. 13.39 [Colour online] Evolution of the radius of gyration of the three-soliton configuration during the heating and cooling cycle. (b) Evolution of the RMSD from the initial configuration. The grey [red] line is the average value over all configurations, and the light-grey zone marks the extent of one standard deviation from the average value. The Monte Carlo steps are displayed in multiples of 10^6.

Figure 13.40 shows results for a representative simulation with 1500 complete heating and cooling cycles; an increase in the number of cycles does not have a qualitative effect on the result. The figure shows the distribution of the final snapshot conformations, grouped according to their radius of gyration versus end-to-end distance. The final conformations form clusters, and we identify the six major clusters that we observe in our simulations. By construction, the clusters correspond to local extrema of the energy function tht we have constructed to model 2L86: the average conformation of each cluster can be identified with a particular snapshot state $|s\rangle$ in the expansion (13.152), (13.156). Five of the clusters, denoted 2–6 in Fig. 13.40, have an apparent spread. This implies that the energy has a flat conformational direction around its extremum. Clusters 3 and 5 are also somewhat more scattered than clusters 2, 4, and 6. Finally, cluster 1 is a localized one: this cluster corresponds to a sharply localized snapshot state $|s\rangle$. Note that the initial conformation, marked with a [red online] triangle in Fig. 13.40, does not appear among the final configurations. It is apparently an unstable extremum of the energy, stabilized by the micelles.

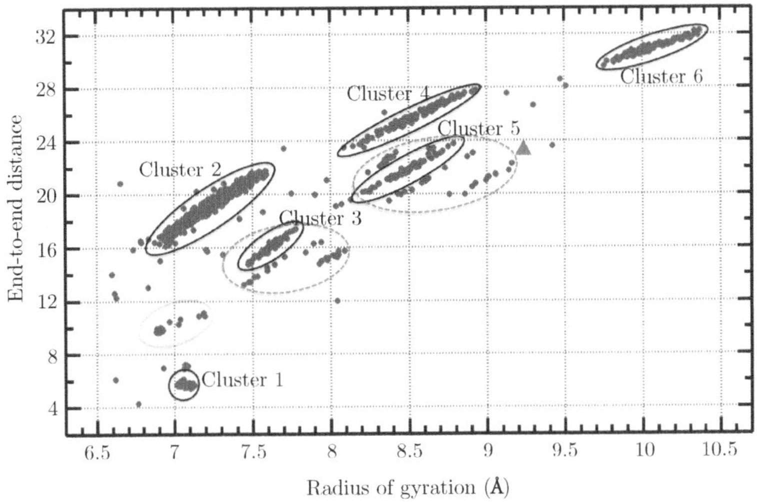

Fig. 13.40 [Colour online] Distribution of all final configurations in a run with 1500 full heating and cooling cycles, classified in terms of the radius of gyration and end-to-end distance of the final configuration. Each [blue] dot represents a single final configuration. The six major clusters are each encircled by a black ellipse; the wider grey ellipses around clusters 3 and 5 include some nearby scattered states. The [red] triangle identifies the initial configuration, entry 1 in 2L86. Note also the presence of a cluster encircled [in yellow] between clusters 1 and 2.

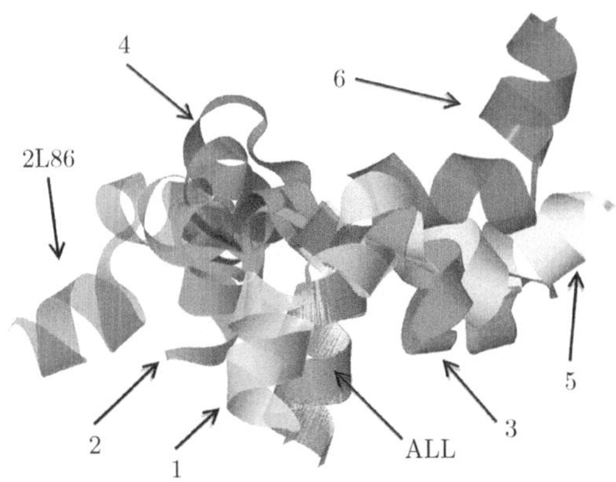

Fig. 13.41 Superposition of all six major clusters from Fig. 13.40, interlaced with each other and with PDB entry 2L86.

In Fig. 13.41, we display the average conformations in each of the six clusters, interlaced with each other and the initial 2L86 configuration. In this figure, the first two C_α atoms from the N-terminus are made to coincide. We have maximized the alignment of the subsequent C_α atoms, to the extent that this is possible. The figure reveals the presence of substantial conformational differences among the clusters. The totality

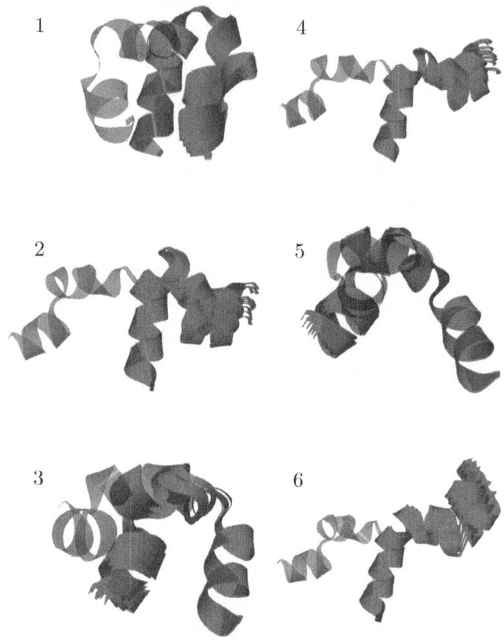

Fig. 13.42 [Colour online] Superposition of ten representative conformations (dark grey [red]) in each of the six clusters, as marked, together with PDB entry 2L86 (light grey [blue]).

of the conformations shown in Fig. 13.41 can be given an interpretation in terms of the dynamical hIAPP. It is a long-time-period average picture of the Hartree state (13.152), (13.156) that is a linear combination of the various snapshot conformations $|s_i\rangle$ $(i = 1, \ldots, 6)$.

In Fig. 13.42, we compare the individual clusters with the initial 2L86 configuration (light grey [blue online]). In each of these comparisons, we show ten representative entries in each of the clusters (dark grey [red online]), to visualize the extent of conformational fluctuations within each cluster. We observe that the conformational spread within each of the six clusters is not very large.

In conclusion, we have found that the three-soliton configuration that models the C_α backbone of the human islet amyloid polypeptide is quite unsettled: its low-temperature limit comes endowed with six different conformational clusters. This is a marked contrast with the properties of a multisoliton configuration that models a protein that is known to possess a unique folded native state, such as myoglobin. The low-temperature clustering of hIAPP is in full accord with the intrinsically disordered character of the protein: the different clusters can be viewed as instantaneous snapshot conformations, between which the dynamic hIAPP swings and sways in an apparently unsettled manner that is characteristic of an intrinsically disordered protein.

Only cluster 1 appears different. This cluster has a much more localized conformational distribution than the other five clusters, and the posture comprises two antiparallel helices. This suggests to us that cluster 1 is a good candidate to trigger the formation of hIAPP fibrils and amyloidosis that correlates with type 2 diabetes.

My advice to all my students is—be careful with your lifestyle to keep your hIAPP folds under good control.

13.7 Beyond C_α

Thus far, we have analysed protein structure and dynamics in terms of the C_α atoms only. We have argued that the virtual C_α backbone bond and torsion angles form a complete set of local order parameters to describe the protein backbone conformation.

The C_α atoms play a central role in X-ray crystallography, where the experimental determination of protein structure often starts with a skeletonization of the electron density map. From Figs. 13.1 and 13.7 we observe that the C_α atoms are located centrally, forming the vertices that connect the peptide planes and coinciding with the branch points between the backbone and the side chain. Thus, the C_α atoms are subject to stringent stereochemical constraints. Accordingly, the first step in experimental model building is the initial identification of the skeletal C_α trace.

The central role of the C_α atoms is widely exploited in structural classification schemes like CATH (http://www.cathdb.info/) and SCOP (http://scop.mrc-lmb.cam.ac.uk/scop/), in various homology modelling techniques [27–29], in *de novo* approaches [3], and in the development of coarse-grained energy functions for folding prediction [44, 45]. The so-called C_α-trace problem has been formalized and has been the subject of extensive investigations [92–94]. The aim is to construct an accurate main chain and/or all-atom model of the folded protein from knowledge of the positions of the central C_α atoms only. Both knowledge-based approaches such as *MaxSprout* (http://www.ebi.ac.uk/Tools/structure/maxsprout/) and *de novo* methods like *PULCHRA* (http://cssb.biology.gatech.edu/PULCHRA) have been developed for this purpose. In the case of the backbone atoms, various geometric algorithms can be utilized. For the side-chain atoms, most approaches rely either on a statistical approach or on a conformer rotamer library in combination with steric constraints, complemented by an analysis based on diverse scoring functions. For the final fine-tuning of the model, all-atom molecular dynamics simulations can be utilized.

The Ramachandran map shown in Fig. 13.10 is used widely both in various analyses of the protein structures and as a tool in protein visualization. It describes the statistical distribution of the two dihedral angles ϕ and ψ that are adjacent to the C_α carbons along the protein backbone. In the case of side-chain atoms, visual analysis methods similar to the Ramachandran map have been introduced. For example, there is the Janin map that can be used to compare observed side-chain dihedrals in a given protein against their statistical distribution in a manner that is analogous to the Ramachandran map. Crystallographic refinement and validation programs like *PHENIX* (http://www.phenix-online.org/), *REFMAC* (http://www2.mrc-lmb.cam.ac.uk/groups/murshudov/), and many others, utilize the statistical data obtained from libraries such as the Engh and Huber library [12] that are built using small molecular structures that have been determined with a very high resolution. At the level of entire proteins, side-chain restraints are commonly derived from analysis of high-resolution PDB crystallographic structures [95]. Backbone-independent rotamer libraries that make no reference to backbone conformation and both secondary-structure-dependent and backbone-dependent rotamer libraries have been developed. According to [95] the information content in the secondary-structure-dependent libraries and the backbone-independent libraries essentially coincide, and both are often used during crystallographic protein structure model building and refinement. But for the prediction of side-chain conformations, for example in the case of homology modelling and protein design, it is often an advantage to use the more revealing backbone-dependent rotamer libraries.

13.7.1 'What-you-see-is-what-you-have'

We shall present a short introductory outline of how the C_α Frenet frames can be utilized to develop a new generation of visualization techniques for protein structure analysis, refinement, and validation. Our outline is based on [96–98].

Despite the availability of various three-dimensional visualization tools such as the Java-based viewer on the PDB website, thus far the visualization of proteins has not yet taken *full* advantage of modern visualization techniques. The commonly available three-dimensional viewers present the protein in the 'laboratory' frame, and as such they provide mainly an *external* geometry-based characterization of protein structure. On the other hand, the method that we describe is based on internal, *comoving* framing of the protein backbone—watching a roller-coaster is not the same as taking the ride. As such, our approach provides complementary visual information. Starting from the positions of the C_α atoms, we aim for a three-dimensional *what-you-see-is-what-you-have* type of visual map of the all-atom structure. Indeed, the visualization of a three-dimensional discrete framed curve is an important and widely studied topic in computer graphics, from the association of ribbons and tubes to the determination of camera gaze directions along trajectories.

In lieu of the backbone dihedral angles that appear as coordinates in the Ramachandran map and correspond to a toroidal topology, we use the geometry of virtual 2-spheres that surround each heavy atom. For this, we employ the geometric interpretation of the virtual C_α backbone bond and torsion angles in terms of latitude and longitude on the surface of a sphere $\mathbb{S}^2$. We shall outline how the approach works in the case of the backbone N and C atoms, and the side-chain C_β atoms. The approach can easily be extended to visually describe all the higher-level side-chain atoms on the surface of the sphere, level-by-level along the backbone and side chains [96–98]. The outcome is a three-dimensional visual map that describes the backbone and side-chain atoms exactly in the manner they are seen by an imaginary, geometrically determined and C_α-based miniature observer who roller-coasts along the backbone: at each C_α atom, the observer orients herself consistently according to the purely geometrically determined C_α-based discrete Frenet frames. Thus, the visualization of all the other atoms depends only on the C_α geometry—there is no reference to the other atoms in the initialization of the construction. The other atoms—including subsequent C_α atoms along the backbone chain—are all mapped on the surface of a sphere that surrounds the observer, as if these atoms were stars in the sky; the construction proceeds along the ensuing side chain, until the positions of all the heavy atoms have been determined. This provides purely geometric and equitable direct visual information on the statistically expected all-atom structure in a given protein, based entirely on the C_α trace.

We start with the bond and torsion angles (13.83) and (13.84), and we choose each bond angle to take values $\kappa \in [0, \pi]$. We identify the bond angle with the latitude angle of a two-sphere centred at the C_α carbon. We orient the sphere so that the north pole where $\kappa = 0$ is in the direction of $\mathbf{t}$. The torsion angle $\tau \in [-\pi, \pi)$ is the longitudinal angle of the sphere. It is defined so that $\tau = 0$ on the great circle that passes both through the north pole and through the tip of the normal vector $\mathbf{n}$. The longitude angle increases in the counterclockwise direction around the vector $\mathbf{t}$. We also find it useful to introduce the stereographic projection of the sphere onto the plane. The standard stereographic projection from the south pole of the sphere to the plane with coordinates (x, y) is given by

$$x + iy \equiv re^{i\tau} = \tan\left(\frac{1}{2}\kappa\right) e^{i\tau}. \tag{13.158}$$

This maps the north pole where $\kappa = 0$ to the origin $(x, y) = (0, 0)$. The south pole where $\kappa = \pi$ is sent to infinity; see Fig. 13.43.

If need be, the visual effects of the projection can be enhanced by sending

$$\kappa \to f(\kappa), \tag{13.159}$$

where $f(\kappa)$ is a properly chosen function of the latitude angle κ.

13.7.1.1 The C_α map

We first explain how to visually describe the C_α trace in terms of the C_α Frenet frames (13.80)–(13.82). Consider the virtual miniature observer who roller-coasts the backbone by moving between the C_α atoms. At the location of each C_α, the observer has an orientation that is determined by the Frenet frames (13.80)–(13.82). The base of the ith tangent vector $\mathbf{t}_i$ is at the position $\mathbf{r}_i$. The tip of $\mathbf{t}_i$ is a point on the surface of the sphere (κ, τ) that surrounds the observer and points towards the north pole. The vectors $\mathbf{n}_i$ and $\mathbf{b}_i$ determine the orientation of the sphere. These vectors define

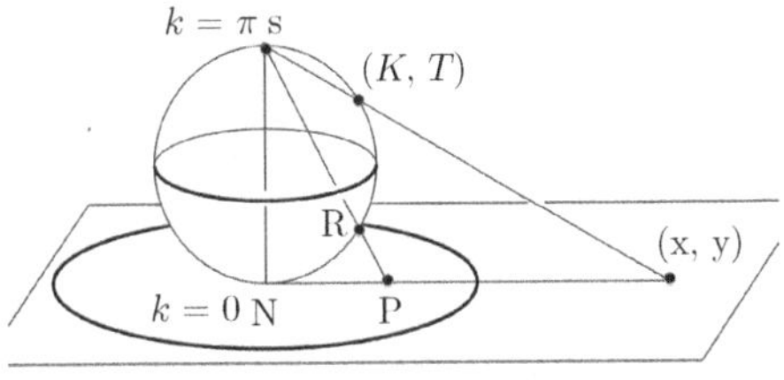

Fig. 13.43 Stereographic projection of the 2-sphere $\mathbb{S}^2$ on the plane $\mathbb{R}^2$ from the south pole.

a frame on the normal plane to the backbone trajectory, as shown in Fig. 13.21. The observer maps the various atoms in the protein chain on the surface of the surrounding two-sphere, as if the atoms were stars in the sky.

The map of the C_α backbone is constructed as follows. The observer first translates the centre of the sphere from the location of the ith C_α all the way to the location of the $(i+1)$th C_α with *no rotation* of the sphere with respect to the ith Frenet frames. The observer then identifies the direction of $\mathbf{t}_{i+1}$, i.e. the direction towards the site $\mathbf{r}_{i+2}$ to which she proceeds from the next C_α carbon, as a point on the surface of the sphere. This determines the corresponding coordinates (κ_i, τ_i). After this, the observer redefines her orientation so that it matches the Frenet framing at the $(i+1)$th central carbon, and then proceeds by repeating the construction, in exactly the same manner. The ensuing map, over the entire backbone, gives an instruction to the observer at each point $\mathbf{r}_i$ how to turn at site $\mathbf{r}_{i+1}$ to reach the $(i+2)$th C_α carbon at the point $\mathbf{r}_{i+2}$.

In Fig. 13.44, we show the C_α Frenet-frame backbone map. It describes the statistical distribution that we obtain when we plot all PDB structures that have been measured with better than 2.0 Å resolution and using the stereographic projection (13.158). For our observer, who always fixes her gaze position towards the north pole of the surrounding 2-sphere at each C_α, i.e. towards the small filled [red online] dot at the centre of the annulus, the intensity of shading [of colour online] in this map reveals the probability of the direction at position $\mathbf{r}_i$ where the observer will turn at the next C_α carbon when she moves from $\mathbf{r}_{i+1}$ to $\mathbf{r}_{i+2}$. In this way, the map is in a direct visual correspondence with the way in which the Frenet-frame observer perceives the backbone geometry. Note how the probability distribution is concentrated within an annulus, roughly between the latitude angle values $\kappa \sim 1$ and $\kappa \sim 3/2$. The exterior of the annulus is a sterically excluded region, while the entire interior is in

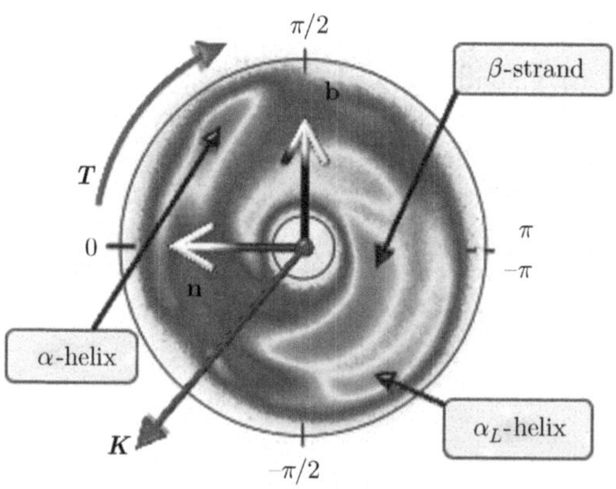

Fig. 13.44 [Colour online] The stereographically projected Frenet-frame map of backbone C_α atoms, with major secondary structures identified. Also shown is the direction of the Frenet-frame normal vector $\mathbf{n}$; the vector $\mathbf{t}$ corresponds to the small solid [red] circle at the centre and it points away from the viewer. The map is constructed using all PDB structures that have been measured with better than 2.0 Å resolution.

principle sterically allowed but very rarely occupied in the case of folded proteins. In the figure, we identify the four major secondary structure regions, according to the PDB classification; α-helices, β-strands, left-handed α-helices, and loops.

13.7.1.2 The backbone C and N and the side-chain C_β maps

Consider our imaginary miniature observer, located at the position of a C_α atom and oriented according to the discrete Frenet frames. She proceeds to observe and record the backbone heavy atoms N and C and the side-chain C_β atoms that are covalently bonded to C_α. These atoms form the covalently bonded heavy-atom corners of the C_α centred sp^3-hybridized tetrahedron. In Fig. 13.45, we show the ensuing density distributions on the surface of the C_α-centred sphere, the way in which they are seen by the miniature observer. These figures are constructed from all the PDB entries that have been measured using diffraction data with better than 1.0 Å resolution.

As visible in Fig. 13.45 in the C_α-centred Frenet frames, the C_β, C, and N atoms each oscillate in a manner that depends on the local secondary structure. Note that in the case of both C_β and N, the left-handed α region (L-α) is distinctly detached from the rest. But in the case of C, the L-α region is connected with the other regions. On the other hand, also in the case of C, the *cis*-prolines form a clearly detached and localized region, which is not similarly visible in the case of C and C_β.

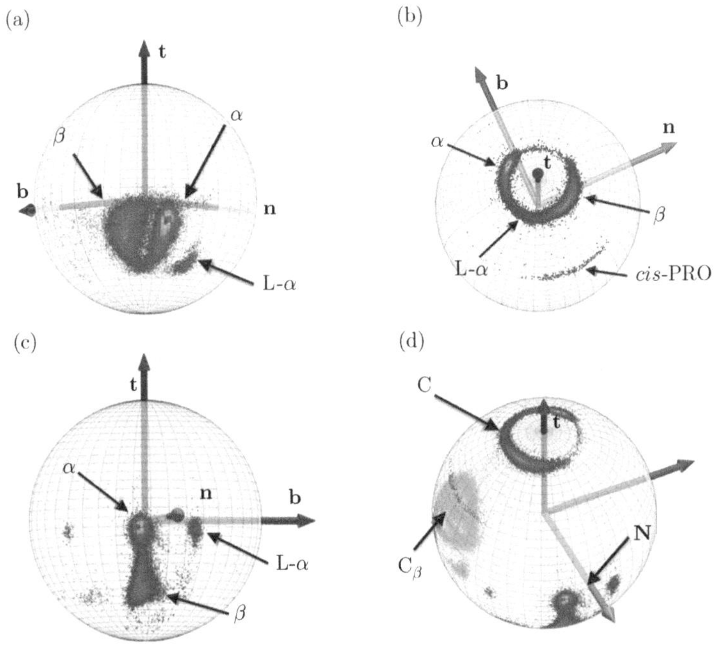

Fig. 13.45 [Colour online] Distributions of (a) C_β atoms, (b) backbone C atoms, and (c) backbone N atoms in the C_α-centred Frenet frames in PDB structures measured with better than 1.0 Å resolution. The three major domains, namely α-helices, β-strands, and left-handed α-helices, are indicated. (d) A combination of the distributions in (a)–(c). Note that *cis*-prolines are also clearly identifiable in the C-atom distribution (b).

We now consider the three bond angles

$$\begin{aligned} \vartheta_{\text{NC}} &\simeq \text{N–C}_\alpha\text{–C}, \\ \vartheta_{\text{N}\beta} &\simeq \text{N–C}_\alpha\text{–C}_\beta \\ \vartheta_{\beta\text{C}} &\simeq \text{C}_\beta\text{–C}_\alpha\text{–C}. \end{aligned} \tag{13.160}$$

The angle ϑ_{NC} relates to the backbone only, while the definitions of the other two involve the side-chain C_β. In experimental protein structure validation, these three angles are often presumed to have their ideal values. For example, the deviation of the C_β atom from its ideal position is among the validation criteria used in *MolProbity* (http://molprobity.biochem.duke.edu/), which uses it in identifying potential backbone distortions around C_α.

In Fig. 13.46, we show the distribution of the three tetrahedral bond angles in our 1.0 Å PDB data set. We find that in the case of the two side-chain-related angles $\vartheta_{\text{N}\beta}$ and $\vartheta_{\beta\text{C}}$, the distribution displays a single peak that is compatible with the ideal values (the isolated small peak in Fig. 13.46(b) is due to *cis*-prolines). But in the case of the backbone-specific angle ϑ_{NC}, we find that this is not the case. The PDB data show a clear correlation between the ϑ_{NC} distribution and the backbone secondary structure. We recall that ϑ_{NC} pertains to the relative orientation of the two peptide planes that are connected by the C_α.

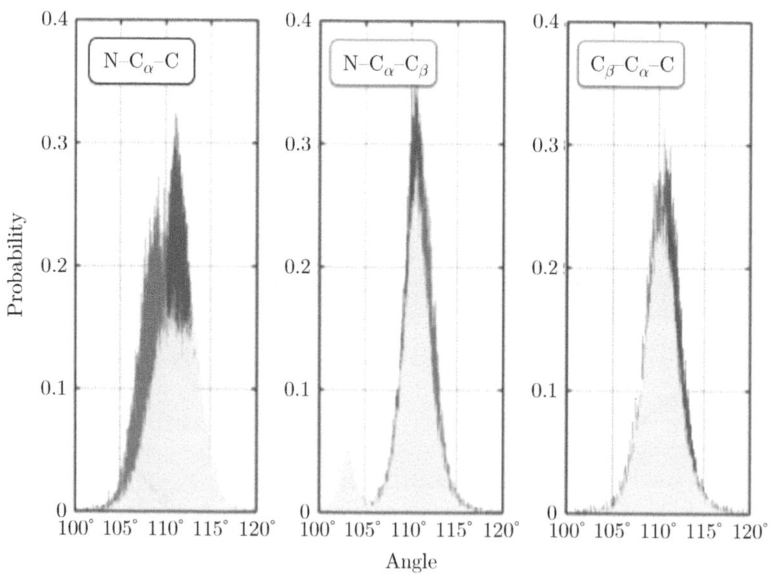

Fig. 13.46 [Colour online] Distribution of the three angles (13.160) according to secondary structures. Blue is for α-helices, red is for β-strands, and yellow is for loops; the small (yellow) peak in N–C_α–C_β at an angle around 103° is due to prolines.

The two Ramachandran angles (ϕ, ψ) in Fig. 13.10 are directly adjacent to the given C_α, and each is specific to a *single* peptide plane; see Fig. 13.7. But this is not the case for ϑ_{NC}, which connects two peptide planes. This angle contributes to the bending of the backbone. Consequently, a systematic secondary structure dependence *should* be present in this angle. Owing to the very rigid structure of the C_α-centred sp^3-hybridized covalent tetrahedron, one then expects that the secondary structure dependence should also become visible in the side-chain-specific angles $\vartheta_{N\beta}$ and $\vartheta_{\beta C}$. The lack of any observable secondary-structure dependence in the experimental data on these two angles suggests that existing validation methods distribute the refinement tension entirely to ϑ_{NC}.

Research project 13.10 Can you explain in detail why secondary-structure dependence is absent from PDB data on $\vartheta_{N\beta}$ and $\vartheta_{\beta C}$?

The construction that we have presented can be continued to visualize all the higher-level side-chain atoms in a protein, beyond C_β [96–98]. It turns out that these higher-level atoms also display a highly organized, systematic pattern akin to those shown in Fig. 13.45. In particular, the underlying geometry of the C_α backbone is clearly visible in these side-chain atoms; there is a very strong coupling between the C_α backbone geometry and the side-chain geometry. Accordingly, it becomes possible, in principle, to determine with high accuracy the all-atom structure of the protein from knowledge of the C_α atom positions only. This enables us to extend our C_α-based approach that builds on (13.103) to construct the full all-atom structure of the entire protein: the C_α positions can be computed from the energy function (13.103), and the remaining atoms are located by s statistical analysis.

According to Fig. 13.45(a), in the intrinsic, purely geometric Frenet-frame coordinate system, the directions of the C_β atoms are subject to only small fluctuations. At the level of C_β, the side chains all point in essentially the same direction. Thus, the crystallographic protein structures are like a spin chain where the side chains are the spin variables, and the collapsed phase bears a resemblance to a ferromagnetic phase.

Research project 13.11 See how far you can get using the spin-chain analogy of folded proteins, with side chains as the spin variables.

That's all, folks

Acknowledgements

I wish to thank my many collaborators, postdocs, and students from whom I have learned so much on the subject of these lectures. They include Si Chen, Alireza Chenani, Maxim Chernodub, Jin Dai, Ulf Danielsson, Jan Davidsson, Thomas Garandel, Ivan Gordeliy, Jianfeng He, Yanzhen Hou, Konrad Hinsen, Shuangwei Hu, Theodora Ioannidou, Ying Jiang, Gerald Kneller, Andrei Krokhotine, Nevena Litova, Jiaojia Liu, Adam Liwo, Martin Lundgren, Gia Maisuradze, Nora Molkenthin, Alexander Molochov, Alexander Nasedkin, Daniel Neiss, Xuan Nguyen, Stam Nicolis, Xubiao Peng, Fan Sha, Harold Scheraga, Adam Sieradzan, Ann Sinelnikova, Maxim Ulybushev, Yifan Zhou, and many, many others (to whom I apologize if they are not mentioned above). I also thank Frank Wilczek for making me curious about Anderson's *больше*.

My research has been supported by a CNRS PEPS grant, a Region Centre Recherche d'Initiative Academique grant, the Sino-French Cai Yuanpei Exchange Program (Partenariat Hubert Curien), Vetenskapsrådet, Carl Trygger's Stiftelse för vetenskaplig forskning, and a Qian Ren Grant. I thank the organizers of the Les Houches Summer School on Topological Aspects of Condensed Matter Physics, Claudio Chamon, Mark Goerbig, and Roderich Moessner, for giving me the opportunity to present these lectures. I thank the International Institute of Physics in Natal, Brazil for hospitality during the writing of these notes. Last, but not least, my thanks go to the great audience at Les Houches!

References

[1] E. Schrödinger. *'What is Life?' The Physical Aspect of the Living Cell.* Cambridge University Press, Cambridge (1948)

[2] P. W. Anderson. More is different. Broken symmetry and the nature of the hierarchical structure of science. *Science* **177**, 393–396 (1972)

[3] K. Dill, S. B. Ozkan, T. R. Weikl, J. D. Chodera, V. A. Voelz. The protein folding problem: when will it be solved? *Curr. Opin. Struct. Biol.* **17**, 342 (2007).

[4] K. A. Dill and J. K. MacCallum. The protein-folding problem, 50 years on. *Science* **338**, 1042–1046 (2012).

[5] F. Chiti and C. Dobson. Protein misfolding, functional amyloid, and human disease. *Annu. Rev. Biochem.* **75**, 333 –366 (2006).

[6] J. Davies. Bacteria on the rampage. *Nature* **383**, 219 (1996).

[7] C. T. Walsh. Molecular mechanisms that confer antibacterial drug resistance. *Nature* **406**, 775 (2000).

[8] M. A. Fischbach and C.T. Walsh. Antibiotics for emerging pathogens. *Science* **325**, 1089 –1093 (2009).

[9] B. Alberts, A. Johnson, J. Lewis, D. Morgan, M. Raff, K. Roberts, and P. Walter. *Molecular Biology of the Cell*, 6th edn. Garland Science, New York (2014).

[10] L. Frappat, P. Sorba, and A. Sciarrino. A crystal base for the genetic code. *Phys. Lett. A* **250**, 214 –221 (1998).

[11] R. J. Read et al. A new generation of crystallographic validation tools for the Protein Data Bank. *Structure* **19**, 1395 –1412 (2011).

[12] R. A. Engh and R. Huber. Structure quality and target parameters. In *International Tables for Crystallography*, Vol. F (ed. M. G. Rossmann and E. Arnold), pp. 382–392. Kluwer, Dordrecht (2001).

[13] R. Joosten et al. PDB_REDO: automated re-refinement of X-ray structure models in the PDB. *J. Appl. Crystallogr.* **42**, 376–384 (2009)

[14] K. Sanderson. New protein structures replace the old. *Nature* **459**, 1038–1039 (2009).

[15] Z. Dauter, A. Wlodawer, W. Minor, M. Jaskolski, and B. Rupp. Avoidable errors in deposited macromolecular structures: an impediment to efficient data mining. *IUCrJ* **1**, 179–193 (2014).

[16] P. G. De Gennes. *Scaling Concepts in Polymer Physics.* Cornell University Press, Ithaca, NY (1979).

[17] L. Schäfer. *Excluded Volume Effects in Polymer Solutions: As Explained by the Renormalization Group.* Springer-Verlag, Berlin (1999).

[18] L. P. Kadanoff. Scaling laws for Ising models near T_c. *Physics* **2**, 263 (1966).

[19] K. G. Wilson, Renormalization group and critical phenomena. I. Renormalization group and Kadanoff scaling picture. *Phys. Rev. B* **4**, 3174 (1971).

[20] K. G. Wilson and J. Kogut. The renormalization group and the ϵ expansion, *Phys. Rept.* **12**, 75–200 (1974).

[21] M. L. Huggins. Solutions of long chain compounds. *J. Chem. Phys.* **9**, 440 (1941).

[22] P.J. Flory, Thermodynamics of high polymer solutions. *J. Chem. Phys.* **10**, 51–61 (1942).

[23] B. Li, N. Madras, and A. Sokal. Critical exponents, hyperscaling, and universal amplitude ratios for two- and three-dimensional self-avoiding walks. *J. Stat. Phys.* **80**, 661–754 (1995).

[24] P. G. De Gennes. Exponents for the excluded volume problem as derived by the Wilson method. *Phys. Lett. A* **38**, 339–340 (1972).

[25] J. C. LeGuillou and J. Zinn-Justin. Critical exponents from field theory. *Phys. Rev. B* **21**, 3976–3998 (1980).

[26] K. Hinsen, S. Hu, G.R. Kneller, and A.J. Niemi. A comparison of reduced coordinate sets for describing protein structure. *J. Chem. Phys.* **139**, 124115 (2013).

[27] M. A. Marti-Renom, A. C. Stuart, A. Fiser, R. Sánchez, F. Melo, and A. Sali. Comparative protein structure modelling of genes and genomes. *Annu. Rev.Biophys. Biomol. Struct.* **29** 291–325 (2000).

[28] T. Schwede, J. Kopp, N. Guex, and M. C. Peitsch. SWISS-MODEL: an automated protein homology-modeling server *Nucleic Acids Res.* **31**, 3389 (2003).

[29] Y. Zhang. Protein structure prediction: when is it useful? *Curr. Opin. Struct. Biol.* **19**, 145–155 (2009).

[30] C. Chothia. Proteins. One thousand families for the molecular biologist. *Nature* **357** 543–544 (1992).

[31] E. Ott, *Chaos in Dynamical Systems*, 2nd edn. Cambridge University Press, Cambridge (2002).

[32] V. E. Korepin, N.M. Bogoliubov, A.G. Izergin, *Quantum Inverse Scattering Method and Correlation Functions.* Cambridge University Press, Cambridge (1997)

[33] D. E. Shaw et al. Anton, a special-purpose machine for molecular dynamics simulation. *Commun. ACM* **51.7**, 91–97 (2008).

[34] D. E. Shaw et al. Millisecond-scale molecular dynamics simulations on Anton. In *Proceedings of the Conference on High Performance Computing Networking, Storage and Analysis, Portland, OR, 14–20 November 2009.* IEEE, New York (2009).

[35] K. Lindorff-Larsen, S. Piana, R. O. Dror, and D. E. Shaw. How fast-folding proteins fold. *Science* **334**, 517–520 (2011).

[36] G. Gallavotti. *Nonequilibrium and Irreversibility.* Springer-Verlag, Cham (2014). (Also arXiv:1311.6448v4 [cond-mat.stat-mech].)

[37] S. Nosé. A unified formulation of the constant temperature molecular dynamics methods. *J. Chem. Phys.* **81**, 511–519 (1984).

[38] W. G. Hoover, Canonical dynamics: equilibrium phase-space distributions. *Phys. Rev. A* **31**, 1695–1697 (1985).

[39] G. J. Martyna, M. L. Klein, and M. Tuckerman. Nosé–Hoover chains: the canonical ensemble via continuous dynamics. *J. Chem. Phys.* **97**, 2635–2643 (1992).

[40] Y. Liu and M. E. Tuckerman. Generalized Gaussian moment thermostatting: a new continuous dynamical approach to the canonical ensemble *J. Chem. Phys.* **112** 1685–1700 (2000).

[41] A. J. Niemi. Gauge fields, strings, solitons, anomalies, and the speed of life. *Theor. Math. Phys.* **181**, 1235–1262 (2014).

[42] H. J. C. Berendsen, J. P. M. Postma, W. F. van Gunsteren, A. R. H. J. DiNola, and J. R. Haak. Molecular dynamics with coupling to an external bath. *J. Chem. Phys.* **81**, 3684–3690 (1984).

[43] A. Liwo, S. Oλdziej, M. R. Pincus, R. J. Wawak, S. Rackovsky, and H. A. Scheraga. A united-residue force field for off-lattice protein-structure simulations. I. Functional forms and parameters of long-range side chain interaction potentials from protein crystal data. *J. Comput. Chem.* **18**, 849–873 (1997).

[44] H. A. Scheraga, M. Khalili, and A. Liwo. Protein-folding dynamics: overview of molecular simulation techniques. *Annu. Rev. Phys. Chem.* **58**, 57–83 (2007).

[45] A. Liwo, A. J. Niemi, X. Peng, and A. K. Sieradzan. A novel coarse-grained description of protein structure and folding by UNRES force field and discrete nonlinear Schrödinger equation. In *Frontiers in Computational Chemistry,* Vol. 1 (ed. Z. Ul-Haq and J. D. Madura), pp. 257–289. Bentham Science Publishers, Sharjah, UAE (2015).

[46] N. N. Gō, Theoretical studies of protein folding. *Annu. Rev. Biophys. Bioeng.* **12**, 183–210 (1983).

[47] T. Haliloglu, I. Bahar, and B. Erman,. Gaussian dynamics of folded proteins. *Phys. Rev. Lett.* **79**, 3090–3093 (1997).

[48] C. K. Chiang, C. R. Fincher Jr, Y. W. Park, A. J. Heeger, H. Shirakawa, E. J. Louis, S. C. Gau, and A. G. MacDiarmid. Electrical conductivity in doped polyacetylene. *Phys. Rev. Lett.* **39**, 1098–2001 (1977).

[49] R. Jackiw and J.R. Schrieffer. Solitons with fermion number $\frac{1}{2}$ in condensed matter and relativistic field theories. *Nucl. Phys. B* **190**, 253–265 (1981).

[50] A. J. Niemi and G. W. Semenoff. Fermion number fractionization in quantum field theory. *Phys. Rept.* **135**, 99–193 (1986).

[51] G. Baskaran, Z. Zou, and P. W. Anderson. The resonating valence bond state and high-T_c superconductivity. A mean field theory. *Solid State Commun.* **63**, 973–976 (1987).

[52] B. J. Kim, H. Koh, E. Rotenberg, S.-J. Oh, H. Eisaki, N. Motoyama, S. Uchida, T. Tohyama, S. Maekawa, Z.-X. Shen, and C. Kim. Distinct spinon and holon dispersions in photoemission spectral functions from one-dimensional $SrCuO_2$. *Nat. Phys.* **2**, 397–401 (2006).

[53] S. Coleman and E. Weinberg, Radiative corrections as the origin of spontaneous symmetry breaking. *Phys. Rev. D* **7**, 1888–1910 (1973).

[54] M. N. Chernodub, L. D. Faddeev, and A. J. Niemi. Non-Abelian supercurrents and de Sitter ground state in electroweak theory *JHEP* **12**, 014 (2008).

[55] A. J. Niemi. Phases of bosonic strings and two dimensional gauge theories. *Phys. Rev. D* **67**, 106004 (2003)

[56] R. L. Bishop, There is more than one way to frame a curve. *Am. Math. Monthly* **82**, 246–251 (1975).

[57] S. Hu, M. Lundgren, and A. J. Niemi. Discrete Frenet frame, inflection point solitons, and curve visualization with applications to folded proteins. *Phys. Rev. E* **83**, 061908 (2011).

[58] L. D. Faddeev and L. A. Takhtajan. *Hamiltonian Methods in the Theory of Solitons.* Springer-Verlag, Berlin (1987).

[59] M. J. Ablowitz, B. Prinari, and A. D. Trubatch. *Discrete and Continuous Nonlinear Schrödinger Systems.* London Mathematical Society Lecture Note Series, No. 302. Cambridge University Press, Cambridge (2003).

[60] A. M. Polyakov. Fine structure of strings. *Nucl. Phys. B* **268**, 406–412 (1986).

[61] O. Kratky and G. Porod. Röntgenuntersuchung gelöster Fadenmoleküle. *Recl. Trav. Chim. Pays-Bas* **68**, 1106–1123 (1949).

[62] P. G. Kevrekidis, *The Discrete Nonlinear Schrödinger Equation: Mathematical Analysis, Numerical Computations and Physical Perspectives.* Springer-Verlag, Berlin (2009).

[63] N. Manton and P. Sutcliffe. *Topological Solitons.* Cambridge University Press, Cambridge (2004).

[64] S. Hu, Y. Jiang, and A. J. Niemi. Energy functions for stringlike continuous curves, discrete chains, and space-filling one dimensional structures. *Phys. Rev. D* **87**, 105011 (2013).

[65] T. Ioannidou, Y. Jiang, and A. J. Niemi. On spinors, strings, integrable models and decomposed Yang–Mills theory. *Phys. Rev. D* **90**, 025012 (2014).

[66] V. I. Arnold. *Singularities of Caustics and Wave Fronts.* Kluwer, Dordrecht (1990).

[67] V. I. Arnold. The geometry of spherical curves and the algebra of quaternions. *Russ. Math. Surv.* **50**, 1–68 (1995).
[68] V. I. Arnold, On the number of flattening points on space curves. *Am. Math. Soc. Transl.* **171**, 11 (1996).
[69] F. Aicardi, Self-linking of spatial curves without inflections and its applications. *Funct. Anal. Appl.* **34**, 79–85 (2000).
[70] R. Uribe-Vargas. On singularities, 'perestroikas' and differential geometry of space curves. *Enseign. Math.* **50**, 69–101 (2004).
[71] U. H. Danielsson, M. Lundgren, and A. J. Niemi. Gauge field theory of chirally folded homopolymers with applications to folded proteins. *Phys. Rev. E* **82**, 021910 (2010).
[72] M. N. Chernodub, S. Hu, and A. J. Niemi. Topological solitons and folded proteins. *Phys. Rev. E* **82** 011916 (2010).
[73] N. Molkenthin, S. Hu, and A. J. Niemi. Discrete nonlinear Schrödinger equation and polygonal solitons with applications to collapsed proteins. *Phys. Rev. Lett.* **106**, 078102 (2011).
[74] S. Hu, A. Krokhotin, A. J. Niemi, and X. Peng. Towards quantitative classification of folded proteins in terms of elementary functions. *Phys. Rev. E* **83**, 041907 (2011).
[75] A. Krokhotin, A. J. Niemi, and X. Peng. Soliton concepts and protein structure. *Phys. Rev. E* **85** 031906 (2011).
[76] A. Krokhotin, M. Lundgren, and A. J. Niemi. Solitons and collapse in the λ-repressor protein. *Phys. Rev. E* **86**, 021923 (2012).
[77] A. Krokhotin, A. J. Niemi, and X. Peng. On the role of thermal backbone fluctuations in myoglobin ligand gate dynamics. *J. Chemical Phys.* **138**, 175101 (2013).
[78] A. Krokhotin, M. Lundgren, A. J. Niemi, and X. Peng. Soliton driven relaxation dynamics and protein collapse in the villin headpiece. *J. Phys. Condens. Matter* **25**, 325103 (2013).
[79] R. J. Glauber. Time-dependent statistics of the Ising model. *J. Math. Phys.* **4**, 294–307 (1963).
[80] A. B. Bortz, M. H. Kalos, J. L. Lebowitz. A new algorithm for Monte Carlo simulation of Ising spin systems. *J. Comput. Phys.* **17**, 10–18 (1975).
[81] J. Skolnick, A. K. Arakaki, Y. L. Seung, and M. Brylinski. The continuity of protein structure space is an intrinsic property of proteins. *Proc. Natl Acad. Sci. USA* **106**, 15690 (2009).
[82] M. Ptashne, *A Genetic Switch,: Phage Lambda Revisited*, 3rd edn. Cold Spring Harbor Laboratory Press, Cold Spring Harbor, NY (2004).
[83] M. E. Gottesman and R. A. Weisberg. Little lambda, who made thee? *Microbiol. Mol. Biol. Rev.* **68**, 796–813 (2004).
[84] P. A. Jennings and P. E. Wright. Formation of a molten globule intermediate early in the kinetic pathway of apomyoglobin. *Science* **262**, 892–896 (1993).
[85] Y. Moriyama, and K. Takeda, Critical temperature of secondary structural change of myoglobin in thermal denaturation up to 130°C and effect of sodium dodecyl sulfate on the change. *J. Phys. Chem. B* **114**, 2430 (2010).

[86] Y. Ochia, Y. Watabane , H. Ozawa, S. Ikegami, N. Uchida, and S. Watabe, Thermal denaturation profiles of tuna myoglobin. *Biosci. Biotechnol. Biochem.* **74**, 1673–1679 (2010).

[87] M. N. Chernodub, M. Lundgren, and A. J. Niemi. Elastic energy and phase structure in a continuous spin Ising chain with applications to chiral homopolymers. *Phys. Rev. E* **83**, 011126 (2010).

[88] K. N. Woods. Using THz time-scale infrared spectroscopy to examine the role of collective, thermal fluctuations in the formation of myoglobin allosteric communication pathways and ligand specificity. *Soft Matter* **10**, 4387–4402 (2014).

[89] P. Westermark, A. Andersson, and G. T. Westermark. Islet amyloid polypeptide, islet amyloid, and diabetes mellitus. *Physiol. Rev.* **91**, 795–826 (2011).

[90] K. Pillay and P. Govender. Amylin uncovered: a review on the polypeptide responsible for type II diabetes. *BioMed Res. Int.* **2013**, 826706 (2013).

[91] J. He, J. Dai, J. Li, X. Peng, and A. J. Niemi. Aspects of structural landscape of human islet amyloid polypeptide. *J. Chem. Phys.* **142**, 045102 (2015).

[92] L. Holm and C. Sander. Database algorithm for generating protein backbone and side-chain co-ordinates from a C^{α} trace: application to model building and detection of co-ordinate errors. *J. Mol. Biol.* **218** 183–194 (1991).

[93] S. C. Lovell, I. W. Davis, W. B. Arendall III, P. I. W. de Bakker, J. M. Word, M. G. Prisant, J. S. Richardson, and D. C. Richardson, Structure validation by C_α geometry: ϕ, ψ and C_β. *Proteins* **50**, 437–450 (2003).

[94] P. Rotkiewicz and J. Skolnick. Fast procedure for reconstruction of full atom protein models from reduced representations. *J. Comput. Chem.* **29** 1460–1465 (2008).

[95] R. L. Dunbrack Jr. Rotamer libraries in the 21st century. *Curr. Opin. Struct. Biol.* **12**, 431–440 (2002).

[96] M. Lundgren, A. J. Niemi and F. Sha. Protein loops, solitons, and side chain visualization with applications to the left-handed helix region *Phys. Rev. E* **85**, 061909 (2012).

[97] M. Lundgren and A. J. Niemi. Correlation between protein secondary structure, backbone bond angles, and side chain orientations. *Phys. Rev. E* **86**, 021904 (2013).

[98] X. Peng, A. Chenani, S. Hu, Y. Zhou, and A. J. Niemi. A three dimensional visualisation approach to protein heavy-atom structure reconstruction. *BMC Struct. Biol.* **14**, 27 (2014).

The manufacturer's authorised representative in the EU for product safety is Oxford University Press España S.A. of El Parque Empresarial San Fernando de Henares, Avenida de Castilla, 2 - 28830 Madrid (www.oup.es/en or product.safety@oup.com). OUP España S.A. also acts as importer into Spain of products made by the manufacturer.
Printed and bound by CPI Group (UK) Ltd, Croydon, CR0 4YY
10/07/2026
02163887-0001